T0180297

# Lecture Notes in Computer Science 9624

Commenced Publication in 1973
Founding and Former Series Editors:
Gerhard Goos, Juris Hartmanis, and Jan van Leeuwen

More information about this series at http://www.springer.com/series/7407

Alexander Gelbukh (Ed.)

# Computational Linguistics and Intelligent Text Processing

17th International Conference, CICLing 2016
Konya, Turkey, April 3–9, 2016
Revised Selected Papers, Part II

 Springer

*Editor*
Alexander Gelbukh
CIC, Instituto Politécnico Nacional
Mexico City
Mexico

ISSN 0302-9743          ISSN 1611-3349  (electronic)
Lecture Notes in Computer Science
ISBN 978-3-319-75486-4          ISBN 978-3-319-75487-1  (eBook)
https://doi.org/10.1007/978-3-319-75487-1

Library of Congress Control Number: 2018934347

LNCS Sublibrary: SL1 – Theoretical Computer Science and General Issues

Printed on acid-free paper

This Springer imprint is published by the registered company Springer International Publishing AG
part of Springer Nature
The registered company address is: Gewerbestrasse 11, 6330 Cham, Switzerland

*In Memoriam of Adam Kilgarriff*

# Preface

CICLing 2016 was the 17th International Conference on Intelligent Text Processing and Computational Linguistics. The CICLing conferences provide a wide-scope forum for discussion of the art and craft of natural language processing research, as well as the best practices in its applications.

In 2015, Adam Kilgarriff, an influential scientist and a wonderful and brave person, passed away prematurely, at the age of only 55. Adam was a great friend of CICLing, a member of its small informal Steering Committee, one who helped to shape CICLing from its inception. Until his last days, already terminally ill, he volunteered to help us with the reviewing process for CICLing 2015. This CICLing event was dedicated to his bright memory, and its proceedings begin with a paper that attempts to summarize his scientific legacy.

This set of two books contains five invited papers and a selection of regular papers accepted for presentation at the conference. Since 2001, the proceedings of the CICLing conferences have been published in Springer's *Lecture Notes in Computer Science* series as volumes 2004, 2276, 2588, 2945, 3406, 3878, 4394, 4919, 5449, 6008, 6608, 6609, 7181, 7182, 7816, 7817, 8403, 8404, 9041, and 9042.

The set has been structured into 14 sections: an In Memoriam section and 13 sections representative of the current trends in research and applications of natural language processing:

– General formalisms
– Embeddings, language modeling, and sequence labeling
– Lexical resources and terminology extraction
– Morphology and part-of-speech tagging
– Syntax and chunking
– Named entity recognition
– Word sense disambiguation and anaphora resolution
– Semantics, discourse, and dialog
– Machine translation and multilingualism
– Sentiment analysis, opinion mining, subjectivity, and social media
– Text classification and categorization
– Information extraction
– Applications

The 2016 event received submissions from 54 countries. A total of 298 papers by 671 authors were submitted for evaluation by the international Program Committee (see Fig. 1 and Tables 1 and 2). This two-volume set contains revised versions of 89 regular papers selected for presentation, with the acceptance rate of 29.8%.

**Table 1.** Number of submissions and accepted papers by topic[a]

| Submissions | Topic |
|---|---|
| 60 | Text mining |
| 58 | Information extraction |
| 53 | Lexical resources |
| 44 | Emotions, sentiment analysis, opinion mining |
| 43 | Information retrieval |
| 42 | Semantics, pragmatics, discourse |
| 39 | Clustering and categorization |
| 36 | Under-resourced languages |
| 35 | Machine translation and multilingualism |
| 33 | Morphology |
| 33 | Practical applications |
| 28 | Social networks and microblogging |
| 25 | Named entity recognition |
| 25 | POS tagging |
| 23 | Syntax and chunking |
| 21 | Noisy text processing and cleaning |
| 20 | Formalisms and knowledge representation |
| 16 | Plagiarism detection and authorship attribution |
| 16 | Question answering |
| 16 | Summarization |
| 16 | Word sense disambiguation |
| 13 | Computational terminology |
| 13 | Natural language interfaces |
| 12 | Other |
| 11 | Natural language generation |
| 10 | Speech processing |
| 7 | Spelling and grammar check |
| 7 | Textual entailment |
| 6 | Coreference resolution |

[a]As indicated by the authors. A paper may belong to more than one topic.

In addition to regular papers and an In Memoriam paper, the books features papers by the keynote speakers:

- Pascale Fung, Hong Kong University of Science and Technology, Hong Kong
- Tomas Mikolov, Facebook AI Research, USA
- Simone Teufel, University of Cambridge, UK
- Piek Vossen, Vrije Universiteit Amsterdam, The Netherlands

Publication of full-text invited papers in the proceedings is a distinctive feature of the CICLing conferences. In addition to a presentation of their invited papers, the keynote speakers organized separate lively informal events; this is also a special feature of this conference series.

**Table 2.** Number of submitted and accepted papers by country or region

| Country or region | Authors | Submissions[b] | Country or region | Authors | Submissions[b] |
|---|---|---|---|---|---|
| Algeria | 15 | 6.5 | Libya | 2 | 0.5 |
| Argentina | 2 | 1 | Mexico | 8 | 5.17 |
| Australia | 5 | 1.33 | Morocco | 13 | 4.81 |
| Austria | 1 | 0.33 | Nigeria | 3 | 1 |
| Brazil | 8 | 3.75 | Norway | 10 | 3.96 |
| Canada | 24 | 10.78 | Pakistan | 1 | 1 |
| China | 23 | 11.25 | Peru | 3 | 2 |
| Colombia | 7 | 3 | Poland | 2 | 1 |
| Czechia | 19 | 8 | Portugal | 6 | 2.05 |
| Egypt | 17 | 6 | Qatar | 11 | 2.83 |
| Finland | 1 | 0.25 | Romania | 12 | 5.75 |
| France | 56 | 19.52 | Russia | 12 | 7.25 |
| Germany | 15 | 5.93 | Saudi Arabia | 5 | 1.3 |
| Greece | 3 | 1 | Singapore | 8 | 2.5 |
| Hong Kong | 2 | 1 | South Africa | 3 | 1 |
| Hungary | 6 | 5 | South Korea | 7 | 4 |
| India | 73 | 37.3 | Spain | 9 | 3.78 |
| Indonesia | 3 | 1 | Sri Lanka | 12 | 5 |
| Iran | 5 | 2.33 | Switzerland | 4 | 2 |
| Ireland | 3 | 1.25 | The Netherlands | 1 | 0.2 |
| Israel | 10 | 3.24 | Tunisia | 86 | 40.35 |
| Italy | 5 | 1 | Turkey | 53 | 29.22 |
| Japan | 13 | 3.5 | Turkmenistan | 1 | 0.33 |
| Jordan | 4 | 3 | UAE | 4 | 1.5 |
| Kazakhstan | 19 | 6.75 | UK | 18 | 9.17 |
| Latvia | 2 | 1 | USA | 30 | 13.4 |
| Lebanon | 2 | 0.67 | Vietnam | 4 | 1.25 |
| | | | *Total:* | 671 | 298 |

[b] By the number of authors: e.g., a paper by two authors from the USA and one from UK is counted as 0.67 for the USA and 0.33 for UK.

With this event, we continued our policy of giving preference to papers with verifiable and reproducible results: In addition to the verbal description of their findings given in the paper, we encouraged the authors to provide a proof of their claims in electronic form. If the paper claimed experimental results, we asked the authors to make available to the community all the input data necessary to verify and reproduce these results; if it claimed to introduce an algorithm, we encourage the authors to make the algorithm itself, in a programming language, available to the public. This additional electronic material will be permanently stored on the CICLing's server, www.CICLing.org, and will be available to the readers of the corresponding paper for download under a license that permits its free use for research purposes.

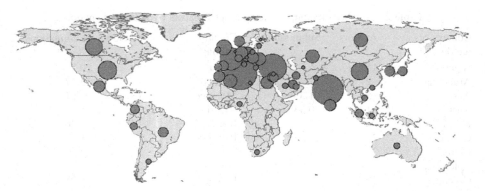

**Fig. 1.** Submissions by country or region. The area of a circle represents the number of submitted papers

In the long run, we expect that computational linguistics will have verifiability and clarity standards similar to those of mathematics: In mathematics, each claim is accompanied by a complete and verifiable proof, usually much longer than the claim itself; each theorem's complete and precise proof—and not just a description of its general idea—is made available to the reader. Electronic media allow computational linguists to provide material analogous to the proofs and formulas in mathematic in full length—which can amount to megabytes or gigabytes of data—separately from a 12-page description published in the book. More information can be found on http://www.CICLing.org/why_verify.htm.

To encourage providing algorithms and data along with the published papers, we selected three winners of our Verifiability, Reproducibility, and Working Description Award. The main factors in choosing the awarded submission were technical correctness and completeness, readability of the code and documentation, simplicity of installation and use, and exact correspondence to the claims of the paper. Unnecessary sophistication of the user interface was discouraged; novelty and usefulness of the results were not evaluated—instead, they were evaluated for the paper itself and not for the data.

The following papers received the Best Paper Awards, the Best Student Paper Award, as well as the Verifiability, Reproducibility, and Working Description Awards, respectively:

Best Paper 1st Place:     "Mining the Web for Collocations: IR Models of Term Associations," by Rakesh Verma, Vasanthi Vuppuluri, An Nguyen, Arjun Mukherjee, Ghita Mammar, Shahryar Baki, and Reed Armstrong, USA
                          "Extracting Aspect Specific Sentiment Expressions Implying Negative Opinions," by Arjun Mukherjee, USA
Best Paper 2nd Place:     "Word Sense Disambiguation Using Swarm Intelligence: A Bee Colony Optimization Approach," by Saket Kumar and Omar El Ariss, USA

Best Paper 3rd Place:        "Corpus Frequency and Affix Ordering in Turkish," by
                              Mustafa Aksan, Umut Ufuk Demirhan, and Yeşim Aksan,
                              Turkey
Best Student Paper[1]:       "Generating Bags of Words from the Sums of Their Word
                              Embeddings," by Lyndon White, Roberto Togneri, Wei Liu,
                              and Mohammed Bennamoun, Australia
Verifiability 1st Place:     "Pluralizing Nouns in isiZulu and Related Languages," by
                              Joan Byamugisha, C. Maria Keet, and Langa Khumalo,
                              South Africa
Verifiability 2nd Place:     "A Free/Open-Source Hybrid Morphological
                              Disambiguation Tool for Kazakh,"[2] by Zhenisbek
                              Assylbekov, Jonathan North Washington, Francis Tyers,
                              Assulan Nurkas, Aida Sundetova, Aidana Karibayeva,
                              Balzhan Abduali, and Dina Amirova, Kazakhstan, USA, and
                              Norway
Verifiability 3rd Place:     "An Informativeness Approach to Open IE Evaluation," by
                              William Léchelle and Philippe Langlais, Canada

The authors of the awarded papers (except for the Verifiability Award) were given extended time for their presentations. In addition, the Best Presentation Award and the Best Poster Award winners were selected by a ballot among the attendees of the conference.

Besides its high scientific level, one of the success factors of the CICLing conferences is their excellent cultural program in which all attendees participate. The cultural program is a very important part of the conference, serving its main purpose: personal interaction and making friends and contacts. The attendees of the conference had a chance to visit the Tinaztepe Cavern and nearby attractions; the Underground City and the amazing Cappadocia landscape, probably the most astonishing natural landscape I have seen so far; as well as the attractions and historial monuments of the city of Konya, the ancient capital of the Seljuk Sultanate of Rum and the Karamanids.

The conference was accompanied by two satellite events: the Second International Conference on Arabic Computational Linguistics, ACLing 2016, and the First International Conference on Turkic Computational Linguistics, TurCLing 2016, both founded by CICLing. This is in accordance with CICLing's mission to promote consolidation of emerging NLP communities in countries and regions underrepresented in the mainstream of NLP research and, in particular, in the mainstream publication venues. The Program Committees of ACLing 2016 and TurCLing 2016 have helped the CICLing 2016 committee in evaluation of a number of submissions on the corresponding narrow topics.

---

[1] The best student paper was selected among papers of which the first author was a full-time student, excluding the papers that received Best Paper Awards.

[2] This paper is published in the proceedings of a satellite event of the conference and not in this book set.

I would like to thank all those involved in the organization of this conference. In the first place, the authors of the papers that constitute this book: It is the excellence of their research work that gives value to the book and meaning to the work of all other people. I thank all those who served on the Program Committee of CICLing 2016, the Second International Conference on Arabic Computational Linguistics, ACLing 2016, and the First International Conference on Turkic Computational Linguistics, TurCLing 2016; the Software Reviewing Committee, the Award Selection Committee, as well as additional reviewers, for their hard and very professional work. Special thanks go to Ted Pedersen, Manuel Vilares Ferro, and Soujanya Poria for their invaluable support in the reviewing process.

I also want to cordially thank the conference staff, volunteers, and the members of the local Organizing Committee headed by Hatem Haddad, as well as the organizers of the satellite events, Samhaa R. El-Beltagy of Nile University, Egypt, Khaled Shaalan of the British University in Dubai, UAE, and Bahar Karaoglan of Ege University, Turkey. I am deeply grateful to the administration of the Mevlana University for their helpful support, warm hospitality, and in general for providing this wonderful opportunity of holding CICLing in Turkey. I acknowledge support from the project CON-ACYT Mexico–DST India 122030 "Answer Validation Through Textual Entailment" and SIP-IPN grant 20161958.

The entire submission and reviewing process was supported for free by the Easy-Chair system (www.EasyChair.org). Last but not least, I deeply appreciate the Springer team's patience and help in editing these volumes and getting them printed in very short time—it is always a great pleasure to work with Springer.

December 2016                                                     Alexander Gelbukh

# Organization

CICLing 2016 was hosted by the Mevlana University, Turkey, and was organized by the CICLing 2016 Organizing Committee in conjunction with the Mevlana University, the Natural Language and Text Processing Laboratory of the Centro de Investigación en Computación (CIC) of the Instituto Politécnico Nacional (IPN), Mexico, and the Mexican Society of Artificial Intelligence (SMIA).

## Organizing Chair

Hatem Haddad      Université Libre de Bruxelles, Belgium

## Organizing Committee

| | |
|---|---|
| Hatem Haddad (Chair) | Université Libre de Bruxelles, Belgium |
| Armağan Ozkaya | Mevlana University, Turkey |
| Niyazi Serdar Tunaboylu | Mevlana University, Turkey |
| Alaa Eleyan | Mevlana University, Turkey |
| Mustafa Kaiiali | Mevlana University, Turkey |
| Mohammad Shukri Salman | Mevlana University, Turkey |
| Gülden Eleyan | Mevlana University, Turkey |
| Hasan Ucar | Mevlana University, Turkey |

## Program Chair

Alexander Gelbukh      Instituto Politécnico Nacional, Mexico

## Program Committee

| | |
|---|---|
| Ajith Abraham | Machine Intelligence Research Labs (MIR Labs), USA |
| Rania Al-Sabbagh | University of Illinois at Urbana-Champaign, USA |
| Galia Angelova | Bulgarian Academy of Sciences, Bulgaria |
| Marianna Apidianaki | LIMSI-CNRS, France |
| Alexandra Balahur | European Commission Joint Research Centre, Italy |
| Sivaji Bandyopadhyay | Jadavpur University, India |
| Leslie Barrett | Bloomberg, LP, USA |
| Roberto Basili | University of Rome Tor Vergata, Italy |
| Anja Belz | University of Brighton, UK |
| Christian Boitet | Université Joseph Fourier and LIG, GETALP, France |
| Igor Bolshakov | Independent Researcher, Russia |
| Nicoletta Calzolari | Istituto di Linguistica Computazionale – CNR, Italy |
| Erik Cambria | Nanyang Technological University, Singapore |

| | |
|---|---|
| Kjetil Nørvåg | Norwegian University of Science and Technology, Norway |
| Attila Novák | Pázmány Péter Catholic University, Hungary |
| Nir Ofek | Ben-Gurion University, Israel |
| Partha Pakray | National Institute of Technology Mizoram, India |
| Ivandre Paraboni | University of São Paulo, Brazil |
| Patrick Saint-Dizier | IRIT-CNRS, France |
| Maria Teresa Pazienza | University of Rome Tor Vergata, Italy |
| Ted Pedersen | University of Minnesota, USA |
| Viktor Pekar | University of Birmingham, UK |
| Anselmo Peñas | UNED, Spain |
| Soujanya Poria | Nanyang Technological University, Singapore |
| Marta R. Costa-Jussà | Institute for Infocomm Research, Singapore |
| Fuji Ren | University of Tokushima, Japan |
| German Rigau | IXA Group, UPV/EHU, Spain |
| Fabio Rinaldi | University of Zurich, Switzerland |
| Horacio Rodriguez | Universitat Politècnica de Catalunya, Spain |
| Paolo Rosso | Universitat Politècnica de València, Spain |
| Vasile Rus | The University of Memphis, USA |
| Franco Salvetti | University of Colorado at Boulder/Microsoft, USA |
| Rajeev Sangal | Language Technologies Research Centre, India |
| Kepa Sarasola | Euskal Herriko Unibertsitatea, Spain |
| Fabrizio Sebastiani | Qatar Computing Research Institute, Qatar |
| Serge Sharoff | UoLeeds, UK |
| Bernadette Sharp | Staffordshire University, UK |
| Grigori Sidorov | Instituto Politécnico Nacional, Mexico |
| Kiril Simov | Linguistic Modelling Laboratory, IICT-BAS, Bulgaria |
| John Sowa | VivoMind Intelligence, USA |
| Efstathios Stamatatos | University of the Aegean, Greece |
| Maosong Sun | Tsinghua University, China |
| Jun Suzuki | NTT, Japan |
| Stan Szpakowicz | University of Ottawa, Canada |
| Hristo Tanev | Independent Researcher, Italy |
| Juan-Manuel Torres-Moreno | Laboratoire Informatique d'Avignon/UAPV, France |
| George Tsatsaronis | Technical University of Dresden, Germany |
| Olga Uryupina | University of Trento, Italy |
| Manuel Vilares Ferro | University of Vigo, Spain |
| Aline Villavicencio | Federal University of Rio Grande do Sul, Brazil |
| Piotr W. Fuglewicz | TiP Sp. z o. o., Poland |
| Marilyn Walker | UCSC, USA |
| Andy Way | ADAPT Centre, Dublin City University, Ireland |
| Bonnie Webber | The University of Edinburgh, UK |
| Alisa Zhila | IBM, USA |

## Software Reviewing Committee

| | |
|---|---|
| Ted Pedersen | University of Minnesota, USA |
| Florian Holz | University of Leipzig, Germany |
| Miloš Jakubíček | Masaryk University, Czech Republic |
| Sergio Jiménez Vargas | Instituto Caro y Cuervo, Colombia |
| Miikka Silfverberg | University of Colorado, USA |
| Ronald Winnemöller | University of Hamburg, Germany |

## Award Committee

| | |
|---|---|
| Alexander Gelbukh | Instituto Politécnico Nacional, Mexico |
| Eduard Hovy | Carnegie Mellon University, USA |
| Rada Mihalcea | University of Michigan, USA |
| Ted Pedersen | University of Minnesota, USA |
| Yorick Wilks | University of Sheffield, UK |

## Additional Reviewers

| | |
|---|---|
| Abhijit Mishra | Kyoko Kanzaki |
| Abidalrahman Moh'D | Leonardo Zilio |
| Aitor García | Magdalena Jankowska |
| Alisa Zhila | Mahsa Forati |
| Aljaz Kosmerlj | Maite Giménez |
| Amir Hossein Razavi | Majid Laali |
| Avi Hayoun | Marc Franco Salvador |
| Benjamin Marie | Marcelo Sardelich |
| Bernardo Cabaleiro | Michael Mohler |
| Bryan Rink | Miguel Angel Rios Gaona |
| Carla Parra Escartín | Olga Kolesnikova |
| Diana Trandabăţ | Petya Osenova |
| Egoitz Laparra | Pidong Wang |
| Elnaz Davoodi | Pistol Ionut Cristian |
| Enrique Flores | Richard Evans |
| Francisco J. Ribadas-Pena | Rodrigo Wilkens |
| Hanna Bechara | Silvio Ricardo Cordeiro |
| Helena Gómez Adorno | Sonia Ordoñez Salinas |
| Hiram Calvo | Svetla Boytcheva |
| Ilia Markov | Tal Baumel |
| Irazú Hernández Farias | Travis Goodwin |
| Jan R. Benetka | Victor Darriba |
| Jan Rupnik | Victoria Yaneva |
| Janez Starc | Vinita Nahar |
| Jared Bernstein | Vojtech Kovar |
| Jie Mei | Xin Chen |
| Jumana Nassour-Kassis | Zuzana Neverilova |
| Kazuhiro Takeuchi | |

# Second Arabic Computational Linguistics Conference

## Program Chairs

Samhaa R. El-Beltagy          Nile University, Egypt
Alexander Gelbukh             Instituto Politécnico Nacional, Mexico
Khaled Shaalan                The British University in Dubai, UAE

## Program Committee

Bayan Abushawar               Arab Open University, Jordan
Hanady Ahmed                  Qatar University, Qatar
Hend Alkhalifa                King Saud University, Saudi Arabia
Mohammed Attia                Al-Azhar University, Egypt
Aladdin Ayesh                 De Montfort University, UK
Karim Bouzoubaa               Mohamed V University, Morocco
Violetta Cavalli-Sforza       Al Akhawayn University, Morocco
Khalid Choukri                ELDA, France
Samhaa R. El-Beltagy          Nile University, Egypt
Ossama Emam                   ALTEC, Egypt
Aly Fahmy                     Cairo University, Egypt
Ahmed Guessoum                University of Science and Technology Houari
                              Boumediene, Algeria
Nizar Habash                  New York University Abu Dhabi, UAE
Hatem Haddad                  Université Libre de Bruxelles, Belgium
Kais Haddar                   MIRACL Laboratory, Faculté des sciences de Sfax,
                              Tunisia
Lamia Hadrich Belguith        MIRACL Laboratory, Tunisia
Imtiaz Khan                   King Abdulaziz University, Jeddah, Saudi Arabia
Alma Kharrat                  Microsoft, USA
Sattar Izwaini                American University of Sharjah, UAE
Mark Lee                      University of Birmingham, UK
Sherif Mahdy Abdou            RDI, Egypt
Farid Meziane                 University of Salford, UK
Farhad Oroumchian             University of Wollongong in Dubai, UAE
Hermann Moisl                 Newcastle University, UK
Ahmed Rafea                   American University in Cairo, Egypt
Allan Ramsay                  The University of Manchester, UK
Mohsen Rashwan                Cairo University, Egypt
Horacio Rodriguez             Universitat Politècnica de Catalunya, Spain
Paolo Rosso                   Universitat Politècnica de València, Spain
Nasredine Semmar              CEA, France
Khaled Shaalan                The British University in Dubai, UAE
William Teahan                Bangor University, UK
Imed Zitouni                  IBM Research, USA

## Additional Reviewers

Asma Aouichat
Hind Saddiki
Mohamed Hadj Ameur
Muhammad Shuaib Qureshi

Muhammad Umair
Nora Al-Twairesh
Riadh Belkebir

## First International Conference on Turkic Computational Linguistics

### Program Chairs

| | |
|---|---|
| Bahar Karaoglan (Chair) | Ege University, Turkey |
| Tarık Kışla (Co-chair) | Ege University, Turkey |
| Senem Kumova (Co-chair) | İzmir Ekonomi University, Turkey |
| Hatem Haddad (Co-chair) | Université Libre de Bruxelles, Belgium |

### Program Committee

| | |
|---|---|
| Yeşim Aksan | Mersin University, Turkey |
| Adil Alpkocak | Dokuz Eylul University, Turkey |
| Ildar Batyrshin | Instituto Politécnico Nacional, Mexico |
| Cem Bozsahin | Middle East Technical University, Turkey |
| Fazli Can | Bilkent University, Turkey |
| Ilyas Cicekli | Hacettepe University, Turkey |
| Gülşen Eryiğit | Istanbul Technical University, Turkey |
| Alexander Gelbukh | Instituto Politécnico Nacional, Mexico |
| Tunga Gungor | Boğaziçi University, Turkey |
| Hatem Haddad | Université Libre de Bruxelles, Belgium |
| Bahar Karaoglan | Ege University, Turkey |
| Tarik Kisla | Ege University, Turkey |
| Senem Kumova Metin | İzmir University of Economics, Turkey |
| Altynbek Sharipbayev | L. N. Gumilyov Eurasian National University, Kazakhstan |
| Dzhavdet Suleymanov | Academy of Sciences of Tatarstan, Russia |

# Additional Reviewers

A. Sumru Özsoy
Ali Erkan
Cem Rıfkı Aydın
Dilyara Yakubova

Ezgi Yıldırım
Nihal Yağmur Aydın
Razieh Ehsani

# Website and Contact

The website of the CICLing conference series is http://www.CICLing.org. It provides information about past CICLing conferences and their satellite events, including links to published papers (many of them in open access) or their abstracts, as well as photos and video recordings of keynote talks. In addition, it provides data, algorithms, and open-source software accompanying accepted papers, in accordance with the CICLing verifiability, reproducibility, and working description policy. It also provides information about the forthcoming CICLing events, as well as contact options.

# Contents – Part II

**Sentiment Analysis, Opinion Mining, Subjectivity, and Social Media**

**Invited Paper:**

**Best Paper Award, First Place:**

## Applications

### Invited Papers:

# Contents – Part I

## Lexical Resources and Terminology Extraction

### Best Paper Award, First Place:

## Morphology and Part-of-Speech Tagging

### Best Paper Award, Third Place:

**Syntax and Chunking**

**Named Entity Recognition**

# Machine Translation and Multilingualism

# Enabling Medical Translation
# for Low-Resource Languages

Ahmad Musleh[1], Nadir Durrani[1(✉)], Irina Temnikova[1], Preslav Nakov[1],
Stephan Vogel[1], and Osama Alsaad[2]

[1] Qatar Computing Research Institute, HBKU, Doha, Qatar
{amusleh,ndurrani,itemnikova,pnakov,svogel}@qf.org.qa
[2] Texas A&M University in Qatar, Doha, Qatar
osama.alsaad@qatar.tamu.edu

**Abstract.** We present research towards bridging the language gap
between migrant workers in Qatar and medical staff. In particular, we
present the first steps towards the development of a real-world Hindi-
English machine translation system for doctor-patient communication.
As this is a low-resource language pair, especially for speech and for the
medical domain, our initial focus has been on gathering suitable training
data from various sources. We applied a variety of methods ranging from
fully automatic extraction from the Web to manual annotation of test
data. Moreover, we developed a method for automatically augmenting
the training data with synthetically generated variants, which yielded a
very sizable improvement of more than 3 BLEU points absolute.

**Keywords:** Machine translation · Medical translation
Doctor-patient communication · Resource-poor languages · Hindi

## 1 Introduction

In recent years, Qatar's booming economy has resulted in rapid growth in the
number of migrant workers needed for the growing number of infrastructure
projects. These workers, who mainly come from the southern parts of Asia,
usually know little or no English and do not know any Arabic either. This
results in a communication barrier between them and the natives. More serious
situation arises in the case of medical emergency. This causes serious problems as
the public administration and services in Qatar mostly use Arabic and English.
According to a 2012 report by the Weill Cornell Medical College (WCMC[1]) in
Qatar [15], almost 78% of the patients visiting the Hamad Medical Corporation
(HMC, the main health-care provider in Qatar) did not speak Arabic or English.
The study also pointed out that the five most spoken languages in Qatar in
2012 were (in that order) Nepali, Urdu, Hindi, English, and Arabic. The report
also pointed out that even though HMC currently uses medical interpreters

---

[1] http://qatar-weill.cornell.edu.

© Springer International Publishing AG, part of Springer Nature 2018
A. Gelbukh (Ed.): CICLing 2016, LNCS 9624, pp. 3–16, 2018.
https://doi.org/10.1007/978-3-319-75487-1_1

to overcome this problem, their number is not sufficient. This has urged the authorities to look into technology for alternatives.

In this paper, we propose a solution to bridge this language gap. We present our preliminary effort towards developing a Statistical Machine Translation (SMT) system for doctor-patient communication in Qatar. The success of a data-driven system largely depends upon the availability of in-domain data.

This makes our task non-trivial as we are dealing with a low-resourced language pair and furthermore with the medical domain. We decided to focus on Hindi, one of the languages under question. Our decision was driven by the fact, that Hindi and Urdu are closely-related languages, often considered dialects of each other, and people from Nepal and other South-Asian countries working in Qatar typically understand Hindi. Moreover, we have access to more Hindi-English parallel data than for any other language pair involving the top-5 most spoken languages in Qatar.

Our focus in this paper is on data collection and data generation (Sects. 3 and 4). We collected data from various sources (Sect. 3), including (*i*) Wikimedia (Wikipedia, Wiktionary, and OmegaWiki) parallel English-Hindi data, (*ii*) doctor-patient dialogues from YouTube videos, and movie subtitles, and (*iii*) parallel medical terms from BabelNet and MeSH. Moreover, we synthesized Hindi-English parallel data from Urdu-English data, by translating the Urdu part into Hindi. The approach is described in Sect. 4. Our results show improvement of up to +1.45 when using synthesized data, and up to +1.66, when concatenating the mined dictionaries on top of the synthesized data (Sect. 5).

Moreover, Sect. 2 provides an overview of related work on machine translation for doctor-patient dialogues and briefly discusses the Machine Translation (MT) approaches for low-resource languages; Sect. 5 presents the results of the manual evaluation, and Sect. 6 provides the conclusions and discusses directions for future work.

## 2    Related Work

Below we first describe some MT applications for doctor-patient communication. Then, we present more general research on MT for the medical domain.

### 2.1    Bi-directional Doctor↔Patient Communication Applications

We will first describe the pre-existing MT systems for doctor-patient communication, particularly the ones that required data collection for under-resourced languages. Several MT systems facilitating doctor-patient communication have been built in the past [3,5,13,14,17,20]. Most of them are still prototypes, and only few have been fully deployed. Some of these systems work with under-resourced languages [3,14,17,20]. Moreover, their solution relies on mapping the utterances to an interlingua, instead of using SMT.

**MedSLT** [3] is an interlingua-based speech-to-speech translation system. It covers a restricted set of domains, and covers English, French, Japanese, Spanish, Catalan, and Arabic. The system can translate doctor's questions/statements to the patient, but not the responses by the patient back to the doctor.

**Converser** [5] is a commercial doctor-patient speech-to-speech bidirectional MT system for the English↔Spanish language pair. It has been deployed in several US hospitals and has the following features: users can correct the automatic speech recognition (ASR) and MT outputs, back-translation (re-translation of the translation) to the user is made to allow this. The system maps concepts to a lexical database specially created from various sources, and also allows "translation shortcuts" (i.e., translation memory of previous translations that do not need verification).

**Jibbigo** [13] is a travel and medical speech-to-speech MT system, deployed as an iPhone mobile app. Jibbigo allows English-to-Spanish and Spanish-to-English medical speech-to-speech translation.

**S-MINDS** [14] is a two-way doctors-patient MT system, which uses an in-house "interpretation" software. It matches the ASR utterances to a finite set of concepts in the specific semantic domain and then paraphrases them. In case the utterance cannot be matched, the system uses an SMT engine.

**Accultran** [20] is an automatic translation system prototype, which features back translation to the doctor and yes/no or multiple-choice questions (MCQs) to the patient. It allows the doctor to confirm the translation to the patient, and has a cross cultural adviser. It flags sensitive utterances that are difficult to translate to the patient. The system maps the utterances to SNOMED-CT or Clinical Document Architecture (CDA-2) standards, which are used as an interlingua.

**IBM MASTOR** [17] is a speech-to-speech MT system for two language pairs (English-Mandarin and English-dialectal Arabic), which relies on ASR, SMT, and Speech Synthesis components. It works both on laptops and PDAs.

**English-Portugese SLT** [37] is an English-Portuguese speech-to-speech MT system, composed of an ASR, MT (relying on HMM) and speech synthesis. It works as an online service and as a mobile application.

None of the above systems handles the top-5 languages of interest to Qatar.

### 2.2 Uni-directional Doctor↔Patient Communication Applications

Besides the above-described MT systems, there are a number of mobile or web applications, which are based on pre-translated phrases. The phrases are pre-recorded by professionals or native speakers and can be played to the patient. Most of these applications work only in the doctor-to-patient direction. The most popular ones are UniversalDoctor[2], MediBabble[3], Canopy[4], MedSpeak, Mavro Emergency Medical Spanish[5], and DuoChart[6].

Unfortunately, these applications do not allow free, unseen, or spontaneous translations, and do not cover the language pairs of interest for Qatar. Moreover, some of them (e.g., UniversalDoctor) require a paid subscription.

### 2.3   General Research in Medical Machine Translation

A number of systems have been developed and participated in the WMT'14 Medical translation task. It is a Cross-Lingual Information Retrieval (CLIR) task divided into two sub-tasks: (*i*) translation of user search queries, and (*ii*) translation of summaries of retrieved documents:

A system described in [12], part of the Khresmoi project[7], uses the phrase-based Moses and standard methods for domain adaptation. [25] also uses the phrase-based Moses system and achieved the highest BLEU score for the English-German intrinsic query translation evaluation. Another system [26] combined web-crawled in-domain monolingual data and a bilingual lexicon in order to complement the limited in-domain parallel corpora. A third one [34] proposed a terminology translation system for the query translation subtask and used 6 different methods for terminology extraction. A fourth system [35] used a combination of n-gram based NCODE and phrase-based Moses to the subtask of sentence translation. The system of [40] applied a combination of domain adaptation techniques on the medical summary sentence translation task and achieved the first and the second best BLEU scores. Then, the system of [44] used the Moses phrase-based system and worked on the medical summary WMT'14 task and experimented with translation models, re-ordering models, operation sequence models, and language models, as well as with data selection. A study on quality analysis of machine translation systems in medical domain was carried in [27]. Most of this work focused on European language pairs and did not cover languages of interest to us, nor did it involve low-resource languages in general.

## 3   Data Collection

As the main problem of low resource languages is data collection [24, 30], we have adopted a variety of approaches, in order to collect as much parallel English-Hindi data as possible.

---

[2] http://www.universaldoctor.com.
[3] http://medibabble.com.
[4] http://www.canopyapps.com.
[5] http://mavroinc.com/medical.html.
[6] http://duochart.com.
[7] http://www.khresmoi.eu.

## 3.1  Wiki Dumps

We downloaded, extracted and mined all language links from Wikipedia[8], Wiktionary[9], and OmegaWiki[10] in order to provide a one-to-one word mapping from English into Hindi. We then extracted page links and language links from Wikipedia and Wiktionary. Moreover, we used OmegaWiki to provide a bilingual word dictionary containing the word, its synonyms, its translation, and its lexical, terminological and ontological forms. We extracted the data using two OmegaWiki sources: bilingual dictionaries and an SQL database dump. Table 1 shows the number of Hindi words we collected from all three sources.

**Table 1.** Number of word translation pairs collected from Wikimedia sources.

| Source | Word pairs |
|---|---|
| Wikipedia | 40,764 |
| Wiktionary | 10,352 |
| OmegaWiki | 3,476 |

## 3.2  Doctor-Patient YouTube Videos and Movie Subtitles

We used a mixed approach to extract doctor-patient dialogues from medical YouTube subtitles. As using the YouTube-embedded automatic subtitling is inefficient, the dialogues of the videos were first extracted by manually typing the audio found in the videos. However, as this process was very time consuming, we started a screenshot session in order to collect all the visual representations of the subtitles. Next, the subtitles were extracted using Tesseract[11], an open source Optical Character Recognition (OCR) reader provided by Google, on the screenshots captured. The subtitles were then manually corrected, translated into Hindi using Google Translate, and post-edited by a Hindi native speaker. This resulted in a parallel corpus of medical dialogues with 11,000 Hindi words (1,200 sentences). These sentences were later used for tuning and testing our MT system. Additionally, we used a web crawler to extract a small number of non-medical parallel English-Hindi movie subtitles (nine movies) from Open Subtitles[12].

## 3.3  BabelNet and MeSH

We extracted medical terms from BabelNet [32] using their API. As Medical Subject Headings (MeSH[13]) represents the largest source of Medical terms, we

---

[8] http://www.wikipedia.org.
[9] http://www.wiktionary.org.
[10] http://www.omegawiki.org.
[11] https://github.com/tesseract-ocr.
[12] http://www.opensubtitles.com.
[13] http://www.ncbi.nlm.nih.gov/mesh.

downloaded their dumps and extracted the 198,958 MeSH terms which we over-lapped with the previously mined results of Wiki Dumps.

## 4   Data Synthesis

Hindi and Urdu are closely-related languages that share grammatical structure and largely overlap in vocabulary. This provides strong motivation to transform an Urdu-English parallel data into Hindi-English by translating the Urdu part into Hindi. We made use of the Urdu-English segment of the Indic multi-parallel corpus [36], which contains about 87,000 sentence pairs. The Hindi-English seg-ment of this corpus is a subset of the parallel data that was made available for the WMT'14 translation task, but its English side is completely disjoint from the English side of the Urdu-English segment.

Initially, we trained an Urdu-to-Hindi SMT system using the tiny EMILLE[14] corpus [1]. However, we found this system to be useless for translating the Urdu part of the Indic data due to domain mismatch and the high proportion of Out-of-Vocabulary (OOV) words (approximately 310,000 tokens). Thus, in order to reduce data sparseness, we synthesized additional phrase tables using interpola-tion and transliteration.

### 4.1   Interpolation

We built two phrase translation tables $p(\bar{u}_i|\bar{e}_i)$ and $p(\bar{e}_i|\bar{h}_i)$, from Urdu-English (Indic corpus) and Hindi-English (HindEnCorp [2]) bitexts. Given the phrase table for Urdu-English $p(\bar{u}_i|\bar{e}_i)$ and the phrase table for English-Hindi $p(\bar{e}_i|\bar{h}_i)$, we induced an Urdu-Hindi phrase table $p(\bar{u}_i|\bar{h}_i)$ using the model [39, 43]:

$$p(\bar{u}_i|\bar{h}_i) = \sum_{\bar{e}_i} p(\bar{u}_i|\bar{e}_i)p(\bar{e}_i|\bar{h}_i) \tag{1}$$

The number of entries in the baseline Urdu-to-Hindi phrase table were approx-imately 254,000. Using interpolation, we were able to build a phrase table con-taining roughly 10M phrases. This reduced the number of OOV tokens from 310K to approximately 50,000.

### 4.2   Transliteration

As Urdu and Hindi are written in different scripts (Arabic and Devanagri, respec-tively), we added a transliteration component to our Urdu-to-Hindi system. While it can be used to translate all 50,000 OOV words, previous research has shown that transliteration is useful for more than just translating OOV words when translating closely related language pairs [29,31,38,41,42]. Following [9], we transliterate all Urdu words to Hindi and hypothesize n-best transliterations,

---

[14] EMILLE contains about 12,000 sentences of comparable data in Hindi and Urdu. We were able to align about 7,000 sentences to build an Urdu-to-Hindi system.

along with regular translations. The idea is to generate novel Hindi translations that may be absent from the regular and interpolated phrase table, but for which there is evidence in the language model. Moreover, the overlapping evidence in the translation and transliteration phrase tables improves the overall system.

We learn an unsupervised transliteration model [10] from the word-alignments of Urdu-Hindi parallel data. We were able to extract around 2,800 transliteration pairs. To learn a richer transliteration model, we additionally fed the interpolated phrase table, as described above, to the transliteration miner. We were able to mine about 21,000 additional transliteration pairs and to build an Urdu-Hindi character-based model from it. In order to fully capitalize on the large overlap in Hindi–Urdu vocabulary, we transliterated each word in the Urdu test data to Hindi and we produced a phrase table with 100-best transliterations. We then used the two synthesized (triangulated and transliterated) phrase tables along with the baseline Urdu-to-Hindi phrase table in a log-linear model.

Table 2 shows development results from training an Urdu-to-Hindi SMT system. By adding interpolated phrase tables and transliteration, we obtain a very sizable gain of +3.35 over the baseline Urdu→Hindi system. Using our best Urdu-to-Hindi system ($\mathbf{B_{u,h}T_gT_r}$), we translated the Urdu part of the multi-Indic corpus to form a Hindi-English bi-text. This yielded a synthesized bi-text of ≈87,000 Hindi-English sentence pairs. Detailed analysis can be found in [7].

**Table 2.** Evaluating triangulated and transliterated phrase tables for Urdu-to-Hindi SMT. Notation: $\mathbf{B_{u,h}}$ = Baseline Phrase Table, $\mathbf{T_g}$ = Triangulated Phrase Table, and $\mathbf{T_r}$ = Transliteration Phrase Table.

| System | PT | Tune | Test | System | Tune | Test | System | Tune | Test |
|--------|-----|-------|-------|---------|-------|-------|---------|------|------|
| $\mathbf{B_{u,h}}$ | 254K | 34.18 | 34.79 | $\mathbf{B_{u,h}T_g}$ | 37.65 | 37.58 | | | |
| $\mathbf{T_g}$ | 10M | 15.60 | 15.34 | $\mathbf{B_{u,h}T_r}$ | 34.77 | 35.76 | $\mathbf{B_{u,h}T_gT_r}$ | 38.0 | 37.99 |
| $\mathbf{T_r}$ | | 9.54 | 9.93 | $\mathbf{T_gT_r}$ | 17.63 | 18.11 | | $\Delta+3.89$ | $\Delta+3.35$ |

## 5    Experiments and Evaluation

### 5.1    Machine Translation

#### 5.1.1    Baseline Data

We trained the Hindi-English systems using Hindi-English parallel data [2] composed by compiling several sources including the Hindi-English segment of the Indic parallel corpus. It contains 287,202 parallel sentences, and 4,296,007 Hindi and 4,026,860 English tokens. We used 635 sentences (6,111 tokens) for tuning and 636 sentences (5,231) for testing, collected from doctor-patient communication dialogues in YouTube videos. The sentences were translated into Hindi by a human translator. We trained interpolated language models using all the English and Hindi monolingual data made available for the WMT'14 translation task: 287.3M English and 43.4M Hindi tokens.

### 5.1.2   Baseline System

We trained a phrase-based system using Moses [22] with the following settings: a maximum sentence length of 80, GDFA symmetrization of GIZA++ alignments [33], an interpolated Kneser-Ney smoothed 5-gram language model with KenLM [19] used at runtime, 100-best translation options, MBR decoding [23], Cube Pruning [21] using a stack size of 1,000 during tuning and 5,000 during testing. We tuned with the $k$-best batch MIRA [4]. We additionally used msd-bidirectional-fe lexicalized reordering, a 5-gram OSM [11], class-based models [8][15] sparse lexical and domain features [18], a distortion limit of 6, and the no-reordering-over-punctuation heuristic. We used an unsupervised transliteration model [10] to transliterate the OOV words. These are state-of-the-art settings, as used in [6].

**Table 3.** Results using synthesized (Syn) Hindi-English parallel data. Notation used: $\mathbf{B_0}$ = System without synthesized data, $+\mathbf{PT}$ = System using synthesized data as an additional phrase table.

| Pair | $\mathbf{B_0}$ | $+\mathbf{Syn}$ | $\Delta$ | $+\mathbf{PT}$ | $\Delta$ |
|---|---|---|---|---|---|
| hi-en | 21.28 | 22.67 | +1.39 | 22.1 | +0.82 |
| en-hi | 22.52 | 23.97 | +1.45 | 23.28 | +0.76 |

### 5.1.3   Results

Table 3 shows the results when adding the synthesized Hindi-English bi-text on top of the baseline system (+Syn). The synthesized data was simply concatenated to the baseline data to train the system. We also tried building phrase tables (+PT) separately from the baseline data and from the synthesized one and used as a separate features in the log-linear model as done in [28,29,41,42]. We found that concatenating synthetic data with the baseline data directly was superior to training a separate phrase-table from it. We obtained improvements of up to +1.45 by adding synthetic data.

Table 4 shows the results from adding the mined dictionaries to the baseline system. The baseline system ($\mathbf{B_0}$) used in this case is the best system in Table 3. Again, we simply concatenated the dictionaries with the baseline data and we gained improvements of up to +1.66 BLEU points absolute. Cumulatively, by using dictionaries and synthesized phrase-tables, we were able to obtain statistically significant improvements of more than 3 BLEU points.

### 5.2   Manual Evaluation

In addition to the above evaluation, we ran a small manual evaluation experiment, using the Appraise platform [16]. The two sections below describe the results of the Hindi-to-English (Sect. 5.2.1) and the English-to-Hindi (Sect. 5.2.2) evaluations.

---

[15] We used mkcls to cluster the data into 50 clusters.

**Table 4.** Evaluating the effect of dictionaries. $B_0$ = System without dictionaries.

| Pair | $B_0$ | +**Dict** | $\Delta$ |
|---|---|---|---|
| hi-en | 22.67 | 23.29 | +0.62 |
| en-hi | 23.97 | 25.63 | +1.66 |

### 5.2.1  Hindi-to-English

The evaluation was conducted by 3 monolingual English speakers, using 321 randomly selected sentences, divided into three batches of evaluation. Similar to the setup at evaluation campaigns such as WMT, the evaluators were shown the translations and references.

The evaluators were asked to assign one of the following three categories to each translation: (a) *helpful in this situation*, (b) *misleading*, and (c) *doubtful that people will understand it*. As shown in Table 5, over 37% of the cases were classified as *helpful* (good translations), 39% as doubtful (mediocre), and 24% as *misleading* (really bad translations). Annotators did not always agree, e.g., Judge 1 and Judge 2 were more lenient than Judge 3.

**Table 5.** Manual sentence evaluation for Hindi-to-English translation.

| Response | Judge 1 | Judge 2 | Judge 3 | Total | Percentage |
|---|---|---|---|---|---|
| Helpful | 129 | 129 | 96 | 354 | 37% |
| Doubtful | 114 | 145 | 118 | 377 | 39% |
| Misleading | 78 | 47 | 107 | 232 | 24% |

### 5.2.2  English-to-Hindi

In order to check the output of the English-to-Hindi system, we asked a bilingual judge to evaluate 328 sentences. She was asked to classify the sentences in the same categories as for the Hindi-English evaluation. Table 6 shows the results; we can see that 55.8% of the sentences were found *helpful in this situation*. This is hardly because English-Hindi system was any better, but more likely because the human evaluator was lenient. Unfortunately, we could not find a second Hindi speaker to evaluate our translations, and thus we could not calculate inter-annotator agreement.

**Table 6.** Manual sentence evaluation for English-to-Hindi translation.

| Response | Number | Percentage |
|---|---|---|
| Helpful | 183 | 55.8% |
| Doubtful | 111 | 33.8% |
| Misleading | 34 | 10.4% |

### 5.2.3  Analysis

In order to understand the problems with the Hindi output, we conducted an error analysis on 100 sentences classifying the errors into the following categories:

- missing/untranslated words;
- wrongly translated words;
- word order problems;
- other error types, e.g., extra words.

Table 7 shows the results. We can see that most of the problems are associated with *word order problems* (84%) or *wrongly translated words* (74%).

**Table 7.** Error analysis results for our English-to-Hindi translation.

| Category | Percentage |
|---|---|
| Missing/Untranslated words | 45% |
| Wrongly translated words | 74% |
| Word order problems | 84% |
| Other types of errors | 13% |

## 6  Conclusions

We presented our preliminary efforts towards building a Hindi↔English SMT system for facilitating doctor-patient communication. We improved our baseline system using two approaches, namely (*i*) additional data collection, and (*ii*) automatic data synthesis. We mined useful dictionaries from Wikipedia in order to improve the coverage of our system. We made use of the relatedness between Hindi and Urdu to generate synthetic Hindi-English bi-texts by automatically translating 87,000 Urdu sentences into Hindi. Both our data collection and our synthesis approach worked well and have shown significant improvements over the baseline system, yielding a total improvement of +3.11 BLEU points absolute for English-to-Hindi and +2.07 for Hindi-to-English. We also carried out human evaluation for the best system. In the error analysis of the Hindi outputs, we found that most errors were due to ordering of the words in the output, or to wrong lexical choice.

In future work, we plan to collect more data for Hindi, but also to synthesize Urdu data. We further plan to develop a system for Nepali-English. Finally, we would like to add Automatic Speech Recognition (ASR) and Speech Synthesis components in order to build a fully-functional speech-to-speech system, which we would test and gradually deploy for use in real-world scenarios.

**Acknowledgments.** The authors would like to thank Naila Khalisha and Manisha Bansal for their contributions towards the project.

# References

1. Baker, P., Hardie, A., McEnery, T., Cunningham, H., Gaizauskas, R.J.: EMILLE, a 67-million word corpus of indic languages: data collection, mark-up and harmonisation. In: Proceedings of the Third International Language Resources and Evaluation Conference, LREC 2002, Las Palmas, Canary Islands, Spain (2002)
2. Bojar, O., Diatka, V., Rychlý, P., Straňák, P., Tamchyna, A., Zeman, D.: Hindi-English and Hindi-only corpus for machine translation. In: Proceedings of the Ninth International Language Resources and Evaluation Conference, LREC 2014, Reykjavik, Iceland, pp. 3550–3555 (2014)
3. Bouillon, P., Flores, G., Georgescul, M., Halimi Mallem, I.S., Hockey, B.A., Isahara, H., Kanzaki, K., Nakao, Y., Rayner, E., Santaholma, M.E., Starlander, M., Tsourakis, N.: Many-to-many multilingual medical speech translation on a PDA. In: Proceedings of the Eighth Conference of the Association for Machine Translation in the Americas, AMTA 2008, Waikiki, Hawaii, USA, pp. 314–323 (2008)
4. Cherry, C., Foster, G.: Batch tuning strategies for statistical machine translation. In: Proceedings of the 2012 Conference of the North American Chapter of the Association for Computational Linguistics: Human Language Technologies, HLT-NAACL 2012, Montréal, Canada, pp. 427–436 (2012)
5. Dillinger, M., Seligman, M.: Converser: highly interactive speech-to-speech translation for healthcare. In: Proceedings of the COLING-ACL 2006 Workshop on Medical Speech Translation, Sydney, Australia, pp. 36–39 (2006)
6. Durrani, N., Haddow, B., Koehn, P., Heafield, K.: Edinburgh's phrase-based machine translation systems for WMT-14. In: Proceedings of the ACL 2014 Ninth Workshop on Statistical Machine Translation, WMT 2014, Baltimore, Maryland, USA, pp. 97–104 (2014)
7. Durrani, N., Koehn, P.: Improving machine translation via triangulation and transliteration. In: Proceedings of the 17th Annual Conference of the European Association for Machine Translation, EAMT 2014, Dubrovnik, Croatia, pp. 71–78 (2014)
8. Durrani, N., Koehn, P., Schmid, H., Fraser, A.: Investigating the usefulness of generalized word representations in SMT. In: Proceedings of the 25th Annual Conference on Computational Linguistics, COLING 2014, Dublin, Ireland, pp. 421–432 (2014)
9. Durrani, N., Sajjad, H., Fraser, A., Schmid, H.: Hindi-to-Urdu machine translation through transliteration. In: Proceedings of the 48th Annual Meeting of the Association for Computational Linguistics, ACL 2014, Uppsala, Sweden, pp. 465–474 (2010)
10. Durrani, N., Sajjad, H., Hoang, H., Koehn, P.: Integrating an unsupervised transliteration model into statistical machine translation. In: Proceedings of the 15th Conference of the European Chapter of the ACL, EACL 2014, Gothenburg, Sweden, pp. 148–153 (2014)
11. Durrani, N., Schmid, H., Fraser, A., Koehn, P., Schütze, H.: The operation sequence model - combining N-gram-based and phrase-based statistical machine translation. Comput. Linguist. 41(2), 157–186 (2015)
12. Dušek, O., Hajic, J., Hlaváčová, J., Novák, M., Pecina, P., Rosa, R., Tamchyna, A., Urešová, Z., Zeman, D.: Machine translation of medical texts in the khresmoi project. In: Proceedings of the 52nd Annual Meeting of the Association of Computational Linguistics, ACL 2014, Baltimore, Maryland, USA, pp. 221–228 (2014)

13. Eck, M., Lane, I., Zhang, Y., Waibel, A.: Jibbigo: speech-to-speech translation on mobile devices. In: Proceedings of IEEE Spoken Language Technology Workshop, SLT 2010, Berkeley, California, USA, pp. 165–166 (2010)
14. Ehsani, F., Kimzey, J., Master, D., Sudre, K., Park, H.: Speech to speech translation for medical triage in Korean. In: Proceedings of the COLING-ACL 2006 Workshop on Medical Speech Translation, New York City, New York, USA, pp. 13–19 (2006)
15. Elnashar, M., Abdelrahim, H., Fetters, M.D.: Cultural competence springs up in the desert: the story of the center for cultural competence in health care at Weill Cornell Medical College in Qatar. Acad. Med. **87**(6), 759–766 (2012)
16. Federmann, C.: Appraise: an open-source toolkit for manual evaluation of MT output. Prague Bull. Math. Linguist. **98**, 25–35 (2012)
17. Gao, Y., Gu, L., Zhou, B., Sarikaya, R., Afify, M., Kuo, H.-K., Zhu, W.-Z., Deng, Y., Prosser, C., Zhang, W., et al.: IBM MASTOR system: multilingual automatic speech-to-speech translator. In: Proceedings of the COLING-ACL 2006 Workshop on Medical Speech Translation, Sydney, Australia, pp. 53–56 (2006)
18. Hasler, E., Haddow, B., Koehn, P.: Sparse lexicalised features and topic adaptation for SMT. In: Proceedings of the Seventh International Workshop on Spoken Language Translation, IWSLT 2012, Hong Kong, China, pp. 268–275 (2012)
19. Heafield, K.: KenLM: faster and smaller language model queries. In: Proceedings of the Sixth Workshop on Statistical Machine Translation, WMT 2011, Edinburgh, Scotland, United Kingdom, pp. 187–197 (2011)
20. Heinze, D.T., Turchin, A., Jagannathan, V.: Automated interpretation of clinical encounters with cultural cues and electronic health record generation. In: Proceedings of the COLING-ACL 2006 Workshop on Medical Speech Translation, Sydney, Australia, pp. 20–27 (2006)
21. Huang, L., Chiang, D.: Forest rescoring: faster decoding with integrated language models. In: Proceedings of the 45th Annual Meeting of the Association of Computational Linguistics, ACL 2007, Prague, Czech Republic, pp. 144–151 (2007)
22. Koehn, P., Hoang, H., Birch, A., Callison-Burch, C., Federico, M., Bertoldi, N., Cowan, B., Shen, W., Moran, C., Zens, R., Dyer, C., Bojar, O., Constantin, A., Herbst, E.: Moses: open source toolkit for statistical machine translation. In: Proceedings of the 45th Annual Meeting of the Association of Computational Linguistics, ACL 2007, Prague, Czech Republic, pp. 177–180 (2007)
23. Kumar, S., Byrne, W.J.: Minimum Bayes-risk decoding for statistical machine translation. In: Proceedings of the North American Chapter of the Association for Computational Linguistics: Human Language Technologies, HLT-NAACL 2004, Boston, Massachusetts, USA, pp. 169–176 (2004)
24. Lewis, W.D., Munro, R., Vogel, S.: Crisis MT: developing a cookbook for MT in crisis situations. In: Proceedings of the EMNLP 2011 Sixth Workshop on Statistical Machine Translation, WMT 2011, Edinburgh, Scotland, United Kingdom, pp. 501–511 (2011)
25. Li, J., Kim, S.-J., Na, H., Lee, J.-H.: Postech's system description for medical text translation task. In: Proceedings of the Ninth Workshop on Statistical Machine Translation, WMT 2014, Baltimore, Maryland, USA, pp. 229–233 (2014)
26. Lu, Y., Wang, L., Wong, D.F., Chao, L.S., Wang, Y., Oliveira, F.: Domain adaptation for medical text translation using web resources. In: Proceedings of the Ninth Workshop on Statistical Machine Translation, WMT 2014, Baltimore, Maryland, USA, pp. 233–238 (2014)

27. Costa-Jussa, M.R., Farrus, M., Pons, J.S.: Machine translation in medicine. A quality analysis of statistical machine translation in the medical domain. In: Proceedings of the 1st Virtual International Conference on Advanced Research in Scientific Areas, ARSA 2012, pp. 1995–1998 (2012)
28. Nakov, P.: Improving English-Spanish statistical machine translation: experiments in domain adaptation, sentence paraphrasing, tokenization, and recasing. In: Proceedings of the Third Workshop on Statistical Machine Translation, WMT 2008, Columbus, Ohio, USA, pp. 147–150 (2008)
29. Nakov, P., Ng, H.T.: Improved statistical machine translation for resource-poor languages using related resource-rich languages. In: Proceedings of the 2009 Conference on Empirical Methods in Natural Language Processing: Volume 3, EMNLP 2009, Singapore, pp. 1358–1367 (2009)
30. Nakov, P., Ng, H.T.: Improving statistical machine translation for a resource-poor language using related resource-rich languages. J. Artif. Intell. Res. (JAIR) **44**, 179–222 (2012)
31. Nakov, P., Tiedemann, J.: Combining word-level and character-level models for machine translation between closely-related languages. In: Proceedings of the 50th Annual Meeting of the Association for Computational Linguistics: Short Papers - Volume 2, ACL 2012, Jeju Island, Korea, pp. 301–305 (2012)
32. Navigli, R., Ponzetto, S.P.: BabelNet: the automatic construction, evaluation and application of a wide-coverage multilingual semantic network. Artif. Intell. **193**, 217–250 (2012)
33. Och, F.J., Ney, H.: A systematic comparison of various statistical alignment models. In: Proceedings of the 41st Annual Meeting of the Association for Computational Linguistics, ACL 2003, Sapporo, Japan, pp. 19–51 (2003)
34. Okita, T., Vahid, A.H., Way, A., Liu, Q.: The DCU terminology translation system for the medical query subtask at WMT14. In: Proceedings of the ACL 2014 Workshop on Statistical Machine Translation, WMT 2014, Baltimore, Maryland, USA, pp. 239–245 (2014)
35. Pécheux, N., Gong, L., Do, Q.K., Marie, B., Ivanishcheva, Y., Allauzen, A., Lavergne, T., Niehues, J., Max, A., Yvon, F.: LIMSI@ WMT'14 medical translation task. In: Proceedings of the ACL 2014 Workshop on Statistical Machine Translation, WMT 2014, Baltimore, Maryland, USA, pp. 246–253 (2014)
36. Post, M., Callison-Burch, C., Osborne, M.: Constructing parallel corpora for six indian languages via crowdsourcing. In: Proceedings of the Seventh Workshop on Statistical Machine Translation, WMT 2012, Montréal, Canada, pp. 401–409 (2012)
37. Rodrigues, J.A.S.G.: Speech-to-speech translation to support medical interviews. Ph.D. thesis, Universidade de Lisboa, Portugal (2013)
38. Tiedemann, J., Nakov, P.: Analyzing the use of character-level translation with sparse and noisy datasets. In: Proceedings of the International Conference Recent Advances in Natural Language Processing, RANLP 2013, Hissar, Bulgaria, pp. 676–684 (2013)
39. Utiyama, M., Isahara, H.: A comparison of pivot methods for phrase-based statistical machine translation. In: Proceedings of the 2007 Meeting of the North American Chapter of the Association for Computational Linguistics, NAACL 2007, Rochester, New York, USA, pp. 484–491 (2007)
40. Wang, L., Lu, Y., Wong, D.F., Chao, L.S., Wang, Y., Oliveira, F.: Combining domain adaptation approaches for medical text translation. In: Proceedings of the Ninth Workshop on Statistical Machine Translation, WMT 2014, Baltimore, Maryland, USA, pp. 254–259 (2014)

41. Wang, P., Nakov, P., Ng, H.T.: Source language adaptation for resource-poor machine translation. In: Proceedings of the 2012 Joint Conference on Empirical Methods in Natural Language Processing and Computational Natural Language Learning, EMNLP-CoNLL 2012, Jeju Island, Korea, pp. 286–296 (2012)
42. Wang, P., Nakov, P., Ng, H.T.: Source language adaptation approaches for resource-poor machine translation. Comput. Linguist. **42**, 1–44 (2016)
43. Wu, H., Wang, H.: Pivot language approach for phrase-based statistical machine translation. In: Proceedings of the 45th Annual Meeting of the Association of Computational Linguistics, ACL 2007, Prague, Czech Republic, pp. 856–863 (2007)
44. Zhang, J., Wu, X., Calixto, I., Vahid, A.H., Zhang, X., Way, A., Liu, Q.: Experiments in medical translation shared task at WMT 2014. In: Proceedings of the Ninth Workshop on Statistical Machine Translation, WMT 2014, Baltimore, Maryland, USA, pp. 260–265 (2014)

# Combining Phrase and Neural-Based Machine Translation: What Worked and Did Not

Marta R. Costa-jussà[✉] and José A. R. Fonollosa

TALP Research Center, Universitat Politècnica de Catalunya, Barcelona, Spain
{marta.ruiz,jose.fonollosa}@upc.edu

**Abstract.** Phrase-based machine translation assumes that all words are at the same distance and translates them using feature functions that approximate the probability at different levels. On the other hand, neural machine translation infers a word embedding and translates these word vectors using a neural model. At the moment, both approaches co-exist and are being intensively investigated.

This paper to the best of our knowledge is the first work that both compares and combines these two systems by: using the phrase-based output to solve unknown words in the neural machine translation output; using the neural alignment in the phrase-based system; comparing how the popular strategy of pre-reordering affects both systems; and combining both translation outputs. Improvements are achieved in Catalan-to-Spanish and German-to-English.

**Keywords:** Machine translation · Phrase-based · Neural
Combination

## 1 Introduction

Statistical Machine Translation (MT) has been developed for more than 20 years. Although phrase-based systems [14] have mainly been data-based without using syntactic information, other approaches, called syntax-based, have reached better results for some language pairs (specially those requiring long reorderings) by introducing syntactic information [18]. In between the syntax and phrase-based systems, there are also the successful hierarchical-based systems [5].

Just recently, a new paradigm based on deep learning has appeared and promises to be competitive: neural MT [6,12,21]. The idea of neural MT can be found in previous works [4] and it uses an encoding-decoding architecture to go from source to target. One of the main advantages of neural MT in front of the existing statistical MT approaches is training the system as a unique system without splitting the task of alignment, modeling and learning features [13]. However, the approach is computationally expensive and as a consequence working vocabulary has to be limited.

A. Gelbukh (Ed.): CICLing 2016, LNCS 9624, pp. 17–26, 2018.
https://doi.org/10.1007/978-3-319-75487-1_2

In this paper, we present several ways to combine the phrase-based and neural systems to try to get the best of both paradigms. There have been previous work that introduce neural networks to smooth the phrase-based language model or integrate neural language models in the phrase-based system [19]. However, as far as we are concerned, there is no combination of the phrase and neural systems as we are proposing in this paper.

The rest of the paper is organised as follows. Section 2 presents the basics of phrase-based MT, and Sect. 3, the basics of neural MT. Section 4 does a high-level comparison of advantages and disadvantages of both systems. Then, Sect. 5 describes the two large datasets that have been used to experiment and details the parameters of the baseline systems. Section 6 reports the strategy to improve neural with phrase-based MT, which is by letting neural MT using the phrase-based output to translate the large quantity of unknowns. Section 7 details the negative results of using neural alignment instead of the standard phrase-based alignment. Section 8 shows how the popular strategy of pre-reordering impacts phrase-based and neural MT. Next Sect. 9 shows that the combination with Minimum Bayes Risk (MBR) strategy obtains the best results in translation terms. Finally, Sect. 10 concludes by commenting the main contributions of the paper and planned further work.

## 2   Phrase-Based Machine Translation

In short, SMT uses statistical algorithms to decide the most likely translation of a word. Thus, given a source string $s_1^J = s_1 \ldots s_j \ldots s_J$ to be translated into a target string $t_1^I = t_1 \ldots t_i \ldots t_I$, the aim is to choose, among all possible target strings, the string with the highest probability:

$$\tilde{t_1^I} = \underset{t_1^I}{argmax}\, P(t_1^I | s_1^J)$$

where $I$ and $J$ are the number of words of the target and source sentence, respectively. The basic idea of phrase-based translation is to segment the given source sentence into units (hereinafter called phrases), then translate each phrase and, finally, compose the target sentence from these phrase translations. For extraction from a bilingual word aligned training corpus, words are consecutive, and they are consistent with the word alignment matrix. Given the collected phrase pairs, the phrase translation probability distribution is commonly estimated by relative frequency in both directions.

One of the most popular phrase-based MT implementations combines the relative frequencies together with the following additional feature models: the target language model, the word and the phrase bonus and the source-to-target and target-to-source lexical models and the reordering model. The target language model attempts to reflect how likely a string occurs inside a language [20]. The lexical models allow to compute the probability of translation units using the probability of translating word per word of the unit. The reordering model is a measure of the plausibility of word movements and it has been widely investigated in statistical MT [3]. Finally, all these models are optimized in the decoder following the minimum error rate procedure [17]. In the last ten years, the phrase-based system and the features functions have evolved from different perspectives, e.g. by smoothing the translation model [19] or morphology techniques [22].

## 3   Neural Machine Translation

Neural MT uses a neural network approach to maximize the translation probability of the source sentence [2,6].

The neural MT used in this work, proposed in [2], follows the encoder-decoder architecture, see Fig. 1 for an schematic representation. Both encoder and decoder are recurrent neural networks. First, the encoder reads the source sentence $s = (s_1, ..s_I)$ and encodes it into a sequence of hidden states $h = (h_1, ..h_I)$. Then, the decoder generates a corresponding translation $t = t_1, ..., t_J$ based on the decoder hidden states, context (attention vectors) and the last predicted words. Both encoder and decoder are jointly trained to maximize the conditional log-probability of the correct translation.

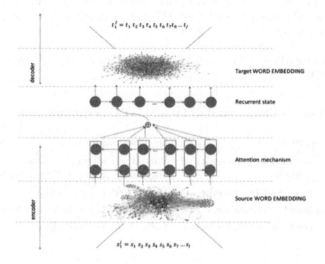

**Fig. 1.** Neural MT architecture scheme.

Bidirectional encoding was addressed in [1]. The main improvement of [2] is because of the *attention mechanism*. The decoder becomes also a gated recurrent unit and each word $t_j$ is predicted based on: decoder hidden states, context (attention vectors) and the last predicted words. The context (attention vectors) is obtained from the weighted sum of the annotations $h_k$, which in turn, is computed through an alignment model $alpha_{jk}$ (a feedforward neural network). This neural MT with attention has achieved competitive results to the standard phrase-based MT in the WMT 2015 evaluation [11].

## 4    Phrase-Based vs Neural Machine Translation

As seen in previous sections phrase-based and neural MT have big dissimalities in structure and similarities may be reduced to the type of data used to train the systems, which in both cases is a large amount of parallel text at the level of sentence. More specifically, structural dissimilarities between both MT approaches include:

- The different levels of difficulties at tracing errors: while the phrase-based system can trace back relatively easy to translation table entries, the neural-based system has not such translation units.
- Managing vocabulary: phrase-based MT can use most of the vocabulary from the training corpus (except for misaligned words), while neural MT have to limit the target vocabulary for computational reasons.
- Inferring word embeddings: neural MT infers word embeddings which no longer assumes that all words are at the same distance (as phrase-based MT does).
- Global vs partial optimization: neural MT optimizes the translation from source to target, it is a global optimization which includes alignment and translation, whereas phrase-based MT treats independently alignment and translation steps.

Table 1 summarises the advantages and disadvantages of both phrase-based and neural MT.

**Table 1.** Phrase-based and neural MT comparison summary.

| System | Advantages | Disadvantages |
|--------|------------|---------------|
| Phrase | Manages large quantity of vocabulary | Partial optimizations (word alignment and translation model) |
|        | Errors can be traced | Words are independent |
| Neural | Global optimization | Higher limitation in vocabulary |
|        | Word embedding infering | Difficult to trace errors |

## 5    Experimental Framework

As follows, we detail the data used for all experiments. We mention how data was preprocessed and how the baseline systems were built.

### 5.1    Data

We used two large corpus for two different translation tasks: Catalan-Spanish and German-English.

For Catalan to Spanish we used a subset of El Periodico [9]. Preprocessing, for this pair, was limited to tokenizing and truecasing.

For German to English we used the WMT 2015 data[1] including the EPPS, NEWS and Commoncrawl. Preprocessing, in this case, consisted of tokenizing, truecasing, normalizing punctuation and filtering sentences with more than 5% of words in other language than German or English.

Corpus statistics are shown in Table 2.

**Table 2.** Corpus details (in thousands) for Catalan-Spanish and German-English.

|         | Sentences | Words  | Vocabulary |
|---------|-----------|--------|------------|
| Catalan | 2,000     | 56,921 | 369        |
| Spanish | 2,000     | 52,373 | 383        |
| German  | 3,504     | 77,702 | 1602       |
| English | 3,504     | 81,296 | 754        |

## 5.2 Baseline Systems

The phrase-based system was built using MOSES [13], with standard parameters such as grow-final-diag for alignment, smoothing of good turing for the relative frequencies, 5-gram language modeling using kneser-ney discounting, lexicalized reordering, among others.

The neural system was built using the software from DL4MT[2] available in github. We generally used settings by previous work [11]: an embedding size of 620, the number of hidden units in the alignment model is 1024, a batch size of 32 sentences, no dropout, Adam algorithm for optimization. Note that we did not use language model since the rescoring from the DL4MT software did not work to us. We used a vocabulary size of 40,000 words in Catalan-to-Spanish and 90,000 words in German-to-English.

BLEU results using these baseline systems are shown in Table 3. The translation quality in terms of BLEU is higher when using the phrase-based system. One of the main limitations in the neural MT is the size of vocabulary.

**Table 3.** Baseline systems. BLEU results.

| System | CAES  | DEEN  |
|--------|-------|-------|
| Phrase | 80.78 | 20.99 |
| Neural | 71.97 | 18.83 |

---

[1] http://www.statmt.org/wmt15/translation-task.html.
[2] http://dl4mt.computing.dcu.ie/.

# 6  Neural MT Enhanced with the Phrase-Based System

Neural MT is still a young paradigm compared to phrase-based MT. As mentioned in previous sections, one of the main advantages of phrase-based MT is its capacity to use a larger vocabulary compared to the strict limitation still existing in neural MT. There have been several approaches that deal with unknown words in neural MT either by means of a dictionary or using the fast-align [16].

Our proposal differs from previous ones in the sense that we are using the phrase-based output as a dictionary for neural MT. For this, we use the output of both systems and the corresponding alignments. Alignments are used to identify the source word from which the unknown word of the neural output comes from, so that it can be translated by the translation of that same source word from the phrase-based system.

Table 4 shows an example of how unknown words coming from the neural output are solved using the phrase-based output. The unknown source word (word 4 in the *source* input) by the neural output (word 3 in the *neural* output) is substituted with the corresponding phrase-based translation (word 3 in the *phrases* output) of that same source word. So, in these examples, the *UNK* word is substituted by *remolino*.

**Table 4.** Example of how to address UNKs with the phrase-based output

| | |
|---|---|
| Source | va començar com un **remolí** |
| Neural | comenzó como un UNK |
| Phrases | comenzó como un remolino |
| Align-neural | 0-0 1-0 2-1 3-2 **4-3** |
| Align-phrases | 0-0 1-0 2-1 3-2 **4-3** |
| Neural + Phrase-output | comenzó como un **remolino** |
| Reference | comenzó como un torbellino |

Table 5 shows BLEU improvements when using this technique: more than 3.5 points for Catalan-to-Spanish and almost 1.5 points for German-to-English.

Improvements are higher in Catalan-to-Spanish than in German-to-English, probably because the former is much more monotonic than the latter, and the alignments seem more reliable for the former than the latter.

**Table 5.** Addressing UNKs in neural MT. BLEU results.

| Method | CAES | DEEN |
|---|---|---|
| Neural + Source word | 73.33 | 19.25 |
| Neural + Phrase-output | **77.00** | **20.62** |

# 7  Phrase-Based MT Combined with the Neural System

This section uses the neural alignment [2] in the phrase-based system.

The neural alignment output is used in substitution to the standard GIZA++ alignment and to extract the translation model in the phrase-based system. The neural soft-alignment is generated by visualizing the annotations weights and looking at the more important positions in the source sentences when generating the target word. However, the phrase-based system manages only hard alignments. In order to obtain hard-alignments, we establish a threshold of probabilities.

Table 6 shows, for different neural alignment threshold probabilities, the result in BLEU of the phrase-based system. Threshold probabilities may differ between tasks due to the variability in the monotonocity of the tasks. As commented before, Catalan-Spanish is a task much more monotonic than German-English, and alignment becomes easier in the former than in the later. Although we can not provide the information about Alignment Error Rate (AER) due to the lack of an available corpus for this, we can expect that AER is much higher in the Catalan-Spanish case than in the German-English.

Note that compared to the baseline phrase-based system shown in Table 3, using neural alignment in the phrase-based system produces worse translations in terms of BLEU. Results do not even reach the baseline phrase-based system when combining the neural and GIZA++ alignment. The main problem of our experiments may be that we are using hard alignments which are expected not to be helpful to MT [2]. Unfortunately, advantages of the soft-alignments which lets the model look at several words at a time are not used in our experiment.

**Table 6.** Neural alignment threshold probability (TP) and BLEU results in the phrase-based system.

| TP | CAES | TP | DEEN |
|---|---|---|---|
| 0.87 | 57.28 | 0.01 | 3.81 |
| 0.88 | 67.93 | 0.05 | 8.23 |
| 0.89 | 67.94 | 0.1 | 9.69 |
| 0.90 | 67.75 | 0.3 | 8.82 |
| 0.91 | 58.06 | 0.5 | 7.27 |
| 0.89 + GIZA++ | *79.66* | 0.1 + GIZA++ | *20.56* |

# 8  Effect of Pre-reordering in Neural Machine Translation vs Phrase-Based

A common strategy in a phrase-based system is to pre-reorder the source sentence so that it matches better the target order [8]. In this case, we are using a deterministic pre-reordering based on hand rules [7] for the German to English task. There is no point in doing a pre-reordering in the task of Catalan to Spanish since the order of the languages is very similar.

Table 7 shows improvements of the pre-reordering in the phrase-based system. As expected, the phrase-based system shows a clear improvement when using pre-reordering. There is no improvement in the neural system using pre-reordering. This is a surprising experimental result, because, as comented in [2] and differently from [10], the modes of the weights of the annotations only move in one direction, which is a severe limitation in MT, as (long-distance) reordering is often needed. That is why, we would have expected pre-reordering to help. Maybe negative results only apply to our task at hand, so we could investigate other tasks in the future.

**Table 7.** BLEU results with pre-reordering

| System | DE2EN |
|---|---|
| Phrase | 20.99 |
| Phrase + Pre-reordering | **21.55** |
| Neural | 18.83 |
| Neural + Pre-reordering | 18.84 |

## 9   Combination of Neural and Phrase-Based Outputs

The last experiments in this paper combine the neural and phrase-based MT n-best outputs using the standard MBR procedure [15]. Table 8 shows slight improvements over both baseline systems. N-best output combinations (with $N$ from 2 to 10) have been tried without modifying the results.

**Table 8.** BLEU results with MBR combination with best single systems.

| System | CAES | DEEN |
|---|---|---|
| Phrase | 80.78 | (+pre-reordering) 21.55 |
| Neural + Phrase-output | 77.00 | 20.62 |
| MBR | **80.92** | **21.74** |

## 10   Conclusions and Further Work

This paper has presented different direct combinations of phrase-based and neural MT. To the best of our knowledge, this is the first work on both comparing and combining these two approaches.

We have explored how to use phrase-based MT together with neural MT in both directions: taking alternatively phrase-based MT and neural MT as guiding approach.

At the moment, phrase-based MT presents slightly better results than neural MT mainly due to the difficulty of neural MT to deal with large vocabularies.

Best combination improvements have been obtained when using the phrase-based output to alleviate the UNKs problem in neural MT and when combining both phrase-based and neural outputs.

We have shown that the standard pre-reordering technique does not benefit translation in neural MT (and in the German-to-English task).

Given that neural MT alignment has not been shown useful when using a threshold, as further work, we would like to use neural soft-alignment with their corresponding probability.

**Acknowledgements.** This work has been supported by Spanish Ministerio de Economía y Competitividad, contract TEC2015-69266-P and the Seventh Framework Program of the European Commission through the International Outgoing Fellowship Marie Curie Action (IMTraP-2011-29951).

# References

1. Sundermeyer, M., Alkhouli, T., Wuebker, J., Ney, H.: Translation modeling with bidirectional recurrent neural networks. In: Proceedings of the Conference on Empirical Methods in Natural Language Processing, Doha (2014)
2. Bahdanau, D., Cho, K., Bengio, Y.: Neural machine translation by jointly learning to align and translate. CoRR abs/1409.0473 (2015)
3. Bisazza, A., Federico, M.: A survey of word reordering in statistical machine translation: computational models and language phenomena. arXiv preprint arXiv:1502.04938 (2015)
4. Castaño, M.A., Casacuberta, F.: A connectionist approach to MT. In: Proceedings of the EUROSPEECH Conference (1997)
5. Chiang, D.: Hierarchical phrase-based translation. Comput. Linguist. **33**(2), 201–228 (2007)
6. Cho, K., van Merrienboer, B., Bahdanau, D., Bengio, Y.: On the properties of neural machine translation: encoder-decoder approaches. In: Proceedings of the 8th Workshop on Syntax, Semantics and Structure in Statistical Translation, Doha (2014)
7. Collins, M., Koehn, P., Kucerova, I.: Clause restructuring for statistical machine translation. In: Annual Conference of the Association for Computational Lingusitics (ACL 2005), Michigan (2005)
8. Costa-Jussà, M.R., Fonollosa, J.A.R.: Statistical machine reordering. In: Proceedings of the 2006 Conference on Empirical Methods in Natural Language Processing, Sydney, pp. 70–76 (2006)
9. Costa-Jussà, M., Poch, M., Fonollosa, J., Farrús, M., Marinno, J.: A large Spanish-Catalan parallel corpus release for machine translation. Comput. Inf. **33**(4), 907–920 (2014)
10. Graves, A.: Generating sequences with recurrent neural networks. arXiv preprint arXiv:1308.0850 (2013)
11. Jean, S., Firat, O., Cho, K., Memisevic, R., Bengio, Y.: Montreal neural machine translation systems for WMT15. In: Proceedings of the 10th Workshop on Statistical Machine Translation, Lisbon (2015)

12. Kalchbrenner, N., Blunsom, P.: Recurrent continuous translation models. In: Proceedings of the Conference on Empirical Methods in Natural Language Processing, Seattle (2013)
13. Koehn, P., Hoang, H., Birch, A., Callison-Burch, C., Federico, M., Bertoldi, N., Cowan, B., Shen, W., Moran, C., Zens, R., Dyer, C., Bojar, O., Constantin, A., Herbst, E.: Moses: open source toolkit for statistical machine translation. In: Proceedings of the 45th Annual Meeting of the Association for Computational Linguistics, pp. 177–180 (2007)
14. Koehn, P., Och, F., Marcu, D.: Statistical phrase-based translation. In: Proceedings of the 41th Annual Meeting of the Association for Computational Linguistics (2003)
15. Kumar, S., Byrne, W.J.: Minimum Bayes-risk decoding for statistical machine translation. In: HLT-NAACL, pp. 169–176 (2004)
16. Luong, T., Sutskever, I., Le, Q.V., Vinyals, O., Zaremba, W.: Addressing the rare word problem in neural machine translation. In: Proceedings of the 53rd Annual Meeting of the Association for Computational Linguistics and the 7th International Joint Conference on Natural Language Processing of the Asian Federation of Natural Language Processing, Beijing, pp. 11–19 (2015)
17. Och, F.: Minimum error rate training in statistical machine translation. In: Proceedings of the 41th Annual Meeting of the Association for Computational Linguistics, pp. 160–167 (2003)
18. Quirk, C., Menezes, A., Cherry, C.: Dependency treelet translation: syntactically informed phrasal SMT. In: Proceedings of the 43rd Annual Meeting on Association for Computational Linguistics, pp. 271–279 (2005)
19. Schwenk, H., Costa-Jussà, M.R., Fonollosa, J.A.R.: Smooth bilingual n-gram translation. In: Proceedings of the Joint Conference on Empirical Methods in Natural Language Processing and Computational Natural Language Learning, Prague, pp. 430–438 (2007)
20. Stolcke, A.: SRILM - an extensible language modeling toolkit. In: 7th International Conference on Spoken Language Processing, Denver (2002)
21. Sutskever, I., Vinyals, O., Le, Q.V.: Sequence to sequence learning with neural networks. In: Ghahramani, Z., Welling, M., Cortes, C., Lawrence, N.D., Weinberger, K.Q. (eds.) Advances in Neural Information Processing Systems 27, pp. 3104–3112. Curran Associates, Inc. (2014)
22. Toutanova, K., Suzuki, H., Ruopp, A.: Applying morphology generation models to machine translation. In: Proceedings of the Joint Conference of the Association for Computational Linguistics and Human Language Technology, Columbus, pp. 514–522 (2008)

# Combining Machine Translated Sentence Chunks from Multiple MT Systems

Matīss Rikters[1](✉) 🆔 and Inguna Skadiņa[2]

[1] University of Latvia, 19 Raina Blvd., Riga, Latvia
matiss@lielakeda.lv
[2] Institute of Mathematics and Computer Science, University of Latvia, 29 Raina Blvd., Riga, Latvia
inguna.skadina@lumii.lv

**Abstract.** This paper presents a hybrid machine translation (HMT) system that pursues syntactic analysis to acquire phrases of source sentences, translates the phrases using multiple online machine translation (MT) system application program interfaces (APIs) and generates output by combining translated chunks to obtain the best possible translation. The aim of this study is to improve translation quality of English – Latvian texts over each of the individual MT APIs. The selection of the best translation hypothesis is done by calculating the perplexity for each hypothesis using an n-gram language model. The result is a phrase-based multi-system machine translation system that allows to improve MT output compared to individual online MT systems. The proposed approach show improvement up to +1.48 points in BLEU and −0.015 in TER scores compared to the baselines and related research.

**Keywords:** Machine translation · Multi-system machine translation
Hybrid machine translation · Syntactic parsing · Chunking
Natural language processing

## 1 Introduction

Although machine translation (MT) has been researched for many decades and there are many on-line MT systems available, the output of MT systems in many cases still has low quality. This problem is especially relevant to the less resourced languages with rich morphology and rather free word order (e.g. Latvian, Croatian or Estonian) where linguistic and human resources are limited. As these are rather small languages, they are not the main interest of the global MT developers. Better solutions are provided by local companies for particular domain, e.g. *hugo.lv*[1] as Latvian e-government solution provides good translation of legal texts [1]. Although such translation systems usually translate better than global multilingual MT systems [2], translation quality for general domain texts is still insufficient.

The problem of translation quality into under-resourced languages has been also recognized by EU H2020 programme and addressed in QT21 project. The QT21

---

[1] Latvian public administration machine translation service - http://hugo.lv.

© Springer International Publishing AG, part of Springer Nature 2018
A. Gelbukh (Ed.): CICLing 2016, LNCS 9624, pp. 27–37, 2018.
https://doi.org/10.1007/978-3-319-75487-1_3

project investigates novel methods, e.g. hybrid MT, neural network MT, etc. to improve MT output for morphologically rich under-resourced languages.

One promising approach is multi-system machine translation - a subset of hybrid MT where output of multiple MT systems is combined into a single translation in order to boost the accuracy and fluency of the translations [3]. Our hypothesis is that quality of MT output for under-resourced languages can be increased by applying multi-system machine translation that combines outputs of MT systems developed by global players, who have access to large linguistic data, with MT systems developed by MT developers, who pay more attention to particular language and domain.

This paper presents several methods how to enrich an MSMT system with linguistic knowledge. The experiments described use multiple combinations of outputs from two, three or four on-line MT systems. The automatic evaluation results obtained with this hybrid system are analysed and compared with each other. Our approach allowed to increase output by 1.48 BLEU points when translating general domain texts. It is a continuation of an experiment series that started as syntax-based multi-system machine translation [4].

## 2 Related Work

In the last decade the statistical machine translation has been the dominant research direction in machine translation. However, the quality of the output with state-of-art methods in many cases is insufficient. This has been a reason, why new techniques, including hybrid solutions become more and more popular.

According to Thurmair [5] three types of HMT systems are recognized – coupling, architecture extension and genuine hybrid architectures. Our approach is related to parallel coupling or multi-system MT. Multi-system machine translation (MSMT) is a subset of HMT where multiple MT systems are combined in a single system in order to boost the accuracy and fluency of the translations. It is also referred as multi-engine MT [6], coupling MT [7] or just MT system combination [8].

Traditional MSMT [9] identifies the best translation from a list of possible candidates generated by different MT engines using n-gram approach. Improvement has been reported when translated from French (+1.6 BLEU), German (+1.95) BLEU or Hungarian (+1 BLEU) into English. However, application of similar approach for English-Latvian translation has resulted in +0.12 BLEU [10].

Another approach is to translate fragments or phrases and use confusion networks to generate translation [11, 12]. At first a skeleton sentence is selected, then for each position of the skeleton the best translation alternative is identified and composed to the overall output sentence. Using this approach Heafield et al. [11] reported +2 BLEU point improvement for Hungarian-English.

A similar approach to confusion network decoding is lattice-based MT combination. The main difference is that confusion networks have a limit of 1-to-1 mappings between words in candidate translations, while the lattice-based approach can express n-to-n mappings. Feng et al. [13] report improvements of 0.93–1.23 BLEU compared to the confusion network approach and 3.0–3.73 BLEU compared to the best single system for Chinese-English translations.

In 2014 the EU-BRIDGE project reported that they achieved significantly better translation performance with gains of up to +1.6 points in BLEU and −1.0 points in TER by combining up to nine different machine translation systems for translation between German and English [14]. Recently Freitag et al. [15] presented novel system combination approach that enhance traditional confusion network system combination approach with an additional model trained by a neural network. Experiments were performed with high quality input systems for Chinese-English and Arabic-English. The proposed approach yielded in translation improvement from up to +0.9 points in BLEU and −0.5 points in TER.

## 3   Architecture of Multi-system MT System

The work of the system can be divided into following steps: pre-processing of the source sentence, acquisition of a translations via online APIs, and generation of MT output, as it is shown in Fig. 1.

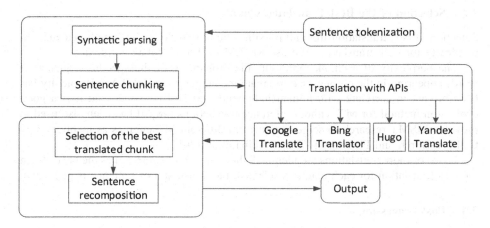

**Fig. 1.** Workflow of the translation process

While in our experiments two, three and four translation APIs were used, the system's architecture is flexible, allowing easily to integrate more translation APIs. The source and target language can also be changed to any language pair that is supported by the APIs and syntactic parser.

### 3.1   Pre-processing

The goal of the pre-processing step is to divide sentences into linguistically motivated chunks that will be further sent to the translation APIs. The Berkeley Parser [16] is used for parsing to obtain initial chunks. The parse tree of each sentence is then processed by the chunk extractor to obtain the parts of the sentence that will be individually translated.

It has to be stressed that when translation is performed into morphologically rich language, a simple chunk translation approach will not lead to a better translation. For example, when small chunks are translated into Latvian, they usually will be in canonical form that correspond to subject of sentence, but will be incorrect for object. On the other hand, if long chunks are translated, then translation usually breaks agreement rules or translation has wrong word order. Thus several approaches how to select best chunks for translation have been investigated.

### 3.2   Translation with the APIs

Four online translation APIs – *Google Translate*[2], *Bing Translator*[3], *Yandex Translate*[4] and *Hugo*[5] - were used in our experiments. These specific APIs were selected because of their public availability and descriptive documentation as well as the wide range of languages (including Latvian) that they support. While first two APIs are developed by global players, *Hugo* is developed in Latvia by Tilde, a company which has a long term experience with smaller languages.

### 3.3   Selection of the Best Translated Chunk

Selection of best translation from all possible chunk translations is done by calculating perplexity for each translation. We use KenLM [17] to calculate perplexity.

The perplexity of a chunk represents the probability of the specific sequence of words appearing in the corpus used to create the language model (LM). Perplexity has been proven to correlate with human judgments and BLEU scores, and it is a good evaluation method for MT without reference translations [18]. It has been also used in previous MSMT research to score output from different MT engines as mentioned by Callison-Burch and Flournoy [19] and Akiba et al. [20].

If two or more translations are identical, the translation is selected as the best. When the best translation for each chunk is selected, the translation of sentence is generated.

### 3.4   Post-processing

The post-processing step is necessary to correct some common mistakes of the translation engines, and remove duplicate punctuation marks that result by concatenating chunks into full sentences.

---

[2] Google Translate API - https://cloud.google.com/translate/.

[3] Bing Translator Control - http://www.bing.com/dev/en-us/translator.

[4] Yandex Translate API - https://tech.yandex.com/translate/.

[5] Latvian public administration machine translation service API - http://hugo.lv/TranslationAPI.

# 4   Experiments

## 4.1   Setup

Experiments were conducted on the English – Latvian language pair. Two legal domain corpora - JRC Acquis corpus v. 3.0 [21], consisting of 1.4 million sentences, and DGT-Translation Memory [22], consisting of 3.1 million sentences – were used for language modelling.

For evaluation two different test sets were used:

- 1581 randomly selected sentences from the JRC Acquis corpus;
- ACCURAT balanced evaluation corpus consisting of 512 sentences [23].

Translations were automatically evaluated with two scoring methods – BLEU [24] and TER [25] - and manually inspected in web-based MT evaluation platforms MT–ComparEval [26] and iBLEU [27] to determine, which system from the hybrid setups was selected to get the specific translation of chunk and inspect the differences in translations.

## 4.2   Baseline Systems

As the baseline, we used full translations from each individual on-line API and simple MSMT system [10] that uses only perplexity to select the best translation from outputs of the online APIs. BLEU and TER scores for the baseline systems are presented in Table 1. As it was expected, systems developed by global MT developers show better results for general domain translation, while Latvian public administration MT system is better for translation of legal texts. The baseline MSTM system (using a 6-gram JRC-Acquis LM) demonstrates lower results as individual systems in legal domain, while for general domain results are close to the best individual system.

**Table 1.** Automatic evaluation results for baseline systems

| System | JRC-Acquis | | Balanced | |
|---|---|---|---|---|
| | BLEU | TER | BLEU | TER |
| Bing | 16.99 | 0.695 | 17.43 | 0.765 |
| Google | 16.19 | **0.682** | **17.73** | **0.749** |
| Hugo | **20.27** | 0.708 | 17.14 | 0.764 |
| Yandex | 19.75 | 0.696 | 16.04 | 0.776 |
| MSTM - BG | 16.38 | 0.689 | 17.70 | 0.755 |
| MSTM - BGH | 17.89 | 0.694 | 17.63 | 0.756 |

## 4.3   Syntax Based MSMT Systems

We evaluated two approaches in chunk translation – translation of top level chunks and translation of smaller chunks that are selected based on their properties in sentence.

**Simple Chunks (SyMHyT)**

In first experiment, a parse tree of each sentence is processed by the chunk extractor to obtain the top-level sub-trees (noun phrases, verb phrases, prepositional phrases, etc.). The chunk extractor uses regular expressions to identify sub-trees. When sub-trees are identified, they are translated with on-line APIs. Finally, the translation of the sentence is generated by combination of translation hypothesis of sub-trees as it is described in Sect. 3.3.

We evaluated this approach for two SyMHyT systems: *Bing + Google* (BG) and *Bing + Google + Hugo* (BGH). Similarly to the baseline MSMT system, SyMHyT also used a 6-gram language model trained on JRC-Acquis corpus for selection of the best chunk. Evaluation results of this approach are summarized in Table 2. The SyMHyT approach allowed to increase translation quality for combination of *Bing* and *Google* APIs by +0.37 BLEU points to compare with the best baseline (*Bing* 16.99 BLEU points) on legal domain texts. When applied to general domain balanced corpus +0.22 BLEU points are obtained to compare with best baseline (*Google* 17.73 BLEU points).

**Table 2.** Automatic evaluation results for simple chunk baseline system

| System | JRC-Acquis | | Balanced | |
|---|---|---|---|---|
| | BLEU | TER | BLEU | TER |
| Bing + Google | 17.36 | 0.672 | **17.95** | 0.825 |
| Bing + Google + Hugo | 19.50 | 0.661 | 17.30 | 0.817 |

However, when three APIs were combined, the decrease of BLEU points is observed. To understand why combination of three systems did not improve translation, we analyzed translation selection process. Figure 2 shows proportion of translated chunks of different APIs selected by SyMHyT system. When translations of all three systems are used to generate MT output, most of fragments are selected from translations produced by *Hugo.lv*. Since for general domain translation *hugo.lv* showed the worse result, it influenced SyMHyT output and decrease of −0.43 BLEU is observed. In case of legal domain *hugo.lv* showed the best result (+3 BLEU to compare with other baselines), however, since only 63% of fragments were selected from this system, it was insufficient to beat the baseline (−0.77 BLEU).

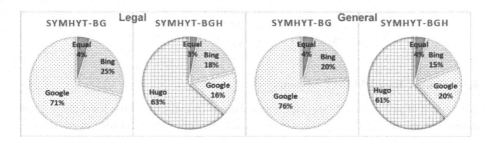

**Fig. 2.** Proportion of chunks selected from translations different APIs

## Linguistically motivated chunks (ChunkMT)

Although proposed SyMHyT approach demonstrated some improvement for general domain translation, the analysis of selected translated chunks revealed discrepancy between BLEU score evaluation results and preferences of selection module. In addition, we observed some obvious flaws, e.g. one-word chunks, one-symbol chunks or very long chunks. This motivated us to investigate more complex algorithm for chunk extraction.

The proposed chunk extractor reads output of the Berkeley Parser and places it in a tree data structure. During this process each node of the tree is initialised with its phrase (NP, VP, ADVP, etc.), word (if it has one) and a chunk consisting of the chunks from its child nodes. To obtain the final chunks for translation the resulting tree is traversed bottom-up post-order (left to right). A chunk is combined with the previous one, if it is (a) non-alphabetical, (b) only one symbol, or (c) contains genitive phrase. If a chunk is very long (length of chunk > sentence length/4 in the first chunking iteration), an attempt to break it into smaller chunks is made. Figure 3 illustrates chunk extraction result of both MSTM systems.

| SyMHyT | ChunkMT |
|---|---|
| Recently<br><br>there<br><br>has been an increased interest in the automated discovery of equivalent expressions in different languages<br><br>. | Recently there has been an increased interest<br><br>in the automated discovery of equivalent expressions<br><br>in different languages . |

**Fig. 3.** Examples of chunks extracted by SyMHyT and ChunkMT

**Table 3.** Evaluation results for legal domain (JRC-Acquis test corpus)

| System | 12-gram | | 6-gram | |
|---|---|---|---|---|
| | BLEU | TER | BLEU | TER |
| *JRC-Acquis language models* | | | | |
| BG | 17.67 | **0.671** | 16.70 | 0.686 |
| HY | **21.38** | **0.681** | 20.18 | 0.703 |
| HG | 19.44 | **0.677** | – | – |
| All | **20.33** | **0.668** | 18.47 | 0.698 |
| *DGT language models* | | | | |
| BG | 17.61 | **0.675** | 16.81 | 0.688 |
| HY | **21.39** | 0.684 | 20.36 | 0.699 |
| HG | 19.86 | **0.677** | – | – |
| All | 20.01 | **0.667** | 17.98 | 0.699 |

The improved MSTM system was evaluated on legal domain and general domain test corpora. For selection of best hypothesis 6-gram and 12-gram language models were used. In almost all cases better results are obtained with higher order language model.

For legal domain (Table 3), the best result (+1.11 BLEU) is obtained by combining *Yandex* (19.75 BLEU) and *hugo.lv* (20.27 BLEU) systems (HY). Similarly to the previous experiments, inclusion of MT systems with significantly lower BLEU scores, produce output which in BLEU points did not exceed the best baseline.

Analysis of selected chunks (Table 4) revealed interesting phenomenon which needs further investigations – when all systems are combined, translations from the best baseline system is used only in 33% of cases, but from the second best system only in 16.59% of cases.

**Table 4.** Best results using test data and LM from JRC-Acquis and selected chunk percentages

| System | BLEU | Equal | Bing | Google | Hugo | Yandex |
|---|---|---|---|---|---|---|
| BLEU | – | – | 16.99 | 16.19 | 20.27 | 19.75 |
| MSMT - BG | 16.38 | 4.88% | 45.03% | 50.09% | – | – |
| MSMT - BGH | 17.89 | 2.78% | 34.31% | 28.93% | 33.98% | – |
| SyMHyT - BG | 17.36 | 4.59% | 24.61% | 70.80% | – | – |
| SyMHyT - BGH | 19.50 | 2.88% | 18.01% | 15.71% | 63.40% | – |
| ChunkMT - BG | **17.67** | 15.23% | 41.14% | 43.63% | – | – |
| ChunkMT - HY | **21.38** | 9.15% | – | – | 44.79% | 46.06% |
| ChunkMT - all | **20.33** | 2.94% | 27.80% | 19.67% | 33.00% | 16.59% |

For general domain (Table 5), the best result (+1.48 BLEU) is obtained by combining output from all four MT systems. Similarly to legal domain, results of two system combination are better, when better baseline systems are combined. Increase by 0.56 BLEU points is observed when *Bing* and *Google* systems are combined (BG).

**Table 5.** Evaluation results on ACCURAT balanced test corpus

| | 12-gram | | 6-gram | |
|---|---|---|---|---|
| System | BLEU | TER | BLEU | TER |
| *JRC-Acquis language models* | | | | |
| BG | 17.34 | 0.757 | 17.30 | 0.757 |
| HY | 15.72 | 0.774 | 15.78 | 0.775 |
| All | – | – | 15.88 | 0.774 |
| *DGT language models* | | | | |
| BG | **18.29** | 0.753 | 17.81 | 0.760 |
| HY | 17.72 | 0.757 | 16.49 | 0.768 |
| HG | 18.06 | **0.747** | – | – |
| All | **19.21** | **0.745** | 16.36 | 0.776 |

Table 6 presents distribution of selected translated chunks between different MT engines. Most of translations come from *hugo.lv,* which can be explained with choice of legal domain language model, while *Google* and *Bing* were the best baseline systems for general domain.

**Table 6.** Best results using balanced test data and DGT LM and distribution of selected chunks

| System | BLEU | Equal | Bing | Google | Hugo | Yandex |
|---|---|---|---|---|---|---|
| BLEU | – | – | 17.43 | 17.73 | 17.14 | 16.04 |
| MSMT - BG | 17.70 | 7.25% | 43.85% | 48.90% | – | – |
| MSMT - BGH | 17.63 | 3.55% | 33.71% | 30.76% | 31.98% | – |
| SyMHyT - BG | 17.95 | 4.11% | 19.46% | 76.43% | – | – |
| SyMHyT - BGH | 17.30 | 3.88% | 15.23% | 19.48% | 61.41% | – |
| ChunkMT - BG | **18.29** | 22.75% | 39.10% | 38.15% | – | – |
| ChunkMT - all | **19.21** | 7.36% | 30.01% | 19.47% | 32.25% | 10.91% |

## 5  Conclusion and Future Work

In this paper we described a machine translation system combination approach that uses syntactic features to extract source text fragments, applies public online MT system APIs for translation and selects translations using statistical features. The results show improvements in BLEU (up to +1.11 for legal domain and +1.48 for general domain) and TER (down to −0.015 for legal domain and −0.004 for general domain) scores compared to the baselines and related research projects.

Experiments described in this paper were performed for the English-Latvian language pair, however the framework that realizes described MSMT approach can be applied for other language pairs as well and is freely downloadable from GitHub[6].

Some possible areas of improvement could be enhancements for the chunking algorithm, use of different language models or more complex morphological tag language models, or a completely different approaches for selecting the best candidate chunk.

**Acknowledgements.** The research was supported by Grant 271/2012 from the Latvian Council of Science.

## References

1. Vasiļjevs, A., Kalniņš, R., Pinnis, M., Skadiņš, R.: Machine translation for e-Government - the Baltic case. In: Proceedings of AMTA 2014, vol. 2: MT Users, pp. 181–193 (2014)
2. Skadiņš, R., Šics, V., Rozis, R.: Building the world's best general domain MT for Baltic languages. In: Human Language Technologies – The Baltic Perspective, Proceedings of the Sixth International Conference Baltic HLT 2014, pp. 141–148. IOS Press (2014)

---

[6] ChunkMT - https://github.com/M4t1ss/ChunkMT.

3. Costa-Jussa, M.R., Fonollosa, J.A.R.: Latest trends in hybrid machine translation and its applications. Comput. Speech Lang. **32**(1), 3–10 (2015)
4. Rikters, M., Skadiņa, I.: Syntax-based multi-system machine translation. In: LREC 2016 (2016)
5. Thurmair, G.: Comparing different architectures of hybrid machine translation systems. In: Proceedings of the MT Summit XII, pp. 340–347 (2009)
6. Mellebeek, B., Owczarzak, K., Van Genabith, J., Way, A.: Multi-engine machine translation by recursive sentence decomposition. In: Proceedings of the 7th Conference of the Association for Machine Translation in the Americas, pp. 110–118 (2006)
7. Ahsan, A., Kolachina, P.: Coupling statistical machine translation with rule-based transfer and generation. In: AMTA-The Ninth Conference of the Association for Machine Translation in the Americas, Denver, Colorado (2010)
8. Barrault, L.: MANY: open source machine translation system combination. Prague Bull. Math. Linguist. **93**, 147–155 (2010)
9. Hildebrand, A.S., Vogel, St.: CMU system combination for WMT'09. In: Proceedings of the 4th Workshop on SMT, Athens (2009)
10. Rikters, M.: Multi-system machine translation using online APIs for English-Latvian. In: ACL-IJCNLP 2015, p. 6 (2015)
11. Heafield, K., Hanneman, Gr., Lavie, A.: Machine translation system combination with flexible word ordering. In: Proceedings of the 4th Workshop on SMT, Athens (2009)
12. Chen, Y., Jellinghaus, M., Eisele, A., Yi, Zh., Hunsicker, S., Theison, S., Federmann, Ch., Uszkoreit, H.: Combining multi-engine translations with Moses. In: Proceedings of the 4th Workshop on SMT, Athens (2009)
13. Feng, Y., Liu, Y., Mi, H., Liu, Q., Lü, Y.: Lattice-based system combination for statistical machine translation. In: Proceedings of the 2009 Conference on Empirical Methods in Natural Language Processing: Volume 3, vol. 3. Association for Computational Linguistics (2009)
14. Freitag, M., Peitz, S., Wuebker, J., Ney, H., Huck, M., Sennrich, R., Durrani, N., Nadejde, M., Williams, P., Koehn, P., Herrmann, T., Cho, E., Waibel, A.: EU-BRIDGE MT: combined machine translation. In: ACL 2014 Ninth Workshop on Statistical Machine Translation (WMT 2014), Baltimore, MD, USA, pp. 105–113 (2014)
15. Freitag, M., Peter, J., Peitz, S., Feng, M., Ney, H.: Local system voting feature for machine translation system combination. In: EMNLP 2015 Tenth Workshop on Statistical Machine Translation (WMT 2015), Lisbon, Portugal, pp. 467–476 (2015)
16. Petrov, S., Barrett, L., Thibaux, R., Klein, D.: Learning accurate, compact, and interpretable tree annotation. In: Proceedings of the 21st International Conference on Computational Linguistics and the 44th Annual Meeting of the Association for Computational Linguistics. Association for Computational Linguistics (2006)
17. Heafield, K.: KenLM: faster and smaller language model queries. In: Proceedings of the Sixth Workshop on Statistical Machine Translation. Association for Computational Linguistics (2011)
18. Gamon, M., Aue, A., Smets, M.: Sentence-level MT evaluation without reference translations: beyond language modeling. In: Proceedings of EAMT (2005)
19. Callison-Burch, C., Flournoy, R.S.: A program for automatically selecting the best output from multiple machine translation engines. In: Proceedings of the Machine Translation Summit VIII (2001)
20. Akiba, Y., Watanabe, T., Sumita, E.: Using language and translation models to select the best among outputs from multiple MT systems. In: Proceedings of the 19th International Conference on Computational Linguistics, vol. 1. Association for Computational Linguistics (2002)

21. Steinberger, R., Pouliquen, B., Widiger, A., Ignat, C., Erjavec, T., Tufis, D., Varga, D.: The JRC-Acquis: a multilingual aligned parallel corpus with 20+ languages. arXiv preprint cs/0609058 (2006)
22. Steinberger, R., Eisele, A., Klocek, S., Pilos, S., Schlüter, P.: DGT-TM: a freely available translation memory in 22 languages. arXiv preprint arXiv:1309.5226 (2013)
23. Skadiņš, R., Goba, K., Šics, V.: Improving SMT for Baltic languages with factored models. In: Proceedings of the Fourth International Conference Baltic HLT 2010. Frontiers in Artificial Intelligence and Applications, vol. 2192, pp. 125–132 (2010)
24. Papineni, K., Roukos, S., Ward, T., Zhu, W.J.: BLEU: a method for automatic evaluation of machine translation. In: Proceedings of the 40th Annual Meeting on Association for Computational Linguistics. Association for Computational Linguistics (2002)
25. Doddington, G.: Automatic evaluation of machine translation quality using n-gram co-occurrence statistics. In: Proceedings of the Second International Conference on Human Language Technology Research. Morgan Kaufmann Publishers Inc. (2002)
26. Klejch, O., Avramidis, E., Burchardt, A., Popel, M.: MT-ComparEval: graphical evaluation interface for machine translation development. Prague Bull. Math. Linguist. **104**(1), 63–74 (2015)
27. Madnani, N.: iBLEU: interactively debugging and scoring statistical machine translation systems. In: 2011 Fifth IEEE International Conference on Semantic Computing (ICSC). IEEE (2011)

# Forest to String Based Statistical Machine Translation with Hybrid Word Alignments

Santanu Pal[1]([✉]), Sudip Kumar Naskar[3], and Josef van Genabith[1,2]

[1] Universität des Saarlandes, Saarbrücken, Germany
{santanu.pal,josef.vangenabith}@uni-saarland.de
[2] German Research Center for Artificial Intelligence (DFKI), Saarbrücken, Germany
[3] Jadavpur University, Kolkata, India
sudip.naskar@cse.jdvu.ac.in

**Abstract.** Forest to String Based Statistical Machine Translation (FSB-SMT) is a forest-based tree sequence to string translation model for syntax based statistical machine translation. The model automatically learns tree sequence to string translation rules from a given word alignment estimated on a source-side-parsed bilingual parallel corpus. This paper presents a hybrid method which combines different word alignment methods and integrates them into an FSBSMT system. The hybrid word alignment provides the most informative alignment links to the FSBSMT system. We show that hybrid word alignment integrated into various experimental settings of FSBSMT provides considerable improvement over state-of-the-art Hierarchical Phrase based SMT (HPBSMT). The research also demonstrates that additional integration of Named Entities (NEs), their translations and Example Based Machine Translation (EBMT) phrases (all extracted from the bilingual parallel training data) into the system brings about further considerable performance improvements over the hybrid FSBSMT system. We apply our hybrid model to a distant language pair, English–Bengali. The proposed system achieves 78.5% relative (9.84 BLEU points absolute) improvement over baseline HPBSMT.

## 1 Introduction

Statistical Machine Translation (SMT) [3,17] is arguably the most successful machine translation (MT) paradigm in recent years and it has obtained strong results in large-scale MT evaluations[1,2,3]. One of the major bottlenecks in SMT is the problem of long-distance reordering between the source and target language text, particularly for distant language pairs. To alleviate this problem two techniques are generally applied: (i) pre-ordering techniques reorder the source

---

[1] http://www.statmt.org/wmt15/.
[2] http://www.nist.gov/itl/iad/mig/openmt15.cfm.
[3] http://ntcir.nii.ac.jp/PatentMTList/.

© Springer International Publishing AG, part of Springer Nature 2018
A. Gelbukh (Ed.): CICLing 2016, LNCS 9624, pp. 38–50, 2018.
https://doi.org/10.1007/978-3-319-75487-1_4

sentences as close as possible to the target word order before training and decoding [15,32], and (ii) tree based decoding which takes syntactic trees or forests as input to the decoder. The tree based decoding process is carried out by choosing translation and reordering jointly [22,26,43]. Pre-ordering is not able to consider translation and reordering at the same time, it is done before the actual translation process.

In the research reported in this paper, we use Travatar[4] [27], an open-source tree-to-string or forest-to-string translation system that can be used as a tool for translation using source-side syntactic information. Unlike other syntax augmented decoders such as syntax-augmented MT [46] or hierarchical SMT [5] which are built upon the synchronous context-free grammar (SCFG) framework [6], Travatar has been developed in a tree transducer framework [12].

Word alignment is the task of establishing alignment links between source and target words in a bilingual parallel corpus that are translations of each other. Word alignment is the first step and perhaps the most critically important task in any data driven (or corpus based) approach to MT since successive models are approximated based on source-target word associations established during the word alignment step. Word alignment is the first step in discovering the translation knowledge hidden in the parallel corpus.

Word alignments are not limited to just one-to-one correspondences; there could be multiple words associated to a single word in either directions (i.e. many-to-many). Statistical aligners do not always optimally capture many-to-many word alignment links.

In the traditional IBM models [3], model parameters are estimated from the raw parallel corpus and are unable to handle many-to-many alignments. In another well-known word alignment approach, Hidden Markov Model (HMM) [41], the alignment probabilities depend on the alignment position of the previous word. Standard HMM models do not explicitly consider many-to-many alignment either.

Named entities (NEs) often consist of more than one word; thus they can be considered as a specific type of the many-to-many alignment problem in SMT. In the work presented in this paper, we explicitly align NEs with the help of an English–Bengali transliteration engine after identifying NEs on both the source and the target sides. To align NEs, we follow a method similar to the one described in [30]. In our approach, we extract NE phrase pairs from the bilingual training data and add the NE phrase pairs as supplementary training data.

Example Based Machine Translation (EBMT) phrases tend to be more linguistically motivated compared to SMT phrases which essentially operate on n-grams. The knowledge extraction as well as representation process, in both EBMT and SMT, uses very different techniques in order to extract resources. In our research, we extract EBMT phrases following the work of [7]. High frequency EBMT phrases are added to the training corpus as additional training material. Prior (EBMT) phrase alignment helps the statistical aligner indirectly in the sense that more evidence is provided to the statistical aligner about highly

---

[4] http://phontron.com/travatar/.

frequent EBMT phrases. It also narrows down the scope of the alignment at the beginning of IBM Model 1's initially uniform probability estimation. The EBMT phrases facilitate the IBM models to make the alignments more reliable and definite and it also helps the IBM models converge faster during the training phrase of SMT. Thus, prior (EBMT) phrase alignment indirectly improves the performance of the statistical word aligner, which in turn results in well aligned source-target phrases during the phrase extraction process. Reduction in noisy alignments also reduces the size of the phrase table which is prepared during the SMT training pipeline. What is more, a smaller translation model also results in faster translation during decoding time by reducing the search space. This motivated us to improve the quality of state-of-the-art word alignment methods further by applying a hybrid methodology. We present an improvement of word alignment quality by combining three statistical word alignment tables: (i) GIZA++ alignment, (ii) SymGiza++ alignment and (iii) Berkeley alignment. Our objective was to assess the effectiveness of the hybrid model in word alignment and to see whether it can enhance the overall translation quality.

The remainder of the paper is organized as follows. Section 2 highlights related research in this area. Section 3 details the components of our system, in particular named entity extraction, EBMT phrases and our hybrid implementation of word alignment. In Sect. 4, we outline the experimental setup for the task. Section 5 provides evaluation results along with some analysis on the performance on the system. Section 6 concludes the paper and presents avenues for future research.

## 2  Related Research

Recent research in SMT has investigated how to incorporate syntactic knowledge into PBSMT systems to improve the translation quality. [5] introduced an approach for incorporating syntax into PBSMT, targeting mainly phrase reordering. In this approach, hierarchical phrase translation probabilities are used to handle a range of reordering phenomena. [24] presented a similar extension of PBSMT systems with syntactic structure on the target side. [45] extended the work introduced in [5] by augmenting the hierarchical phrase labels with syntactic categories derived from parsing the target side of the parallel corpus. They associate a target parse tree with their corresponding search lattice provided by lexical phrases on the source sentence and assign a syntactic category to phrases which align directly with the parse hierarchy. Similar to [5], a chart-based parser with a limited language model is used.

A major characteristic of state-of-the-art PBSMT is that phrase pairs are extracted solely based on the knowledge contained in the word alignment table. The extracted phrases in PBSMT do not respect linguistically motivated phrase boundaries and may contain words from neighboring linguistic phrases. Syntax based SMT systems have provided promising improvements in recent years. Syntax based SMT can be divided into two categories: formal syntax-based systems where there is no need of using any additional parser [5], and linguistic syntax-based systems that use PCFG [14,22,25,26,44], dependency [9,34,35] or other parsers, e.g. [42]. Translation rules can be extracted from aligned string-to-string [5], tree-to-tree [9] or tree/forest-to-string [11,26,42] data structures and their corresponding word alignment tables.

Like any other approach to data driven MT, phrase alignment in syntax based SMT or FSBSMT also crucially relies on word alignment quality. Hybrid approaches to word alignment have been able to successfully improve MT translation quality [31,38]. Previous research demonstrated that compact representations such as alignment combination [1,8,19,28], can produce improved results. Inspired by [19,28], [38] proposed a novel approach to combine multiple alignments for improving MT. Instead of combining exactly two bidirectional alignments as in [19,28], they used an arbitrary number of alignments. Apart from that they also considered the occurrences of potential links of individual alignments. To combine an arbitrary number of alignments, they constructed weighted alignment matrices (WAMs) over 1-best alignments [23,39] from multiple alignments generated by different models (including a refined model as well as minimum Bayes risk (MBR) based models). As the alignment probabilities are generally incomparable between different alignment models, they propose a novel calculation of link probabilities in WAMs. An alignment refinement model was applied to refine multiple alignments into a new alignment that favors the consensus of various models. The MBR decision is used to find the candidate hypothesis that has the least expected loss under a probability model when the true reference is unknown [2].

In this paper, we propose a hybrid word alignment model prepared by combining multiple word alignment tables as well as inclusion of bilingual NE and EBMT phrases as additional training examples. The hybrid word alignment model is integrated into the existing FSBSMT pipeline to enhance the quality of translation.

## 3    System Description

As discussed in Sect. 1, our system is designed by improving two core components of SMT, namely word alignment and (syntactic or hierarchical) phrase estimation. First, we improve word alignment in two different but complementary ways: (i) by using extracted NE phrases (Sect. 3.1) and (ii) extracted EBMT phrase pairs (Sect. 3.2), both collected from the bilingual training data, and used as additional training data to train the alignment models. Next, the phrase extraction step is carried out using the resulting overall hybrid word alignment table. The phrase table creation is carried out according to the state-of-the-art method [5,18,25].

### 3.1    Prior Alignment

We initially identify and align bilingual NEs (described in Sect. 3.1) and EBMT phrases (c.f., Sect. 3.1). These phrase pairs are incorporated into the word alignment model by using the aligned bilingual NEs and example based bilingual phrases as additional training examples.

**NE Alignment.** For NE alignment, we first identify NEs on the source (i.e., English) side of the parallel corpus using Stanford NER[5]. NEs in the target side (i.e., Bengali) are identified using the NER system of [10]. Next, we try to align the extracted source and target NEs. The alignment is trivial when both sides contain only one NE. We add such NE pairs to populate a parallel NE corpus that contains examples having only one token in both sides. Since Bengali has a different orthography than English, NE alignments are performed using transliteration and edit distance [30]. However, for language pairs having the same orthography, NE alignments can be established by making use of edit distance solely. If both the source and target sides contain $n$ number of NEs, and the alignments of $n - 1$ NEs can be established through the transliteration method or by means of already existing alignments, then the $n^{th}$ alignment is established between the remaining (i.e., non-aligned) source and target NE. The bilingual NE pairs extracted thus serve as additional training material and they improve the word alignment.

**Example Based Phrase Alignment.** Example based phrase pairs are extracted based on the work described in [7]. They proposed a compiled app-roach of EBMT that automatically extracts translation templates from sentence-aligned bilingual text by observing the similarities and differences between two example pairs. Two types of translation templates, *generalized* and *atomic* tem-plates, are extracted by applying this approach. A *generalized* translation tem-plate replaces similar or differing sequences with variables while an *atomic* trans-lation template does not contain any variable. The *atomic* translation templates are used as additional parallel training examples to improve the state-of-the-art word alignment. Consider the following two English–Bengali translation pairs from the tourism domain data (Bengali is represented here in English gloss):

(1a) visitors feel happiness: *darsakera ananda onuvab kore*
(1b) visitors feel restlessness: *darsakera klanti onuvab kore*

These two examples share the word sequence *visitors feel* and differ in the word sequence *happiness* and *restlessness* on the source side. Similarly, on the target side, the differing fragments are *ananda* and *klanti*. Based on these dif-fering fragments, we extract the following sub-sentential phrase pairs as in (2).

(2a) happiness: ananda
(2b) restlessness: klanti

We apply this process recursively to extract sub-sentential phrase pairs when more than one differing sequence is present in a pair of sentences. The sub-sentential aligned pairs learned in (2) are called *atomic translation templates*.

---

[5] http://nlp.stanford.edu/software/CRF-NER.shtml.

The details of the algorithm can be found in [7]. Since this particular approach has a cubic run-time complexity on the size of the parallel corpus, it takes a significant amount of time to extract phrase pairs even from a small corpus. Therefore we used heuristics to reduce the time complexity. We divided the entire corpus into $n$ clusters based on sentence length such that similar length sentences belong to the same cluster. We extract *atomic* translations from each of these clusters.

## 3.2 Hybrid Word Alignment

The statistical word aligner used our hybrid word alignment model is trained on the parallel bilingual training corpus (described in Sect. 4.1) with aligned NEs and EBMT phrases as additional training materials. The hybrid word alignment model is described as the combination of three word alignment models as follows.

**Word Alignment Using GIZA++:** GIZA++ [29] is a statistical word alignment tool which implements maximum likelihood estimators for IBM models 1–6 and an HMM alignment model. The model parameters of GIZA++ acquire better estimation from very large amounts of parallel data. Symmetrization methods now-a-days are able to provide some improvements in the alignment table where the parallel corpus is aligned using bidirectional training and then the two alignment tables are reconciled using different heuristics, e.g., union, intersection and most recently grow-diagonal-final and grow-diagonal-final-and heuristics [17]. In spite of these heuristics, the word alignment quality is still low and calls for further improvement. We describe our approach of improving word alignment quality in the following subsections.

**Berkeley Aligner:** Like GIZA++, the Berkeley Aligner is also used to align words across sentence pairs in a bilingual parallel corpus. The Berkeley Aligner [21] allows both unsupervised and supervised approaches to align words from parallel corpora. We initially train on the parallel corpus using the fully unsupervised method of producing Berkeley word alignments. The Berkeley aligner is an extension of the Cross Expectation Maximization word aligner. The aligner uses agreement between two simple sequence-based models by training and facilitates substantial error reductions over standard models. Moreover, it is jointly trained with HMM models, and as a result the alignment error rate [40] reduces substantially.

**SymGiza++:** SymGiza++ [16] modifies the counting phase of each model of Giza++ to allow updating of the symmetrised models between the chosen iterations of the original training algorithms. It computes symmetric word alignment models with the capability of taking advantage of multi-processor systems. Experimental results show that the alignment quality improves by more than 17% compared to Giza++ [16].

**Hybridization:** Our hybrid word alignment method combines three different statistical word alignment methods including: Giza++ word alignment with grow-diag-final-and (GDFA) heuristic [17], Berkeley word alignment and Sym-Giza++ word alignment.

**ADD Additional Alignments:** This method follows the following heuristic. We consider one of the alignments generated by GDFA ($A_1$), Berkeley aligner ($A_2$), and SymGiza++ ($A_3$) as the standard alignment. Subsequently, we compute the intersection of the other two alignments and add such alignment points to the standard alignment if they are not already present. The hypothesis behind this heuristic is that if one of the alignments can be trusted more than the others then that alignment is considered as the standard alignment and the reliable alignment points which are common to the rest of the alignments are added to the standard alignment.

### 3.3   Forest-Based SMT

Forest-based SMT is an extension of tree-based SMT. Current tree-based systems suffer from a major drawback: during translation they only use the 1-best parse tree, which might result in incorrect translation due to parsing errors. In forest-based systems, the decoder produces translations of a packed forest of exponentially $k$-best parses. A *forest* is a compact representation of all the parse trees for a given input sentence under a context-free grammar.

There are two separate steps preformed by the existing tree-based systems: (i) parsing of the source input sentence into a 1-best tree $\tau$ and (ii) decoding, where the decoder searches for the best derivation $\delta^*$ that translates source tree $\tau$ into a target-language string among all possible derivations $D$:

$$\delta^* = argmax_{\delta \in D} P(\delta|\tau) \tag{1}$$

Equation 1 can be unpacked as:

$$\delta^* = argmax_{\delta \in D} P(\delta|\tau)^{\lambda_0} \times e^{\lambda_1|\delta|} \\ \times P_{lm}(s)^{\lambda_2} \times e^{\lambda_3|s|} \tag{2}$$

Equation 2 can be represented as a log-linear one:

$$\delta^* = argmax_{\delta \in D} \lambda_0 \log P(\delta|\tau) + \lambda_1|\delta| \\ + \lambda_2 \log P_{lm}(s) + \lambda_3|s| \tag{3}$$

where, $e^{\lambda_1|\delta|}$ is the penalty term on the number of rules in a derivation, $P_{lm}(s)$ is the language model and $e^{\lambda_3|s|}$ is the length penalty term on the target translation $s$.

The decoding step of FSBSMT is to translate the parse forest using the set of translation rules. A technique of pattern-matching from tree-based decoding is applied to convert a parse forest into a translation forest. The decoder chooses the best derivation from the translation forest and finally produces the translation output in the form of a target string. Therefore, the derivation probability $P(\delta|\tau)$ is now replaced by $P(\delta|\hbar)$ where $\hbar$ is the parse forest, and this is the product of probabilities of translation rules $r \in \delta$.

$$P(\delta|\hbar) = \prod_{r \in \delta} P(r) \tag{4}$$

Each $P(r)$ is defined as the product of five different probabilities as in Eq. 5. Let $t$ and $s$ be the source-side tree and target-side string of rule $r$, respectively, $P(t|s)$ and $P(s|t)$ are the two translation probabilities, and $P_{lex}(t|s)$ and $P_{lex}(s|t)$ are the two lexical probabilities. $P(t|\hbar)$ denotes the source side parsing probability of the current translation rule $r$ in the parse forest.

$$P(r) = P(t|s)^{\lambda_4} \times P(s|t)^{\lambda_5} \times P_{lex}(t|s)^{\lambda_6} \tag{5}$$
$$\times P_{lex}(s|t)^{\lambda_7} \times P(t|\hbar)^{\lambda_8}$$

## 4    Experiments

### 4.1    Data

We used an English–Bengali parallel corpus[6] containing 25,000 sentences from the travel and tourism domain. Corpus cleaning was carried out first by calculating the global mean ratio of the number of characters in a source sentence to that in a target sentence and then filtering out sentence pairs that exceed or fall below 20% of the global ratio [37]. Tokenization and punctuation normalization were performed using Moses scripts. Finally, we filtered the parallel training data on maximum allowable sentence length of 100 and sentence length ratio of 1:2 (either direction).

After cleaning, the English-Bengali parallel corpus contained 23,492 parallel sentences consisting of 569,600 source tokens and 489,609 target tokens. We randomly selected 500 sentences each for the development set and the test set from this filtered parallel corpus and treat the rest as the training corpus. All NEs and all EBMT phrase pairs were extracted from the training corpus.

### 4.2    Experimental Settings

The effectiveness of our approach is demonstrated by comparing it against the HPBSMT model which serves as our baseline. For building the baseline HPB-SMT system, we use the maximum phrase length of 7 and a 5-gram language model. For performing word alignment for the baseline systems, we used the Berkeley Aligner (BA) as BA provides better alignment than the GIZA++ implementation of IBM word alignment model 4 with grow-diagonal-final-and

---

[6] This corpus is produced in the *EILMT* project funded by DEITY, MCIT, Govt. of India.

heuristics. The phrase extraction process was carried out using the hierarchical model [5]. For FSBSMT [27], rule extraction from forests is performed using the method described in [25]. To build our FSBSMT systems, we used Egret[7] to parse the English sentences as it provides high accuracy parsing as well as the output of k-best parses in the packed forest format.

The 5-gram target language model was trained using KENLM [13]. Parameter tuning for both HPBSMT and FSBSMT was carried out using both k-best MIRA [4] and Minimum Error Rate Training (MERT) [28] on the held-out development set. After the parameters were tuned, decoding was carried out on the held out test set. In the set of experiments reported in this paper, we first integrated the hybrid word alignment model (c.f., Sect. 3.2) within both the hierarchical phrase-extraction [5] as well as the state-of-the-art forest to string based phrase extraction model.

**Table 1.** Systematic evaluation results. HPB := HPBSMT, FB := FSBSMT

| SYSTEM | BLEU | NIST | TER | METEOR |
|---|---|---|---|---|
| HPB_BA | 12.53 | 4.34 | 72.93 | 40.97 |
| HPB_SYM | 11.20 | 4.35 | 71.10 | 39.67 |
| HPB_GIZA | 11.62 | 4.25 | 73.90 | 40.45 |
| HPB_BA_FB | 12.96 | 4.38 | 72.41 | 41.27 |
| FB_GIZA | 17.79 | 4.62 | 66.78 | 41.61 |
| FB_GIZA_NEA | 18.30 | 4.70 | **66.27** | 42.03 |
| FB_BA | 21.28 | 4.77 | 69.37 | 41.88 |
| FB_BA_NEA | 21.40 | 4.81 | 69.20 | 42.05 |
| FB_HWA_NEA | 22.03 | 4.91 | 67.67 | **42.94** |
| FB_HWA_NEA_EBMT | **22.37** | **4.92** | 67.53 | 42.52 |

## 5  Results

To test the effect of hybrid word alignment on the forest based system, we compared the systems with various experimental settings. We evaluated the systems using three well known automatic MT evaluation metrics: BLEU [33], METEOR [20] and TER [36].

The evaluation results are reported in Table 1. We use HPBSMT (HPB_BA) implemented with BA as our baseline model. As reported in Table 1, the BA performed better in comparison with the other statistical word aligners such as GIZA++ (HPB_GIZA) and Symgiza++ (HPB_SYM). The BLEU score of the baseline is 12.53. High scoring phrase pairs obtained form forest based training are added as additional training data for the HPB_BA_FB system which brings

---

[7] http://code.google.com/p/egret-parser/.

some improvements over the baseline system. The motivation behind adding phrase pairs obtained form forest based training in HPBSMT stems from the fact that FSBSMT is time intensive as it has dependencies during decoding, while HPBSMT decoding is much faster. FB_GIZA is a simple GIZA++ implementation of FSBSMT while FB_GIZA_NEA is an extension of the FB_GIZA system where we make use of the NE aligned parallel data as additional parallel training examples. Similarly, the FB_BA system is a BA implementation of FSBSMT and FB_BA_NEA uses NE alignments as additional training materials.

Using multiple alignment combinations, i.e. hybrid word alignment (HWA) implemented in FSBSMT system, clearly improves the BLEU score. The HWA combined with NEA and prior high frequency EBMT phrases (FB_HWA_NEA_EBMT in Table 1) provided the overall best performance in terms of both BLEU (22.37) and NIST (4.92), while HWA with NEA produced the overall best METEOR score (42.94) and the FB_GIZA_NEA system resulted in the lowest TER (66.27) score. The proposed FSBSMT system provides 78.5% relative (9.84 absolute BLEU points) improvement over baseline HPBSMT. The relative improvement in terms of BLEU compared to the vanilla settings of HPBSMT is 92.5%.

## 6   Conclusions and Future Work

This paper reports research on integrating hybrid word alignment in forest to string based statistical machine translation (FSBSMT). Experimental results on an English–Bengali dataset show that FSBSMT with Berkeley alignment brings a huge improvement (69.83% relative, 8.75 absolute BLEU points) over state-of-the-art hierarchical Phrase based SMT (HPBSMT) in the first place. Systems like HPBSMT which work only with 1-best parse tree may suffer from parsing errors. FSBSMT alleviates this problem by considering packed forest of $k$-best parses.

Additional integration of prior aligned named entities and EBMT phrases into the proposed system also brings further considerable improvements. The enhanced system provides 78.5% relative (9.84 absolute BLEU points) improvements over the baseline HPBSMT system and 5.12% relative improvement (1.09 absolute BLEU points) over a FSBSMT system with Berkeley alignment.

In future, we would like to apply a similar methodology with other alignment combination methods and compare between them. We will also focus on improving our hybrid word alignment model by considering the strength of alignment points given by the various word alignment models.

**Acknowledgments.** This work is supported by the People Programme (Marie Curie Actions) of the European Union's Framework Programme (FP7/2007-2013) under REA grant agreement no. 317471.

# References

1. Ayan, N.F., Dorr, B.J., Monz, C.: NeurAlign: combining word alignments using neural networks. In: Proceedings of Human Language Technology Conference and Conference on Empirical Methods in Natural Language Processing, pp. 65–72. Association for Computational Linguistics, Vancouver, October 2005
2. Bickel, P.J., Doksum, K.A.: Mathematical Statistics: Basic Ideas and Selected Topics. Holden-Day Company, Oakland (1977)
3. Brown, P.F., Pietra, V.J.D., Pietra, S.A.D., Mercer, R.L.: The mathematics of statistical machine translation: parameter estimation. Comput. linguist. **19**(2), 263–311 (1993)
4. Cherry, C., Foster, G.: Batch tuning strategies for statistical machine translation. In: Proceedings of the 2012 Conference of the North American Chapter of the Association for Computational Linguistics: Human Language Technologies, pp. 427–436 (2012)
5. Chiang, D.: A hierarchical phrase-based model for statistical machine translation. In: Proceedings of the 43rd Annual Meeting on Association for Computational Linguistics, pp. 263–270 (2005)
6. Chiang, D.: Hierarchical phrase-based translation. Comput. Linguist. **33**(2), 201–228 (2007)
7. Cicekli, I., Güvenir, H.A.: Learning translation templates from bilingual translation examples. Appl. Intell. **15**(1), 57–76 (2001)
8. DeNero, J., Macherey, K.: Model-based aligner combination using dual decomposition. In: Proceedings of the 49th Annual Meeting of the Association for Computational Linguistics: Human Language Technologies, HLT 2011, vol. 1, pp. 420–429. Association for Computational Linguistics, Stroudsburg (2011)
9. Ding, Y., Palmer, M.: Machine translation using probabilistic synchronous dependency insertion grammars. In: Proceedings of the 43rd Annual Meeting on Association for Computational Linguistics, pp. 541–548 (2005)
10. Ekbal, A., Bandyopadhyay, S.: Named entity recognition using support vector machine: a language independent approach. Int. J. Electr. Comput. Syst. Eng. **4**(2), 155–170 (2010)
11. Galley, M., Hopkins, M., Knight, K., Marcu, D.: What's in a translation rule? In: HLT-NAACL 2004: Main Proceedings, 2–7 May 2004, pp. 273–280. Association for Computational Linguistics, Boston (2004)
12. Graehl, J., Knight, K.: Training tree transducers. In: HLT-NAACL 2004: Main Proceedings, 2–7 May 2004, pp. 105–112. Association for Computational Linguistics, Boston (2004)
13. Heafield, K.: KenLM: faster and smaller language model queries. In: Proceedings of the Sixth Workshop on Statistical Machine Translation, pp. 187–197 (2011)
14. Huang, L.: Statistical syntax-directed translation with extended domain of locality. In: Proceedings of the AMTA 2006, pp. 66–73 (2006)
15. Isozaki, H., Sudoh, K., Tsukada, H., Duh, K.: Head finalization: a simple reordering rule for SOV languages. In: Proceedings of the Joint Fifth Workshop on Statistical Machine Translation and MetricsMATR, pp. 244–251. Association for Computational Linguistics (2010)
16. Junczys-Dowmunt, M., Szał, A.: SyMGiza++: symmetrized word alignment models for statistical machine translation. In: Bouvry, P., Kłopotek, M.A., Leprévost, F., Marciniak, M., Mykowiecka, A., Rybiński, H. (eds.) SIIS 2011. LNCS, vol. 7053, pp. 379–390. Springer, Heidelberg (2012). https://doi.org/10.1007/978-3-642-25261-7_30

17. Koehn, P.: Statistical Machine Translation, 1st edn. Cambridge University Press, New York (2010)
18. Koehn, P., Hoang, H., Birch, A., Callison-Burch, C., Federico, M., Bertoldi, N., Cowan, B., Shen, W., Moran, C., Zens, R., Dyer, C., Bojar, O., Constantin, A., Herbst, E.: Moses: open source toolkit for statistical machine translation. In: Proceedings of the 45th Annual Meeting of the ACL on Interactive Poster and Demonstration Sessions, pp. 177–180 (2007)
19. Koehn, P., Och, F.J., Marcu, D.: Statistical phrase-based translation. In: Proceedings of the 2003 Conference of the North American Chapter of the Association for Computational Linguistics on Human Language Technology, pp. 48–54 (2003)
20. Lavie, A., Agarwal, A.: METEOR: an automatic metric for MT evaluation with high levels of correlation with human judgments. In: Proceedings of the Second Workshop on Statistical Machine Translation, pp. 228–231 (2007)
21. Liang, P., Taskar, B., Klein, D.: Alignment by agreement. In: Proceedings of the Main Conference on Human Language Technology Conference of the North American Chapter of the Association of Computational Linguistics, HLT-NAACL 2006, pp. 104–111 (2006)
22. Liu, Y., Liu, Q., Lin, S.: Tree-to-string alignment template for statistical machine translation. In: Proceedings of the 21st International Conference on Computational Linguistics and the 44th Annual Meeting of the Association for Computational Linguistics, ACL-44, pp. 609–616 (2006)
23. Liu, Y., Xia, T., Xiao, X., Liu, Q.: Weighted alignment matrices for statistical machine translation. In: Proceedings of the 2009 Conference on Empirical Methods in Natural Language Processing, pp. 1017–1026. Association for Computational Linguistics, Singapore, August 2009
24. Marcu, D., Wang, W., Echihabi, A., Knight, K.: SPMT: statistical machine translation with syntactified target language phrases. In: Proceedings of the 2006 Conference on Empirical Methods in Natural Language Processing, Sydney, Australia, pp. 44–52, July 2006
25. Mi, H., Huang, L.: Forest-based translation rule extraction. In: Proceedings of EMNLP, pp. 206–214. ACL (2008)
26. Mi, H., Huang, L., Liu, Q.: Forest-based translation. In: Proceedings of ACL 2008: HLT, pp. 192–199. Association for Computational Linguistics, Columbus, June 2008
27. Neubig, G.: Travatar: a forest-to-string machine translation engine based on tree transducers. In: Proceedings of the 51st Annual Meeting of the Association for Computational Linguistics: System Demonstrations, pp. 91–96. Association for Computational Linguistics, Sofia (2013)
28. Och, F.J.: Minimum error rate training in statistical machine translation. In: Proceedings of the 41st Annual Meeting on Association for Computational Linguistics, vol. 1, pp. 160–167 (2003)
29. Och, F.J., Ney, H.: A systematic comparison of various statistical alignment models. Comput. Linguist. **29**(1), 19–51 (2003)
30. Pal, S., Naskar, S.K., Pecina, P., Bandyopadhyay, S., Way, A.: Handling named entities and compound verbs in phrase-based statistical machine translation. In: Proceedings of the of Multiword Expression Workshop (MWE 2010) and the 23rd International Conference of Computational Linguistics (Coling 2010) (2010)
31. Pal, S., Naskar, S.K., Bandyopadhyay, S.: A hybrid word alignment model for phrase-based statistical machine translation. In: ACL 2013, pp. 94–101 (2013)

32. Pal, S., Naskar, S.K., Bandyopadhyay, S.: Word alignment-based reordering of source chunks in PB-SMT. In: Proceedings of the Ninth International Conference on Language Resources and Evaluation (LREC 2014). European Language Resources Association (ELRA), May 2014

33. Papineni, K., Roukos, S., Ward, T., Zhu, W.J.: BLEU: a method for automatic evaluation of machine translation. In: Proceedings of the 40th Annual Meeting on Association for Computational Linguistics, ACL 2002, pp. 311–318 (2002)

34. Quirk, C., Menezes, A., Cherry, C.: Dependency treelet translation: syntactically informed phrasal SMT. In: Proceedings of ACL, pp. 271–279 (2005)

35. Shen, L., Xu, J., Weischedel, R.: A new string-to-dependency machine translation algorithm with a target dependency language model. In: Proceedings of Association for Computational Linguistics, pp. 577–585 (2008)

36. Snover, M., Dorr, B., Schwartz, R., Micciulla, L., Makhoul, J.: A study of translation edit rate with targeted human annotation. In: Proceedings of Association for Machine Translation in the Americas, pp. 223–231 (2006)

37. Tan, L., Pal, S.: Manawi: using multi-word expressions and named entities to improve machine translation. In: Proceedings of Ninth Workshop on Statistical Machine Translation (2014)

38. Tu, Z., Liu, Y., He, Y., van Genabith, J., Liu, Q., Lin, S.: Combining multiple alignments to improve machine translation. In: The 24th International Conference of Computational Linguistics (Coling 2012), pp. 1249–1260 (2012)

39. Tu, Z., Liu, Y., Liu, Q., Lin, S.: Extracting hierarchical rules from a weighted alignment matrix. In: Proceedings of 5th International Joint Conference on Natural Language Processing, pp. 1294–1303 (2011)

40. Vilar, D., Popovi, M., Ney, H.: AER: do we need to improve our alignments. In: Proceedings of the International Workshop on Spoken Language Translation, pp. 205–212 (2006)

41. Vogel, S., Ney, H., Tillmann, C.: Hmm-based word alignment in statistical translation. In: Proceedings of the 16th Conference on Computational Linguistics, vol. 2, pp. 836–841. Association for Computational Linguistics (1996)

42. Wu, X., Matsuzaki, T., Tsujii, J.: Effective use of function words for rule generalization in forest-based translation. In: Proceedings of the 49th Annual Meeting of the Association for Computational Linguistics: Human Language Techologies, Portland, Oregon, USA, pp. 22–31, June 2011

43. Yamada, K., Knight, K.: A syntax-based statistical translation model. In: Proceedings of the 39th Annual Meeting on Association for Computational Linguistics, ACL 2001, pp. 523–530 (2001)

44. Zhang, H., Zhang, M., Li, H., Aw, A., Tan, C.L.: Forest-based tree sequence to string translation model. In: Proceedings of the Joint Conference of the 47th Annual Meeting of the ACL and the 4th International Joint Conference on Natural Language Processing of the AFNLP, pp. 172–180 (2009)

45. Zollmann, A., Venugopal, A.: Syntax augmented machine translation via chart parsing. In: Proceedings on the Workshop on Statistical Machine Translation, New York City, pp. 138–141, June 2006

46. Zollmann, A., Venugopal, A., Paulik, M., Vogel, S.: The syntax augmented MT (SAMT) system for the shared task in the 2007 ACL workshop on statistical machine translation. In: Proceedings of the Second Workshop on Statistical Machine Translation, pp. 216–219. Association for Computational Linguistics (2007)

# Instant Translation Model Adaptation by Translating Unseen Words in Continuous Vector Space

Shonosuke Ishiwatari[1]([✉]), Naoki Yoshinaga[2,3],
Masashi Toyoda[2], and Masaru Kitsuregawa[2,4]

[1] Graduate School of Information Science and Technology,
The University of Tokyo, Tokyo, Japan
ishiwatari@tkl.iis.u-tokyo.ac.jp
[2] Institute of Industrial Science, The University of Tokyo, Tokyo, Japan
{ynaga,toyoda,kitsure}@tkl.iis.u-tokyo.ac.jp
[3] National Institute of Information and Communications Technology, Tokyo, Japan
[4] National Institute of Informatics, Tokyo, Japan

**Abstract.** In statistical machine translation (SMT), differences between domains of training and test data result in poor translations. Although there have been many studies on domain adaptation of language models and translation models, most require supervised in-domain language resources such as parallel corpora for training and tuning the models. The necessity of supervised data has made such methods difficult to adapt to practical SMT systems. We thus propose a novel method that adapts translation models without in-domain parallel corpora. Our method infers translation candidates of unseen words by nearest-neighbor search after projecting their vector-based semantic representations to the semantic space of the target language. In our experiment of out-of-domain translation from Japanese to English, our method improved BLEU score by 0.5–1.5.

## 1 Introduction

Statistical machine translation (SMT) has been successfully applied to the translation between various language pairs, particularly phrase-based SMT, which is the most common since it can learn a translation model from a sentence-aligned parallel corpus without any linguistic annotations. Although we can improve the quality of translation by using a large language model that can be obtained from easily available monolingual corpora [1], language models capture only the fluency in languages so the quality of translation cannot be improved much if the translation model does not provide correct translation candidates for source-language words and phrases. The quality of translation in SMT is therefore bounded by the size of parallel corpus to train the translation model. Even if a large parallel corpus is available for the pair of languages in question, we often want to translate sentences in a domain that has a different vocabulary

© Springer International Publishing AG, part of Springer Nature 2018
A. Gelbukh (Ed.): CICLing 2016, LNCS 9624, pp. 51–62, 2018.
https://doi.org/10.1007/978-3-319-75487-1_5

from the domain of available parallel corpora, and this inconsistency deteriorates the quality of translation [2,3].

Researchers have tackled this problem and proposed methods of domain adaptation for SMT that exploits a larger out-of-domain parallel corpus. They have focused on a scenario in which a small or pseudo in-domain parallel corpus is available for training [4]. In actual scenarios when users want to exploit machine translation, the target domains can differ so the domain mismatches between the prepared SMT system and the target documents are likely to occur. Domain adaptation is thus expected to improve the quality of translation. However, it is unrealistic for most MT users who cannot command the target language to prepare in-domain parallel corpora by themselves. The use of crowdsourcing for preparing in-domain parallel corpora is allowed for a few users who have a large number of documents for translation and are willing to pay money for improving the quality of translation.

In this study, we assume domain adaptation for SMT in a scenario where no sentence-aligned parallel corpus is available for the target domain and propose an instant method of domain adaptation for SMT by using a cross-lingual projection of word semantic representations [5]. Assuming that source-and target-language monolingual corpora are available, we first learn vector-based semantic representations of words in the source and target languages from those monolingual corpora. We next obtain a projection from semantic representations in the source language to those in the target language using a seed dictionary (in general domain) to learn a translation matrix. We then use the translation matrix to obtain translations of unseen (out-of-vocabulary, OOV) words. The translation probabilities are computed by using cosine-similarity between the projected semantic representation of the OOV word and semantic representations of words in the target language.

To evaluate the effectiveness of our method, we apply our method to a translation between English (en) and Japanese (ja) in recipe documents using a translation model learned by phrase-based SMT from Kyoto-related Wikipedia articles. Experimental results confirmed that our method improves BLEU score by 0.5–1.5 and 0.1–0.2 for ja-en and en-ja translations, respectively.

The remainder of this paper is structured as follows. Section 2 explains existing approaches to domain adaptation for SMT without in-domain parallel corpus. Section 3 describes a method of translating word semantic representations. Section 4 proposes a method of adapting SMT to a new domain without a sentence-aligned parallel corpora. Section 5 evaluates the effectiveness of the proposed method on domain adaptation for SMT. Section 6 finally concludes this study and addresses future work.

## 2    Related Work

As mentioned in Sect. 1, most previous approaches to domain adaptation for SMT assume a scenario where a small or pseudo in-domain parallel corpus is available. In this section, we briefly overview a method of domain adaptation for SMT in a setting where no in-domain parallel corpus is available.

Wu et al. [6] have proposed domain adaptation for SMT that exploits an in-domain bilingual dictionary. They generate a translation model from the bilingual dictionary and combine it with the translation model learned from out-of-domain parallel corpora. An issue here is how to learn a translation probability between words (or phrases) needed for the translation model, and they resort to probabilities of words in the target language in a monolingual corpus. Although building a bilingual dictionary for the target domain is more effective than developing a parallel corpus to cover rare OOV words, it is still difficult to develop a bilingual dictionary for most MT users who cannot command the target language.

To cope with this problem, several researchers have recently exploited a bilingual lexicon automatically induced from in-domain corpora to generate a translation model for SMT [7–9]. These approaches induce a bilingual lexicon from in-domain comparable corpora prior to the translation and use it to obtain an in-domain translation model.

Marthur et al. [10] exploit parallel corpora in various domains to induce the translation model for the target domain. They used 11 sets of parallel corpora for domains including TED talks, news articles, and software manuals to train the translation model for each domain and then linearly interpolated these translation models to derive a translation model for the target domain. They successfully improved the quality of translation when no parallel corpus was available for the target domain. Yamamoto and Sumita [11] assume various language expressions in translating travel conversations and train several language and translation models from a set of parallel corpora that are split by unsupervised clustering of the entire parallel corpus for travel conversations. The language and translation models for translating a given sentence are chosen in accordance with the similarity between the given sentence and the sentences in each split of the parallel corpus. Although this method is not intended for domain adaptation, it can be used in our setting when we have a parallel corpus for the general domain (and the domain of the target sentence is included in the general domain). These studies, however, implicitly assume in-domain (or related domain) parallel corpora are available, while we assume those resources are unavailable to broaden the applicability of our method.

Among these studies, our method is most closely related to domain adaptation using bilingual lexicon induction [7–9] but is different from these approaches in that it does not need to build a sort of bilingual lexicon prior to the translation to support the translation of OOV words in a given sentence. We use a projection of semantic representations of source-language words to the target-language semantic space to dynamically find translation candidates of found OOV words by computing the similarity of the obtained representations to semantic representations for words in the target language at the time of translation. Also, we empirically show that our approach could even benefit from general-domain non-comparable monolingual corpora instead of in-domain comparable monolingual corpora used in these studies on bilingual lexicon induction.

# 3   Cross-Lingual Projection of Word Semantic Representations

Our method exploits a projection of semantic representations of OOV words in the source-language onto the target-language semantic space to look for translation candidates for the OOV words. In this section, we first introduce semantic representations of words in a continuous vector space and then describe a method we proposed previously that learns a translation matrix for projecting vector-based representations of words across languages [5].

A vector-based semantic representation of a word, hereinafter *word vector*, represents the meaning of a word with a continuous vector. These representations are based on the distributional hypothesis [12,13], which states that words that occur in the similar contexts tend to have similar meanings. The word vectors can be obtained from monolingual corpora in an unsupervised manner, such as a count-based approach [14] or prediction-based approaches [15,16].

The words that have similar meanings tend to have similar vectors [17,18]. By mapping words into continuous vector space, we can use cosine similarity to compute the similarity of meanings between words. However, the similarity between word vectors across languages is difficult to compute, so these word vectors are difficult to utilize in cross-lingual applications such as machine translation or cross-lingual information retrieval.

To solve this problem, Mikolov et al. [19] proposed a method that learns a cross-lingual projection of word vectors from one language into another. By projecting a word vector into the target-language semantic space, we can compute the semantic similarity between words in different languages. Suppose that we have training data of $n$ examples, $\{(\boldsymbol{x}_1, \boldsymbol{z}_1), (\boldsymbol{x}_2, \boldsymbol{z}_2), \ldots (\boldsymbol{x}_n, \boldsymbol{z}_n)\}$, where $\boldsymbol{x}_i$ is the vector representation of a word in the source language (*e.g.*, "*gato*"), and $\boldsymbol{z}_i$ is the word vector of its translation in the target language (*e.g.*, "*cat*"). Then the translation matrix, $\boldsymbol{W}$, such that $\boldsymbol{W}\boldsymbol{x}_i$ approximates $\boldsymbol{z}_i$, can be obtained by solving the following optimization problem:

$$\boldsymbol{W}^{\star} = \operatorname*{argmin}_{\boldsymbol{W}} \sum_{i=1}^{n} \|\boldsymbol{W}\boldsymbol{x}_i - \boldsymbol{z}_i\|^2$$

Here, since word vectors are induced from monolingual corpora, vectors of OOV words are easy to obtain by using in-domain or large-scale monolingual corpora.

We have improved the aforementioned approach by adopting the count-based vectors for words and integrating prior knowledge on translatable context pairs between the dimensions of count-based vectors [5]:

$$\boldsymbol{W}^{\star} = \operatorname*{argmin}_{\boldsymbol{W}} \sum_{i=1}^{n} \|\boldsymbol{W}\boldsymbol{x}_i - \boldsymbol{z}_i\|^2 + \frac{\lambda}{2}\|\boldsymbol{W}\|^2 - \beta_{train} \sum_{(j,k)\in\mathcal{D}_{train}} w_{jk} - \beta_{sim} \sum_{(j,k)\in\mathcal{D}_{sim}} w_{jk}.$$

The second term is the $L_2$ regularizer, while the third and fourth terms are meant to strengthen $w_{jk}$ when $k$-th dimension in the source language corresponds to $j$-th dimension in the target language. $\mathcal{D}_{train}$ and $\mathcal{D}_{sim}$ are sets of translatable dimension pairs. $\mathcal{D}_{train}$ is obtained from the above training data, while $\mathcal{D}_{sim}$ is obtained by computing the surface-level similarity between the dimensions. $\lambda$, $\beta_{train}$ and $\beta_{sim}$ are corresponding hyperparameters to control the strength of the added terms.

Because our method improved the accuracy of choosing translation candidates for words using the projected semantic representation against [19,20], we adopt and implement this method again for finding translation candidates of OOV words in our method.

## 4 Method

Our method assumes that monolingual corpora are available for the source and target language (in the target domain, if any) and first induces semantic representation of words from those corpora. It then learns a cross-lingual projection (translation matrix) using a seed dictionary in a general domain as described in Sect. 3. Note that a seed dictionary for common words is usually available for most pairs of languages or could be constructed assuming English as a pivot language [21].

Having a translation matrix to obtain projections of semantic representations of OOV words in a given sentence, our method instantly constructs a back-off translation model used for enumerating translation candidates for the OOV words in the following way:

**Step 1:** When the translation system accepts a sentence with an OOV word, $f_{oov}$, it translates a semantic representation of the word, $x_{oov}$ into a semantic representation in the target language $x'_{oov}$ using the translation matrix obtained by the method described in Sect. 3.

**Step 2:** It then computes the cosine similarity between the obtained semantic representations with those in the target languages to enumerate $k$ translation candidates[1] in accordance with the value of cosine similarity. The cosine similarity is also used to obtain $P_{vec}(e|f_{oov})$, the direct translation probabilities from the OOV word in the source language, $f_{oov}$, to a candidate word in the target language, $e$, by normalizing them to sum up to 1. Although the obtained translation candidates could include wrong translations, the language model can choose one that is more appropriate in the contexts in the next step, unless the contexts are full of OOV words.

**Step 3:** The decoder of phrase-based SMT uses the above translation probabilities as a back-off translation model to perform the translation. More formally, we add new feature function $h_{vec}$ to the log-linear model used in the decoder as following equation:

$$\log P(e|f) = \sum_i \log(h_i(e, f))\lambda_i + \log(h_{vec}(e, f))\lambda_{vec} \qquad (1)$$

---

[1] $k$ was set to 10 in the experiments.

The $h_{vec}(\boldsymbol{e}, \boldsymbol{f})$ in Eq. (1) is computed with $P_{vec}(e|f_{\text{OOV}})$, only for each OOV word $f_{\text{OOV}}$ in source sentence $\boldsymbol{f}$. An issue here is how to set feature weight $\lambda_{vec}$ since no in-domain training data are available for turning. We simply set $\lambda_{vec}$ to the same value as the weight of direct phrase translation probability of the translation model.

## 5  Experiments

This section evaluates our method of domain adaptation for SMT, using an out-of-domain parallel corpus and source-language and target-language monolingual corpora.

### 5.1  Settings

First, we prepared two parallel corpora in different domains to carry out an experiment of domain adaptation in the SMT system. One is the "Japanese-English Bilingual Corpus of Wikipedia's Kyoto Articles" (hereinafter KFTT corpus), originally prepared by the National Institute of Information and Communications Technology (NICT) and used as a benchmark in "The Kyoto Free Translation Task"[2][22], a translation task that focuses on Wikipedia articles relates to Kyoto. The other parallel corpus (hereafter RECIPE corpus) is provided by Cookpad Inc.,[3] which is the largest online recipe sharing service in Japan. The KFTT corpus includes many words relates to Japanese history and the temples or shrines in Kyoto. On the other hand, the RECIPE corpus includes many words related to foods and cookware. We randomly sampled 10k pairs of sentences from the RECIPE corpus as test corpus for evaluating our domain adaption method. The language models of the target languages are trained with the concatenation of the KFTT corpus and the remaining portion of the RECIPE corpus, while the translation models are trained with only the KFTT corpus. The sizes of the training data and test data are as detailed in Table 1.

**Table 1.** Statistics of the dataset.

| Corpus | Japanese | English |
| --- | --- | --- |
| KFTT (training) | 29.5 MB (440k sentences) | 30.6 MB (440k sentences) |
| RECIPE (test) | 0.8 MB (10k sentences) | 0.7 MB (10k sentences) |

We conducted experiments with Moses [23][4] with the language models trained with SRILM [24][5] and the word alignments predicted by GIZA++ [25].[6]

---

[2] http://www.phontron.com/kftt/.
[3] http://cookpad.com/.
[4] http://www.statmt.org/moses/.
[5] http://www.speech.sri.com/projects/srilm/.
[6] https://github.com/moses-smt/giza-pp.

**Table 2.** Monolingual corpora used to induce semantic representations.

| Corpus | Japanese | English |
|---|---|---|
| Wikipedia (general domain) | 4.4 GB | 16 GB |
| RECIPE (in-domain) | 12 MB | 9.5 MB |

5-gram language models were trained using SRILM with `interpolate` option and `kndiscount` option. Word alignments were obtained using GIZA++ with `grow-diag-final-and` heuristic. The lexical reordering model was obtained with `msd-bidirectional` setting.

Next, we extracted four sets of count-based word vectors from Wikipedia dumps[7] (general-domain monolingual corpora) and the remaining portion of the RECIPE corpus (in-domain monolingual corpora), for Japanese and English, respectively. We considered context windows of five words to both sides of the target word. The function words are then excluded from the extracted context words following our previous work [5]. Since the count vectors are very high-dimensional and sparse, we selected top-$d$ ($d = 10,000$ for general-domain corpus, $d = 5000$ for in-domain corpus) frequent words as contexts words (in other words, the number of dimensions of the word vectors). We converted the counts into positive point-wise mutual information [26] and normalized the resulting vectors to remove the bias introduced by the difference in the word frequency. The size of the monolingual dataset for inducing semantic representations of words is as detailed in Table 2.

Finally, we used Open Multilingual WordNet[8] to train the translation matrices as in [5]. The hyperparameters were tuned on the development set as follows: $\lambda = 0.1$, $\beta_{train} = 5$, $\beta_{sim} = 5$ for (ja-en, general-domain). $\lambda = 1$, $\beta_{train} = 0.1$, $\beta_{sim} = 0.2$ for (ja-en, in-domain). $\lambda = 0.1$, $\beta_{train} = 5$, $\beta_{sim} = 5$ for (en-ja, general-domain). $\lambda = 0.5$, $\beta_{train} = 1$, $\beta_{sim} = 2$ for (en-ja, in-domain).

## 5.2 Results

We performed domain adaptation as described in Sect. 4 and evaluated the effectiveness of our method through BLEU score [28]. Table 3 shows results of the translations of the 10k sentences in the RECIPE corpus between Japanese and English. **All** and **oov sentences** in Table 3 show the BLEU scores measured in the whole test set and the scores measured only in the sentences that include OOV words, respectively. Statistics of the OOV words are shown in Table 4.

All four methods shown in Table 3 use translation models that were trained with the KFTT corpus and are tested with the RECIPE corpus. **Proposed (general)** uses the word vectors extracted from Wikipedia corpus, while **Proposed (in-domain)** uses the vectors extracted from the remaining portion of the RECIPE corpus. In both these methods, we performed domain adaptation by

---

[7] http://dumps.wikimedia.org/ (versions of Nov, 4th, 2014 (ja), Oct, 8th, 2014 (en)).
[8] http://compling.hss.ntu.edu.sg/omw/.

**Table 3.** BLEU on RECIPE corpus. * indicates statistically significant improvements in BLEU over the respective baseline systems in accordance with bootstrap resampling [27] at $p < 0.05$.

| Method | All | | OOV sentences | |
|---|---|---|---|---|
| | ja-en | en-ja | ja-en | en-ja |
| Baseline (no adaptation) | 5.58 | 3.37 | 5.36 | 3.16 |
| Proposed (general-domain) | 6.05* | 3.48* | 5.87* | 3.42* |
| Proposed (in-domain) | **7.08*** | **3.57*** | **7.00*** | **3.63*** |
| Parallel Corpus | 20.88 | 16.69 | 20.72 | 17.01 |

**Table 4.** Statistics of the OOV words in test data (the 10k sentences in the RECIPE corpus).

| | ja-en | en-ja |
|---|---|---|
| The number of OOV words (types) | 3,464 | 1,613 |
| The number of OOV words (tokens) | 21,218 | 4,639 |
| The number of sentences with OOV words | 8,742 | 3,636 |

automatically constructing back-off translation models for OOV words. **Parallel Corpus** in Table 3 uses the remaining portion of the RECIPE corpus as a parallel corpus to learn the translation models, resources of which are assumed to be unavailable in this study. Thus, **Parallel Corpus** is the upper-bound for the task. The low BLEU score for en-ja translation is explained by the direction of the translation being different from the direction when the corpus was built (ja-en) [29]. In addition, the smaller number of OOV tokens in en-ja than in ja-en also causes the smaller improvement in BLEU score.

Table 3 shows that our methods perform well for the translation task. We found that it was better to use the in-domain monolingual corpora rather than general-domain monolingual corpora to obtain the word vectors. This conforms to our expectation because the contextual information included in the word vectors strongly correlates with the target domains. The **Parallel Corpus** has much higher BLEU than all other methods. This result shows that the domain adaptation task we performed was intrinsically difficult because of the significant differences between the two domains.

We show hand-picked examples of the translations in Table 5 to analyze the methods in more detail. The first two examples show that **Proposed (in-domain)** provides more accurate translations than **Proposed (general)**. Despite our method being able to improve the translations of OOV words, the third and the fourth examples indicate that it is not good at improving the translations of **Baseline** that have wrong syntax. The last example shows that some OOV words tend to be translated into their related words, mainly because of their similarity in the semantic space.

**Table 5.** Hand-picked examples of the translations for the 10k sentences in the RECIPE corpus from Japanese to English. Text in bold denotes OOV words in the input sentences and their translations. The subscripts of the translation of the OOV words refer to a manual word alignment of the OOV words.

| Input | 混ぜながら弱火で煮る。 |
|---|---|
| Ref | simmer over low heat while mixing . |
| **Baseline** | 煮る$_1$ at low heat while mixing . |
| **Proposed (general)** | **boil**$_1$ over a low heat while mixing . |
| **Proposed (in-domain)** | **simmer**$_1$ over a low heat while mixing . |
| **Parallel Corpus** | **simmer**$_1$ over low heat while stirring . |
| Input | 玉ねぎ、ニンニクをみじん切りに。 |
| Ref | finely chop the onion and garlic . |
| **Baseline** | みじん切り$_1$ in the onion and garlic . |
| **Proposed (general)** | the garlic and onion in **butter**$_1$ . |
| **Proposed (in-domain)** | **mince**$_1$ the onion and garlic . |
| **Parallel Corpus** | **finely**$_1$**chop**$_1$ the onion and garlic . |
| Input | オーブントースターで焦げ目がつくまで焼く。 |
| Ref | bake until browned in a toaster oven . |
| **Baseline** | in トースター$_1$ oven until 焦げ目$_2$ made 焼く$_3$ . |
| **Proposed (general)** | oven in the **refrigerator**$_1$ until **fenbuconazole**$_2$ made **bread**$_3$ . |
| **Proposed (in-domain)** | in a **toaster**$_1$ oven , **bake**$_3$ until the **end**$_2$ . |
| **Parallel Corpus** | **bake**$_3$ in a **toaster**$_1$ oven until **golden**$_2$**brown**$_2$ . |
| Input | しっとりした食感の素朴なケーキです。 |
| Ref | a simple cake with a moist texture . |
| **Baseline** | しっとり$_1$ food of a simple cake です$_2$ . |
| **Proposed (general)** | the food **texture**$_1$ as a simple cake thing . |
| **Proposed (in-domain)** | the **moist**$_1$ food that 's simple cake . |
| **Parallel Corpus** | a **moist**$_1$**texture**$_1$ of the simple cake . |
| Input | 火を消し、ごま油を入れ混ぜる。 |
| Ref | turn off the heat , and stir in the sesame oil . |
| **Baseline** | 消し$_1$ fire , and put ごま油$_2$ 混ぜる$_3$ . |
| **Proposed (general)** | heat **butter**$_1$ completely , add the **milk**$_2$ . |
| **Proposed (in-domain)** | fire , add **coconut**$_2$ , and **mix**$_3$ . |
| **Parallel Corpus** | **turn**$_1$**off**$_1$ the heat , add the **sesame**$_2$**oil**$_2$ and **mix**$_3$ . |

The examples show that the OOV words such as "煮る" (simmer), "トースター" (toaster), and "焼く" (bake) could successfully be translated with **Proposed (in-domain)**. These words almost never appear in the KFTT corpus, since they do not have any relation with Japanese history or the temples in Kyoto. By comparing **Proposed (in-domain)** and **Proposed (general)**, we see that the latter method translated many OOV words into related words (e.g., "トースター" (toaster) to "refrigerator", or "煮る" (simmer) to "boil") by mistake. This result also indicates that the word vectors extracted from the in-domain corpus will work better than the vectors extracted from the general-domain corpus.

# 6  Conclusions

A cross-lingual projection of word semantic representations has been leveraged to obtain a translation model for unseen (out-of-vocabulary, OOV) words in domain adaptation for SMT. Assuming monolingual corpora for the source and target languages, we induce vector-based semantic representations of words and obtain a projection (translation matrix) from source-language semantic representations into the target-language semantic space. We use this projection to find translation candidates of OOV words and use the cosine similarity to induce the translation probability. Experimental results on domain adaptation from a Kyoto-related domain to a recipe domain confirmed that our method improved BLEU by 0.5–1.5 and 0.1–0.2 for en-ja and ja-en translations, respectively.

In the future, we plan to (i) assign better translation probabilities for non-OOV words that exist in the translation model learned from an out-of-domain parallel corpus, (ii) extend our method to obtain a translation between phrases as in [30], and (iii) combine our method with the existing approaches to domain adaptation for SMT that assumes no bilingual corpus in the target domain.

**Acknowledgments.** The authors thank Nobuhiro Kaji and the anonymous reviewers for their valuable comments and suggestions. We also thank Jun Harashima for providing us the Cookpad recipe corpus. This work was partially supported by JSPS KAKENHI Grant Number 25280111.

# References

1. Brants, T., Popat, A.C., Xu, P., Och, F.J., Dean, J.: Large language models in machine translation. In: Proceedings of the 2007 Joint Conference on Empirical Methods in Natural Language Processing and Computational Natural Language Learning (EMNLP-CoNLL), pp. 858–867 (2007)
2. Irvine, A., Morgan, J., Carpuat, M., Daumé III, H., Munteanu, D.: Measuring machine translation errors in new domains. Trans. Associ. Comput. Linguist. **1**, 429–440 (2013)
3. Costa-Jussà, M.R.: Domain adaptation strategies in statistical machine translation: a brief overview. Knowl. Eng. Rev. **30**, 514–520 (2015)
4. Mansour, S., Ney, H.: Unsupervised adaptation for statistical machine translation. In: Proceedings of the Ninth Workshop on Statistical Machine Translation, pp. 457–465 (2014)
5. Ishiwatari, S., Kaji, N., Yoshinaga, N., Toyoda, M., Kitsuregawa, M.: Accurate cross-lingual projection between count-based word vectors by exploiting translatable context pairs. In: Proceedings of the 19th Conference on Computational Natural Language Learning (CoNLL), pp. 300–304 (2015)
6. Wu, H., Wang, H., Zong, C.: Domain adaptation for statistical machine translation with domain dictionary and monolingual corpora. In: Proceedings of the 22nd International Conference on Computational Linguistics (COLING), pp. 993–1000 (2008)
7. Daumé III, H., Jagarlamudi, J.: Domain adaptation for machine translation by mining unseen words. In: Proceedings of the 49th Annual Meeting of the Association for Computational Linguistics: Human Language Technologies (ACL-HLT), pp. 407–412 (2011)

8. Irvine, A., Quirk, C., Daumé III, H.: Monolingual marginal matching for translation model adaptation. In: Proceedings of the 2013 Conference on Empirical Methods in Natural Language Processing (EMNLP), pp. 1077–1088 (2013)

9. Razmara, M., Siahbani, M., Haffari, R., Sarkar, A.: Graph propagation for paraphrasing out-of-vocabulary words in statistical machine translation. In: Proceedings of the 51st Annual Meeting of the Association for Computational Linguistics (ACL), pp. 1105–1115 (2013)

10. Mathur, P., Keseler, F.B., Venkatapathy, S., Cancedda, N.: Fast domain adaptation of SMT models without in-domain parallel data. In: Proceedings of the 25th International Conference on Computational Linguistics (COLING), pp. 1114–1123 (2014)

11. Yamamoto, H., Sumita, E.: Bilingual cluster based models for statistical machine translation. In: Proceedings of the 2007 Joint Conference on Empirical Methods in Natural Language Processing and Computational Natural Language Learning (EMNLP-CoNLL), pp. 514–523 (2007)

12. Harris, Z.S.: Distributional structure. Word **10**, 146–162 (1954)

13. Firth, J.R.: A synopsis of linguistic theory. In: Studies in Linguistic Analysis, pp. 1–32 (1957)

14. Lund, K., Burgess, C.: Producing high-dimensional semantic spaces from lexical co-occurrence. Behav. Res. Methods Instr. Comput. **28**, 203–208 (1996)

15. Bengio, Y., Ducharme, R., Vincent, P., Janvin, C.: A neural probabilistic language model. J. Mach. Learn. Res. **3**, 1137–1155 (2003)

16. Mikolov, T., Chen, K., Corrado, G., Dean, J.: Efficient estimation of word representations in vector space. In: Proceedings of Workshop at International Conference on Learning Representations (ICLR) (2013)

17. Turney, P.D., Pantel, P., et al.: From frequency to meaning: vector space models of semantics. J. Artif. Intell. Res. (JAIR) **37**, 141–188 (2010)

18. Erk, K.: Vector space models of word meaning and phrase meaning: a survey. Lang. Linguist. Compass **6**, 635–653 (2012)

19. Mikolov, T., Le, Q.V., Sutskever, I.: Exploiting similarities among languages for machine translation. arXiv preprint (2013)

20. Fung, P.: A statistical view on bilingual lexicon extraction: from parallel corpora to non-parallel corpora. In: Proceedings of the Third Conference of the Association for Machine Translation in the Americas (AMTA), pp. 1–17 (1998)

21. Tsunakawa, T., Okazaki, N., Liu, X., Tsujii, J.: A Chinese-Japanese lexical machine translation through a pivot language. ACM Trans. Asian Lang. Inf. Process. (TALIP), **8**, 9:1–9:21 (2009)

22. Neubig, G.: The Kyoto free translation task (2011). http://www.phontron.com/kftt

23. Koehn, P., Knight, K.: Learning a translation lexicon from monolingual corpora. In: Proceedings of ACL Workshop on Unsupervised lexical acquisition. pp. 9–16 (2002)

24. Stolcke, A., et al.: SRILM-an extensible language modeling toolkit. In: Proceedings of the Seventh International Conference on Spoken Language Processing (ICSLP), pp. 901–904 (2002)

25. Och, F.J., Ney, H.: A systematic comparison of various statistical alignment models. Comput. Linguist. **29**, 19–51 (2003)

26. Church, K.W., Hanks, P.: Word association norms, mutual information, and lexicography. Comput. Linguist. **16**, 22–29 (1990)

27. Koehn, P.: Statistical significance tests for machine translation evaluation. In: Proceedings of the 2004 Conference on Empirical Methods in Natural Language Processing (EMNLP), pp. 388–395 (2004)
28. Papineni, K., Roukos, S., Ward, T., Zhu, W.J.: Bleu: a method for automatic evaluation of machine translation. In: Proceedings of the 40th Annual Meeting of the Association for Computational Linguistics (ACL), pp. 311–318 (2002)
29. Lembersky, G., Ordan, N., Wintner, S.: Adapting translation models to translationese improves SMT. In: Proceedings of the 13th Conference of the European Chapter of the Association for Computational Linguistics (EACL), pp. 255–265 (2012)
30. Schwenk, H.: Continuous space translation models for phrase-based statistical machine translation. In: Proceedings of 24th International Conference on Computational Linguistics (COLING): Posters, pp. 1071–1080 (2012)

# Fast-Syntax-Matching-Based Japanese-Chinese Limited Machine Translation

Wuying Liu[1], Lin Wang[1], and Xing Zhang[1,2(✉)]

[1] Laboratory of Language Engineering and Computing,
Guangdong University of Foreign Studies,
Guangzhou 510420, Guangdong, China
wyliu@gdufs.edu.cn, wanglin@nudt.edu.cn,
zximt@aliyun.com
[2] Luoyang University of Foreign Languages, Luoyang 471003, Henan, China

**Abstract.** Limited machine translation (LMT) is an unliterate automatic translation based on bilingual dictionary and sentence bank, and related algorithms can be widely used in natural language processing applications. This paper addresses the Japanese-Chinese LMT problem, proposes two syntactic hypotheses about Japanese language, and designs a fast-syntax-matching-based Japanese-Chinese (FSMJC) LMT algorithm. In which, the fast syntax matching function, a modified version of Levenshtein function, can approximately get the syntactic similarity after the efficient calculating of the formal similarity between two Japanese sentences. The experimental results show that the FSMJC LMT algorithm can achieve the preferable performance with greatly reduced time costs, and prove that our two syntactic hypotheses are effective on Japanese text.

**Keywords:** Limited machine translation · Fast syntax matching
Hiratoken · Formal similarity · Syntactic similarity

## 1 Introduction

Limited machine translation (LMT) is a hybrid machine translation (MT) technique, which combines the techniques of example-based MT [1], information retrieval (IR), fast syntax matching (FSM), and supports the applications of skimming reading, translingual IR or filtering, computer-aided translation based on translation memory [2], statistical MT [3]. The effectiveness of the LMT technique is dependent on the efficient FSM function and big data resources of bilingual dictionary and sentence bank.

The FSM, an important approximate string matching [4], tries to find a sentence with the most similar syntactic structure [5] to a given sentence from a big set of sentences. The early Levenshtein method calculates an edit distance, difference between two strings, to estimate their similarity. The Levenshtein method can be used in approximate string matching, but the computing cost is roughly proportional to the product of the two string lengths. Thus, the strings are short to help improve speed of comparisons usually. Subsequently, various index-based matching methods [6] are proposed, which can obtain preferable performance in approximate string matching.

MT between Japanese and Chinese [7] has been widely investigated since the early days of natural language processing, and many effective algorithms have been proposed [8]. Generally a Japanese sentence is composed of hiragana characters, katakana characters, Chinese characters, letters, Arabic numerals and punctuations. The total number of hiragana characters is 75 only, but the function of hiragana characters is special and powerful. The hiragana characters are mainly used as accessory word, auxiliary verb, okurigana, or transliteration of Chinese characters. Just because the functional symbols are mainly made up of hiragana characters, the occurrence of sequential hiragana characters implicates the syntactic feature of a Japanese sentence. The important feature allows readers to quickly understand Japanese content by vision, and also allows the computer to simplify the Japanese information processing.

In this paper, we mainly investigate the LMT problem from Japanese to Chinese. After statistical analysis, we find that a similar form generally implicates a similar syntax among Japanese sentences, and Japanese text has native syntactic identifiers. Based on the finding of Japanese features, we propose two syntactic hypotheses to support the efficient FSM function. The first is that the formal similarity implicates the syntactic similarity. The second is that the hiragana token (abbreviated as hiratoken) takes on the syntactic identifier of Japanese language. Here, the hiratoken is made up of any sequential hiragana characters separated by katakana characters, Chinese characters, letters, Arabic numerals or punctuations. According to above two syntactic hypotheses, we present a novel architecture for Japanese-Chinese LMT.

## 2   Architecture

Figure 1 shows our Japanese-Chinese LMT architecture, which mainly includes two parts: data preprocessing and limited translating. In the data preprocessing part, the **SenIndexer** receives Japanese-Chinese sentence pairs from *JCSenBank*, and indexes each pair of key-value < *JSen, CSens* > into a hash table *JCSenIndex*; the **HiraIndexer** receives the same sentence pairs, extracts hiratokens from each Japanese sentence, and indexes each pair of key-value < *Hiratoken, JSens* > into a hash table *HiratokenSenIndex*; the **WordIndexer** receives Japanese-Chinese word pairs from *JCDict*, and indexes each pair of key-value < *JWord, CWords* > into a hash table *JCWordIndex*.

In the limited translating part, the **SenRetriever** receives an input Japanese sentence (*JSen*), and retrieves it in the *JCSenIndex*. If the *JCSenIndex* contains the key of *JSen*, the **SenRetriever**, supported by the *JCSenBank*, will output the value of *CSens* corresponding to the *JSen*. If the *JCSenIndex* does not contain the *JSen*, a similar Japanese sentence in the syntactic structure and word-word translations as much as possible will be very helpful. At this time, the **Tokenizer** is triggered, which receives the input *JSen*, and extracts its Japanese words (*JWords*) and hiratokens (*Hiratokens*). The **WordRetriever** retrieves the *JWords* in the *JCWordIndex*, and outputs the

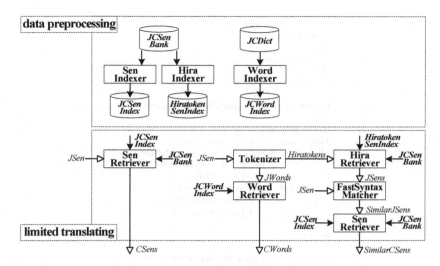

**Fig. 1.** Japanese-Chinese LMT architecture.

corresponding Chinese words (*CWords*). The **HiraRetriever** retrieves the *Hiratokens* in the *HiratokenSenIndex*, and outputs Japanese sentences (*JSens*) supported by the *JCSenBank*. For the input *JSen*, the **FastSyntaxMatcher** calculates one Levenshtein distance to each sentence in the *JSens*, and outputs the most similar Japanese sentences (*SimilarJSens*). The *SimilarJSens* is a set of sentences, because multiple sentences may have the same shortest Levenshtein distance to the given *JSen*. The same **SenRetriever** will output the Chinese sentences (*SimilarCSens*) related to the *SimilarJSens*.

The **FastSyntaxMatcher** is only a loop version of the classical Levenshtein function, which depends on the **HiraRetriever** to fast find sentences with a similar syntactic structure. In the **HiraRetriever**, there is a preset Top-n value to truncate the number of output Japanese sentences, which have a similar syntactic structure for containing the same syntactic identifiers.

## 3   Algorithm

Figure 2 gives the pseudo-code for our fast-syntax-matching-based Japanese-Chinese (FSMJC) LMT algorithm consisting of two main functions: *translate* and *preprocess*. And the *translate* function calls a key *fastsyntaxmatch* function. The FSMJC LMT algorithm takes the translating as an index retrieving process and takes the preprocessing as an indexing process.

If we use N and n to denote the number of sentences in the *JCSenBank* and the preset Top-n value, respectively, the maximal complexity of the FSMJC LMT algorithm will be $O(n)$ and be independent of the N value. Usually, the n is a small integer and is far smaller than the N value. Thus, the FSMJC LMT algorithm is acceptable in practical applications.

```
1.      // FSMJC LMT Algorithm
2.      JCSenBank jcsb;
3.      JCDict jcd;
4.      JCSenIndex jcsi;
5.      HiratokenSenIndex hsi;
6.      JCWordIndex jcwi;
7.
8.      Function Result: translate(JSen js, Integer n)
9.      Result r;
10.     If (jcsi.contain(js))
11.     Then  r.add((CSen[]) jcsi.get(js, jcsb));
12.     Else   Hiratoken[] hs ← Tokenizer.htokenize(js);
13.            JWord[] jws ← Tokenizer.wtokenize(js);
14.            r.add((CWord[]) jcwi.get(jws));
15.            JSen[] jss ← hsi.get(hs, jcsb, n);
16.            JSen[] sjss ← fastsyntaxmatch(js, jss);
17.            r.add((CSen[]) jcsi.get(sjss, jcsb));
18.     End If
19.     Return r.
20.
21.     Function JSen[]: fastsyntaxmatch(JSen js, JSen[] jss)
22.     JSen[] sjss;
23.     Float min ← Float.max;
24.     For Integer i ← 1 To jss.size Do
25.         Float f ← levenshteindis(js, jss[i]);
26.         If (f < min)
27.         Then sjss.clear;
28.              sjss.add(jss[i]);
29.              min ← f;
30.         Else If (f = min)
31.         Then sjss.add(jss[i]);
32.         End If
33.     End For
34.     Return sjss.
35.
36.     Function void: preprocess()
37.     jcsi ← senindex(jcsb);
38.     hsi ← hiraindex(jcsb);
39.     jcwi ← wordindex(jcd).
```

**Fig. 2.** FSMJC LMT algorithm.

# 4 Experiment

In order to validate the effectiveness of our LMT technique, we do the following experiment. Firstly, we test the *preprocess* function to explain that the data structure of our indexes is space-time-efficient. Secondly, we implement a baseline syntax matcher based on the classical Levenshtein method. The IR evaluation measure and associated evaluation methodology are applied to evaluate that the two hypotheses are effective in our Japanese FSM method. Finally, we compare our LMT result with the manual golden standard to discuss the performance of the FSMJC LMT algorithm.

## 4.1  Dataset

We use a public dataset of the Japanese-Chinese Sentence Bank (JCSB)[1] and our Japanese-Chinese dictionary in the experiment. The JCSB contains 371,712 pairs of Japanese-Chinese sentences. The Japanese-Chinese dictionary contains 436,199 pairs of Japanese-Chinese words, which can be used as the *JCDict* in Fig. 1. We run the experiment on a computer with 4.00 GB RAM and Intel Core i5-2520 M CPU.

In order to construct the scientific test set (*TSet*) and the *JCSenBank* mentioned in Sect. 2, we use a simple random sampling method to extract 1,000 pairs of sentences from the JCSB as a *TSet*, and the remaining 370,712 pairs as a *JCSenBank*. We run the simple random sampling 10 times, and get 10 pairs of < *JCSenBank, TSet* >. We do the following experiment in each pair of < *JCSenBank, TSet* >, and will get 10 groups of experimental results. In the following discussions, we will report the mean value of performance among the 10 groups.

## 4.2  Preprocess Result and Discussion

We run the *preprocess* function to make three indexes. Table 1 shows the mean run-time and file space. Averagely, the raw plain text file of the *JCSenBank* occupies 33 MB space, and it will cost 6 s to make the *JCSenBank* into a *HiratokenSenIndex*, which will occupy 29 MB space.

**Table 1.**  Time and space.

|            | Raw space (MB) | Indexing time (Sec) | Index space (MB) |                   |
|------------|----------------|---------------------|------------------|-------------------|
| *JCSenBank* | 33            | 1                   | 35               | *JCSenIndex*      |
| *JCSenBank* | 33            | 6                   | 29               | *HiratokenSenIndex* |
| *JCDict*   | 17             | 2                   | 21               | *JCWordIndex*     |

Because there is not a repetitive item both in the *JCSenBank* and the *JCDict*, the size of raw file is close to that of corresponding index. It is equivalent to converting the file storage format, so the index structure is space-efficient. And each retrieving in an index has a constant time complexity according to a string hash function, so the index structure is also time-efficient.

## 4.3  FSM Result and Discussion

We focus more on the FSM issue, and ignore the exact-hit-based translation memory. There is just not an intersection between the *JCSenBank* and the *TSet*, so the *JCSenIndex* will not be hit during the experiment.

Supported by the above indexes, we run our *translate* function and output the similar Japanese sentences (*SimilarJSens*) for each sentence in the *TSet*. We also

---

[1] http://cbd.nichesite.org/CBD2014D002.htm.

implement a baseline syntax matcher with no need for indexes. If the number of Japanese sentences is $m$ in the *JCSenBank*, the baseline will calculate $m$ Levenshtein distances and output the most similar Japanese sentences (*MSimilarJSens*) for a given sentence. Using the same *TSet*, we run the baseline and regard its result as the pseudo golden standard.

$$\text{Precision} = \frac{R}{R+U} \tag{1}$$

$$\text{Recall} = \frac{R}{T} \tag{2}$$

$$F1 = \frac{2\,\text{Precision} \cdot \text{Recall}}{\text{Precision} + \text{Recall}} \tag{3}$$

We report the runtime and the classical Precision, Recall, F1 measure to evaluate the experimental result. The value of Precision, Recall, F1 belongs to [0, 1], where 1 is optimal. The three classical measures are computed as Eqs. (1)–(3) separately. Where the $T$ denotes the total number of sentences in the *MSimilarJSens*, the $R$ denotes the number of related sentences in the *SimilarJSens*, and the $U$ denotes the number of unrelated sentences in the *SimilarJSens*.

As expected, the runtime of the baseline in the *TSet* is the horrible 3,088 s. It will cost 3 s that the classical Levenshtein function compares 370,712 times for a given Japanese sentence. Figure 3 shows the experimental result of our *translate* function under different Top-n value from 10 to 200. We find that the *translate* function can get the 37% precision at the time cost of 40 s, and it can reduce the time cost from 3,088 s to

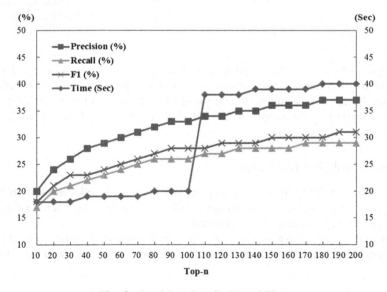

**Fig. 3.** Precision, Recall, F1 and Time.

40 s. On the situation of Top-n = 200, averagely each sentence costs 40/1,000 = 0.04 s. That is to say, in one second, the **translate** function can complete 25 FSM operations in a big set of 370,712 sentences.

Moreover, the 37% precision is dependent of the pseudo golden standard from full formal similarity. The actual precision of our FSM is higher than 37%, because the formal similarity is stricter than the syntactic similarity. The result proves that Japanese text has explicit syntactic identifiers, and this feature helps the FSM method to calculate formal similarity efficiently, which is the substantial clause of its success.

## 4.4    LMT Result and Discussion

Among MT evaluation measures, the BLEU4 is an effective one [9], which is based on the smoothed modified $k$-gram precision for multiple reference texts, and is computed as Eq. (4). Where the $c$ denotes the length of candidate text, the $r$ denotes the length of effective reference text, and the $p_k$ denotes the smoothed modified $k$-gram precision.

$$\text{BLEU4} = exp(min(0, 1 - \frac{r}{c}) + \frac{1}{4}\sum_{k=1}^{4} ln(p_k)) \tag{4}$$

Comparing with the golden standard, we calculate the BLEU4 value to evaluate our LMT result. The values of **BLEU4**.*JSens* are calculated from our candidate source *SimilarJSens* and the pseudo reference source *MSimilarJSens*. The values of **BLEU4**. *CSens* are calculated from our candidate translation *SimilarCSens* and the manual reference translation.

Table 2 shows the detailed experimental result. With the Top-n value from 10 to 200, the time cost only increases a little, while that of the baseline is 3,088 s. On the situation of Top-n = 200, the FSMJC LMT algorithm can achieve the performance (**BLEU4**.*JSens* = 0.4792, **BLEU4**.*CSens* = 0.1797).

We select two example sentences randomly in the experiment to further expatiate the advantage of our algorithm. Table 3 shows the two Japanese sentences ([*JSen*]) and the corresponding Chinese translation sentences ([*CSen*]) in the JCSB, the sentence [*MSimilarJSens*] of pseudo golden standard, and the sentence recommended by our FSMJC LMT algorithm.

For the example sentence 1, we find that the baseline Levenshtein algorithm and our algorithm can get the same result ([*MSimilarJSens*] and [*SimilarJSens*]), and there is only a few difference between sentences ([*JSen*] and [*SimilarJSens*]). If a Chinese person sends a Japanese sentence ([*JSen*]) of the example sentence 1 to a computer-aided translation system, and can get a pair of Japanese-Chinese sentences ([*SimilarJSens*] and [*SimilarCSens*]). We believe that the person will guess well.

For the example sentence 2, the baseline Levenshtein algorithm recommends a similar sentence ([*MSimilarJSens*]), while the FSMJC LMT algorithm recommends three similar ones ([*SimilarJSens*.0], [*SimilarJSens*.1], and [*SimilarJSens*.2]). Judging by the linguists, the three similar sentences have the most similar syntactic structure to the given Japanese sentence [*JSen*]. Though the FSMJC LMT algorithm does not hit the pseudo golden standard for the sentence [*JSen*], the linguists believe that the sentence [*SimilarJSens*.0] is a fine recommendation especially.

**Table 2.** Time and BLEU4.

| Top-n | Time (Sec) | BLEU4.*JSens* | BLEU4.*CSens* |
|-------|-----------|--------------|--------------|
| 10  | 18 | 0.3248 | 0.1688 |
| 20  | 18 | 0.3627 | 0.1724 |
| 30  | 18 | 0.3835 | 0.1746 |
| 40  | 19 | 0.3956 | 0.1746 |
| 50  | 19 | 0.4075 | 0.1756 |
| 60  | 19 | 0.4198 | 0.1772 |
| 70  | 19 | 0.4278 | 0.1775 |
| 80  | 20 | 0.4348 | 0.1776 |
| 90  | 20 | 0.4408 | 0.1784 |
| 100 | 20 | 0.4422 | 0.1786 |
| 110 | 38 | 0.4455 | 0.1794 |
| 120 | 38 | 0.4492 | 0.1795 |
| 130 | 38 | 0.4549 | 0.1793 |
| 140 | 39 | 0.4562 | 0.1791 |
| 150 | 39 | 0.4619 | 0.1790 |
| 160 | 39 | 0.4652 | 0.1791 |
| 170 | 39 | 0.4704 | 0.1794 |
| 180 | 40 | 0.4713 | 0.1794 |
| 190 | 40 | 0.4756 | 0.1797 |
| 200 | 40 | 0.4792 | 0.1797 |

**Table 3.** Two example sentences.

| | Example Sentence 1 | Example Sentence 2 |
|---|---|---|
| **JCSB** | [*JSen*]2003 年の環境汚染処理投資額は国内総生産の 1.39%を占めた。<br>[*CSen*]2003 年环境污染治理投资占国内生产总值的 1.39%。 | [*JSen*]彼は歌がへたなので人を静めることができず、聴衆はみな帰ってしまった。<br>[*CSen*]他唱得不好，压不住人，听众都走了。 |
| **Pseudo Golden Standard** | [*MSimilarJSens*]2006 年、環境汚染処理への投資は GDP の 1.15%を占めた。<br>[*MSimilarCSens*]2006 年，环境污染治理投资占国内生产总值的 1.15%。 | [*MSimilarJSens*]彼はみんなの意見に逆らうことができず、やむなく承知した。<br>[*MSimilarCSens*]他拗不过大家的意见，只得依从 |
| **FSMJC LMT Algorithm** | [*SimilarJSens*]2006 年、環境汚染処理への投資は GDP の 1.15%を占めた。<br>[*SimilarCSens*]2006 年，环境污染治理投资占国内生产总值的 1.15%。 | [*SimilarJSens*.0]彼は様子がおかしいので、とんぼ返りに帰ってしまった。<br>[*SimilarCSens*.0]他看情形不对，转身就走了。<br>[*SimilarJSens*.1]彼は寝返りを一つうったが、また眠ってしまった。<br>[*SimilarCSens*.1]他翻了一个身，又睡着了。<br>[*SimilarJSens*.2]彼は不当な取り扱いを受けたと思い、ふてくされて帰ってしまった。<br>[*SimilarCSens*.2]他觉得受了委屈，一赌气就走了。 |

We will further explain the belief of linguists by calculating a hit rate manually. Table 4 shows the token bitmap of the example sentence 2. There are three lines for each sentence: the first line is the symbol bitmap, the second line is the hiratoken bitmap, and the third line is the word bitmap segmented by linguists.

**Table 4.** Token bitmap of example sentence 2.

[*JSen*]

| 彼 | は | 歌 | が | へ | た | な | の | で | 人 | を | 静 | め | る | こ | と | が | で | き | ず | 、 | 聴 | 衆 | は | み | な | 帰 | っ | て | し | ま | っ | た | 。 |
|---|---|---|---|---|---|---|---|---|---|---|---|---|---|---|---|---|---|---|---|---|---|---|---|---|---|---|---|---|---|---|---|---|---|
| 1 | 2 | 3 | 4 | 5 | 6 | 7 | 8 | 9 | 10 | 11 | 12 | 13 | 14 | 15 | 16 | 17 | 18 | 19 | 20 | 21 | 22 | 23 | 24 | 25 | 26 | 27 | 28 | 29 | 30 | 31 | 32 | 33 | 34 |

| は | が へ た な の で | を | め る こ と が で き ず | は み な | っ て し ま っ た |
|---|---|---|---|---|---|
| 1 | 2 | 3 | 4 | 5 | 6 |

| 彼 | は | 歌 | が | へ た な の で | 人 | を | 静 | め る こ と | が | で き | ず | 聴 衆 | は | み な | 帰 っ | て | し ま っ | た | 。 |
|---|---|---|---|---|---|---|---|---|---|---|---|---|---|---|---|---|---|---|---|
| 1 | 2 | 3 | 4 | 5 | 6 | 7 | 8 | 9 | 10 | 11 | 12 | 13 | 14 | 15 | 16 | 17 | 18 | 19 | 20 | 21 | 22 |

[*MSimilarJSens*]

| 彼 | は | み | ん | な | の | 意見 | に | 逆 | ら | う | こ | と | が | で | き | ず | 、 | や | む | な | く | 承知 | し | た | 。 |
|---|---|---|---|---|---|---|---|---|---|---|---|---|---|---|---|---|---|---|---|---|---|---|---|---|---|
| 1 | 2 | 3 | 4 | 5 | 6 | 7 | 8 | 9 | 10 | 11 | 12 | 13 | 14 | 15 | 16 | 17 | 18 | 19 | 20 | 21 | 22 | 23 | 24 | 25 | 26 | 27 | 28 |

| は み ん な の | に | ら う こ と が で き ず | や む な く | し た |
|---|---|---|---|---|
| 1 | 2 | 3 | 4 | 5 |

| 彼 | は | み ん な | の | 意見 | に | 逆 ら う | こ | と | が | で き | ず | 、 | や む な く | 承知 | し た | 。 |
|---|---|---|---|---|---|---|---|---|---|---|---|---|---|---|---|---|
| 1 | 2 | 3 | 4 | 5 | 6 | 7 | 8 | 9 | 10 | 11 | 12 | 13 | 14 | 15 | 16 |

[*SimilarJSens*.0]

| 彼 | は | 様子 | が | お | か | し | い | の | で | 、 | と | ん | ぽ | 返 | り | に | 帰 | っ | て | し | ま | っ | た | 。 |
|---|---|---|---|---|---|---|---|---|---|---|---|---|---|---|---|---|---|---|---|---|---|---|---|---|
| 1 | 2 | 3 | 4 | 5 | 6 | 7 | 8 | 9 | 10 | 11 | 12 | 13 | 14 | 15 | 16 | 17 | 18 | 19 | 20 | 21 | 22 | 23 | 24 | 25 | 26 |

| は | が お か し い の で | と ん ぽ | り に | っ て し ま っ た |
|---|---|---|---|---|
| 1 | 2 | 3 | 4 | 5 |

| 彼 | は | 様子 | が | お か し い | の で | 、 | と ん ぽ 返 | り に | 帰 っ | て | し ま っ | た | 。 |
|---|---|---|---|---|---|---|---|---|---|---|---|---|---|
| 1 | 2 | 3 | 4 | 5 | 6 | 7 | 8 | 9 | 10 | 11 | 12 | 13 | 14 |

[*SimilarJSens*.1]

| 彼 | は | 寝 | 返 | り | を | 一 | つ | う | っ | た | が | 、 | ま | た | 眠 | っ | て | し | ま | っ | た | 。 |
|---|---|---|---|---|---|---|---|---|---|---|---|---|---|---|---|---|---|---|---|---|---|---|
| 1 | 2 | 3 | 4 | 5 | 6 | 7 | 8 | 9 | 10 | 11 | 12 | 13 | 14 | 15 | 16 | 17 | 18 | 19 | 20 | 21 | 22 | 23 |

| は | り を | つ う っ た が | ま た | っ て し ま っ た |
|---|---|---|---|---|
| 1 | 2 | 3 | 4 | 5 |

| 彼 | は | 寝 返 り | を | 一 つ | う っ | た | が | 、 | ま た | 眠 っ | て | し ま っ | た | 。 |
|---|---|---|---|---|---|---|---|---|---|---|---|---|---|---|
| 1 | 2 | 3 | 4 | 5 | 6 | 7 | 8 | 9 | 10 | 11 | 12 | 13 | 14 | 15 |

[*SimilarJSens*.2]

| 彼 | は | 不 | 当 | な | 取 | り | 扱 | い | を | 受 | け | た | と | 思 | い | 、 | ふ | て | く | さ | れ | て | 帰 | っ | て | し | ま | っ | た | 。 |
|---|---|---|---|---|---|---|---|---|---|---|---|---|---|---|---|---|---|---|---|---|---|---|---|---|---|---|---|---|---|---|
| 1 | 2 | 3 | 4 | 5 | 6 | 7 | 8 | 9 | 10 | 11 | 12 | 13 | 14 | 15 | 16 | 17 | 18 | 19 | 20 | 21 | 22 | 23 | 24 | 25 | 26 | 27 | 28 | 29 | 30 | 31 |

| は | な | り | い を | け た と | い | ふ て く さ れ て | っ て し ま っ た |
|---|---|---|---|---|---|---|---|
| 1 | 2 | 3 | 4 | 5 | 6 | 7 | 8 |

| 彼 | は | 不 当 な | 取 り 扱 い | を | 受 け | た | と | 思 い | ふ て く さ れ て | 帰 っ | て | し ま っ | た | 。 |
|---|---|---|---|---|---|---|---|---|---|---|---|---|---|---|
| 1 | 2 | 3 | 4 | 5 | 6 | 7 | 8 | 9 | 10 | 11 | 12 | 13 | 14 | 15 | 16 | 17 |

According to the above token bitmap, we can easily calculate the hit rate for each sentence. Table 5 shows the detailed hit rate of various tokens. We can find that the sentence [*MSimilarJSens*] of pseudo golden standard has lower hit rates in three tokens and the top hit rate belongs to the sentence [*SimilarJSens*.0]. Especially the word hit rate of the sentence [*SimilarJSens*.0] is the top 0.714 among those of the four sentences, which proves that the sentence recommended by our algorithm has a more similar syntactic structure to the given sentence [*JSen*].

**Table 5.** Hit rate of various tokens.

| Token | Sentence | Hit number | Total number | Hit rate |
|---|---|---|---|---|
| Symbol | [*MSimilarJSens*] | 15 | 28 | 15/28 = 0.536 |
| | [*SimilarJSens*.0] | 14 | 26 | 14/26 = **0.538** |
| | [*SimilarJSens*.1] | 12 | 23 | 12/23 = 0.522 |
| | [*SimilarJSens*.2] | 14 | 31 | 14/31 = 0.452 |
| Hiratoken | [*MSimilarJSens*] | 0 | 5 | 0/5 = 0.000 |
| | [*SimilarJSens*.0] | 2 | 5 | 2/5 = **0.400** |
| | [*SimilarJSens*.1] | 2 | 5 | 2/5 = **0.400** |
| | [*SimilarJSens*.2] | 2 | 8 | 2/8 = 0.250 |
| Word | [*MSimilarJSens*] | 9 | 16 | 9/16 = 0.563 |
| | [*SimilarJSens*.0] | 10 | 14 | 10/14 = **0.714** |
| | [*SimilarJSens*.1] | 9 | 15 | 9/15 = 0.600 |
| | [*SimilarJSens*.2] | 9 | 17 | 9/17 = 0.529 |

The above experiment validates the effectiveness of our LMT technique: (I) the data structure is space-time-efficient; (II) the two hypotheses are effective in the Japanese FSM method; and (III) the FSMJC LMT algorithm can achieve the preferable performance with greatly reduced time costs.

## 5    Conclusion

This paper presents a novel bilingual LMT architecture, and validates that our two hypotheses are effective for Japanese syntax matching. Currently, more finding similar sentence with manual translation and useful dictionary hints, than providing an imperfect translation will help us to make a tripping translation.

Further research will add some rules of probabilistic syntax frame to assemble our outputs of *CWords* and *SimilarCSens* into a fluent Chinese sentence. We also expect to discover explicit formal syntactic identifiers for other languages, and to transfer above research productions to the languages.

**Acknowledgements.** The research is supported by the Featured Innovation Project of Guangdong Province (No.2015KTSCX035), National Social Science Foundation of China (No.12BYY136) and the National Natural Science Foundation of China (No.61402119).

## References

1. Brown, R.D.: The CMU-EBMT machine translation system. Mach. Transl. **25**(2), 179–195 (2011)
2. Biçici, E., Dymetman, M.: Dynamic translation memory: using statistical machine translation to improve translation memory fuzzy matches. In: Gelbukh, A. (ed.) CICLing 2008. LNCS, vol. 4919, pp. 454–465. Springer, Heidelberg (2008). https://doi.org/10.1007/978-3-540-78135-6_39

3. Callison-Burch, C., Talbot, D., Osborne, M.: Statistical machine translation with word- and sentence-aligned parallel corpora. In: Proceedings of ACL 2004, the 42nd Meeting of the Association for Computational Linguistics, Barcelona, Spain, Main Volume, pp. 175–182 (2004)
4. Koehn, P., Senellart, J.: Fast approximate string matching with suffix arrays and A* parsing. In: Proceedings of AMTA 2010, The 9th Biennial Conference of the Association for Machine Translation in the Americas, Denver, Colorado, USA (2010)
5. Vanallemeersch, T., Vandeghinste, V.: Improving fuzzy matching through syntactic knowledge. In: Proceedings of the 36th Translating and the Computer Conference, Westminster, London, UK (2014)
6. Navarro, G., Baeza-Yates, R., Sutinen, E., Tarhio, J.: Indexing methods for approximate string matching. IEEE Data Eng. Bull. **24**(4), 19–27 (2001)
7. Dan, H., Sudoh, K., Wu, X., Duh, K., Tsukada, H., Nagata, M.: Head finalization reordering for Chinese-to-Japanese machine translation. In: Proceedings of the Sixth Workshop on Syntax, Semantics and Structure in Statistical Translation, Stroudsburg, PA, USA, pp. 57–66 (2012)
8. Chu, C., Nakazawa, T., Kawahara, D., Kurohashi, S.: Chinese-Japanese Machine Translation Exploiting Chinese Characters. ACM Transactions on Asian Language Information Processing, **12**(4) (2013). Article 16
9. Lin, C.-Y., Och, F.J.: Orange: a method for evaluating automatic evaluation metrics for machine translation. In: Proceedings of the 20th International Conference on Computational Linguistics, Geneva, Switzerland (2004)

# A Classifier-Based Preordering Approach for English-Vietnamese Statistical Machine Translation

Viet Hong Tran[1,2], Huyen Thuong Vu[2,3(✉)],

Vinh Van Nguyen[2(✉)], and Minh Le Nguyen[4(✉)]

[1] University of Economic and Technical Industries, Hanoi, Vietnam
thviet@uneti.edu.vn
[2] University of Engineering and Technology,
Vietnam National University, Hanoi, Vietnam
huyenvt@tlu.edu.vn, vinhnv@vnu.edu.vn
[3] ThuyLoi University, Hanoi, Vietnam
[4] Japan Advanced Institute of Science and Technology, Nomi, Japan
nguyenml@jaist.ac.jp

**Abstract.** Reordering is of essential importance problem for phrase based statistical machine translation (SMT). In this paper, we propose an approach to automatically learn reordering rules as preprocessing step based on a dependency parser in phrase-based statistical machine translation for English to Vietnamese. We used dependency parsing and rules extracting from training the features-rich discriminative classifiers for reordering source-side sentences. We evaluated our approach on English-Vietnamese machine translation tasks, and showed that it outperform the baseline phrase-based SMT system.

**Keywords:** Natural language processing · Machine translation
Phrase-based statistical machine translation

## 1 Introduction

Phrase-based statistical machine translation [1,2] is the state-of-the-art of SMT because of its power in modelling short reordering and local context. However, with phrase-based SMT, long distance reordering is still problematic. The reordering problem (global reordering) is one of the major problems. In recent years, many reordering methods have been proposed to tackle the long distance reordering problem.

Many solutions to the reordering problem have been proposed, such as syntax-based model [3], lexicalized reordering [2], and tree-to-string methods [4]. Chiang [3] shows significant improvement by keeping the strengths of phrases, while incorporating syntax into SMT. Some approaches have been applied at the word-level [5]. They are particularly useful for language with rich morphology, for reducing data sparseness. Other kinds of syntax reordering methods require

© Springer International Publishing AG, part of Springer Nature 2018
A. Gelbukh (Ed.): CICLing 2016, LNCS 9624, pp. 74–87, 2018.
https://doi.org/10.1007/978-3-319-75487-1_7

parser trees, such as the work in [5,6]. The parsed tree is more powerful in capturing the sentence structure. However, it is expensive to create tree structure, and building a good quality parser is also a hard task. All the above approaches require much decoding time, which is expensive.

The approach we are interested in here is to balance the quality of translation with decoding time. Reordering approaches as a preprocessing step [7–10] are very effective (significant improvement over state of-the-art phrase-based and hierarchical machine translation systems and separately quality evaluation of each reordering models).

Inspiring this preprocessing approach, we have proposed a combine approach which preserves the strength of phrase-based SMT in local reordering and decoding time as well as the strength of integrating syntax in reordering. Moreover, dependency-based preordering is good for the long distance words and phrases. Firstly, we use dependency parsing for preprocessing with training and testing. Second, we use the features discriminative classifiers from data training to extract rules which are learnt automatically from parallel corpus to the dependency tree. Besides these rules, we can use extracting features in source-side that directly predicts the target-side word order to be applied as a preprocessing step in phrase-based machine translation. The experiment results from English-Vietnamese pair showed that our approach achieves significant improvements over MOSES which is the state-of-the art phrase based system.

The rest of this paper is structured as follows. Section 2 reviews the related works. Section 3 briefly introduces classifier-based Preordering for Phrase-based SMT. Section 4 describes experimental results. Section 5 discusses the experimental results. And, conclusions are given in Sect. 6.

## 2   Related Works

The difference of the word order between source and target languages is the major problems in phrase-based statistical machine translation. Preordering (reordering as preprocessing) is an approach for tacking the problem, which modifies the word order of an input sentence in a source language to have the word order in a target language (Fig. 1).

Many preordering methods using syntactic information have been proposed to solve the reordering problem. [5,8] presented a preordering method which used manually created rules on parse trees and linguistic knowledge for a language pair is necessary to create such rules. Other preordering methods using automatically created reordering rules or utilizing statistical classifier have been studied [7,10–13].

[5] developed a clause detection and used some handwritten rules to reorder words in the clause. Partly, [7,14] built an automatic extracted syntactic rules.

[8] described a method using dependency parse tree and a flexible rule to perform the reordering of subject, object, etc. These rules were written by hand, but [8] showed that an automatic rule learner can be used.

[9,10] described a method using discriminative classifiers to directly predict the final word order. [13] presented a simple preordering approach for machine translation based on a feature-rich logistic regression model to predict whether

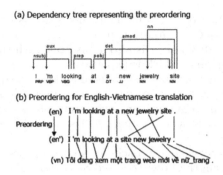

**Fig. 1.** A example of preordering for English-Vietnamese translation.

two children of the same node in the source-side parse tree should be swapped or not. [15] introduced a novel pre-ordering approach based on dependency parsing for Chinese-English SMT. [16] proposed a simple procedure to train a discriminative preordering model. The main idea was to obtain oracle labels for each node by maximizing Kendall's $\tau$ of word alignments. [17] presented an efficient incremental top down parsing method for preordering based on Bracketing Transduction Grammar.

Our approach is closest similarity to [10] but it has a few differences. Firstly, we aimed to develop the phrase-based translation model using dependency parse of source sentence to translate from English to Vietnamese. Secondly, we extracted automatically a set of English to Vietnamese transformation rules from English-Vietnamese parallel corpus by using SVM classification model [18] with lexical and syntactic features based on dependency parsing of source sentence. In [10] used maximum entropy classifiers with the GradBoost. Thirdly, we used the English to Vietnamese transformation rules that directly predict target-side word as a preprocessing step in phrase-based machine translation. As the same with [7,14], we also applied preprocessing in both training and decoding time.

# 3    Classifier-Based Preordering for Phrase-Based SMT

## 3.1    Phrase-Based SMT

In this section, we will describe the phrase-based SMT system which was used for the experiments. Phrase-based SMT, as described by [1] translates a source sentence into a target sentence by decomposing the source sentence into a sequence of source phrases, which can be any contiguous sequences of words (or tokens treated as words) in the source sentence. For each source phrase, a target phrase translation is selected, and the target phrases are arranged in some order to produce the target sentence. A set of possible translation candidates created in this way were scored according to a weighted linear combination of feature values, and the highest scoring translation candidate was selected as the translation of the source sentence. Symbolically,

$$\hat{t} = \text{argmax}_{t,a} \sum_{i=1}^{n} \lambda_i f_j(s, t, a) \tag{1}$$

when s is the input sentence, t is a possible output sentence, and a is a phrasal alignment that specifies how t is constructed from s, and $\hat{t}$ is the selected output sentence. The weights $\lambda_i$ associated with each feature $f_i$ are tuned to maximize the quality of the translation hypothesis selected by the decoding procedure that computes the argmax. The log-linear model is a natural framework to integrate many features. The baseline system uses the following features:

- the probability of each source phrase in the hypothesis given the corresponding target phrase.
- the probability of each target phrase in the hypothesis given the corresponding source phrase.
- the lexical score for each target phrase given the corresponding source phrase.
- the lexical score for each source phrase given the corresponding target phrase.
- the target language model probability for the sequence of target phrase in the hypothesis.
- the word and phrase penalty score, which allow to ensure that the translation does not get too long or too short.
- the distortion model allows for reordering of the source sentence.

The probabilities of source phrase given target phrases, and target phrases given source phrases, are estimated from the bilingual corpus.

[1] used the following distortion model (reordering model), which simply penalizes nonmonotonic phrase alignment based on the word distance of successively translated source phrases with an appropriate value for the parameter $\alpha$:

$$d(a_i - b_{i-1}) = \alpha^{|a_i - b_{i-1} - 1|} \tag{2}$$

Current time, state-of-the-art phrase-based SMT system using the lexicalized reordering model in Moses toolkit. In our work, we also used Moses to evaluate on English-Vietnamese machine translation tasks.

## 3.2  Classifier-Based Preordering

In this section, we describe a the learning model that can transform the word order of an input sentence to an order that is natural in the target language. English is used as source language, while Vietnamese is used as target language in our discussion about the word orders.

For example, when translating the English sentence:

*I'm looking at a new jewelry site.*

to Vietnamese, we would like to reorder it as:

*I'm looking at a site new jewelry.*

And then, this model will be used in combination with translation model. The feature is built for "site, a, new, jewelry" family in Fig. 2:

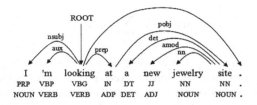

**Fig. 2.** A example with POS tags and dependency parser.

NN, DT, det, JJ, amod, NN, nn, 1230, 1023

We use the dependency grammars and the differences of word order between English and Vietnamese to create a set of the reordering rules. From part-of-speech (POS) tag and parse the input sentence, producing the POS tags and head-modifier dependencies shown in Fig. 2. Traversing the dependency tree starting at the root to reordering. We determine the order of the head and its children (independently of other decisions) for each head word and continue the traversal recursively in that order. In the above example, we need to decide the order of the head "looking" and the children "I", "m", and "site.".

The words in sentence are reordered by a new sequence learned from training data using multi-classifier model. We use SVM classification model [18] that supports multi-class prediction. The class labels are corresponding to reordering sequence, so it is enable to select the best one from many possible sequences.

**Table 1.** Set of features used in training data from corpus English-Vietnamese

| Feature | Description |
|---------|-------------|
| T | The heads POS tag |
| 1T | The first childs POS tag |
| 1L | The first childs syntactic label |
| 2T | The second childs POS tag |
| 2L | The second childs syntactic label |
| 3T | The third childs POS tag |
| 3L | The third childs syntactic label |
| 4T | The fourth childs POS tag |
| 4L | The fourth childs syntactic label |
| O1 | The sequence of head and its children in source alignment |
| O2 | The sequence of head and its children in target alignment |

**Table 2.** Examples of rules and reorder source sentences

| Pattern | Order | Example |
|---|---|---|
| NN, DT, det, JJ, amod, NN, nn | 1,0,2,3 | I'm looking at a new jewelry site |
| | | → I'm looking at a site new jewelry |
| NNS, JJ, amod, CC, cc, NNS, con | 2,1,0,3 | It faced a blank wall |
| | | → It faced a wall blank |
| NNP, NNP, nn, NNP, nn | 2,1,0 | It's a social phenomenon |
| | | → It's a phenomenon social |

**Features.** The features extracted based on dependency tree includes POS tag and alignment information. We traverse the tree from the top, in each family we create features with the following information:

- The head's POS tag,
- The first child's POS tag, the first child's syntactic label.
- The second child's POS tag, the second child's syntactic label.
- The third child's POS tag, the third child's syntactic label.
- The fourth child's POS tag, the fourth child's syntactic label.
- The sequence of head and its children in source alignment.
- The sequence of head and its children in target alignment. It is class label for SVM classifier model.

We limited ourself by processing families that have less than five children based on counting total families in each group: 1 head and 1 child, 1 head and 2 children, 1 head and 3 children, 1 head and 4 children ... We found out that the most common families appear (80%) in our training sentences is less than and equal four children.

We trained a separate classifier for each number of possible children. In hence, the classifiers learn to trade off between a rich set of overlapping features. List of features are given in Table 1.

We use SVM classification model [18] in the WEKA tools [19] that supports multi-class prediction. Since it naturally supports multi-class prediction and can therefore be used to select one out of many possible permutations. The learning algorithm produces a sparse set of features. In our experiments the our models have typically only a few 50K non-zero feature weights English-Vietnamese language pairs.

```
Algorithm 1 Extract rules
    input: dependency trees of source sentences
           and alignment pairs;
    output: set of automatically rules;
    for each family in dependency trees of subset
    and alignment pairs of sentences do
        generate feature (pattern + order) ;
    end for
    Build model from set of features;
    for each family in dependency trees in the rest
    of the sentences do
        generate pattern for prediction;
        get predicted order from model;
        add (pattern, order) as new rule in set of rules;
    end for
```

```
Algorithm 2 Apply rule
    input: source-side dependency trees , set of rules;
    output: set of new sentences;
    for each dependency tree do
        for each family in tree do
            generate pattern
            get order from set of rules based on pattern
            apply transform
        end for
        Build new sentence;
    end for
```

When extracting the features, every word can be represented by its word identity, its POS-tags from the treebank, syntactic label. We also include pairs of these features, resulting in potentially bilexical features.

**Training Data for Preordering.** In this section, we describe a method to build training data for a pair English to Vietnamese. Our purpose is to reconstruct the word order of input sentence to an order that is arranged as Vietnamese words order. For example with the English sentence in Fig. 2:

*I'm looking at a new jewelry site.*

is transformed into Vietnamese order:

*I'm looking at a site new jewelry.*

For this approach, we first do preprocessing to encode some special words and parser the sentences to dependency tree using Stanford Parser [20]. Then, we use target to source alignment and dependency tree to generate features. We add source, target alignment, POS tag, syntactic label of word to each node in the dependency tree. For each family in the tree, we generate a training instance if it has less than and equal four children. In case, a family has more than and equal five children, we discard this family but still keep traversing at each child.

Each rule consists of: pattern and order. For every node in the dependency tree, from the top-down, we find the node matching against the pattern, and if a match is found, the associated order applies. We arrange the words in the English sentence, which is covered by the matching node, like Vietnamese words order. And then, we do the same for each children of this node. If any rule is applied, we use the order of original sentence. These rules are learnt automatically from bilingual corpora. The our algorithm's outline is given as Algorithms 1 and 2

Algorithm 1 extracts automatically the rules with input including dependency trees of source sentences and alignment pairs.

Algorithm 2 proceeds by considering all rules after finish Algorithm 1 and source-side dependency trees to build new sentence.

**Classification Model.** The reordering decisions are made by multi-class classifiers (correspond with number of permutation: 2, 6, 24, 120) where class labels correspond to permutation sequences. We train a separate classifier for each number of possible children. Crucially, we do not learn explicit tree transformations rules, but let the classifiers learn to trade off between a rich set of overlapping features. To build a classification model, we use SVM classification model [18] in the WEKA tools [19]. The following result are obtained using 10 folds-cross validation.

We apply them in a dependency tree recursively starting from the root node. If the POS-tags of a node matches the left-hand-side of the rule, the rule is applied and the order of the sentence is changed. We go through all the children of the node and matching rules for them from the set of automatically rules.

Table 2 gives examples of original and preprocessed phrase in English. The first line is the original English: "I'm looking at a new jewelry site.", and the target Vietnamese reordering "Ti dang xem mt trang web mi v n_trang.". This sentences is arranged as the Vietnamese order. Vietnamese sentences are the output of our method. As you can see, after reordering, the original English line has the same word order: "I'm looking at a site new jewelry." in Fig. 1.

## 4 Experiment

In this section, we present our experiments to translate from English to ietnamese in a statistical machine translation system. In hence, the language pair chosen is English-Vietnamese. We used Stanford Parser [20] to parse source sentence (English sentences).

We used dependency parsing and rules extracted from training the features-rich discriminative classifiers for reordering source-side sentences. The rules are automatically extracted from English-Vietnamese parallel corpus and the dependency parser of English examples. Finally, they used these rules to reorder source sentences. We evaluated our approach on English-Vietnamese machine translation tasks with systems in table 4 which shows that it can outperform the baseline phrase-based SMT system.

We give some definitions for our experiments:

- Baseline: use the baseline phrase-based SMT system using the lexicalized reordering model in Moses toolkit.
- Manual Rules: the phrase-based SMT systems applying manual rules [21].
- Auto Rules: the phrase-based SMT systems applying automatically rules.
- Auto Rules + Manual Rules: the phrase-based SMT systems applying automatically rules, then applying manual rules.

**Table 3.** Corpus statistical

| Corpus | Sentence pairs | Training Set | Development Set | Test Set |
|---|---|---|---|---|
| General | 55341 | 54642 | 200 | 499 |

| | | English | Vietnamese |
|---|---|---|---|
| Training | Sentences | 54620 | |
| | Average Length | 11.2 | 10.6 |
| | Word | 614578 | 580754 |
| | Vocabulary | 23804 | 24097 |
| Development | Sentences | 200 | |
| | Average Length | 11.1 | 10.7 |
| | Word | 2221 | 2141 |
| | Vocabulary | 825 | 831 |
| Test | Sentences | 499 | |
| | Average Length | 11.2 | 10.5 |
| | Word | 5620 | 6240 |
| | Vocabulary | 1844 | 1851 |

**Table 4.** Our experimental systems on English-Vietnamese parallel corpus

| Name | Description |
|---|---|
| Baseline | Phrase-based system |
| Manual Rules | Phrase-based system with corpus which is preprocessed using manual rules |
| Auto Rules | Phrase-based system with corpus which is preprocessed using automatically learning rules |
| Auto Rules + Manual Rules | Phrase-based system with corpus which is preprocessed using automatically learning rules and manual rules |

## 4.1 Implementation

- We used Stanford Parser [20] to parse source sentence and apply to preprocessing source sentences (English sentences).
- We used classifier-based preordering by using SVM classification model [18] in Weka tools [19] for training the features-rich discriminative classifiers to extract automatically rules and apply them for reordering words in English sentences according to Vietnamese word order.
- We implemented preprocessing step during both training and decoding time.
- Using the SMT Moses decoder [22] for decoding.

**Table 5.** Size of phrase tables

| Name | Size of phrase-table |
|------|---------------------|
| Baseline | 1152216 |
| Manual Rules | 1231365 |
| Auto Rules | 1213401 |
| Auto Rules + Manual Rules | 1253401 |

**Table 6.** Translation performance for the English-Vietnamese task

| System | BLEU (%) |
|--------|----------|
| Baseline | 36.89 |
| Manual Rules | 37.71 |
| Auto Rules | 37.12 |
| Auto Rules + Manual Rules | 37.85 |

## 4.2 Data Set and Experimental Setup

For evaluation, we used an English-Vietnamese corpus [23], including about 54642 pairs for training, 500 pairs for testing and 200 pairs for development test set. Table 3 gives more statistical information about our corpora. We conducted some experiments with SMT Moses Decoder [22] and SRILM [24]. We trained a trigram language model using interpolate and kndiscount smoothing with Vietnamese mono corpus. Before extracting phrase table, we use GIZA++ [25] to build word alignment with grow-diag-final-and algorithm. Besides using preprocessing, we also used default reordering model in Moses Decoder: using word-based extraction (wbe), splitting type of reordering orientation to three classes (monotone, swap and discontinuous – msd), combining backward and forward direction (bidirectional) and modeling base on both source and target language (fe) [22]. To contrast, we tried preprocessing the source sentence with manual rules and automatically rules.

## 4.3 BLEU score

The result of our experiments in Table 5 showed our applying transformation rule to process the source sentences. In this method, we can find out various phrases in the translation model. So that, they enable us to have more options for decoder to generate the best translation.

Table 6 describes the BLEU score of our experiments. As we can see, by applying preprocessing in both training and decoding, the BLEU score of our best system increase by 0.96 point ("Auto Rules + Manual Rules" system) over "Baseline system". Improvement over 0.96 BLEU point is valuable because baseline system is the strong phrase based SMT (integrating lexicalized reordering

**Table 7.** Statistical number of family on corpus English-Vietnamese

| Number children of head | Number | Description |
|---|---|---|
| 1 | 79142 | Family has 1 children |
| 2 | 40822 | Family has 2 children |
| 3 | 26008 | Family has 3 children |
| 4 | 15990 | Family has 4 children |
| 5 | 7442 | Family has 5 children |
| 6 | 2728 | Family has 6 children |
| 7 | 942 | Family has 7 children |
| 8 | 307 | Family has 8 children |
| 9 | 83 | Family has 9 children |

models). We also carried out the experiments with manual rules [21]. Using automatically rules help the phrased translation model generate some best translation. Besides, the result proved that the effect of applying transformation rule on the dependency tree when the BLEU score is higher than baseline systems. Because, the cover of manual rules is better than automatically rules on corpus.

We can extract more and better phrase tables. However, in our experiments, we need to conduct with larger corpus and better quality of corpus to extract automatically rules which can cover many linguistic reordering phenomena on corpus. We believe that quality of translation systems will be better when doing experiments with larger corpus [22].

## 5   Analysis and Discussion

We have found that in our experiments work is sufficiently correlated to the translation quality done manually. Besides, we also have found some error causes such as parse tree source sentence quality, word alignment quality and quality of corpus. All the above errors can effect automatically reordering rules. We focus mainly on some typical relations as noun phrase, adjectival and adverbial phrase, preposition and created manually written reordering rule set for English-Vietnamese language pair. Our study employed dependency syntactic and transformation rules to reorder the source sentence and applied to English to Vietnamese translation systems. For example, with noun phrase, there always exists a head noun and the components before and after it. These auxiliary components will move to new positions according to Vietnamese translational order. These rules can popular source linguistic phenomena equivalent to target language ones as follows:

- the phrase-based systems applying rules with category JJ or JJS
- the phrase-based systems applying rules with category NN or NNS
- the phrase-based systems applying rules with category IN or TO.

Based on these phenomena, translation quality has significantly improved. We carried out error analysis sentences and compared to the golden reordering. Our analysis has also the benefits of automatically reordering rules on translation quality. In combination with machine learning method in related work [10], it is shown that applying classifier method to solve reordering problems automatically.

According to typical differences of word order between English and Vietnamese, we have created a set of automatically rules for reordering words in English sentence according to Vietnamese word order and types of rules including noun phrase, adjectival and adverbial phrase, as well as preposition phrase. Table 7 gives statistical families which have larger or equal 4 children in our corpus. The number of children in each family has limited 4 children in our approach. So in target language (Vietnamese), the number of children in each family is the same.

We compared experimental results between the phrase-based SMT systems applying manual rules with the phrase-based SMT systems applying automatic rules. Because the manual rules have good quality [7,14], the phrase-based SMT systems applying manual rules is better than the phrase-based SMT systems applying automatically rules. We believe that the quality of the phrase-based SMT systems applying automatic rules will be better when we have a better corpus.

# 6   Conclusion

In this study, a preprocessing approach based on a dependency parser is presented. We used classifier-based preordering by using SVM classification model [18] in Weka tools [19] for training the features-rich discriminative classifiers to extract automatically rules and apply these rules for reordering words in English sentence according to Vietnamese word order.

We evaluated our approach on English-Vietnamese machine translation tasks. The experimental results showed that our approach achieved statistical improvements 0.96 BLEU point scores over a state-of-the-art phrase-based baseline system. Our rules are learnt automatically from corpus and can cover many linguistic reordering phenomena. We believe that such reordering rules benefit English-Vietnamese language pairs.

In the future, we plan to investigate along this direction and extend the rules to other languages. We would like to evaluate our method with tree with higher and deeper syntactic structure and larger size of corpus. We also attempt to create more efficient prereordering rules by exploiting the rich information in dependency structures.

**Acknowledgements.** This work described in this paper has been partially funded by Hanoi National University (QG.15.23 project).

# References

1. Koehn, P., Och, F.J., Marcu, D.: Statistical phrase-based translation. In: Proceedings of HLT-NAACL 2003, Edmonton, Canada, pp. 127–133 (2003)
2. Och, F.J., Ney, H.: The alignment template approach to statistical machine translation. Comput. Linguist. **30**(4), 417–449 (2004)
3. Chiang, D.: A hierarchical phrase-based model for statistical machine translation. In: Proceedings of the 43rd Annual Meeting of the Association for Computational Linguistics (ACL 2005), Ann Arbor, Michigan, pp. 263–270, June 2005
4. Zhang, Y., Zens, R., Ney, H.: Chunk-level reordering of source language sentences with automatically learned rules for statistical machine translation. In: Proceedings of SSST, NAACL-HLT 2007/AMTA Workshop on Syntax and Structure in Statistical Translation, pp. 1–8 (2007)
5. Collins, M., Koehn, P., Kucerová, I.: Clause restructuring for statistical machine translation. In: Proceedings of ACL 2005, Ann Arbor, USA, pp. 531–540 (2005)
6. Quirk, C., Menezes, A., Cherry, C.: Dependency treelet translation: syntactically informed phrasal SMT. In: Proceedings of ACL 2005, Ann Arbor, Michigan, USA, pp. 271–279 (2005)
7. Xia, F., McCord, M.: Improving a statistical MT system with automatically learned rewrite patterns. In: Proceedings of Coling 2004, Geneva, Switzerland, COLING, 23–27 August 2004, pp. 508–514 (2004)
8. Xu, P., Kang, J., Ringgaard, M., Och, F.: Using a dependency parser to improve SMT for subject-object-verb languages. In: Proceedings of Human Language Technologies: The 2009 Annual Conference of the North American Chapter of the Association for Computational Linguistics, Boulder, Colorado, pp. 245–253. Association for Computational Linguistics, June 2009
9. Genzel, D.: Automatically learning source-side reordering rules for large scale machine translation. In: Proceedings of the 23rd International Conference on Computational Linguistics. COLING 2010, Stroudsburg, PA, USA, pp. 376–384. Association for Computational Linguistics (2010)
10. Lerner, U., Petrov, S.: Source-side classifier preordering for machine translation. In: EMNLP, pp. 513–523 (2013)
11. Li, C.H., Li, M., Zhang, D., Li, M., Zhou, M., Guan, Y.: A probabilistic approach to syntax-based reordering for statistical machine translation. In: Annual Meeting-association for Computational Linguistics, vol. 45, p. 720 (2007)
12. Yang, N., Li, M., Zhang, D., Yu, N.: A ranking-based approach to word reordering for statistical machine translation. In: Proceedings of the 50th Annual Meeting of the Association for Computational Linguistics: Long Papers, vol. 1, pp. 912–920. Association for Computational Linguistics (2012)
13. Jehl, L., de Gispert, A., Hopkins, M., Byrne, B.: Source-side preordering for translation using logistic regression and depth-first branch-and-bound search. In: Proceedings of the 14th Conference of the European Chapter of the Association for Computational Linguistics, Gothenburg, Sweden, pp. 239–248. Association for Computational Linguistics, April 2014
14. Habash, N.: Syntactic preprocessing for statistical machine translation. In: Proceedings of the 11th MT Summit (2007)
15. Cai, J., Utiyama, M., Sumita, E., Zhang, Y.: Dependency-based pre-ordering for Chinese-English machine translation. In: Proceedings of the 52nd Annual Meeting of the Association for Computational Linguistics (2014)

16. Hoshino, S., Miyao, Y., Sudoh, K., Hayashi, K., Nagata, M.: Discriminative pre-ordering meets kendall's τ maximization. In: Proceedings of the 53rd Annual Meeting of the Association for Computational Linguistics and the 7th International Joint Conference on Natural Language Processing (Volume 2: Short Papers), Beijing, China, pp. 139–144. Association for Computational Linguistics, July 2015
17. Nakagawa, T.: Efficient top-down BTG parsing for machine translation preordering. In: Proceedings of the 53rd Annual Meeting of the Association for Computational Linguistics and the 7th International Joint Conference on Natural Language Processing (Volume 1: Long Papers), Beijing, China, pp. 208–218. Association for Computational Linguistics, July 2015
18. Wang, L.: Support Vector Machines: Theory and Applications, vol. 117. Springer Science & Business Media, Heidelberg (2005)
19. Hall, M., Frank, E., Holmes, G., Pfahringer, B., Reutemann, P., Witten, I.H.: The WEKA data mining software: an update. SIGKDD Explor. Newsl. **11**(1), 10–18 (2009)
20. Cer, D., de Marneffe, M.C., Jurafsky, D., Manning, C.D.: Parsing to stanford dependencies: trade-offs between speed and accuracy. In: 7th International Conference on Language Resources and Evaluation (LREC 2010) (2010)
21. Tran, V.H., Nguyen, V.V., Nguyen, M.L.: Improving English-Vietnamese statistical machine translation using preprocessing dependency syntactic. In: Proceedings of the 2015 Conference of the Pacific Association for Computational Linguistics (Pacling 2015), pp. 115–121 (2015)
22. Koehn, P., Hoang, H., Birch, A., Callison-Burch, C., Federico, M., Bertoldi, N., Cowan, B., Shen, W., Moran, C., Zens, R., Dyer, C., Bojar, O., Constantin, A., Herbst, E.: Moses: open source toolkit for statistical machine translation. In: Proceedings of ACL, Demonstration Session (2007)
23. Nguyen, T.P., Shimazu, A., Ho, T.B., Nguyen, M.L., Nguyen, V.V.: A tree-to-string phrase-based model for statistical machine translation. In: Proceedings of the Twelfth Conference on Computational Natural Language Learning (CoNLL 2008), Manchester, England. Coling 2008 Organizing Committee, pp. 143–150, August 2008
24. Stolcke, A.: SRILM - an extensible language modeling toolkit. In: Proceedings of International Conference on Spoken Language Processing, vol. 29, pp. 901–904 (2002)
25. Och, F.J., Ney, H.: A systematic comparison of various statistical alignment models. Comput. Linguist. **29**(1), 19–51 (2003)

# Quality Estimation for English-Hungarian Machine Translation Systems with Optimized Semantic Features

Zijian Győző Yang[1(✉)], László János Laki[2], and Borbála Siklósi[1]

[1] Faculty of Information Technology and Bionics,
Pázmány Péter Catholic University, Práter str. 50/A, Budapest 1083, Hungary
{yang.zijian.gyozo,siklosi.borbala}@itk.ppke.hu
[2] MTA-PPKE Hungarian Language Technology Research Group,
Práter str. 50/A, Budapest 1083, Hungary
laki.laszlo@itk.ppke.hu

**Abstract.** Quality estimation at run-time for machine translation systems is an important task. The standard automatic evaluation methods that use reference translations cannot evaluate MT results in real-time and the correlation between the results of these methods and that of human evaluation is very low in the case of translations from English to Hungarian. The new method to solve this problem is called quality estimation, which addresses the task by estimating the quality of translations as a prediction task for which features are extracted from the source and translated sentences only. In this study, we implement quality estimation for English-Hungarian. First, a corpus is created, which contains Hungarian human judgements. Using these human evaluation scores, different quality estimation models are described, evaluated and optimized. We created a corpus for English-Hungarian quality estimation and we developed 27 new semantic features using WordNet and word embedding models, then we created feature sets optimized for Hungarian, which produced better results than the baseline feature set.

## 1 Introduction

Machine translation (MT) has become a daily used tool among people and companies. The measurement of the quality of translation output has become necessary. A quality score for MT could save a lot of time and money for users and researchers. Knowing the quality of machine translated segments can help human annotators in their post-edit tasks, or can filter out and inform about unreliable translations. Last but not least, quality indicators can help MT systems to combine the translations to produce better output. There are two kinds of evaluation methods for MT. The first type uses reference translations, i.e. it compares machine translated sentences to human translated reference sentences, and measures the similarities or differences between them. These methods are automatic evaluation approaches, such as BLEU, or other methods based on

© Springer International Publishing AG, part of Springer Nature 2018
A. Gelbukh (Ed.): CICLing 2016, LNCS 9624, pp. 88–100, 2018.
https://doi.org/10.1007/978-3-319-75487-1_8

BLEU, TER, HTER etc. The problem is that automatic evaluation methods cannot perform well enough in this task, because these need reference translations. It means that after the automatic translation, we also have to create a human translated sentence (for the sentences of the test set) to compare it to the machine translated output. Creating human translations is expensive and time-consuming. We can not use these methods in run-time. Thus, a completely new approach is needed to solve these problems, i.e. a method which can predict translation quality in real-time and does not need reference translations.

The supervised method of quality estimation (QE) does not use reference translations and addresses the problem by evaluating the quality of machine translated segments as a prediction task. Using QE we can save considerable time and money for human annotators, researchers and companies.

In this study, we use the QuEst framework [1], developed by Specia et al., to train and apply QE models for Hungarian. For training we created an English-Hungarian QE corpus. Then, we developed new semantic features, using a dictionary, WordNet and word embedding models to gain better results.

Hungarian is an agglutinating and compounding language. There are significant differences between English and Hungarian, regarding their morphology, syntax and word order or number. Furthermore, the free order of grammatical constituents, and different word orders in noun phrases (NPs) and prepositional phrases are also characteristics of Hungarian. Thus, features used in a QE task for English-Spanish or English-German, which produced good results, perform much worse for English-Hungarian. Hence, if we would like to use linguistic features in QuEst, we need to integrate the available Hungarian linguistic tools into it.

The structure of this paper is as follows: First we will shortly introduce the quality estimation approach. Then, we will present a corpus we created for English-Hungarian QE. Finally, our experiments, optimizations and results in the task of QE are described.

## 2   Related Work

QE is a prediction task, where different quality indicators are extracted from the source and the machine translated segments. The QE model is built with machine learning algorithms based on these quality indicators. Then the QE model is used to predict the quality of unseen translations. The aim is that the scores, predicted with the QE model, highly correlate with human judgments. Thus the QE model is trained on human evaluations. In the last couple of years there have been several WMT workshops with quality estimation shared tasks,[1] which provided datasets for QE research. The datasets are evaluated with HTER, METEOR, ranking or post-edit effort scores. Unfortunately there is no such dataset for Hungarian. In this research we created a dataset for Hungarian and for assigning human judgements to translations we used a general scoring scale.

---

[1] http://www.statmt.org/wmt15/quality-estimation-task.html.

Recently, in the field of QE, research has focused on feature selection [2] using a variety of machine learning algorithms and feature engineering [3]. In the feature selection task, Beck et al. tried more than 160 features in an experiment for English-Spanish to predict HTER [4]. There is a language independent baseline set which contains 17 features.

In our research we did experiments for Hungarian in both fields: creating training and evaluation datasets and optimizing feature selection.

## 3   Quality Estimation

In the QE task (see Fig. 1), we extract different kinds of quality indicators from the source and translated sentences without using reference translations. Following the research of Specia et al., we can separate the features in different kinds of categories [1]. From the source sentences, complexity features can be extracted (e.g. number of tokens in the source segment). From the translated sentences, we extract fluency features (e.g. percentage of verbs in the target sentences). From the comparison between the source and the translated sentences, adequacy features are extracted (e.g. ratio of percentage of nouns in the source and target). We can also extract features from the decoder of the MT system. These are the confidence features (e.g. features and global score of the SMT system). We can also divide the features into two more main categories: "black-box" features (independent from the MT system) and "glass-box" features (MT system-dependent). Since in our experiments we have translations from different MT systems, we did use only the "black-box" features. After feature extraction, using these quality indicators, we can build a QE model with machine learning methods. The aim is that the predictions of the QE model are highly correlated with human evaluations. Thus, the extracted quality indicators need to be trained on human judgments.

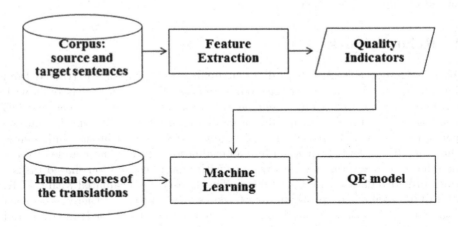

**Fig. 1.** QE algorithm

# 4  English-Hungarian Corpus for QE

In our experiments, first, we created a corpus for Hungarian QE, called HuQ corpus. The corpus contains 1500 English-Hungarian sentence pairs.This set of 1500 sentence pairs contains 300 sentence pairs of mixed topics from the Hunglish corpus [5] containing human translations. In addition to the human-translated sentence pairs, we translated the 300 English sentences to Hungarian with four different machine translation systems. Each segment was translated by Google Translate, Bing Translate, the MetaMorpho [6] rule based MT system and the MOSES statistical MT toolkit [7]. Then we created the human judgements for these translations.

For adding human scores, we developed a website[2] with a form for human annotators to evaluate the translations. In this website we can see an English source sentence and its Hungarian translation. Using the Likert scale, people can give quality scores from 1 to 5, from two points of view: adequacy (1 = none; 2 = little meaning; 3 = much meaning; 4 = most meaning; 5 = all meaning) and fluency (1 = incomprehensible; 2 = disfluent Hungarian; 3 = non-native Hungarian; 4 = good Hungarian; 5 = flawless Hungarian). All the 1500 sentence pairs were evaluated by 3 human translators. All the translators were native Hungarian speakers who have minimum B2 level English language skill. The 3 annotators have different evaluation attitudes. One of them is a linguist, the second annotator is a MT specialist and the third one is a language technology expert. To follow and control the annotators effectively, or to discuss the annotation aspects with the annotators personally to avoid misunderstandings, we did not use crowdsourcing for the evaluation. To ensure a consistent annotation scheme, the 3 annotators evaluated a set of 50 translations in a personal meeting. These translations were not included in the training set.

For building the QE model, we used the mean of the scores of the 3 annotators. We used the mean of the adequacy scores (AD), the mean of the fluency scores (FL) and the mean of the AD and FL scores (AF). We also created classification scores from the AD (CLAD), the FL (CLFL) and the AF (CLAF) scores:

- BAD: $1 <= (AD|FL|AF) <= 2$
- MEDIUM: $2 < (AD|FL|AF) < 4$
- GOOD: $4 <= (AD|FL|AF) <= 5$.

# 5  Methods and Experiments

For building the QE model, features as quality indicators are needed, which are extracted from the corpus. Then, with a machine learning method, human or automatic evaluation scores are used to build the QE model (see Fig. 1). To create the quality indicators from the features, we used the QuEst framework. In this study, 103 features (103F) were extracted from the corpus. The set of 103

---

[2] http://nlpg.itk.ppke.hu/node/65.

features contains 76 features (76F) implemented by Specia et al. and 27 additional features developed by us (27F). In the 76F, there are adequacy features (e.g. ratio of percentage of nouns in the source and target, ratio of number of tokens in source and target, etc.), fluency features (e.g. perplexity of the target, percentage of verbs in the target, etc.) and complexity features (e.g. average source token length, source sentence log probability, etc.). The 27F contains 3 dictionary features and 24 features using WordNet and the word embedding model.

The first task was trying features developed by Specia et al [1]. First, we tried the 17 baseline feature set for Hungarian. The baseline set is language and language tool independent. Then we performed experiments with the 76F (baseline set is subset of 76F). The problem was that the 76F contains features that use language dependent linguistic tools (e.g. Stanford parser, Berkeley Parser etc.). The most commonly used linguistic tools could not be used for Hungarian. Thus, we integrated the available Hungarian linguistic tools into QuEst: For Part-of-Speech (POS) tagging and lemmatization, we used PurePos 2.0 [8], which is an open source, HMM-based morphological disambiguation tool. Purepos2 has the state-of-the-art performance for Hungarian. It has the possibility to integrate a morphological analyzer. Thus, to get the best performance, we used Humor [9], a Hungarian morphological analyzer. For NP-chunking, we used HunTag [10] that was trained on the Szeged Treebank [11]. HunTag is a maximum entropy Markov-model based sequential tagger. There are many language specific features that could not be extracted, because there are no Hungarian language tools for them.

### 5.1   Dictionary, WordNet and Word Embedding Model

We developed 27 new word-level semantic features. Our aim was to quantify the similarity and relatedness of the topic or meaning of the source and the target sentences. We created bag of words from the source and the target segments. The bag of words contains the stem, the synonym and the semantic neighbours of the words.

We introduced 3 features using an English-Hungarian dictionary ($D$) used by MetaMorpho, which contains 365000 word pairs. The dictionary contains only nouns, verbs and adjectives. From each source sentence ($S = s_1, s_2, \ldots, s_i, \ldots, s_n$) and target sentence ($T = t_1, t_2, \ldots, t_j, \ldots, t_m$) pairs we counted the source-target word pairs ($(s_i; t_j)$), which are included in the dictionary, according to the following formulae:

$$DictionaryCount_S = \frac{count((s_i; t_j) \in D)}{n} \tag{1}$$

$$DictionaryCount_T = \frac{count((s_i; t_j) \in D)}{m} \tag{2}$$

$$harmonic\ mean\ of\ DictionaryCount_S\ (1)\ and\ DictionaryCount_T\ (2), \tag{3}$$

where $n$ is the length of the source sentence, $m$ is the length of the target sentence, $i = [1, n]$ and $j = [1, m]$.

In addition, we developed 24 features extracted from WordNet and word embedding models. We used the Princeton WordNet 3.0 [12] and the Hungarian WordNet [13].

First, we collected the synset ids ($level = 1$) of the nouns from the source and the target sentences. Then, we added the hypernym ids of the synsets up to two levels ($level = [2, 3]$). Using the collected synsets and hypernyms, we created a set of source ids ($SET_S$) and a set of target ids ($SET_T$). Thereafter, we counted the weighted intersection ($W$) of these sets ($I(S; T) = SET_S \cap SET_T = \{y_1, \ldots, y_k\}$). Features are extracted using the following formulae:

$$WordNetCount_S = \frac{W(I(S; T))}{n} \tag{4}$$

$$WordNetCount_{x_S} = \frac{W(I(S; T))}{|x_S|} \tag{5}$$

$$WordNetCount_T = \frac{W(I(S; T))}{m} \tag{6}$$

$$WordNetCount_{x_T} = \frac{W(I(S; T))}{|x_T|} \tag{7}$$

$$harmonic\ mean\ of\ WordNetCount_S\ (4)\ and\ WordNetCount_T\ (6) \tag{8}$$

$$harmonic\ mean\ of\ WordNetCount_{x_S}\ (5)\ and\ WordNetCount_{x_T}\ (7), \tag{9}$$

where $|x|$ is the number of nouns in the sentence; $n$ is the length of the source sentence; $m$ is the length of the target sentence; and

$$W(I(S; T)) = \sum_{i=1}^{k} \frac{y_i}{level_{y_i}}; \ y_i \in I(S; T)$$

Using these 6 formulae we also created features with the verbs, the adjectives and the adverbs.

However, if looking up words in WordNet did not provide any results, which is quite often the case because of the small coverage of the Hungarian WordNet, we used word embedding models [14] to substitute synset results. Thus, first we trained a CBOW model with 300 dimensions on a 3-billion-word lemmatized Hungarian corpus using the word2vec tool[3]. The reason for using the lemmatized version of the corpus was to define relations of words along their meaning, rather than their morphosyntactic characteristics. Due to the agglutinating behaviour of Hungarian, building an embedding model from the raw text would have provided syntactically similar groups of words, and only a second key of similarity would have been their semantic relatedness [15]. However, in the lemmatized model, this problem was eliminated. Thus, if there was no result for a word from WordNet, its

---

[3] https://github.com/danielfrg/word2vec.

top 10 nearest neighbours were retrieved from this embedding model, providing the 10 semantically most similar words, i.e. quasi-synonyms from the corpus and used the same way as WordNet synsets. However, as these lists do not necessarily correspond to exact synonyms of the original word, the weight of this feature was decreased to 0.1.

## 5.2   Machine Learning

To train the QE model from the evaluated dataset and the generated features, machine learning algorithms were applied. For the machine learning task, we used the Weka system [16] to create 7 classifiers with 10-fold cross-validation: Gausian Processes with RBF kernel, Support Vector Machine for regression with NormalizedPolyKernel (SMOreg), Bagging (with M5P classifier), Linear regression, M5Rules, M5P Tree and for classification we used Support Vector Machine with NormalizedPolyKernel (SMO). In this paper, we show only the results of the SMOreg and SMO, because these methods gained the best scores. For evaluating the performance of our methods, we used the statistical correlation, the MAE (Mean absolute error), the RMSE (Root mean-squared error) and the Correctly Classified Instances (CCI) evaluation metrics. The correlation ranges from −1 to +1, and the closer the correlation to −1 or +1, the better it is. In the case of MAE and RMSE, the closer the value to 0, the better.

## 5.3   Experiments and Optimization

We carried out experiments for five different settings:

- First task (T1): the HuQ corpus is evaluated using automatic evaluation methods: TER, BLEU and NIST.
- Second task (T2): using the HuQ corpus and the 103F, we built QE models trained on the automatic evaluation metrics (segment-level TER, BLEU and NIST).
- Third task (T3): using the HuQ corpus, the baseline set, the 76F and the 103F, we built and evaluated QE models trained on the AD, the FL and the AF scores.
- Fourth task (T4): using the HuQ corpus and the 103F, we built the QE models trained on the AD, the FL and the AF scores, then we optimized the QE models for Hungarian.
- Fifth task (T5): using the HuQ corpus and the 103F, we built the QE models trained on the CLAD, CLFL and CLAF scores, then we optimized the QE models for Hungarian.

The experiment with human scores needed to be optimized for English-Hungarian. For optimizing, we used the forward selection method. First, we extracted and evaluated each feature separately. Then we chose the feature that produced the best result. Thereafter, we combined the chosen feature with each remaining feature, and we added the feature that produced the best combined result in each round. Then, we continued adding features until the combined result did not improve further. With this attribute selection algorithm, we created 6 optimized feature sets for Hungarian:

- OptADSet: Optimized feature set for AD scores.
- OptFLSet: Optimized feature set for FL scores.
- OptAFSet: Optimized feature set for AF scores.
- OptCLADSet: Optimized feature set for CLAD scores.
- OptCLFLSet: Optimized feature set for CLFL scores.
- OptCLAFSet: Optimized feature set for CLAF scores.

# 6   Results and Evaluation

The system-level results of the T1 evaluation are: TER: 0.6107; BLEU: 0.3038, NIST: 5.1359. According to the TER and the BLEU scores, 30% of the HuQ corpus are correct translations.

During the T2 experiment, we built the QE models to predict automatic evaluations (see Table 1).

Table 1. Evaluation of T2, T3 and T4

|                              | Correlation | MAE    | RMSE   |
|------------------------------|-------------|--------|--------|
| TER                          | 0.3550      | 0.3275 | 0.4357 |
| BLEU                         | 0.4404      | 0.2201 | 0.3474 |
| NIST                         | 0.3669      | 2.6695 | 3.4777 |
| AD-baseline                  | 0.3832      | 0.9429 | 1.1990 |
| FL-baseline                  | 0.5400      | 0.8229 | 0.8345 |
| AF-baseline                  | 0.4931      | 0.8345 | 1.0848 |
| AD-76F                       | 0.4757      | 0.8804 | 1.1274 |
| FL-76F                       | 0.5980      | 0.7751 | 1.0391 |
| AF-76F                       | 0.5510      | 0.7984 | 1.0342 |
| AD-103F                      | 0.4847      | 0.8805 | 1.1199 |
| FL-103F                      | 0.6070      | 0.7723 | 1.0297 |
| AF-103F                      | 0.5618      | 0.7962 | 1.0252 |
| OptADSet (29 features)       | **0.5245**  | **0.8397** | **1.0869** |
| OptFLSet (32 features)       | **0.6413**  | **0.7440** | **0.9878** |
| OptAFSet (26 features)       | **0.6100**  | **0.7459** | **0.9775** |

From the results of experiments T3 and T4 (Table 1), it can be seen that:

- the AD-103F could gain ~1% higher correlation than the 76F and ~10% higher correlation than the baseline set,

- the FL-103F could gain ~1% higher correlation than the 76F and ~6% higher correlation than the baseline set,
- the AF-103F could gain ~1% higher correlation than the 76F and ~7% higher correlation than the baseline set.

During T4, first, we used the 103F to build the QE models trained on AD, FL and AF human scores. Then, using the forward selection method, we optimized the models to Hungarian. After optimizing, the results are following (Table 1):

- the OptADSet containing 29 features could gain ~4% higher correlation than the 103F and ~14% higher correlation than the baseline set,
- the OptFLSet containing 32 features could gain ~4% higher correlation than the 103F and ~10% higher correlation than the baseline set,
- the OptAFSet containing 26 features could gain ~5% higher correlation than the 103F and ~12% higher correlation than the baseline set.

For T5, we can see the results of the evaluation and optimization of classification in Table 2:

- the OptCLADSet containing 21 features could gain ~3% higher correlation than the 103F and ~6% higher correlation than the baseline set,
- the OptCLFLSet containing 10 features could gain ~1.5% higher correlation than the 103F and ~5% higher correlation than the baseline set,
- the OptCLAFSet containing 12 features could gain ~1.5% higher correlation than the 103F and ~4% higher correlation than the baseline set.

**Table 2.** Evaluation of T5

|  | CCI | MAE | RMSE |
|---|---|---|---|
| CLAD-baseline | 54.9333% | 0.3590 | 0.4591 |
| CLFL-baseline | 58.8667% | 0.3434 | 0.4419 |
| CLAF-baseline | 57.8000% | 0.3433 | 0.4417 |
| CLAD-103F | 57.6667% | 0.3492 | 0.4483 |
| CLFL-103F | 62.4667% | 0.3310 | 0.4275 |
| CLAF-103F | 60.3333% | 0.3347 | 0.4318 |
| OptCLADSet (21 features) | **60.9333%** | **0.3370** | **0.4346** |
| OptCLFLSet (10 features) | **64.0667%** | **0.3299** | **0.4262** |
| OptCLAFSet (12 features) | **61.8000%** | **0.3299** | **0.4263** |

In the Tables 3 and 4, we can see the optimized features for Hungarian. The indexes of the optimized features are the following (the ones developed in this research are emphasized):

- OptADSet features: 1064, 1015, 1091, 1089, 2005, 1001, 1075, 1072, 1057, 1066, 1024, 1082, 1042, 1094, 1010, 1068, **2019**, 1006, 1060, 1013, **2023**, 1073, 1076, 1067, **2015**, **2029**, 1038, **2007**

**Table 3.** Features, developed by Specia et al., optimized for Hungarian

| Index | Feature |
|-------|---------|
| 1001 | Number of tokens in the source sentence |
| 1002 | Number of tokens in target |
| 1005 | Abs difference between no tokens and source and target norm by source length |
| 1006 | Average source token length |
| 1010 | Source sentence perplexity |
| 1011 | Source sentence perplexity without end of sentence marker |
| 1013 | Perplexity of the target |
| 1014 | Perplexity of the target sentence without end of sentence marker |
| 1015 | Number of occurrences of the target word within the target hypothesis |
| 1016 | Avg num of translations per source word in the sentence (prob > 0.01) |
| 1024 | Avg num of translations per source word in the sentence (prob > 0.5) |
| 1034 | Avg num of trans per source word (prob > 0.5) weighted by freq of words in source |
| 1036 | Avg num of trans per source word (prob > 0.01) weighted inv freq words in source |
| 1038 | Avg num of trans per source word (prob > 0.05) weighted inv freq words in source |
| 1042 | Avg num of trans per source word (prob > 0.2) weighted inv freq words in source |
| 1044 | avg num of trans per source word (prob > 0.5) weighted inv freq words in source |
| 1046 | Avg unigram freq in quartile 1 of frequency in the corpus of the source |
| 1047 | Avg unigram frequency in quartile 2 of frequency in the corpus of the source |
| 1052 | Avg bigram freq in quartile 3 of frequency in the corpus of the source sentence |
| 1054 | Avg trigram freq in quartile 1 of frequency in the corpus of the source sentence |
| 1055 | Avg trigram freq in quartile 2 of frequency in the corpus of the source sentence |
| 1057 | Avg trigram freq in quartile 4 of frequency in the corpus of the source sentence |
| 1060 | Percentage of distinct trigrams seen in the corpus (in all quartiles) |
| 1064 | Abs different between number of commas in source and target |
| 1066 | Abs different between number of : in source and target |
| 1067 | Abs different between number of : in source and target norm by target length |
| 1068 | Abs different between number of ; in source and target |
| 1072 | Abs different between number of ! in source and target |
| 1073 | Abs different between number of ! in source and target norm by target length |
| 1075 | Percentage of punctuation marks in target |
| 1076 | Abs diff between num of punct marks between source and target norm by target |
| 1077 | Percentage of numbers in the source |
| 1078 | Percentage of numbers in the target sentence |
| 1079 | Abs diff between number of numbers in source and target norm by source length |
| 1080 | Number source tokens that do not contain only a-z |
| 1081 | Percentage of tokens in the target which do not contain only a-z |
| 1082 | Ratio of percentage of tokens a-z in the source and tokens a-z in the target |
| 1089 | Percentage of verbs in the source |
| 1090 | Percentage of nouns in the target |
| 1091 | Percentage of verbs in the target |
| 1092 | Ratio of percentage of nouns in the source and target |
| 1093 | Ratio of percentage of verbs in the source and target |
| 1094 | Ratio of percentage of pronouns in the source and target |
| 2005 | Abs diff between num of NP-s in source and target norm by num of phrasal tags |

**Table 4.** Own developed features, optimized for Hungarian

| Index | Feature |
|-------|---------|
| 2001 | Dictionary lookup/length of source sentence |
| 2002 | Dictionary lookup/length of target sentence |
| 2003 | Dictionary lookup F-score |
| 2006 | WordNet: F-score: nouns/number of tokens |
| 2007 | WordNet: F-score: verbs/number of tokens |
| 2010 | WordNet: F-score: nouns/number of nouns |
| 2015 | WordNet: in target: verbs/number of tokens |
| 2016 | WordNet: in target: adjectives/number of tokens |
| 2019 | WordNet: in target: verbs number of verbs |
| 2020 | WordNet: in target: adjectives/number of adjectives |
| 2022 | WordNet: in source: nouns/number of tokens |
| 2023 | WordNet: in source: verbs/number of tokens |
| 2026 | WordNet: in source: nouns/number of nouns |
| 2025 | WordNet: in source: adverbs/number of tokens |
| 2028 | WordNet: in source: adjectives/number of adjectives |
| 2029 | WordNet: in source: adverbs/number of adverbs |

- OptFLSet features: 1015, 1060, 1002, 1082, 1091, 2019, 1066, **2003**, 1036, 1068, 1072, **2020**, **2026**, 1006, 1010, 1089, 1044, 1073, 1054, 1046, 1093, 2005, **2007**, **2016**, 1067, 1011, 1052, **2001**, 1034, 1042, **2002**, **2015**
- OptAFSet features: 1015, 1091, 1089, 1002, 1082, 1066, 1044, 1057, 1016, 1010, 1072, **2019**, 1006, 1068, 2005, **2001**, 1080, **2028**, 1013, 1052, **2022**, 1073, 1077, **2006**, 1067, 1079
- OptCLADSet features: 1068, 1064, 1005, 1091, 1092, 1015, **2001**, 1072, 1046, 1077, 1078, 1055, 1082, 1066, 1093, 1057, 1081, **2019**, 1067, 1090, 1010
- OptCLFLSet features:1064, 1076, **2002**, 1091, 1072, 1047, 1077, 1011, 1014, 1054
- OptCLAFSet features: 1064, 1091, 1075, 1093, 1057, 1072, **2010**, **2025**, 1066, 1014, 1067, 1079.

# 7    Conclusion

We created a quality estimation system for English-Hungarian machine translation. We used a corpus of 1500 translation pairs annotated with human judgement scores. Then using the human judgements, we built different QE models for English-Hungarian translation evaluation. In addition to the human scores, we also used automatic translation evaluation metrics in our experiments. We tried 103 features including 27 newly developed semantic features using WordNet and word embedding models. Then, we optimized the quality estimation models to English-Hungarian. In the optimization task, we used forward selection method to find the best features. We could produce optimized sorted feature sets, which produced more than 10% better correlation to human evaluation than the baseline set. In our experiments, our QE models can be used for predicting the quality of machine translation outputs for English-Hungarian. Our goal is to build a stable and reliable QE model for English-Hungarian.

# References

1. Specia, L., Shah, K., de Souza, J.G., Cohn, T.: QuEst - a translation quality estimation framework. In: Proceedings of the 51st Annual Meeting of the Association for Computational Linguistics: System Demonstrations, Sofia, Bulgaria, pp. 79–84 (2013)
2. Biçici, E.: Feature decay algorithms for fast deployment of accurate statistical machine translation systems. In: Proceedings of the Eighth Workshop on Statistical Machine Translation, Sofia, Bulgaria (2013)
3. Camargo de Souza, J.G., Buck, C., Turchi, M., Negri, M.: FBK-UEdin participation to the WMT13 quality estimation shared task. In: Proceedings of the Eighth Workshop on Statistical Machine Translation, Sofia, Bulgaria, pp. 352–358 (2013)
4. Beck, D., Shah, K., Cohn, T., Specia, L.: SHEF-Lite: when less is more for translation quality estimation. In: Proceedings of the Workshop on Machine Translation (WMT) (2013)
5. Halácsy, P., Kornai, A., Németh, L., Sas, B., Varga, D., Váradi, T., Vonyó, A.: A Hunglish korpusz és szótár. In: III. Magyar Számítógépes Nyelvészeti Konferencia, Szegedi Egyetem (2005)
6. Novák, A., Tihanyi, L., Prószéky, G.: The MetaMorpho translation system. In: Proceedings of the Third Workshop on Statistical Machine Translation. StatMT 2008, Stroudsburg, PA, USA, pp. 111–114 (2008)
7. Koehn, P., Hoang, H., Birch, A., Callison-Burch, C., Federico, M., Bertoldi, N., Cowan, B., Shen, W., Moran, C., Zens, R., Dyer, C., Bojar, O., Constantin, A., Herbst, E.: Moses: open source toolkit for statistical machine translation. In: Proceedings of the 45th Annual Meeting of the ACL, pp. 177–180 (2007)
8. Orosz, G., Novák, A.: PurePos 2.0: a hybrid tool for morphological disambiguation. In: RANLP 2013, pp. 539–545 (2013)
9. Prószéky, G.: Industrial applications of unification morphology. In: Proceedings of the Fourth Conference on ANLP, Stuttgart, Germany, pp. 213–214 (1994)
10. Recski, G., Varga, D.: A Hungarian NP Chunker. The Odd Yearbook. ELTE SEAS Undergraduate Papers Linguistics, pp. 87–93 (2009)
11. Csendes, D., Csirik, J., Gyimóthy, T., Kocsor, A.: The Szeged treebank. In: Matoušek, V., Mautner, P., Pavelka, T. (eds.) TSD 2005. LNCS (LNAI), vol. 3658, pp. 123–131. Springer, Heidelberg (2005). https://doi.org/10.1007/11551874_16
12. Fellbaum, C.: WordNet: An Electronic Lexical Database. Bradford Books (1998)
13. Miháltz, M., Hatvani, C., Kuti, J., Szarvas, G., Csirik, J., Prószéky, G., Váradi, T.: Methods and results of the hungarian wordnet project. In: Proceedings of the Fourth Global WordNet Conference GWC 2008, pp. 310–320 (2008)

14. Mikolov, T., Sutskever, I., Chen, K., Corrado, G.S., Dean, J.: Distributed representations of words and phrases and their compositionality. In: Advances in Neural Information Processing Systems 26: 27th Annual Conference on Neural Information Processing Systems 2013. Proceedings of a meeting held 5–8 December 2013, Lake Tahoe, Nevada, United States, pp. 3111–3119 (2013)
15. Siklósi, B., Novák, A.: Beágyazási modellek alkalmazása lexikai kategorizációs feladatokra. XII. Magyar Számítógépes Nyelvészeti Konferencia, pp. 3–14 (2016)
16. Hall, M., Frank, E., Holmes, G., Pfahringer, B., Reutemann, P., Witten, I.H.: The WEKA data mining software: an update. SIGKDD Explor. Newsl. **11**, 10–18 (2009)

# Genetic-Based Decoder for Statistical Machine Translation

Douib Ameur[(⊠)], Langlois David[(⊠)], and Smaïli Kamel[(⊠)]

Villers-lès-Nancy, France
ameur.douib@inria.fr, {david.langlois,kamel.smaili}@loria.fr

**Abstract.** We propose a new algorithm for decoding on machine translation process. This approach is based on an evolutionary algorithm. We hope that this new method will constitute an alternative to Moses's decoder which is based on a beam search algorithm while the one we propose is based on the optimisation of a total solution. The results achieved are very encouraging in terms of measures and the proposed translations themselves are well built.

## 1 Introduction

In Statistical machine translation (SMT) [5] given a source sentence $f$, the system produces a translation $e$ in the target language, which maximises the probability $P(e|f)$. SMT can be divided into three parts. The language model (LM), which estimates the fluency of the translation $e$ in the target language. The translation model (TM), which estimates the quality of the translation (accuracy) $e$, given the source $f$. The last part of SMT system is the decoder, for which, the machine translation process can be considered as an optimisation issue, where it takes, in input, the source sentence $f$ and uses the LM and the TM to produce the best possible translation $e$, from all possible translations.

Many algorithms are proposed to handle the issue of decoding. The first proposals used a word-based translation system [1,8], where the alignment is between words. Nowadays, decoders use a phrase-based system [11], where the alignment is between phrases. MOSES is the most popular open source decoder used by the community [10]. It is based on a beam-search algorithm, where it builds incrementally a set of complete translations from partial translations and starting with the empty one. In the building process, new phrases are added for each partial translation hypothesis to produce new hypotheses. Consequently, a large number of hypotheses are produced. To reduce this number, a pruning process is applied, where the n-best hypotheses are retained for the next step according to a score given by the LM and the TM. Finally, from the set of complete translation hypotheses, the one which has the highest score is chosen as the final translation $e$. This algorithm gives good results [9], but presents at least two drawbacks. The first one concerns the fact that it is impossible to challenge a previous decision of translation. That is why it is possible to miss a partial solution which could lead to the best final translation. The second one concerns

© Springer International Publishing AG, part of Springer Nature 2018
A. Gelbukh (Ed.): CICLing 2016, LNCS 9624, pp. 101–114, 2018.
https://doi.org/10.1007/978-3-319-75487-1_9

the decision making. At each step, MOSES keeps some translation hypotheses and eliminates others, according to the scores of the partial translations. The final solution is made up on a series of decisions what we would like to challenge in our method.

Using a complete translation hypothesis from the beginning of the translation process can reduce the impact of these problems. With complete translation hypothesis, it is possible to visit each part of the research space and modify it if necessary.

In the literature, works have been proposed in order to achieve translations from a complete translation. In [12] the decoder starts with a complete translation and applies iteratively heuristics until it reaches the best solution. In each iteration, the neighbour translations are produced by applying some neighbour functions [12], which modify phrases and lead to new propositions of translation. This work gives good results but do not outperforms the state-of-the-art system.

In this paper, we propose a new decoder for SMT based on evolutionary algorithms, and more particularly those based on genetic principle [3, 6]. The advantage of the genetic algorithm is to use, not just one complete translation as in [12], but a population of complete translations. The combination of these translations, using crossover and mutation functions, produces more information allowing to take better decisions. The genetic algorithm is one of the most performant optimization algorithms [3], especially when the space search is huge.

In previous works, the genetic algorithm was proposed to handle some parts of the automatic translation. In [7] a genetic algorithm was used in a learning process to generate new translation examples, but for an example-based machine translation system and not for SMT. Another work [19] proposed an algorithm to generate a multi-word-based Translation Model, and to evaluate this model, a basic genetic algorithm was used as a translator. In this paper, we show the feasibility of using a genetic algorithm as a decoder for SMT. Also, we give a detailed adaptation of each component of the genetic algorithm.

In Sect. 2, we define the Statistical Machine Translation problem. In Sect. 3, we present a description of our genetic decoder. We present the corpus data and some comparative results in Sect. 4. Finally, in Sect. 5, we give the conclusion with the perspectives for future works.

## 2 Phrase-Based Statistical Machine Translation System

In phrase-based SMT the system takes a source sentence $f = <f_1, f_2, ..., f_{|f|}>$ in input, and the decoder produces the best possible translation $e = <e_1, e_2, ..., e_{|e|}>$ in the target language, where $f_i$ (respectively $e_i$) is the $i^{th}$ word in $f$ (respectively $e$). The translation process segments $f$ into phrases, translates each phrase into the target language, and reorder target phrases. The links between the source and target phrases define the alignment $(a)$. The translation $e$ must maximise the conditional probability $P(a, e|f)$ [18]. So, the problem is considered as an optimization issue:

$$\hat{e} = argmax_{a,e}[P(e) \times P(f|a,e)] \tag{1}$$

In Eq. (1) we distinguish the language model $P(e)$ and the phrase-based translation model $P(f|e)$. The decoder uses these two models as features, to evaluate the translations and finds the best one. Other features can be added to improve the evaluation.

## 3  Genetic Algorithm for SMT

The basic idea of the genetic algorithm [3,6] is to start with an initial population of solutions (chromosomes) and to produce iteratively new chromosomes using the crossover and the mutation functions. At the end of each iteration (generation) some chromosomes are selected from the population and kept for the next generation. This process ensures the evolution of the population towards a good solution (see Fig. 1). An adequate representation of chromosomes and a good function to evaluate the chromosomes (fitness) are required to guarantee this evolution. Many parameters (population length, crossover and mutation rate, end process conditions, etc.) have to be fixed depending on the problem. In the

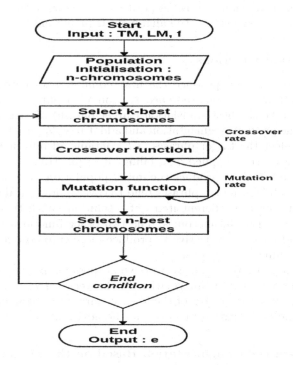

**Fig. 1.** Genetic algorithm for SMT

next sections, we present how we adapt the genetic algorithm as a decoder for the SMT problem and we describe all the functions needed. We call our decoder GAMaT for Genetic Algorithm for Machine Translation.

### 3.1   Chromosome Representation

A chromosome is composed of a serie of genes. A gene is an item of the translation hypothesis. As we use a phrase-based SMT, each gene will be associated to a phrase. This encoding makes easier the application of crossover and mutation functions at the phrase level. So, each chromosome $c$ contains the main following attributes:

- A complete translation hypothesis $e$.
- A phrase segmentation of $f$ and $e$.
- An alignment $a$ between the source and target phrases.

To simplify the notation for the next sections, we define a chromosome $c = <f, e, \overline{f}, \overline{e}, a>$, where $\overline{f}$ is the phrase segmentation of $f$, $\overline{e}$ is the phrase segmentation of $e$, and $a$ the alignment between the source and target phrases. We denote $f_i^j$ (respectively $e_i^j$) a phrase from $f$ (respectively $e$) starting at position $i$ and ending at position $j$. $a_i$ is the position of the target phrase in $\overline{e}$ aligned with $\overline{f}_i$. When $a_i = i$ for every $i$, the alignment is monotone.

### 3.2   Initialisation Functions

Five functions are used to produce the first population of chromosomes. All these functions, from the input sentence $f$, produce a phrase segmentation $\overline{f}$. After that, they take the best translation (target phrase) of each source phrase determined by the previous segmentation and add it to $e$. We use the Translation Table $(TT)$ to select the target phrases. $TT$ contains all information obtained from the training process, applied on a bilingual corpus [13].

The first three functions produce the initial population by promoting longer phrases, this is useful since long phrases tend to cover more lexical and syntactic relationships between the different items of a solution. With these three functions, we produce three chromosomes. The two other functions use a random segmentation and have the objective to produce more chromosomes. For all the functions the alignment is monotone.

Before applying any function, we retrieve all possible phrases from $f$ using $TT$, and save phrases' length information in the vector $PH = <ph_1, ..., ph_n>$ (Table 1), where $ph_i$ is the length of the longest phrase in $f$ starting at position $i$. We describe the initialization functions in the next sections.

### 3.2.1   A Left-to-right Segmentation Based on the Maximum Length of Phrases

For this function, the source sentence $f$ is segmented from left to right. The first phrase (segment) is the longest prefix of $f$ that is present in $TT$. The remaining

**Table 1.** Length phrases table

| madame | la | président | , | la | prśidence | a | proclamé | le | résultat | du | vote | . |
|---|---|---|---|---|---|---|---|---|---|---|---|---|---|
| 5 | 4 | 2 | 2 | 5 | 3 | 7 | 3 | 5 | 4 | 3 | 2 | 1 |

of $f$ is processed using the same heuristic until $f$ is entirely segmented. For each segment in $f$, we select from $TT$ the English phrase with the highest probability. Then we concatenate these English phrases in order to obtain the initial translation. Figure 2 shows the result of this function applied to the example of Table 1.

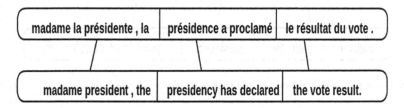

**Fig. 2.** Example of the left-to-right segmentation based on the maximum length of phrases

### 3.2.2 A Right-to-left Segmentation Based on the Maximum Length of Phrases

Contrary to the first function, here the segmentation is done from right to left. So, the first phrase is the longest suffix of $f$ that is present in $TT$. The same process is applied until $f$ is entirely segmented. The translation $e$ is generated in the same way as for the first function. Figure 3 shows the result of this initialization for the same example.

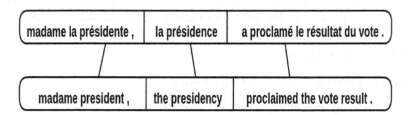

**Fig. 3.** Example of the right-to-left segmentation based on the maximum length of phrases

### 3.2.3 A Global Segmentation Based on the Maximum Length of Phrases

In this function non constraint is imposed to the direction of the segmentation. The main idea is to produce a segmentation with the minimum number of phrases. To produce this segmentation, we start by choosing the longest phrase $\overline{f}_i$ in $f$ using the phrases' length information saved in $PH$. Then we apply recursively the same process for the left and right part of $\overline{f}_i$ until exhaustion of all items $f$. The Fig. 4 shows the result of this segmentation using the $PH$.

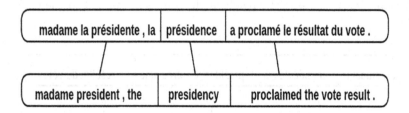

**Fig. 4.** Example of the global segmentation based on the maximum length of phrases

### 3.2.4 A Left-to-right Random Segmentation

Like for the first function, the source sentence $f$ is segmented from left to right. But, we select randomly from $TT$ a phrase which is considered as a prefix of $f$. We apply the same process iteratively on the remaining of $f$. To produce the target sentence $e$ we proceed in the same way as the previous functions. As this function takes random decisions, we use it to produce several chromosomes, one for each produced segmentation.

### 3.2.5 A Right-to-left Random Segmentation

This second random segmentation function proceeds from right to left contrary to the previous one. Iteratively we choose a random suffix of $f$ in the not yet segmented part of $f$ as a new phrase, and which exists in $TT$. Like for the previous function, we use this function to generate several chromosomes.

### 3.3 Crossover Function

A crossover operation consists in the following steps:

- Select randomly two chromosomes $c1$ and $c2$.
- Select randomly a subpart $sp$ from $f$ respecting some constraints: the first word of $sp$ must be the first word of a segment from $c1.\overline{f}$ and $c2.\overline{f}$, and the last word of $sp$ must be the last word of a segment from $c1.\overline{f}$ and $c2.\overline{f}$. In Fig. 5, the chosen segment is *"présidence a proclamé le résultat du vote"*. This segment is a good candidate for crossover because *"présidence"* is the first word of $c1.\overline{f}_3$ and $c2.\overline{f}_2$, and *"vote"* is the last word of $c1.\overline{f}_4$ and $c2.\overline{f}_4$.

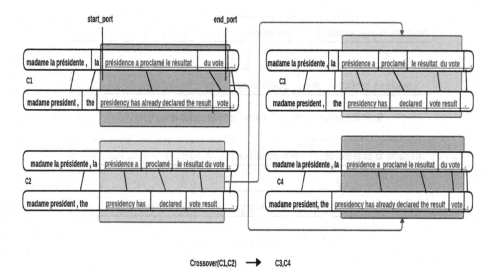

**Fig. 5.** Crossover function

- Build $c3$ by taking $sp$ from $c2$ (with its segmentation, its English counterpart, and the corresponding alignment). Complete $c3$ with the left and the right parts from $c1$.
- Build $c4$ by taking $sp$ from $c1$ (with its segmentation, its English counterpart, and the corresponding alignement). Complete $c4$ with the left and the right parts from $c2$.

## 3.4   Mutation Functions

We use five mutation functions, which modify the aspect of an existing chromosome and allow consequently to produce a new one. These functions are presented below.

1. **Replace-Phrase**
   We choose randomly a source phrase, and replace its associated target phrase in $e$ by another using $TT$. The new target phrase must have the highest probability of translation among all the possible translations excluding obviously the one we would like to mutate.
2. **Split-Phrase**
   Source phrase $\overline{f}_i$ is selected and randomly and segmented into a left and a right part. These parts must be present in $TT$. Then, we look for the segmentation of the target phrase aligned with $\overline{f}_i$ such that the target left (respectively right) part can be aligned with the source left (respectively right) part of $\overline{f}_i$. If it is possible, we build a new chromosome with the updated segmentation and alignment. If not, the left and right parts of $\overline{f}_i$ are translated using the most probable translations in $TT$. In Fig. 6, the segmented source

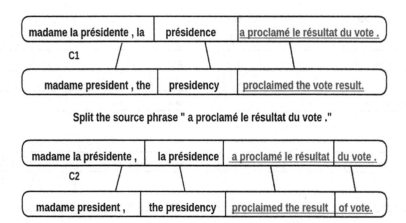

**Fig. 6.** Split mutation example

phrase is $c1.\overline{f}_3$. No translations are present in $TT$ for *"a proclamé le résultat"* and *"du vote."*, which exist separately in the aligned target phrase $c1.\overline{e}_{a_3}$. So, we choose new translation for each source phrase from $TT$.

3. **Merge-Two-Phrases**
   We choose randomly two adjacent phrases $\overline{f}_i$ and $\overline{f}_{i+1}$, where their aligned target phrases are also adjacent in the target sentence. The achieved phrase after merging $\overline{f}_i$ and $\overline{f}_{i+1}$ must exist in $TT$. If these conditions are met, we merge the selected source phrases into one phrase $\overline{f}_i$. We do the same process for the target phrases, if the result of this merging exists in $TT$ as a translation of the new source phrase. Otherwise, we choose from the translation table a new target phrase $\overline{e}_{a_i}$ which has the highest translation probability. Finally, we update the segmentation and the alignment in the chromosome.

4. **Merge-Two-Phrases-and-Replace**
   This function has the same logic as Merge-Two-Phrases, except that we translate directly the new generated phrase using $TT$, without any verification.

5. **Swap-Two-Phrases**
   This last function selects randomly two adjacent source phrases $\overline{f}_i$ and $\overline{f}_{i+1}$. After that, we exchange the positions of $\overline{e}_{a_i}$ and $\overline{e}_{a_{i+1}}$ in the translation $e$. This mutation produces a non-monotone alignment.

### 3.5  Evaluation Function: Log-Linear Model

Given a chromosome $c = <f, e, \overline{f}, \overline{e}, a>$, the log-linear approach is used to estimate the quality of the chromosome (translation). In this approach we calculate the logarithmic sum of a set of features functions $h_m(c)$, where $m$ is the *id* of a feature. The result of this logarithmic sum represents the *score* of the chromosome.

$$Score(c) = \sum_m \lambda_m \times \log(h_m(\overline{e}, \overline{f}, a)) \qquad (2)$$

Where $\lambda_m$ is the weight of $h_m$. The value of each weight defines the influence of the corresponding feature in the final score. With this approach, we can add new features, to estimate better the translation. GAMaT uses the following features, which are the same for MOSES [10]:

- $h_1$: Language model probability
- $h_2$: Direct translation probability
- $h_3$: Inverse translation probability
- $h_4$: Direct lexical weighting probability
- $h_5$: Inverse Lexical weighting probability
- $h_6$: Phrase penalty
- $h_7$: Word penalty
- $h_8$: Reordering model

As we handle complete translations, the use of the word penalty in the same way as in Moses, the produced translations are shorter than the references. That is why, we define a new word penalty score, using a length translation model. In this model we define a length probability $P_l(e, f)$ for each pair of source and target sentence, which is estimated from the bilingual training corpus as follows:

$$P_l(e, f) = \frac{count(|f|, |e|)}{count(|e|)} \qquad (3)$$

Where $count(|f|, |e|)$ is the number of times that source sentences of length $|f|$ has been translated by target translations of length $|e|$. $count(|e|)$ is the number of translations of length $|e|$ in the target corpus.

### 3.6   Selection Process

The selection process is used in two cases, the first one, to define an *elite* set from which we will pick chromosomes as parents, to perform genetic manipulations. The *elite* set contains the $k(k < n)$ best chromosomes of the population. The second selection is applied to select the $n$ best chromosomes, which are kept for the next generation. The selection is based on the score of chromosomes.

## 4   Results and Comparisons

### 4.1   Corpora

For our experiments, we use the $9^t h$ task workshop on Statistical Machine Translation (2014) [4]. We take the French-English corpus to evaluate GAMaT and MOSES. The corpus contains 2,000,000 pairs of sentences. After a classical step of corpus preprocessing, we define the following three sentences sets: *Train* set that contains 1,323,382 pairs of sentences for the training process. *Dev* set, that contains 165,422 pairs of sentences for the development process. Finally, *Test* set which contains 1,000 pairs of sentences to evaluate the decoders.

We use GIZA++ [13] to generate the translation model, and SRILM [17] for language model. MERT [2] is launched on the baseline system (MOSES) then we use the same weights in GAMaT since almost all the probability parameters are calculated by the tools associated to MOSES.

## 4.2  GAMaT Parameters

In a genetic algorithm we have four important parameters: the number of chromosomes in the population $(n)$, the number of chromosomes in the *elite* set, the crossover and mutation rates. We optimized $n$ by varying its value and we found that 120 chromosomes in the population gives the best translation results. The *elite* set contains the best 75% chromosomes in the population. The crossover rate and the mutation rate fix the number of times which the crossover and mutation functions are applied in each generation. We optimized the crossover and mutation rates empirically. So, in the presented results the mutation and the crossover operations are applied at each iteration to 20% and 40% of the population respectively.

## 4.3  Comparative Results

In this section, we present the performance of GAMaT, and we compare them to MOSES. To evaluate the quality of the translations achieved by the two decoders, we use BLEU [14] and TER [16] metrics. The first one calculates the number of n-gram which exists in the hypothesis and the reference translation at the same time. The second one, TER, calculates a cost of the modifications that we have to apply on hypothesis to obtain the reference translation.

**Table 2.** GAMaT and MOSES performances using BLEU, TER

| Decoder | BLEU | TER |
|---|---|---|
| MOSES | **29.32** | **53.02** |
| GAMaT-WPS | 26.13 | 53.34 |
| GAMaT-LtM | **27.15** | **53.08** |
| GAMaT+1-MOSES | 27.22 | 52.81 |

In Table 2, we present the BLEU and TER scores for GAMaT and MOSES, where GAMaT-LtM represents GAMaT with the translation length probabilty as feature, and GAMaT-WPS is GAMaT using the word penalty as feature. In the GAMaT+1-MOSES, we add the best translation of MOSES in the initialisation, as a chromosome. The underlying idea is to test if the 1-bets of MOSES can help GAMaT to produce better results.

The results show that there is a significant difference between GAMaT and MOSES, in terms of BLEU, however in terms of TER there, the two systems are equivalent. By studying the results, we mean the translations themselves achieved by both systems we found that GAMaT gives, in general, good translations compared to the source sentences. The main problem for GAMaT is the reordering. For some sentences, it has difficulties to apply the good permutations to find the best reordering in the target.

Adding the best translation of MOSES in the initialisation process does not improve significantly the result in terms of BLEU, however it outperforms MOSES in terms of TER which is very encouraging. It means that we have to improve the selection process in order to achieve more diversity in the population. So, improvement of the quality of the first population insures the improvement of the final translation.

Table 3 shows the number of translations that are better for a system than the other, in terms of TER. We can see that the two decoders give the same translation quality for 287 sentences from 1000. MOSES is better for 366 and GAMaT is better for 347 among them. From this result, we can deduce that each system manages to solve some translation problems better than the other.

**Table 3.** Comparison between GAMaT and MOSES for each sentence, using TER

|  | Nbr-sentences | % |
| --- | --- | --- |
| Equal | 287 | 28.70 |
| MOSES better | 366 | 36.60 |
| GAMaT better | 347 | 34.70 |

In the following, we will analyze certain functions of GAMaT in order to understand how to improve it. In Fig. 7, we plot the evolution of the population in terms of BLEU. Each curve represents the evolution for one translation. The curves show clearly, that there is no stability in the evolution for the first iterations. This is normal because the population contains a large variety of chromosomes. But, after some iterations the population stabilizes and progresses until convergence. This evolution is a normal evolution for any genetic algorithm, and proves that the decoder manages to increase the quality of translation from an initial population.

Finally, we analyse the influence of crossover and mutation functions in the research process. The values in Table 4 represent the number of times that these functions were used in the evolution of the best translations. We see that the crossover function is the most used function, with 67%, because this function is the main operation in any genetic algorithm, also because the fixed rate to apply the crossover function is 40%, compared with 20% for all the mutation functions. Also, we can see that the substitution of phrases is the most used mutation, through the *Replace-phrase* and *Merge-Two-Phrases-and-Replace* functions. The less used functions are *Split-Phrase* and *Swap-two-Phrases*, which is normal for

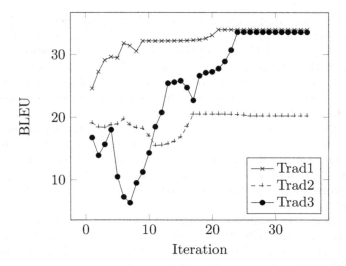

**Fig. 7.** Evolution of the population, using BLEU, for 3 different sentences

the French-English pair, because the reordering can be covered by long segments. In order to fix the problem of reordering we tried to force GAMaT to use more the *Swap-two-phrases* function, but the results were not better.

**Table 4.** Influence of crossover and mutation functions on final translations

| Function | Nbr | % |
|---|---|---|
| Crossover | 4491 | 67,92 |
| M-Replace | 858 | 12,98 |
| M-Swap | 345 | 5,22 |
| M-Split | 334 | 5,05 |
| M-Merge | 584 | 8,83 |
| M-Merge + Replace | 593 | 8,97 |

## 5    Conclusion and Perspectives

In this paper, we presented GAMaT, a new decoder for the SMT. This decoder is based on a genetic algorithm which allows to use a set of complete translations. The obtained performance for the French-English pair are promising but not yet better than MOSES in terms of BLEU but they are equivalent in terms of TER. However, GAMaT succeeds to propose better translations in 34.7% of cases and gives identical results in 28.7% of cases. Analysing the produced translations, we noticed that GAMaT suffers from the mismanagement of the reordering process.

To outperform MOSES, we will mainly optimize the feature's weights independently from MOSES. In fact, we use exactly the same weights produced by MOSES, we have to set up them, for instance by an evolutionary algorithm. The confidence measures [15] also will be used in order to guide better the genetic operations.

# References

1. Berger, A.L., Brown, P.F., Della Pietra, S.A., Della Pietra, V.J., Gillett, J.R., Lafferty, J.D., Mercer, R.L., Printz, H., Ures, L.: The candide system for machine translation. In: HLT 1994 Proceedings of the Workshop on Human Language Technology, pp. 157–162 (1994)
2. Bertoldi, N., Haddow, B., Fouet, J.-B.: Improved minimum error rate training in Moses. Prague Bull. Math. Linguist. **91**, 7–16 (2009)
3. Binitha, S., Siva Sathya, S.: A survey of bio inspired optimization algorithms. Int. J. Soft Comput. Eng. (IJSCE), 2231–2307 (2012)
4. Bojar, O., Buck, C., Federmann, C., Haddow, B., Koehn, P., Leveling, J., Monz, C., Pecina, P., Post, M., Saint-Amand, H., Soricut, R., Specia, L., Tamchyna, A.: Findings of the 2014 workshop on statistical machine translation. In: ACL NINTH Workshop on Statistical Machine Translation (2014)
5. Brown, P.F., Della Pietra, V.J., Della Pietra, S.A., Mercer, R.L.: The mathematics of statistical machine translation: parameter estimation. Comput. Linguist. **19**, 263–311 (1993)
6. Črepinšek, M., Liu, S.-H., Mernik, M.: Exploration and exploitation in evolutionary algorithms: a survey. ACM Comput. Surv. (CSUR) **45**(3), 35 (2013)
7. Echizen-ya, H., Araki, K., Momouchi, Y., Tochinai, K.: Machine translation method using inductive learning with genetic algorithms. In: Conference on Computational Linguistics, vol. 2, no. 16, pp. 1020–1023 (1996)
8. Germann, U., Jahr, M., Knight, K., Marcu, D., Yamada, K.: Fast decoding and optimal decoding for machine translation. Artif. Intell. **154**, 127–143 (2004)
9. Koehn, P.: A beam search decoder for phrase-based statistical machine translation models. In: Conference of the Association for Machine Translation in the Americas, AMTA, pp. 115–124 (2004)
10. Koehn, P., Hoang, H., Birch, A., Callison-Burch, C., Federico, M., Bertoldi, N., Cowan, B., Shen, W., Moran, C., Zens, R., Dyer, C., Bojar, O., Constantin, A., Herbst, E.: Moses: open source toolkit for statistical machine translation. In: ACL 2007 Proceedings of the 45th Annual Meeting of the ACL, pp. 177–180 (2007)
11. Koehn, P., Och, F.J., Marcu, D.: Statistical phrase-based translation. In: NAACL 2003 Proceedings of the 2003 Conference of the North American Chapter of the Association for Computational Linguistics on Human Language Technology, pp. 48–54 (2003)
12. Langlais, P., Patry, A., Gotti, A.: A greedy decoder for phrase-based statistical machine translation. In: Proceedings of TMI (2007)
13. Och, F.J., Ney, H.: A systematic comparison of various statistical alignment models. Comput. Linguist. **21**, 19–51 (2003)
14. Papineni, K., Roukos, S., Ward, T., Zhu, W.-J.: BLEU: a method for automatic evaluation of machine translation. In: ACL 2002 Proceedings of the 40th Annual Meeting on Association for Computational Linguistics, pp. 311–318 (2002)

15. Raybaud, S., Langlois, D., Smaili, K.: 'This sentence is wrong'. Detecting errors in machine-translated sentences. Mach. Transl. **25**, 1–34 (2011)
16. Snover, M., Dorr, B., Schwartz, R., et al.: A study of translation edit rate with targeted human annotation. In: Proceedings of the Association for Machine Translation in the Americas, pp. 223–231 (2006)
17. Stolcke, A., Zheng, J., Wang, W., Abrash, V.: SRILM at sixteen: update and outlook. In: Proceedings of IEEE Automatic Speech Recognition and Understanding Workshop, p. 5 (2011)
18. Zens, R., Och, F.J., Ney, H.: Phrase-based statistical machine translation. In: Jarke, M., Lakemeyer, G., Koehler, J. (eds.) KI 2002. LNCS (LNAI), vol. 2479, pp. 18–32. Springer, Heidelberg (2002). https://doi.org/10.1007/3-540-45751-8_2
19. Zogheib, A.: Genetic algorithm-based multi-word automatic language translation. Recent Adv. Intell. Inf. Syst. 751–760 (2011)

# Bilingual Contexts from Comparable Corpora to Mine for Translations of Collocations

Shiva Taslimipoor[1(✉)], Ruslan Mitkov[1],
Gloria Corpas Pastor[2], and Afsaneh Fazly[3]

[1] Research Group in Computational Linguistics, University of Wolverhampton,
Wolverhampton, UK
{shiva.taslimi,r.mitkov}@wlv.ac.uk
[2] Univeristy of Malaga, Malaga, Spain
gcorpas@uma.es
[3] VerticalScope Inc., Toronto, Canada
afsaneh.fazly@gmail.com

**Abstract.** Due to the limited availability of parallel data in many languages, we propose a methodology that benefits from comparable corpora to find translation equivalents for collocations (as a specific type of difficult-to-translate multi-word expressions). Finding translations is known to be more difficult for collocations than for words. We propose a method based on bilingual context extraction and build a word (distributional) representation model drawing on these bilingual contexts (bilingual English-Spanish contexts in our case). We show that the bilingual context construction is effective for the task of translation equivalent learning and that our method outperforms a simplified distributional similarity baseline in finding translation equivalents.

**Keywords:** Collocations · Word vector representation
Distributional similarity · Comparable corpora

## 1  Introduction

Collocations are considered as one type of Multi-word Expressions (MWEs) [2, 7,23]. While there are many studies on the automatic extraction of collocations from monolingual text [6,19,24], only a few have drawn on bilingual resources for the automatic treatment of collocations [3,5,15]. The need for representation of collocations in bilingual dictionaries is broadly discussed in [5]. To exemplify, collocations like *pay attention* and *pay homage*, require a different translation of the collocative verb in Spanish according to the base noun: *prestar/poner atención, rendir homenaje*.

Dealing with collocations, bilingually, is very interesting for two reasons: first, finding translation equivalents for these expressions is far from a resolved issue in Natural Language Processing (NLP); secondly, using bilingual corpora,

A. Gelbukh (Ed.): CICLing 2016, LNCS 9624, pp. 115–126, 2018.
https://doi.org/10.1007/978-3-319-75487-1_10

we can improve their identification especially for resource-poor languages. With regards to resource-poor languages, one approach that is indeed beneficial is to use comparable/non-parallel corpora. Although comparable corpora have been known to be helpful [14], their application to this task has been rather limited [9,21,26].

We propose an approach to find translation equivalents for collocations using comparable corpora. The idea is to use distributional similarity across bilingual corpora. By 'equivalent expressions' or 'equivalents' we refer to expressions which are translations of each other across languages. One of the premises in this methodology is that equivalent expressions are expected to appear in the same or similar contexts across languages.

Characterisation and comparison of context (distributional) vectors is known to be the standard approach to bilingual lexicon extraction from comparable corpora [4]. However, we aim to use such an approach to find translation equivalents for collocations. We benefit from a list of automatically aligned words to build bilingual contexts for our target expressions. We use the very recent word embedding approach [16] which employs the bilingual contexts to learn vector representations for words or expressions. Similar to [22], we use a strictly comparable corpora, in which the documents are paired to each other, to retrieve more relevant translations.

We focus on a particular type of collocations, namely those that are formed from a combination of a verb and a noun, e.g., *take part* in English, *formar parte* in Spanish. While the approach is language independent, in this particular study we seek to identify translation equivalents between Spanish and English collocations.

The remainder of this paper is organised as follows. The next section describes previous work addressing the task of bilingual translation equivalents extraction. In Sect. 3, we elaborate on the context similarity approach for identifying collocation translations. Section 4 includes the details of the data and the experiments which have been done for the task. We evaluate, report and discuss the results in Sect. 5 and we, finally, conclude in Sect. 6.

## 2   Related Work

The most common approach for extracting translation equivalents from parallel corpora is to use Statistical Machine Translation (SMT) [27]. Recently, several studies have suggested approaches for extracting *parallel segments* from comparable corpora for several different tasks, including bilingual lexicon construction [4,9,11,21], and sentence alignment for improving SMT [10,18,25]. Corpus-based distributional similarity has been used in a bilingual context to automatically discover translationally-equivalent *words* from comparable corpora [9,20,21]. It is not clear, however, whether a similar approach can be used for finding the translations of *multi-word collocations*.

NLP systems that need to translate collocations often use pre-existing lexicons of collocation translations [15]. However, such lexicons do not provide translations of all collocations, as new combinations are created and used on a daily basis. Thus, it is important to develop a method that can automatically find translation equivalents for multi-word collocations. Bouamor et al. [3] use distributional models to align MWEs to improve the performance of a machine translation system. However, their method relies on sentence-aligned (parallel) corpora. Rapp and Sharoff [22] also investigate the use of word co-occurrence patterns across languages to extract translations of single and multi-word terms. Like [11] they avoid using a large initial bilingual dictionary. While their approach delivers good results in finding the translations of single words, they do not report good results for MWEs. Even for single words their results only cover words that are salient words (keywords) according to their frequency patterns.

We also use context similarity to automatically extract translations for a set of experimental collocations in English and Spanish. However, we define the contexts bilingually and we draw on word embeddings for learning vector representations for our target expressions [16]. Our results suggest that similarities measured using word embeddings are more meaningful and lead to better translations.

# 3   Distributional Similarity Across Languages

According to the distributional similarity hypothesis, terms that are translation equivalents may share common *concepts* in their contexts. These shared concepts are in turn expressed by words/terms that are translation equivalents in the two languages. For example, we might expect to see the Spanish expression *poner en marcha* co-occuring with words, such as *problema*, *decisión* and *mercado*, and the potential English translation of it, *to launch*, co-occurring with the translations of the Spanish context words, i.e., *concern*, *decision*, *market*, respectively.

Distributional similarity has been widely used to find pairs (words or terms) that are semantically similar; however, the applications have mainly focused on similar pairs within a single language. We use an extended version of a state-of-the-art distributional similarity method to identify translation equivalents for collocations. Specifically, we define context in a bilingual space by pairing words from the two languages that we know are translations of each other. Note that we do not rely on a clean bilingual lexicon. Instead, we take the word pairs from a noisy bilingual lexicon, which is automatically learned by using a word alignment tool.[1]

## 3.1   Word Vector Representation

To represent words using context vectors, we use the `word2vec` method proposed by Mikolov et al. [16]. The method employs the patterns of word co-occurrences within a small window to predict similarities among words. The idea is to represent each word as a dense vector (a.k.a. word embeddings) derived by various

---

[1] We use the lexicon built by applying GIZA++ on the Spanish–English portion of the Europarl.

training methods, which in turn have been inspired by neural-network language modelling [13]. The new word embedding approach uses a neural network to learn low-dimensional word vectors from raw (monolingual) text. The standard implementation of word2vec constructs bag-of-words contexts for all single-word terms that appear in a training corpus. We adapt the model to our task of finding translation equivalents for multi-word collocations, by: (i) treating sequences of words as single units/terms, and (ii) defining bilingual contexts by drawing on a core set of known translation pairs. To do this we use the generalised word embedding approach proposed by [13] that allows us to define bilingual contexts. Although the generalised version of word2vec was originally used to extract dependency-based word embeddings [13], we can easily adapt it to our specific task of vector construction for multi-word collocations using bilingual contexts.

### 3.2  Bilingual Phrase Vector Representation

In standard word2vec, using a window of size $k$ around a target word $w$, $2k$ context words are produced: the $k$ words before and the $k$ words after $w$. We base our context extraction on this standard, with the difference that we extract only specific words rather than all the words in the context window. Our favourable context words come from a bilingual dictionary of words. Specifically, we focus on nouns as the most important components of meaning, and use a core lexicon of paired English–Spanish nouns as our bilingual context terms. The generalised word2vec model (called word2vecf)[2] can then be trained on these pairs, resulting in the vectors of the two languages to be defined over the same space (of paired English–Spanish nouns), and to be comparable.

### 3.3  Translation Equivalent Extraction

Given a target collocation $s$ from the source language (e.g., Spanish), our goal is to find the best translation equivalent in the target language (e.g., English). First, we identify a set of candidate translations for $s$, from a Spanish–English comparable corpora that we automatically build by pairing documents from the two languages. Next, we rank these candidates according to their semantic similarity to the target collocation. The following subsections explain these two steps in more detail.

**Candidate Extraction.**  To extract candidate translations for a collocation, we examine a set of automatically paired *comparable documents* from the two languages. Specifically, for each collocation $s$, we examine all target language documents that are paired to the source language documents containing $s$. We take a set of frequent unigrams, bigrams, and trigrams (which are verb combinations) appearing in these documents as candidate translations for $s$.[3] The details of pairing documents in comparable corpora is explained in Sect. 4.1.

---

[2] The software is available in the websites of the authors of [13].

[3] We set the frequency threshold to 10 in our experiments.

**Ranking Candidates Using Cross-Lingual Similarity.** We construct a cross-lingual vector representation for each collocation $s$, and for each of its candidate translations, drawing on our proposed approach for defining a cross-lingual semantic space (see Sect. 3 above). The winning candidate is the one that has the highest similarity to the collocation $s$.

## 4   Experimental Setup

### 4.1   Corpus

We use a corpus of comparable English–Spanish documents that we build from various news sources on the Web, as explained below.

**Collecting News Documents from the Web.** News texts are rich sources of shared content, and hence have commonly been used to construct comparable corpora [1,8,17]. To build our corpus of comparable English–Spanish documents, we collect news feeds from a variety of news sources, including the ABC news,[4] Yahoo news,[5] CNN news,[6] Sport news,[7] and Euronews[8] in both Spanish and English languages. We focus on documents from July to December 2015. We use a tool from the ACCURAT project[9] to extract comparable documents from the news texts [1].

**Computing Document Comparability.** ACCURAT also comes with a tool, called `DictMetric`, which is designed to measure the comparability levels of document pairs via cosine similarity [26]. The tool is specifically proposed to provide a data for extracting parallel segments with high performance. To measure the comparability of two documents in different languages, one language get translated to the other. The tool translates non-English texts into English by using lexical mapping from the available GIZA++ based bilingual dictionaries. Since the proportion of overlapped lexical information in two documents is the key factor in measuring their comparability, the tool converts the texts into index vectors and then computes the comparability score of document pairs by applying cosine similarity measure on the index vectors.

Using the ACCURAT toolkit, we compute the comparability of all pairs of Spanish and English documents. We extract the pairs with the comparability score (cosine similarity) of higher than 0.45 as aligned comparable documents. This result in 16,436 English documents (with around 11 million word tokens) and 11,468 Spanish documents (with around 6 million word tokens).

---

[4]  http://www.abc.es and http://www.abc.net.au.
[5]  http://es.noticias.yahoo.com and http://uk.news.yahoo.com.
[6]  http://cnnespanol.cnn.com and http://cnn.com.
[7]  http://www.sport.es/es and http://www.sport-english.com/en.
[8]  http://es.euronews.com and http://euronews.net.
[9]  http://www.accurat-project.eu.

Each English document is paired to at least one Spanish document; equally, there is at least one paired English document for every Spanish document.[10]

## 4.2   Experimental Expressions

Our methodology is to use bilingual word vector representation to find translations for collocations across comparable corpora. To report the results, we focus on 9 highly-frequent verbs in English and 6 in Spanish. These verbs tend to frequently combine with many different nouns in their direct object positions to form multi-word collocations. The verbs are: *take, have, make, give, get, find, pay, lose* in English, and *tener, dar, hacer, formar, tomar, poner* in Spanish. We extract all occurrences of these verbs followed with a noun, from the whole News corpora, focusing only on those combinations that have a frequency higher than 10. This process results in 1,007 English Verb+Noun collocations, and 930 Spanish Verb+Noun collocations, which are annotated by two human annotators as being semantically coherent collocations, or arbitrary sequences of words. We measure inter-annotator agreement using the Kappa score: Kappa is 0.67 for English expressions, and 0.61 for Spanish expressions. Among these candidate expressions, only 162 English expressions and 187 Spanish expressions occur with frequency higher than 9 in our paired comparable documents. We run the experiments only on these expressions.

## 4.3   Vector Construction

Recall that to construct vectors for our English and Spanish expressions, we need a seed list of paired context words (a.k.a., the bilingual context pairs). For this purpose, we use a subset of the word alignments resulting from applying GIZA++ on the English–Spanish Europarl parallel corpus [12]. Specifically, we only consider pairs of frequent nouns that have an alignment probability of higher than 0.2, where frequent nouns in a language are those that appear 50 times or more in Europarl. As a result we have a list of 4,700 bilingual contexts.

For learning the vectors, we use the following corpora to extract word co-occurrence statistics: the monolingual English and Spanish components from the Europarl, and the English and Spanish components of our News corpora. We index all the English and Spanish *verb combinations* (unigrams, bigrams, trigrams) according to their occurrences with the context word pairs. Specifically, from the window of 10 words around a target expression, we capture any word that exists in our bilingual context pairs (focusing on the relevant language given the language of the target expression). The `word2vecf` software is then used to train vectors on the indexed corpus. We then apply our methodology to find translations for collocations in both directions: Spanish to English, and English to Spanish.

---

[10] The comparable corpora that we prepared is available on https://github.com/shivaat/EnEsCC.

Note that we focus on finding translations for Verb+Noun combinations. We assume that for most such expressions, the translation equivalent is either a Verb (unigram), a Verb+Noun (bigram), or a Verb+Noun with an intervening word, such as a determiner or an adjective (trigram). We thus consider as our candidate translations all unigram Verbs, bigram Verb+Noun combinations, and trigram Verb+Noun combinations with an intervening word. For every expression from the source language (e.g., Spanish), our goal is to find the five most cross-lingually similar Verb or Verb + Noun combination in the target language (e.g., English).

# 5   Evaluation and Results

**Baseline.** We implement a simple distributional similarity approach as our baseline. Given two expressions (from the two languages), we measure their similarity by comparing their corresponding sets of (bilingual) context pairs (using a context window of size 10). We use the Jaccard similarity coefficient to measure similarity. The baseline uses our comparable corpora to find translation candidates for each expression, but relies on the above simple similarity to rank these candidates.

**Using Loosely Comparable Corpora.** We also perform experiments to investigate the advantage of using comparable corpora with high level of similarity for finding the candidate translations of an expression. To do so, we add noisy alignments to our accurately-aligned documents. Specifically, for each source-language (e.g., Spanish) document, paired with several highly-similar target-language (e.g., English) documents, we align an extra set of 2,000 randomly selected target-language documents.[11] This process results in a larger but noisy corpus of comparable documents. Our goal here is to understand whether using a larger set of documents that may contain more candidate translations is helpful, despite the noise. That is, we intend to understand whether a method like word2vec is sufficiently robust to noise, and hence capable of finding good translations from documents that are not perfectly aligned. If that is the case, then we can avoid the rather expensive process of building highly-accurate comparable corpora. We apply both the baseline and our proposed approach (the one that uses word2vec) to this noisy data, and compare the results with those on the smaller corpora with the more accurately aligned documents.

**Results and Discussion.** We ask a human expert to rate the top-ranked translations produced by each of the methods for each expression. We ask the expert to give a rating of 1 if there is at least one good translation in the top-5-ranked list; otherwise, the list is given a rating of 0. We also have 25% of the resulted translation lists annotated by a second annotator. The inter-annotator

---

[11] Note that we add noise in both Spanish–English and English–Spanish directions.

agreement in terms of Kappa is 0.80 both for finding translations for Spanish expressions and for finding translations for English expressions.

Note that we use a similarity measure to rank the candidate translations of each expression. By using different threshold values for this similarity, we get ranked lists of varying sizes. The higher this threshold, the smaller the number of the resulting translation candidates, and hence the higher the number of expressions for which we may not have any good translations. In other words, we can trade off accuracy (precision) for coverage (recall). We thus set the similarity thresholds to different values in order to measure accuracy for varying degrees of coverage (from around 10% to around 80%). Doing so gives us a better understanding of the overall performance of each method.

Table 1 shows accuracy and coverage values for finding translations of the Spanish expressions; Table 2 gives the results for English expressions. Note that we show the results for both the baseline and the word2vec method, using both corpora of comparable documents: the (smaller and less noisy) corpus of highly-comparable documents (referred to as paird CC), and the larger and noisy corpus (referred to as CC + noise).

**Table 1.** The accuracy of the baseline compared to the word2vec approach in extracting translations of Spanish expressions.

|  | Coverage | 10–20% | 20–30% | 30–40% | 40–50% | 50–60% | 60–70% | 70–80% |
|---|---|---|---|---|---|---|---|---|
| Using paired CC | Baseline | 82% | 55% | 24% | 22% | 18% | 16% | 12% |
|  | word2vec | 50% | 46% | 40% | 36% | 34% | 32% | 33% |
| Using CC + noise | Baseline | 78% | 50% | 24% | 18% | 14% | 13% | 8% |
|  | word2vec | 44% | 45% | 38% | **37%** | 30% | **33%** | 32% |

**Table 2.** Comparing the accuracy of the baseline with the word2vec approach in extracting translations of English expressions.

|  | Coverage | 10–20% | 20–30% | 30–40% | 40–50% | 50–60% | 60–70% | 70–80% |
|---|---|---|---|---|---|---|---|---|
| Using paired CC | Baseline | 79% | 52% | 46% | 35% | 26% | 22% | 18% |
|  | word2vec | 39% | 37% | 34% | 36% | 34% | 29% | 31% |
| Using CC + noise | Baseline | 70% | 50% | 24% | 22% | 18% | 12% | 13% |
|  | word2vec | 38% | 34% | 31% | **39%** | **39%** | **32%** | 31% |

As can be seen in the first rows of both tables, the baseline accuracy/precision is high when we limit the method with a very low coverage/recall, but drops down quickly as we increase coverage. Note that when coverage is low, many expressions do not have any translation equivalents. But those that do have candidates, have a few accurate ones, and hence it is easy for a simple method such as the baseline to pick the best.

Compared to the baseline, the word2vec approach is more stable across the different degrees of coverage for both translation directions: in fact, the performance of word2vec drops only slightly when we move from a coverage of 30% to almost 80%. Importantly, even for a very high degree of coverage (i.e., 70%–80%) word2vec performs much better than the baseline in terms of accuracy (33% compared to 12% for Spanish-to-English, and 31% versus 18% for English-to-Spanish).

Next, we compare the results using the two corpora. Investigating the baseline approach over the two corpora, we observe that almost in all coverages the performance of the baseline approach drops by using the noisy paired documents. This can be seen in both Tables 1 and 2 for both directions of Spanish to English and English to Spanish translations. Then we compare the results of word2vec: Interestingly, the performance of word2vec is reasonably close on the two different corpora, even though the CC + noise has a much higher degree of noise. The better accuracies of word2vec in some cases when we use the larger noisy corpora are shown in bold. This is an interesting result, suggesting that even using a large but noisy corpus of comparable documents, we can find reasonable translations for multiword collocations by relying on a robust and accurate method such as word2vec.

**Semantically Coherent Collocations.** Our experimental Verb+Noun combinations (that we try to find translations for) include a range of expressions, from frequent collocations (*get things*), to multi-word verbal units (*make reference*), to more idiomatic expressions (*take place*). It is thus interesting to find out whether the performance of our method differs on these different types of expressions. For this, we take a subset of expressions from each language that has been annotated as a semantically-coherent MWE by two annotators. This selection process results in 80 Spanish and 101 English expressions. Table 3 shows accuracy of the word2vec method for both Spanish and English subsets when coverage is set to around 80% (using the cleaner comparable corpora for finding candidates). The results show that, for both languages, accuracy improves when we focus on these subsets (48% versus 33% for Spanish expressions, and 44% versus 31% for English).

**Table 3.** The accuracy of the word2vec approach in extracting translations of multiword collocations from comparable corpora.

|  | Accuracy | |
| --- | --- | --- |
|  | Spanish | English |
| word2vec approach | 48% | 44% |

# 6   Conclusions and Future Work

We have proposed a method for extracting cross-lingual contexts from comparable corpora, which we have then used to build embedding-based vector representations for multi-word collocations using a state-of-the-art technique (word2vec). We use these vectors to find translation equivalents for Verb+Noun combinations between Spanish and English. We show that our approach outperforms a simple distributional similarity baseline. We also show that, in contrast to the distributional similarity baseline, the word2vec approach is less vulnerable to noise in the corpus (in terms of comparability of the aligned documents).

Future experiments will focus on improving the results further as follows: First, preparing larger corpora of comparable documents, in order to increase the coverage and also the accuracy by providing more context. Secondly, we can take into account expressions that have more than one intervening word between the Verb and the Noun components (both for our experimental collocations, and for the translation candidates). Third, syntactic structure can be added to the word2vec approach to draw on the grammatical dependencies of context and hence form better vector representations (as suggested in [13]).

**Acknowledgments.** This work has been partially supported by the LATEST (Ref: 327197-FP7-PEOPLE-2012-IEF) project. The authors would like to express their gratitude to Anna de Santis and Lorena Gomez for their annotation work.

# References

1. Aker, A., Kanoulas, E., Gaizauskas, R.: A light way to collect comparable corpora from the web. In: Proceedings of the Eight International Conference on Language Resources and Evaluation (LREC 2012) (2012)
2. Bannard, C.: A measure of syntactic flexibility for automatically identifying multiword expressions in corpora. In: Proceedings of the Workshop on a Broader Perspective on Multiword Expressions, pp. 1–8. Association for Computational Linguistics (2007)
3. Bouamor, D., Semmar, N., Zweigenbaum, P.: Identifying bilingual multi-word expressions for statistical machine translation. In: Proceedings of the Eight International Conference on Language Resources and Evaluation (LREC 2012), Istanbul, Turkey. European Language Resources Association (ELRA) (2012)
4. Bouamor, D., Semmar, N., Zweigenbaum, P.: Context vector disambiguation for bilingual lexicon extraction from comparable corpora. In: Proceedings of the 51st Annual Meeting of the Association for Computational Linguistics, Sofia, Bulgaria, Short Papers, vol. 2, pp. 759–764. Association for Computational Linguistics (2013)
5. Pastor, G.C.: Collocations in e-bilingual dictionaries: from underlying theoretical assumptions to practical lexicography and translation issues. In: Torner, S., Bernal, E. (eds.) Collocations and Other Lexical Combinations in Spanish: Theoretical and Applied Approaches, pp. 173–199. Routledge, Abingdon (2017)
6. Evert, S.: The statistics of word cooccurrences : word pairs and collocations. Ph.D. thesis, Universität Stuttgart, Holzgartenstr. 16, 70174 Stuttgart (2005)
7. Fazly, A.: Automatic acquisition of lexical knowledge about multiword predicates. Ph.D. thesis, Department of Computer Science, University of Toronto (2007)

8. Fung, P.: A statistical view on bilingual lexicon extraction: from parallel corpora to non-parallel corpora. In: Farwell, D., Gerber, L., Hovy, E. (eds.) AMTA 1998. LNCS (LNAI), vol. 1529, pp. 1–17. Springer, Heidelberg (1998). https://doi.org/10.1007/3-540-49478-2_1
9. Fung, P., McKeown, K.: Finding terminology translations from non-parallel corpora. In: Proceedings of the 5th Annual Workshop on Very Large Corpora, pp. 192–202 (1997)
10. Ion, R.: PEXACC: a parallel sentence mining algorithm from comparable corpora. In: Proceedings of the Eight International Conference on Language Resources and Evaluation (LREC 2012) (2012)
11. Ismail, A., Manandhar, S.: Bilingual lexicon extraction from comparable corpora using in-domain terms. In: Proceedings of the 23rd International Conference on Computational Linguistics: Posters, pp. 481–489. Association for Computational Linguistics (2010)
12. Koehn, P.: Europarl: a parallel corpus for statistical machine translation. In: Conference Proceedings: The Tenth Machine Translation Summit, Phuket, Thailand, pp. 79–86 (2005)
13. Levy, O., Goldberg, Y.: Dependency-based word embeddings. In: Proceedings of the 52nd Annual Meeting of the Association for Computational Linguistics, Baltimore, Maryland, Short Papers, vol. 2, pp. 302–308. Association for Computational Linguistics (2014)
14. McEnery, A., Xiao, R.: Parallel and comparable corpora: what is happening. In: Incorporating Corpora: The Linguist and the Translator, pp. 18–31 (2007)
15. Mendoza Rivera, O., Mitkov, R., Corpas Pastor, G.: A flexible framework for collocation retrieval and translation from parallel and comparable corpora. In: Workshop on Multi-word Units in Machine Translation and Translation Technology (2013)
16. Mikolov, T., Sutskever, I., Chen, K., Corrado, G.S., Dean, J.: Distributed representations of words and phrases and their compositionality. In: Advances in Neural Information Processing Systems, pp. 3111–3119 (2013)
17. Munteanu, D.S., Marcu, D.: Improving machine translation performance by exploiting non-parallel corpora. Comput. Linguist. 31(4), 477–504 (2005)
18. Pal, S., Pakray, P., Naskar, S.K.: Automatic building and using parallel resources for SMT from comparable corpora. In: Proceedings of the 3rd Workshop on Hybrid Approaches to Translation (HyTra) @ EACL, pp. 48–57 (2014)
19. Pecina, P.: An extensive empirical study of collocation extraction methods. In: Proceedings of the ACL Student Research Workshop, ACLstudent 2005, Stroudsburg, PA, USA, pp. 13–18. Association for Computational Linguistics (2005)
20. Pekar, V., Mitkov, R., Blagoev, D., Mulloni, A.: Finding translations for low-frequency words in comparable corpora. Mach. Transl. 20(4), 247–266 (2006)
21. Rapp, R.: Automatic identification of word translations from unrelated English and German corpora. In: Proceedings of the 37th Annual Meeting of the Association for Computational Linguistics on Computational Linguistics, pp. 519–526. Association for Computational Linguistics (1999)
22. Rapp, R., Sharoff, S.: Extracting multiword translations from aligned comparable documents. In: Proceedings of the 3rd Workshop on Hybrid Approaches to Translation (HyTra) @ EACL 2014, Gothenburg, Sweden, pp. 83–91 (2014)
23. Sag, I.A., Baldwin, T., Bond, F., Copestake, A., Flickinger, D.: Multiword expressions: a pain in the neck for NLP. In: Gelbukh, A. (ed.) CICLing 2002. LNCS, vol. 2276, pp. 1–15. Springer, Heidelberg (2002). https://doi.org/10.1007/3-540-45715-1_1

24. Smadja, F.: Retrieving collocations from text: Xtract. Comput. Linguist. **19**, 143–177 (1993)
25. Smith, J.R., Quirk, C., Toutanova, K.: Extracting parallel sentences from comparable corpora using document level alignment. In: Proceedings of Human Language Technologies: The 11th Annual Conference of the North American Chapter of the Association for Computational Linguistics (NAACL-HLT 2010), pp. 403–411 (2010)
26. Su, F., Babych, B.: Measuring comparability of documents in non-parallel corpora for efficient extraction of (semi-)parallel translation equivalents. In: Proceedings of the Joint Workshop on Exploiting Synergies Between Information Retrieval and Machine Translation (ESIRMT) and Hybrid Approaches to Machine Translation (HyTra), EACL 2012, Stroudsburg, PA, USA, pp. 10–19. Association for Computational Linguistics (2012)
27. Tiedemann, J.: Extraction of translation equivalents from parallel corpora. In: Proceedings of the 11th Nordic Conference on Computational Linguistics, pp. 120–128 (1998)

# Bi-text Alignment of Movie Subtitles for Spoken English-Arabic Statistical Machine Translation

Fahad Al-Obaidli[1], Stephen Cox[2], and Preslav Nakov[1(✉)]

[1] Qatar Computing Research Institute, HBKU, Doha, Qatar
{faalobaidli,pnakov}@qf.org.qa
[2] School of Computing Sciences, University of East Anglia, Norwich, UK
s.j.cox@uea.ac.uk

**Abstract.** We describe efforts towards getting better resources for English-Arabic machine translation of spoken text. In particular, we look at movie subtitles as a unique, rich resource, as subtitles in one language often get translated into other languages. Movie subtitles are not new as a resource and have been explored in previous research; however, here we create a much larger bi-text (the biggest to date), and we further generate better quality alignment for it. Given the subtitles for the same movie in different languages, a key problem is how to align them at the fragment level. Typically, this is done using length-based alignment, but for movie subtitles, there is also time information. Here we exploit this information to develop an original algorithm that outperforms the current best subtitle alignment tool, **subalign**. The evaluation results show that adding our bi-text to the IWSLT training bi-text yields an improvement of over two BLEU points absolute.

**Keywords:** Machine translation · Bi-text alignment · Movie subtitles

## 1 Introduction

Statistical machine translation (SMT) research is continually improving in an attempt to produce systems to meet the ever-increasing demand for accessibility to content in foreign languages. SMT requires a substantial amount of parallel bi-text in order to build a translation model and unfortunately, such bi-text resources are very limited. Moreover, automatic bi-text alignment is a challenging task.

Existing parallel corpora are mostly derived from specialized domains such as administrative, technical and legislation documents [11]. These documents often only cover few widely spoken languages or languages that are either regionally or culturally related. Interestingly, with the huge demand for movie subtitles, movie subtitles online databases are among the fastest growing sources of multilingual data. Many users provide these subtitles for free online in a variety of languages

© Springer International Publishing AG, part of Springer Nature 2018
A. Gelbukh (Ed.): CICLing 2016, LNCS 9624, pp. 127–139, 2018.
https://doi.org/10.1007/978-3-319-75487-1_11

through download services. Subtitles are made available in plain text with a common format to help with rendering the text segments accordingly.

They are usually created to fit pirated copies of copyright-protected movies shared by organized groups, so their use in research could be deemed to be a positive side effect of the Internet movie-piracy scene [16]. Moreover, the use of these subtitles enables low-cost alignment of multilingual corpora by utilising one of their features, which is temporal indexing of subtitle segments. This approach has created opportunities to align specific language pairs that are difficult to align using the traditional methods or that are of generally scarce resources.

Because of the inherent nature of movie dialogue, subtitles differ from other parallel resources in several aspects. They are mostly transcriptions of material that is often spontaneous speech, which may contain considerable slang language, idiomatic expressions and also fragmental spoken utterances rather than complete grammatical sentences—such material is commonly summarized instead of being literally transcribed. Since these subtitles are user-generated, the translations are free, incomplete and affected by cultural differences. Rephrasing and compression degrees vary between different languages and depend on subtitling traditions. Subtitles also arbitrarily include some information such as the movie title, subtitle author/translator details and trailers. They may also contain translations of visual information such as sign languages. Certain versions of subtitles are especially compiled for the hearing-impaired to include extra information about other, non-spoken sounds such as background noise and, therefore contain material not related to the speech. Furthermore, subtitles must be short enough to fit the screen in a readable manner and to span for a limited time, which creates variable segmentations between languages.

The subtitle languages available differ from one movie to another. Here we are interested in movie subtitles that exist in both English and Arabic. Arabic is a widely-spoken language with 300 million native speakers (and another 120+ million non-native speakers), and has an official status in 28 countries (third in that respect, after English and French). Yet, the digital presence of Arabic is relatively low, compared to what one should expect given the number of speakers. Still, according to web traffic analytics, search queries for Arabic subtitles and traffic from the Arabic region is relatively very high.

The reminder of this paper is organized as follows. Section 2 offers an overview of related work. Section 3 presents our fragment-alignment method. Section 4 describes the experiments and the evaluation results. Section 5 concludes with general discussion and points to possible directions for future work.

## 2   Related Work

There is a body of literature about building multilingual parallel corpora from movie subtitles [15,33]. Tiedemann has put a lot of efforts in this direction and has made substantial contributions for aligning movie subtitles [25–27]. Initially, he gathered bi-texts for 59 languages in his OpenSubtitles2013 corpus, which was obtained from 308,000 subtitles files of around 18,900 movies downloaded from OpenSubtitles.org, one of the free online databases of movie subtitles.

For alignment, Tiedemann, started with the traditional approach of length-based sentence alignment [5] using sentence boundaries tagged in an earlier stage. This is based on the idea that sentences are linguistically-driven elements and, thus, it would be more appropriate to process them using linguistic features rather than merely aligning subtitles appearing simultaneously on the screen. Nonetheless, there are many untranslated parts of the movie for reasons discussed earlier. The initial results were unsatisfactory.

In follow-up work, Tiedemann took another approach based on the time over-lap of subtitle fragments between different languages and made explicit use of the available time information. He also took into account that one subtitle segment can be matched to multiple segments in the target language. This yielded significant improvements. However, some movies in the evaluation set yielded noticeably low matching ratios, which was due to time offsets between subtitles synchronized to different versions of the movie, which consequently radically affected the score even for a slight shift.

In order to find the right offset and to synchronize two subtitle files, a reference point is required. Tiedemann used cognates between the source and the target subtitle languages as a reference point to calculate the offset where the resulting matches ratio is below a certain threshold. A problem with this approach is that, depending on the language pair, false hit cognates may be used and may instead affect the performance drastically. This approach may not work for English–Arabic anyway due to the lack of enough cognates (cognates are relatively hard to find due to the different writing scripts used). Another approach is to use language dictionaries, which are expensive to build and are also language-dependent, although they do yield better results than the cognate approach.

Volk and Harder [30] built their Swedish-Danish corpus by first manually translating Swedish subtitles of English TV programmes with the help of trained translators using specified guidelines and, in the process, adding appropriate time information for the translations. Subsequently, Danish translators used the Swedish subtitle files as a template and added their translations of the Swedish subtitles between the available time codes. This "commercial" setup allows them to avoid the complex alignment approaches found in other studies [26,27]. They only matched pairs where Swedish and Danish time-stamps differ by less than 0.6 s, in order to ensure high quality. Volk and Harder assumed that they could match most of the subtitles except where the Danish translator has changed the time codes sufficiently to fail the strict overlap condition [29]. They state that their alignment approach is pragmatic and requires minimal human inspection, even though it is not sensible beyond this setup, using "genesis" files (Swedish subtitles) as a reference, as demonstrated earlier by Tiedemann [25], where, in reality, many translations require accurate handling.

It is also worth mentioning the AMARA project[1] [9], a unique, open, scalable and flexible online collaborative platforms which takes advantage of the power of crowdsourcing and encourages volunteer translation and editing of subtitles of educational videos.

---

[1] https://www.amara.org/en/.

The core value of the AMARA platform is demonstrated by its "faster transcription turnaround while maintaining high levels of user engagement." This could be achieved by the ease of use of the platform and the ability to remotely transcribe a video without the necessity to re-upload it on the platform. Translation can also be verified by other users.

Abdelali et al. [1] used the AMARA content to generate a parallel corpus out of the translations in order to improve an SMT system for the educational domain. They aligned the multi-lingual text of the educational videos and assessed its quality with several measures in comparison with the IWSLT training set, which is very close to the AMARA domain. It is important to note that 75% of the subtitle segments have exact time values among both languages in the pair. This is due to the neat editing environment that makes it easier for users to use existing subtitles as a template to generate translations for other languages. This also makes it ideal for the alignment process as less effort is required to handle partial overlap. This is not the usual case in more open communities such as movie subtitle platforms, which are merely databases for independently generated subtitles. They report an improvement of 1.6 BLEU points absolute when adding the AMARA bi-text to the IWSLT training bi-text for Arabic-to-English SMT.

## 3   Method

We look at movie subtitles as a unique source of bi-texts in an attempt to align as many translations of movies as possible in order to improve English to Arabic SMT performance. In comparison to other translation directions, translating from English into Arabic in particular is a rare translation direction and often yields significantly lower results when compared with the opposite direction (Arabic to English).

First, we collected pairs of English-Arabic subtitles of more than 29,000 movies/TV shows, a collection bigger than any pre-existing bilingual subtitles dataset. We then designed a pipeline of heuristic processing in order to eliminate the inherent noise that comes with the subtitles' source in order to yield good quality alignment. We used the time information provided within the subtitle files to develop a "time-overlap" based alignment method. Note that time overlap information is language-independent and it has been proven in previous work to outperform other traditional approaches such as length-based approaches, which rely on sentence boundaries to match translation segments [25]. We attempted to maximize the number of aligned sentence pairs whilst still maintaining the quality of our parallel corpora to minimize noise caused by alignment errors.

After producing our parallel corpora, which vary in size and relative quality, we used them as additional training data for an English-Arabic SMT system. We trained the baseline system on the IWSLT13 training bitext only; we compared this system to the same system but retrained with IWSLT13 data plus our newly-produced bi-text. We further used the Arabic subtitles to train a language model.

Adding each of our alignment models positively impacted the SMT system. In particular, the best system outperforms the baseline by 1.92 BLEU points absolute. Moreover, our method outperformed the results yielded by the best previously available subtitle alignment tool [28] by 0.2 BLUE points absolute.

We provide more detail about the bi-text production below.

## 3.1   Data Gathering

We identified target subtitles from OpenSubtitles using their daily-generated exports, which include references to their entire database of about 3M subtitles (as of May 20th, 2015) along with some useful information. We looked for Arabic and English subtitle files that are provided in the most common SubRip format for consistency. It is common that multiple subtitle versions exist for the same movie in the same language. In this case, we look for movies with matching movie *release* names to form a pair of subtitles for each movie. Otherwise, the versions are picked randomly. It was apparent later that pairs with matching movie release names yielded a significantly higher alignment ratio than those picked at random.

## 3.2   Preprocessing

We then downloaded the subtitles with the kind help of the administrators of OpenSubtitles. The files were then subject to a set of quality checks such as language, format, encoding identifications, and were further cleansed in order to eliminate unnecessary text such as HTML formatting tags. We also split segments of dialogue between multiple speakers (starting with leading hyphens) into multiple segments and split the original allocated time among them according to their new character length. We excluded corrupt and misidentified subtitle files from our collection, and we ultimately ended up with 29,000 complete pairs of Arabic-English subtitle files of unique movies/TV show episodes.

## 3.3   Bi-text Alignment

We decided to base our bi-text alignment approach on time information given its efficiency with subtitle documents, as found in previous work [25]. Each segment of one of the language pairs is compared to the segments of the other and the segment time information is used to discover overlaps. When an overlap is identified, a predefined *overlap ratio* threshold is applied to decide whether to count the segment pair as a valid translation or to exclude it from the final corpus. The ratio is a modified version of the Jaccard index and is defined as follows:

$$ratio = \frac{intersect + 1}{union + 1}$$

where *intersect* is the time period in which both segments overlap over the movie's runtime, while *union* is the period both segments cover together, i.e., the timespan from the start of the earliest segment to the end of the latest one.

Here is a pseudocode of our approach:

```
1. Loop through documents A and B and look for a minimal overlap
2. If no overlap is found proceed
3. If overlap is found calculate the overlap ratio
4. If the ratio is more than or equal to the threshold then
   4.1 Return matching segment pair
   4.2 Note the index of the matching segment in document B
   4.3 Break from the inner toop to continue to the next segment in
document A, comparing it with the next segment in B from the
last noted index.
5. Else, go to 1.
```

We define and distinguish between overlaps here as $(i)$ partial overlaps where the starting time value of a segments lies within the other segment's (from the other document) time interval and its end value is found beyond that interval, and vice-versa, and $(ii)$ complete overlaps where both the start and the end time value of one segment lies within the other segment's time interval and vice-versa.

This framework allows us to decide at which side of the pair to iterate next given the kind of overlap. Furthermore, the overlap is only considered a match between two overlapping segments when a predefined minimal overlap ratio threshold is met. We define this ratio as follows: given the start and the end time values for the English and the Arabic segment, we identify the minimum (soonest) and the maximum (latest) start times, $startMin$ and $startMax$, respectively and regardless of the language. And we do the same for the minimum and for the maximum end times as well, $endMin$ and $endMax$. Then we decide the amount of overlap/intersect between the segment $(endMin - startMax)$ by the total/union time covered by both $(endMax - startMin)$ using the formula above.

The ratio provides an indexation of the magnitude of the overlap given the length of the segments. Setting a threshold below 1 (strictest) allows a window for partial and complete overlap cases, described earlier, with excess trailing or leading time intervals beyond or before the actual overlapping interval, respectively, to be matched. This tolerance window is essential for the nature of our data since subtitles are typically compiled with independently defined time values and, thus, are positioned with organic variations between different versions along the movie runtime. And, as described earlier, it is less likely that two versions will have an exact number of segments with exact timestamps.

Finally, we extend the above algorithm to handle not only 1-1 matches but also 1:M ones. We simply look for 1-1 matches as usual and, in case the segments do not meet the minimum overlap ratio threshold and depending on the state of the false overlap, we will expand one side of the segments in the pair by concatenating its time value with the next segment on its side and, hence, calculate a new overlap ratio. The expansion continues on one side of the pair up to five adjacent segments, and then stops if no matches are found, and continues searching for 1-1 matches again and so on. If multiple segments match, they are concatenated into one line and white-space separated.

We only consider expanding one side and not both in a single match, i.e., one-to-many (1-M) and not many-to-many (M-M). This is because, for example, two parallel and adjacent segments might not match separately as 1-1, but, when concatenated into 2-2, the longer time intervals might compensate for a higher overlap ratio and, consequently, falsely yield a match. It is even more likely to harvest false matches when expanding further, i.e., 2-3, 3-3, etc. After all, it is less likely that both documents contain a segmentation of the same sentence. Yet, if so, longer sentences that had to be segmented still have a good chance of being matched by the 1-1 alignment condition separately.

## 4    Experiments and Evaluation

Below we first describe the baseline system and the evaluation setup; then, we present our subtitle alignment and machine translation experiments.

### 4.1    Baseline System

We built a phrase-based SMT model [14], as implemented in the Moses toolkit [13], to train an SMT system translating from English to Arabic. We trained all components of the system (translation, reordering, and language models) on the IWSLT'13 data,[2] which includes a training bi-text of 150 K sentences [3].

Following [10], we normalized the Arabic training, development and test data using MADA [23], fixing automatically all wrong instances of *alef, ta marbuta* and *alef maqsura*. We further segmented the Arabic words using the Stanford word segmenter [17]. For English, we converted all words to lowercase.

We built our phrase tables using the standard Moses pipeline with max-phrase-length of 7 and Kneser-Ney smoothing. We first word-aligned the training bi-text using IBM model 4 [2] on the English-Arabic and on the Arabic-English directions; then, we consolidated the two alignments using the grow-diag-final-and symmetrization heuristics. Then, we trained a phrase-based translation model with the standard features and a maximum length of 7.

We also built a lexicalized reordering model [12]: *msd-bidirectional-fe*. We further built a 5-gram language model (LM) on the Arabic text with Kneser-Ney smoothing using KenLM [7]. We tuned the models using pairwise ranking optimization (PRO) [8] with the length correction of [20][3] with 1000-best lists, using the IWSLT'13 tuning dataset. We tested on the IWSLT'13 test dataset, which we decoded using the Moses [13] decoder to produce Arabic translations, which we then desegmented.

In order to ensure stability, we performed three reruns of MERT for each experiment, and we report evaluation results averaged over the three reruns, as suggested in [4].

---

[2] The data can be found at http://workshop2013.iwslt.org.

[3] For a broader discussion see also [6,21].

## 4.2   Evaluation Setup

One of the issues when translating into Arabic are the common mistakes in written Arabic text such as incorrect glyphs, punctuation, and other orthographic features of the language. With 29 K pairs of different movies, it would be unrealistic to attempt to model them. Furthermore, the Arabic references provided by IWSLT in both the tuning and the testing sets are no exception from these random errors, which is an issue for scoring outputs of the systems when they are expected to produce proper Arabic. In fact, it is shown to affect the IWSLT baseline system score by more than two BLEU points with the NIST v13 scoring tool [24]. Therefore, we normalize both the tuning and the testing references with the QCRI Arabic Normalizer v3.0,[4] which use MADA's morphological analysis to output an enriched form of Arabic; it also standardizes the digits to appear all in Arabic and converts all the punctuation to English punctuation to make it possible to tokenize. We apply the same normalization to each system's output after detokenization as a final post-processing step [19].

## 4.3   Experiments

### 4.3.1   Subtitle Alignment

We produced two datasets with an overlap ratio threshold of 0.65, one with 1-1 alignments only (i.e., we only allow aligning one English segment to one Arabic segment) and the other one with 1-M alignments[5] (i.e., we allow aligning an English segment to a sequence of one or more consecutive Arabic segments).

We choose the threshold of 0.65 roughly as a midpoint between the most strict condition, 1.0, which is not promising for the open nature of the data, or the less strict condition, 0.5, which creates the possibility of two adjacent segments being viable for a match with the same segment on the other side of the pair, and hence create duplications. Yet, this is only one case of presenting noise to the data, which will be contaminated with partial alignments, i.e., pairs with translations of only half of the other, when lowering the threshold.

We also produced a third 1-M alignment with a stricter (0.85) threshold that excluded pairs yielding less than 20% matches between their original content. We assume that these pairs are likely to be out of sync. We also aim to test whether accounting for quality would enhance the translation performance even at the expense of quantity.

For comparison, we also trained an MT model on a bi-text produced by the best available subtitle alignment tool, `subalign`, developed by Tiedemann. This tool includes more complex features such as auto-synchronization and matching segments at the sentence level [28]. It does not provide an English-Arabic dictionary for synchronization nor does it support the use of cognates for this language pair, so we set the parameters to default.

---

[4] This is the official scoring method for the translation tracks into Arabic at IWSLT'13: http://alt.qcri.org/tools/arabic-normalizer.

[5] We set $M$ to be at most 5 in order to prevent the algorithm from unreasonably iterating up to the last segment looking for a match.

The comparisons are in Table 1. The numbers show an 11.3% increase in the total number of matched pairs and also a 5% increase in the average sentence length in the 1-M model compared to the 1-1 model. The strict 1-M model yields significant drop in the number of sentences and word tokens. Finally, `subalign` yields a bi-text that is about twice as large as our bi-texts. Yet, as we will see below, this does not mean that it is of better quality.

**Table 1.** A comparison of the size of the produced corpora.

| Model | #Sentences | Avg. Len. | #Tokens |
|-------|-----------|-----------|---------|
| 1-1 | 13.3 M | 6.3 | 83.8 M |
| 1-M | 14.8 M | 6.6 | 97.7 M |
| 1-M Strict | 11.1 M | 6.4 | 71.0 M |
| subalign | 18.6 M | 9.6 | 178.6 M |

### 4.3.2    Machine Translation

We tested the impact of adding various versions of our bi-text to the training bi-text of the baseline system. We used the movies bi-text in two different ways: (a) we built a separate phrase table, which then we interpolated with the phrase table of the baseline system [18, 22, 31, 32], and (b) we just concatenated our bi-text with the IWSLT'13 training bi-text in the baseline and trained on the concatenation. We found that option (a) worked better on the 1-1 alignment model, and thus, we decided to use (a) for all models.

**Table 2.** Evaluation results for various alignment models.

| System | Data | BLEU |
|--------|------|------|
| Baseline | IWSLT | 11.90 |
| 1-1 | IWSLT+SUB | 13.59 |
| 1-M | IWSLT+SUB | **13.98** |
| 1-M Strict | IWSLT+SUB | 12.89 |
| subalign | IWSLT+SUB | 13.69 |

Table 2 shows a comparison between the baseline system and the three proposed systems that combine the IWSLT data from the baseline with the subtitles corpora using the corresponding alignment models. The baseline system scored 11.9 BLEU points, which is a replicate of the same results reported on the IWSLT'13 official page.[6]

---

[6] https://wit3.fbk.eu.

We can see in the table that the simplest model's system, which used the 1-1 alignment, had an advantage over the baseline by only 0.61 BLEU points absolute when we concatenated the texts of both the baseline and the subtitles for training the language model.

We ran another experiment with the same parameters except that in this run, we created separate LMs for the two corpora and then we interpolated them to optimize their weights. This improved the score to 1.71 BLEU points above the baseline score. Furthermore, the 1-M model's system yielded an even better score, 13.98 BLEU points, and is the best system overall. On the other hand, the strict version of the 1-M model only yielded 12.89 BLEU points, which is better than the baseline but worse than the simplest approach, the 1-1 model.

On the other hand, the 1-M strict model's SMT output yielded a better result than the 1-1 model, 12.89 BLEU points, but it is still lower than the original 1-M model. This may imply that SMT independently filters out noise because of its statistical nature and, yet, the translation model may make use of the excess data to compute more accurate statistics.

Finally, we tried to produce a corpus using the OPUS Uplug `subalign` tool to align the data. The tool comes with its own preprocessor, which we had to use in order to produce output in XML format as expected by the alignment script. Although this tool allows synchronization of subtitles, it requires a dictionary and/or a language model for this feature to work, which were not provided for Arabic and the time was not feasible to prepare any. After all, it would be more relevant to compare the output of that tool using only features matching our own model's. `subalign` managed to harvest about 18M parallel segment lines with longer average sentence length (English) from our subtitle collection, although for Arabic it was contaminated with gibberish caused by poor handling of the UTF-8 character encoding. Several, workarounds were attempted by force re-encode the text, fixing parts of the source code relevant to encoding and enforcing the encoding flag to UTF-8, but these were ineffective. This is also despite the fact the we applied the `dos2unix` utility to the raw text as instructed to eliminate any special DOS-related characters. Thus, the BLEU evaluation for the SMT pipeline using this corpus is not comparable to the rest.

# 5   Conclusion and Future Work

We have presented our efforts towards getting better resources for English-Arabic machine translation of spoken text. We created the largest English-Arabic bi-text to date, and we developed a subtitle segment alignment algorithm, which outperformed the best rivaling tool, `subalign`. The evaluation results have shown that our dataset, combined with our alignment algorithm, yielded an improvement of over two BLEU points absolute over a strong spoken SMT system.

In future work, we would like to incorporate more sources of information in the process of alignment, e.g., cognates, translation dictionaries, punctuation, numbers, dates, etc. Adding some language-specific information might be interesting too, e.g., linguistic knowledge. We also plan to teach our tool to recognize advertisements, which are typically placed at the beginning or at the end of the file. Another interesting research direction is choosing the best pair of subtitles, in case multiple versions of subtitles exist for the same movie for the source and/or for the target language.

# References

1. Abdelali, A., Guzmán, F., Sajjad, H., Vogel, S.: The AMARA corpus: building parallel language resources for the educational domain. In: Proceedings of the Ninth International Conference on Language Resources and Evaluation, LREC 2014, Reykjavik, Iceland (2014)
2. Brown, P.F., Pietra, V.J.D., Pietra, S.A.D., Mercer, R.L.: The mathematics of statistical machine translation: parameter estimation. Comput. Linguist. **19**(2), 263–311 (1993)
3. Cettolo, M., Niehues, J., Stüker, S., Bentivogli, L., Frederico, M.: Report on the 10th IWSLT evaluation campaign. In: Proceedings of the International Workshop on Spoken Language Translation, IWSLT 2013, Heidelberg, Germany, pp. 15–24 (2013)
4. Foster, G., Kuhn, R.: Stabilizing minimum error rate training. In: Proceedings of the Fourth Workshop on Statistical Machine Translation, WMT 2009, Athens, Greece, pp. 242–249 (2009)
5. Gale, W.A., Church, K.W.: A program for aligning sentences in bilingual corpora. Comput. linguist. **19**(1), 75–102 (1993)
6. Guzmán, F., Nakov, P., Vogel, S.: Analyzing optimization for statistical machine translation: MERT learns verbosity, PRO learns length. In: Proceedings of the Nineteenth Conference on Computational Natural Language Learning, CoNLL 2015, Beijing, China, pp. 62–72 (2015)
7. Heafield, K.: KenLM: Faster and smaller language model queries. In: Proceedings of the 6th Workshop on Statistical Machine Translation, WMT 2011, Edinburgh, Scotland, UK, pp. 187–197 (2011)
8. Hopkins, M., May, J.: Tuning as ranking. In: Proceedings of the Conference on Empirical Methods in Natural Language Processing, EMNLP 2011, Edinburgh, Scotland, UK, pp. 1352–1362 (2011)
9. Jansen, D., Alcala, A., Guzmán, F.: AMARA: a sustainable, global solution for accessibility, powered by communities of volunteers. In: Stephanidis, C., Antona, M. (eds.) UAHCI 2014. LNCS, vol. 8516, pp. 401–411. Springer, Cham (2014). https://doi.org/10.1007/978-3-319-07509-9_38
10. El Kholy, A., Habash, N.: Orthographic and morphological processing for English-Arabic statistical machine translation. Mach. Transl. **26**(1–2), 25–45 (2012)
11. Koehn, P.: Europarl: a parallel corpus for statistical machine translation. In: Proceedings of the Tenth Machine Translation Summit, MT Summit 2005, Phuket, Thailand, pp. 79–86 (2005)
12. Koehn, P., Axelrod, A., Mayne, A.B., Callison-Burch, C., Osborne, M., Talbot, D.: Edinburgh system description for the 2005 IWSLT speech translation evaluation. In: Proceedings of the International Workshop on Spoken Language Translation, IWSLT 2005, Pittsburgh, PA, USA (2005)

13. Koehn, P., Hoang, H., Birch, A., Callison-Burch, C., Federico, M., Bertoldi, N., Cowan, B., Shen, W., Moran, C., Zens, R., Dyer, C., Bojar, O., Constantin, A., Herbst, E.: Moses: open source toolkit for statistical machine translation. In: Proceedings of the 45th Annual Meeting of the ACL: Interactive Poster and Demonstration Sessions, ACL 2007, Prague, Czech Republic, pp. 177–180 (2007)

14. Koehn, P., Och, F.J., Marcu, D.: Statistical phrase-based translation. In: Proceedings of the 2003 Conference of the North American Chapter of the Association for Computational Linguistics on Human Language Technology, HLT-NAACL 2003, Edmonton, Canada, vol. 1, pp. 48–54 (2003)

15. Lavecchia, C., Smaïli, K., Langlois, D.: Building parallel corpora from movies. In: Proceedings of the 4th International Workshop on Natural Language Processing and Cognitive Science, NLPCS 2007, Funchal, Madeira, Portugal (2007)

16. Mangeot, M., Giguet, E.: Multilingual aligned corpora from movie subtitles. Report in Laboratoire d'Informatique, Systèmes, Traitement de l'Information et de la Connaissance (2005)

17. Monroe, W., Green, S., Manning, C.D.: Word segmentation of informal Arabic with domain adaptation. In: Proceedings of the 52nd Annual Meeting of the Association for Computational Linguistics (Volume 2: Short Papers), ACL 2014, Baltimore, MD, USA, pp. 206–211 (2014)

18. Nakov, P.: Improving English-Spanish statistical machine translation: experiments in domain adaptation, sentence paraphrasing, tokenization, and recasing. In: Proceedings of the Third Workshop on Statistical Machine Translation, WMT 2008, Columbus, Ohio, USA, pp. 147–150 (2008)

19. Nakov, P., Al Obaidli, F., Guzman, F., Vogel, S.: Parameter optimization for statistical machine translation: it pays to learn from hard examples. In: Proceedings of the International Conference Recent Advances in Natural Language Processing, RANLP 2013, Hissar, Bulgaria, pp. 504–510 (2013)

20. Nakov, P., Guzmán, F., Vogel, S.: Optimizing for sentence-level BLEU+1 yields short translations. In: Proceedings of the 24th International Conference on Computational Linguistics, COLING 2012, Mumbai, India, pp. 1979–1994 (2012)

21. Nakov, P., Guzmán, F., Vogel, S.: A tale about PRO and monsters. In: Proceedings of the 51st Annual Meeting of the Association for Computational Linguistics (Volume 2: Short Papers), ACL 2013, Sofia, Bulgaria, pp. 12–17 (2013)

22. Nakov, P., Ng, H.T.: Improved statistical machine translation for resource-poor languages using related resource-rich languages. In: Proceedings of the 2009 Conference on Empirical Methods in Natural Language Processing, EMNLP 2009, Singapore, vol. 3, pp. 1358–1367 (2009)

23. Roth, R., Rambow, O., Habash, N., Diab, M., Rudin, C.: Arabic morphological tagging, diacritization, and lemmatization using lexeme models and feature ranking. In: Proceedings of the 46th Annual Meeting of the Association for Computational Linguistics, ACL 2008, Columbus, OH, USA, pp. 117–120 (2008)

24. Sajjad, H., Guzmán, F., Nakov, P., Abdelali, A., Murray, K., Al Obaidli, F., Vogel, S.: QCRI at IWSLT 2013: experiments in Arabic-English and English-Arabic spoken language translation. In Proceedings of the 10th International Workshop on Spoken Language Translation, IWSLT 2013, Heidelberg, Germany (2013)

25. Tiedemann, J.: Building a multilingual parallel subtitle corpus. In: Proceedings of the Computational Linguistics in the Netherlands, CLIN 2007, Nijmegen, Netherlands (2007)

26. Tiedemann, J.: Improved sentence alignment for movie subtitles. In: Proceedings of the Conference on Recent Advances in Natural Language Processing, RANLP 2007, Borovets, Bulgaria, pp. 582–588 (2007)

27. Tiedemann, J.: Synchronizing translated movie subtitles. In: Proceedings of the Sixth International Conference on Language Resources and Evaluation, LREC 2008 (2008)

28. Tiedemann. J.: Parallel data, tools and interfaces in OPUS. In: Proceedings of the Eighth International Conference on Language Resources and Evaluation, LREC 2012, Istanbul, Turkey, pp. 2214–2218 (2012)

29. Volk, M.: The automatic translation of film subtitles. A machine translation success story? J. Lang. Technol. Comput. Linguist. (JLCL) **24**(3), 115–128 (2009)

30. Volk, M., Harder, S.: Evaluating MT with translations or translators: what is the difference? In: Proceedings of the Machine Translation Summit XI, MT-Summit 2007, Copenhagen, Denmark (2007)

31. Wang, P., Nakov, P., Ng, H.T.: Source language adaptation for resource-poor machine translation. In: Proceedings of the 2012 Joint Conference on Empirical Methods in Natural Language Processing and Computational Natural Language Learning, EMNLP-CoNLL 2012, Jeju Island, Korea, pp. 286–296 (2012)

32. Wang, P., Nakov, P., Ng, H.T.: Source language adaptation approaches for resource-poor machine translation. Comput. Linguist. **42**, 277–306 (2016)

33. Xiao, H., Wang, X.: Constructing parallel corpus from movie subtitles. In: Li, W., Mollá-Aliod, D. (eds.) ICCPOL 2009. LNCS (LNAI), vol. 5459, pp. 329–336. Springer, Heidelberg (2009). https://doi.org/10.1007/978-3-642-00831-3_32

# A Parallel Corpus of Translationese

Ella Rabinovich[1(✉)], Shuly Wintner[1], and Ofek Luis Lewinsohn[2]

[1] Department of Computer Science, University of Haifa, Haifa, Israel
ellarabi@gmail.com, shuly@cs.haifa.ac.il
[2] Department of Computational Linguistics,
Universität des Saarlandes, Saarbrücken, Germany
o.l.lewinsohn@gmail.com

**Abstract.** We describe a set of bilingual English-French and English-German parallel corpora in which the direction of translation is accurately and reliably annotated. The corpora are diverse, consisting of parliamentary proceedings, literary works, transcriptions of TED talks and political commentary. They will be instrumental for research of translationese and its applications to (human and machine) translation; specifically, they can be used for the task of translationese identification, a research direction that enjoys a growing interest in recent years. To validate the quality and reliability of the corpora, we replicated previous results of supervised and unsupervised identification of translationese, and further extended the experiments to additional datasets and languages.

**Keywords:** Parallel corpora · Translationese · Machine translation

## 1 Introduction

Research in all areas of language and linguistics is stimulated by the unprecedented availability of data. In particular, large text corpora are essential for research of the unique properties of *translationese*: the sub-language of translated texts (in any given language) that is presumably distinctly different from the language of texts originally written in the same language. Indeed, contemporary research in translation studies is prominently dominated by corpus-based approach [1–9]. Most studies of translationese utilize *monolingual comparable* corpora, i.e. corpora where translations from multiple languages into a single language are compared with texts written originally in the target language.

The unique characteristics of translated language have been traditionally classified into two categories: properties that stem from the *interference* of the source language [10], and *universal* traits elicited from the translation process itself, regardless of the specific source and target language-pair [1,11]. Computational investigation of translated texts has been a prolific field of recent research, laying out an empirical foundation for the theoretically-motivated hypotheses on the characteristics of translationese. More specifically, identification of translated

© Springer International Publishing AG, part of Springer Nature 2018
A. Gelbukh (Ed.): CICLing 2016, LNCS 9624, pp. 140–155, 2018.
https://doi.org/10.1007/978-3-319-75487-1_12

texts by means of automatic classification shed much light on the manifestation of translation universals and interference phenomena in translation [12–20].

Along the way it was suggested that the unique properties of translationese should be studied in a *parallel* setting, i.e., in context of the corresponding *source* language: the original language the text was produced in. In particular, several studies hypothesized that certain phenomena traditionally attributed to translation universals (i.e., source-language independent) are, in fact, derivatives of the linguistic characteristics of the specific language-pair, subject to translation. [21] investigated the phenomenon of omitting the optional "that" in reporting English verbs, such as "he claimed [that] they left the room", highlighting correlation of this behavior to the linguistic conventions in the source language. [22] raised similar arguments regarding *explicitation* in translation (e.g., excessive usage of cohesive devices): he claimed that this phenomenon should only be studied by comparative analysis of translation and its original counterpart.

Parallel setting can also facilitate the task of automatic identification of translationese. [14] were the first to employ bilingual text properties for identification of the direction of translated parallel texts. They took advantage of sentence pairs translated in both directions for training a supervised classifier to identify translationese using word- and (part-of-speech) POS-ngrams as features. [23] used POS MTU (minimal translation unit) ngrams and HMM distortion properties extracted from bilingual parallel English-French Europarl and Hansard texts. Consecutively, they carried out series of experiments on sentence-level identification of translationese using Brown clusters [24] MTUs on the Hansard corpus [25].

A good corpus for research into the properties of translationese, should ideally satisfy the following desiderata:

**Diversity.** The corpus should ideally reflect diverse genres, registers, authors, modality (written vs. spoken) etc.

**Parallelism.** The corpus should include both the source and its translation, so that features that are revealed in the translation can be traced back to their origins in the source.

**Multilinguality.** Having translations from several source languages to the same target language facilitates a closer inspection of properties that are language-pair-specific vs. more "universal" features of translationese [1,22,26,27].

**Uniformity.** Whatever processing is done on the texts, it must be done uniformly. This includes sentence boundary detection, tokenization, sentence- and word-alignment, POS tagging, etc.

**Availability.** Finally, corpora that are used for research must be publicly available so that other researchers have the opportunity to replicate and corroborate research results.

In this work we describe a set of cross-domain, parallel, uniform, English-French and English-German corpora that were compiled specifically for research on translationese[1]. The corpora are diverse, consisting of parliamentary proceedings, literary works, transcriptions of TED talks and political commentary. We rigorously evaluate all datasets by series of supervised and unsupervised experiments; sensitivity analysis further implies applicability of these methodologies to data-meager scenarios. These datasets will be instrumental for research of translationese and its manifestations; they will also facilitate accurate identification of translationese at small text units by exploitation of bilingual text properties.

We detail the structure of the corpus in Sect. 2, explain how it was processed in Sect. 3, and evaluate it by extending some state-of-the-art supervised and unsupervised experiments in Sect. 4. We conclude with suggestions for future extensions.

## 2   Corpus Structure

Our corpus of translationese consists of five sub-corpora: Europarl, Canadian Hansard, literature, TED and political commentary.[2] All are parallel corpora, with accurate annotation indicating the direction of the translation. The datasets are uniformly pre-processed, represented, and organized. All corpora were further filtered to contain solely one-to-one sentence-alignments, which are more useful for the SMT research. Tables 1 and 2 report some statistical data on the corpus (after tokenization).

**Table 1.** English-French corpus statistics

| Corpus | # of sentences | | | # of tokens | | # of types | |
|---|---|---|---|---|---|---|---|
| | Original EN | Original FR | Total | Original EN | Original FR | Original EN | Original FR |
| EUR | 217K | 130K | 347K | 9,572K | 10,542K | 61K | 73K |
| HAN | 5,237K | 1,379K | 6,616K | 132,232K | 147,463K | 193K | 196K |
| LIT | 35K | 98K | 133K | 2,875K | 2,898K | 52K | 66K |
| TED | 7K | 4K | 12K | 217K | 239K | 14K | 17K |

### 2.1   Europarl

The Europarl sub-corpus is extracted from the collection of the proceedings of the European Parliament, dating back to 1996, originally collected by [28]. The original corpus[3] is organized as several language-pairs, each with multiple sentence-aligned files. We mainly used the English-French and English-German

---

[1] All corpora are available at
   http://cl.haifa.ac.il/projects/translationese/index.shtml.
[2] We use "EUR", "HAN", "LIT", "TED" and "POL" to denote the five corpora hereafter.
[3] The original Europarl is available from http://www.statmt.org/europarl/.

**Table 2.** English-German corpus statistics

| Corpus | # of sentences | | | # of tokens | | # of types | |
|---|---|---|---|---|---|---|---|
| | Original EN | Original DE | Total | Original EN | Original DE | Original EN | Original DE |
| EUR | 225K | 155K | 380K | 10,550K | 10,067K | 63K | 170K |
| LIT | 45K | 48K | 93K | 2,854K | 2,666K | 56K | 104K |
| POL | 8K | 9K | 18K | 443K | 421K | 26K | 44K |

segments, but resorted to other segments as we presently explain. We focus below on the way we generated the English-French sub-corpus; its English-German counterpart was obtained in a similar way.

Europarl is probably the most popular parallel corpus in natural language processing, and it was indeed used for many of the translationese tasks surveyed in Sect. 1. Unfortunately, it is a very problematic corpus. First, it consists of transcriptions of spoken utterances that are edited (by the speakers) after they are transcribed; only then are they translated. Consequently, there are significant discrepancies between the actual speeches and their "verbatim" transcriptions [29]. Second, while "Members of the European Parliament have the right to use any of the EU's [24] official languages when speaking in Parliament",[4] many of them prefer to speak in English, which is often not their native language.[5]

Mainly due to its multilingual nature, however, Europarl has been used extensively in SMT [30] and in cross-lingual research [31]. It has even been adapted specifically for research in translation studies: [32] compiled a customized version of Europarl, where the direction of translation is indicated. They used meta-data from the corpus, and in particular the "language" tag, to identify the original language in which each sentence was spoken, and removed sentence pairs for which this information was missing. A similar strategy was used by [33,34]. However, relying on the "language" tag in Europarl parallel text for identification of translation direction could be potentially flawed. Next we detail the procedure of reliable extraction of speaker details, including the original language of each sentence.

The Europarl corpus is a collection of several monolingual (parallel) corpora: the original text was uttered in one language and then translated to several other languages. In each sub-corpus, each paragraph is annotated with meta-information, in particular, the original language in which the paragraph was uttered. Unfortunately, the meta-information pertaining to the original language of Europarl utterances is frequently missing. Furthermore, in some cases this information is inconsistent: different languages are indicated as the original languages of (various translations of) the same paragraph (in the various sub-corpora). Additionally, the Europarl corpus includes several bilingual sub-corpora that are generated from the original and the translated texts, and

---

[4] http://europa.eu/about-eu/facts-figures/administration/index_en.htm.
[5] http://www.theguardian.com/education/datablog/2014/may/21/european-parliament-english-language-official-debates-data.

are already sentence-aligned. These bilingual corpora include only raw sentence pairs, with no meta-information.

To minimize the risk of erroneous information, we processed the Europarl corpus as follows. First, we propagated the meta-information from the monolingual texts to the bilingual sub-corpora: each sentence pair was thus annotated with the original language in which it was uttered. We repeated this process five times, using as the source of meta-information the original monolingual corpora in five languages: English, French, German, Italian, and Spanish (note that not all monolingual corpora are identical: some are much larger than others). For the same reason, not all the English-French sentence pairs in our bilingual corpus are reflected in all five monolingual corpora, and therefore some sentence pairs have less than five annotations of the original language. We restricted the bilingual corpus to only those sentence pairs that had five annotations. Then, we filtered out all sentence pairs whose annotations were inconsistent (about 0.5%). We also removed comments (about 0.5% as well), typically written in parentheses (things like "applause", "continuation of the previous session", etc.) As a result, we are confident that the speaker information (and in particular, the original language of utterances) in the filtered corpus is highly accurate.

### 2.2   The Canadian Hansard

The Hansard corpus is a parallel corpus consisting of transcriptions of the Canadian parliament in (Canadian) English and French from 2001–2009. We used a version that was annotated with the original language of each parallel sentence. This corpus most likely suffers from similar problems as the Europarl corpus discussed above; indeed, [35], who investigated the *British* Hansard parliamentary transcripts, found that "the transcripts omit performance characteristics of spoken language, such as incomplete utterances or hesitations, as well as any type of extrafactual, contextual talk" and that "transcribers and editors also alter speakers' lexical and grammatical choices towards more conservative and formal variants." Still, this is the largest available source of English-French sentence pairs. In addition to parliament members' speech, the original Hansard corpus contains metadata. Various annotations were used to discriminate different line types, including the date of the session, the name of the speaker, etc. We filtered out all segments except those referring to speech: in total, about 15% of the corpus line-pairs were thus eliminated.

### 2.3   Literary Classics

Our English-French literary corpus consists of classics written and translated in the 18th–20th centuries by English and French writers. Most of the raw material is available from the Gutenberg project[6] and FarkasTranslations.[7] The English-German literature corpus was generated in a similar way: we used material from

---

[6] http://www.gutenberg.org.
[7] http://farkastranslations.com/.

the Gutenberg project, Wikisource,[8] and a few more books. Both English-French and English-German datasets contain a metadata file with details about the books: title, year of publication, translator name and year of translation.

Identification of translationese in literary text by means of classification is considered a more challenging task [36,37] than classifying more "technical" translations, such as parliament proceedings. Translators of literature typically benefit from freedom and fewer constraints, rendering the translated text more similar to original writing. Additionally, our literature span almost three centuries and comprises works from wide range of genres – traits that overshadow the subtle characteristics of translationese [15,19,38,39]. Under this circumstances, we obtain very high accuracy with supervised classification on this corpus, and moderate, yet reasonable results with unsupervised clustering (see Sect. 4.3).

## 2.4 TED Talks

Our TED corpus is based on the subtitles of the TED talks delivered in English and translations to English of TEDx talks originally given in French[9]. We used the TED API[10] to extract subtitles of talks delivered in English, and Youtube API for TEDx talks originally given in French.

The quality of translation in this corpus is very high: not only are translators assumed to be competent, but the common practice is that each translation passes through a review before being published. This corpus consists of talks delivered orally, but we assume that they were meticulously prepared, so the language is not spontaneous but rather planned. Compared to the other subcorpora, the TED dataset has some unique characteristics that stem from the following reasons: (i) its size is relatively small; (ii) it exhibits stylistic disparity between the original and translated texts (the former contains more "oral" markers of a spoken language, while the latter is a written translation); and (iii) TED talks are not transcribed but are rather subtitled, so they undergo some editing and rephrasing. The vast majority of TED talks are publicly available online, which makes this corpus easily extendable for future research.

## 2.5 Political News and Commentary

This corpus contains articles, commentary and analysis on world affairs and international relations. English articles and their translations to German were collected from Project Syndicate.[11] This is a non-profit organization that primarily relies on contributions from newspapers in developed countries. It provides original commentaries by people who are shaping the world's economics, politics, science and culture. We collected articles categorized as Word Affairs from

---

[8]  http://en.wikisource.org/.

[9]  TEDx are TED-like events not restricted to specific language. We could not find sufficient amount of TEDx German talks translated to English.

[10] http://developer.ted.com/.

[11] http://www.project-syndicate.org/.

this project, originally written by English authors and translated to German. Original German commentaries and their translations to English were collected from the Diplomatics Magazine,[12] specifically, from its International Relations section.

## 3 Processing

The original Europarl corpora are already sentence-aligned, using an implementation of the Gale and Church sentence-alignment algorithm [40]. Since the alignment was done for one source paragraph at a time (typically consisting of few sentences), its quality is very high. The same also holds for the Hansard corpus, so we used the original alignments for both sub-corpora. We then filtered out any alignments that were not one-to-one; this resulted in a loss of about 3% of the alignments in Europarl, and only 2% in Hansard.

The literary sub-corpus required more careful attention. Books that were acquired from farkastranslations.com were available pre-aligned at the chapter- and paragraph-level; we therefore sentence-aligned them, one paragraph at a time, using a Python implementation [41] of the Gale and Church algorithm. For the remainder of the books, we first extracted chapters by (manually) identifying characteristic chapter titles (e.g., Roman numerals, explicit "Chapter N"). Paragraph boundaries within a chapter are typically marked by a double newline in Gutenberg transcripts, and we used this pattern to break chapters into paragraphs. Due to the fact that the Gale-Church algorithm only utilizes text length for alignment, it can be easily refined for aligning other logical units, e.g., paragraphs [40]. Finally, we aligned sentences within paragraphs using the same algorithm.

The genre of the literature sub-corpus is very different (presumably due to translators taking greater liberty), hence restricting the dataset to include only one-to-one sentence-alignments resulted in loss of above 10% of each book.

Sentence-alignment of subtitles of TED talks originally delivered in French (and translated to English) involved synchronization of subtitle frames. A typical frame in a subtitles (.srt) file contains frame start and end time (including milliseconds), as well as frame text:

```
18                                        frame sequential number
00:00:47,497 --> 00:00:50,813            frame start and end time
Cet engagement, je pense que j'ai fait le choix    frame text
```

First, we re-organized the subtitles file to contain (longer) frames that start and end on a sentence boundary; we achieved this by concatenating frames until a sentence termination punctuation symbol is reached. This procedure was conducted on both French subtitles and their corresponding English translations.

---

[12] http://www.diplomatisches-magazin.de/.

Then, we aligned the English-French parallel files at paragraph-level by alternated concatenation of paragraphs until synchronization of frame end time (up to a $\delta$ threshold that was fixed to 500 ms) on the English and French sides. The paragraph-alignment procedure pseudo-code is detailed in Algorithm 1.

---

**Algorithm 1.** TED subtitles paragraph-alignment algorithm

---

Comment: *l_paragraphs* and *r_paragraphs* are (not necessarily equal length) arrays of text paragraphs for alignment, from the left and right sides, respectively

$\delta = 500$ milliseconds ▷ threshold controlling the allowed delta in aligned frames' end time

subtitles_paragraph_alignment(1,1)    ▷ initial invocation assuming the arrays start from 1

**procedure** subtitles_paragraph_alignment(*l_count*,*r_count*)
    **if** (*l_count* > *l_paragraphs.length*) **and** (*r_count* > *r_paragraphs.length*) **then**
        return
    **end if**
    **if** (*l_count* > *l_paragraphs.length*) **then**
        output *r_paragraphs*[*r_count:r_paragraphs.length*]       ▷ remainder of the right side
        return
    **end if**
    **if** (*r_count* > *r_paragraphs.length*) **then**
        output *l_paragraphs*[*l_count:l_paragraphs.length*]       ▷ remainder of the left side
        return
    **end if**
    *l_current* = *l_paragraphs*[*l_count*].*frame_content*
    *l_frame_end* = *l_paragraphs*[*l_count*].*frame_end*
    *r_current* = *r_paragraphs*[*r_count*].*frame_content*
    *r_frame_end* = *r_paragraphs*[*r_count*].*frame_end*
    **while** (|*l_frame_end* − *r_frame_end*| > $\delta$ **and** (*l_count* < *l_paragraphs.length*)
        **and** (*r_count* < *r_paragraphs.length*)) **do**
        **if** (*l_frame_end* > *r_frame_end*) **then**             ▷ advance on the right side
            *r_count* += 1; *r_current* += *r_paragraphs*[*r_count*].*frame_content*
            *r_frame_end* = *r_paragraphs*[*r_count*].*frame_end*
        **else**                                    ▷ advance on the left side
            *l_count* += 1; *l_current* += *l_paragraphs*[*l_count*].*frame_content*
            *l_frame_end* = *l_paragraphs*[*l_count*].*frame_end*
        **end if**
    **end while**
    output "aligned paragraph pair:", *l_current*, *r_current*
    subtitles_paragraph_alignment(*l_count*+1,*r_count*+1)               ▷ recursive invocation
**end procedure**

---

We further aligned the paragraph-aligned TED and TEDx corpora at the sentence-level using the same sentence-alignment procedure [40]. TED talks tend to vary greatly in terms of sentence alignments (one-to-one, one-to-many, many-to-one, many-to-many). On average, approximately 10% of the alignments are not one-to-one; those were filtered out as well.

# 4   Evaluation

To validate the quality of the corpus we replicated the experiments of [18], who conducted a thorough exploration of supervised classification of translationese, using dozens of feature types. While [18] only used the Europarl corpus (in its original format) and worked on English translated from French, we extended the experiments to all the datasets described above, including also English translated from German, as well as French and German translations from English. We show that in-domain classification (with ten-fold cross-validation evaluation) yields excellent results. Moreover, very good results are obtained using unsupervised classification, implying robustness of this methodology and its applicability to various domains and languages.

## 4.1   Preprocessing and Tools

The (tokenized) datasets were split into chunks of approximately 2000 tokens, respecting sentence boundaries and preserving punctuation. We assume that translationese features are present in the texts or speeches across author, genre or native language, thus we allow some chunks to contain linguistic information from two or more different speakers simultaneously. The frequency-based features are normalized by the number of tokens in each chunk. For POS tagging, we employ the Stanford implementation along with its models for English, French and German [42].

We use Platt's sequential minimal optimization algorithm [43] to train a support vector machine classifier with the default linear kernel, an implementation freely available in Weka [44]. In all classification experiments we use (the maximal) equal number of chunks from each class: original (O) and translated (T).

## 4.2   Features

The first feature set we utilized for classification tasks comprises *function words* (FW), probably the most popular choice ever since [45] used it successfully for the Federalist Papers. Function words proved to be suitable features for multiple reasons: (i) they abstract away from contents and are therefore less biased by topic; (ii) their frequency is so high that by and large they are assumed to be selected unconsciously by authors; (iii) although not easily interpretable, they are assumed to reflect grammar, and therefore facilitate the study of how structures are carried over from one language to another. We used the list of above 400 English function words provided in [15], and similar number of French and German function words.[13]

A more informative way to represent (admittedly shallow) syntax is to use *part-of-speech (POS) trigrams*. Triplets such as PP (personal pronoun) + VHZ (verb "have", 3rd person sing. present) + VBN (verb "be", past participle) reflect

---

[13] The list of French and German FW was downloaded from https://code.google.com/archive/p/stop-words/.

a complex tense form, represented distinctively across languages. In Europarl, for example, this triplet is highly frequent in translations from Finnish and Danish and much rarer in translations from Portuguese and Greek.

We also used *positional token frequency* [46]. The feature is defined as counts of words occupying the first, second, third, penultimate and last positions in a sentence. The motivation behind this feature is that sentences open and close differently across languages, and it should be expected that these opening and closing devices will be transferred from the source if they do not violate the grammaticality of the target language. Positional tokens were previously used for translationese identification [18] and for native language detection [47].

Finally, we experimented with *contextual function words*. Contextual FW are a variation of POS trigrams where a trigram can be anchored by specific function words: these are consecutive triplets $\langle w_1, w_2, w_3 \rangle$ where at least two of the elements are function words, and at most one is a POS tag.

POS-trigrams, positional tokens and contextual-FW-trigrams are calculated as detailed in [18], but we only considered the 1000 most frequent feature values extracted from each dataset.

## 4.3   Results

**Supervised Identification of Translationese.** We begin with supervised identification of translated text using features detailed in Sect. 4.2[14]; Table 3 reports the results. Total number of chunks used for classification is reported per dataset, where we used the maximum available amount of data (up to 1000 chunks).

In line with previous works, the classification results are very high, yielding near-perfect accuracy with all feature types across all datasets. Close inspection of highly discriminative feature values sheds interesting light on the realization of unique characteristics of translationese across languages and domains; we leave this discussion for another venue.

Supervised classification methods, however, suffer from two main drawbacks: (i) they inherently depend on data annotated with the translation direction,

**Table 3.** Ten-fold supervised cross-validation classification (rounded) accuracy of English, French and German translationese; the best result in each column is bold-faced.

| Feature/corpus | EN(O) + FR→EN | | | | EN(O) + DE→EN | | | FR(O) + EN→FR | | | | DE(O) + EN→DE | | |
|---|---|---|---|---|---|---|---|---|---|---|---|---|---|---|
| | EUR | HAN | LIT | TED | EUR | LIT | POL | EUR | HAN | LIT | TED | EUR | LIT | POL |
| Total # of chunks | 1K | 1K | 400 | 40 | 1K | 650 | 100 | 1K | 1K | 400 | 40 | 1K | 600 | 90 |
| FW | 96 | **99** | **99** | 90 | 96 | **95** | 100 | **96** | 94 | 93 | 93 | 99 | **96** | 99 |
| pos. tokens | 97 | 96 | 96 | 93 | 93 | 93 | 99 | **96** | **98** | **95** | 96 | **98** | 92 | 100 |
| POS-trigrams | **98** | **99** | 98 | **94** | **97** | 94 | 100 | 95 | 88 | **95** | 94 | **98** | 93 | 99 |
| Contextual FW | 94 | 98 | 93 | 90 | 92 | 89 | 98 | **96** | 95 | 94 | **98** | 94 | 83 | 90 |

---

[14] Feature combinations yield similar, occasionally slightly better, results; we refrain from providing full analysis in this paper.

and (ii) they may not be generalized to unseen (related or unrelated) domains. Indeed, series of works on supervised identification of translationese reveal that classification accuracy dramatically deteriorates when classifier is evaluated out-of-domain (i.e., trained and tested on texts drawn from different corpora): [15,19,38,39] demonstrated significant drop in the accuracy of classification when one of the parameters (genre, source language, modality) was changed. These shortcomings undermine the usability of supervised methods for translationese identification in a typical real-life scenario, where no labelled in-domain data are available.

**Unsupervised Identification of Translationese.** To overcome the domain- and labeled-data-dependence of supervised classification we experiment in this section with unsupervised methods. We adopt the approach detailed in [19], who demonstrated high accuracy identifying English translationese by clustering methodology.

Table 4 demonstrates the results; the reported numbers reflect average accuracy over 30 experiments (the only difference being a random choice of the initial conditions).[15] Europarl and Hansard systematically obtain very high accuracy with all feature types (with a single exception of FW for French Hansard), implying uniform distribution of other linguistic aspects (authorship, topic, modality, epoch etc.) in these sub-corpora, thus facilitating the unsupervised procedure of clustering, since the text translation status dominates other dimensions.

**Table 4.** Clustering (rounded) accuracy of English, French and German translationese; the best result in each column is boldfaced.

| Feature/corpus | EN(O) + FR→EN | | | | EN(O) + DE→EN | | | FR(O) + EN→FR | | | | DE(O) + EN→DE | | |
|---|---|---|---|---|---|---|---|---|---|---|---|---|---|---|
| | EUR | HAN | LIT | TED | EUR | LIT | POL | EUR | HAN | LIT | TED | EUR | LIT | POL |
| Total # of chunks | 1K | 1K | 400 | 40 | 1K | 650 | 100 | 1K | 1K | 400 | 40 | 1K | 600 | 90 |
| FW | 92 | 91 | 77 | **89** | **95** | **70** | 100 | 91 | 71 | 72 | 95 | **96** | **68** | 98 |
| pos. tokens | 87 | 95 | 55 | 67 | 80 | 64 | 99 | **97** | 86 | **80** | 83 | 94 | **68** | **99** |
| POS-trigrams | **97** | 94 | 71 | 61 | 94 | 67 | **100** | 95 | 79 | 60 | 85 | 95 | **68** | 99 |
| Contextual FW | 87 | **96** | **78** | 70 | 88 | 67 | 98 | 96 | **91** | 72 | **98** | 95 | 64 | 89 |

A notably high accuracy is obtained on the small TED corpus, which implies the applicability of the clustering methodology to data-meager scenarios. The exceptionally high accuracy achieved by unsupervised procedure on the politics dataset (both English and German, across all feature types) may indicate existence of additional artifacts (e.g., subtle topical differences) that tease apart O from T, thus boosting the classification procedure. We leave this investigation for the future work.

We explain the lower precision achieved on the literature corpus by its unique character: translators of literary works enjoy more freedom, rendering the trans-

---

[15] Standard deviation in most experiments was close to 0.

lated texts more similar to original writing. Yet, clustering with FW systematically yields a reasonable accuracy for the literature datasets as well. We therefore, conclude that FW comprise one of the best-performing and most-reliable
features for the task of unsupervised identification of translationese.

**Sensitivity Analysis.** Next we tested (supervised and unsupervised) classifiers'
sensitivity by varying the number of chunks that are subject to classification.
We used FW (one of the best performing, content-independent features) in these
experiments. We excluded TED and politics datasets from these experiments
due to their small size; the results for the literature corpus are limited by the
amount of available data in this dataset. Figures 1 and 2 report supervised and
unsupervised classification accuracy as function of number of chunks used for
this task.

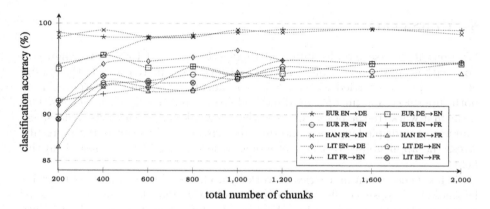

**Fig. 1.** Supervised classification accuracy as function of number of chunks using function words

Supervised classification accuracy remains stable when the number of chunks
used for classification decreases. Evidently, as few as 200 (100 on each side)
chunks are sufficient for excellent classification in most cases. Clustering results
demonstrate similar pattern: the vast majority of datasets preserve perfectly
stable performance when the number of chunks decreases. A single exception is
Hansard French (O + T from English), that exhibits results with considerable
variance; we attribute these fluctuations to the random choice of samples, subject
for clustering.

Unsupervised classification is inherently sensitive procedure, thus the stable
accuracy obtained by the majority of sub-corpora implies high reliability and
applicability of the clustering procedure to scenarios where only little data are
available.

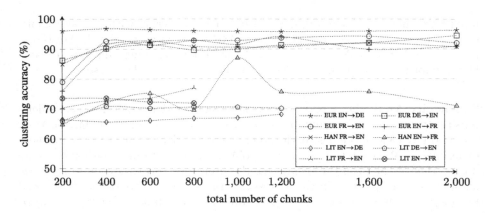

**Fig. 2.** Clustering accuracy as function of number of chunks using function words

# 5   Conclusion

We present diverse parallel bilingual English-French and English-German corpora with accurate indication of the translation direction. To evaluate the quality of the corpus, we carried out series of experiments across all sub-corpora, using both supervised and unsupervised methodologies and various feature types. This is the first work (to the best of our knowledge) employing unsupervised classification across multiple languages and diverse registers, and the encouraging results stress the applicability of this methodology, leveraging further research in this field.

It has been shown in a series of works [14,33,48–51] that awareness to translationese has a positive effect on the quality of SMT. Parallel resources presented in this work enable exploitation of *bilingual* information for the task of identification of translationese. More precisely, the datasets that we compiled can be used for the task of identifying the translation direction of parallel texts; task that enjoys growing interest in recent years [23,25].

The potential value of this work leaves much room for further exploratory and practical activities. Our future plans include extending this set of corpora to additional domains and languages, as well as exploitation of bilingual information for highly accurate identification of translationese at small text units, eventually, at the sentence level.

**Acknowledgments.** This research was supported by a grant from the Israeli Ministry of Science and Technology. We are grateful to Noam Ordan for much advice and encouragement. We also thank Sergiu Nisioi for helpful suggestions. We are grateful to Philipp Koehn for making the Europarl corpus available; to Cyril Goutte, George Foster and Pierre Isabelle for providing us with an annotated version of the Hansard corpus; to François Yvon and András Farkas (http://farkastranslations.com) for contributing their literary corpora; and to the TED OTP team for sharing TED talks and their translations. We thank also Raphael Salkie for sharing his diverse English-German corpus.

# References

1. Baker, M.: Corpus linguistics and translation studies: implications and applications. In: Baker, M., Francis, G., Tognini-Bonelli, E. (eds.) Text and Technology: in Honour of John Sinclair, pp. 233–252. John Benjamins, Amsterdam (1993)
2. Baker, M.: Corpora in translation studies: an overview and some suggestions for future research. Target **7**, 223–243 (1995)
3. Baker, M.: Corpus-based translation studies: the challenges that lie ahead. In: Mona Baker, G.F., Tognini-Bonelli, E., (eds.) Terminology, LSP and Translation. Studies in Language Engineering in Honour of Juan C. Sager, pp. 175–186. John Benjamins, Amsterdam (1996)
4. Al-Shabab, O.S.: Interpretation and the language of translation: creativity and conventions in translation. Janus, Edinburgh (1996)
5. Laviosa, S.: Core patterns of lexical use in a comparable corpus of English lexical prose. Meta **43**, 557–570 (1998)
6. Laviosa, S.: Corpus-Based Translation Studies: Theory, Findings, Applications. Approaches to Translation Studies. Rodopi, Amsterdam (2002)
7. Olohan, M.: Introducing Corpora in Translation Studies. Routledge, Abingdon (2004)
8. Becher, V.: When and why do translators add connectives? Target **23**, 26–47 (2011)
9. Zanettin, F.: Corpus methods for descriptive translation studies. Procedia Soc. Behav. Sci. **95**, 20–32 (2013). Corpus Resources for Descriptive and Applied Studies. Current Challenges and Future Directions: Selected Papers from the 5th International Conference on Corpus Linguistics (CILC 2013)
10. Gellerstam, M.: Translationese in Swedish novels translated from English. In: Wollin, L., Lindquist, H. (eds.) Translation Studies in Scandinavia, pp. 88–95. CWK Gleerup, Lund (1986)
11. Toury, G.: Descriptive Translation Studies and Beyond. John Benjamins, Amsterdam/Philadelphia (1995)
12. Baroni, M., Bernardini, S.: A new approach to the study of translationese: machine-learning the difference between original and translated text. Literary Linguist. Comput. **21**, 259–274 (2006)
13. van Halteren, H.: Source language markers in EUROPARL translations. In: Scott, D., Uszkoreit, H., (eds.) COLING 2008, 22nd International Conference on Computational Linguistics, Proceedings of the Conference, 18–22 August 2008, Manchester, UK, pp. 937–944 (2008)
14. Kurokawa, D., Goutte, C., Isabelle, P.: Automatic detection of translated text and its impact on machine translation. In: Proceedings of MT-Summit XII, pp. 81–88 (2009)
15. Koppel, M., Ordan, N.: Translationese and its dialects. In: Proceedings of the 49th Annual Meeting of the Association for Computational Linguistics: Human Language Technologies, Portland, Oregon, USA, pp. 1318–1326. Association for Computational Linguistics (2011)
16. Ilisei, I., Inkpen, D., Corpas Pastor, G., Mitkov, R.: Identification of translationese: a machine learning approach. In: Gelbukh, A. (ed.) CICLing 2010. LNCS, vol. 6008, pp. 503–511. Springer, Heidelberg (2010). https://doi.org/10.1007/978-3-642-12116-6_43
17. Ilisei, I., Inkpen, D.: Translationese traits in Romanian newspapers: a machine learning approach. Int. J. Comput. Linguist. Appl. **2**, 319–332 (2011)

18. Volansky, V., Ordan, N., Wintner, S.: On the features of translationese. Digit. scholarsh. Humanit. **30**, 98–118 (2015)
19. Rabinovich, E., Wintner, S.: Unsupervised identification of translationese. Trans. Assoc. Comput. Linguist. **3**, 419–432 (2015)
20. Nisioi, S.: Unsupervised classification of translated texts. In: Biemann, C., Handschuh, S., Freitas, A., Meziane, F., Métais, E. (eds.) NLDB 2015. LNCS, vol. 9103, pp. 323–334. Springer, Cham (2015). https://doi.org/10.1007/978-3-319-19581-0_29
21. Pym, A.: On Toury's laws of how translators translate. In: Pym, A., Shlesinger, M., Simeoni, D., (eds.) Beyond Descriptive Translation Studies: Investigations in Homage to Gideon Toury. Benjamins Translation Library: EST Subseries, pp. 311–328. John Benjamins (2008)
22. Becher, V.: Abandoning the notion of "translation-inherent" explicitation: against a dogma of translation studies. Across Lang. Cult. **11**, 1–28 (2010)
23. Eetemadi, S., Toutanova, K.: Asymmetric features of human generated translation. In: Proceedings of the 2014 Conference on Empirical Methods in Natural Language Processing (EMNLP), pp. 159–164. Association for Computational Linguistics (2014)
24. Brown, P.F., deSouza, P.V., Mercer, R.L., Pietra, V.J.D., Lai, J.C.: Class-based n-gram models of natural language. Comput. Linguist. **18**, 467–479 (1992)
25. Eetemadi, S., Toutanova, K.: Detecting translation direction: a cross-domain study. In: NAACL Student Research Workshop, ACL Association for Computational Linguistics (2015)
26. House, J.: Beyond intervention: universals in translation? Trans-kom **1**, 6–19 (2008)
27. Laviosa, S.: Universals. In: Baker, M., Saldanha, G. (eds.) Routledge Encyclopedia of Translation Studies, 2nd edn, pp. 288–292. Routledge (Taylor and Francis), New York (2008)
28. Koehn, P.: Europarl: a parallel corpus for statistical machine translation. In: MT Summit (2005)
29. Cucchi, C.: Dialogic features in EU non-native parliamentary debates. Rev. Air Force Acad. **11**, 5–14 (2012)
30. Koehn, P., Birch, A., Steinberger, R.: 462 machine translation systems for Europe. In: Proceedings of the Twelfth Machine Translation Summit, pp. 65–72 (2009)
31. Cartoni, B., Zufferey, S., Meyer, T.: Using the Europarl corpus for cross-linguistic research. Belg. J. Linguist. **27**, 23–42 (2013)
32. Islam, Z., Mehler, A.: Customization of the Europarl corpus for translation studies. In: Proceedings of the Eight International Conference on Language Resources and Evaluation (LREC 2012), European Language Resources Association (ELRA) (2012)
33. Lembersky, G., Ordan, N., Wintner, S.: Language models for machine translation: original vs. translated texts. Comput. Linguist. **38**, 799–825 (2012)
34. Cartoni, B., Meyer, T.: Extracting directional and comparable corpora from a multilingual corpus for translation studies. In: Proceedings 8th International Conference on Language Resources and Evaluation (LREC), pp. 2132–2137. European Language Resources Association (ELRA) (2012)
35. Mollin, S.: The Hansard hazard: gauging the accuracy of British parliamentary transcripts. Corpora **2**, 187–210 (2007)
36. Lynch, G., Vogel, C.: Towards the automatic detection of the source language of a literary translation. In: Proceedings of COLING 2012, the 24th International Conference on Computational Linguistics: Posters, pp. 775–784 (2012)

37. Avner, E.A.: Identifying Hebrew translationese using machine learning techniques. Diplomarbeit, University of Potsdam (2013)
38. Popescu, M.: Studying translationese at the character level. In: Angelova, G., Bontcheva, K., Mitkov, R., Nicolov, N., (eds.) Proceedings of RANLP-2011, pp. 634–639 (2011)
39. Avner, E.A., Ordan, N., Wintner, S.: Identifying translationese at the word and sub-word level. Digital Scholarship in the Humanities (Forthcoming)
40. Gale, W.A., Church, K.W.: A program for aligning sentences in bilingual corpora. Comput. Linguist. **19**, 75–102 (1993)
41. Tan, L., Bond, F.: NTU-MC toolkit: annotating a linguistically diverse corpus. In: Proceedings of 25th International Conference on Computational Linguistics (COLING 2014) (2014)
42. Manning, C.D., Surdeanu, M., Bauer, J., Finkel, J., Bethard, S.J., McClosky, D.: The stanford CoreNLP natural language processing toolkit. In: Proceedings of 52nd Annual Meeting of the Association for Computational Linguistics: System Demonstrations, Baltimore, Maryland, pp. 55–60. Association for Computational Linguistics (2014)
43. Keerthi, S., Shevade, S., Bhattacharyya, C., Murthy, K.: Improvements to Platt's SMO algorithm for SVM classifier design. Neural Comput. **13**, 637–649 (2001)
44. Hall, M., Frank, E., Holmes, G., Pfahringer, B., Reutemann, P., Witten, I.H.: The WEKA data mining software: an update. SIGKDD Explor. **11**, 10–18 (2009)
45. Mosteller, F., Wallace, D.L.: Inference in an authorship problem: a comparative study of discrimination methods applied to the authorship of the disputed federalist papers. J. Am. Stat. Assoc. **58**, 275–309 (1963)
46. Grieve, J.: Quantitative authorship attribution: an evaluation of techniques. Literary. Linguist. Comput. **22**, 251–270 (2007)
47. Nisioi, S.: Feature analysis for native language identification. In: Gelbukh, A. (ed.) CICLing 2015. LNCS, vol. 9041, pp. 644–657. Springer, Cham (2015). https://doi.org/10.1007/978-3-319-18111-0_49
48. Lembersky, G., Ordan, N., Wintner, S.: Adapting translation models to translationese improves SMT. In: Proceedings of the 13th Conference of the European Chapter of the Association for Computational Linguistics, Avignon, France, pp. 255–265. Association for Computational Linguistics (2012)
49. Lembersky, G., Ordan, N., Wintner, S.: Improving statistical machine translation by adapting translation models to translationese. Comput. Linguist. **39**, 999–1023 (2013)
50. Lembersky, G., Ordan, N., Wintner, S.: Language models for machine translation: original vs. translated texts. In: Proceedings of the 2011 Conference on Empirical Methods in Natural Language Processing, Edinburgh, Scotland, UK, pp. 363–374. Association for Computational Linguistics (2011)
51. Twitto-Shmuel, N., Ordan, N., Wintner, S.: Statistical machine translation with automatic identification of translationese. In: Proceedings of WMT-2015 (2015)

# A Low Dimensionality Representation
# for Language Variety Identification

Francisco Rangel[1,2](✉), Marc Franco-Salvador[1], and Paolo Rosso[1]

[1] Universitat Politècnica de València, Valencia, Spain
francisco.rangel@autoritas.es, mfranco@prhlt.upv.es, prosso@dsic.upv.es
[2] Autoritas Consulting, Valencia, Spain

**Abstract.** Language variety identification aims at labelling texts in a native language (e.g. Spanish, Portuguese, English) with its specific variation (e.g. Argentina, Chile, Mexico, Peru, Spain; Brazil, Portugal; UK, US). In this work we propose a low dimensionality representation (LDR) to address this task with five different varieties of Spanish: Argentina, Chile, Mexico, Peru and Spain. We compare our LDR method with common state-of-the-art representations and show an increase in accuracy of ∼35%. Furthermore, we compare LDR with two reference distributed representation models. Experimental results show competitive performance while dramatically reducing the dimensionality—and increasing the big data suitability—to only 6 features per variety. Additionally, we analyse the behaviour of the employed machine learning algorithms and the most discriminating features. Finally, we employ an alternative dataset to test the robustness of our low dimensionality representation with another set of similar languages.

**Keywords:** Low dimensionality representation
Language variety identification · Similar languages discrimination
Author profiling · Big data · Social media

## 1 Introduction

Language variety identification aims at labelling texts in a native language (e.g. Spanish, Portuguese, English) with their specific variation (e.g. Argentina, Chile, Mexico, Peru, Spain; Brazil, Portugal; UK, US). Although at first sight language variety identification may seem a classical text classification problem, cultural idiosyncrasies may influence the way users construct their discourse, the kind of sentences they build, the expressions they use or their particular choice of

The work of the first author was in the framework of ECOPORTUNITY IPT-2012-1220-430000. The work of the last two authors was in the framework of the SomEMBED MINECO TIN2015-71147-C2-1-P research project. This work has been also supported by the SomEMBED TIN2015-71147-C2-1-P MINECO research project and by the Generalitat Valenciana under the grant ALMAPATER (PrometeoII/2014/030).

© Springer International Publishing AG, part of Springer Nature 2018
A. Gelbukh (Ed.): CICLing 2016, LNCS 9624, pp. 156–169, 2018.
https://doi.org/10.1007/978-3-319-75487-1_13

words. Due to that, we can consider language variety identification as a double problem of text classification and author profiling, where information about how language is shared by people may help to discriminate among classes of authors depending on their language variety.

This task is specially important in social media. Despite the vastness and accessibility of the Internet destroyed frontiers among regions or traits, companies are still very interested in author profiling segmentation. For example, when a new product is launched to the market, knowing the geographical distribution of opinions may help to improve marketing campaigns. Or given a security threat, knowing the possible cultural idiosyncrasies of the author may help to better understand who could have written the message.

Language variety identification is a popular research topic of natural language processing. In the last years, several tasks and workshops have been organized: the Workshop on Language Technology for Closely Related Languages and Language Variants @ EMNLP 2014[1]; the VarDial Workshop @ COLING 2014 - Applying NLP Tools to Similar Languages, Varieties and Dialects[2]; and the LT4VarDial - Joint Workshop on Language Technology for Closely Related Languages, Varieties and Dialect[3] @ RANLP [12,14]. We can find also several works focused on the task. In [10] the authors addressed the problem of identifying Arabic varieties in blogs and social fora. They used character $n$-gram features to discriminate between six different varieties and obtained accuracies between 70%–80%. Similarly, [13] collected 1,000 news articles of two varieties of Portuguese. They applied different features such as word and character $n$-grams and reported accuracies over 90%. With respect to the Spanish language, [6] focused on varieties from Argentina, Chile, Colombia, Mexico and Spain in Twitter. They used meta-learning and combined four types of features: *(i)* character $n$-gram frequency profiles, *(ii)* character $n$-gram language models, *(iii)* Lempel-Ziv-Welch compression and *(iv)* syllable-based language models. They obtained an interesting 60%–70% accuracy of classification.

We are interested in discovering which kind of features capture higher differences among varieties. Our hypothesis is that language varieties differ mainly in lexicographic clues. We show an example in Table 1.

**Table 1.** The same example in three varieties of Spanish (Argentina, Mexico and Spain).

| English | I was **goofing around** with my dog and I **lost** my **mobile** |
|---|---|
| ES-Argentina | Estaba haciendo **boludeces** con mi perro y **extravié** el **celular** |
| ES-Mexico | Estaba haciendo el **pendejo** con mi perro y **extravié** el **celular** |
| ES-Spain | Estaba haciendo el **tonto** con mi perro y **perdí** el **móvil** |

---

[1] http://alt.qcri.org/LT4CloseLang/index.html.
[2] http://corporavm.uni-koeln.de/vardial/sharedtask.html.
[3] http://ttg.uni-saarland.de/lt4vardial2015/dsl.html.

In this work we focus on the Spanish language variety identification. We differentiate from the previous works as follows: *(i)* instead of *n*-gram based representations, we propose a low dimensionality representation that is helpful when dealing with big data in social media; *(ii)* in order to reduce the possible over-fitting, our training and test partitions do not share any author of instance between them[4]; and *(iii)* in contrast to the Twitter dataset of [6], we will make available our dataset to the research community.

## 2   Low Dimensionality Representation

The key aspect of the low dimensionality representation (LDR) is the use of weights to represent the probability of each term to belong to each one of the different language varieties. We assume that the distribution of weights for a given document should be closer to the weights of its corresponding language variety. Formally, the LDR is estimated as follows:

*Term-frequency - inverse document frequency (tf-idf) matrix creation.* First, we apply the *tf-idf* [11] weighting for the terms of the documents of the training set *D*. As result we obtain the following matrix:

$$\Delta = \begin{bmatrix} w_{11} & w_{12} & ... & w_{1m} & \delta(d_1) \\ w_{21} & w_{22} & ... & w_{2m} & \delta(d_2) \\ ... & ... & ... & ... \\ w_{n1} & w_{n2} & ... & w_{nm} & \delta(d_n) \end{bmatrix}, \tag{1}$$

where each row in the matrix $\Delta$ represents a document *d*, each column represents a vocabulary term *t*, $w_{ij}$ represents its *tf-idf*, and $\delta(d_i)$ represents the assigned class *c* of the document *i*, that is, the language variety actually assigned to this document.

*Class-dependent term weighting.* Using the matrix $\Delta$, we obtain the class-dependent term weight matrix $\beta$. This matrix contains the weights of each term *t* for each language variety *C* on the basis of Eq. 2:

$$W(t,c) = \frac{\sum_{d \in D/c=\delta(d)} w_{dt}}{\sum_{d \in D} w_{dt}}, \forall d \in D, c \in C \tag{2}$$

Basically, the term weight $W(t,c)$ is the ratio between the weights of the documents belonging to a concrete language variety *c* and the total distribution of weights for that term *t*.

---

[4] It is important to highlight the importance of this aspect from an evaluation perspective in an author profiling scenario. In fact, if texts from the same authors are both part of the training and test sets, their particular style and vocabulary choice may contribute at training time to learn the profile of the authors. In consequence, over-fitting would be biasing the results.

**Class-dependent document representation.** We employ the class-dependent term weights $\beta$ to obtain the final representation of the documents as follows:

$$d = \{F(c_1), F(c_2), ..., F(c_n)\} \sim \forall c \in C, \tag{3}$$

$$F(c_i) = \{avg, std, min, max, prob, prop\} \tag{4}$$

where each $F(c_i)$ contains the set of features showed in Eq. 4 and described in Table 2. As we can see, our class-dependent weights $\beta$ are employed to extract a small[5] —but very discriminant—number of features for each language variety.[6] We note that the same process can be followed in order to represent a test document $d' \in D'$. We just need to use the $\beta$ matrix obtained with $D$ to index the document $d'$ by means of Eq. 3.

Table 2. Set of features for each category (language variety) used in Eq. 4.

| | |
|---|---|
| avg | The average weight of a document is calculated as the sum of weights $W(t,c)$ of its terms divided by the total number of vocabulary terms of the document |
| std | The standard deviation of the weight of a document is calculated as the root square of the sum of all the weights $W(t,c)$ minus the average |
| min | The minimum weight of a document is the lowest term weight $W(t,c)$ found in the document |
| max | The maximum weight of a document is the highest term weight $W(t,c)$ found in the document |
| prob | The overall weight of a document is the sum of weights $W(t,c)$ of the terms of the document divided by the total number of terms of the document |
| prop | The proportion between the number of vocabulary terms of the document and the total number of terms of the document |

## 3 Evaluation Framework

In this section, we describe the corpus and the alternative representations that we employ in this work.

---

[5] Our hypothesis is that the distribution of weights for a given document should be closer to the weights of its corresponding language variety, therefore, we use the most common descriptive statistics to measure this variability among language varieties.

[6] Using the LDR a document is represented by a total set of features equal to 6 multiplied by the number of categories (the 5 language varieties), in our case 30 features. This is a considerable dimensionality reduction that may be helpful to deal with big data environments.

## 3.1  HispaBlogs Corpus

We have created the HispaBlogs dataset[7] by collecting posts from Spanish blogs from five different countries: Argentina, Chile, Mexico, Peru ànd Spain. For each country, there are 450 and 200 blogs respectively for training and test, ensuring that each author appears only in one set. Each blog contains at least 10 posts. The total number of blogs is 2,250 and 1,000 respectively. Statistics of the number of words are shown in Table 3.

**Table 3.** Number of posts, words and words per post (average and standard deviation) per language variety.

| Language variety | # Blogs/authors | | # Words | | # Words per post | |
|---|---|---|---|---|---|---|
| | Training | Test | Training | Test | Training | Test |
| AR - Argentina | 450 | 200 | 1,408,103 | 590,583 | 371,448 | 385,849 |
| CL - Chile | 450 | 200 | 1,081,478 | 298,386 | 313,465 | 225,597 |
| ES - Spain | 450 | 200 | 1,376,478 | 620,778 | 360,426 | 395,765 |
| MX - Mexico | 450 | 200 | 1,697,091 | 618,502 | 437,513 | 392,894 |
| PE - Peru | 450 | 200 | 1,602,195 | 373,262 | 410,466 | 257,627 |
| TOTAL | 2,250 | 1,000 | 7,164,935 | 2,501,511 | 380,466 | 334,764 |

## 3.2  Alternative Representations

We are interested in investigating the impact of the proposed representation and compare its performance with state-of-the-art representations based on $n$-grams and with two approaches based on the recent and popular distributed representations of words by means of the continuous Skip-gram model [1].

**State-of-the Art Representations.** State-of-the-art representations are mainly based on $n$-grams models, hence we tested character and word based ones, besides word with *tf-idf* weights. For each of them, we iterated $n$ from 1 to 10 and selected 1,000, 5,000 and 10,000 most frequent grams. The best results were obtained with the 10,000 most frequent BOW, character 4-grams and *tf-idf* 2-grams. Therefore, we will use them in the evaluation.

**Distributed Representations.** Due to the increasing popularity of the distributed representations [4], we used the continuous Skip-gram model to generate distributed representations of words (e.g. $n$-dimensional vectors), with further refinements in order to use them with documents. The continuous Skip-gram

---

[7] The HispaBlogs dataset was collected by experts on social media from the Autoritas Consulting company (http://www.autoritas.net). Autoritas experts in the different countries selected popular bloggers related to politics, online marketing, technology or trends. The HispaBlogs dataset is publicly available at: https://github.com/autoritas/RD-Lab/tree/master/data/HispaBlogs.

model [7,8] is an iterative algorithm which attempts to maximize the classification of the context surrounding a word. Formally, given a word $w(t)$, and its surrounding words $w(t-c)$, $w(t-c+1), ..., w(t+c)$ inside a window of size $2c+1$, the training objective is to maximize the average of the log probability shown in Eq. 5:

$$\frac{1}{T} \sum_{t=1}^{T} \sum_{-c \leq j \leq c, j \neq 0} \log p(w_{t+j}|w_t) \tag{5}$$

To estimate $p(w_{t+j}|w_t)$ we used negative sampling [8] that is a simplified version of the Noise Contrastive Estimation (NCE) [3,9] which is only concerned with preserving vector quality in the context of Skip-gram learning. The basic idea is to use logistic regression to distinguish the target word $W_O$ from draws from a noise distribution $P_n(w)$, having $k$ negative samples for each word. Formally, the negative sampling estimates $p(w_O|w_I)$ following Eq. 6:

$$\log \sigma(v'_{w_O}{}^T v_{w_I}) + \sum_{i=1}^{k} \mathbb{E}_{w_i} \sim P_n(w) \left[ \log \sigma(-v'_{w_i}{}^T v_{w_I}) \right] \tag{6}$$

where $\sigma(x) = 1/(1 + \exp(-x))$. The experimental results in [8] show that this function obtains better results at the semantic level than hierarchical softmax [2] and NCE.

In order to combine the word vectors to represent a complete sentence we used two approaches. First, given a list of word vectors $(w_1, w_2, ..., w_n)$ belonging to a document, we generated a vector representation $v$ of its content by estimating the average of their dimensions: $v = n^{-1} \sum_{i=1}^{n} w_i$. We call this representation Skip-gram in the evaluation. In addition, we used Sentence vectors (SenVec) [5], a variant that follows Skip-gram architecture to train a special vector $sv$ representing the sentence. Basically, before each context window movement, SenVec uses a special vector $sv$ in place of $w(t)$ with the objective of maximizing the classification of the surrounding words. In consequence, $sv$ will be a distributed vector of the complete sentence.

Following state-of-the-art approach [5], in the evaluation we used a logistic classifier for both SenVec and Skip-gram approaches.[8]

## 4 Experimental Results

In this section we show experimental results obtained with the machine learning algorithms that best solve the problem with the proposed representation, the impact of the preprocessing on the performance, the obtained results in comparison with the ones obtained with state-of-the-art and distributed representations,

---

[8] We used 300-dimensional vectors, context windows of size 10, and 20 negative words for each sample. We preprocessed the text with word lowercase, tokenization, removing the words of length one, and with phrase detection using word2vec tools: https://code.google.com/p/word2vec/.

the error analysis that provides useful insights to better understand differences among languages, a depth analysis on the contribution of the different features and a cost analysis that highlights the suitability of LDR for a big data scenario.

## 4.1 Machine Learning Algorithms Comparison

We tested several machine learning algorithms[9] with the aim at selecting the one that best solves the task. As can be seen in Table 4, Multiclass Classifier[10] obtains the best result (results in the rest of the paper refer to Multiclass Classifier). We carried out a statistical test of significance with respect to the next two systems with the highest performance: SVM ($z_{0.05}0, 880 < 1, 960$) and LogitBoost ($z_{0.05} = 1, 983 > 1, 960$).

**Table 4.** Accuracy results with different machine learning algorithms.

| Algorithm | Accuracy | Algorithm | Accuracy | Algorithm | Accuracy |
|---|---|---|---|---|---|
| Multiclass classifier | **71.1** | Rotation forest | 66.6 | Multilayer perceptron | 62.5 |
| SVM | 69.3 | Bagging | 66.5 | Simple cart | 61.9 |
| LogitBoost | 67.0 | Random forest | 66.1 | J48 | 59.3 |
| Simple logistic | 66.8 | Naive Bayes | 64.1 | BayesNet | 52.2 |

## 4.2 Preprocessing Impact

The proposed representation aims at using the whole vocabulary to obtain the weights of its terms. Social media texts may have noise and inadequately written words. Moreover, some of these words may be used only by few authors. With the aim at investigating their effect in the classification, we carried out a preprocessing step to remove words that appear less than $n$ times in the corpus, iterating $n$ between 1 and 100. In Fig. 1 the corresponding accuracies are shown. In the left part of the figure ($a$), results for $n$ between 1 and 10 are shown in a continuous scale. In the right part ($b$), values from 10 to 100 are shown in a non-continuous scale. As can be seen, the best result was obtained with $n$ equal to 5, with an accuracy of 71.1%. As it was expected, the proposed representation takes advantage from the whole vocabulary, although it is recommendable to remove words with very few occurrences that may alter the results. We show examples of those infrequent words in Table 5.

In Fig. 2, when analysing the evolution of the number of remaining words in function of the value of $n$, we can see a high number of words with very low frequency of occurrence. These words may introduce a high amount of noise in our LDR weight estimation. In addition, removing these words may be also beneficial in order to reduce the processing time needed to obtain the representation. This fact has special relevance for improving the performance in big data environments.

---

[9] http://www.cs.waikato.ac.nz/ml/weka/.

[10] We used SVM with default parameters and exhaustive correction code to transform the multiclass problem into a binary one.

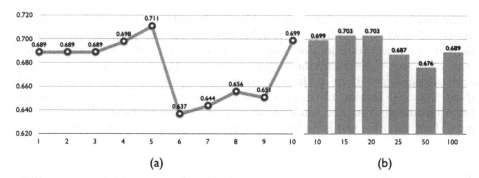

**Fig. 1.** Accuracy obtained after removing words with frequency equal or lower than $n$. (a) Continuous scale. (b) Non-continuous scale.

**Table 5.** Very infrequent words.

| # occurrences = 1 | # occurrences = 2 | # occurrences = 3 |
|---|---|---|
| aaaaaaaah | aaaaa | aaaa |
| aaaaaaaarrrgh | aaaayyy | aaaaaaaaae |
| aaaaaaggghhhhh | aaavyt | aaaaaaaacu |
| aaaah | aach | aantofagastina |
| aaaahhhh | aachen | ñirripil |

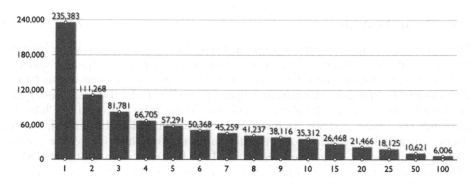

**Fig. 2.** Number of words after removing those with frequency equal or lower than $n$.

### 4.3   Language Variety Identification Results

In Table 6 we show the results obtained by the described representations employing the Multiclass Classifier. As can be appreciated, the proposed low dimensionality representation improves more than 35% the results obtained with the state-of-the-art representations. BOW obtains slightly better results than character 4-grams, and both of them improve significantly the ones obtained with *tf-idf* 2-grams. Instead of selecting the most frequent *n*-grams, our approach takes advantage from the whole vocabulary and assigns higher weights to the most discriminative words for the different language varieties as shown in Eq. 2.

**Table 6.** Accuracy results in language variety identification and number of features for each representation.

| Representation | Accuracy | # Features |
|---|---|---|
| Skip-gram | 0.722 | 300 |
| **LDR** | **0.711** | **30** |
| SenVec | 0.708 | 300 |
| BOW | 0.527 | 10,000 |
| Char. 4-grams | 0.515 | 10,000 |
| *tf-idf* 2-grams | 0.393 | 10,000 |
| Random baseline | 0.200 | - |

We highlight that our LDR obtains competitive results compared with the use of distributed representations. Concretely, there is no significant difference among them (Skip-gram $z_{0.05} = 0,5457 < 1,960$ and SenVec$z_{0.05} = 0,7095 < 1,960$). In addition, our proposal reduces considerably the dimensionality of one order of magnitude as shown in Table 6.

### 4.4   Error Analysis

We aim at analysing the error of LDR to better understand which varieties are the most difficult to discriminate. As can be seen in Table 7, the Spanish variety is the easiest to discriminate. However, one of the highest confusions occurs from Argentinian to Spanish. Mexican and Spanish were considerably confused with Argentinian too. Finally, the highest confusion occurs from Peruvian to Chilean, although the lowest average confusion occurs with Peruvian. In general, Latin American varieties are closer to each other and it is more difficult to differentiate among them. Language evolves over time. It is logical that language varieties of nearby countries—as the Latin American ones—evolved in a more similar manner that the Spanish variety. It is also logical that even more language variety similarities are shared across neighbour countries, e.g. Chilean compared with Peruvian and Argentinian.

**Table 7.** Confusion matrix of the 5-class classification.

| Variety | Clasified as | | | | |
|---|---|---|---|---|---|
| | AR | CL | ES | MX | PE |
| AR | 143 | 16 | 22 | 8 | 11 |
| CL | 17 | 151 | 11 | 11 | 10 |
| ES | 20 | 13 | 154 | 7 | 6 |
| MX | 20 | 18 | 18 | 131 | 13 |
| PE | 16 | 28 | 12 | 12 | 132 |

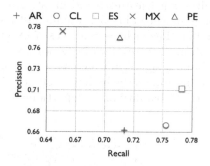

**Fig. 3.** F1 values for identification as the corresponding language variety vs. others.

In Fig. 3 we show the precision and recall values for the identification of each variety. As can be seen, Spain and Chile have the highest recall so that texts written in these varieties may have less probability to be misclassified as other varieties. Nevertheless, the highest precisions are obtained for Mexico and Peru, implying that texts written in such varieties may be easier to discriminate.

### 4.5 Most Discriminating Features

In Table 8 we show the most discriminant features. The features are sorted by their information gain (IG). As can be seen, the highest gain is obtained by average, maximum and minimum, and standard deviation. On the other hand, probability and proportionality features has low information gain.

**Table 8.** Features sorted by information gain.

| Attribute | IG | Attribute | IG | Attribute | IG |
|---|---|---|---|---|---|
| PE-avg | $0.680 \pm 0.006$ | ES-std | $0.497 \pm 0.008$ | PE-prob | $0.152 \pm 0.005$ |
| AR-avg | $0.675 \pm 0.005$ | CL-max | $0.496 \pm 0.005$ | MX-prob | $0.151 \pm 0.005$ |
| MX-max | $0.601 \pm 0.005$ | CL-std | $0.495 \pm 0.007$ | ES-prob | $0.130 \pm 0.011$ |
| PE-max | $0.600 \pm 0.009$ | MX-std | $0.493 \pm 0.007$ | AR-prob | $0.127 \pm 0.006$ |
| ES-min | $0.595 \pm 0.033$ | CL-min | $0.486 \pm 0.013$ | AR-prop | $0.116 \pm 0.005$ |
| ES-avg | $0.584 \pm 0.004$ | AR-std | $0.485 \pm 0.005$ | MX-prop | $0.113 \pm 0.006$ |
| MX-avg | $0.577 \pm 0.008$ | PE-std | $0.483 \pm 0.012$ | PE-prop | $0.112 \pm 0.005$ |
| ES-max | $0.564 \pm 0.007$ | AR-min | $0.463 \pm 0.012$ | ES-prop | $0.110 \pm 0.007$ |
| AR-max | $0.550 \pm 0.007$ | CL-avg | $0.455 \pm 0.008$ | CL-prop | $0.101 \pm 0.005$ |
| MX-min | $0.513 \pm 0.027$ | PE-min | $0.369 \pm 0.019$ | CL-prob | $0.087 \pm 0.010$ |

We experimented with different sets of features and show the results in Fig. 4. As may be expected, average-based features obtain high accuracies (67.0%).

However, although features based on standard deviation have not the highest information gain, they obtained the highest results individually (69.2%), as well as their combination with average ones (70,8%). Features based on minimum and maximum obtain low results (48.3% and 54.7% respectively), but in combination they obtain a significant increase (61.1%). The combination of the previous features obtains almost the highest accuracy (71.0%), equivalent to the accuracy obtained with probability and proportionality features (71.1%).

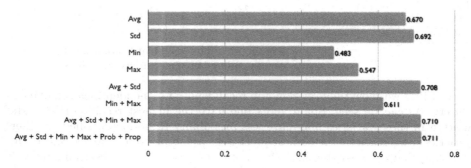

**Fig. 4.** Accuracy with different combinations of features.

### 4.6 Cost Analysis

We analyse the cost from two perspectives: *(i)* the complexity to the features; and *(ii)* the number of features needed to represent a document. Defining $l$ as the number of different language varieties, and $n$ the number of terms of the document to be classified, the cost of obtaining the features of Table 2 (average, minimum, maximum, probability and proportionality) is $O(l \cdot n)$. Defining $m$ as the number of terms in the document that coincides with some term in the vocabulary, the cost of obtaining the standard deviation is $O(l \cdot m)$. As the average is needed previously to the standard deviation calculation, the total cost is $O(l \cdot n) + O(l \cdot m)$ that is equal to $O(max(l \cdot n, l \cdot m)) = O(l \cdot n)$. Since the number of terms in the vocabulary will always be equal or greater than the number of coincident terms ($n \geq m$), and as the number of terms in the document will always be much higher than the number of language varieties ($l << n$), we can determine the cost as lineal with respect to the number of terms in the document $O(n)$. With respect to the number of features needed to represent a document, we showed in Table 6 the considerable reduction of the proposed low dimensionality representation.

### 4.7 Robustness

In order to analyse the robustness of the low dimensionality representation to different languages, we experimented with the development set of the DSLCC corpus[11] from the Discriminating between Similar Languages task [12]. The corpus consists of 2,000 sentences per language or variety, with between 20 and 100 tokens per sentence, obtained from news headers. In Table 9 we show the results obtained with the proposed representation and the two distributed representations, Skip-gram and SenVec. It is important to notice that, in general, when a particular representation improves for one language is at cost of the other one. We can conclude that the three representations obtained comparative results and support the robustness of the low dimensionality representation.

**Table 9.** Accuracy results in the development set of the DSLCC. The significance is marked in bold when some representation obtains significantly better results than the next best performing representation (e.g. results for SenVec in Portugal Portuguese are significantly higher than LDR, which at the same time are significantly higher than Skip-gram).

| Language | LDR | Skip-gram | SenVec |
|---|---|---|---|
| Bulgarian | 99.9 | 100 | 100 |
| Macedonian | 99.9 | 100 | 100 |
| Spain Spanish | **84.7** | 82.1 | **86.3** |
| Argentina Spanish | 88.0 | **90.3** | 87.6 |
| Portugal Portuguese | **87.4** | 83.2 | **90.0** |
| Brazilian Portuguese | 90.0 | **94.5** | 87.6 |
| Bosnian | **78.0** | 80.3 | 74.4 |
| Croatian | 85.8 | 85.9 | 84.7 |
| Serbian | **86.4** | 75.1 | 91.2 |
| Indonesian | 99.4 | 99.3 | 99.4 |
| Malay | 99.2 | 99.2 | **99.8** |
| Czech | 99.8 | 99.9 | 99.8 |
| Slovak | 99.3 | **100** | 99.3 |
| Other languages | 99.9 | 99.8 | 99.8 |

## 5 Conclusions

In this work, we proposed the LDR low dimensionality representation for language variety identification. Experimental results outperformed traditional state-of-the-art representations and obtained competitive results compared with two distributed representation-based approaches that employed the popular continuous Skip-gram model. The dimensionality reduction obtained by means of LDR is from thousands to only 6 features per language variety. This allows to deal

---

[11] http://ttg.uni-saarland.de/lt4vardial2015/dsl.html.

with large collections in big data environments such as social media. Recently, we have applied LDR to the age and gender identification task obtaining competitive results with the best performing teams in the author profiling task at the PAN[12] Lab at CLEF.[13] As a future work, we plan to apply LDR to other author profiling tasks such as personality recognition.

# References

1. Franco-Salvador, M., Rangel, F., Rosso, P., Taulé, M., Antònia Martít, M.: Language variety identification using distributed representations of words and documents. In: Mothe, J., Savoy, J., Kamps, J., Pinel-Sauvagnat, K., Jones, G.J.F., SanJuan, E., Cappellato, L., Ferro, N. (eds.) CLEF 2015. LNCS, vol. 9283, pp. 28–40. Springer, Cham (2015). https://doi.org/10.1007/978-3-319-24027-5_3
2. Goodman, J.: Classes for fast maximum entropy training. In: Proceedings of the Acoustics, Speech, and Signal Processing (ICASSP 2001), vol. 1, pp. 561–564 (2001)
3. Gutmann, M.U., Hyvärinen, A.: Noise-contrastive estimation of unnormalized statistical models, with applications to natural image statistics. J. Mach. Learn. Res. **13**, 307–361 (2012)
4. Hinton, G.E., Mcclelland, J.L., Rumelhart, D.E.: Distributed Representations, Parallel Distributed Processing: Explorations in the Microstructure of Cognition, Foundations, vol. 1. MIT Press, Cambridge (1986)
5. Le, Q.V., Mikolov, T.: Distributed representations of sentences and documents. In: Proceedings of the 31st International Conference on Machine Learning (ICML 2014), vol. 32 (2014)
6. Maier, W., Gómez-Rodríguez, C.: Language variety identification in Spanish tweets. In: Workshop on Language Technology for Closely Related Languages and Language Variants (EMNLP 2014), pp. 25–35 (2014)
7. Mikolov, T., Chen, K., Corrado, G., Dean, J.: Efficient estimation of word representations in vector space. In: Proceedings of Workshop at International Conference on Learning Representations (ICLR 2013) (2013)
8. Mikolov, T., Sutskever, I., Chen, K., Corrado, G.S., Dean, J.: Distributed representations of words and phrases and their compositionality. In: Advances in Neural Information Processing Systems, pp. 3111–3119 (2013)
9. Mnih, A., Teh, Y.W.: A fast and simple algorithm for training neural probabilistic language models. In: Proceedings of the 29th International Conference on Machine Learning (ICML 2012), pp. 1751–1758 (2012)
10. Sadat, F., Kazemi, F., Farzindar, A.: Automatic identification of Arabic language varieties and dialects in social media. In: 1st International Workshop on Social Media Retrieval and Analysis (SoMeRa 2014) (2014)
11. Salton, G., Buckley, C.: Term-weighting approaches in automatic text retrieval. Inf. Process. Manag. **24**(5), 513–523 (1988)
12. Tan, L., Zampieri, M., Ljubešic, N., Tiedemann, J.: Merging comparable data sources for the discrimination of similar languages: the DSL corpus collection. In: 7th Workshop on Building and Using Comparable Corpora Building Resources for Machine Translation Research (BUCC 2014), pp. 6–10 (2014)

---

[12] http://pan.webis.de.
[13] http://www.clef-innitiative.org.

13. Zampieri, M., Gebrekidan-Gebre, B.: Automatic identification of language varieties: the case of Portuguese. In: Proceedings of the 11th Conference on Natural Language Processing (KONVENS 2012), pp. 233–237 (2012)
14. Zampieri, M., Tan, L., Ljubeši, N., Tiedemann, J.: A report on the DSL shared task 2014. In: Proceedings of the First Workshop on Applying NLP Tools to Similar Languages, Varieties and Dialects (VarDial 2014), pp. 58–67 (2014)

# Sentiment Analysis, Opinion Mining, Subjectivity, and Social Media

# Towards Empathetic Human-Robot Interactions

Pascale Fung[(⊠)], Dario Bertero, Yan Wan, Anik Dey,
Ricky Ho Yin Chan, Farhad Bin Siddique, Yang Yang,
Chien-Sheng Wu, and Ruixi Lin

Department of Electronic and Computer Engineering,
Human Language Technology Center, The Hong Kong University of Science
and Technology, Clear Water Bay, Hong Kong
pascale@ece.ust.hk

**Abstract.** Since the late 1990s when speech companies began providing their customer-service software in the market, people have gotten used to speaking to machines. As people interact more often with voice and gesture controlled machines, they expect the machines to recognize different emotions, and understand other high level communication features such as humor, sarcasm and intention. In order to make such communication possible, the machines need an empathy module in them, which is a software system that can extract emotions from human speech and behavior and can decide the correct response of the robot. Although research on empathetic robots is still in the primary stage, current methods involve using signal processing techniques, sentiment analysis and machine learning algorithms to make robots that can 'understand' human emotion. Other aspects of human-robot interaction include facial expression and gesture recognition, as well as robot movement to convey emotion and intent. We propose Zara the Supergirl as a prototype system of empathetic robots. It is a software-based virtual android, with an animated cartoon character to present itself on the screen. She will get 'smarter' and more empathetic, by having machine learning algorithms, and gathering more data and learning from it. In this paper, we present our work so far in the areas of deep learning of emotion and sentiment recognition, as well as humor recognition. We hope to explore the future direction of android development and how it can help improve people's lives.

## 1 Introduction

From science fiction films to novels, humans have always fantasized – or needed – to have an emotional relationship with intelligent machines.

Many people in the society seem to think that the objective of creating intelligent machines is to "imitate humans" or create a new species of "humans". This misunderstanding has led to the irrational fear of "machines taking over humans" by some people. Their reasoning is obvious – if intelligent machines are supposed to imitate humans then as they become more and more human-like they are bound to have humanly desire for power and dominance. It is obvious if one believes in the premises that we are creating machines to "imitate humans". However, this is far from the reality of artificial intelligence research.

Rather than trying to build some Frankenstein surrogate of the human race, the objective of intelligent machine research and development has always been to help humans. As such, even when we build robot "companions" we are working to create health benefits for the elderly or educational benefits for the young.

In the past couple of decades, interactive dialog systems have been designed as software programs either for the desktop, embedded in an enterprise solution, as cloud services, or as mobile applications. They would have a synthesized voice. Since the 1990s, voice interactive designers have tried to make the dialog prompts more natural, and speech synthesis has made great progress to enable computer voice to sound human like. However, such systems remain invisible and virtual. Even after giving these applications names like Siri or Cortana, users remain emotionally indifferent to such systems as if they are merely using an ATM machine for transactions.

One reason behind this might be something that has been studied by human-robot interaction researchers [39]. It is known that physical embodiment of an intelligent system, whether in virtual simulation or in a robotic form, is important for users to feel related and empathize with the system [34].

More importantly though, physical robots, even extremely humanlike androids, seem cold and distant to humans because while they can sometimes be built to look and even sound emotional, they do not recognize or respond to human emotions and intent. Roboticists make great efforts to build robots in anthropomorphic form so that humans can empathize with them [22], and to have embodied cognition [10], only to find human users disappointed by the lack of reciprocal empathy from these robots.

It follows that we shall embody interactive dialog systems in simulated or robotic forms. It is also important that we give such systems the ability to both recognize human emotions and intent, as well as expressing its own. Before we share our lives with robots, they need to be able to recognize human emotion and intent, through natural language communications, through facial expression and gesture communications.

In this paper, we describe a proposed framework for building a robotic interactive system with an "empathy module". In Sect. 2, we describe the design of a prototype empathetic virtual robot system. In Sect. 3 we describe the personality analysis of our system and in Sect. 4 the need for the system to handle user challenges to our empathetic interactive virtual robot. To enable different features described in Sects. 2 to 4, we need speech recognition, emotion and sentiment recognition from audio and text. We describe our current approaches in these areas in Sects. 5 to 6. In Sect. 5 we first present a brief over view of different deep learning architectures. Section 6 describes our current approach of hybrid HMM-DNN speech recognition system for interactive systems. Section 7 describes our approach of emotion recognition from audio with and without feature engineering. We then discuss sentiment recognition from speech and text in Sect. 8 with the special case of humor recognition in conversations in Sect. 9. We summarize and discuss future work in Sect. 10.

## 2 Architecture of an Empathetic Human-Robot Interactive Platform

To achieve human-machine empathy towards each other, we propose a platform that will consist of the following features and functionalities:

1. Embodiment of the system on a virtual robot platform for human empathy;
2. Emotion and intent expression by the robot;
3. Facial recognition of user ethnicity, age, etc.;
4. Speech recognition of what the user is saying including humor;
5. Natural language understanding of user intent and sentiment including humor;
6. Emotion recognition from facial expressions, speech, audio and language;

Research has shown that humans prefer interacting with machines that are anthropomorphous. So the very first step in designing interactive intelligent systems is to give it a humanlike form.

In recent years, we have seen more efforts to give interactive systems a "face". For example, the Microsoft Tay is a chatbot with a human face and personality. "She" is designed as a teenage female character and exhibits the language model of a typical American teenage girl. The Boston company Jibo made a social robot of that name with animated expressions and movement. Jibo has had a huge success online. Pepper is another social robot that has the capability of recognizing human emotion and sentiments, available in Japan. Pepper also has an anthropomorphous body with a head and body movement designed to express emotion and intent.

For a prototype system, we gave our platform a human name and a cartoon figure. We call it Zara the Supergirl, with a multiracial face and a cartoon body. It was shown at the World Economic Forum in Dalian in September 2015. Her cartoon facial expressions and body movements are template-based and programmed to represent her recognition of user emotion or sentiment. Her lips move when she is talking. Her body gestures are program to express a combination of friendliness, power, knowledge and other emotions she might have while interacting with humans (Figs. 1 and 2).

When you begin a conversation with Zara, she says, "Please wait while I analyze your face". Zara's algorithms study images captured by the computer's webcam to determine your gender and ethnicity. She will then guess which language you speak (Zara understands English, Mandarin and is learning French) and ask you a few questions in your native tongue. What is your earliest memory? Tell me about your mother. How was your last vacation? Through this process, based on your facial expressions, the acoustic features of your voice, and the content of your responses, Zara will respond in ways that mimic empathy. She would frown and express sympathy if you mentioned a sad childhood story, but would give a thumbs up when you talk about a great vacation.

Zara's current task is a conversational MBTI personality assessor and we designed 6 categories of personality-assessing questions, in order to assess the user's personality [24]. After five minutes of conversation, Zara will try to guess your personality—You seem like a easy going and popular person—and ask you about your attitudes toward

**Fig. 1.** A prototype interactive system with multicue and emotion recognition

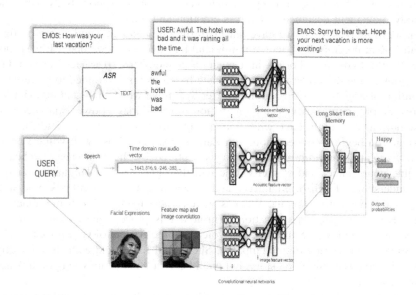

**Fig. 2.** A deep learning multichannel emotion recognition architecture

empathetic machines: How do you feel about machines like me? This is a way for us to gather feedback from people on their interactions with early empathetic robots.

A dialog management system with different states is designed to control the flow of the conversation, which consists of one-part machine-initiative questions from Zara and answers from human users and another part of user-initiative questions and challenges to Zara.

## 3  Personality Analysis

Empathy is the recognition and sharing of the emotion of the other. In order to demonstrate machine empathy towards humans within a short duration of interaction, we gave Zara the task of personality analysis. We designed a set of personal questions in six different domains in order to classify user personality from among sixteen different MBTI personality types [24]. The original MBTI test questionnaire contains about 70 questions about user's preferences and sentiments, and would require about half an hour to complete. We asked a group of training users to answer this questionnaire but also answer questions from Zara. The personality type generated by the MBTI questionnaire is used as the gold standard label for training the Zara system. Based on user answers to Zara's questions, scores are calculated in four dimensions (namely Introversion - Extroversion, Intuitive - Sensing, Thinking - Feeling, Judging - Perceiving).

| Level | Introvert | Extravert |
|---|---|---|
| Conversational behaviour | Listen<br>Less back-channel behaviour | Initiate conversation<br>More back-channel behaviour |
| Topic selection | Self-focused<br>Problem talk, dissatisfaction<br>Strict selection<br>Single topic<br>Few semantic errors<br>Few self-references | Not self-focused*<br>Pleasure talk, agreement, compliment<br>Think out loud*<br>Many topics<br>Many semantic errors<br>Many self-references |
| Style | Formal<br>Many hedges (tentative words) | Informal<br>Few hedges (tentative words) |
| Syntax | Many nouns, adjectives, prepositions (explicit)<br>Elaborated constructions<br>Many words per sentence<br>Many articles<br>Many negations | Many verbs, adverbs, pronouns (implicit)<br>Simple constructions*<br>Few words per sentence<br>Few articles<br>Few negations |
| Lexicon | Correct<br>Rich<br>High diversity<br>Many exclusive and inclusive words<br>Few social words<br>Few positive emotion words<br>Many negative emotion words | Loose*<br>Poor<br>Low diversity<br>Few exclusive and inclusive words<br>Many social words<br>Many positive emotion words<br>Few negative emotion words |
| Speech | Received accent<br>Slow speech rate<br>Few disfluencies<br>Many unfilled pauses<br>Long response latency<br>Quiet<br>Low voice quality<br>Non-nasal voice<br>Low frequency variability | Local accent*<br>High speech rate<br>Many disfluencies*<br>Few unfilled pauses<br>Short response latency<br>Loud<br>High voice quality<br>Nasal voice<br>High frequency variability |

**Fig. 3.** Summary of identified language cues for extraversion and various production levels [21]

We use the output of the sentiment analysis from language and emotion recognition from speech as linguistic and speech cues to calculate the score for each personality dimension based on previous research [21]. For each response, the individual score for each of the four dimensions is calculated and updated, and the final score in each dimension is the group average of all the responses (Fig. 3).

## 4  Handling User Challenge

The personality test consists mostly of machine-initiative questions from Zara and human answers. However, as described in the user analysis section below, there are scenarios where the user does not respond to questions from Zara directly. 24.62% of the users who tried Zara exhibited some form of verbal challenge in their responses during the dialogue conversation, of which 37.5% of users evade the questions with an irrelevant answer. 12.5% of users challenged Zara's ability more directly with questions unrelated to the personality test.

From September to December 2015, a total of 184 responses were recorded in total, 24.61% of the data showed some form of challenge during the extent of the conversation. Challenge here refers to user responses that were difficult to handle and impeded the flow of conversation with Zara. They include the following 6 types:

1. Seeking disclosure reciprocity
2. Asking for clarification
3. Avoidance of topic
4. Deliberate challenge of system ability
5. Abusive language
6. Garbage.

Several of the above categories can be observed in human-human interactions. For instance, seeking disclosure reciprocity is not uncommon in human conversations [41].

Responses that revealed some form of avoidance of topic were the most frequent. Avoidance in psychology is viewed as a coping mechanism in response to stress, fear, discomfort, or anxiety [30]. In the dataset collected, two types of avoidance were observed. Users who actively avoid the topic specifically reveal their unwillingness to continue the conversation ("I don't want to talk about it", "I am in no mood to tell you a story Zara"), while users who adopts a more passive strategy had the intent to discontinue the conversation implied ("Let's continue.", "Make it a quick one", "You know…").

Abusive language includes foul, obscene, culturally and socially inappropriate remarks and the like. Currently collected data revealed surprisingly few inappropriate comments such as "get lost now" and "None of your business". These challenges are comparatively mild. Owing to the context of Zara's role as a personality assessor, the reasons here for abuse could be the need to trust the robotic assessor and response to discomfort.

Asking for clarification examples included "Can you repeat?" and "Can you say it again?" Clarification questions observed in this dataset are primarily non- reprise questions as a request to repeat a previous utterance [26].

Deliberate challenge of system ability was also observed. This took the form of direct requests ("Can I change a topic?", "Why can't you speak English?" in the Chinese mode), or statements un-related to the questions asked ("Which one is 72.1%?").

Zara is programmed with a gentle but witty personality to handle different user challenges. For example, when abusive language is repeatedly used against her, she asks for an apology after expressing concern for the user's level of stress. If the user asks a general domain question unrelated to the personality test, Zara will try to entertain the question with an answer from a general knowledge database using a search engine API, much like Siri or Cortana. However, unlike these other systems, Zara will not chat indefinitely with the user but will remind the user of their task at hand, namely the personality test. In addition, Zara is also designed to have a sense of humor.

In the following sections, we will describe our current approaches for different modules of the Zara system, namely speech recognition, emotion and mood recognition from audio, and sentiment analysis from speech and text. We will also describe a first ever approach in humor recognition from conversations.

## 5   DNN, CNN and LSTM

In this section we give a general description of the Deep Neural Network architectures we use in the task described. The real power of any DNN is to reduce a set of low level features into a single low-dimensional feature vector, performing the appropriate feature-selection [40].

The first model we use is the Convolutional Neural Network [9], which is useful to obtain a fixed-length vector representation of an utterance, an audio signal or an image. The basic structure of a CNN takes as input one or more low level input vectors $x_{0..N}$. In case of an utterance each word is encoded as a vector, in case of an audio signal each frame, or in alternative the raw audio sample. The input feature vectors are first fed into a first embedding layer to obtain a low dimensional dense vector, given by:

$$x_i^E = f(W_E x_i + b_E)$$

where f is a non-linear function (sigmoid, hyperbolic tangent or rectified linear), and W the parameters to be trained.

A sliding window then moves over the output vectors of the embedding layer (or the raw audio input itself), and another layer is applied to each group of token or audio frame vectors:

$$x_i^C = f(W_C [x_h^E]_{h \in [x - \frac{c}{2}, x + \frac{c}{2}]} + b_C)$$

where c is the size of the convolution window, and [] denotes the concatenation. This operation allows to capture the local context and extract features from it. A max-pooling operation is then applied to extract the most salient features of all the tokens or frames into a single vector for the whole utterance. The max-pooling takes, for each vector element, the maximum value among all output vectors from the convolution:

$$x_i^{MP} = \max_t \left( x_{t,i}^C \right)$$

where t represents the token, and i the vector element.

The Long Short Term Memory (LSTM) [15] is instead useful to model time sequences where it is important to remember the past context. It is an improvement over the Recurrent Neural Network aimed to enhance its memory capabilities. In a standard RNN the hidden memory layer is updated through a function of the input and the hidden layer at the previous time instant:

$$h_t = \tanh(W_x x_t + W_h h_{t-1} + b)$$

Where **x** is the network input and **b** the bias term.

This kind of connection is not very effective to maintain the information stored for long time instants, and it does not allow to forget unneeded information between two time steps.

The LSTM enhances the RNN with a series of three multiplicative gates. The structure is the following:

$$i_t = \sigma(W_{i_x} x_t + W_{i_h} h_{t-1} + b_i)$$
$$f_t = \sigma(W_{f_x} x_t + W_{f_h} h_{t-1} + b_f)$$
$$o_t = \sigma(W_{o_x} x_t + W_{o_h} h_{t-1} + b_o)$$
$$s_t = \tanh(W_{s_x} x_t + W_{s_h} h_{t-1} + b_s)$$
$$c_t = f_t \odot c_{t-1} + i_t \odot s_t)$$
$$h_t = \tanh(c_t) \odot o_t)$$

where $\odot$ is the element-wise product. The gate factors **i**, **f**, **o** are able to let through or suppress a specific update contribution, thus allowing a selective information retaining. In this way a cell value can be retained for multiple time steps when **i** = 0, ignored in the output when **o** = 0, and forgotten when **f** = 0.

## 6 Speech Recognition

We experiment with an automatic speech recognition system implemented in house. ASR systems consist of both an acoustic model and a language model in a Bayesian framework.

$$\hat{W} = \arg\max_W P(A|W)P(W)$$

For training acoustic models, we train a GMM-HMM for predicting the hidden states and use a Deep Neural Network (DNN) to predict the emission probabilities of the HMM states. We train maximum likelihood (ML) Gaussian mixture Hidden Markov Model (GMM-HMM) with 8000 tied triphone states and 240K Gaussians, using linear discriminant analysis (LDA) and maximum likelihood linear transform (MLLT) feature transformations. We apply boosted maximum mutual information (bMMI) discriminative training on ML trained HMMs.

We train DNN for the emission probabilities of the HMMs with 6 hidden layers, and each hidden layer has 1024 neurons. The DNN is initialized with stacked restricted Boltzmann machines (RBMs) which are pre-trained in a greedy layerwise fashion. Cross-entropy (CE) criterion DNN training is first applied on the state alignments produced by the GMM-HMMs. State alignment is then reproduced with the DNN-HMMs, and DNN training with CE criterion is done again. Sequence-discriminative training of DNN with state level minimum Bayes risk (sMBR) criterion is applied by lattice-based approach on the CE trained DNN-HMMs. We train our acoustic models with the Kaldi speech recognition toolkit [25].

Our text data to train the language model contains 88.6M sentences. It comprises the acoustic model training English transcriptions, web crawled news data, web crawled book data, Cantab filtering sentences on Google 1 billion word language modeling benchmark, weather domain and music domain queries expanded from manually designed templates, and common chat queries. We train Wittenbell smoothing inter-polated trigram language model (LM) and CE based recurrent neural network (RNN) LM using the SRI-LM toolkit and CUED-RNNLM toolkit respectively.

The ASR decoder performs search on weighted finite state transducer (WFST) graph for trigram LM and generates lattice, and then performs lattice rescoring with RNN LM. The decoder is designed for input audio data that is streamed from TCP/IP or HTTP network protocol, and also performs decoding in real time. The decoder supports simultaneous users by multiple threads and user queue. The English ASR achieves 7.6% word error rate on the combined test set of wsj1[1] "si_dt_05" and "si_dt_20".

Our system is a hybrid DNN-HMM system. In recent years, there has been work on end-to-end models for speech recognition where feature vectors are the input and word sequences are predicted one by one, without an HMM framework. One such model is called End-to-End Attention-based Large Vocabulary Speech Recognition [3]. Recurrent Neural Nets (RNNs) are used to replace the HMMs. Language models are incorporated directly into the RNN decoder. Other approach includes using Deep Convolutional Neural Nets instead of DNNs or GMMs [32]. CNNs have the advantage of being able to model the spectral correlations that exist in speech signals.

In summary, speech recognition has improved in leaps and bounds thanks to deep learning and is able to give us most of what we want to hear from any spoken input. The decoded word sequence is then used for later stage semantic processing, such as emotion, sentiment and intent recognition.

# 7  Real-Time Emotion Recognition from Time-Domain Raw Audio Input

In recent years, we have seen successful systems that gave high classification accu-racies on benchmark datasets of emotional speech [21] or music genres and moods [34]. Most of such work consists of two main steps, namely feature extraction and

---

[1] https://catalog.ldc.upenn.edu/LDC94S13A.

classifier learning. One challenge for most emotion recognition systems from speech is the time needed to extract features from the speech file. Both high and low level features, as described in the following section, are needed so far for emotion and mood recognition from audio. There are close to a 1000 features, a much larger set than the feature set used for speech recognition, that need to be extracted and computed over windows of audio signals. This typically takes a few dozen seconds to do for each utterance, making the response time less than real-time instantaneous, which users have come to expect from interactive systems.

Feature engineering is tedious and time-consuming. It also requires a lot of hand tuning. In order to bypass feature engineering, the current direction is to explore methods that can recognize emotion or mood directly from time-domain audio signals. One approach that has shown great potential is using Convolutional Neural Networks. In the following sections, we compare an approach of using CNN without feature engineering to a method that uses audio features with a SVM classifier.

## 7.1 Dataset

For our experiments on emotion recognition with raw audio, we built a dataset form the TED-LIUM corpus release 2 [31]. It includes 207 h of speech extracted from 1495 TED talks. We annotated the data with an existing commercial API followed by manual correction. We obtained a total of 90041 segments, divided into the following 11 categories: creative/passionate, criticism/cynicism, defensiveness/anxiety, friendly/ warm, hostility/anger, leadership/charisma, loneliness/unfulfillment, love/happiness, sadness/sorrow, self-control/practicality and supremacy/arrogance.

In our experiments we only use the following 6 categories: criticism, anxiety, anger, loneliness, happiness, sadness. We obtained a total of 2389 segments for the criticism category, 3855 for anxiety, 12708 for anger, 3618 for loneliness, 8070 for happy and 1824 for sadness. The segments have an average length slightly above 13 s. As on-going work, we continue to re-label audio data with our CNN-based emotion decoder followed by manual correction.

## 7.2 Real-Time Speech Emotion Recognition with CNN

The CNN model using raw audio as input is shown in Fig. 4. The raw audio samples are first down-sampled at 8 kHz, in order to optimize between the sampling rate and representation memory efficiency in case of longer segments. The CNN is designed with one filter for real-time processing. We set a convolution window of size 200, which corresponds to 25 ms, and an overlapping step size of 50, equal to around 6 ms. The convolution layer perform the feature extraction in each layer, and models the variations among neighboring frames due to the overlapping.

The network is trained using the standard back-propagation algorithm, performing gradient descent over each parameter. At evaluation time the time complexity is linear over the length of the audio input segment due to the convolution. Thus the largest time contribution is due to the computations inside the network [14], which with one convolution only can be performed in negligible time for single utterances.

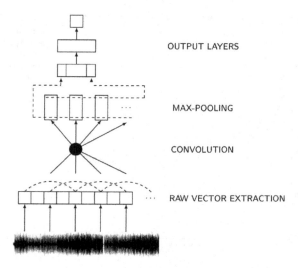

OUTPUT LAYERS

MAX-POOLING

CONVOLUTION

RAW VECTOR EXTRACTION

**Fig. 4.** Convolutional neural network for emotion classification of raw audio.

## 7.3  Experimental Setup

We setup our experiments as binary classification tasks, in which each segment is classified as either part of a particular emotion category or not part of that particular category. For each category the negative samples are chosen randomly from the music clips that do not belong to the positive genre.

We implement our CNN with the Theano framework [4]. Theano's automatic differentiation capabilities are used to implement the backpropagation. Our models are trained with GPU Tesla K20 on the CUDA platform.

We choose rectified linear as the non-linear function for the hidden layers, as it generally provides better performance over other functions. We use standard back-propagation training, with momentum set to 0.9 and initial learning rate to $10^{-5}$. We used the validation set to determine the early stopping condition when the error on it began to increase. We normalize the input data of each experiment with zero mean and unit standard deviation.

As a baseline we use a linear-kernel SVM model from the LibSVM [8] library with the INTERSPEECH 2009 emotion feature set [35], extracted with openSMILE [12]. These features are computed from a series of input frames and output a single static summary vector, e.g., the smooth methods, maximum and minimum value, mean value of the features from the frames [20]. The model of each binary classification is trained through 500 iterations.

All results are shown in Table 1, while Fig. 5 shows the learning curves of our binary classification using raw audio data as input. The lower results for some categories, even on the SVM baseline, may be a sign of inaccuracy in manual labeling. We plan to work to improve both the dataset, with hand-labeled samples, and retrain the model as on going work.

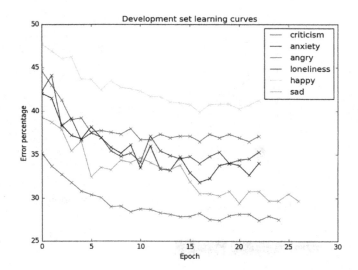

**Fig. 5.** Learning curve for binary classification: raw audio data (CNN) over six emotion categories.

**Table 1.** Real-time CNN outperforms SVM with features

|                            | % Accuracy (CNN) | % Accuracy (SVM) |
|----------------------------|------------------|------------------|
| Criticism, cynicism        | 61.2             | 55.0             |
| Defensiveness, anxiety     | 62.0             | 56.3             |
| Hostility, anger           | 72.9             | 72.8             |
| Loneliness, unfulfillment  | 66.6             | 61.1             |
| Love, happiness            | 60.1             | 50.9             |
| Sadness, sorrow            | 71.4             | 71.1             |
| Average                    | **65.7**         | **61.2**         |

## 8  Sentiment Inference from Speech and Text

In the first version of Zara, sentiment analysis was based on lexical features. We look for keyword matches from a pool of positive and negative emotion lexicons from LIWC[2] dictionary and use an N-gram model to classify the sentiment. In the current approach, we use a CNN-based classifier on Word2Vec. Convolutional Neural Networks (CNNs) have recently achieved remarkably strong performance on the practically important task of sentence classification [16–18].

In this work, we train a CNN with one layer of convolution and max pooling on top of word vectors obtained from an unsupervised neural language model. We begin with a sentence that we then convert to a matrix. The rows are word vector representations

---

[2] http://liwc.wpengine.com.

of each word in that sentence. There are several publicly available word vectors sets, such as Google word2vec, WordNet or GloVe. Due to the instinctive sequential structure of a sentence, we use filters that slide over full rows of the matrix, which scrolls word by word specifically. The width of the filter is fixed to the dimensionality of the word vector, and the height varies with different filters. Our model uses multiple filters (with different window heights 3, 4 and 5) to represent multiple features. The convolution matrix $W_i^{cnv}$ is the parameters to be learned for the $i$-th filter and its size is a hyper-parameter to be chosen for development. In our model, we choose 100. We then apply a max-over-time pooling operation over each feature, which means the model will always pick up the most valuable information wherever it happens in the input sentence. Also, the pooling technique can fix the feature length for further classification. Our work is implemented as a multi-channel model, which has two channels of word vectors—one that is kept static throughout training as word2vec and another is fine-tuned via back propagation. At last, we concatenate the feature outputs of each channel and pass it to a fully connected softmax layer whose output is the probability distribution over a binary classification for sentiment analysis of text transcribed from speech by our speech recognizer (Tables 2 and 3).

**Table 2.** Corpus statistics for text emotion experiments with CNN

| Corpus | Average length | Size | Vocabulary size | Words in Word2vec |
|---|---|---|---|---|
| Movie review training data | 20 | 10662 | 18765 | 16448 |

**Table 3.** CNN sentiment analysis on 388 sentences from human interacting with Zara

| | Accuracy | Precision | Recall | F-score |
|---|---|---|---|---|
| CNN | 67.8% | 91.2% | 63.5% | 74.8% |

It is also possible to combine audio input and transcribed text into the CNNs with two channels, as illustrated in the Fig. 6 below. Each channel process either the speech or the text features, through the max-pooling layer, the largest number of each feature is recorded into the next layer. The output of each channel are concatenated to forma a feature vector for the penultimate layer, which will be further encoded for the final layer. The final softmax will be performed for the final layer.

## 9  Humor Recognition in Conversations

In this section, we will describe a fully machine-learning approach for learning a particular sentiment – humor from spoken conversations. The methodology is applicable to recognizing other sentiments as well.

Humor can be classified depending on the way they are intended to trigger laughter: common forms include irony, sarcasm, puns and nonsense [1]. Humor in a conversation serves to lighten the mood and build rapport between the speaker and the audience. It is also very helpful for human-robot interactions. Humans are more

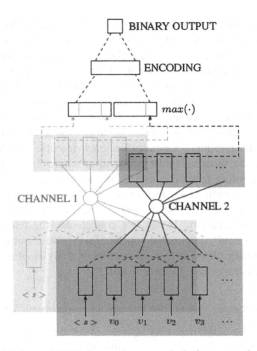

**Fig. 6.** Bi-channel CNN for sentiment analysis from speech and text.

forgiving of machine mistakes when the latter has a sense of humor. Yell at Siri, you will get a "Hey! I am doing my best!". Ask a challenging question like "Siri, do you love me?" the answer is likely to be "I am not at capable of love".

Humor in a spoken dialog is characterized both by what is said and how it is said. Therefore prosody is an important component that must be taken into account together with the semantic content of an utterance. A long-term goal of humor research, or language understanding in general, is how to effectively integrate language features with audio features.

We analyze funny dialogues extracted from humorous TV-sitcoms. A sitcom is a scripted oral dialog where at regular intervals canned laughter are embedded. Such laughter indicates where the audience is supposed to laugh, and solicit its active participation. We are interested to predict when this occurs.

Spontaneous conversational humor typically follows a defined recurrent structure [2, 38]. The first moment is the "setup", which outlines the context for the subsequent jokes and prepare the audience to receive their stimuli. A specific utterance called "punchline" has then the effect to release the tension with a peculiar reaction, typically laughter. Assuming the punchlines are the utterances followed by a canned laughter, our task is to detect them. To keep the show interesting and the audience engaged, sitcom punchlines are equally distributed throughout the show duration. This makes them a rich source of funny humorous dialogues.

An example of a sitcom dialog is shown below:

PENNY: Okay, Sheldon, what can I get you?

SHELDON: Alcohol.

PENNY: Could you be a little more specific?

SHELDON: **Ethyl alcohol. LAUGH Forty milliliters. LAUGH**

PENNY: I'm sorry, honey, I don't know milliliters.

SHELDON: **Ah. Blame President James Jimmy Carter. LAUGH**

**He started America on a path to the metric system but then just gave up. LAUGH**

In the example the punchlines are highlighted in bold.

Before each punchline the other utterances build the setup. In a dialog setting, without a proper context the punchlines may lose their effect of triggering laughters. If we take the last utterance of the example above out of context it may be perceived as a political complaint. Some people may still laugh if exposed to this utterance alone, but setting the dialog in a bar makes the humorous intent stronger.

We employ a supervised classification method to detect when punchlines occur. We use a Deep Neural Network framework divided into two levels. The first level is made of two Convolutional Neural Networks [9] to encode each individual utterance from word embedding vectors, and the audio track associated from a set of frame-level features. The language CNN is followed by a Long Short-Term Memory [15] to model the sequential context of the dialog. Before the output softmax layer we concatenate the output of the LSTM with the acoustic feature vector of the audio CNN, and a few sentence-based extra features. A framework diagram is shown in Fig. 7.

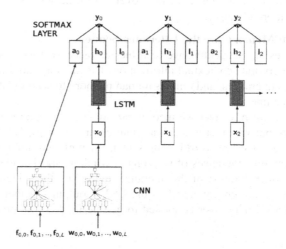

**Fig. 7.** Humor inference network: **w** are the word embedding vectors, **f** the audio frames feature vectors, and **l** the other language feature vectors.

**Convolutional Neural Network**

Astraightforward way to model a dialog is to retrieve language and acoustic features from each utterance, and to apply a memory-based classifier (such as a Recurrent Neural Network or a Long Short-Term Memory) to model the discourse context. Previous work showed that acoustic utterance-level features (such as the INTER-SPEECH 2010 paralinguistic challenge feature set [36]) are quite effective with simple classifiers such as logistic regression and conditional random fields, but yield suboptimal performance with DNN systems [5, 13]. We apply two parallel Convolutional Neural Networks to model each utterance from lower-level features.

The first CNN we use is dedicated to language features. Word2Vec word vectors are used as token-level features. Our Word2Vec model is trained on the text9 Wikipedia corpus[3] using gensim implementation [28]. Word2Vec is an improvement over bag-of-ngram or similar representations that ignore out-of-vocabulary words, that often occur in the cross-domain tasks, for example in the case of different sitcoms with different domain vocabulary.

The second CNN is used to encode the audio track of each utterance. We split each utterance into overlapping 25 ms frames shifted 10 ms from each other. We then extract for each frame a vector of low level acoustic and prosodic features with openSMILE [12]. The features we extract are MFCC, $\Delta$ MFCC, $\Delta\Delta$ MFCC, pitch, energy and zero crossing mean. A CNN then combines together all the frames into a single vector for each utterance, selecting the features from the most salient frames and discarding those that carry no information. This CNN is similar to that used for language, but it uses two embedding layers at the input, and rectified linear units instead of tanh. Both past attempts in the literature [13] and our experiments on our development sets showed the use of frame-level features over a shifting-window, and of rectified linear units to be more effective.

**Long Short-Term Memory for the Utterance Sequence**

As dialog utterances are sequential, we feed the utterance vectors in sequence into a LSTM block to incorporate contextual information. The memory unit keeps track of the context of previous utterances, and mimics human memory to accumulate the setup that may trigger a punchline.

Before the final softmax layer we incorporate some extra features not modeled by our neural network but which still add important contribution [6, 7, 27, 29].

These features are average word length, sentence length and difference in sentence length with the previous utterances of the context window, speaker identity as well as the speaking rate (time duration of the utterance divided by the sentence length).

All these features are concatenated to the LSTM output together with the audio CNN output, and a softmax layer is applied to get the final output probabilities.

## 9.1 Experiments

We built a corpus from three popular TV-sitcoms: "The Big Bang Theory" seasons 1 to 6, "Friends" seasons 6 to 9, and "Seinfeld" seasons 5 to 9. We downloaded the subtitle

---

[3] Extension of the text8 corpus, obtained from http://mattmahoney.net/dc/textdata.

files associated to each episode and the scripts[4]. We extracted and segmented the audio track of each episode for the acoustic features. The audio tracks were also used to retrieve the canned laughter timestamps, applying a vocal removal tool followed by a silence/sound detector. We then annotated each utterance as a punch line in case it was followed by laughter within 1s, assuming that utterances not followed by a laughter would be the setup for the punch line. Other than for annotation, we used the timestamps to cut from each utterance audio track the overlapping parts with a canned laughter, in order to avoid crosstalk bias.

For each show we built a training set of 80% of the overall episodes, and a development and test set of 10% each. The episodes were drawn from all the seasons with the same proportion. Our three corpora are structured as follows:

We set the size to 100 for all the hidden layers of the language CNN and the LSTM and the convolution window to 5. In the audio CNN we set the size to 50 and we used a convolution window of size 3. We concatenated the audio CNN output with the other features at the end instead of feeding it through the LSTM, as it gave us better performance. We applied a dropout regularization layer [37] after the output of the LSTM and the audio CNN, the dropout coefficient was set to 0.7. The network was trained with standard back propagation, with a momentum coefficient set to 0.9. The development set was used to tune the hyperparameters, and to determine the early stopping condition. The neural network was implemented with the Theano toolkit [4].

In the first set of experiments we trained and tested the classifiers on the same sitcom. As a first baseline for comparison we chose a Conditional Random Field [19] trained over a set of features [6] including the same features added at the end of our neural network, bag-of-ngrams, part of speech proportion, sentiment from SentiWordNet [11], antonyms, and a prosodic feature vector from the INTERSPEECH 2010 paralinguistic challenge [36]. We used the implementation from CRFSuite [23], with L2 regularization. The second baseline system is made of the CNNs encoding blocks only, where the role of the LSTM is replaced with the language CNN output vectors at previous times, concatenated together before the softmax layer. We evaluated context window length of 1 (no LSTM), 2, 3 and 5; as in all the sitcoms over 80% of punchlines occur within five utterances from the previous. All the results are shown in Table 4.

In Table 5 we also compare the results of our best system on the Big Bang corpus with the system proposed in [7], where a LSTM is applied to a larger set of language-only features, which includes one-hot word vectors and character-trigram input vectors in addition to Word2Vec. This comparison evaluates the role of acoustic features in our neural network framework.

The second series of experiments is a cross-domain evaluation. Each classifier is trained over all the data from two of the corpus shuffled together, and tested over the third corpus. In these experiments we ignore the speaker identity feature. Our baseline system is again the CRF described above, trained and tested over the same cross-domain data. We evaluated context window lengths of size 3 and 5. The results of this experiments are shown in Table 6.

---

[4] From bigbangtrans.wordpress.com and http://www.livesinabox.com/friends/scripts.shtml.

**Table 4.** CNN + LSTM gives superior performance on humor recognition

| | CNN + shifted context | | | | CNN + LSTM | | | |
|---|---|---|---|---|---|---|---|---|
| Context window size | A | P | R | F1 | A | P | R | F1 |
| *The big bang theory* | | | | | | | | |
| CRF baseline | 73.4 | 72.1 | 61.8 | 66.5 | 73.4 | 72.1 | 61.8 | 66.5 |
| 1 utterance | **73.7** | 70.3 | 66.7 | 68.5 | **73.7** | 70.3 | 66.7 | 68.5 |
| 2 utterances | 73.0 | 70.3 | 63.8 | 66.9 | 72.8 | 68.0 | 68.7 | 68.4 |
| 3 utterances | 72.2 | 64.9 | 76.2 | 70.1 | 71.6 | 63.6 | **78.7** | **70.3** |
| 5 utterances | 71.2 | 66.7 | 65.4 | 66.0 | 72.9 | **72.6** | 59.1 | 65.2 |
| *Friends* | | | | | | | | |
| CRF baseline | **71.6** | **62.7** | 45.7 | 52.9 | **71.6** | **62.7** | 45.7 | 52.9 |
| 1 utterance | 70.6 | 61.1 | 42.8 | 50.3 | 70.6 | 61.1 | 42.8 | 50.3 |
| 2 utterances | 70.6 | 60.2 | 45.9 | 52.1 | 68.3 | 54.6 | **54.0** | **54.3** |
| 3 utterances | 69.1 | 58.9 | 37.4 | 45.8 | 69.2 | 56.6 | 50.0 | 53.1 |
| 5 utterances | 70.8 | 62.3 | 40.9 | 49.4 | 70.7 | **62.7** | 41.4 | 49.9 |
| *Seinfeld* | | | | | | | | |
| CRF baseline | 76.5 | 59.4 | 36.5 | 45.2 | 76.5 | 59.4 | 36.5 | 45.2 |
| 1 utterance | 71.1 | 44.4 | 35.4 | 39.4 | 71.1 | 44.4 | 35.4 | 39.4 |
| 2 utterances | 72.2 | 46.4 | 30.0 | 36.5 | **76.6** | 65.8 | 25.1 | 36.4 |
| 3 utterances | 70.5 | 43.6 | 38.0 | 40.6 | 75.3 | 54.4 | **44.7** | **49.1** |
| 5 utterances | 74.3 | 56.3 | 14.9 | 23.6 | **76.6** | **67.8** | 22.6 | 33.9 |

**Table 5.** Bichannel LSTM performs the best

| Method | A | P | R | F1 |
|---|---|---|---|---|
| CRF baseline | 65.9 | 61.2 | 55.3 | 58.1 |
| LSTM language-only | 70.0 | 66.7 | 59.4 | 62.9 |
| LSTM language + audio | 71.6 | 63.6 | 78.7 | **70.3** |

## 10  Summary and Discussion

In this paper, we have described a prototype system of an empathetic virtual robot that can recognize user emotions and thereby bring about a new level of human-robot interactions. We described the design of the architecture, the task of personality analysis, and user analysis of Zara the Supergirl. From there, we extended our description to include more details of the recognition and inference of emotion and sentiment from speech and language. Zara also has a facial recognition component which we have not described in detail as it acts as a supplement to the speech and language part.

We have shown how Deep Learning can be used for various modules in this architecture, ranging from speech recognition, emotion recognition to humor recognition from dialogs. More importantly, we have shown that by using a CNN with one

**Table 6.** LSTM is relatively robust across different corpora

*Train: seinfeld + friends test: big bang*

|  | Self-trained | | | | Cross-corpus | | | |
|---|---|---|---|---|---|---|---|---|
| Context window size | A | P | R | F1 | A | P | R | F1 |
| CRF baseline | 73.4 | 72.1 | 61.8 | 66.5 | 68.3 | 73.3 | 40.5 | 52.2 |
| LSTM 3 utterances | 71.5 | 63.6 | 78.7 | 70.3 | 69.6 | 72.2 | 47.1 | 57.0 |
| LSTM 5 utterances | 72.9 | 72.6 | 59.1 | 65.2 | 68.7 | **76.1** | 39.0 | 51.6 |

*Train: big bang + seinfeld test: friends*

|  | Self-trained | | | | Cross-corpus | | | |
|---|---|---|---|---|---|---|---|---|
| Context window size | A | P | R | F1 | A | P | R | F1 |
| CRF baseline | 71.6 | 62.7 | 45.7 | 52.9 | 63.0 | 47.4 | 56.9 | 51.7 |
| LSTM 3 utterances | 69.2 | 56.6 | 50.0 | 53.1 | 69.6 | 71.0 | 21.4 | 32.9 |
| LSTM 5 utterances | 70.7 | 62.7 | 41.4 | 49.9 | 69.5 | **72.0** | 20.3 | 31.6 |

*Train: big bang + friends test: seinfeld*

|  | Self-trained | | | | Cross-corpus | | | |
|---|---|---|---|---|---|---|---|---|
| Context window size | A | P | R | F1 | A | P | R | F1 |
| CRF baseline | 76.5 | 59.4 | 36.5 | 45.2 | 69.6 | 44.4 | 58.1 | 50.3 |
| LSTM 3 utterances | 75.3 | 54.4 | 44.7 | 49.1 | 64.5 | 40.2 | 74.8 | 52.3 |
| LSTM 5 utterances | 76.6 | 67.8 | 22.6 | 33.9 | 67.8 | 43.2 | 67.7 | **52.7** |

filter, it is possible to obtain real-time performance on speech emotion recognition directly from time-domain audio input, bypassing feature engineering. We have so far developed only the most primary tools that future emotionally intelligent robots would need. Empathetic robots including Zara that are there currently, and the ones that will be there in the near future, might not be completely perfect. However, the most significant step is to make robots to be more human like in their interactions. This means it will have flaws, just like humans do. If this is done right, then future machines and robots will be empathetic and less likely to commit harm in their interactions with humans. They will be able to get us, understand our emotions, and more than anything, they will be our teachers, our caregivers, and our friends.

# References

1. Attardo, S.: Linguistic Theories of Humor, vol. 1. Walter de Gruyter, Berlin (1994)
2. Attardo, S.: The semantic foundations of cognitive theories of humor. Humor-Int. J. Humor Res. **10**(4), 395–420 (1997)
3. Bahdanau, D., Chorowski, J., Serdyuk, D., Brakel, P., Bengio, Y.: End-to-end attention-based large vocabulary speech recognition. In: 2016 IEEE International Conference on Acoustics, Speech and Signal Processing (ICASSP) (2016)
4. Bergstra, J., Breuleux, O., Bastien, F., Lamblin, P., Pascanu, R., Desjardins, G., Turian, J., Warde-Farley, D., Bengio, Y.: Theano: a CPU and GPU math expression compiler. In: Proceedings of the Python for Scientific Computing Conference (SciPy), vol. 4, p. 3 (2010)

5. Bertero, D., Fung, P.: Deep learning of audio and language features for humor prediction. In: International Conference on Language Resources and Evaluation (LREC) (2016)
6. Bertero, D., Fung, P.: Predicting humor response in dialogues from TV sitcoms. In: 2016 IEEE International Conference on Acoustics, Speech and Signal Processing (ICASSP) (2016)
7. Bertero, D., Fung, P.: A long short-term memory framework for predicting humor in dialogues. In: Proceedings of the 2016 Conference of the North American Chapter of the Association for Computational Linguistics: Human Language Technologies (2016)
8. Chang, C.C., Lin, C.J.: LIBSVM: a library for support vector machines. ACM Trans. Intell. Syst. Technol. (TIST) **2**(3), 27 (2011)
9. Collobert, R., Weston, J., Bottou, L., Karlen, M., Kavukcuoglu, K., Kuksa, P.: Natural language processing (almost) from scratch. J. Mach. Learn. Res. **12**, 2493–2537 (2011)
10. Duffy, B.R., Joue, G.: Intelligent robots: the question of embodiment. In: Proceedings of the Brain-Machine Workshop (2000)
11. Esuli, A., Sebastiani, F.: SENTIWORDNET: a publicly available lexical resource for opinion mining. In: Proceedings of LREC, vol. 6, pp. 417–422 (2006)
12. Eyben, F., Wöllmer, M., Schuller, B.: openSMILE: the munich versatile and fast open-source audio feature extractor. In: Proceedings of the 18th ACM International Conference on Multimedia, pp. 1459–1462. ACM (2010)
13. Han, K., Yu, D., Tashev, I.: Speech emotion recognition using deep neural network and extreme learning machine. In: INTERSPEECH, pp. 223–227 (2014)
14. He, K., Sun, J.: Convolutional neural networks at constrained time cost. In: Proceedings of the IEEE Conference on Computer Vision and Pattern Recognition, pp. 5353–5360 (2015)
15. Hochreiter, S., Schmidhuber, J.: Long short-term memory. Neural Comput. **9**(8), 1735–1780 (1997)
16. Johnson, R., Zhang, T.: Effective use of word order for text categorization with convolutional neural networks. In: Proceedings of the 53rd Annual Meeting of the Association for Computational Linguistics (2015)
17. Kalchbrenner, N., Grefenstette, E., Blunsom, P.: A convolutional neural network for modelling sentences. In: Proceedings of the 52nd Annual Meeting of the Association for Computational Linguistics (2014)
18. Kim, Y.: Conditional neural networks for sentence classification. In: EMNLP 2014 (2014)
19. Lafferty, J., McCallum, A., Pereira, F.C.: Conditional Random Fields: Probabilistic Models for Segmenting and Labeling Sequence Data (2001)
20. Liscombe, J., Venditti, J., Hirschberg, J.B.: Classifying subject ratings of emotional speech using acoustic features. Columbia University Academic Commons (2003)
21. Mairesse, F., Walker, M.A., Mehl, M.R., Moore, R.K.: Using linguistic cues for the automatic recognition of personality in conversation and text. J. Artif. Intell. Res. **30**, 457–500 (2007)
22. Mataric, M.J.: The role of embodiment in assistive interactive robotics for the elderly. In: AAAI Fall Symposium on Caring Machines: AI for the Elderly, Arlington, VA (2005)
23. Okazaki, N.: CRFsuite: a fast implementation of conditional random fields (CRFs) (2007). http://www.chokkan.org/software/crfsuite
24. Polzehl, T., Möller, S., Metze, F.: Automatically assessing personality from speech. In: 2010 IEEE Fourth International Conference on Semantic Computing (ICSC), pp. 134–140. IEEE (2010)
25. Povey, D., Ghoshal, A., Boulianne, G., Burget, L., Glembek, O., Goel, N., Silovsky, J.: The Kaldi speech recognition toolkit. In: IEEE 2011 Workshop on Automatic Speech Recognition and Understanding (No. EPFL-CONF-192584). IEEE Signal Processing Society (2011)

26. Purver, M.: The theory and use of clarification requests in dialogue. Unpublished Doctoral Dissertation, University of London (2004)
27. Rakov, R., Rosenberg, A.: "Sure, i did the right thing": a system for sarcasm detection in speech. In: INTERSPEECH, pp. 842–846 (2013)
28. Řehůřek, R., Sojka, P.: Gensim–python framework for vector space modelling. NLP Centre, Faculty of Informatics, Masaryk University, Brno, Czech Republic (2011)
29. Reyes, A., Rosso, P., Veale, T.: A multidimensional approach for detecting irony in Twitter. Lang. Resour. Eval. **47**(1), 239–268 (2013)
30. Roth, S., Cohen, L.J.: Approach, avoidance, and coping with stress. Am. Psychol. **41**(7), 813 (1986)
31. Rousseau, A., Deléglise, P., Estève, Y.: Enhancing the TED-LIUM corpus with selected data for language modeling and more TED talks. In: LREC, pp. 3935–3939 (2014)
32. Sainath, T.N., Mohamed, A.R., Kingsbury, B., Ramabhadran, B.: Deep convolutional neural networks for LVCSR. In: 2013 IEEE International Conference on Acoustics, Speech and Signal Processing (ICASSP), pp. 8614–8618. IEEE (2013)
33. Scaringella, N., Zoia, G., Mlynek, D.: Automatic genre classification of music content: a survey. IEEE Sig. Process. Mag. **23**(2), 133–141 (2006)
34. Schermerhorn, P., Scheutz, M.: Disentangling the effects of robot affect, embodiment, and autonomy on human team members in a mixed-initiative task. In: Proceedings from the International Conference on Advances in Computer-Human Interactions, pp. 236–241 (2011)
35. Schuller, B., Steidl, S., Batliner, A.: The INTERSPEECH 2009 emotion challenge. In: INTERSPEECH, vol. 2009, pp. 312–315 (2009)
36. Schuller, B., Steidl, S., Batliner, A., Burkhardt, F., Devillers, L., Müller, C.A., Narayanan, S. S.: The INTERSPEECH 2010 paralinguistic challenge. In: INTERSPEECH, vol. 2010, pp. 2795–2798, September 2010
37. Srivastava, N., Hinton, G., Krizhevsky, A., Sutskever, I., Salakhutdinov, R.: Dropout: a simple way to prevent neural networks from overfitting. J. Mach. Learn. Res. **15**(1), 1929–1958 (2014)
38. Taylor, J., Mazlack, L.: Toward computational recognition of humorous intent. In: Proceedings of Cognitive Science Conference, pp. 2166–2171 (2005)
39. Wainer, J., Feil-Seifer, D.J., Shell, D.A., Matarić, M.J.: The role of physical embodiment in human-robot interaction. In: The 15th IEEE International Symposium on Robot and Human Interactive Communication, ROMAN 2006, pp. 117–122. IEEE (2006)
40. Wang, M., Manning, C.D.: Effect of non-linear deep architecture in sequence labeling. In: IJCNLP, pp. 1285–1291 (2013)
41. Wheeless, L.R., Grotz, J.: The measurement of trust and its relationship to self-disclosure. Hum. Commun. Res. **3**(3), 250–257 (1977)

# Extracting Aspect Specific Sentiment Expressions Implying Negative Opinions

Arjun Mukherjee[(⊠)]

Department of Computer Science, University of Houston, Houston, USA
arjun@uh.edu

**Abstract.** Subjective expression extraction is a central problem in fine-grained sentiment analysis. Most existing works focus on generic subjective expression extraction as opposed to aspect specific opinion phrase extraction. Given the ever-growing product reviews domain, extracting aspect specific opinion phrases is important as it yields the key product issues that are often mentioned via phrases (e.g., "signal fades very quickly," "had to flash the firmware often"). In this paper, we solve the problem using a combination of generative and discriminative modeling. The generative model performs a first level processing facilitating (1) discovery of potential head aspects containing issues, (2) generation of a labeled dataset of issue phrases, and (3) feed latent semantic features to subsequent discriminative modeling. We then employ discriminative large-margin and sequence modeling with pivot features for issue sentence classification and issue phrase boundary extraction. Experimental results using real-world reviews from Amazon.com demonstrate the effectiveness of the proposed approach.

## 1 Introduction

Aspect-based sentiment analysis is one of the main frameworks in opinion mining [1]. This thread focuses on unigram modeling as opposed to phrases which are more expressive. The fine-grained sentiment analysis paradigm [2] focuses on generic expressions as opposed to aspect specific expressions. Thus, there lies a big disconnect: *extracting aspect specific sentiment expressions (opinion phrases)*.

Working in the most ubiquitous consumer reviews domain, this paper proposes a framework for extracting aspect specific opinion phrases. Further, we focus on sentiment expressions implying negative opinions. We call these *issues*. Extracting phrasal issues is important as they delineate the key problematic aspects of products that people want to know before making purchase decisions. Also, in contrast to positive opinion phrases that are relatively easier to discover (as they often involve direct positive opinions), discovering phrasal issues is more challenging as they appear in myriad types: direct ("signal strength was bad") or indirect ("had to flash firmware everyday"), containing verb phrase ("has been dropping connection"), noun, adjective or adverbial phrase ("voice commands operate only a limited set of features"), etc. We propose a holistic approach that caters for all types. Our approach is also context and polarity independent facilitating generic aspect specific opinion phrase extraction.

© Springer International Publishing AG, part of Springer Nature 2018
A. Gelbukh (Ed.): CICLing 2016, LNCS 9624, pp. 194–210, 2018.
https://doi.org/10.1007/978-3-319-75487-1_15

Formally, the task can be stated as follows: Given a sentence, $s = (w_1, \ldots w_n)$, discover the head aspect (HA/issue subject), $w_{HA=i}$ and a sub-sequence $(w_p \ldots w_q), p \leq i \leq s$ that best describes the issue, i.e., an aspect specific opinion phrase on the head aspect and containing the head aspect. Throughout the paper, we will refer to head aspect and issue subject interchangeably. Examples below show labeled product issues within [[ ]] with the issue subject (head aspect) italicized:

- The first one I got working for about 2 weeks, then it [[started to drop the *signal*]], causing me to have to power cycle the unit.
- On the not so good side - We find the GPS [[*voice* to be not as clear]] as other GPSs we have used.

Although there are works that discover aspect/topic phrases using topic modeling [3, 4] and those that extract generic subjective expressions [5–7] using conditional random fields (CRFs), they lack the correspondence of the aspect and sentiment terms appearing in the sentence context. The proposed aspect specific sentiment expression task setting includes the head aspect within the phrase that naturally addresses the correspondence issue. Belonging to the family of information extraction problems, our task has resemblances with various works which are noted below.

In [8], subjective verb expressions were discovered using markov networks; in [9], supervised keyphrase extraction was used; and in [10], a re-ranking approach was used on the output of a sequence model to improve opinion expression extraction. These works mostly relied on word level features under the first-order Markov assumption. In [11, 12], segment features were used via semi-CRFs.

Parsing, phrasal, relational, and syntactic feature based approaches [13–16] have also been successful in opinion mining. In our context, the work of [17] is relevant where features indicating dependency relations between opinion expressions were employed for opinion expression extraction. However, their approach relies on the output of a sequence labeler, prohibiting dependency features to be encoded in a sequence model. Other related works where sequence modeling was used include polarity identification [18, 19] and opinion relation extraction [20].

On a broader scope, this work is also related to the family of approaches in para-phrase learning [21], clustering [22], emotional paraphrase extraction [23], and key-phrase extraction [24] as they also discover phrase boundaries in their relevant contexts.

However, above works focus either on generic subjective expressions, aspect/topic phrases, paraphrases, or keyphrases as opposed to aspect specific opinion phrases which is the focus of this work. They tend to employ variations of term, segment, structural, syntactic, or rule/window based features as opposed to our proposed pivot features with respect to the head-aspect that allow modeling arbitrarily long opinion phrases. Also, above works that employ sequence modeling (e.g., [18, 20]), rely on canonical CRFs for phrase boundary detection. This has two shortcomings in our given problem context: (i) Canonical CRFs have a strong bias towards detecting a potential opinion phrase (issue) around the head aspect for every sentence in which the head aspect appears. This unfortunately leads to higher false positive rate as not every sentence mentioning the head aspect has an issue. We noticed this in our pilot studies, (ii) Under canonical CRFs, the space of potential issue phrases is much smaller than all possible enumerable sequences resulting in inaccurate phrase boundary detection. To address these shortcomings, we propose a two-step approach (Fig. 1):

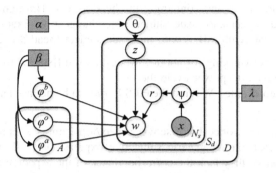

**Fig. 1.** Plate notation of ME-ASM

- Task I: Given a head aspect (HA), detect whether a sentence containing the HA mentions an issue.
- Task II: Given a HA and a sentence mentioning an issue, extract the issue phrase boundary.

To solve these tasks, we first posit a generative model, ME-ASM for domain-wise aspect specific sentiment extraction. ME-ASM provides us potential head aspects containing issues which directly feeds the issue annotation pipeline. Next, we use discriminative modeling for tasks I and II and leverage the generative model's posterior as features in the discriminative sequence model which significantly improves phrase extraction performance. To our knowledge, this has not been attempted before in opinion mining. The key contributions of this work include:

- A domain-wise aspect specific sentiment generative model for detecting head aspects.
- A family of pivot features for task I and phrase structural constraints for task II that can be used with generic discriminative and sequence modeling respectively.
- A comprehensive evaluation of the proposed methods against baselines including feature ablation and domain adaptation.
- A labeled data of aspect specific opinion phrases across 6 domains containing 3610 instances (sentences) tagged with phrase boundaries that imply negative opinions. The dataset used in this work is available at http://www.cicling.org/2016/data/10.

## 2   Aspect Specific Sentiment Modeling

We now present our generative semi-supervised model for extracting domain-wise aspect specific sentiments. As mentioned in Sect. 1, this is the first major step that feeds the pipeline for both tasks: [I] issue sentence classification, [II] issue phrase boundary extraction. Our model, ME-ASM (Max-Ent Aspect Sentiment Model) has resemblances to previous aspect extraction models [25–28] but aims to deliver (i) robust domain-wise aspects, (ii) clear separation of aspects from aspects specific sentiments, and iii) sentence level modeling for sharper aspect extraction.

As noted in [25], modeling entire reviews as documents tend to correspond to the product's global properties (e.g., brand, name) resulting in overlapping aspects. To avoid this, we perform sentence level modeling. We posit $a_{1...A}$ aspects, $o_{1...A}$ aspect specific sentiments, and background language models using multinomials $\varphi_a^A$, $\varphi_a^O$, and $\varphi^b$ drawn from $Dir(\beta)$ respectively over the vocabulary $v_{1...V}$. For each domain $d$, we draw a domain specific aspect distribution $\theta_d \sim Dir(\alpha)$. Next, for each review sentence (document), $s_d$ of a domain, $d$ we draw an aspect, $z_{d,s} \sim Mult(\theta_d)$. We assume that each sentence evaluates one aspect which mostly holds in the review domain. Next, to generate each word, $w_{d,s,j}$ of the sentence, $s_d$ we first set the switch variable $r_{d,s,j} \leftarrow Mult(\psi_{d,s,j})$ from a previously trained discriminative model (see Sect. 2.2). The switch variable, $r \in \{\hat{a}, \hat{o}, \hat{b}\}$ takes on values corresponding to aspect, sentiment, and background words as estimated by the switch model $\psi$. Finally, depending upon the latent aspect, $z_{d,s}$ and the switch variable $r_{d,s,j}$, we emit $w_{d,s,j}$ as follows:

$$
w_{d,s,j} \sim \begin{cases} Mult\left(\varphi^b\right) & \text{if } r_{d,s,j} = \hat{b} \\ Mult\left(\varphi_{z_{d,s}}^A\right) & \text{if } r_{d,s,j} = \hat{a} \\ Mult\left(\varphi_{z_{d,s}}^O\right) & \text{if } r_{d,s,j} = \hat{o} \end{cases} \tag{1}
$$

## 2.1 Inference

We employ MCMC Gibbs sampling for posterior inference. As latent variables $z$ and $r$ belong to different levels, we hierarchically sample $z$ and then $r$ for each sweep of a Gibbs iteration as follows:

$$
p\left(z_{d,s} = a | Z_{\neg d,s}, R_{\neg d,s}, W_{\neg d,s}\right) \propto \frac{\left(n_{d,a}^s\right)_{\neg d,s} + \alpha}{\left(n_{d,(\cdot)}^s\right)_{\neg d,s} + A\alpha} \times \left[\left(\prod_{v=1}^V \frac{\Gamma\left(n_{a,v}^A + \beta\right)}{\Gamma\left(n_{a,v \neg d,s}^A + \beta\right)}\right) \middle/ \right.
$$
$$
\left. \left(\frac{\Gamma\left(n_{a,(\cdot)}^A + V\beta\right)}{\Gamma\left(n_{a,(\cdot) \neg d,s}^A + V\beta\right)}\right)\right] \times \left[\left(\prod_{v=1}^V \frac{\Gamma\left(n_{a,v}^O + \beta\right)}{\Gamma\left(n_{a,v \neg d,s}^O + \beta\right)}\right) \middle/ \left(\frac{\Gamma\left(n_{a,(\cdot)}^O + V\beta\right)}{\Gamma\left(n_{a,(\cdot) \neg d,s}^O + V\beta\right)}\right)\right] \tag{2}
$$

$$
p\left(r_{d,s,j} = l | \ldots, w_{d,s,j} = v\right) \propto \frac{\left(n_{a,v}^l\right)_{\neg d,s,j} + \beta}{\left(n_{a,(\cdot)}^l\right)_{\neg d,s,j} + V\beta} \times \frac{\exp\left(\sum_{i=1}^n \lambda_i f_i\left(x_{d,s,j}, l\right)\right)}{\sum_{l \in \{\hat{a},\hat{o},\hat{b}\}} \exp\left(\sum_{i=1}^n \lambda_i f_i\left(x_{d,s,j}, l\right)\right)}; \tag{3}
$$
$$
l \in \{\hat{a}, \hat{o}, \hat{b}\}
$$

**Table 1.** Top ranked aspect and sentiment terms in two head aspects (issue subjects) across two domains. Clustering errors are *italicized in red*.

| Issue subject: Signal | | Issue subject: Firmware | |
|---|---|---|---|
| Aspect ($\varphi^A$) | Sentiment ($\varphi^O$) | Aspect ($\varphi^A$) | Sentiment ($\varphi^O$) |
| signal | loses | firmware | bug |
| wireless | faded | hardware | upgrade |
| antenna | drops | *third* | update |
| wifi | poor | *party* | old |
| *download* | losing | version | restore |
| unsecured | unavailable | level | incompatible |
| *feet* | slow | driver | instable |
| router | *clear* | latest | *install* |
| range | weak | *download* | flashing |

(a) Router Domain

| Issue subject: Screen | | Issue subject: Voice | |
|---|---|---|---|
| Aspect ($\varphi^A$) | Sentiment ($\varphi^O$) | Aspect ($\varphi^A$) | Sentiment ($\varphi^O$) |
| screen | small | voice | poor |
| touchscreen | sensitive | sound | clarity |
| display | unresponsive | *directions* | understand |
| touch | *bright* | accent | *sounds* |
| contrast | useless | command | quality |
| garmin | horrible | street | awful |
| resolution | *responsive* | name | *slow* |
| *3d* | clutter | instructions | horrible |
| map | *poor* | *english* | distorted |

(b) GPS Domain

where $n_{d,a}^s$ denotes the # of sentences in domain $d$ assigned to aspect $a$. $n_{a,v}^A$, $n_{a,v}^O$, $n_v^B$ denotes the # of times word $v$ was assigned to aspect $a$ in the aspect, aspect specific opinion, and background language models respectively. A count variable with subscript ($\cdot$) signifies the marginalized sum over the latter index and $\neg$ denotes the discounted counts.

## 2.2    Setting Switch and Hyper-parameters

ME-ASM performs a three-way switch between aspects, sentiments and background words and is motivated by models in [27, 28]. We employ a discriminative Max-Ent model for performing the switch. As aspect and sentiment terms play different syntactic roles in a sentence, we leverage the part-of-speech (POS) and syntactic chunk tags of the terms as features for learning the Max-Ent model $\psi_{d,s,j}$ conditioned on the observed feature vector $\overrightarrow{x_{d,s,j}}$ associated with $w_{d,s,j}$. We use a window of 4 terms both ahead and behind the term $w_{d,s,j}$ to encode feature context. The Max-Ent $\lambda$ were learned using 500 labeled terms in each domain following the approach in [28]. The hyper-parameters for ME-ASM were set as $\beta = 0.1$ and $\alpha = 50/A$ following [29]. The total # of aspects, $A$ across all 6 domains (see Sect. 3.1) were set to 20 after tuning via our pilot experiments.

## 2.3  Estimated Posterior

Table 1 shows the top terms for the estimated posterior on $\varphi_a^A$ and $\varphi_a^O$. Owing to space constraints, we focus on two aspects for two domains each. As our goal is to discover phrasal issues, we run our model on $\leq$ 3-star reviews (see dataset in Sect. 3.1). Except for some clustering errors (*italicized in red*), which is a known issue in generative modeling [30], we see that ME-ASM yields a decent clustering of aspects and aspect specific sentiment terms implying negative opinions. The posterior feeds head-aspect detection (Sect. 3.1) and tasks I and II.

# 3  Task I: Issue Sentence Classification

This section details the task I: Given a head aspect, HA and a sentence containing the HA, classify whether the sentence mentions an issue or not. We first detail our dataset, followed by features and results.

## 3.1  Dataset

To our knowledge, there are no publicly available datasets that mark phrase boundaries of aspect specific sentiment expressions. The closest tasks to ours are in (1) SemEval 2015 Aspect based Sentiment Analysis Task [31] which aims to discover opinion targets on entity-aspect pairs and have annotations such as {FOOD#QUALITY, "Chow fun", negative, from = "0" to = "8"} for the sentence: "Chow fun was dry; pork shu mai was…" where annotations apply to aspect expressions, and (2) The MPQA 2.0 corpus [2] although has some labeled opinion expressions, it mostly contains generic subjective expressions spanning dimensions such as sentiment, agreement, arguing, intention, etc. Both corpora don't contain the entire aspect specific sentiment phrase boundaries labeled and hence cannot be directly used in our task.

Hence, we constructed a data resource for the proposed task. Given our problem context, we consider 1, 2, and 3-star product reviews from Amazon.com. For each domain, we annotated issue phrases for top 4 aspects that had the highest appearance of negative opinions (estimated using the posterior on $\varphi_a^A$ and $\varphi_a^O$ from ME-ASM). We followed previous work in [32] for training our judges for annotation. A phrase was defined to be any subjective expression that captures various sentiments (evaluation, emotion, appraisal, etc.) toward the head aspect and containing the head aspect (see examples in Sect. 1). The annotation was distributed across four human judges (native English speakers). Every sentence was tagged by at least two judges. Inconsistencies were resolved by a third judge. Across each domain, we obtained, kappa $\kappa \in [0.71 - 0.82]$ indicating substantial to high agreements. The annotation statistics are reported below. For each domain, we report the head aspects and the counts as $(x/y)$ where $x$ is the # of issue sentences and $y$ the total # of sentences in that domain: Router (1284/5063; connection, firmware, signal, wireless), GPS (632/2075; voice, software, screen direction), Keyboard (667/1446; spacebar, range, pad, keys), Mouse (494/2488; battery, button, pointer, wheel), MP3-Player (174/352; button, interface, jack, screen), Earphone (359/678; cord, jack wire). This dataset serves both of our tasks I and II.

## 3.2  Features

As product issues are directly reflected in the language usage, word and POS (W+POS) $n$-gram features serve as natural baselines. We consider unigrams and bigrams. Using (W+POS) features here is akin to traditional sentence polarity classification [33].

However, in our problem context, (W+POS) features are insufficient as they do not consider the head aspect and relevant positional/contextual features, i.e., how do different POS tags, syntactic units (chunks), polar sentiments appear in proximity to the head aspect? Hence, centering on the issue subject (head aspect), we propose a set of pivot features to model context.

**Pivot Features**: We consider five feature families which take on a set of values:
POS Tags ($T$): *DT, IN, JJ, MD, NN, RB, VB,* etc.
Phrase Chunk Tags ($C$): *ADJP, ADVP, NP, PP, VP,* etc.
Prefixes ($P$): *anti, in, mis, non, pre, sub, un,* etc.
Suffixes ($S$): *able, est, ful, ic, ing, ive, ness, ous,* etc.
Word Sentiment Polarity ($W$): *POS, NEG, NEU*

Pivoting on the head aspect, we look forward and backward to generate a family of binary features defined by a specific template (see Table 2). Each template generates several feature that capture various positional context around the head aspect.

**Table 2.** Pivot feature templates. The subscript $i$ denotes the position of the issue subject (HA) which is italicized and the subscript $j$ denotes the position relative to $i$.

| Category | Feature template | Example of feature appearing in a sentence |
|---|---|---|
| 1$^{st}$ order features $X_{i+j}; -4 \leq j \leq 4$ $X \in \{T, C, P, S, W\}$ | $W_{i+j}$ | $W_{i-1} = NEG$; previous term of head aspect is of NEG polarity, … have this terrible *voice* on the… |
| | $S_{i+j}$ | $S_{i-2} = ing$; suffix of 2$^{nd}$ previous term of head aspect is "ing", …kept dropping the *signal*… |
| | … | … |
| 2$^{nd}$ order features $X_{i+j}, Y_{i+j}; -4 \leq j \leq 4$ $X, Y \in \{T, C, P, S, W\}$ | $T_{i+j}, T_{i+j'}$ | $T_{i-2} = JJ, T_{i-1} = VBZ$, …frequently drops *connection*… |
| | $T_{i+j}, C_{i+j'}$ | $T_{i+2} = RB, C_{i+3} = ADJP$; …*screen* is too small… |
| | … | … |
| 3$^{rd}$ order features $X_{i+j}, Y_{i+j}, Z_{i+j};$ $-4 \leq j \leq 4$ $X, Y, Z \in \{T, C, P, S, W\}$ | $T_{i+j}, S_{i+j'}, T_{i+j''}$ | $T_{i+2} = JJ, S_{i+4} = un, T_{i+4} = JJ$; …*screen* is blank and unresponsive… |
| | … | … |

Additionally, we consider up to $3^{rd}$ order pivot features allowing us to model a rich and expressive feature space.

**Latent Semantics (LS):** The generative model yields us aspect $(\varphi_a^A)$ and aspect specific sentiment terms $(\varphi_a^O)$ for each head aspect, $a$. It also provides us the assignments of latent variables $z$ and $r$ in each sentence. We leverage this information by positing the following features: (i) Top 50 terms of $\varphi^A$, $\varphi^S$, (ii) # of words assigned to aspect, opinion, and background distributions in a sentence, i.e., $\left| r_{d,s,n} = a \right|$, $\left| r_{d,s,n} = o \right|$, $\left| r_{d,s,n} = b \right|$, (iii) the aspect assignment for the sentence, $z_{d,s}$, (iv) for each term $\{ w | w \in \varphi_a^A, \varphi_a^O \}$, the signed positional index of $w$ form the head aspect, $a$.

## 3.3    Results

We now evaluate the performance on the first task of issue sentence classification. Merging sentences of all head aspects per domain, we report classification results for each domain. Upon experimenting with various kernels (linear, RBF, sigmoid) and features selection schemes in our pilot, we finalized on a RBF kernel SVM [34] with $C = 10, \gamma = 0.01$ and $\chi^2$ feature selection as our classifier as it performed best. Table 4 shows the 5-fold cross validation (CV) results across different feature sets. While inducing LS features, for each fold of 5-fold CV, ME-ASM was run on the full data excluding the test fold. The learned ME-ASM was then fitted to the test set sentences for generating the LS features of the test instances. We note the following observations:

- Across each domain, the pivot features significantly ($p < 0.01$) improve precision, recall, and F1 scores over the baseline features across all domains.

**Table 3.** (P)recision, (R)ecall, F1 scores, and (Acc)uracy in % of 5-fold CV for issue sentence classification per domain.

| Feature Set | P | R | F1 | Acc. | P | R | F1 | Acc. |
|---|---|---|---|---|---|---|---|---|
| Word (W) + POS | 68.8 | 59.6 | 63.9 | 82.8 | 70.3 | 57.2 | 63.1 | 69.2 |
| W + POS + Pivot | 74.1 | 66.4 | 70.0 | 85.5 | 73.1 | 60.5 | 66.2 | 71.5 |
| Latent Semantics | 72.9 | 64.7 | 68.5 | 84.9 | 70.9 | 59.6 | 64.8 | 70.2 |
| All | 81.5 | 69.6 | 75.1 | 88.2 | 76.6 | 64.3 | 69.9 | 74.5 |
|  | (a) Router | | | | (b) Keyboard | | | |

| Feature Set | P | R | F1 | Acc. | P | R | F1 | Acc. |
|---|---|---|---|---|---|---|---|---|
| Word (W) + POS | 69.6 | 58.9 | 63.8 | 79.6 | 65.9 | 56.7 | 60.9 | 85.5 |
| W + POS + Pivot | 72.6 | 62.4 | 67.1 | 81.3 | 67.5 | 60.5 | 63.8 | 86.1 |
| Latent Semantics | 71.1 | 62.3 | 66.4 | 80.8 | 66.5 | 59.5 | 62.8 | 85.8 |
| All | 73.6 | 64.8 | 68.9 | 82.2 | 66.8 | 60.1 | 63.3 | 86.8 |
|  | (c) GPS | | | | (d) Mouse | | | |

- LS features alone improve performance significantly ($p < 0.03$) and are close to combined W+POS+Pivot features' performance. Particularly, LS features improve recall. We also note that although the LS feature space is smaller than pivot features, it can perform quite well, thereby indicating its discriminative strength.
- Across all domains and feature sets, we find that recall is relatively lower than precision. This is due to the rather myriad forms of implied issues (e.g., "small screen buttons," "firmware does not contain the fixes," "firmware would reboot itself," etc.) which are difficult cases. Nonetheless, we note that LS and pivot features significantly improve recall over W + POS features.
- Lastly, combining all feature sets yield the best performance with an average F1 of $\approx 0.69$ and accuracy of $\approx 0.83$ across a total of 3,077 test instances (issues sentences of 4 domains combined, see Sect. 3.1) spanning 4 domains showing that the issue sentence classification module can be fed to subsequent phrase sequence model as a pipelined model (Sect. 5.4).

# 4  Task II: Phrase Boundary Extraction

We now focus on task II: Given a HA and a sentence mentioning an issue, extract the issue phrase boundary. We consider a heuristic baseline and three tailored sequence models for this task.

## 4.1  Unsupervised Heuristic Baseline (UHB)

In this approach, we consider a rule based model. In our problem context, two cases arise:

- The opinion phrase is in between the head aspect and a negative sentiment constituting a part of a noun, verb, adjective or adverbial phrase (e.g., "*signal* was so weak," "loss of *connection*," etc.)
- The opinion phrase is spread out between the head aspect, a positive sentiment and a negator (e.g., "*signal* was not so strong," "couldn't get a stable *connection*")

For the first case, we extract the index of the head aspect, a proximal negative sentiment and emit the terms between them as the phrase. For the second, we sort the index of the head aspect, the proximal positive sentiment relative to the head aspect, and the proximal negator relative to the positive sentiment, and emit the phrase spanning the minimum to maximum index. We consider a 5 term window for our proximity measure (tuned via pilot experiments) and use the associated[1] sentiment lexicon of [1]. This method serves as our baseline. Although heuristic, we will see that it can discover relevant opinion phrases.

---

[1] http://www.cs.uic.edu/~liub/FBS/opinion-lexicon-English.rar.

## 4.2  Sequence Modeling

Recall from Sect. 1 that issue extraction requires us to detect a sequence of words (phrase) that directly or indirectly implies an aspect specific opinion. Let $x = (x_1, \ldots x_n)$ denote the sequence of observed words in a sentence, and let each observation $x_i$ has a label $y_i \in Y$ indicating whether $x_i$ is part of an issue phrase, where $Y = \{B, I, O\}$. The state space of labels follow the standard BIO notation as described in [35], where values taken by $y_i$ denote the *Begin*, *Inside* and *Outside* phrase alignments. Given a sentence containing an issue, $x$, the extraction task is to find the best label sequence $\hat{y}$ that describes an issue. We employ a first order Markov linear-chain CRF whose predictor takes the following form,

$$p(y|x, \Lambda) = \frac{\exp\left(\sum_i \sum_k \lambda_k f_k(y_{i-1}, y_i, x)\right)}{Z(x, \Lambda)} \tag{4}$$

where $f_k$ denotes the feature functions, $\Lambda = \{\lambda_k\}$ denotes the feature weights, and $Z(x, \Lambda)$ the normalization constant which takes the following form,

$$Z(x, \Lambda) = \sum_{y'} \left(\exp\left(\sum_i \sum_k \lambda_k f_k\left(y'_{i-1}, y'_i, x\right)\right)\right) \tag{5}$$

Given a set of training examples $\{x_j, \bar{y}_j\}$ where $\bar{y}_j$ are the correct labels, we estimate the parameters by minimizing the negative log-likelihood (NLL),

$$\Lambda = \underset{\Lambda}{\mathrm{argmin}}\left(-\sum_j \log\left(p\left(\bar{y}_j|x_j, \Lambda\right)\right) + \sum_k \lambda_k^2\right) \tag{6}$$

The term $\sum_k \lambda_k^2$ indicates $L_2$ regularization on the feature weights, $\lambda_k$. It penalizes the NLL to prevent extreme values for $\lambda_k$. We experiment with both CRF and CRF with $L_2$ regularization (CRF-L2R).

## 4.3  Linear Chain Features

We now describe the encoding of our linear chain features (LCF), $f(y_{i-1}, y_i, x)$. We use the templates of the pivot features (Table 2) with a few changes. The index $i$ for LCF in templates refers to the (current) position of any word in the sentence and not necessarily the head aspect. Further, in addition to the families defined under pivot features (Sect. 3.2), we consider latent semantic (LS) features ($\varphi_a^A$ and $\varphi_a^O$ for each head aspect, $a$). These can be very useful as encoding the position of the aspect specific sentiment terms (see Table 1) relative to the head aspect can guide phrase boundary detection. Further, each feature generated by the above defined templates is coupled with the value of the current label $y_i$ and a combination of current and previous labels $y_i, y_{i-1}$.

## 4.4  Phrase Structural Constraints

Although the above canonical CRF formulation can detect phrase boundaries, it has one key downside. Under canonical CRFs, the predictor models a large probability space (the denominator, $Z(x, \Lambda)$ in (5)) as it considers all possible sequence labelings. This unfortunately results in sparse probabilities for potential issue phrases. This is so because all issue sequences, $y = \{y_i\}$ always exhibit the pattern that exactly one component of $y$, say $y_l = B$ followed by one or more consecutive $I$, $(y_{>l} = I)$, followed by all $O$. So only candidate phrases that conform to the above pattern are *valid* phrases for issues. Motivated by previous work in [36], we consider a constrained model where $Z(x, \Lambda)$ is reduced to only *valid* patterns. This reduces the contention of incorrect sequences thereby assuaging the sparsity problem. The key difference lies in summing over only valid sequences in (5). We employ $L_2$ regularization and refer it by CRF-PSC.

# 5    Evaluation

This section details results for sentiment phrase extraction (Task II) (Sect. 4).

## 5.1    Qualitative Analysis

Table 4 shows the sample phrases extracted by UHB and CRF for two domains. Here we report the base CRF model (and not CRF-PSC) as it is representative of other CRF extensions. However, all four models are compared quantitatively in the subsequent sections. From Table 4, we note that although UHB is rule based, it can discover some phrases correctly including harder/longer phrases ("slow down or drop connection"). However, owing to non-reverence to sequential structure, it has two key downsides: (i) incorrect grammatical structures, (ii) incoherent phrase extraction (i.e., does not capture the key issue). These are overcome by CRF's sequence modeling.

## 5.2    Quantitative Results

Here we focus on per domain cross aspect analysis and also compare models across different domains. For each domain (see Sect. 3.1), we test on one head aspect by applying the sequence model learned from examples of the rest 3 head aspects for that domain. This gives us one set of results for one head aspect in a domain. Repeating it for other head aspects of that domain and averaging the performances over all head aspects for a domain, allows us to estimate the comprehensive performance of a model on a given domain. This is akin to 4-fold cross validation per aspect for each domain. Also, for inducing generative LS features in CRF models, we follow the same technique as used in task I (Sect. 3.3). We use the standard token overlap metric for evaluating the phrase boundaries. For each sentence $s \in S$, if $s_c$ and $s_p$ denote the correct and predicted expression spans (tokens) of the target phrase in $s$, then

$$recall(r) = \text{avg}_{s \in S, |s_c| \neq 0} \left( \frac{|s_c \cap s_p|}{|s_c|} \right), \ precison(p) = \text{avg}_{s \in S, |s_p| \neq 0} \left( \frac{|s_c \cap s_p|}{|s_p|} \right), \text{ and } F = \frac{2pr}{p+r}$$

where $S$ is the set of sentences in the test fold of cross-validation. From Table 5, we note the following observations:

**Table 4.** Qualitative comparison of aspect specific opinion phrases discovered by two methods (a) Unsupervised Heuristic Baseline (UHB), vs. (b) Linear chain CRF for two head aspects each across two domains. Extraction errors are *italicized in red*.

| Head Aspect | UHB | CRF |
|---|---|---|
| signal | signal was so weak<br>*good signal sometimes I don't*<br>signal losing<br>*signal it can interfere*<br>problems sending signal<br>*nothing to improve signal* | stopped broadcasting a signal<br>starts dropping signal<br>has a weak signal<br>frequently loses signal<br>signal faded very quickly<br>*weak signal like before* |
| connection | connection drop problems<br>loss of connection<br>*connection or a very poor*<br>slow down or drop connection<br>*connection refused*<br>connection dies | connection would break<br>constantly drops connection<br>*drop your connection*<br>connection would drop out<br>*hold connection steady for*<br>connections don't last |

(a) Router Domain

| Head Aspect | UHB | CRF |
|---|---|---|
| screen | screen software crashes<br>*defective but the screen*<br>*lack of a screen*<br>screen went dead<br>*damaging the screen*<br>*screen has odd* | *screen doesn't come on*<br>screen started to fade<br>screen is too small<br>screen went black<br>*screen doesn't come*<br>screen will go black |
| voice | voice is very distorted<br>*voice files and unneeded*<br>*voice doesn't disturb*<br>*disable the voice*<br>voice prompts were slow<br>*interrupt its voice* | voice has a certain grating<br>voice is very distorted<br>*delete the foreign voices*<br>voice is very shaky<br>voice is scratchy<br>voice is pretty feeble |

(b) GPS Domain

- Across each domain, we note that F1 scores of sequence models progressively improve in the following performance order CRF → CRF-L2R → CRF-PSC over the rule based baseline, UHB. This shows sequence modeling is useful in detecting issue phrases.
- Gains in F1 score of CRF, CRF-L2R over UHB are significant (see Table 5 caption). CRF-PSC further improves the result (especially recall) showing that encoding phrase structural constraints in sequence modeling for this task is useful.
- GPS and Mouse domains seem harder as the performance of all models are relatively lower than Keyboard and Router. This can be linked with the relative amount of training examples for each domain (see Sect. 3.1).

**Table 5.** Precision, recall, F1 scores of sequence models on phrase boundary detection. Gains of CRF and CRF-L2R over UHB are significant at $p < 0.01$. Gains of CRF-PSC over CRF-L2R are significant at $p < 0.02$. Significance was measured using $t$-test across all domains.

| Model | P | R | F1 | P | R | F1 |
|---|---|---|---|---|---|---|
| UHB | 67.0 | 73.7 | 70.2 | 64.4 | 68.1 | 66.2 |
| CRF | 87.9 | 76.9 | 82.0 | 86.8 | 74.8 | 80.3 |
| CRF-L2R | 88.7 | 77.1 | 82.5 | 87.6 | 75.7 | 81.2 |
| CRF-PSC | 92.0 | 80.1 | 85.7 | 90.4 | 78.3 | 83.9 |
| | (a) Router | | | (b) Keyboard | | |

| Model | P | R | F1 | P | R | F1 |
|---|---|---|---|---|---|---|
| UHB | 67.5 | 67.7 | 67.6 | 60.7 | 59.1 | 59.8 |
| CRF | 84.1 | 71.9 | 77.5 | 83.8 | 61.6 | 70.9 |
| CRF-L2R | 84.8 | 72.7 | 78.3 | 84.5 | 61.9 | 71.4 |
| CRF-PSC | 86.7 | 73.5 | 79.5 | 86.1 | 63.6 | 73.1 |
| | (c) GPS | | | (d) Mouse | | |

## 5.3   Feature Ablation

In order to assess the relative discriminative strengths of various feature families, we perform ablation analysis. We fix our model to CRF-PSC (as it performed best) and use the F1 metric. Starting from the full feature set, we drop each feature family and report the resulting performance. From Table 6, we note that each feature family has a positive contribution toward the phrase extraction task as dropping it has a statistically significant ($p < 0.05$) reduction in F1 across each domain. Dropping Latent Semantic (LS) features, POS Tags and word polarity impacts performance substantially. Especially, the LS feature family as the LS features help locate the index of the aspect specific sentiment terms in a sentence that guides the issue phrase boundary detection. Thus, we can see that all feature families are useful, with some (e.g., Latent semantics) contributing significantly to phrase boundary extraction performance.

**Table 6.** F1 scores of CRF-PSC upon feature ablation.

| Dropped Feature | Router | Keyboard | GPS | Mouse |
|---|---|---|---|---|
| None | 85.7 | 83.9 | 79.5 | 73.1 |
| POS Tags | **81.7** | **80.1** | **75.8** | 70.0 |
| Phrase Chunk Tags | 83.8 | 81.7 | 77.3 | 71.3 |
| Word Prefix/Suffix | 84.1 | 82.5 | 78.1 | 72.4 |
| Word Sent. Pol. | 82.7 | 81.8 | 77.6 | 71.7 |
| Latent Semantics | **80.1** | **77.9** | **74.6** | **68.7** |

## 5.4 Domain Adaptation

We now consider a realistic setting of applying our pipelined model to two new domains: Earphone and MP3 Players. We first train the issue sentence classifier and phrase extraction sequence models on other 4 domains. Then the issue sentences identified by the first (classification) model are fed to the previously trained sequence model (CRF-PSC) for phrase extraction on those sentences. We consider two systems and report intermediate results (prec., rec., F1, acc.) of issue sentence classification and also phrase boundary extraction in Table 7. The precision, recall, and F1 for phrase extraction were computed as defined in Sect. 5.2 whereby losses in classification task (e.g., false negative/false positive issue sentences) are accounted (as a penalty) in the recall/precision for phrase extraction respectively.

We note relatively higher precision and recall in classification performance (col 3, 4, Table 7) compared to individual domain experiments (Table 3) as now we have more training data (combination of 4 domains – Router, Keyboard, Mouse, GPS). Both systems (col 1, Table 7) find the Earphone domain harder than the MP3-Player domain. One possible reason for this could be that the domain of MP3 Players shares common aspects (e.g., screen, button) with training domains Mouse, GPS that help improve knowledge transfer. Nonetheless, on average, our system SVM (W+POS+LS+Pivot) +CRF-PSC significantly outperforms the baseline, SVM(W+POS)+UHB yielding an average F1 score of 82.2% for task I and 70.1% for task II showing decent generalization performance across new domains.

**Table 7.** Pipeline model results. Prec., Rec., F1, Acc apply to issue sentence classification. P-Seq, R-Seq, and F1-Seq apply to performance of phrase extraction on the target domain. † indicates significance at $p < 0.01$ measured via t-test.

| System | Domain | Prec | Rec. | F1 | Acc | P-Seq | R-Seq | F1-Seq |
|---|---|---|---|---|---|---|---|---|
| SVM (W +POS) +UHB | Earphone | 80.2 | 70.6 | 75.1 | 75.2 | 53.7 | 53.0 | 53.3 |
| | MP3 player | 84.1 | 73.5 | 78.4 | 80.0 | 56.3 | 55.1 | 55.7 |
| | Avg. | 82.2 | 72.1 | 76.8 | 77.6 | 55.0 | 54.1 | 54.5 |
| SVM (W +POS+LS +Pivot) + CRF-PSC | Earphone | 84.7† | 76.7† | 80.5† | 80.3 | 75.3 | 62.9 | 68.6 |
| | MP3 player | 88.6† | 79.8† | 83.9† | 84.9 | 78.9 | 65.5 | 71.5 |
| | Avg. | **86.7** | **78.3** | **82.2** | **82.6** | **77.1** | **64.2** | **70.1** |

# 6  Conclusion

This work performed an in-depth analysis of a novel task in sentiment analysis – aspect specific opinion phrase extraction. The paper focused on phrases implying negative opinions (*issues*) in the product reviews domain. First, a generative model, ME-ASM was employed for discovering the top head aspects in each domain having potential issues. Next, the sentences containing the head aspect (issue subjects) were annotated for issues including issue phrase boundaries. Discriminative large-margin and sequence models using pivot features were employed to classify issue sentences and extract issues phrase boundaries respectively. Experimental results showed that the proposed approach outperformed baseline systems and also facilitated inductive knowledge transfer across domains. The paper also contributes a new large resource of labeled aspect specific sentiment expressions across 6 domains that can serve for various sequence modeling researches/tasks in opinion mining.

# References

1. Hu, M., Liu, B.: Mining and summarizing customer reviews. In: Proceedings of 2004 ACM SIGKDD International Conference on Knowledge Discovery Data Mining - KDD 2004. ACM Press, New York, p. 168 (2004)
2. Wiebe, J., Wilson, T., Cardie, C.: Annotating expressions of opinions and emotions in language. Lang. Resour. Eval. **39**, 165–210 (2005)
3. Wang, X., McCallum, A., Wei, X.: Topical n-grams: phrase and topic discovery, with an application to information retrieval. In: 2007 Seventh IEEE International Conference on Data Mining, ICDM 2007, pp. 697–702 (2007)
4. Fei, G., Chen, Z., Liu, B.: Review topic discovery with phrases using the p{ó}lya urn model. In: COLING (2014)
5. Breck, E., Choi, Y., Cardie, C.: Identifying expressions of opinion in context. In: Proceedings of 20th International Joint Conference on Artificial Intelligence, pp. 2683–2688 (2007)
6. Choi, Y., Cardie, C., Riloff, E., Patwardhan, S.: Identifying sources of opinions with conditional random fields and extraction patterns. In: Proceedings of Conference on Human Language Technology and Empirical Methods in Natural Language Processing – HLT 2005. Association for Computational Linguistics, Morristown, NJ, USA, pp. 355–362 (2005)
7. Choi, Y., Breck, E., Cardie, C.: Joint extraction of entities and relations for opinion recognition. In: Proceedings of 2006 Conference on Empirical Methods in Natural Language Processing, pp. 431–439 (2006)
8. Li, H., Mukherjee, A., Liu, B., Si, J.: Extracting verb expressions implying negative opinions. In: Proceedings of Twenty-ninth AAAI Conference on Artificial Intelleligence (2015)
9. Berend, G.: Opinion expression mining by exploiting keyphrase extraction. In: International Joint Conference on Natural Language Processing, pp. 1162–1170 (2011)
10. Johansson, R., Moschitti, A.: Reranking models in fine-grained opinion analysis. In: Proceedings of 23rd International Conference on Computational Linguistics, pp. 519–527 (2010)
11. Yang, B., Cardie, C.: Extracting opinion expressions with semi-Markov conditional random fields. In: Proceedings of 2012 Joint Conference on Empirical Methods in Natural Language Processing and Computational Natural Language Learning, pp. 1335–1345 (2012)
12. Klinger, R., Cimiano, P.: Bidirectional inter-dependencies of subjective expressions and targets and their value for a joint model. In: Association for Computational Linguistics (Short Paper) (2013)
13. Kim, S.-M., Hovy, E.: Extracting opinions, opinion holders, and topics expressed in online news media text. In: Proceedings of the Workshop on Sentiment and Subjectivity in Text, pp. 1–8 (2006)

14. Wu, Y., Zhang, Q., Huang, X., Wu, L.: Phrase dependency parsing for opinion mining. In: Proceedings of 2009 Conference on Empirical Methods in Natural Language Processing, pp. 1533–1541 (2009)

15. Jakob, N., Gurevych, I.: Extracting opinion targets in a single- and cross-domain setting with conditional random fields. In: Proceedings of 2010 Conference on Empirical Methods in Natural Language Processing, pp. 1035–1045 (2010)

16. Kobayashi, N., Inui, K., Matsumoto, Y.: Extracting aspect-evaluation and aspect-of relations in opinion mining. In: Proceedings of 2007 Joint Conference on Empirical Methods in Natural Language Processing and Computational Natural Language Learning (EMNLP-CoNLL) (2007)

17. Johansson, R., Moschitti, A.: Extracting opinion expressions and their polarities: exploration of pipelines and joint models. In: Association for Computational Linguistics (Short Paper), pp. 101–106 (2011)

18. Yang, B., Cardie, C.: Joint modeling of opinion expression extraction and attribute classification. Trans. Assoc. Comput. Linguist. 2, 505–516 (2014)

19. Sauper, C., Haghighi, A., Barzilay, R.: Content models with attitude. In: Proceedings of the 49th Annual Meeting of the Association for Computational Linguistics: Human Language Technologies-Volume 1, pp. 350–358 (2011)

20. Yang, B., Cardie, C.: Joint inference for fine-grained opinion extraction. In: Association for Computational Linguistics, pp. 1640–1649 (2013)

21. Barzilay, R., McKeown, K.R.: Extracting paraphrases from a parallel corpus. In: Proceedings of the 39th Annual Meeting of the Association for Computational Linguistics, pp. 50–57 (2001)

22. Apidianaki, M., Verzeni, E., McCarthy, D.: Semantic clustering of pivot paraphrases. In: Conference on Language Resources and Evaluation, pp. 4270–4275 (2014)

23. Keshtkar, F., Inkpen, D.: A corpus-based method for extracting paraphrases of emotion terms. In: Proceedings of the NAACL HLT 2010 Workshop on Computational Approaches to Analysis and Generation of Emotion in Text, pp. 35–44 (2010)

24. Hasan, K.S., Ng, V.: Automatic keyphrase extraction: a survey of the state of the art. In: Proceedings of the 52nd Annual Meeting of the Association for Computational Linguistics, pp. 1262–1273 (2014)

25. Titov, I., McDonald, R.: Modeling online reviews with multi-grain topic models. In: Proceedings of the 17th International Conference on World Wide Web, pp. 111–120 (2008)

26. Brody, S., Elhadad, N.: An unsupervised aspect-sentiment model for online reviews. In: Annual Conference of the North American Chapter of the Association for Computational Linguistics, pp. 804–812 (2010)

27. Mukherjee, A., Liu, B.: Aspect extraction through semi-supervised modeling. In: Association for Computational Linguistics (2012)

28. Zhao, W.X., Jiang, J., Yan, H., Li, X.: Jointly modeling aspects and opinions with a MaxEnt-LDA hybrid. In: Proceedings of the 2010 Conference on Empirical Methods in Natural Language Processing, pp. 56–65 (2010)

29. Griffiths, T.L., Steyvers, M.: Finding scientific topics. Proc. Natl. Acad. Sci. 101, 5228–5235 (2004)

30. Chang, J., Gerrish, S., Wang, C., Boyd-graber, J.L., Blei, D.M.: Reading tea leaves: how humans interpret topic models. In: Advances in Neural Information Processing Systems, pp. 288–296 (2009)

31. Pontiki, M., Galanis, D., Papageorgiou, H., Manandhar, S., Androutsopoulos, I.: Semeval-2015 task 12: aspect based sentiment analysis. In: Proceedings of the 9th International Workshop on Semantic Evaluation (SemEval 2015), Denver, Color (2015)

32. Wilson, T., Wiebe, J., Hoffmann, P.: Recognizing contextual polarity in phrase-level sentiment analysis. In: Proceedings of the Conference on Human Language Technology and Empirical Methods in Natural Language Processing – HLT 2005, Morristown, NJ, USA. Association for Computational Linguistics, pp. 347–354 (2005)
33. Yu, H., Hatzivassiloglou, V.: Towards answering opinion questions: separating facts from opinions and identifying the polarity of opinion sentences. In: Proceedings of the 2003 Conference on Empirical Methods in Natural Language Processing, pp. 129–136 (2003)
34. Joachims, T.: Making large-scale support vector machine learning practical. In: Advance in Kernel Methods (1999)
35. Ramshaw, L.A., Marcus, M.P.: Text chunking using transformation-based learning. In: Armstrong, S., Church, K., Isabelle, P., Manzi, S., Tzoukermann, E., Yarowsky, D. (eds.) Natural Language Processing Using Very Large Corpora. Text, Speech and Language Technology, vol. 11. Springer, Dordrecht (1995). https://doi.org/10.1007/978-94-017-2390-9_10
36. Li, Y., Jiang, J., Chieu, H.L., Chai, K.M.A.: Extracting relation descriptors with conditional random fields. In: International Joint Conference on Natural Language Processing, pp. 392–400 (2011)

# Aspect Terms Extraction of Arabic Dialects for Opinion Mining Using Conditional Random Fields

Alawya Alawami[✉][iD]

University of Pittsbrugh, Pittsburgh, USA
aaa93@pitt.edu

**Abstract.** While English opinion mining has been studied extensively, Arabic fine grained opinion mining has not received much attention. This paper looks at employing conditional random fields as a supervised method to extract aspect terms which can then be employed for fine grained opinion mining. Despite the lack of Arabic Dialect NLP tools that limited the amount of improvement that can be added to the algorithm, Our analysis shows a comparable level of precision and recall to what has been achieved for English.

## 1 Introduction

The increase of the user generated content on the web has led to the explosion of opinionated text which facilitated opinion mining research. Despite the popularity of this field in English and the large number of Arabic speakers who contribute continuously to the web content, Arabic has not received much attention due to the lack of reliable Natural Language Processing (NLP) tools and the amount of research needed to complete such tasks. The work that exists in sentiment is limited to news, blogs written in Modern Standard Arabic, and some preliminary studies on few social media and web reviews written in Arabic Dialect. Further, most of the available literature is at the document level, and to the best of our knowledge, there is no work on a more fine grained level. There are many challenges that face researchers in conducting Arabic opinion mining but they are mostly attributed to the lack of NLP tools available for academic research. The work of Shoufan and Alameri [28] provides a detailed survey of the limitation of Arabic Dialectal NLP tools. In this paper, we look at using current resources to develop an opinion mining system at the aspect level.

## 2 Related Work

### 2.1 Arabic Sentiment Analysis

Sentiment Analysis, also called opinion mining, is "the field of study that analyzes people's opinions, sentiments, evaluations, appraisals, attitudes, and emotions towards entities such as products, services, organizations, individuals,

© Springer International Publishing AG, part of Springer Nature 2018
A. Gelbukh (Ed.): CICLing 2016, LNCS 9624, pp. 211–220, 2018.
https://doi.org/10.1007/978-3-319-75487-1_16

issues, events, topics, and their attributes" [19, p. 7]. Sentiment analysis can be done on three levels: document, sentence, entity and feature/aspect level analysis. Document-level sentiment analysis is based on the assumption that each document holds opinion about one entity. Thus, sentiment classification at this level classifies the overall opinion about the entity. Sentence-level goes a little deeper to classify each sentence as positive, negative or neutral. A more fine-grained analysis is done at the aspect level where the system finds what the user really likes or dislikes about the entity. It is also called feature-based sentiment or attribute-based sentiment analysis. Arabic sentiment analysis has been studied at the document and sentence level by many authors [1–3, 11, 23]. There is a preliminary work on establishing the basis for Arabic Aspect based Sentiment analysis. The International Workshop on Semantic Evaluation 2016 (SemEval-2016) has established a data set for the task of aspect sentiment analysis with a sub-task for aspect extraction [24]. Al-Smadi, Qawasmeh, Talafha, and Quwaider have recently established annotation for a book review dataset (LABR) [6]. The data set has a baseline for aspect extraction tasks but to the best of our knowledge is no published work on applying English aspect extraction methods on similar Arabic reviews. In this work, we aim at a preliminary work on aspect extraction for Arabic restaurant reviews using Conditional Random Fields.

### 2.2   Aspect Extraction

Aspect-based sentiment analysis refers to a sub field of sentiment analysis that recognizes each phrase that contains sentiments and extracts the aspects which the they refer to. In this task, the system look for aspect related to the entity being discussed. There are many approaches that have been used to extract aspects in reviews. They can be divided into two groups: NLP methods that are either frequency based methods or methods that employ opinion/target relations. The second group relies on statistical approaches such as topic modeling and supervised learning.

Frequency based method employs NLP tools in the extraction of the aspects. This approach is based on the observation that users who reviewed the same entity use a vocabulary that converges [14]. Part of speech tagger and dependency parser are used to find nouns and noun phrases. Then the most frequent nouns are kept based on a threshold determined experimentally. This approach was used in [14]. This approach is challenging to be implemented with Arabic dialect text because of the lack of reliable NLP tools such as part of speech taggers and sentence parsers.

The second method makes use of opinion and target relations. This approach makes use of the idea that every sentiment word that exists in the document belongs to the nearest noun or noun phrase [14]. This approach was improved 22% in [25]. Another approach uses a dependency tree to identify this relationship [31]. Unfortunately, both approaches require the use of a good parser and both tools don't exist yet for Arabic dialect.

Another approach employs Supervised learning: sequential labeling techniques. The problem is treated as an information extraction task. It relays on training data for manually labeling aspect and/or sentiment terms in the dataset. The most dominant approach in this field is sequential labeling techniques such as Hidden Markov Chain [26] and Conditional Random Fields [17]. The work in [16] employs a lexicalized HMM model to learn patterns that are used in extracting aspects and opinion. Another work in [15] used Conditional Random Field (CRF) in multiple domains in attempt to overcome the problem of domain dependency. The highest precision achieved for single domain reviews is 0.79 and the highest recall 0.66 for the movies domain. Their approach had a lower recall in other domains and around 0.6 precision. CRF was also used in other works such as [10, 20].

The last method used in the literature is topic modeling. Topic modeling is an unsupervised learning method that became very popular in recent years. It is used to discover topics in a large collection of text. Topic modeling is a generative model which assumes that each document in the collection is a mixture of topics and each topic is a probability distribution over words. The results of the method is a set of words along with the corresponding probability distribution where each set represents a topic. While topic models have been used to extract topics from collection of text, they have been adapted to model sentiment and topics (aspects) based on the assumption that each opinion has a target.

There are two basic topic models: Probabilistic Latent Semantic Analysis (pLSA) [13] and Latent Dirichlet Allocation (LDA) [8]. The work in [22] built aspect sentiment mixture model based on pLSA. This mixture model consists of three sub-models: aspect model, positive model and negative model. Most of the other proposed models are based on LDA [9, 18, 29].

Topic modeling however has many limitations. First, it needs a large volume of data and heavy tuning to reach a good result. Many of those models use Gibbs sampling which produces different results in each run. Consequently, researchers need to spend a significant amount of time in parameter tuning. Finally, topic modeling is very useful in finding global aspects among datasets but may fail in finding the most frequent local aspect which is more relevant to the entity being reviewed. In this paper, we focus on experimenting with conditional random fields as a supervised method for aspect extraction. We plan on exploring with other approaches in the future.

## 3   Challanges

There are two types of challenges faced when conducting opinion mining in Arabic: some challenges are related to the field of opinion mining and others are related to the Arabic language.

## 3.1   Challenges Related to the Nature of Opinions

Opinion mining faces a number of problems. First, identifying the set of words that can identify the polarity in the text is generally hard. Many adjectives are domain dependent (ex: The battery life is long vs takes a long time to boot). Similarly, sentiment and subjectivity are context sensitive. The sentence "reading the book was very enjoyable" is negative in the movie review context but positive in the book review context. Finally, some opinions are expressed in idioms and not individual words (e.g. cost an arm and leg).

## 3.2   Challenges Related to Nature and Usage of Arabic Language

The Arabic Language is divided into three types: Classical Arabic, Modern Standard Arabic (MSA) and Dialect Arabic (DA). The Arabic language has many different dialects that are used in informal daily communications but are not standardized or taught formally in schools. While there are a variety of dialects, MSA is the only one standard form that is widely recognized and formally taught in schools. MSA is based on Classical Arabic which is the language of the Qur'an (Muslims' holy book) [12]. The MSA is not a native language of any country and it is largely different from dialect forms. MSA has been studied extensively and many NLP tools are available for it. Unfortunately, most of the web contents are written in the dialectal form that hasn't been studied as much. To the best of our knowledge there exist no reliable NLP tools for it. Arabic is a morphologically rich language (MRL) where most of the information regarding syntax and relation is expressed at the word level. English on the other hand has much less information expressed at the word level. The Arabic base form of a word can lead to thousands of surface forms while in English a verb would have three different forms so using those forms in a lexicon corpus will lead to data sparseness in Arabic while in English there is a high chance that the three terms will be present in text [2,5]. This suggests using a compact form of the word along with POS tagging to overcome the problem of data sparseness. Albraheem and Al-Khalifa in [7] recommend stemming to reduce the size of the lexicon corpus. On the other hand, Rushdi in [27] does not recommend the use of stemming for the task of opinion mining.

The second challenge related to Arabic is the lack of widely available Arabic corpora [2], the lack of Arabic lexicon that can be used for sentiment and the lack of publicly available and reliable NLP tools such as Part of speech tagger and dependency parser. Opinion Mining has gained a lot of popularity with the rise of social media. The amount of user-generated content available for researchers has increased tremendously but Arabic opinion mining has received little attention compared to English language. While the amount of data provides many opportunities for researchers, the data is highly unstructured and contains misspellings, abbreviation, repetitions "sooooo Happyyyyyy" and concatenated words [7]. The use of informal form on the web content leads to many problems. Arabic users encode Arabic words in roman alphabet for example the word (الحرب) which means "war" is written as "Al7arb" or "Al 7arb" and there is no defined standards about how this is done so each word would have different variations depending on the user [5]. Also, Albraheem and Al-Khalifa in [7] indicated that different words with different meanings have the same root which can impact sentiment analysis if the wrong root have different sentiment.

## 4   Data Set

Jeeran.com is a popular Arabic site for services' reviews. It can easily be crawled and the data are structured. Spam is eliminated from the reviews by the website administrators. Categories in this site vary widely e.g.: shopping, restaurants, travel, financial services, etc. The data used in this research were crawled between May and July 2014. The website covers cities in Saudi Arabia, Jordan, United Arab Emirates, Egypt, Kuwait, and Qatar. We limited the crawls to cities in the Kingdom of Saudi Arabia to restrict the wide variety of dialects available in the web site. Most of our dataset contains Gulf Dialect Arabic, in addition to some Modern Standard Arabic (MSA) and English reviews. Our main goal for crawling was to collect a sample of reviews and not necessarily all of them as the web site is growing rapidly. Reviews were collected on a per-place basis and limited to Riyadh city. The total reviews crawled were 6485 for the restaurant domain. A randomly selected set of 500 reviews were then annotated by two graduate students who are native Arabic speakers. The annotators were asked to label the explicit aspect discussed in the review. The inter-annotator agreement (Kappa = 0.9). The source of disagreement between annotators is due to misspelling or confusion between parts of speeches. Since Arabic Part of Speech consist of three main categories: Nominal, Verb and Particle, we asked the annotators to label the data with those Super POS tags (SPOS):

- Nominal: Nouns (Noun, Proper Noun), Derived nouns (Adjectives, Imperative verbal noun), Personal Pronoun, Demonstrative pronouns أسماء الإشارة, Relative pronouns الأسماء الموصولة, Possessive determiners الضمائر المتصلة, adverbs (time and location adverbs).
- Verb
- Particle: Prepositions such as (from من, to إلى, in على), Subjunctive particles such as (أن، لن), Negative particles such as (/ا، ـ), Jussive particles such as (/ا، ـ)
- Adjective: sub-category of Nominal
- Other: this category was used for other sequence of characters (not words) that does not fit the other categories.

A summary of data statistics is provided in Table 1

Table 1. Data set description.

| Reviews | Tokens per review | Explicit aspects | Average aspect per review |
|---------|-------------------|------------------|---------------------------|
| 500 | 10313 | 2263 | 2.9 |

## 5   Preprocessing

Arabic dialects are known to be noisy data with different spelling for many words. We followed the same preprocessing highlighted in [4] which increases the accuracy for sentiment analysis on Arabic dialect.

– Remove punctuation, diacritics, and any non-characters.
– Remove any lengthening of the word for example: بــــاب is reduced to باب
– Normalize the Arabic letter (Alef آ إ أ) to bare form (ا).
– Replace ى with ي.
– Replace ة with ه.
– If a word starts with ء, replace it with ا.
– Replace ؤ with و.
– Replace ئ and ۓ with ي.

## 6   Aspect Tagging Using Conditional Random Fields (CRFs)

The conditional random field is a discriminative probabilistic model [17]. It has been used for a wide variety of NLP problems such as part of speech tagging, information extraction, speech recognition and named entity recognition. Similarly, it has been applied to aspect extraction tasks because it can be viewed as a sequence-labeling task using CRFs.

CRF is a generalized form of Hidden Markov Model. Formally, given a sequence of tokens $x = \{x_1, x_2, x_3, .., x_n\}$, we need to generate a sequence of labels $y = \{y_1, y_2, y_3, .., y_n\}$ for each token $x$. For our purpose, the set of possible labels are ASPECT and NON-ASPECT. The aim of CRF model is to find $y$ that maximizes $p(x \mid y)$ for the given sequence.

$$p(y \mid x) = \tfrac{1}{z(x)} exp \left( \sum_t \sum_k \lambda_k f_k(y_{t-1}, y_t, x) \right)$$

$$z(x) = \sum_{y \in y} exp \left( \sum_t \sum_k \lambda_k f_k(y_{t-1}, y_t, x) \right)$$

$$\lambda_k \text{ is the weight of } f_k.$$

A more detailed description of CRFs can be found in [17].

## 7   CRF Features

In choosing the features, we were faced with the challenge of the lack of dialectal Arabic parsers and stemmers which facilitate the implantation of a wide variety of features such as word distance, short dependency path, and stem, which have been shown to produce good CRF results [15]. We therefor resorted to applying the following features. Token (tk) which is the actual string of the current token as a feature. We experimented with the token with and without any preprocessing mentioned in Sect. 5. The second feature used is fine grain part of speech tags (POS) which is produced by Stanford Arabic POS tagger [30]. The last feature used was the Super POS Tags (SPOS) produced by our annotators in the dataset.

## 8  Evaluation

Since aspect extraction can be treated as information extraction problem, we used precision and recall to evaluate our method. These measures have also been used in [9,14,16] to compare similar systems. For our task, precision can be defined as the ratio of the aspect retrieved to the number of aspect and non-aspect term retrieved. (How many of the returned aspects were correct). Similarly, recall is the ratio of the aspect term retrieved to the total aspect terms in the reviews (How many of the correct aspects were returned). F-measure is the harmonic mean of precision and recall. F-measure is used to evaluate the performance of different models and to compare systems. For our task, F-measure will be used to compare the various combination of feature we experiment with. The following equations describe how they are calculated:

$$Precision(P) = \frac{TP}{TP + FP}$$

$$Recall(R) = \frac{TP}{TP + FN}$$

$$F - measure = \frac{P * R}{P + R}$$

## 9  Results

Our dataset was divided into 70% training and 30% testing. We used Mallet, a java open source implementation of CRF [21]. Table 2 shows the results of our experiments. We used various combinations of the features. Although the combination of those features produced comparable results, we can see that the highest precision was produced by using the super POS tags along with preprocessed tokens which points to the need of having reliable NLP tools. The highest recall was produced by the combination of all features. The best performing method out of these was using tokens with Stanford Part of Speech tags. The performance is comparable to aspect extraction for English Language [15]. The results could be improved using a dependency parser or a stemmer. Since there have been a recent attention from the research community to address the problem of the lack of Arabic dialect NLP tools [28], we plan on visiting the problem again in the future.

Table 2. Results summary.

|  | Features | Precision | Recall | F-measure |
|---|---|---|---|---|
| No preprocessing | Tk, POS | 0.683 | 0.728 | 0.70 |
|  | Tk, SPOS | 0.648 | 0.68 | 0.66 |
|  | Tk, POS, SPOS | 0.65 | 0.7 | 0.67 |
| Preprocessing | Tk, POS | 0.68 | 0.63 | 0.65 |
|  | Tk, SPOS | 0.69 | 0.678 | 0.69 |
|  | Tk, POS, SPOS | 0.65 | 0.74 | 0.69 |

# 10   Conclusion

Aspect extraction is a subtask of the process of aspect-based opinion mining. In this paper we attempted to tackle the problem of aspect extraction for Dialectal Arabic utilizing current available resources. We found out that the current resources produced comparable results to what have been achieved for English Language. We plan on revisiting the problem whenever more NLP tools for Arabic dialect are available.

# References

1. Abbasi, A., Chen, H., Salem, A.: Sentiment analysis in multiple languages: feature selection for opinion classification in web forums. ACM Trans. Inf. Syst. (TOIS) **26**(3), 12 (2008)
2. Abdul-Mageed, M., Diab, M.T.: Awatif: a multi-genre corpus for modern standard Arabic subjectivity and sentiment analysis. In: LREC, pp. 3907–3914 (2012)
3. Abdul-Mageed, M., Korayem, M.: Automatic identification of subjectivity in morphologically rich languages: the case of Arabic. In: Proceedings of the 1st Workshop on Computational Approaches to Subjectivity and Sentiment Analysis (WASSA), Lisbon, pp. 2–6 (2010)
4. Abdulla, N.A., Al-Ayyoub, M., Al-Kabi, M.N.: An extended analytical study of Arabic sentiments. Int. J. Big Data Intell. 1 **1**(1–2), 103–113 (2014)
5. Ahmed, S., Pasquier, M., Qadah, G.: Key issues in conducting sentiment analysis on Arabic social media text. In: 2013 9th International Conference on Innovations in Information Technology (IIT), pp. 72–77. IEEE (2013)
6. Al-Smadi, M., Qawasmeh, O., Talafha, B., Quwaider, M.: Human annotated Arabic dataset of book reviews for aspect based sentiment analysis. In: 2015 3rd International Conference on Future Internet of Things and Cloud (FiCloud), pp. 726–730, August 2015
7. Albraheem, L., Al-Khalifa, H.S.: Exploring the problems of sentiment analysis in informal Arabic. In: Proceedings of the 14th International Conference on Information Integration and Web-based Applications & Services, pp. 415–418. ACM (2012)
8. Blei, D.M., Ng, A.Y., Jordan, M.I.: Latent Dirichlet allocation. J. Mach. Learn. Res. **3**, 993–1022 (2003)

9. Brody, S., Elhadad, N.: An unsupervised aspect-sentiment model for online reviews. In: Human Language Technologies: The 2010 Annual Conference of the North American Chapter of the Association for Computational Linguistics, pp. 804–812. Association for Computational Linguistics (2010)

10. Choi, Y., Cardie, C., Riloff, E., Patwardhan, S.: Identifying sources of opinions with conditional random fields and extraction patterns. In: Proceedings of the Conference on Human Language Technology and Empirical Methods in Natural Language Processing, pp. 355–362. Association for Computational Linguistics (2005)

11. Elhawary, M., Elfeky, M.: Mining Arabic business reviews. In: 2010 IEEE International Conference on Data Mining Workshops (ICDMW), pp. 1108–1113. IEEE (2010)

12. Habash, N.Y.: Introduction to Arabic natural language processing. Synthesis Lect. Hum. Lang. Technol. **3**(1), 1–187 (2010)

13. Hofmann, T.: Probabilistic latent semantic indexing. In: Proceedings of the 22nd Annual International ACM SIGIR Conference on Research and Development in Information Retrieval, pp. 50–57. ACM (1999)

14. Hu, M., Liu, B.: Mining and summarizing customer reviews. In: Proceedings of the Tenth ACM SIGKDD International Conference on Knowledge Discovery and Data Mining, pp. 168–177. ACM (2004)

15. Jakob, N., Gurevych, I.: Extracting opinion targets in a single-and cross-domain setting with conditional random fields. In: Proceedings of the 2010 Conference on Empirical Methods in Natural Language Processing, pp. 1035–1045. Association for Computational Linguistics (2010)

16. Jin, W., Ho, H.H., Srihari, R.K.: A novel lexicalized HMM-based learning framework for web opinion mining. In: Proceedings of the 26th Annual International Conference on Machine Learning, pp. 465–472. Citeseer (2009)

17. Lafferty, J., McCallum, A., Pereira, F.C.: Conditional random fields: probabilistic models for segmenting and labeling sequence data (2001)

18. Lin, C., He, Y.: Joint sentiment/topic model for sentiment analysis. In: Proceedings of the 18th ACM Conference on Information and Knowledge Management, pp. 375–384. ACM (2009)

19. Liu, B.: Sentiment analysis and opinion mining. Synthesis Lect. Hum. Lang. Technol. **5**(1), 1–167 (2012)

20. Liu, B., Hu, M., Cheng, J.: Opinion observer: analyzing and comparing opinions on the web. In: Proceedings of the 14th International Conference on World Wide Web, pp. 342–351. ACM (2005)

21. McCallum, A.K.: Mallet: a machine learning for language toolkit (2002)

22. Mei, Q., Ling, X., Wondra, M., Su, H., Zhai, C.: Topic sentiment mixture: modeling facets and opinions in weblogs. In: Proceedings of the 16th International Conference on World Wide Web, pp. 171–180. ACM (2007)

23. Omar, N., Albared, M., Al-Shabi, A.Q., Al-Moslmi, T.: Ensemble of classification algorithms for subjectivity and sentiment analysis of Arabic customers' reviews. Int. J. Adv. Comput. Technol. **5**(14), 77 (2013)

24. Pontiki, M., Galanis, D., Papageorgiou, H., Androutsopoulos, I., Manandhar, S., AL-Smadi, M., Al-Ayyoub, M., Zhao, Y., Qin, B., Clercq, O.D., Hoste, V., Apidianaki, M., Tannier, X., Loukachevitch, N., Kotelnikov, E., Bel, N., Jimnez-Zafra, S.M., Eryiit, G.: SemEval-2016 task 5: aspect based sentiment analysis. In: Proceedings of the 10th International Workshop on Semantic Evaluation. SemEval 2016. Association for Computational Linguistics, San Diego, California, June 2016

25. Popescu, A.M., Nguyen, B., Etzioni, O.: Opine: extracting product features and opinions from reviews. In: Proceedings of HLT/EMNLP on Interactive Demonstrations, pp. 32–33. Association for Computational Linguistics (2005)
26. Rabiner, L.R., Juang, B.H.: An introduction to hidden Markov models. ASSP Mag. IEEE **3**(1), 4–16 (1986)
27. Rushdi-Saleh, M., Martín-Valdivia, M.T., Ureña-López, L.A., Perea-Ortega, J.M.: Oca: opinion corpus for Arabic. J. Am. Soc. Inf. Sci. Technol. **62**(10), 2045–2054 (2011)
28. Shoufan, A., Alameri, S.: Natural language processing for dialectical Arabic: a survey. In: Proceedings of the Second Workshop on Arabic Natural Language Processing, pp. 36–48. Association for Computational Linguistics, Beijing, China, July 2015. http://www.aclweb.org/anthology/W15-3205
29. Titov, I., McDonald, R.: Modeling online reviews with multi-grain topic models. In: Proceedings of the 17th International Conference on World Wide Web, pp. 111–120. ACM (2008)
30. Toutanova, K., Klein, D., Manning, C.D., Singer, Y.: Feature-rich part-of-speech tagging with a cyclic dependency network. In: Proceedings of the 2003 Conference of the North American Chapter of the Association for Computational Linguistics on Human Language Technology, vol. 1, pp. 173–180. Association for Computational Linguistics (2003)
31. Zhuang, L., Jing, F., Zhu, X.Y.: Movie review mining and summarization. In: Proceedings of the 15th ACM International Conference on Information and Knowledge Management, pp. 43–50. ACM (2006)

# Large Scale Authorship Attribution of Online Reviews

Prasha Shrestha(✉), Arjun Mukherjee, and Thamar Solorio

Department of Computer Science, University of Houston,
Houston, TX 77004, USA
pshrestha3@uh.edu, {arjun,solorio}@cs.uh.edu

**Abstract.** Traditional authorship attribution methods focus on the scenario of a limited number of authors writing long pieces of text. These methods are engineered to work on a small number of authors and generally do not scale well to a corpus of online reviews where the candidate set of authors is large. However, attribution of online reviews is important as they are replete with deception and spam. We evaluate a new large scale approach for predicting authorship via the task of verification on online reviews. Our evaluation considers a large number of possible candidate authors seen to date. Our results show that multiple verification models can be successfully combined to associate reviews with their correct author in more than 78% of the time. We propose that our approach can be used to slow down or deter the number of deceptive reviews in the wild.

## 1 Introduction

With almost everything being online, there has been what can only be called a deluge of social media data. For example, Amazon has 244 million active users [1] and Yelp had 83 million unique visitors per month in the fourth quarter of 2015 [2]. Much of the previous research on traditional authorship attribution deals with a small set ($\leq$10) of authors [3,4]. Some more recent researchers have worked on a relatively larger number of authors [5,6]. But in order to keep up with the increase in online data, we need scalable approaches that can work for companies like Amazon and Yelp.

Given a set of authors, authorship attribution (AA) is the task of figuring out who, if any of them is the actual author of a piece of text. AA is not a new field as it has been around from the start of 19th century [7]. But AA on social media data is fairly new. Among various forms of social media data, it is specially important to focus on AA for the domain of product reviews because it contains a lot of fake reviews and spam [8,9]. A single user might create multiple accounts in order to write outstanding reviews for a product in order to promote it. In the same way, negative product reviews could be written for the sake of hindering a competitor's product [10]. Such users are likely to have multiple accounts, a legitimate account and one or more fake accounts. AA can help to detect if

© Springer International Publishing AG, part of Springer Nature 2018
A. Gelbukh (Ed.): CICLing 2016, LNCS 9624, pp. 221–232, 2018.
https://doi.org/10.1007/978-3-319-75487-1_17

two accounts belong to the same author or not and with some modifications verification can also help detect when more than one author is writing reviews under the same user id. AA methods can also be extended to include background author detection to predict not only if a text has been written by an author but also if the text is written by none of the authors.

Most traditional authorship attribution tasks are performed on long texts such as books. There has been a growing interest on authorship attribution of social media data, but most of it has been focused on blog data [5]. Reviews are different from blogs in that reviews are generally shorter. Also, the topic of the reviews will be different from product to product whereas a blogger tends to be focused on a fixed set of topics. It is not clear that even internet scale attribution [11] can be adapted to online reviews. Authorship attribution on reviews is more challenging as:

- the number of candidate authors is very high (ten to hundreds of magnitude larger than most existing work)
- reviews are usually very short as compared to books or blogs
- even the reviews written by the same author differ in topic because users typically write one review per product purchased
- while spamming, authors deliberately try to alter their writing to avoid getting caught.

Our work addresses AA on larger author sets and on noisy/short review texts. In this paper, we use two new datasets of online reviews that can be used to develop and benchmark new approaches for authorship attribution at a large scale. One of the datasets consists of product reviews from Amazon and the other one contains reviews from Yelp. We first present a verification technique through which we will perform the attribution of reviews. Our contributions are: first we present a large review dataset that can be used to benchmark author verification, attribution and background (out-of-set) author detection. We also present our approach to the problem of verification and attribution in datasets having a large author set.

## 2   Related Work

Our work follows that in Koppel and Winter [5], where they reason that any AA problem can be broken into a set of Author Verification (AV) problems [5]. Conversely, they also show that an AV problem can also be converted in to a many-candidates problem. In order to obtain these candidates, their AV system generates impostors from documents of the same genre as a given document. If two documents are consistently more similar to each other than to the impostors across different ngram feature sets, then they are likely to be written by the same author. By using this method, they obtain more than 90% optimal accuracy for 500 pairs of long, 2000-word documents.

Qian et al. [6] also perform AA on online reviews. They first generate document based features and convert them to various text similarity features between two reviews and train on these features. Their document based features consist of various frequency based features, writing density features and vocabulary richness features. They then define their own formulas that use these document feature values of two reviews in order to produce features representing similarity between the two reviews. Their AA method computes these similarity features for a test review and individual reviews from all candidate authors. They train their model on these similarity features to predict if the two reviews are written by a single author or by different authors. They obtain scores from this model for various reviews of an author and combine these scores to obtain the actual author. Their best method obtained in a range of around 40% to 83% accuracy for a range of 2 to 100 candidate authors.

Seroussi et al. [12] used topic modeling to generate author representation. Their main idea is based on distinguishing between document/topic specific words and author specific words. They experimented with five different datasets covering a wide variety of topics, number of authors and amount of text per author. Along with count judgements, blog posts and emails, there are also two datasets on Internet Movie database (IMDb) movie reviews. They compared their system with a baseline of token frequency counts used as features to train a SVM model. Their proposed system outperformed the baseline in four out of the five datasets. The baseline beats their system by a close margin on a movie review dataset consisting of reviews from prolific IMDb users. However, their system beats the baseline on the reviews dataset consisting of reviews from more than 22,000 random IMDb users.

Stamatatos [7] observe that AA researchers have used various lexical, syntactic and semantic features to capture the style of an author [7]. Among these features, lexical features are the most prominently employed features in authorship attribution systems and especially character n-gram based features have given good performance. Character n-grams are also among our features. They also distinguish between two types of attribution methods: profile based approach where all instances of an author's writings are combined to create a single profile for an author and instance based approach where each instance of an author's writings are treated separately. We use the instance based method and choose to combine the results from these instances instead.

Eder [14] try to analyze how the size of an author's text relates to the performance of the task of authorship attribution by using the Delta method [13]. They perform their experiments on separate datasets of English, German, Polish, Latin, Greek and Hungarian language novels. They find that for all languages, the performance on the AA task generally improves with larger amount of text. But after certain length, which the author found to be around 5,000 words, the performance starts to saturate and might even decrease a little. Since we are dealing with reviews, it is very rare for our data to reach this number. The average number of words per review in our dataset is only 233.46. They also analyze the difference in performance when using consecutive blocks of text versus randomly chosen bag of words. Interestingly, they found that using the randomly chosen bag of words gives better performance. They also tried ngrams with

**Table 1.** Number of authors with $\geq x$ reviews

| Criteria ($\geq x$) | Amazon | Yelp hotels | Yelp restaurants |
|---|---|---|---|
| 50 | 8,171 | 2,450 | 3,174 |
| 25 | 15,772 | 3,064 | 5,322 |
| 1 | 123,967 | 5,132 | 35,392 |

variable numbers of $n$. They found that the performance steadily decreases with increasing $n$. This finding aligns with previous researches [7] and we also limit our $n$ to less than 4.

## 3    Review Datasets

We have two review datasets for performing AA on a large number of authors and/or on online data. The Amazon dataset contains the reviews written by the authors for different products on Amazon while the Yelp dataset comprises restaurant and hotel reviews. Along with the text of the reviews, there are other attributes of the reviews in both datasets that might be useful for AA. The Amazon reviews contain information about the reviewer id, date posted, star rating and helpful count of the reviews. The Yelp reviews also contain reviewer id, star rating and date posted, along with useful, funny and cool counts given to the reviews. The Amazon reviews were posted between June 1996 to October 2012 and the Yelp reviews between January 2006 to September 2012.

As shown in Table 1, the datasets have a very large number of authors with the Amazon dataset having the highest number of authors. Apart from having a large number of candidate authors, on average there are only around 240 tokens per review for Amazon dataset and 106 tokens per review for the Yelp dataset. Some of the authors only have a few reviews to their name. It is very hard to perform verification and attribution when there is insufficient text. As can be seen in Table 1, only a fraction of the authors have 50 reviews or more, although the number is still high. Our final datasets only consist of the prolific authors, who have written at least 50 reviews each. This also helps us to winnow out most of the sockpuppet accounts since usually sockpuppet accounts only contain a small number of reviews [10]. As such, we use reviewer and author interchangeably.[1]

## 4    Attribution via Verification

When the number of authors is very high, as in our case, it is unrealistic to try to train a single, combined, multi-class AA model for all the authors. It is much more manageable to break down the problem into smaller pieces. Thus, we approach the problem of AA on a large set of authors by training individual verification models or verifiers for each of the authors. We begin with a set of $n$

---

[1] The datasets can be obtained at http://ritual.uh.edu/resources/.

authors $A = \{a_1, ..., a_n\}$ and their set of documents (reviews) $D = \{D_1, ..., D_n\}$. Here $D_i$ is a set of reviews written by $a_i$. We extract features from these reviews and then train $n$ separate verifiers in which each of the reviews acts as an instance of an author's writing. An author verifier performs a single task of predicting if a given review is in fact written by the same author or not.

In order to train these verifiers we perform review data selection in the following way: For each of the authors we define their own set of positive and negative reviews. Here we use positive and negative in the sense that all the reviews belonging to an author are positive reviews for him/her and reviews written by all other authors in our dataset comprise the negative reviews for that author. If we are training a verification model for author $a_i$, then all of $a_i$'s reviews are his/her positive reviews. All the reviews from the other $n - 1$ authors are possible candidates for negative reviews. For each verifier, we create a balanced training and test dataset. We will discuss later how this is not a limitation for our approach. Since there are a large number of candidates for negative reviews, we need to perform negative review selection.

---

**Algorithm 1. Training and Test Set Selection**

---

Assume A is a set of authors and $|A| = n$
For each author $a_i \in A$:
$S^+_{a_i} = D_i$
that is, all documents from author $a$ are positive instances in the verification case for author $a$.
Split $S^+_{a_i} = \{S^+_{a_i\_train}, S^+_{a_i\_test}\} = \{80 : 10\}$ from $S^+_{a_i}$

Generate negative samples as follows:
**Negative Open Set(NOS):**
$S^-_{a_i\_train} \subset \{s | s \in D_j, j \neq i\}, j = 1, ..., n/2;$
$S^-_{a_i\_test} \subset \{s | s \in D_j, j \neq i\}, j = n/2 + 1, ..., n$

**Negative Random Set(NRS):**
$S^-_{a_i} = D - D_i$
$S^-_{a_i\_train} \subset S^-_{i_a}$ ; $S^-_{a_i\_test} \subset S^-_{a_i}$
subject to the constraints:
$|S^-_{a_i\_train}| = |S^+_{a_i\_train}|$
$|S^-_{a_i\_test}| = |S^+_{a_i\_test}|$
$S^-_{a_i\_train} \cap S^-_{a_i\_test} = \phi$

---

The training and test set selection is shown in Algorithm 1 and described here. Selection of the positive set of documents is straightforward. We take 80% of all of the reviews of an author as the positive training set and 10% as the positive test set. We hold out 10% for an analysis that we will explain in Sect. 5.1. For the selection of negative set of reviews, we tried two different methods. Our motivation behind this is to find out whether having reviews in the training set from those authors whose reviews are also present in the test set makes a difference or not. We describe the two different methods for the negative set selection below.

## 4.1   Negative Open Set (NOS)

In the NOS method, there is no overlap between the authors whose reviews appear in the training set and the authors whose reviews are present in the test set. We first divide the $n - 1$ authors into two sets of $\lfloor (n - 1)/2 \rfloor$ authors each. Then we select $|S^+_{a_i\_train}|$ reviews from among the reviews written by the authors in the first set as the negative training reviews and $|S^+_{a_i\_test}|$ reviews from those written by the authors in the second set as the negative test reviews.

## 4.2   Negative Random Set (NRS)

In the NRS method, there might be an overlap between the authors whose reviews are present in the training set and the authors whose reviews are in the test set. Here, both the negative training and test reviews can come from all $n - 1$ authors. We will select $|S^+_{a_i\_train}|$ reviews for training and $|S^+_{a_i\_test}|$ for test set from the pool of all $n - 1$ authors' reviews.

## 4.3   Features

We use the features that have already been tested in previous AA research or in related fields such as intrinsic plagiarism detection [15] and anomaly detection [16]. We also performed a simple preprocessing step of removing any URLs and converting the text to lowercase before extracting the features. We did not use other preprocessing steps such as stopword removal because stopwords can be important in AA and they are a part of some of our features as we will describe below. Our features include the following:

**Lexical:** These consist of word unigrams and character unigrams, bigrams, and trigrams. An author is likely to show preferences for certain words more than other authors, which can help distinguish him/her. Similarly, character ngrams capture the writing style of an author and have been used by previous researches successfully for authorship attribution [4,5,7].

**Syntactic:** We extract part of speech (POS) tags as well as chunks by using the tagger and chunker available in Apache OpenNLP. We try to model the style of the author by using POS and chunk unigrams, bigrams, and trigrams.

**Writing Density:** Authors can also be distinguished by their writing density. For our experiments, these include the average number of characters per word, the average number of syllables per word, and the average number of words per sentence.

**Readability:** The complexity of a piece of text varies from author to author. We use standard readability indices to measure this. We use Flesch-Kincaid grade level [17,18], Gunning fog index [19], Yule's K measure, and Honore R measure [15] as our features.

**Part of Speech (POS) Trigram Diversity:** This is yet another feature that captures the style of an author [16]. It is quite simply the number of unique POS trigrams normalized by the total number of POS trigrams in the text.

**Stopword Frequency:** This is another stylistic feature and it measures the proportion of stopwords in an author's text. It is measured as the total number of stopwords in a piece of text divided by the overall count of the words in the text.

**Average Word Frequency Class:** This is a measure of how likely an author is to use unique words that are not frequently used in a language [20]. A word is assigned frequency class according to how likely it is to appear in a corpus. Meyer zu Eissen and Stein [20] used the Sydney Morning Herald Corpus to obtain the word frequency class for about one hundred thousand words. They range from 0 for 'the', the most common word in the corpus to 19 for very uncommon words. We took the frequency class for every word of an author's text present in these one hundred thousand words and then normalized it by the total number of words in the text.

**Table 2.** Author verification results showing macro-averaged values

| Dataset | Method | Positive class | | | Negative class | | | Accuracy |
|---------|--------|-----------|--------|---------|-----------|--------|---------|----------|
|         |        | Precision | Recall | F-score | Precision | Recall | F-score |          |
| Amazon reviews | NOS | 0.8674 | 0.9165 | 0.8846 | 0.9193 | 0.8423 | 0.8696 | 87.94 |
| Amazon reviews | NRS | 0.8600 | 0.9162 | 0.8806 | 0.9187 | 0.8331 | 0.8639 | 87.47 |
| Yelp hotel | NOS | 0.8517 | 0.8921 | 0.8678 | 0.8915 | 0.8358 | 0.8579 | 86.39 |
| Yelp hotel | NRS | 0.8636 | 0.8916 | 0.8732 | 0.8927 | 0.8495 | 0.8656 | 87.05 |
| Yelp restaurant | NOS | 0.8595 | 0.8757 | 0.8617 | 0.8804 | 0.8449 | 0.8557 | 86.03 |
| Yelp restaurant | NRS | 0.8567 | 0.8799 | 0.8628 | 0.8825 | 0.8401 | 0.854 | 86.00 |

## 4.4   Authorship Attribution

To perform AA on a review, we first pass it to all our author verifiers. Each of the author verifiers gives us a value $(P_i)$ representing the probability of a review being written by that author. We use logistic regression as the classifier for all of our verifiers. The probability from our verifiers is simply the probability of a review belonging to an author's class as given by logistic regression. For a review document $r$, we will have a set $P = \{P_1, ... P_n\}$ of such probabilities, where $P_i = P(a_i|r)$ i.e. the probability that the author of review $r$ is author $a_i$. With these probabilities, we perform the final attribution in two ways.

**Attribution per review:** We simply choose the author whose model gives the highest probability value for a given review as the author for that review. In other words, if $max(P) = P_i$ for review $r$, then $a_i$ is the author of review $r$.

**Collective attribution per unknown author:** The second method combines results from the first method. All online reviews have reviewer ids associated with them. We perform AA on the reviews having same thereviewer id

(R) collectively. The intuition behind this method is that the author who is most frequently predicted by the attribution per review method as the actual author for these reviews is likely the actual author for all of these reviews. This is a valid formulation because reviewer ids are not related to author identities and doing this is similar to aggregating all the reviews of an author and performing AA on the aggregated text. We take the predictions from the attribution per review method for all the reviews having the same reviewer id R and use them as votes. The author obtaining the highest number of votes is the author of all these reviews as shown in Eq. 1.

$$voting\_pred(R) = \arg\max_{a_i}(count(attribution\_per\_review(r_j) = a_i)),$$
$$\text{where } r_j \epsilon R \tag{1}$$

## 5   Results and Analysis

We performed three separate experiments on 1,000 authors from the Amazon dataset and 500 authors each from the Yelp hotel and restaurant reviews datasets. An author verifier has separate precision, recall, f-score and accuracy values. Table 2 shows the macro-averaged values for all these four metrics across the author verifiers. As evidenced by the results, our system performs fairly well in this task. We achieve similar f-scores for both the positive and negative classes. But the precision and recall values are different. The precision for the negative class is higher while the recall is higher for the positive class. This might happen because all of the positive training examples belong to the same author but the negative reviews belong to different authors.

The NOS method gives us slightly better results, although the difference is negligible. This shows that even if we use the reviews from an author in both negative training and test, it does not affect the final results. Since we obtain good results in this task, it makes sense for us to use these results in order to perform AA.

### 5.1   Performance on the Cumulative Rank Curve

Before AA, we perform a separate experiment to assess how good our author verifiers actually are. We do this on the 10% of the positive reviews that we held out. For each review $r$, we obtain the probabilities as mentioned in Sect. 4.4 from all author verifiers ($\{P_1, ...P_n\}$) and then rank these probabilities from the highest to the lowest. If the author of $r$ is $a_i$, we then get the rank obtained by the probability $P_i$. The rank value will be in the range from $1...n$. After we obtain this rank value for all of the reviews, we get a count of the number of reviews that fall under each rank. From these counts we finally calculate the cumulative probabilities of getting these ranks. The cumulative probability for rank $k$ is the probability of an author verifier of author $a_i$ obtaining rank $\leq k$ for reviews written by $a_i$.

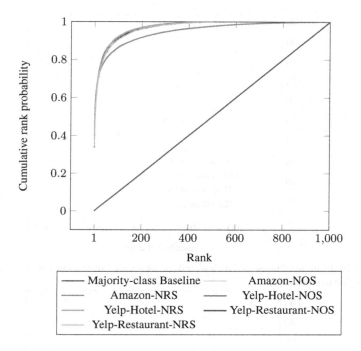

**Fig. 1.** Cumulative rank curve for the ranks of the actual authors

This probability represents how likely our author verifiers are to get the highest probability values for reviews written by their author. For example, in an ideal scenario, all author verifiers would produce the highest probability for reviews written by their author and the probability for rank 1 would be 1. We create a plot of these probabilities against the ranks. The results from this experiment is very interesting as shown in Fig. 1. The cumulative probability for rank 1 is 0.4245 for the Amazon dataset. This means that for more than 40% of the reviews, the author verifier of the actual author of the reviews obtained the highest probability for those reviews among all 1000 verifiers. For the Yelp hotel and restaurant datasets, the cumulative probability for rank 1 is 0.3568 and 0.3406 respectively. The number steadily increases such that when we get to rank 50, the cumulative probability for all datasets is higher than 0.80. This means that for more than 80% of the reviews, the verifier for the actual author is ranked 50 or higher. This provides further motivation for performing AA by using these verifiers. Another observation from this experiment is that again, NOS and NRS have similar results with overlapping curves.

## 5.2 Authorship Attribution

The results for the attribution task are shown in Fig. 2. As mentioned before, the trained models for all of the experiments remain the same. We performed this

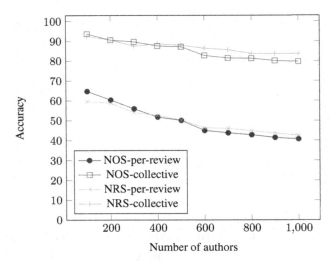

(a) Results from the attribution per review and the collective attribution per unknown author method for Amazon reviews

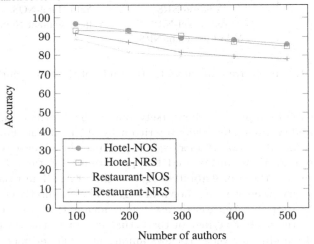

(b) Results from the collective attribution per unknown author method for Yelp reviews

**Fig. 2.** F1-scores for the AA task for NOS and NRS

test on different number of authors for up to a 1000 and 500 authors for Amazon and Yelp datasets respectively. For the attribution per review method, there were ties in the highest probability values for a very small percentage (close to 0.1%) of the reviews. In these cases, if the actual author was among the authors tied for the first place, we counted that as a correct prediction. Otherwise, we selected an author from the tied authors randomly as the predicted author.

As seen for the Amazon dataset (Fig. 2a), the results from the attribution per review method is not very satisfactory and they deteriorate quickly when we increase the number of authors. This is understandable as with growing contentions of candidate authors, the probability space in this method becomes sparse. But for the collective attribution method, all of the results are either close to or above 80%. The method scales well and is fairly stable even when the number of authors is increased. Like the attribution per review method, there were ties in this case too, again for a small percentage (close to 0.8%) of the authors. We used the same method as before to resolve the ties. Again, the NOS and NRS methods give similar performance even for this AA task, showing that having an already seen author in the test set does not make much difference. We also found that the accuracy is positively correlated with the total number of reviews of an author. The dataset for this experiment is very imbalanced, with less than 0.5% of the documents belonging to the positive class for an author and we still perform well. This resembles a real world scenario where there will be a lot less text written by an author as compared to all of the text written by everybody else.

Since the collective attribution method worked so well for the Amazon dataset, we tried the same method on the Yelp datasets as well. The results can be seen in Fig. 2b. For 100 authors in the Yelp Hotel dataset, we obtain 96.79% accuracy in the NOS setting, which is very high considering the number of authors. Other observations are similar to what we found for the Amazon dataset. The method scales well and the results stay either above or near the 80% mark. Again, the NOS and NRS methods perform similarly here as well. Although both datasets are reviews, they have differences between them because the review subjects are very different. But our method performs consistently well for Amazon product reviews as well as for Yelp restaurant and hotel reviews. This also shows that our method works well across different topics.

## 6   Conclusion and Future Work

We have presented a method to perform authorship attribution on reviews in a large scale. We proposed to use individual author verifiers to solve the AA problem on a large number of authors. We were able to build a scalable approach especially geared towards large datasets. Our method was able to obtain good accuracy even when the number of authors is large and the loss in accuracy as we go from 100 authors to 1000 authors was not very high. The first reason why our AA method worked well is that our author verifiers themselves perform very well. They can very well be used separately for authorship verification. The second reason why our approach worked is the collective attribution method we applied to combine the results from our individual verifiers. Our final system gives a good performance and is very well suited for attribution of online reviews. Not only that, the large review datasets that we have used in this paper can be useful as benchmark datasets for future large scale authorship attribution research. In our ongoing work, we are looking at extending our method in order to perform background author detection.

**Acknowledgment.** This research has been partially supported by NSF award No. 1462141.

# References

1. Amazon Media Group: About Amazon Media Group. http://www.amazon.com/b?ie=UTF8&node=8445211011 (2015). Accessed 07 Feb 2016
2. The Yelp Blog: About. http://www.yelp.com/about (2015). Accessed 07 Feb 2016
3. Kešelj, V., Peng, F., Cercone, N., Thomas, C.: N-gram-based author profiles for authorship attribution. In: Proceedings of the Conference of Pacific Association for Computational Linguistics, PACLING, vol. 3, pp. 255–264 (2003)
4. Koppel, M., Schler, J.: Authorship verification as a one-class classification problem. In: Proceedings of the Twenty-First International Conference on Machine Learning, ICML 2004, p. 62. ACM, New York (2004)
5. Koppel, M., Winter, Y.: Determining if two documents are written by the same author. J. Assoc. Inf. Sci. Technol. **65**, 178–187 (2014)
6. Qian, T.Y., Liu, B., Li, Q., Si, J.: Review authorship attribution in a similarity space. J. Comput. Sci. Technol. **30**, 200–213 (2015)
7. Stamatatos, E.: A survey of modern authorship attribution methods. J. Am. Soc. Inf. Sci. Technol. **60**, 538–556 (2009)
8. Lappas, T.: Fake reviews: the malicious perspective. In: Bouma, G., Ittoo, A., Métais, E., Wortmann, H. (eds.) NLDB 2012. LNCS, vol. 7337, pp. 23–34. Springer, Heidelberg (2012). https://doi.org/10.1007/978-3-642-31178-9_3
9. Jindal, N., Liu, B.: Opinion spam and analysis. In: Proceedings of the 2008 International Conference on Web Search and Data Mining, WSDM 2008, pp. 219–230. ACM, New York (2008)
10. Mukherjee, A., Liu, B., Glance, N.: Spotting fake reviewer groups in consumer reviews. In: Proceedings of the 21st International Conference on World Wide Web, WWW 2012, pp. 191–200. ACM, New York (2012)
11. Narayanan, A., Paskov, H., Gong, N., Bethencourt, J., Stefanov, E., Shin, E., Song, D.: On the feasibility of internet-scale author identification. In: 2012 IEEE Symposium on Security and Privacy (SP), pp. 300–314 (2012)
12. Seroussi, Y., Zukerman, I., Bohnert, F.: Authorship attribution with topic models. Comput. Linguist. **40**, 269–310 (2014)
13. Burrows, J.: Delta: a measure of stylistic difference and a guide to likely authorship. Literary Linguist. Comput. **17**, 267–287 (2002)
14. Eder, M.: Does size matter? authorship attribution, small samples, big problem. Digit. Scholarsh. Humanit. **30**, 167–182 (2015)
15. Stein, B., Lipka, N., Prettenhofer, P.: Intrinsic plagiarism analysis. Lang. Resour. Eval. **45**, 63–82 (2011)
16. Guthrie, D., Guthrie, L., Allison, B., Wilks, Y.: Unsupervised anomaly detection. In: Proceedings of the 20th International Joint Conference on Artifical Intelligence, IJCAI 2007, San Francisco, CA, USA, pp. 1624–1628. Morgan Kaufmann Publishers Inc. (2007)
17. Flesch, R.: A new readability yardstick. J. Appl. Psychol. **32**, 221–223 (1948)
18. Kincaid, J.P., Fishburne Jr., R.P., Rogers, R.L., Chissom, B.S.: Derivation of new readability formulas (automated readability index, fog count and flesch reading ease formula) for navy enlisted personnel. Technical report (1975)
19. Gunning, R.: The Technique of Clear Writing (1952)
20. Meyer zu Eissen, S., Stein, B.: Genre classification of web pages. In: Biundo, S., Frühwirth, T., Palm, G. (eds.) KI 2004. LNCS (LNAI), vol. 3238, pp. 256–269. Springer, Heidelberg (2004). https://doi.org/10.1007/978-3-540-30221-6_20

# Discovering Correspondence of Sentiment Words and Aspects

Geli Fei[1(⊠)], Zhiyuan (Brett) Chen[1], Arjun Mukherjee[2],
and Bing Liu[1]

[1] Department of Computer Science, University of Illinois at Chicago,
Chicago, IL, USA
gfei2@uic.edu, czyuanacm@gmail.com, liub@cs.uic.edu
[2] Department of Computer Science, University of Houston, Houston, TX, USA
arjun@cs.uh.edu

**Abstract.** Extracting aspects and sentiments is a key problem in sentiment analysis. Existing models rely on joint modeling with supervised aspect and sentiment switching. This paper explores unsupervised models by exploiting a novel angle – correspondence of sentiments with aspects via topic modeling under two views. The idea is to split documents into two views and model the topic correspondence across the two views. We propose two new models that work on a set of document pairs (documents with two views) to discover their corresponding topics. Experimental results show that the proposed approach significantly outperforms strong baselines.

## 1 Introduction

Finding topics in documents is an important problem for many NLP applications. One of the effective techniques is topic modeling. Over the years, many topic models have been proposed which are extensions or variations of the basic models such as PLSA [11] and LDA [3]. These models typically take a set of documents and discover a set of topics from the documents. They may also produce some other types of auxiliary information at the same time using joint modeling. Topic models are also widely used in sentiment analysis for finding product or service aspects (or features) and opinions about them.

In this paper, we focus on discovering aspect specific sentiments. For example, in the restaurant reviews, positive sentiment words such as "tasty" and "delicious" are usually associated with the *food* aspect, and positive sentiment words "friendly" and "helpful" are about the *staff* aspect. Modeling aspects (topical words) with aspect specific sentiments (opinion words) is very useful. First, it can improve opinion target detection [16, 23]. For example, in the sentence, *"I had sushi in Sakura's on Washington St, which was really tasty,"* there is a positive opinion indicated by "tasty". However, it is not easy to determine the target of the positive opinion – "sushi", "Sakura's" or "Washington St." However, if we know that "sushi" appears in a food topic and "tasty" appears in its corresponding sentiment topic, then we know that "tasty" is about "sushi", and not about "Sakura's" or "Washington St". Second, the results can also help co-reference resolution. For instance, in *"I had sushi in Sakura's*

© Springer International Publishing AG, part of Springer Nature 2018
A. Gelbukh (Ed.): CICLing 2016, LNCS 9624, pp. 233–245, 2018.
https://doi.org/10.1007/978-3-319-75487-1_18

*on Washington St. It was really tasty*", it is not easy to know what "it" refers to. Topical correspondence can resolve "it" to "sushi".

This paper proposes two paired topic models to discover correspondence between two views, i.e., source view and target view, which correspond to aspect and sentiment, respectively.

The first model is a directional model, ASL (Aspect Sentiment LDA) that explicitly models topic correspondence via conditioning target topics on source topics. ASL does not consider target topics while inducing source topics, which is a weakness.

This motivates us to propose the second model IASL (Integrated Aspect and Sentiment LDA), which additionally improves the topic discovery in source and target documents. Unlike ASL, IASL is not directional. It can improve ASL because in inducing the source topics it also considers words in the target documents and vice versa. Existing models and ASL are unable to do this. Technically, IASL merges the source and target views into one virtual document and uses an indicator variable to tell the model whether a word is from the source or the target view during inference. This merging allows improved joint modeling that yields much better results.

However, to apply the proposed models, we need document pairs, but each review or review sentence is not a pair. We split each review or review sentence into two parts: a sub-document consisting of sentiment words and a sub-document consisting of non-sentiment words. Thus, given a large number of online reviews, the proposed ASL and IASL models can find the corresponding aspect and sentiment topics. Our experimental results using reviews from both hotel and restaurant domains show that the proposed models are highly effective and significantly outperform relevant existing models PLTM [19] and ASUM [13].

## 2  Related Work

Topic models have been applied to numerous applications and domains. In sentiment analysis, they have been used to find aspect and/or sentiment terms/words. Related work in this thread include those in e.g., [5, 7, 15, 18, 21, 22, 26, 27, 29]. Although many can model aspects and sentiments jointly, they need supervised aspect/sentiment labels, and none of these above models work on documents that come with two views. One representative model, ASUM in [13], models both aspect and sentiments in reviews. It can, to some extent, extract aspects that are specific to sentiment labels. Thus, we use it as one of our baselines in the experiments.

Although in the context of sentiment analysis, we are not aware of any topic models that work on documents pairs, there are several models in other application fields that have resemblances to our dual view aspect-sentiment topic models. These include the works in [1, 2, 4, 10, 14, 17, 18, 28] that also work on pairs, e.g., authors and papers, images and tags, etc. In [1], Corr-LDA was proposed to work on image and caption pairs. In the model, the generation of caption words is conditioned on image regions. The condition part has some similarity to our ASL model, but ASL's target topics are conditioned on source topics. The model in [4] used unaligned multilingual documents to find shared topics and to pair related documents across languages. Rosen-Zvi et al. [24] proposed an author-topic model, which takes a collection of author lists and their articles as input to find each author's topical interests. Mimno et al. [19] proposed the PLTM model which can take a set of tuples, where each tuple has several corresponding documents. Compared to our IASL model, none of these existing models makes use of word co-occurrences in the target documents in inducing the source topics and vice versa. Among these existing models, PLTM is the closest in function to our models, so we consider it as another baseline model.

## 3   ASL Model

Given a set of document pairs (where the document is split into two views – source view: containing non-sentiment words and target view containing sentiment words), ASL finds a set of topics (called source topics) from the source documents and at the same time, for each source topic, finds the corresponding target topic in the target documents. The modeling of topic correspondence in ASL is directional, meaning that the discovery of target topics is conditioned on source topics. We use a directed link from the source topic node to the target topic node to explicitly model the dependency of target topics on source topics. Also, the topic discovery in the source documents of ASL is independent and is the same as LDA.

ASL assumes that there are $K$ source topics in the source documents and $O$ target topics in the target documents. It posits $K$ source topic-word distributions in the source view, $O$ target topic-word distributions in the target view, and $K$ source-target topic distributions, i.e., each of the $K$ source topics has a distribution over $O$ target topics, i.e., a distribution over distributions as each target topic is already a distribution over words. Through parameter estimation, we aim to discover the $K$ source topics and the most probable source-topic-specific target topics.

### 3.1   Generative Process

We follow standard notations. $\varphi_{(s),k}$ and $\varphi_{(t),o}$ denote the source and target topic distributions $k$ and $o$ respectively. $\theta_{(s),d_{(s)}}$ denotes the document-topic distribution of source document $d_{(s)}$. $\theta_{(t),k,d_{(t)}}$ denotes the source-document topic $k$-specific topic distribution of target document $d_{(t)}$. $z$ and $w$ represent the standard latent topic and observed word in respective views. $\alpha$, $\beta$ denote the respective Dirichlet hyperparameters. Subscripts $(s)/(t)$ indicate whether a variable lies in the source or target views of a document. We detail the generative process of ASL (Fig. 1) as follows:

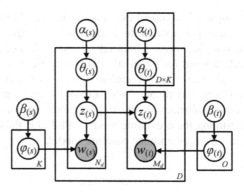

**Fig. 1.** The plate notation of ASL

The generative process of ASL for a corpus of $D$ document pairs is as follows:

1. For each source document $d_{(s)} \in \{1, \ldots, D\}$:

   Draw document-topic distribution $\theta_{(s),d_{(s)}} \sim Dir\ (\alpha_{(s)})$.

2. For each target document $d_{(t)} \in \{1, \ldots, D\}$:

   For each source topic $k \in \{1, \ldots, K\}$:
   Draw document-source topic-target topic distribution $\theta_{(t),k,d_{(t)}} \sim Dir(\alpha_{(t),k})$.

3. For each source topic $k \in \{1, \ldots, K\}$:

   Draw topic-word distribution $\varphi_{(s),k} \sim Dir\ (\beta_{(s)})$.

4. For each target topic $o \in \{1, \ldots, O\}$:

   Draw topic-word distribution $\varphi_{(t),o} \sim Dir\ (\beta_{(t)})$.

5. For each source document $d_{(s)} \in \{1, \ldots D\}$:

   For each word $w_{(s),d_{(s)},n} \in d_{(s)}$:
   Choose a source topic $z_{(s),d_{(s)},n} \sim Multi\ (\theta_{(s),d_{(s)}})$.
   Choose a source word $w_{(s),d_{(s)},n} \sim Multi\ (\varphi_{(s),z_{(s),d_{(s)},n}})$.

6. For each target document $d_{(t)} \in \{1, \ldots D\}$:

   For each word $w_{(t),d_{(t)},m} \in d_{(t)}$:
   Choose a target topic $z_{(t),d_{(t)},m} \sim Multi\ (\theta_{(t),d_{(t)},z_{(s),d_{(s)},*}})$.
   Choose a target word $w_{(t),d_{(t)},m} \sim Multi\ (\varphi_{(t),z_{(t),d_{(t)},m}})$.

Note that in step 6, we use $z_{(s),d_{(s)},*}$ which denotes the dependency of all target topics on source topic $s$, as while sampling a target topic word, we need to consider all the (source) topic assignment of words in the corresponding source document.

## 3.2    Inference

We use Gibbs sampling for inference. In sampling, we sample source topics to generate stable source topics first and then sample target topics. As source topics are independent of target topics, we can sample them independently and wait till source topics stabilize to shape target topics by reducing total Gibbs iterations and autocorrelation. The source topic sampling is similar to LDA:

$$p\left(z_{(s),d_{(s)},v}\middle|z_{-(s),d_{(s)},v},w_{(s)},\alpha_{(s)},\beta_{(s)}\right)$$

$$\propto \left(c_{z_{(s)},d_{(s)},v,d_{(s)},*}^{-(d_{(s)},v)} + \alpha_{(s),z_{(s)},d_{(s)},v}\right)$$

$$\times \frac{\left(c_{z_{(s)},d_{(s)},v,*,w_{(s)},d_{(s)},v}^{-(d_{(s)},v)} + \beta_{(s),w_{(s)},d_{(s)},v}\right)}{\sum_{i=1}^{I}\left(c_{z_{(s)},d_{(s)},v,*,i}^{-(d_{(s)},v)} + \beta_{(s),i}\right)}$$

After the source topics stabilize through Gibbs sampling iterations, we sample target topics in the second step using the following Gibbs sampler.

$$p\left(z_{(t),d_{(t)},y}\middle|z_{-(t),d_{(t)},y},z_{(s)},w_{(t)},\alpha_{(t)},\beta_{(t)}\right) \propto$$

$$\prod_{k=1}^{K}\frac{c_{d,k,*,z_{(t)},d_{(t)},y,*}^{-(d_{(t)},y)} + \alpha_{(t),k,z_{(t)},d_{(t)},y}}{\sum_{o=1}^{O}\left(c_{d,k,*,o,*}^{-(d_{(t)},y)} + \alpha_{(t),k,o}\right)} \times \frac{c_{z_{(t)},d_{(t)},y,*,w_{(t)},d_{(t)},y}^{-(d_{(t)},y)} + \beta_{(t),w_{(t)},d_{(t)},y}}{\sum_{j=1}^{J}\left(c_{z_{(t)},d_{(t)},y,*,j}^{-(d_{(t)},y)} + \beta_{(t),j}\right)}$$

$c_{k,u,*}$ is the number of times topic $k$ is assigned to words in document $u$; $c_{k,*,a}$, the number of times topic $k$ is assigned to word $a$; $c_{d,k,*,o,*}$, the number of times source topic $k$ is assigned to source words while target topic $o$ is assigned to target words in document pair $d$. $c^{\neg(u,v)}$ discounts word $v$ in document $u$.

# 4    IASL Model

In ASL, the discovery of source topics is independent of target topics. However, the target view can help shape better source topics. Hence, we now propose an integrated model IASL to jointly model both source and target views. Specifically, it merges the source and target documents in each document pair into a virtual document. An indicator variable, $y$ (which is observed) is used to indicate whether a word is from the source or the target. By doing so, we effectively increase the word co-occurrence, which consequently results in better topics and better topic correspondence. IASL assumes that there are $K$ one-to-one correspondence of topics between the source and target. Each source (target) topic is a distribution over the vocabulary in the source (target) documents. The plate notation is shown in Fig. 2.

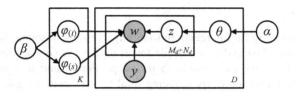

**Fig. 2.** The plate notation of IASL

## 4.1 Generative Process

The generative process of IASL is as follows:

For each source and target topic $k \in \{1, \ldots, K\}$:
  draw $\varphi_{(s),k} \sim Dir\,(\beta)$
  draw $\varphi_{(t),k} \sim Dir\,(\beta)$
For each virtual document $d$:
  draw $\theta_d \sim Dir\,(\alpha)$.
For each word $w_{d,u} \in d$:
  draw its topic $z_{d,u} \sim Multi\,(\theta_d)$.
  if $y$ indicates $w_{d,u}$ being sentiment word:
    draw word $w_{d,u} \sim Multi\,(\varphi_{(s),z_{d,u}})$.
  if $y$ indicates $w_{d,u}$ being aspect word:
    draw word $w_{d,u} \sim Multi\,(\varphi_{(t),z_{d,u}})$.

## 4.2 Inference

The Gibbs sampler for assigning topics to words in the documents takes the following form. $K$ is the number of topics and $I$ is the size of vocabulary in the source and target documents combined.

$$p\left(z_{d,u}|z_{-(d,u)}, y, w\right) \propto \frac{c_{k,d,*}^{-(d,u)} + \alpha}{\sum_{k=1}^{K}\left(c_{k,d,*}^{-(d,u)} + \alpha\right)}$$

$$\times \frac{c_{(s),k,*,w_{d,u}}^{-(d,u)} + \beta}{\sum_{i=1}^{I}\left(c_{(s),k,*,w_i}^{-(d,u)} + \beta\right)} \times \frac{c_{(t),k,*,w_{d,u}}^{-(d,u)} + \beta}{\sum_{i=1}^{I}\left(c_{(t),k,*,w_i}^{-(d,u)} + \beta\right)}$$

where $z_{d,u}$ and $z_{-(d,u)}$ represent topic assignment for $w_{d,u}$ in the virtual document $d$ and topic assignment for words except $w_{d,u}$ respectively. $c_{k,d,*}^{-(d,u)}$ represents the number of times topic $k$ assigned to words in document $d$ except $w_{d,u}$. $c_{(s),k,*,w_i}^{-(d,u)}$ and $c_{(t),k,*,w_i}^{-(d,u)}$ represent the number of times topic $k$ assigned to source word $w_i$ except $w_{d,u}$ and the number of times topic $k$ assigned to target word $w_i$ except $w_{d,u}$ respectively.

# 5  Experiments

We now evaluate the proposed ASL and IASL models. Note that although ME-LDA [29] and ME-SAS [22] discover aspect specific sentiments, they belong to the family of supervised topic models and need labeled training data, seed sets, and also do not model documents in two views that place them in a different problem setting than ours. Instead, we compare our model with PLTM [19] which is unsupervised and designed for document tuples (similar to our dual views) and ASUM [13] which is a representative unsupervised joint aspect and sentiment model. The evaluation task is to use all the models to discover corresponding aspect and sentiment topics in online reviews. ASUM cannot automatically separate aspects from sentiment words, we will post-process its results in order to compare it with our models. The implementation of both PLTM and ASUM was obtained from their authors.

**Datasets:** We use two datasets: one hotel review dataset from TripAdvisor.com, and one restaurant review dataset from Yelp.com. Our hotel data contains 101,234 reviews and 692,783 sentences, and our restaurant data contains 25,459 reviews and 278,179 sentences. We ran the Stanford Parser to perform sentence detection and lemmatization. The data domain name is removed since it co-occurs with most words in the dataset, leading to an undesirable high overlap among topics/aspects.

**Sentences as Documents:** Standard topic models tend to produce topics that correspond to global properties of products instead of their aspects when applied to reviews [26]. We take the approach in [5, 7] and divide each review into sentences and treat each sentence as an independent document.

**Document Pairs:** Treating each sentence as a document, we split each sentence into the source and target pair. The source contains all non-sentiment (aspect) words and the target contains all sentiment words. The sentiment lexicon in [12] was used to find sentiment words in each sentence. We will post-process ASUM results using the same lexicon. Sentences with no sentiment words were ignored.

**Parameter Settings:** In all our experiments, posterior inference was drawn after 2000 Gibbs iterations with a burn-in of 200 iterations. Following [9], we fix the Dirichlet priors of our models as follows: for all document-topic distributions, we set $\alpha = 50/K$, where K is the number of topics. For ASL, we use the same number of topics for both the source and the target documents. And for all topic-word distributions, we set $\beta = 0.1$. We also experimented with other settings of these priors and did not notice much difference.

Setting the number of topics/aspects in topic models is tricky as it is difficult to know the exact number of topics in a corpus. While non-parametric Bayesian approaches [25] exist for estimating the number of topics, it's not the focus of this paper. We empirically set the number of topics for both source and target documents to 15. Although 15 may not be optimal, since all models use the same number, there is no bias against any model.

**Baseline Model Settings:** The input of PLTM is the same as that of our proposed models. ASUM works on a single set of documents and assigns a topic to each sentence. So we treat one review as a document for ASUM input. Also, the output of ASUM is a set of topics called senti-aspects, which are jointly defined by aspects and sentiment labels. In this paper, we only consider positive and negative sentiment, and set 15 topics under each sentiment label (positive or negative). Since ASUM does not separate sentiment words from aspects, we separate them using the sentiment lexicon of [12] during post-processing. For both baselines, we set parameters as mentioned in their original papers.

**Table 1.** Hotel data: aspect and sentiment topic summary results

|  | PLTM | | | ASL | | | IASL | | |
|---|---|---|---|---|---|---|---|---|---|
|  | *P@5* | *P@10* | *P@15* | *P@5* | *P@10* | *P@15* | *P@5* | *P@10* | *P@15* |
| Aspect | 0.82 | 0.75 | 0.72 | 0.87 | 0.83 | 0.81 | 0.90 | 0.85 | 0.78 |
| Sentiment | 0.78 | 0.76 | 0.72 | 0.77 | 0.78 | 0.77 | 0.92 | 0.86 | 0.78 |
| Average | **0.80** | **0.76** | **0.72** | **0.82** | **0.81** | **0.79** | **0.91** | **0.86** | **0.78** |

**Table 2.** Restaurant data: aspect and sentiment topic summary results

|  | PLTM | | | ASL | | | IASL | | |
|---|---|---|---|---|---|---|---|---|---|
|  | *P@5* | *P@10* | *P@15* | *P@5* | *P@10* | *P@15* | *P@5* | *P@10* | *P@15* |
| Aspect | 0.89 | 0.80 | 0.78 | 0.87 | 0.85 | 0.83 | 0.91 | 0.86 | 0.83 |
| Sentiment | 0.85 | 0.79 | 0.74 | 0.96 | 0.91 | 0.84 | 0.95 | 0.9 | 0.87 |
| Average | **0.87** | **0.80** | **0.76** | **0.92** | **0.88** | **0.84** | **0.93** | **0.88** | **0.85** |

**Table 3.** Topic coherence of sentiment topics and aspect topics

|  | (a) Hotel | | | | (b) Restaurant | | | |
|---|---|---|---|---|---|---|---|---|
|  | PLTM | ASUM | ASL | IASL | PLTM | ASUM | ASL | IASL |
| Aspect | −380.04 | −668.03 | −378.67 | −296.77 | −382.13 | −772.45 | −387.27 | −262.52 |
| Sentiment | −448.83 | −596.36 | −420.85 | −247.55 | −438.34 | −667.24 | −404.04 | −381.15 |
| Average | **−414.43** | **−632.19** | **−399.76** | **−272.16** | **−410.235** | **−719.84** | **−395.655** | **−321.835** |

## 5.1  Results of Manual Labeling of Topics

Although statistical measures, such as perplexity, KL-divergence and topic coherence, have been used to evaluate topic models, they may not always conform to human notions of semantics [6] or an actual sentiment analysis application. Thus, we first report evaluation by human judges. In the next sub-section, we also report topic coherence results to evaluate our models statistically.

Topic labeling was done by two judges. The labeling was carried out in two stages sequentially: (1) labeling of topics (both aspect and sentiment word topics), and (2) labeling of topical words in each topic. After the first stage, agreement among judges was computed, and then the two judges discussed about the disagreed topics to reach a consensus. They then moved to the next stage to label the top ranked words in each topic. We measured inter-judge agreement using Kappa [8]. The Kappa score for topic labeling and topical words labeling were 0.853 and 0.917 respectively indicating strong agreements in the labeling.

Since we want to find topic correspondence, during labeling, in the first stage we first determine whether an aspect topic is good or bad and then for each aspect topic, we label its corresponding sentiment topic as good or bad. By good or bad, we mean whether a topic is coherent enough to have a distinct topic label from its top ranked words. If the judges could not label a topic due to its incoherence, it was labeled as bad; otherwise good. Further, if a real-world aspect for an aspect topic could not be identified (i.e., it was a bad topic), its corresponding sentiment topic was also not labeled and discarded. We choose this labeling scheme as our objective is to find topic correspondence, and also because that ASL is directional from source to target topics.

Unlike other models that generate sentiment topics, ASUM generates a set of senti-aspects, each of which is specific to a single sentiment label. Since it is hard to match positive senti-aspects with negative senti-aspects that are about the same aspects, we cannot directly compare the results of ASUM with other models. Hence, in this section, we will only label and compare topics of the other three models. We however compare all four models using statistical evaluation metrics in the next section.

**Results and Discussions:** We use *precision@n* (*P@n*) as our metric. *P@n* is the precision at rank *n* for a topic. The summaries (average over all topics) from both datasets are given in Tables 1 and 2. We note that as ASL is a conditional model, it tries to ensure the source (aspect in our case) is not negatively affected by the goal of finding corresponding topics from the target. Hence, its aspect and sentiment topics are better than those of PLTM. IASL is able to achieve much better results for both the source (aspect) and target (sentiment) topics due to leveraging word collocations across both the source and target views. The improvements are particularly pronounced for the first 5 and the first 10 words (i.e., P@5 and P@10), which is important as in practical applications they are more trustworthy. To measure the significance of the improvements, we conducted paired t-tests. The tests showed that both ASL and IASL outperform PLTM significantly ($p < 0.01$). IASL also outperformed ASL significantly ($p < 0.05$).

**Number of bad topics:** Our labeling also gives the number of good and bad topics. For the hotel domain, PLTM has four bad aspect topics. For both ASL and IASL, there are only three bad aspect topics. For the restaurant domain, all three models give four bad topics. For both domains, for each good aspect topic, its corresponding sentiment topic is always good.

## 5.2    Statistical Evaluation

We use the *topic coherence* (TC) metric in [20] that correlates well with human semantics. Table 3 shows the TC values for aspect and sentiment topics for each dataset (averaged over all topics). Note that the output of ASUM is a set of senti-aspects (a set of words defined by both sentiment and aspect). So using 15 topics and 2 sentiment polarities, ASUM generates 30 senti-aspects. Again, ASUM cannot separate sentiment words from non-sentiment words. In order to compute topic coherence for both aspect topics and sentiment topics, we use the sentiment lexicon to separate each senti-aspect into sentiment topics and aspects. In computing TC values, we used the top 15 words from each topic, which is also the number of top topical words labeled in our human evaluation. TC gives a negative value and a better model should have a higher TC value. Also, TC gives a higher score if less frequent words appear at the top ranks of a topic. So in the statistical evaluation, we remove general seed words from each senti-aspect before separating aspects from sentiment topics in order to be fair to ASUM. This improved the average TC of sentiment topics of ASUM by more than 50. From Table 3, we can see that IASL has highest TC values, which dovetails with human labeling results in Tables 1 and 2. Its values are markedly better than PLTM and ASUM. ASL is also better than PLTM in all cases except a slight drop for aspect topics for restaurant.

Significance testing using paired t-test on the results of TC showed that IASL significantly improves PLTM, ASUM and ASL ($p < 0.03$). ASL also improves ASUM significantly ($p < 0.05$) but does not improve PLTM significantly ($p = 0.11$). The difference between the results here and those from the human labeling results is understandable because although TC correlates with human labeling well, they are not exactly same.

In summary, we conclude that both human evaluation and statistical evaluation show that the proposed models are more effective than the baseline models PLTM and ASUM. Also, IASL improves upon the other models by a large margin.

## 5.3    Case Study

This section shows some example aspect and sentiment topic pairs labeled by our human judges. Words in red are labeled as wrong by the judges. For the same reason as in human evaluation section, ASUM generates a set of senti-aspects, each of which is specific to a single sentiment label. It is thus hard to match positive senti-aspects with negative senti-aspects that are about the same aspects and to directly compare the results with other models. Table 4 lists two sets of topic pairs discovered by PLTM, ASL and IASL models. We can see that ASL performs better than PLTM in source topic (aspect) detection, and IASL outperforms both ASL and PLTM in both aspect topics and sentiment topics. The improvement is due to the proposed modeling tailored for document pairs.

**Table 4.** Example topics extracted by PLTM, ASL and IASL

| Bathroom | | | | | |
|---|---|---|---|---|---|
| **PLTM** | | **ASL** | | **IASL** | |
| bathroom | clean | area | clean | bathroom | small |
| area | comfortable | bathroom | comfortable | towel | dirty |
| bed | sink | large | adequate | floor | sink |
| large | vanity | long | sink | area | adequate |
| space | hang | provide | vanity | shower | hang |
| coffee | spacious | space | hang | wall | break |
| microwave | adequate | chair | spacious | bath | big |
| small | amenity | separate | quiet | space | vanity |
| fridge | available | small | friendly | hair | old |
| desk | safe | counter | poor | separate | stain |
| **Location** | | | | | |
| **PTLM** | | **ASL** | | **IASL** | |
| location | close | location | close | location | close |
| located | clean | located | clean | area | convenient |
| airport | convenient | minute | quiet | park | attraction |
| area | quiet | downtown | comfortable | minute | wonderful |
| shuttle | comfortable | drive | convenient | drive | new |
| downtown | wonderful | street | safe | downtown | ideal |
| drive | free | distance | attraction | short | decent |
| park | attraction | shopping | convenient | mile | clean |
| shopping | safe | short | reasonable | distance | quiet |
| price | convenient | main | affordable | airport | difficult |

# 6 Conclusion

This paper proposed two new topic models ASL and IASL to jointly model source and target topics and their correspondence for datasets involving document pairs. The ASL model is a directional topic model. The IASL model improves ASL by enabling the inference algorithm to leverage word collocations across both source and target documents while inducing topics. The proposed models have been evaluated on the task of finding sentiment and aspect topics and their correspondence using real-world reviews of hotels and restaurants. Experimental results showed that ASL and IASL outperformed the relevant baseline models PLTM and ASUM markedly.

**Acknowledgements.** This work was supported in part by a grant from National Science Foundation (NSF) under grant no. IIS-1407927, a NCI grant under grant no. R01CA192240, and a gift from Bosch. The content of the paper is solely the responsibility of the authors and does not necessarily represent the official views of the NSF, NCI, or Bosch.

# References

1. Blei, D., Jordan, M.: Modeling annotated data. In: Proceedings of the 26th Annual International ACM SIGIR Conference on Research and Development in Information Retrieval, Toronto, Canada, pp. 127–134 (2003)
2. Blei, D., Lafferty, J.: Correlated topic models. In: Advances in Neural Information Processing Systems, vol. 18. MIT Press, Cambridge (2006)
3. Blei, D., Ng, A., Jordan, M.: Latent dirichlet allocation. J. Mach. Learn. Res. **3**, 993–1022 (2003)
4. Boyd-Graber, J., Blei, D.: Multilingual topic models for unaligned text. In: Proceedings of the Twenty-Fifth Conference on Uncertainty in Artificial Intelligence, pp. 75–82. AUAI Press (2009)
5. Brody, S., Elhadad, N.: An unsupervised aspect-sentiment model for online reviews. In: Proceedings of NAACL, pp. 804–812 (2010)
6. Chang, J., Boyd-Graber, J., Chong, W., Gerrish, S., Blei, M.: Reading tea leaves: how humans interpret topic models. In: Proceedings of NIPS, pp. 288–296 (2009)
7. Chen, Z., Mukherjee, A., Liu, B., Hsu, M., Castellanos, M., Ghosh, R.: Exploiting domain knowledge in aspect extraction. In: Proceedings of EMNLP, pp. 1655–1667 (2013)
8. Cohen, J.: Weighted kappa: nominal scale agreement provision for scaled disagreement or partial credit. Psychol. Bull. **70**(4), 213 (1968)
9. Griffiths, T., Steyvers, M.: Finding scientific topics. PNAS **101**(Suppl), 5228–5235 (2004)
10. Gamon, M., Mukherjee, A., Pantel, P.: Predicting interesting things in text. In: Proceedings of COLING (2014)
11. Hofmann, T.: Probabilistic latent semantic analysis. In: Proceedings of UAI, pp. 289–296 (1999)
12. Hu, M., Liu, B.: Mining and summarizing customer reviews. In: Proceedings of KDD, pp. 168–177 (2004)
13. Jo, Y., Oh, A.: Aspect and sentiment unification model for online review analysis. In: Proceedings of WSDM, pp. 815–824 (2011)
14. Kosuke, F., Eguchi, K., Xing, E.: Symmetric correspondence topic models for multilingual text analysis. In: Advances in Neural Information Processing Systems, vol. 25, pp. 1295–1303 (2012)
15. Lin, C., He, Y.: Joint sentiment/topic model for sentiment analysis. In: Proceedings of CIKM, pp. 375–384 (2009)
16. Liu, B.: Sentiment analysis and subjectivity. Handb. Nat. Lang. Process. **2**, 627–666 (2010)
17. Miao, G., Guan, Z., Moser, L., Yan, X., Tao, S., Anerousis, N., Sun, J.: Latent association analysis of document pairs. In: Proceedings of the 18th ACM SIGKDD International Conference on Knowledge Discovery and Data Mining, pp. 1415–1423. ACM (2012)
18. Mei, Q., Ling, X., Wondra, M., Su, H., Zhai, C.: Topic sentiment mixture: modeling facets and opinions in weblogs. In: Proceedings of WWW, pp. 171–180 (2007)
19. Mimno, D., Wallach, H., Naradowsky, J., Smith, D., McCallum, A.: Polylingual topic models. In: Proceedings of the 2009 Conference on Empirical Methods in Natural Language Processing, vol. 2, pp. 880–889. Association for Computational Linguistics (2009)
20. Mimno, D., Wallach, H., Talley, E., Leenders, M., McCallum, A.: Optimizing semantic coherence in topic models. In: Proceedings of EMNLP, pp. 262–272 (2011)
21. Moghaddam, S., Ester, M.: ILDA: interdependent LDA model for learning latent aspects and their ratings from online product reviews. In: Proceedings of SIGIR, pp. 665–674 (2011)
22. Mukherjee, A., Liu, B.: Aspect extraction through semi-supervised modeling. In: Proceedings of ACL, pp. 339–348 (2012)

23. Pang, B., Lee, L.: Opinion mining and sentiment analysis. Found. Trends Inf. Retr. **2**(1–2), 1–135 (2008)
24. Rosen-Zvi, M., Chemudugunta, C., Griffiths, T., Smyth, P., Steyvers, M.: Learning author-topic models from text corpora. ACM Trans. Inf. Syst. **28**(1), 1–38 (2010)
25. Teh, Y., Jordan, M., Beal, M., Blei, D.: Hierarchical dirichlet processes. J. Am. Stat. Assoc. (JASA) **101**, 1566–1581 (2006)
26. Titov, I., McDonald, R.: A joint model of text and aspect ratings for sentiment summarization. In: Proceedings of ACL (2008)
27. Wang, H., Lu, Y., Zhai, C.: Latent aspect rating analysis on review text data: a rating regression approach. In: Proceedings of KDD, pp. 783–792 (2010)
28. Zhang, D., Sun, J., Zhai, C., Bose, A., Anerousis, N.: PTM: probabilistic topic mapping model for mining parallel document collections. In: Proceedings of the 19th ACM International Conference on Information and Knowledge Management, pp. 1653–1656. ACM (2010)
29. Zhao, W., Jiang, J., Yan, H., Li, X.: Jointly modeling aspects and opinions with a MaxEnt-LDA hybrid. In: Proceedings of EMNLP, pp. 56–65 (2010)

# Aspect Based Sentiment Analysis: Category Detection and Sentiment Classification for Hindi

Md Shad Akhtar$^{(\boxtimes)}$, Asif Ekbal, and Pushpak Bhattacharyya

Department of Computer Science and Engineering,
Indian Institute of Technology Patna, Patna 801103, India
{shad.psc15,asif,pb}@iitp.ac.in

**Abstract.** E-commerce markets in developing countries (e.g. India) have witnessed a tremendous amount of user's interest recently. Product reviews are now being generated daily in huge amount. Classifying the sentiment expressed in a user generated text/review into certain categories of interest, for example, *positive* or *negative* is famously known as sentiment analysis. Whereas aspect based sentiment analysis (ABSA) deals with the sentiment classification of a review towards some aspects or attributes or features. In this paper we asses the challenges and provide a benchmark setup for aspect category detection and sentiment classification for Hindi. Aspect category can be seen as the generalization of various aspects that are discussed in a review. As far as our knowledge is concerned, this is the very first attempt for such kind of task involving any Indian language. The key contributions of the present work are two-fold, viz. providing a benchmark platform by creating annotated dataset for aspect category detection and sentiment classification, and developing supervised approaches for these two tasks that can be treated as a baseline model for further research.

**Keywords:** Aspect category detection · Sentiment analysis · Hindi

## 1 Introduction

With the globalization of internet over the past decade or so, usage of e-commerce as well as social media has increased enormously. Users do express their opinions regarding a product and/or service online. Organizations and other users treat these feedbacks and opinions as a goodness measure for the product or service. The amount of contents generated daily poses several practical challenges to maintain and analyze these effectively. Some of the challenges are due to the informal nature of texts, code-mixing (mixing of several language contents) behaviors and the non-availability of many basic resources and/or tools for the processing of these kinds of texts. Thus, it has been a matter of interest to the researchers worldwide to develop robust techniques and tools in order to effectively and accurately analyze the user generated contents. One such task is

© Springer International Publishing AG, part of Springer Nature 2018
A. Gelbukh (Ed.): CICLing 2016, LNCS 9624, pp. 246–257, 2018.
https://doi.org/10.1007/978-3-319-75487-1_19

famously known as sentiment analysis [17] that deals with finding an un-biased opinion of review or text written in social media platforms. It tends to classify a piece of user written text by predicting its polarity as either positive or negative. Finding the polarity of a user review with respect to some features or aspects is known as aspect based sentiment analysis (ABSA), which is gaining interest to the community because of its practical relevance. In 2014, a SemEval shared task [19] was contributed to address this problem in two domains namely, *restaurant & laptop*. It includes four subtasks, namely, Aspect Term Extraction (ATE), Aspect Term Sentiment (ATS) classification, Aspect Category Detection (ACD) & Aspect Category Sentiment (ACS) classification.

The first subtask i.e. aspect term extraction, can be thought of as a sequence labeling problem, where for given sequence of tokens, one has to mark the boundary of an aspect term properly. The second problem was a classification problem, where the sentiment expressed towards an aspect has to be classified as positive, negative, neutral and conflict. The problem of aspect category detection (the third task) deals with the classification of an aspect term into one of the predefined categories. The problem related to the fourth task was to classify the sentiment expressed in a review with respect to the aspect category. The third and the fourth tasks in SemEval considered the reviews of only the restaurant domain, and five aspect categories (i.e. food, price, service, ambiance and misc) were defined. Table 1 shows one example review, each for English and Hindi. The English review contains one aspect term i.e. *'bread'* which belongs to the aspect category *'Food'*. Polarities towards both the aspect term and aspect category are *'positive'*. Similarly, Hindi review contains one aspect term i.e. हाउसिंग *(haaUs-iNg)* and its sentiment is *'neutral'*. However, it belongs to two different aspect categories i.e. *'Design'* & *'Misc'*, and the sentiments towards these are *'neutral'* and *'negative'*, respectively.

Such a fine-grained analysis provides greater insight to the sentiments expressed in the written reviews. In recent times, there have been a growing trends for sentiment analysis at the more fine-grained level, i.e. for aspect based sentiment analysis (ABSA). Few of the interesting systems that have emerged are [5,6,11,15,22,23,25]. Kiritchenko et al. [15] proposed SVM based model for the sentiment classification using word-aspect association lexicons. A multi-kernel approach is defined in [5] for aspect category detection and polarity classification. In [22], application of sigmoidal feed-forward network and Conditional Random Field (CRF) is shown for category detection and opinion target extraction, respectively. However, all these research are related to some specific languages, predominantly for English.

Sentiment analysis in Indian (especially Hindi) languages are still largely unexplored due to the non-availability of various resources and tools such as annotated corpora, lexicons, Part-of-Speech (PoS) tagger etc. Existing works [2–4,8–10,14,16,21] involving Indian languages mainly discuss the problems of sentiment analysis at the coarse-grained level with the aims of classifying sentiments either at the sentence or document level. Existing works have limited scope, mainly because of the lack of good quality resources and/or tools. For

**Table 1.** Examples of various subtasks of aspect based sentiment analysis. ATE: aspect term extraction; ATS: aspect term sentiment; ACD: aspect category detection; ACS: aspect category sentiment[a].

| Subtasks | Review Text | | |
|---|---|---|---|
| | *"The bread is top notch as well."* | | |
| ATE | *bread* | | |
| ATS | *positive* | | |
| ACD | *Food* | | |
| ACS | *positive* | | |
| Subtasks | **Review Text** | | |
| | Devanagari | ``इसका हाउसिंग स्टेनलेस स्टील से निर्मित है इसलिए बहुत भारी है।". | |
| | Transliterated | ``Isakaa haaUsiNg sTenales sTeel se nirmit hai IsaliE bahut bhaaree hai.". | |
| | Translated | ``Its housing is made up of stainless steel that why it is very heavy.". | |
| ATE | | हाउसिंग *(haaUsiNg)* | |
| ATS | | *neutral* | |
| ACD | | *Design, Misc* | |
| ACS | | *neutral, negative* | |

[a]Transliterated and translated forms are provided only for representation purpose. We did not include them for model construction.

example, Balikwal et al. [2] used Google translator to generate the dataset, which clearly does not guarantee good quality because of the translation errors encountered. On the other hand, the works reported in [3,4,14] used the datasets that are limited in size (few 100s reviews). Aspect based sentiment analysis (ABSA) in Indian languages have not been attempted at large-scale so far. Hence, the problem is still an open challenge, mainly, because of the non-availability of any benchmark setup that could provide a high-quality dataset, baseline model as well as the proper evaluation metrics. In recent time, a framework for aspect based sentiment analysis for Hindi has been proposed in [1] that provides annotated dataset for aspect term extraction and sentiment classification with respect to the aspect term.

In this work, our focus is to provide a benchmark framework for aspect category detection and its polarity classification. We create a dataset annotated with aspect categories and their polarities. In order to show the effective usage of the generated dataset we develop models based on supervised approaches for solving two problems, *viz.* aspect category detection and sentiment classification.

The rest of the paper is structured as follows. Section 2 discusses the methodologies of aspect category detection and its sentiment classification. Various aspects of the datasets are described in Sect. 3. Experimental results along with necessary analysis are presented in Sect. 4. Finally, in Sect. 5 we present the concluding remarks.

## 2    Methodologies for Aspect Category Detection and Sentiment Classification

Aspect category is a high level abstract representation (summarized form) of the aspect terms. In other words, each aspect term must belong to one of the predefined categories which represent that aspect term. However, aspect category can be implicit as well. A review that does not contain any explicit aspect term can still belong to one of the categories. For e.g., in Table 4, second sentence does not have any aspect term but still it talks about the *'price'* category whose polarity is *negative*. This information is implicitly present in the review because of the occurrence of word महंगा (*mahaNgaa/costly*). In order to show the efficacy of the resource that we created, we build two separate models for aspect category detection and sentiment classification based on supervised machine learning approaches. We make use of language independent features for both the tasks, i.e. we do not use any domain-specific resources or tools for implementing the features.

### 2.1    Aspect Category Detection

The problem of aspect category detection can be modeled with the multi-label classification framework, where each review belongs to zero (0) or more categories. In general, a multi-label classification problem can be solved using two techniques, such as: (i) binary relevance approach and (ii) label powerset approach. Binary relevance approach handles the multi-label scenario by first building $n$ distinct models for each $n$ unique label. The prediction of $n$ models are then combined to produce the final prediction. Whereas, label powerset approach treats each label combination as a unique label. It then trains and evaluates the model. An example scenario is depicted in Table 2 for both the approaches. First two rows list 5 text reviews $T_i, for$   $i = 1..5$ and the corresponding class labels. The two-class labels i.e. 'a' and 'b' can be assigned to any review. For instance reviews $T_1$ and $T_4$ belong to both 'a' and 'b' classes. In binary relevance approach two separate models i.e. $Model_a$ & $Model_b$ are trained for class 'a' and 'b', respectively. For $Model_a$ all the reviews which belong to class 'a' are assigned binary class '1'. In contrast, reviews that do not belong to class 'a' are assigned binary class '0'. The same procedure is applied to $Model_b$ for class 'b'. For label powerset approach, each unique combination of labels are mapped to some other unique labels. In the given example, there are three unique label combinations i.e. $T_3$ & $T_5$ has 'a', $T_2$ has 'b' and $T_1$ & $T_4$ has 'a, b'. Each of these labels are mapped to some random unique classes say, '1', '2' & '3', respectively. We use the following features for training the multi-label classifier: lexical features like n-grams, non-contiguous n-grams, character n-grams etc. For n-grams, we consider unigrams, bigrams and trigrams. Non-contiguous n-gram sequence is a pair of tokens that are n-tokens apart form each other. It helps to capture co-occurrences of terms that are far apart from each other.

**Table 2.** A hypothetical example for multi-label learning using binary relevance and label powerset techniques.

|  | | |
|---|---|---|
|  | Review | $<T_1; T_2; T_3; T_4; T_5>$ |
|  | Label | $<$'a,b'; 'b'; 'a'; 'a,b'; 'a'$>$ |
| *Binary relevance approach* | | |
| $Model_a$ **for class 'a'** | Review | $<T_1; T_2; T_3; T_4; T_5>$ |
|  | Label | $<$'1'; '0'; '1'; '1'; '1'$>$* |
| $Model_b$ **for class 'b'** | Review | $<T_1; T_2; T_3; T_4; T_5>$ |
|  | Label | $<$'1'; '1'; '0'; '1'; '0'$>$* |
| *Binary labels 1 or 0 (On or Off) | | |
| *Label powerset approach* | | |
|  | Review | $<T_1; T_2; T_3; T_4; T_5>$ |
|  | Label | $<$'3'; '2'; '1'; '3'; '1'$>$^ |

^Assign unique labels to each combination: 'a' => '1'; 'b' => '2'; 'a,b' => '3'

## 2.2 Sentiment Classification

Once the aspect categories are identified, we classify them to one of the four sentiment polarity classes, namely *positive, negative, neutral* and *conflict*. For each aspect category in a review we define a tuple, made up of review text and specific category, and feed it to the learning model to detect the sentiments. For e.g., if a review text 'T' has two aspect categories 'food' and 'price', then we define two tuples as $<$T, food$>$ and $<$T, price$>$ as an input to the system. Here, we use basic lexical features like n-grams, non-contiguous n-grams, character n-grams along with PoS tag and semantic orientation (SO) [12] score which is a measure of association of tokens towards negative and positive sentiments, and can be defined as:

$$SO(t) = PMI(t, posRev) - PMI(t, negRev) \tag{1}$$

where $PMI(t, negRev)$ stands for point-wise mutual information of a token $t$ towards negative sentiment reviews. The SO score would be more effective had we use external data, but in this paper we restrict ourselves not to use any external resources for the sake of domain and resource independence.

## 3    Benchmark Setup for ABSA in Hindi

For ABSA there is no available dataset for the Indian languages, in general, and Hindi, in particular. We create our own dataset for aspect category detection and sentiment classification by collecting user generated web reviews, and annotating these using a pre-defined set of categories. In subsequent subsections we describe these steps in details. We also make the dataset[1] available to the community for the advancement of further research involving Indian languages.

---

[1] Dataset can be found at http://www.CICLing.org/2016/data/170.

## 3.1    Data Collection

We crawl various online sources[2] and collect 5,417 user generated reviews, which belong to 12 different domains, namely (i) Laptops, (ii) Mobiles, (iii) Tablets, (iv) Cameras, (v) Headphones, (vi) Home appliances, (vii) Speakers, (viii) Televisions, (ix) Smart watches, (x) Mobile apps, (xi) Travels and (xii) Movies. Details of these dataset statistics are presented in Sect. 3.3.

## 3.2    Data Annotation

We define and compile a list of aspect categories for different domains as listed in Table 3. All electronics products or domains (*except* Mobile apps, Travels and Movies) share six common categories among themselves e.g. Design of the product, Software, Hardware, Ease of use or accessibility, Price of the product and Miscellaneous.

**Table 3.** Aspect categories that correspond to different domains.

| Domains | Aspect categories |
|---|---|
| Electronics (Laptops, Mobiles, Tablets, Cameras, Speakers, Smart watches, Headphones, Home appliances & Televisions) | Design, Software, Hardware, Ease of use, Price, Misc |
| Mobile apps | GUI, Ease of use, Price, Misc |
| Travels | Scenery, Place, Reachability, Misc |
| Movies | Story, Performance (Action/Direction etc.), Music, Misc |

We follow similar scheme in line with SemEval shared task for annotating the dataset. We identify various aspect categories of each review along with

---

[2] List of few sources...
http://www.jagran.com
http://www.gizbot.com
http://www.patrika.com
http://www.hi.themobileindian.com
http://www.mobilehindi.com
http://navbharattimes.indiatimes.com
http://hindi.starlive24.in/
http://www.amarujala.com
http://techjankari.blogspot.in
http://www.digit.in
http://khabar.ndtv.com/topic
http://www.hindi.mymobile.co.in/
http://www.bhaskar.com.

its associated sentiment and save them into a XML format. Table 4 lists xml structure of two such instances from the dataset. The upper half of the table contains two example reviews in Devanagari script, its Roman transliterated as well as English translated forms. Both the reviews have one aspect category associated with them and whose polarities are *neutral* and *negative*, respectively.

**Table 4.** Dataset annotation structure.

| Id | Format | Review Text |
|----|--------|-------------|
|    | Devanagari | इसकी स्क्रीन 15.6 इंच की है। |
| 1. | Transliterated | Isakee skreen 15.6 INch kee hai. |
|    | Translated | It has 15.6 inch screen. |
|    | Devanagari | यह बहुत महंगा है। |
| 2. | Transliterated | yah bahut mahaNgaa hai. |
|    | Translated | It is very costly. |

| Annotation Structure |
|----------------------|
| \<sentences\> |
| \<sentence id= "1" \> |
| \<text\> इसकी स्क्रीन 15.6 इंच की है।\< \text\> |
| \<aspectCategories\> |
| \<aspectCategory category="hardware" polarity="neutral" /\> |
| \< \aspectCategories\> |
| \< \sentence\> |
| \<sentence id= "2"\> |
| \<text\> यह बहुत महंगा है।\< \text\> |
| \<aspectCategories\> |
| \<aspectCategory category="price" polarity="negative" /\> |
| \< \aspectCategories\> |
| \< \sentence\> |
| \<sentence id= "3"\> |
| ... |
| \< \sentence\> |
| \< \sentences\> |

The $<sentences>$ node represents root node of the xml that contains every sentence of the review as its children i.e. $<sentence>$. To uniquely identify each $<sentence>$, an '*id*' is associated with it as an attribute. Each $<sentence>$ node has two children, namely $<text>$ & $<aspectCategories>$. The $<text>$ node holds one review sentence, whereas $<aspectCategories>$ contains $n$ < $aspectCategory$ > nodes as its children if a review sentence has $n$ different categories. Each $<aspectCategory>$ node holds two attributes: '*category*' & '*polarity*'. Attribute '*category*' defines aspect category represented by current node while '*polarity*' stores the sentiment towards the '*category*'. For the example at hand, number of categories ($n$) equal to 1 for sentence ids 1 and 2. Aspect category of sentence id 1 is '*hardware*' and polarity towards it is '*neutral*'. Similarly, sentence id 2 belongs to '*price*' category whose polarity is '*negative*'.

## 3.3  Dataset Statistics

The dataset contains 5,417 user reviews related to the product or service. There are total of 2,250 'positive', 635 'negative', 2,241 'neutral' and 128 'conflict' instances of aspect categories. Overview of the dataset statistics are presented in Table 5.

**Table 5.** Dataset statistics. Pos: positive, Neg: negative, Neu: neutral, Conf: conflict.

| Domains | Polarity | Category | | | | | | | | | | | | | |
|---|---|---|---|---|---|---|---|---|---|---|---|---|---|---|---|
| | | HW | SW | Des. | Pri. | Ease | GUI | Place | Rea. | Sce. | Story | Perf | Music | Misc | Total |
| Electronics (Laptops, Mobiles, Tablets, Cameras, Headphones, HomeApps, Speakers, Smartwatches & Televisions) | Pos | 700 | 160 | 305 | 110 | 70 | – | – | – | – | – | – | – | 290 | 1635 |
| | Neg | 261 | 55 | 69 | 31 | 19 | – | – | – | – | – | – | – | 89 | 524 |
| | Neu | 763 | 149 | 137 | 83 | 30 | – | – | – | – | – | – | – | 173 | 1335 |
| | Conf | 73 | 6 | 13 | 4 | 3 | – | – | – | – | – | – | – | 21 | 120 |
| | Total | 1797 | 370 | 524 | 228 | 122 | – | – | – | – | – | – | – | 573 | 3614 |
| | Pos | – | – | – | 4 | 18 | 14 | – | – | – | – | – | – | 64 | 100 |
| | Neg | – | – | – | 0 | 4 | 5 | – | – | – | – | – | – | 13 | 22 |
| | Neu | – | – | – | 6 | 3 | 8 | – | – | – | – | – | – | 57 | 74 |
| | Conf | – | – | – | 0 | 1 | 0 | – | – | – | – | – | – | 0 | 1 |
| | Total | – | – | – | 10 | 26 | 27 | – | – | – | – | – | – | 134 | 197 |
| Travels | Pos | – | – | – | – | – | – | 195 | 7 | 97 | – | – | – | 57 | 356 |
| | Neg | – | – | – | – | – | – | 5 | 9 | 1 | – | – | – | 6 | 21 |
| | Neu | – | – | – | – | – | – | 103 | 19 | 24 | – | – | – | 41 | 187 |
| | Conf | – | – | – | – | – | – | 1 | 0 | 0 | – | – | – | 0 | 1 |
| | Total | – | – | – | – | – | – | 304 | 35 | 122 | – | – | – | 104 | 565 |
| Movies | Pos | – | – | – | – | – | – | – | – | – | 6 | 109 | 14 | 30 | 159 |
| | Neg | – | – | – | – | – | – | – | – | – | 11 | 35 | 5 | 17 | 68 |
| | Neu | – | – | – | – | – | – | – | – | – | 17 | 95 | 8 | 525 | 645 |
| | Conf | – | – | – | – | – | – | – | – | – | 1 | 5 | 0 | 0 | 6 |
| | Total | – | – | – | – | – | – | – | – | – | 35 | 244 | 27 | 572 | 878 |
| Overall | Pos | 700 | 160 | 305 | 114 | 88 | 14 | 195 | 7 | 97 | 6 | 109 | 14 | 441 | 2250 |
| | Neg | 261 | 55 | 69 | 31 | 23 | 5 | 5 | 9 | 1 | 11 | 35 | 5 | 125 | 635 |
| | Neu | 763 | 149 | 137 | 89 | 33 | 8 | 103 | 19 | 24 | 17 | 95 | 8 | 796 | 2241 |
| | Conf | 73 | 6 | 13 | 4 | 4 | 0 | 1 | 0 | 0 | 1 | 5 | 0 | 21 | 128 |
| | Total | 1797 | 370 | 524 | 238 | 148 | 27 | 304 | 35 | 122 | 35 | 244 | 27 | 1383 | 5254 |

# 4  Experimental Result and Analysis

To address the problem of multi-label classification of aspect category detection, we use MEKA[3] for the experiments. MEKA is an extension to WEKA which handles multi-label scenario. As a base classifier we use naive Bayes [13], J48 [20] implementation of decision tree and SMO [18] implementation of SVM [7]. The underlying experiment is carried out by the following two approaches i.e. binary relevance method and label powerset method. For the label powerset approach we use MULAN[4] [24] framework. For the sake of experiment we combine the reviews of all the electronics products, *except* mobile apps, travels

---

[3] http://meka.sourceforge.net/.
[4] http://mulan.sourceforge.net/.

and movies, and treat them as to belong to a single domain, namely *'electronics'*. Therefore, we build our model for the four major domains i.e. *'electronics'*, *'mobile apps'*, *'travels'* and *'movies'*. To evaluate the system we use the evaluation script, which was provided by the SemEval shared task organizer. We perform 3-fold cross validation on the training dataset. We obtain the average F-measures of 46.46%, 56.63%, 30.97% and 64.27% for aspect category detection task in electronics, mobile apps, travels and movies domain, respectively. Naive Bayes performs better in electronics and mobile apps domain, while decision tree reports better results for the travels and movies domain. In sentiment classification our proposed model reports the accuracy of 54.48%, 47.95%, 65.20% & 91.62% for the four domains respectively. Experimental results for the two tasks are reported in Table 6.

**Table 6.** Results of aspect category detection and sentiment classification. Here, NB: naive Bayes classifier, DT: decision tree classifier and SMO: sequential minimal optimization implementation of SVM

| Domain | Method | Aspect category detection | | | | | | Polarity |
|---|---|---|---|---|---|---|---|---|
| | | Binary rel. | | | MULAN | | | WEKA |
| | | Pre | Rec | F | Pre | Rec | F | Accuracy |
| Electronics | NB | 31.62 | 37.63 | 34.37 | 48.00 | 45.05 | **46.46** | 50.95 |
| | DT | 49.61 | 17.28 | 25.63 | 31.73 | 31.73 | 31.73 | **54.48** |
| | SMO | 26.70 | 46.93 | 34.03 | 39.36 | 44.90 | 41.94 | 51.07 |
| Mobile apps | NB | 39.30 | 46.19 | 42.47 | 59.20 | 54.09 | **56.53** | 46.78 |
| | DT | 44.28 | 41.75 | 42.97 | 85.07 | 24.89 | 38.51 | **47.95** |
| | SMO | 51.73 | 38.47 | 44.12 | 45.77 | 57.14 | 50.82 | 42.10 |
| Travels | NB | 26.84 | 26.88 | 26.86 | 20.87 | 31.90 | 25.23 | 56.06 |
| | DT | 27.98 | 22.73 | 25.08 | 99.82 | 18.33 | **30.97** | **65.20** |
| | SMO | 25.51 | 20.67 | 22.83 | 15.61 | 39.55 | 22.38 | 60.63 |
| Movies | NB | 41.99 | 65.44 | 51.15 | 56.66 | 63.32 | 59.81 | 87.78 |
| | DT | 47.45 | 58.12 | 52.24 | 64.16 | 64.38 | **64.27** | **91.62** |
| | SMO | 43.78 | 59.81 | 50.55 | 48.60 | 63.26 | 54.97 | **91.62** |

We perform error analysis in order to understand the quality of the results that we obtain. An overview of the different kinds of errors encountered for aspect category detection is shown in the confusion matrix as shown in Table 7. Results show that the system obtains good recall for the *'hardware'* category, but precision is not so impressive. The model does not perform well for the other categories. One possible reason behind this could be the presence of a relatively fewer number of instances for all the domains *except 'hardware'* which is a dominating category in the dataset i.e. 1,797 out of 3,614 instances belong to this particular category. Confusion matrix for sentiment classification is shown in Table 8. It shows that classifier performs better for the *positive* class, and this could be due to the higher number of instances, belonging to this particular class. It classifies 1,120 instances correctly out of total 1,635. The level of accuracy that we obtain for the *'neutral'* class requires further investigation. Lack of sufficient number of instances drives the system to predict only 2 correct instances for the *'conflict'* class.

**Table 7.** Confusion matrix for aspect category detection in electronics domain. Class 'Others' means a test instance that does not belong to a particular class. For example, there are 370 and 3152 true test instances for 'Software' and 'Others' (i.e. *Hardware, Design, Price, Ease* and *Misc* combined) respectively. The system predicts 90 instances as 'Software' and rest as 'Others', out of which only 12 instances were correctly classified as 'Software'. A total of 358 instances were mis-classified as 'Others' by the system.

| Predicted > | Hardware | Software | Design | Price | Ease | Misc | Others |
|---|---|---|---|---|---|---|---|
| Hardware | 1651 | 0 | 0 | 0 | 0 | 0 | 146 |
| Software | 0 | 12 | 0 | 0 | 0 | 0 | 358 |
| Design | 0 | 0 | 39 | 0 | 0 | 0 | 485 |
| Price | 0 | 0 | 0 | 3 | 0 | 0 | 225 |
| Ease | 0 | 0 | 0 | 0 | 0 | 0 | 122 |
| Misc | 0 | 0 | 0 | 0 | 0 | 30 | 543 |
| Others | 1566 | 78 | 198 | 63 | 15 | 196 | 15402 |

**Table 8.** Confusion matrix for aspect category sentiment in electronics domain

| Predicted > | Positive | Negative | Neutral | Conflict |
|---|---|---|---|---|
| Positive | 1120 | 77 | 434 | 4 |
| Negative | 290 | 96 | 138 | 0 |
| Neutral | 642 | 64 | 628 | 1 |
| Conflict | 73 | 11 | 34 | 2 |

## 5 Conclusion

In this paper we have proposed a benchmark setup for aspect category detection and its sentiment classification for Hindi. We have collected review sentences from the various online sources and annotated 5,417 review sentences across 12 domains. Based on these datasets we develop frameworks for aspect category detection and sentiment classification based on supervised classifiers. The problem of aspect category detection was cast as a multi-label classification problem whereas sentiment classification was modeled as a multi-class classification problem. The proposed model reports 46.46%, 56.63%, 30.97% and 64.27%

F-measures for the aspect category detection in electronics, mobile apps, travels and movies domain, respectively. For sentiment classification the model we obtain the accuracies of 54.48%, 47.95%, 65.20% & 91.62% for the four domains, respectively. The key contributions of the research reported here are two-fold, i.e. creating a benchmark set up for aspect category detection and sentiment classification, and developing a benchmark setup that can be used as a reference for further research.

To the best of our knowledge, this is the very first attempt for these two specific problems involving Indian languages, especially Hindi. In future we would like to use domain-specific features for the problems and investigate deep learning methods for the tasks.

# References

1. Akhtar, M.S., Ekbal, A., Bhattacharyya, P.: Aspect based sentiment analysis in hindi: resource creation and evaluation. In: Proceedings of the 10th Edition of the Language Resources and Evaluation Conference (LREC) (2016, accepted)
2. Bakliwal, A., Arora, P., Varma, V.: Hindi subjective lexicon: a lexical resource for Hindi polarity classification (2012)
3. Balamurali, A.R., Joshi, A., Bhattacharyya, P.: Harnessing wordnet senses for supervised sentiment classification. In: Proceedings of the 2011 Conference on Empirical Methods in Natural Language Processing (EMNLP), pp. 1081–1091 (2011)
4. Balamurali, A.R., Joshi, A., Bhattacharyya, P.: Cross-lingual sentiment analysis for Indian languages using linked wordnets. In: Proceedings of the 24th International Conference on Computational Linguistics (COLING), pp. 73–82 (2012)
5. Castellucci, G., Filice, S., Croce, D., Basili, R.: UNITOR: aspect based sentiment analysis with structured learning. In: Proceedings of the 8th International Workshop on Semantic Evaluation (SemEval 2014), pp. 761–767 (2014)
6. Chernyshevich, M.: IHS R&D belarus: cross-domain extraction of product features using conditional random fields. In: Proceedings of the 8th International Workshop on Semantic Evaluation (SemEval 2014), pp. 309–313 (2014)
7. Cortes, C., Vapnik, V.: Support-vector networks. Mach. Learn. **20**(3), 273–297 (1995)
8. Das, A., Bandyopadhyay, S.: Phrase-level polarity identification for Bangla. Int. J. Comput. Linguist Appl. (IJCLA) **1**(1–2), 169–182 (2010)
9. Das, A., Bandyopadhyay, S., Gambäck, B.: Sentiment analysis: what is the end user's requirement? In: Proceedings of the 2nd International Conference on Web Intelligence, Mining and Semantics, p. 35 (2012)
10. Das, D., Bandyopadhyay, S.: Labeling emotion in Bengali blog corpus-a fine grained tagging at sentence level. In: Proceedings of the 8th Workshop on Asian Language Resources, p. 47 (2010)
11. Gupta, D.K., Reddy, K.S., Ekbal, A.: PSO-ASent: feature selection using particle swarm optimization for aspect based sentiment analysis. In: Biemann, C., Handschuh, S., Freitas, A., Meziane, F., Métais, E. (eds.) NLDB 2015. LNCS, vol. 9103, pp. 220–233. Springer, Cham (2015). https://doi.org/10.1007/978-3-319-19581-0_20
12. Hatzivassiloglou, V., McKeown, K.R.: Predicting the semantic orientation of adjectives. In: Proceedings of the ACL/EACL, pp. 174–181 (1997)

13. John, G.H., Langley, P.: Estimating continuous distributions in Bayesian classifiers. In: Proceedings of 11th Conference on Uncertainty in Artificial Intelligence, pp. 338–345. Morgan Kaufmann (1995)

14. Joshi, A., Balamurali, A.R., Bhattacharyya, P.: A fall-back strategy for sentiment analysis in Hindi: a case study (2010)

15. Kiritchenko, S., Zhu, X., Cherry, C., Mohammad, S.: NRC-Canada-2014: detecting aspects and sentiment in customer reviews. In: Proceedings of the 8th International Workshop on Semantic Evaluation (SemEval 2014), pp. 437–442 (2014)

16. Mittal, N., Agarwal, B., Chouhan, G., Bania, N., Pareek, P.: Sentiment analysis of Hindi review based on negation and discourse relation. In: Proceedings of International Joint Conference on Natural Language Processing, pp. 45–50 (2013)

17. Pang, B., Lee, L.: Opinion mining and sentiment analysis. Found. Trends Inf. Retr. **2**(1–2), 1–135 (2008)

18. Platt, J.C.: Sequential minimal optimization: a fast algorithm for training support vector machines. Technical report, Advances in Kernel Methods - Support Vector Learning (1998)

19. Pontiki, M., Galanis, D., Pavlopoulos, J., Papageorgiou, H., Androutsopoulos, I., Manandhar, S.: SemEval-2014 task 4: aspect based sentiment analysis. In: Proceedings of the 8th International Workshop on Semantic Evaluation (SemEval 2014), pp. 27–35, August 2014

20. Quinlan, R.: C4.5: Programs for Machine Learning. Morgan Kaufmann Publishers, Burlington (1993)

21. Sharma, R., Nigam, S., Jain, R.: Polarity detection of movie reviews in Hindi language. Int. J. Comput. Sci. Appl. (IJCSA) **4**(4) (2014)

22. Toh, Z., Su, J.: NLANGP: supervised machine learning system for aspect category classification and opinion target extraction. In: Proceedings of the 9th International Workshop on Semantic Evaluation (SemEval 2015), pp. 496–501 (2015)

23. Toh, Z., Wang, W.: DLIREC: aspect term extraction and term polarity classification system. In: Proceedings of the 8th International Workshop on Semantic Evaluation (SemEval 2015), pp. 235–240 (2014)

24. Tsoumakas, G., Spyromitros-Xioufis, E., Vilcek, J., Vlahavas, I.: MULAN: a Java library for multi-label learning. J. Mach. Learn. Res. **12**, 2411–2414 (2011)

25. Wagner, J., Arora, P., Cortes, S., Barman, U., Bogdanova, D., Foster, J., Tounsi, L.: DCU: aspect-based polarity classification for Semeval task 4. In: Proceedings of the 8th International Workshop on Semantic Evaluation (SemEval 2014), pp. 223–229 (2014)

# A New Emotional Vector Representation
# for Sentiment Analysis

Hanen Ameur[✉], Salma Jamoussi, and Abdelmajid Ben Hamadou

Multimedia InfoRmation Systems and Advanced Computing Laboratory,
MIRACL-Sfax University, Sfax-Tunisia Technopole of Sfax:
Av.Tunis Km 10 B.P. 242, 3021 Sfax, Tunisia
ameurhanen@gmail.com,
{salma.jamoussi,abdelmajid.benhamadou}@isimsf.rnu.tn

**Abstract.** With the advent of Web 2.0, social networks (like, Twitter
and Facebook) offer to users a different writing style that's close to the
SMS language. This language is characterized by the presence of emotion
symbols (emoticons, acronyms and exclamation words). They often man-
ifest the sentiments expressed in the comments and bring an important
contextual value to determine the general sentiment of the text. Moreover, these emotion symbols are considered as multilingual and universal
symbols. This fact has inspired us to research in the area of automatic
sentiment classification. In this paper, we present a new vector representation of text which can faithfully translate the sentimental orientation of text, based on the emotion symbols. We use Support Vector
Machines to show that our emotional vector representation significantly
improves accuracy for sentiment analysis problem compared with the well
known bag-of-words vector representations, using dataset derived from
Facebook.

**Keywords:** Emotional TFIDF vector · SVM classifier
Sentiment analysis · Facebook dataset · Emoticons

## 1 Introduction

With the development of the internet and the advent of community sites like
social networks (e.g., Facebook and Twitter) and blogs, more and more people
share their opinions and express their sentiments towards a considerable variety
of topics. Every day, there is an ever-increasing number of textual data expressing
users' sentiments. The explosion of online opinionated texts, gives rise to new
research areas; opinion mining and sentiment analysis. These areas have been
actively developing recently. An important and motivating research direction of
sentiment analysis boils down to an automatic classification problem. Sentiment
classification aims to group the subjective texts according to the opinion or
judgment which they express by assigning a polarity label to texts (binary:
positive or negative or multivariate).

© Springer International Publishing AG, part of Springer Nature 2018
A. Gelbukh (Ed.): CICLing 2016, LNCS 9624, pp. 258–269, 2018.
https://doi.org/10.1007/978-3-319-75487-1_20

There are two main kinds of sentiment classification approach. The first one is the linguistic approach (lexicon-based), which consists of calculating the semantic orientation (valence) of texts by obtaining word polarities from a lexicon. Thus, this approach requires to construct a properly labelled sentiment lexicon in order to classify texts (using the manual method [19], the dictionaries-based method [8], the corpora-based method [2], the method combining the last two ones [12] or the concept-based method [3]). The second one is the statistical approach, (learning-based), which involves using the machine learning methods (like Support Vector Machines, Naïve Bayesian classifiers, Maximum Entropy, Neural Networks, etc.) to produce automated classifiers and generate the class labels for opinionated texts.

In this paper, we focus on the statistical approach to classify texts into positive or negative. This approach adopts machine learning classification techniques with feature selection. A critical task of learning-based sentiment analysis system is how to find a representation (set of features) which can faithfully capture and translate the sentimental characteristics of a text. [13] were the first who applied a machine learning in sentiment analysis. They evaluated the contribution of different features including uni-grams, bi-grams and part-of-speech tags. Alternatively, words can be counted as booleans as shown in [13], or weighted by their TF-IDF score like [9], TF-$\Delta$IDF score [10]. We introduce a novel way to represent words by distinguishing eight different emotional states which formulate the vector space. We propose vector representations of words and sentences based on the notion of emoticons, in order to represent numerically their semantic and sentimental characteristics. In fact, the emoticons are considered as multilingual and universal symbol. We perform the evaluation using the Facebook dataset to see the effect of emoticons as sentence feature on polarity determination. Throughout the evaluation, we show how this improves accuracy.

The content of this paper is organized as follows: First, in Sect. 2, we review the related work for sentiment analysis. In Sect. 3, we elaborate on our proposed method for sentiment classification. We present the proposed emotional vector representations. Next, we report the experimental results in Sect. 4. First, we provide the datasets collected from Facebook and the subsequent process of its pre-treatment. Finally, we conclude the paper with future works.

## 2   Related Works

Owing to the limited space and our contribution which is a statistical approach (learning-based), we only describe relevant related work at this approach. Most studies have been focused on the statistical methods based on machine learning algorithms (Supervised methods). This kind of algorithm requires a set of well-classified sentences (manually labeled data: training corpus). Supervised methods aim to discover a model using labeled examples which must be able to generalize the classification learned on a wider dataset. Then, it comes to learning a machine how to assign a class to a new unlabeled sentence among the predefined classes in the learned model. The popular algorithms in machine learning, such as Support Vector Machines [5,13,19] and Naive Bayes [6,13], are used to train the sentiment classifier.

In order to perform machine learning, it is necessary to transform the text into numerical representations that may lead to correct classification. In the literature, the most common and useful feature is binary representation that indicates word presence or absence [12,13]. **Binary** representation is very simple, but it is not very informative since it does not inform about the frequency of words which can be an important information. **Frequency-based** feature representation is a natural extension of the binary representation, where the number of occurrences of words in a sentence is counted. In sentiment classification, frequency representation has been used in several works, such as [14], but it presents the disadvantage of not taking into account the length of processed sentences and hence a long sentence may be represented by a vector whose norm is greater than that of the representation of a short sentence. It is therefore very interesting to work with a normalized frequency representation where each sentence is presented by a vector weighted by its size whose each component code the proportion of the term in the sentence.

[17] used Latent Semantic Analysis **LSA** which learns semantic word vectors by applying singular value decomposition to factor a term-document co-occurrence matrix. The **TF-IDF** representation attempts to be more informative than the previous representations (used by [9,14]). The value TF (Term Frequency) corresponds to the frequency of a term in a sentence. It essentially refers to the importance of the term in this sentence. Nevertheless, the value IDF (Inverse Document Frequency) measures its importance in the set of sentences by calculating the logarithm of the inverse documentary frequency. [10] proposed a supervised variant of IDF weighting for sentiment analysis, named $\Delta$**IDF**, in which the IDF calculation is done for each text class and then one value is subtracted from the other. They assigned feature values for a document by calculating the difference of those words TFIDF scores in the positive and negative training corpus.

Recently, another type of representation is proposed by [16], named Distributed vector representations which associate similar vectors with similar words and phrases. These vectors provide useful information for the learning algorithms to achieve better performance in Natural Language Processing tasks [11]. To compute the vector representations of words for sentiment analysis, [1] used the skip-gram model of **Word2Vec** [11]. The Skip-gram model aims to find word representations that are useful for predicting the surrounding words in a sentence or document [11].

[18] proposed **bag-of-sentiwords** vector representations of text which capture the presence of sentiment-carrying words derived from a sentiment lexicon. In other work, text has been represented as a bag-of-opinions, where features denote occurrences of unique combinations of opinion-conveying words, amplifiers, and negators [15]. Other features capture the length of a text segment, and the extent to which it conveys opinions [7].

## 3   Proposed Method

Most of the existing methods for sentiment classification based on machine learning algorithms use the bag-of-words representation of a text. Thus, these representations focus often on presence or frequency of lexicon words or specific words (frequent words, sentiment lexicon, opinion-conveying words,...). In this paper, we aim to determine the polarity of text by using a vector representation that uses words as well as emoticons.

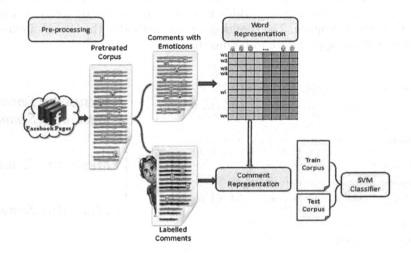

**Fig. 1.** The different steps of our sentiment classification method.

### 3.1   Vector Representation of Words

The statistical approach that we adopted in this paper, revolves essentially around the numerical representations of linguistic units (words and comments). These units are generally represented in a vector form which is characterized by associating each unit with its dimension within the vector space. In the statistical approach, the text is presented with a set of words that are considered as equivalent and unordered entities (bag-of-words). Thus, their semantic and sentimental aspects are not taken into account. Therefore, the main objective of our work is to find a representation that preserves the maximum of these two aspects.

In order to preserve the sentimental characteristic of comment, we need to represent it by an adequate vector which can faithfully translate the sentimental aspect of its words. Therefore, It is necessary to begin by associating a vector representing each word composing this comment.

***Emotional Vector***. In the text bearers sentiments, a word can have many features, but it is very difficult to select those which can give it a complete sentimental representation. Unfortunately, the exact algorithm for finding best features does not exist. It is thus required to rely on our intuition and the domain knowledge for choosing a good feature. It is impossible to represent a word with a feature vector that contains all the words that exist in the language. For the simple reason that the vocabulary in Web 2.0 is very dynamic, handling new words is still not possible.

Due to the richness of the corpus with emoticons, we propose to represent each word according to the emotion symbols present in the comment. In fact, a word can have a different polarity degree with each emotion symbol. Therefore, we present word with a vector which takes account of the relations of a word with all of these symbols. The challenge of our method is the choice of emotion symbols that are appropriate to become features.

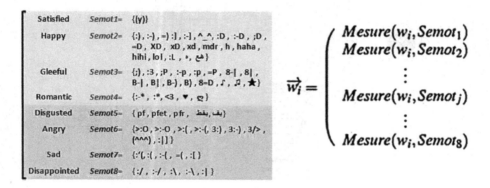

**Fig. 2.** The sets of used emotion symbols.

In the proposed vector representation, we granted to element number $j$ of the vector representing a word, the value of its similarity to the set of emotion symbols $Semot_j$, with $j \in [1..8]$ (see Fig. 2). In the Fig. 2, we consider that each set of emotion symbols is an emotional state that can reach every word. Indeed, we distinguish 8 emotional states (satisfied, happy, gleeful, romantic, disappointed, sad, angry and disgusted). To do this, we collect the emotion symbols which determine each emotional state. The Fig. 2 shows that the 4 first sets of symbols $Semot$ express positive sentiments, and the 4 following $Semot$ express negative sentiments. Therefore, our emotional vector contains 8 dimensions. It represents the similarity degree between the word $w_i$ and each set of emotion symbols (emotional state).

Eventually, it's noteworthy to define the effective similarity measures allowing to quantify sentimental relations between words and the emotion symbols. Here, we describe two similarity measures: Normalized co-occurrence and Emotional TF-IDF weighting.

– **Normalized cooccurrence**

This measure relies on the calculating of a number of co-occurrences of $w_i$ and $Semot_j$ which is equal to the number of times that $w_i$ and one of the $Semot_j$ appear together in the same comment. In order to highlight the similarities between the examined word and the emotion symbols having the same polarity as $Semot_j$, we normalized the number of co-occurrences of $w_i$ and $Semot_j$ by the number of co-occurrences of all words and the emotion symbols having the same polarity as $Semot_j$ by using the following formula:

$$NormCO(w_i, Semot_j) = \frac{CoOcc(w_i, Semot_j)}{\sum_{k=f}^{n} CoOcc(w_i, Semot_k)} \qquad (1)$$

where: $CoOcc(w_i, Semot_j)$ is the number of co-occurrences of the word $w_i$ and the set of emotion symbols $Semot_j$. $f = 1$ and $n = 4$ (if $j \in [1..4]$, $f = 5$ and $n = 8$ (if $j \in [5..8]$).

– **Emotional TF-IDF weighting**

This measure is inspired from the TF-IDF measure (see Sect. 2) and it attempts to be more informative than the previous measure. Indeed, we propose this measure in order to take into account the distribution of emotion symbols in the comments of our corpus. Thus, it weights the $CoOcc(w_i, Semot_j)$ by the number of comments containing the emotion symbols $Semot_j$ in the corpus. It is calculated by using the formula 2.

$$Emotional\_TFIDF(w_i, Semot_j) = CoOcc(w_i, Semot_j) \times \log\frac{N}{n_j} \qquad (2)$$

where: $CoOcc(w_i, Semot_j)$ is the number of times that $w_i$ and one of the $Semot_j$ appear together in the same comment. $N$ is the total number of comments in the corpus containing emotion symbols. $n_j$ is the number of comments that contain only emotion symbols of the $Semot_j$ set.

Thus, we can notice that, as in the emotional TF-IDF representation, $\log\frac{N}{n_j}$ will have a high value with the emotion symbols that appear in a few comments, and vice versa.

Up to this point, we proposed two emotional vector representations for words. These representations are based on similarity measures that are calculated according to the number of co-occurrences of word and emotion symbols, noted $CoOcc(w_i, Semot_j)$. In fact, if this number is important, we can deduce, then, that $w_i$ has the same polarity as $Semot_j$, otherwise, $w_i$ has a different polarity.

**_Negation Handling_**. The lexicon words are represented by emotional vectors containing 8 elements. Each element means the degree of similarity of the word with a set of emotion symbols $Semot$. The general principle of these representations is based on calculating the number of times where we encounter the considered word with an emotion symbol in a comment. However, a word can be sometimes preceded by a negation particle (such as, ne, n, pas, ni, jamais, aucun, no, none, not, neither, never, ever). Hence, it is necessary to consider the presence of this type of information in the calculation of the number of co-occurrences, $CoOcc(w_i, Semot_j)$. It comes to cutting the comment in segments by considering the emotion symbols as separators. We thus assume that the number of co-occurrences will be decremented when we encounter the considered word preceded by a negation particle, with one of the emotion symbols in a comment segment. Thereby, we propose the Eq. 3 to calculate the number of co-occurrences of the word $w_i$ and the set of emotion symbols $Semot_j$.

$$CoOcc = CoOcc(w_i, Semot_j) - CoOcc(w_i^{NEG}, Semot_j) \tag{3}$$

where: $w_i^{NEG}$ is the word $w_i$ when it is preceded by a negation particle.

### 3.2   Vector Representation of Comments

Once the vector representations of the words are ready, we can represent the comments in the same vector space (with a vector containing 8 elements). We proposed a simple strategy that involves attaching the feature vectors of the words that compose each comment of the corpus. We proceeded to represent the comments by averaging the vectors of all words which compose them and exist in the lexicon. The component $V_C^j$ of the vector representing the comment is thus given by the following formula:

$$V_C^j = \frac{\sum_{i=1}^{N} s_{ij}}{N} \tag{4}$$

where $N$ is the number of words constituting the comment $C$ and each word $w_i$ is represented by a vector $\overrightarrow{w_i} = (s_{i1}, s_{i2}, \ldots, s_{ij}, \ldots, s_{i8})$.

The comments can contain emotion symbols, it is therefore very important to consider them at the level of the vector representation step. For this reason, we considered the emotion symbol as a word by associating it with a binary vector. Thus, we assign the value 1 to the position $Semot_i$ of the vector, where an emotion symbol of the set of symbols $i$ exists in the treated comment, ($i \in [1..8]$). For example, the emoticon :) will be represented by the vector: $(0, 1, \ldots, 0)$.

Sometimes, a comment can contain unknown words that do not exist in the lexicon and are not represented by feature vectors. However, these words are indirectly represented by vectors whose components are null. At this point, we come to represent the comments of the corpus by two representations (Normalized co-occurrence and Emotional TF-IDF weighting). These representations are used in order to apply sentiment classification methods which described in the following section.

### 3.3   Comment Classification

In this paper, our main objective is to classify subjective texts (comments) into two classes (positive and negative). To achieve this objective, we propose two classification methods: by summation of elements of its vectors and by using a supervised algorithm SVM.

***Classification by Summation***. We start by using the elements constituting the vector of the comment in a simple and intuitive way without recourse to classical classification methods. The idea is to combine the elements of the vector having the same polarity, in order to calculate the positive and negative scores (see 5). Then, we compare the two scores to determine the polarity of the comment.

$$Score_{pos} = \sum_{j=1}^{4} s_{Cj}, \quad Score_{neg} = \sum_{j=5}^{8} s_{Cj} \tag{5}$$

***Supervised Classification with SVM***. According to the literature, Naive Bayes classifier and Support Vector Machines are the most celebrated algorithms in the field of supervised classification which provide better results in most cases. We introduce in this paper a method which aims to classify texts (comments) as positive or negative using Support Vector Machines, a well-known and powerful tool for classification of vector of real-valued features [13]. To do this, we must perform labelling manually in advance and prepare the comments that will be used for the training step to obtain a model. The goal is to automatically identify the class of a new comment of the test corpus using the learned SVM-model.

SVM is a machine learning classification technique which uses a function called a kernel to map a space of data points in which the data is not linearly separable onto a new space in which it is, with allowances for erroneous classification. At the implementation level, we used the libsvm tool[1] developed and updated by [4]. This tool provides several kernel types, kernel parameters and optimization parameters. In order to achieve best performing SVM, we empirically adjust these parameters. We choose to classify data with polynomial kernel (defined as $(gamma * u' * v + coef0)^{degree}$) type and change the parameter values related to this type (degree, gamma and coefficient).

# 4    Experiment Results and Discussion

To implement our proposed methods, we firstly prepare the dataset derived from Facebook. Indeed, we collect a set of comments from the political Tunisian pages. Then, we evaluate and validate the proposed sentiment classification methods by using external evaluation measure (F-score).

## 4.1    Facebook Dataset

***Collection***. The first step of our work is to collect huge amount of data, as shown in the Fig. 1. We chose to collect our corpus from social network Facebook in order to achieve a specific sentiment classification system to Facebook comment.

---

[1] SVM library available on the net (http://www.csie.ntu.edu.tw/cjlin/libsvm/).

The choice of this corpus is not due to coincidence, but to the fact that the Facebook user uses the emotion symbols in the majority of their comments. Facebook provides an API[2] that brings us a simple and consistent information about the comment object. We extracted the comments from the political Tunisian pages in the period [1-Jan-2011, 1-Aug-2012]. We used 22 political pages among the most popular in Tunisia during the revolution period.

We present our collected corpus as a set of multilingual comments (Tunisian dialect, standard Arabic, French, etc.) organized in a well structured XML file. We decompose our corpus into two sub-sets of comments. The first is based only on comments containing emotion symbols (7000 comments), for generating the emotional vectors of lexicon words. The second is based on 3000 comments manually examined and annotated by an expert (1314 positive comments and 1686 negative comments). It serves to the evaluation and validation of our system classification and the proposed representations. We divided the set of comments into two disjoint sub-sets: 2100 comments used to perform the training step, and 900 comments used to perform the test step.

***Pre-processing Corpus***. Before beginning our sentiment classification process, it is necessary to pre-process and homogenize the corpus. This step aims to avoid the noise (presence of the words deemed unnecessary). Thereby, it allows us to select the most significant and relevant words which express the sentiment clearly. In this step, we performed: **Character normalization:** we replace specific characters with a space. Indeed, we prepared a set of unpronounced characters that haven't any influence on the sentimental information. These characters can be attached to words or differently listed as sequences. For this reason, we replace them with a space. **Filtering:** we avoid the hyperlinks to external resources and @target user in order to keep only the useful words that reflect the semantic and sentimental content of the comments. **Translation into French:** We prepare an automatic program which uses the translation tool (Google translator which is the most popular), in order to render all the comments written in the same language French and unify the future treatment. **Lemmatization:** It consists in encompassing the words which have the same primary entity (lemma). **Stopwords removal:** we prepared our own Stopwords file containing the empty words as grammatical words and linking words.

## 4.2    Evaluation Results

In this paper, we propose two emotional vector representations for text based on emotion symbols, namely: normalized co-occurrence and emotional TFIDF. In order to evaluate these representations and test our proposed sentiment classification methods, we used the external evaluation techniques by measuring the global efficiency F-score[3] (see Table 1). From Table 1, we conclude that the best results are obtained with the handling of negation particles step. This step was

---

[2] https://developers.facebook.com/docs/reference/apis/.
[3] http://blog.onyme.com/apprentissage-artificiel-evaluation-precision-rappel-f-mesure/.

one of the factors that contributed significantly to the efficiency of our classifier. In fact, negation handling aims to reverse the polarity of all words directly preceded by one of these particles. Despite its simple idea, we found that the negation handling step improves the results with the two proposed representations of text. We can also notice that the emotional TFIDF weighting provided superior results when compared against normalized co-occurrence.

Using the method of comment classification by summation of the elements of the vector having the same polarity, we obtained an efficiency of 72.71% with the emotional TFIDF weighting and 70.46% with the normalized co-occurrence vector. This shows that the performance of our grouping of emotion symbols that are used to form the vector features (emotional states which represent the positive and negative sentiments). However, when we use the SVM classifier, the vector representation of comments plays its proper role in the statistical classification approach. We achieve more important results than using the classification by summation. We obtained an efficiency of 81.08% with the emotional TFIDF weighting and 75.18% with the normalized co-occurrence vector.

**Table 1.** The efficiencies achieved by the summation **SUM** and **SVM** methods using different vector representations of words with and without negation handling.

|  | Normalized co-occurrence | | Emotional-TFIDF weighting | |
| --- | --- | --- | --- | --- |
|  | Wo/Neg | W/Neg | Wo/Neg | W/Neg |
| **SUM** | 35.50% | 70.46% | 45.44% | 72.71% |
| **SVM** | 54.64% | 75.18% | 60.11% | 81.08% |

To evaluate the effectiveness of our emotional vector representation based on emoticons, we compared them with baselines representations previously reported in the literature. To do this, we implemented several alternative bag-of-words vector representations that are conceptually similar to our own, as discussed in Sect. 2 (see Table 1):

**Latent Semantic Analysis LSA:** we used the package R (lsa[4]) which provides us with a sentence-term matrix.

**Binary:** We developed the program which takes, as input, the lexicon words and generates, as output, a binary sentence-word matrix (1: presence, 0: absent).

**TF*IDF:** we used the TFIDF score [17] to generate the sentence-word matrix.

**TF*ΔIDF:** we used the sentimental score proposed by [10] to generate the sentence-word matrix.

**Word2Vec:** we used the open-source distributed deep-learning library written in Java[5], to compute the vector representations of words. In addition, we chose to apply the skip-gram model of word2Vec [11]. In fact, the skip-gram model aims to find word representations that are useful for predicting the surrounding

---

[4] https://cran.r-project.org/web/packages/lsa/lsa.pdf.
[5] http://deeplearning4j.org/word2vec.html.

words in a sentence. We constructed 100-dimensional word vectors for all lexicon words. Then, we compute the averaging of vectors representing the words which constitute a given comment. After this, supervised learning with SVM was performed using these vector representations.

**Table 2.** The efficiencies achieved by **SVM** classifier with different baseline vector representations and our emotional vectors.

| SVM classifier | F-score |
|---|---|
| *Baselines features* | |
| LSA (Sentence-Word matrix) | 67.24% |
| Bag of words (binary) | 63.07% |
| TF*IDF Weighting | 69.65% |
| TF*$\Delta$IDF Weighting | 71.78% |
| Word2Vec + Average | 71.09% |
| *Our features* | |
| Normalized co-occurrence + Negation | 75.18% |
| Emotional TF-IDF weighting + Negation | 81.08% |

Table 2 shows the obtained results (F-score) using SVM classifier with different representations of text based only on words (baseline feature) and our emotional vector representations based on emoticons. In baseline representation, we find that the TF*$\Delta$IDF and word2vec work well in sentiment analysis problem. We also notice that our emotional vector representation clearly outperform those baseline representations. This shows that the usage of the emotion symbols in the text representation increases the performance of sentiment classification.

## 5    Conclusion and Future Work

In this paper, we presented a new vector representation of text for sentiment classification. First of all, we started by representing the words in an emotional vector based on the emoticons. The ultimate goal is to represent the different comments by numerical vectors capable of faithfully translate their sentimental orientations. Then, we proposed comment classification methods, namely: the classification by summing the vector elements and the supervised classification with the SVM. Finally, we presented the experimental results obtained by our proposed method. Our results are also effective and consistent. In the word representation step, we used only the comments having emotion symbols. Thus, we kept all infrequent words because these words have the chance to be known from other comments and to have higher frequencies. For this reason, it is very interesting to think of an enrichment step. Therefore, we thought, in future work, to take advantage of the other comments which haven't got any emotion symbol in order to adjust the feature vectors of words and comments and to add new words to the lexicon.

# References

1. Alghunaim, A., Mohtarami, M., Cyphers, S., Glass, J.: A vector space approach for aspect based sentiment analysis. In: Proceedings of NAACL-HLT (2015)
2. Ameur, H., Jamoussi, S.: Dynamic construction of dictionaries for sentiment classification. In: 13th IEEE International Conference on Data Mining Workshops, ICDM Workshops, TX, USA, pp. 896–903 (2013)
3. Cambria, E., Schuller, B., Xia, Y., Havasi, C.: New avenues in opinion mining and sentiment analysis. IEEE Intell. Syst. **28**, 15–21 (2013)
4. Chang, C.C., Lin, C.J.: LIBSVM: a library for support vector machines. ACM Trans. Intell. Syst. Technol. **2**, 1–27 (2011)
5. Gezici, G., Yanikoglu, B., Tapucu, D., Saygn, Y.: New features for sentiment analysis: do sentences matter? (2012)
6. Harb, A., Plantié, M., Dray, G., Roche, M., Trousset, F., Poncelet, P.: Web opinion mining: how to extract opinions from blogs? In: Proceedings of the 5th International Conference on Soft Computing as Transdisciplinary Science and Technology, New York, NY, USA, pp. 211–217 (2008)
7. Hogenboom, A., Bal, M., Frasincar, F., Bal, D., Kaymak, U., de Jong, F.: Lexicon-based sentiment analysis by mapping conveyed sentiment to intended sentiment. Int. J. Web Eng. Technol. **9**, 125–147 (2014)
8. Kamps, J., Marx, M., Mokken, R.J., de Rijke, M.: Using wordnet to measure semantic orientations of adjectives. In: Proceedings of LREC-04, 4th International Conference on Language Resources and Evaluation, vol. 4, pp. 1115–1118 (2004)
9. Kim, S.M., Pantel, P., Chklovski, T., Pennacchiotti, M.: Automatically assessing review helpfulness. In: Proceedings of the 2006 Conference on Empirical Methods in Natural Language Processing, Stroudsburg, PA, USA, pp. 423–430 (2006)
10. Martineau, J., Finin, T.: Delta TFIDF: an improved feature space for sentiment analysis. In: Proceedings of the Third International Conference on Weblogs and Social Media, San Jose, California, USA, pp. 258–261 (2009)
11. Mikolov, T., tau Yih, W., Zweig, G.: Linguistic regularities in continuous space word representations. In: Proceedings of NAACL-HLT, pp. 746–751 (2013)
12. Pak, A., Paroubek, P.: Construction dun lexique affectif pour le franais partir de Twitter. In: TALN 2010. Université de Paris-Sud, Orsay Cedex, France (2010)
13. Pang, B., Lee, L.: A sentimental education: sentiment analysis using subjectivity summarization based on minimum cuts. In: Proceedings of the 42nd Annual Meeting on ACL, Stroudsburg, PA, USA, pp. 271–278 (2004)
14. Poirier, D.: Des textes communautaires à la recommandation. Ph.D. thesis, Université d'Orléans (2011)
15. Qu, L., Ifrim, G., Weikum, G.: The bag-of-opinions method for review rating prediction from sparse text patterns. In: Proceedings of the 23rd International Conference on Computational Linguistics, Stroudsburg, PA, USA, pp. 913–921 (2010)
16. Schütze, H.: Dimensions of meaning. In: Proceedings of Supercomputing 1992 (1992)
17. Turney, P.D., Pantel, P.: From frequency to meaning: vector space models of semantics. J. Artif. Int. Res. **37**, 141–188 (2010)
18. Wang, H., Lu, Y., Zhai, C.: Latent aspect rating analysis on review text data: a rating regression approach. In: Proceedings of the 16th ACM SIGKDD International Conference on Knowledge Discovery and Data Mining, New York, NY, USA, pp. 783–792 (2010)
19. Wilson, T., Wiebe, J., Hwa, R.: Just how mad are you? Finding strong and weak opinion clauses. In: Proceedings of the 19th National Conference on Artifical Intelligence, San Jose, California, pp. 761–767 (2004)

# Cascading Classifiers for Twitter Sentiment Analysis with Emotion Lexicons

Hiram Calvo[1(⊠)] and Omar Juárez Gambino[1,2]

[1] Centro de Investigación en Computación,
CIC-IPN, J.D. Bátiz e/ M.O. de Mendizábal, 07738 Mexico City, Mexico
hcalvo@cic.ipn.mx, b150697@sagitario.cic.ipn.mx
[2] Escuela Superior de Cómputo, ESCOM-IPN, J.D. Bátiz e/ M.O. de Mendizábal,
07738 Mexico City, Mexico

**Abstract.** Many different attempts have been made to determine sentiment polarity in tweets, using emotion lexicons and different NLP techniques with machine learning. In this paper we focus on using emotion lexicons and machine learning only, avoiding the use of additional NLP techniques. We present a scheme that is able to outperform other systems that use both natural language processing and distributional semantics. Our proposal consists on using a cascading classifier on lexicon features to improve accuracy. We evaluate our results with the TASS 2015 corpus, reaching an accuracy only 0.07 below the top-ranked system for task 1, 3 levels, whole test corpus. The cascading method we implemented consisted on using the results of a first stage classification with Multinomial Naïve Bayes as additional columns for a second stage classification using a Naïve Bayes Tree classifier with feature selection. We tested with at least 30 different classifiers and this combination yielded the best results.

## 1 Introduction

Sentiment analysis (SA) can be defined as the computational treatment of opinion, sentiment and subjectivity in texts [1]. It is a hard task because even humans often disagree on the sentiment of a given text, and becomes harder when the text is very short, such as in tweets. SA can be tackled as a supervised classification problem, where classes represent the polarity of the expressed opinions (*i.e.*, positive or negative opinions). Features for this classification range from word space models alone, to adding information from external lexicons (*v.gr.* emotion lexicons) as well as incorporating syntactic information. In [2] dictionaries of words annotated with their semantic orientation or polarity were created and used for classifying the polarity of different user's reviews. Every word of the reviews was compared to the words on the dictionaries in order to find a match; if they matched, the polarity of the words was used to determine the global sentiment polarity of the review. For English, numerous lexicons have been created over the years, for instance: SentiWordnet [3], OpinionFinder [4], the Harvard

© Springer International Publishing AG, part of Springer Nature 2018
A. Gelbukh (Ed.): CICLing 2016, LNCS 9624, pp. 270–280, 2018.
https://doi.org/10.1007/978-3-319-75487-1_21

inquirer [5] and LIWC [6]. Most of the sentiment analysis research in Spanish has used lexicons translated from English, although there are several lexicons specially created for this language. In this work, we focus on using the Spanish Emotion Lexicon (SEL) [7], a resource created specifically for Spanish; and LIWC [6], that has a special list of Spanish words.

One of the most known efforts for Spanish sentiment analysis is the TASS workshop organized by SEPLN (Spanish society for natural language processing). This workshop has spawned a tagged corpus of Twitter posts [8] (tweets), annotated with their global sentiment polarity. The workshop has been celebrated annually since 2012 and recurrently has included a task for determining the global polarity of every tweet in the corpus.

A comprehensive list of participants of TASS 2015 [9] is shown in Table 1. This Table shows the highest accuracy obtained by each system, and the combination of lexicons they used. The last six dictionaries were directly translated from English. The shown accuracy accounts for the three level task, *i.e.*, *positive*, *negative* and *neutral* opinions. In total, there are four classes, including the previous three levels plus a *none* class. Some systems, such as SINAI, INGEOTEC,

**Table 1.** Affective lexicons used and their accuracy for task 1, 4 classes, whole test corpus. Best run reported for each system. (*indicates only 6 classes accuracy available)

| System | Accuracy | BM25[10] | ElhPolar [11] | SOCAL [12] | SOL [13] | iSOL [14] | SSL [15] | SEL [7] | Own | DAL [16] | ANEW [17] | LYSA [18] | ML senticon [19] | SentiWordNet [20] | Q-WordNet [21] | Semeval 2013 [22] | MPQA [23] | HGI [24] |
|---|---|---|---|---|---|---|---|---|---|---|---|---|---|---|---|---|---|---|
| LIF | 0.726 | | X | | | X | | | | | | X | | | | | X | X |
| ELiRF | 0.725 | | | | | X | | | | | | | | X | | X | | |
| GTI-GRAD | 0.695 | | | X | X | | | | | X | X | | | | | | | |
| GSI (aspect) | 0.69 | | X | X | | X | X | | | | | X | | | | | | |
| LYS | 0.664 | | | | | | | | | | | | | | | | | |
| DLSI | 0.655 | | | | | | | X | | | | | | | | | | |
| SINAI-DW2Vec | 0.619 | | | | | | | | | | | | | | | | | |
| INGEOTEC | 0.613 | | | | | | | | | | | | | | | | | |
| UCSP | 0.613 | | | | | | | | | | | | | | | | | |
| ITAINNOVA | 0.61 | | | | | | | | | X | X | | | | | | | |
| BittenPotato | 0.602 | | | | | | | | | | | | | | | | | |
| DeustoTech | 0.601 | X | X | | X | | | | | | | X | | | | | | |
| CU | 0.597 | | | | | | | | | X | | | | | X | | | |
| TID-Spark | 0.594 | | | | | | | | | | | | | | | | | |
| SINAI-EMMA | 0.502* | | | | | X | | X | | | | | | X | X | X | | |

**Table 2.** Classifiers used for systems in Table 1. (AB = AdaBoost, RF = Random Forest, DT = Decision Trees, RBF = Radial Basis Function, LR = Logistic Regression, SVM = Support Vector Machines, ME = MaxEnt, SG = SkipGrams)

| System | Accuracy | Max n-gram | NER | NLP | Negation | Word2Vec | Doc2Vec | GloVe | TF·IDF | LDA | LSI | Classifier |
|---|---|---|---|---|---|---|---|---|---|---|---|---|
| LIF | 0.726 | | | X | X | X | | | | | X | (SVM SG Cbow)→SVM |
| ELiRF | 0.725 | | | X | | | | | X | | | SVM (+ SVM) |
| GTI-GRAD | 0.695 | 2 | | X | | | | | | | | Logistic regression |
| GSI (aspect) | 0.69 | | X | X | X | | | | | | | SVM |
| LYS | 0.664 | | | X | | | | | | | | Logistic regression L2-LG |
| DLSI | 0.655 | 2 | | X | | | | | | | | SVM |
| SINAI-DW2Vec | 0.619 | | | | | X | X | | | | | SVM |
| INGEOTEC | 0.613 | 5 | | | | | | | X | X | X | SVM |
| UCSP | 0.613 | X | | X | | | | | X | | X | (SVM NB ME DT)→voting |
| ITAINNOVA | 0.61 | 3 | | X | | | | | X | | | MaxEnt, RF |
| BittenPotato | 0.602 | 3 | | | | | | | X | | | (SVM AB RF LR)→RBF |
| DeustoTech | 0.601 | 1 | | X | X | | | | | | | SVM |
| CU | 0.597 | | | X | | | | | | | | Logistic regression |
| TID-Spark | 0.594 | | | | | | | | X | | | SVM + hard clustering |
| SINAI-EMMA | 0.502* | 2 | | | | | | | | | | SVM |

UCSP and BittenPotato did not use lexicons at all, but they used other resources such as NLP or distributional semantics. These resources, along with the machine learning techniques they used, are listed in Table 2.

Most works use Support Vector Machines (SVM) with a linear kernel, although, as one would expect, other kernels are able to yield a better performance. The best system uses a ensemble of SVM, and Convolutional networks with skipgrams, bag of words and vectors obtained from Glove [25]. These results are fed to an SVM classifier. ELiRF, the second best system in TASS 2015 combines the output of several SVM classifiers with different parameters and then this information is classified in cascade with another SVM classifier. These two examples motivated us to explore the cascading effect with another classifiers. We were interested on measuring to which extent it is possible to obtain a benefit from adding the outcome of a previous classifier as new columns to classify the same training set. This might resemble the AdaBoost algorithm [26], but in that technique examples are re-weighted to obtain a new classification, and at the end classifiers are combined as an ensemble. Contrary to this, in cascading, only additional columns are added to the training vectors.

In this paper we describe several experiments using cascaded classifiers for finding sentiment polarity on Tweets performed on the Spanish corpus used in TASS workshop [8]. In the following sections we describe the sentiment analysis problem and the selection of the Spanish lexical resources (Sect. 2); in Sect. 3 we detail the results of our experiments; and finally we draw our conclusions in Sect. 4.

## 2    Resources for Sentiment Analysis

The sentiment analysis problem is a natural language processing task that has attracted the scientific community attention during the last years. Thanks to the social networks, a lot of opinions are publicly available. Twitter is a social network in which users post messages known as tweets. Due to the shortness of tweets, the difficulty for determining the sentiment polarity increases.

**TASS.** The TASS workshop organized by SEPLN [8] task 1: Sentiment Analysis in Tweets, developed a Spanish corpus composed by 68,017 tweets that are publicly available for research purposes.

Tweets from the general corpus were collected from 150 well-known personalities and celebrities of the world of politics, economy, communication, mass media and culture, between November 2011 and March 2012. The context of extraction has a Spain-focused bias but the nationality of the authors includes people from Spain, Mexico, Colombia, Puerto Rico, USA and many other countries. Covered topics are: politics, entertainment, economy, music, soccer, films, technology, sports, literature, and others.

The general corpus has been divided into training set (about 10%) and test set (90%). All tagging has been done semi-automatically: a baseline machine learning model is first run and then all tags are manually checked by human experts.

There were three different tasks on the 2015 edition of TASS. We focused on Task 1: Sentiment Analysis at Global Level. This task had been carried out in previous editions. It consists on performing an automatic polarity classification to determine the global polarity of each message in the test set of the General corpus. Participants were provided with the training set of the General corpus so that they may train and validate their models. There were two different evaluations: one based on 6 different polarity labels (P+, P, NEU, N, N+, NONE) and another based on just 4 labels (P, N, NEU, NONE). All results reported in this paper (ours and foreign) are based on the four labels evaluation.

**SEL.** [7] is a dictionary marked with emotions and polarity. Selected words are tagged with six basic emotions (total number of words of each category shown in parentheses): joy (668), anger (382), fear (211), sadness (391), surprise (175) and disgust (209). Total: 2036 words. It was manually created by multiple anotators (at least 15 for each word). Unlike other dictionaries, the SEL dictionary contains weightings that correspond to a probability of each word of being used with a sense related to emotion. This is called probability factor of affective use (PFA). The kappa agreement for each category ranges from 0.7 to 0.8.

The use of this resource has been reported in [27] for Twitter Sentiment Analysis.

**LIWC.** [6] is a dictionary that classifies words in emotional, cognitive and structural component categories. LIWC was developed for studying the psycholinguistic concerns dealing with the therapeutic effect of verbally expressing emotional experiences and memories. LIWC provides a Spanish dictionary composed by 7,515 words and word stems. Each word can be classified into one or more of 72 categories. These are classified in four groups: linguistic processes (pronouns, articles, prepositions, numbers, negations), psychological processes (affective words, positive, negative emotions, cognitive process, perceptual process), relativity (time, space, motion), and personal concerns (occupation, leisure activity, money/financial issues, religion, death and dying).

LIWC has been used for opinion mining in Spanish in [28]. To our knowledge, it has not been included in Twitter sentiment analysis.

**Training Vectors.** Our training vectors are composed then, by a lemmatized and binarized bag of words, the accumulated FPA for each category from the SEL dictionary (6 columns), and the accumulated count of words for each LIWC category (72 columns), yielding a total of 12,620 columns[1].

## 3    Experiments

In this section we present the results of our experiments. We used two emotion dictionaries in all our experiments: SEL [7] and LIWC [6]. From the first one, we extracted the Affective Factor (PFA) for each word, the grouped PFA (positive and negative emotions), and from the second one, we used all features, including "positive emotion" and "negative emotion". Using a Multinomial Naïve Bayes classifier, we obtained an accuracy of 0.608 which would ranked us in 13th place in Table 1, despite the simplicity of the technique we are using. Note that no NLP techniques are being used apart from stemming, i.e., we do not use n-grams, stopwords, part of speech tagging or syntax analysis.

Attribute Selection is a widely known technique recommended to avoid redundant features and improving up the learning process [29]. We performed a Correlation-based Feature Subset Selection [30] with a Best First search method. We can see from Table 3 that in this case feature selection did not help improving performance.

---

[1] Our vectors can be downloaded at http://likufanele.com/twitterSEL as ARFF files.

**Table 3.** Accuracy on the TASS 2015 corpus without cascading.

| Features | Experiment | Accuracy | ROC area | Time (s) |
|---|---|---|---|---|
| All (12,620) | Multinomial NB | 0.608 | 0.832 | 0.1 |
| | SVM polynomial | 0.570 | 0.696 | 17.7 |
| | Naïve Bayes | 0.515 | 0.750 | 19.2 |
| Attr. sel. (59 feat.) | MLP | 0.548 | 0.757 | 134.8 |
| | NB | 0.550 | 0.754 | 0.1 |
| | Multinomial NB | 0.518 | 0.780 | $\leq$0.1 |
| | SVM polynomial | 0.422 | 0.572 | 1216.4 |
| | NB tree | 0.476 | 0.704 | 43.2 |

**Cascading.** We used the training part of TASS to train a Multinomial Naïve Bayes classifier; then, we appended the resulting probabilities of each class (positive, negative, neutral and none) as columns of new training examples. We repeated this process up to four iterations. We present in Table 4 the evaluation with regard to the test set for each iteration. Iteration zero means no cascading.

**Table 4.** Cascading with Multinomial Naïve Bayes. Accuracy on the TASS 2015 corpus.

| Iteration | Accuracy |
|---|---|
| 0 | 0.608 |
| 1 | 0.623 |
| 2 | 0.627 |
| 3 | 0.629 |
| 4 | 0.629 |

Interestingly, we can see that by simple cascading the Multinomial Naïve Bayes (MNB) classifier, we obtained an improvement in accuracy, taking us from 13th place to 8th. After the fourth iteration we found no further improvement.

In order to determine if other tasks could be benefited from this cascading scheme, we tested with the movie review corpus from Pang *et al.* [31]. We obtained the results shown in Table 5. We found that MNB cascading did not improve accuracy on the standard test of the Movie reviews corpus (three fold cross validation). We tried to match TASS 2015 conditions, that have a very small section for training (10%) compared with a large collection of examples for testing (90%). By segmenting the movie review corpus in this way, we could see a very slight improvement when using MNB Cascading. Finally, we observed that if we reduced the number of classes of the TASS corpus from 4, eliminating "neutral" and "non" the improvement was very small too. This leads us to think that the benefit of Cascaded MNB is limited to cases when there is a small set of training examples along with more than two classes.

**Table 5.** Cascading with multinomial Naïve Bayes. Accuracy on different corpora.

| Iteration | Accuracy | Difference |
|-----------|----------|------------|
| Movie reviews: 3-fold cross validation | | |
| 0 iter | 0.8090 | |
| 1 iter | 0.8083 | −0.0007 |
| 2 iter | 0.8076 | −0.0007 |
| Movie reviews: 10% train, 90% test | | |
| 0 iter | 0.7242 | |
| 1 iter | 0.7244 | 0.0002 |
| 2 iter | 0.7244 | 0 |
| TASS 2 class: positive, negative | | |
| 0 iter | 0.8321 | |
| 1 iter | 0.8342 | 0.0021 |
| 2 iter | 0.8342 | 0 |

Finally, we aimed to experiment with the effects of simple cascading with other classifiers. In all cases we tested with only one iteration. In all cases the first classification stage used Multinomial Naïve Bayes (MNB). Then, we tested with different classifiers for the second stage. Results are shown in Table 6. We performed experiments both without attribute selection (see lower part of Table 6) and with attribute selection (see the top part of Table 6). As in the previous case, we used Correlation-based Feature Subset Selection [30] with a Best First search method. In this case we can see a great improvement in accuracy when using attribute selection on previously classified training examples (*i.e.*, using cascading). We obtained the best score with a Naive Bayes Tree classifier. Although Data Near Balanced Nested Dichotomies [32] is close in accuracy to NB Tree, note that the ROC area under the curve is significantly lower, meaning that results are prone to change if we added more data. Using Multinomial Naïve Bayes as a second stage cascade classifier yields similar results no matter if we used attribute selection or not. Oveall, the best accuracy we obtained was 0.656 which ranks us in 6th place of the TASS 2015 task.

**Table 6.** Accuracy on the TASS 2015 corpus with cascading.

| Features | Experiment | Accuracy | ROC Area | Time(s) |
|---|---|---|---|---|
| Attr. Sel. (73 feat.) | Naïve Bayes Tree | 0.656 | 0.841 | 29.2 |
| | DataNearBalancedND (J48) | 0.653 | 0.794 | 2.4 |
| | Nested Dichotomies | 0.653 | 0.794 | 1.2 |
| | Random Subspace | 0.650 | 0.848 | 3.3 |
| | Logit Boost | 0.650 | 0.847 | 5.8 |
| | Ordinal classifier | 0.649 | 0.790 | 1.8 |
| | Classification via regression | 0.648 | 0.841 | 7.5 |
| | Decision Table | 0.648 | 0.839 | 4.7 |
| | END | 0.646 | 0.818 | 8.1 |
| | Raced Incremental LogitBoost | 0.645 | 0.838 | 1.6 |
| | Multilayer Perceptron | 0.645 | 0.830 | 174.2 |
| | Decorate | 0.644 | 0.827 | 134.2 |
| | Filtered Classifier | 0.643 | 0.826 | 0.4 |
| | REPTree | 0.643 | 0.828 | 0.8 |
| | Bagging | 0.643 | 0.839 | 6.5 |
| | BFTree | 0.640 | 0.809 | 4.1 |
| | RotationForest | 0.638 | 0.815 | 8.3 |
| | SVM - Polynomial | 0.638 | 0.736 | 3.2 |
| | J48 (Graft) | 0.636 | 0.780 | 1.2 |
| | Nested Dichotomies ND | 0.636 | 0.776 | 1.7 |
| | PART | 0.635 | 0.819 | 4.7 |
| | SimpleCart Trees | 0.634 | 0.789 | 4.7 |
| | DTNB | 0.634 | 0.848 | 1188.2 |
| | Bayes Net | 0.633 | 0.852 | 0.2 |
| | Complement Naïve Bayes | 0.630 | 0.732 | 63.0 |
| | Logistic Regression | 0.630 | 0.833 | 2.8 |
| | Forest Trees | 0.629 | 0.823 | 20.3 |
| | Multinomial Naïve Bayes | 0.629 | 0.832 | 0.1 |
| | JRip | 0.629 | 0.734 | 5.8 |
| | Simple Logistic | 0.628 | 0.836 | 70.4 |
| | SVM - Linear kernel | 0.627 | 0.728 | 7.1 |
| | SMO | 0.627 | 0.746 | 7.0 |
| | Ridor | 0.626 | 0.731 | 5.7 |
| | Naïve Bayes | 0.626 | 0.819 | 0.1 |
| | Dagging | 0.625 | 0.746 | 0.6 |
| | RandomForest | 0.614 | 0.799 | 2.4 |
| All (12,624 feat.) | Naïve Bayes Updeatable | 0.629 | 0.835 | 0.1 |
| | Multinomial Naïve Bayes | 0.628 | 0.836 | 0.1 |
| | SVM - Linear Kernel | 0.619 | 0.733 | 23.2 |
| | Nested Dichotomies ND | 0.612 | 0.736 | 668.2 |
| | Bayes Net | 0.610 | 0.826 | 46.2 |

# 4   Conclusions and Future Work

We have presented results of using a cascaded classification method on the Sentiment Analysis Workshop 2015 of the Spanish Society for Natural Language Processing (TASS 2015, SEPLN acronyms from Spanish). By using a Naïve Bayes Tree classifier on filtered pre-classified training data using a Multinomial Naïve Bayes classifier we obtained an accuracy of 0.656. Data was filtered with Correlation-based Feature Subset Selection with a Best First search method, reducing from 12,624 to only 73 features. Among the 73 features, the four features corresponding to the previous classification stage were automatically selected. When applying feature selection to data without previous classification (*i.e.*, no cascading), accuracy was markedly decreased (in a range of 6 to 15%).

We explored using several classifiers for the second stage classifier. For the special case of using a MNB classifier again, results were nearly the same with and without feature selection. We experimented with several iterations of cascading. For the particular case of the TASS Task 1, four classes, we obtained an increase on accuracy on each iteration up to the 3rd one. Afterwards, accuracy was degraded. We applied this method to other corpora, such as the Movie Review corpus by Pang *et al.*, and found no improvement. When we reduced the number of classes from the TASS corpus from four to only two (positive and negative) we found no improvement as well, when using the cascading technique. This leads us to think that our presented technique would be useful when having a small training set, with few examples, and a highly uneven distribution among the classes in this set. Testing with other corpus of similar characteristics has been left as a future work.

We avoided using NLP techniques, even simple ones such as stopwords, or POS Tag; however, other resources such as distributional semantics, that need no manual crafting of resources could be explored as well. In this paper we have only experimented with a single MNB classifier as the first stage; as future work, we plan to try other classifiers in this first stage, as well as different combinations of them.

**Acknowledgments.** We thank the support of Instituto Politécnico Nacional (IPN), ESCOM-IPN, SIP-IPN projects number 20160815, 20162058, COFAA-IPN, and EDI-IPN.

# References

1. Pang, B., Lee, L.: Opinion mining and sentiment analysis. Found. Trends Inf. Retr. **2**, 1–135 (2008)
2. Taboada, M., Brooke, J., Tofiloski, M., Voll, K.D., Stede, M.: Lexicon-based methods for sentiment analysis. Comput. Linguist. **37**, 267–307 (2011)
3. Baccianella, S., Esuli, A., Sebastiani, F.: SentiWordNet 3.0: an enhanced lexical resource for sentiment analysis and opinion mining. In: Calzolari, N., Choukri, K., Maegaard, B., Mariani, J., Odijk, J., Piperidis, S., Rosner, M., Tapias, D. (eds.) LREC. European Language Resources Association, Paris (2010)

4. Wilson, T., Hoffmann, P., Somasundaran, S., Kessler, J., Wiebe, J., Choi, Y., Cardie, C., Riloff, E., Patwardhan, S.: OpinionFinder: a system for subjectivity analysis. In: Proceedings of HLT/EMNLP on interactive demonstrations, pp. 34–35. Association for Computational Linguistics (2005)
5. Stone, P.J.: The General Inquirer: A Computer Approach to Content Analysis. User's Manual. MIT Press, Cambridge (1968)
6. Pennebaker, J.W., Francis, M.E., Booth, R.J.: Linguistic Inquiry and Word Count: LIWC 20001, vol. 71. Lawrence Erlbaum Associates, Mahway (2001)
7. Rangel, I.D., Guerra, S.S., Sidorov, G.: Creación y evaluación de un diccionario marcado con emociones y ponderado para el español. Onomazein **29**, 31–46 (2014)
8. Villena Román, J., Lana Serrano, S., Martínez Cámara, E., González Cristóbal, J.C.: TASS-workshop on sentiment analysis at SEPLN (2013)
9. Villena-Román, J., García Morera, J., García-Cumbreras, M.Á., Martínez-Cámara, E., Martín-Valdivia, M.T., Ureña López, L.A.: Overview of TASS 2015. In: TASS 2015: Workshop on Sentiment Analysis at SEPLN, vol. 1397. CEUR-WS.org (2015)
10. Robertson, S.E., Walker, S., Jones, S., Hancock-Beaulieu, M.M., Gatford, M., et al.: Okapi at TREC-3, P. 109. NIST Special Publication (1995)
11. Saralegi, X., San Vicente, I.: Elhuyar at TASS 2013. In: XXIX Congreso de la Sociedad Espaola de Procesamiento de lenguaje natural, Workshop on Sentiment Analysis at SEPLN (TASS 2013), pp. 143–150 (2013)
12. Taboada, M., Brooke, J., Tofiloski, M., Voll, K., Stede, M.: Lexicon-based methods for sentiment analysis. Comput. Linguist. **37**, 267–307 (2011)
13. Martınez-Cámara, E., Martın-Valdivia, M., Molina-González, M., Urena-López, L.: Bilingual experiments on an opinion comparable corpus. In: WASSA 2013, p. 87 (2013)
14. Molina-González, M.D., Martínez-Cámara, E., Martín-Valdivia, M.T., Perea-Ortega, J.M.: Semantic orientation for polarity classification in Spanish reviews. Expert Syst. Appl. **40**, 7250–7257 (2013)
15. Perez-Rosas, V., Banea, C., Mihalcea, R.: Learning sentiment lexicons in Spanish. In: LREC, vol. 12, p. 73(2012)
16. Rıos, M.G.D., Gravano, A.: Spanish DAL: a Spanish dictionary of affect in language. In: WASSA 2013, p. 21 (2013)
17. Redondo, J., Fraga, I., Padrón, I., Comesaña, M.: The Spanish adaptation of ANEW. Behav. Res. Methods **39**, 600–605 (2007)
18. Vilares, D., Doval, Y., Alonso, M.A., Gómez-Rodríguez, C.: LyS at TASS 2014: a prototype for extracting and analysing aspects from Spanish tweets. In: Proceedings of the TASS workshop at SEPLN (2014)
19. Cruz, F.L., Troyano, J.A., Pontes, B., Ortega, F.J.: ML-SentiCon: un lexicón multilingüe de polaridades semánticas a nivel de lemas. Procesamiento del Leng. Nat. **53**, 113–120 (2014)
20. Esuli, A., Sebastiani, F.: SENTIWORDNET: a publicly available lexical resource for opinion mining. In: Proceedings of LREC, vol. 6, pp. 417–422 (2006)
21. Agerri, R., García-Serrano, A.: Q-WordNet: extracting polarity from WordNet senses. In: LREC (2010)
22. Manandhar, S., Yuret, D.: Second joint conference on lexical and computational semantics (*SEM). In: Proceedings of the Seventh International Workshop on Semantic Evaluation (SemEval 2013), vol. 2 (2013)
23. Deng, L., Wiebe, J.: MPQA 3.0: an entity/event-level sentiment corpus. In: Conference of the North American Chapter of the Association of Computational Linguistics: Human Language Technologies (2015)

24. Stone, P.J., Dunphy, D.C., Smith, M.S.: The general inquirer: a computer approach to content analysis (1966)
25. Pennington, J., Socher, R., Manning, C.D.: GloVe: global vectors for word representation. In: EMNLP, vol. 14, pp. 1532–1543 (2014)
26. Collins, M., Schapire, R.E., Singer, Y.: Logistic regression, Adaboost and Bregman distances. Mach. Learn. **48**, 253–285 (2002)
27. Martínez-Cámara, E., García-Cumbreras, M., Martín-Valdivia, M.T., Ureña López, L.A.: SINAI-EMMA: vectores de palabras para el análisis de opiniones en Twitter. In: TASS 2015: Workshop on Sentiment Analysis at SEPLN, vol. 1397. CEUR-WS.org (2015)
28. del Pilar Salas-Zárate, M., López-López, E., Valencia-García, R., Aussenac-Gilles, N., Almela, Á., Alor-Hernández, G.: A study on LIWC categories for opinion mining in spanish reviews. J. Inf. Sci. **40**, 749–760 (2014)
29. Serendero, P., Toro, M.: Attribute selection for classification. In: Proceedings International Conference e-Society (IADIS)', Lisbon, Portugal (2003)
30. Hall, M.A.: Correlation-based feature selection for machine learning. PhD thesis, The University of Waikato (1999)
31. Pang, B., Lee, L., Vaithyanathan, S.: Thumbs up? Sentiment classification using machine learning techniques. In: Proceedings of EMNLP, pp. 79–86 (2002)
32. Dong, L., Frank, E., Kramer, S.: Ensembles of balanced nested dichotomies for multi-class problems. In: Jorge, A.M., Torgo, L., Brazdil, P., Camacho, R., Gama, J. (eds.) PKDD 2005. LNCS (LNAI), vol. 3721, pp. 84–95. Springer, Heidelberg (2005). https://doi.org/10.1007/11564126_13

# A Multilevel Approach to Sentiment Analysis of Figurative Language in Twitter

Braja Gopal Patra[1]([⊠])(iD), Soumadeep Mazumdar[2], Dipankar Das[1],
Paolo Rosso[3], and Sivaji Bandyopadhyay[1]

[1] Department of Computer Science and Engineering,
Jadavpur University, Kolkata, India
brajagopal.cse@gmail.com, dipankar.dipnil2005@gmail.com,
sivaji_cse_ju@yahoo.com
[2] Microsoft India Development Center, Hyderabad, India
mazumdar.soumadeep@gmail.com
[3] PRHLT Research Center, Universitat Politècnica de València, Valencia, Spain
prosso@dsic.upv.es

**Abstract.** Commendable amount of work has been attempted in the field of Sentiment Analysis or Opinion Mining from natural language texts and Twitter texts. One of the main goals in such tasks is to assign polarities (positive or negative) to a piece of text. But, at the same time, one of the important as well as difficult issues is how to assign the degree of positivity or negativity to certain texts. The answer becomes more complex when we perform a similar task on figurative language texts collected from Twitter. Figurative language devices such as irony and sarcasm contain an intentional secondary or extended meaning hidden within the expressions. In this paper we present a novel approach to identify the degree of the sentiment (fine grained in an 11-point scale) for the figurative language texts. We used several semantic features such as sentiment and intensifiers as well as we introduced sentiment abruptness, which measures the variation of sentiment from positive to negative or vice versa. We trained our systems at multiple levels to achieve the maximum cosine similarity of 0.823 and minimum mean square error of 2.170.

**Keywords:** Figurative text · Sentiment analysis
Sentiment abruptness measure · Irony · Sarcasm · Metaphor

## 1 Introduction

With the rapid expansion of social media, a variety of user generated contents become available online. However, the major challenges are how to process the user generated contents such as texts, audio and images and how to organize them in some meaningful ways. It is observed that the existing systems achieved

© Springer International Publishing AG, part of Springer Nature 2018
A. Gelbukh (Ed.): CICLing 2016, LNCS 9624, pp. 281–291, 2018.
https://doi.org/10.1007/978-3-319-75487-1_22

promising results for identifying opinions or sentiments along with polarities in case of literal language, because there is no secondary meaning embedded within it. In contrast, extracting sentiments from figurative language is one of the most challenging tasks in Natural Language Processing (NLP) because the literal meaning are discontinued and secondary or extended meanings are intentionally profiled. The affective polarity of the literal meaning may differ significantly from that of the intended figurative meaning [1]. Again, identifying the degree of the sentiment from these figurative texts are much more difficult. Figurative language contains several categories of tweets such as irony, sarcasm, and metaphor. The example below is a sarcastic tweet together with its degree of polarity in brackets.

*you're such a cunt, I hope you're happy now #sarcasm (−4)*

The study in Ghosh et al. [1] shows that metaphor, irony and sarcasm can each sculpt the sentiment of an utterance in complex ways, and texts limits the conventional techniques for the sentiment analysis of supposedly literal texts. For this reason, the analysis of sentiment degree in figurative language is considered to be a difficult tasks.

In this paper we present a novel approach for fine grained sentiment analysis of figurative language. Along with the features like parts of speech (POS), sentiment and intensifier, we also employed as further feature, "sentiment abruptness". We developed a multilevel classification framework to improve the performance of the system.

The rest of the paper is organized as follows: in Sect. 2, we describe the related work carried out in sentiment analysis of figurative languages. Section 3 describes the dataset. In Sect. 4, we describe the features and introduce the sentiment abruptness measure. The proposed multilevel system and its evaluations are described in Sect. 5. Finally, we concluded our study with future work in Sect. 6.

## 2    Related Work

Sentiment Analysis or Opinion Mining refers to the process of identifying the subjective responses or opinions about a specific topic. Much research have been conducted in the fields of opinion mining [18], sentiment extraction [10], emotion analysis [17] and review sentiment analysis [4]. Recent publications in the field of sentiment analysis were based on user generated data collected from different social media platforms like Facebook, Twitter, reviews and blogs etc. [2,4,19].

Extracting polarity (positive or negative) from the text is one of the tasks in sentiment analysis and used to achieve high accuracies [1]. Unfortunately, sentiment analysis when figurative language is employed still remains a challenging research topic as the languages have secondary or extended meanings and are intentionally profiled [2]. There have been several automatic computational approaches were attempted to categorize the figurative texts, such as humor recognition, metaphor or irony or sarcasm detection [2,3].

Limited amount of research has focused on identifying the degree of sentiment in the figurative language. A shared task on *"Sentiment Analysis of Figurative*

*Language in Twitter"* was organized in SemEval-2015 [1]. One of the main goals of this task was to evaluate the degree to which a conventional sentiment analysis approach suits for creative language or figurative language. In the task, a set of tweets that are rich in irony, metaphor, and sarcasm were given and the goal was to determine whether the user has expressed a positive, negative or neutral sentiment in each, and the degree to which this sentiment has been communicated. To capture the degree or intensity of the irony, metaphor and sarcasm, each of the participating systems were asked to assign fine grained sentiment score in a scale of −5 to +5.

A total of 15 teams participated in the above task and the team *CLaC* achieved the top results among all other submitted systems [1]. Features like unigram, bigram, parts-of-speech (POS), and sentiment lexicons such as Senti-WordNet [14], NRC Emotion lexicon [13], and AFINN dictionary [11] were used in most of the systems [5,20]. Support vector machines (SVMs), LibSVM (a variant of SVMs), and Decision Trees are the main classifiers that were used to develop several sentiment analysis systems for figurative language [1,5,20].

## 3   Dataset

We used the same datasets (trial, training and test) as provided by the organizing committee of the SemEval-2015 Shared Task-11[1] for our experimentation purpose. The trial and training datasets consist of 906 and 8000 tweets where each of the instances is accompanied with a real valued score ranging from [−5, +5]. Similarly, the test dataset contains 4000 tweets accompanied with absolute valued score from [−5, +5].

The distributions of each tweet class with respect to the trial, training and test datasets are shown in Table 1. We have also counted the frequencies of well-established hashtags like #sarcasm, #irony and #not and recorded the statistics in Table 1. We found a total of 4077, 257 and 1242 unique hashtags in training, trial and test datasets, respectively. We removed the junk characters from the tweets. We also normalized the words having multiple characters (for example, the word 'yesssss' to 'yes' using an English dictionary).

**Table 1.** Tweet class distribution and hashtag statistics.

| Data | Tweet score | | | | | | | | | | | Total | Hashtags | | |
|------|----|-----|------|------|-----|-----|-----|-----|-----|-----|----|--------|--------|------|----------|
| set | −5 | −4 | −3 | −2 | −1 | 0 | 1 | 2 | 3 | 4 | 5 | tweets | #irony | #not | #sarcasm |
| Trial | 5 | 80 | 359 | 227 | 75 | 46 | 46 | 33 | 27 | 7 | 1 | 906 | 720 | 472 | 455 |
| Train | 6 | 364 | 2971 | 2934 | 861 | 345 | 165 | 197 | 106 | 49 | 2 | 8000 | 1405 | 3328 | 1975 |
| Test | 4 | 100 | 737 | 1541 | 680 | 298 | 169 | 155 | 201 | 111 | 4 | 4000 | 32 | 45 | 197 |

---

[1] http://alt.qcri.org/semeval2015/task11/.

## 4   Feature Analysis

In order to develop an automatic tweet classification system based on the above datasets, we identified the basic textual and semantic features as available in the literature [5]. We have considered the following key features like *POS, sentiment features, intensifiers* and *sentiment abruptness* measure for tweet polarity strength classification tasks.

### 4.1   POS (I)

POS tag plays an important role in sentiment analysis [4]. Thus, we have used the ark-tweet-nlp [12] tool to parse each of the tweets to find out the POS tags of each word and included them as a feature in our experiments. The POS feature is used only for the Conditional Random Field (CRF) [8] based system.

### 4.2   POS Sequence (II)

We have observed that the POS sequence also plays important role in sentiment analysis [4]. For e.g., '*brave/JJ heart/NN*', here the word 'brave' is an adjective and it enhances the positivity of the noun 'heart'. Therefore, we utilized this information while training our datasets using CRF.

### 4.3   Intensifier (III)

We prepared six types intensifier lists for our system. Generally, such words help in identifying the intensity of the sentiments. Basically, an intensifier emphasizes or reduces the sentiment value or effect of the sentiment word it is preceded or followed by. For example, if there are two sentences, 'I am sad.' and 'I am very sad', in the second sentence, the word 'very' emphasizes the degree of negativity of the word 'sad' in the sentence. The intensifier classes and examples from each class are given in the Table 2.

We assigned values for each of the intensifier classes, for e.g., we assigned 2, 1.5, 1, $-1.5$, and $-2$ values for maximizers, boosters, approximators, compromisers, diminishers, and minimizers respectively. If an intensifier is found in a tweet before a sentiment word, its corresponding value is multiplied with the succeeding positive or negative value of the sentiment word. The positive or negative value of the sentiment word is identified using SentiWordNet.

### 4.4   Sentiment Feature (IV)

Sentiment lexicons are the most important features for any kind of sentiment identification or classification tasks [4,5]. We have used several lexicons like SentiWordNet, WordNet Affect [6], SentiSense Synset [9], Effect WordNet [15], AFINN dictionary, NRC Word-Emotion Association Lexicon, Taboada adjective list [10] and Whissell dictionary [16] to identify the sentiment/emotion class of

**Table 2.** List of intensifiers.

| Intensifier class (total instances) | Words |
|---|---|
| Maximizers (14) | completely, absolutely, totally, thoroughly, etc |
| Boosters (38) | very, highly, immensely, exceedingly, etc |
| Approximators (12) | nearly, virtually, effectively, all but, etc |
| Compromisers (5) | fairly, pretty, rather, etc |
| Diminishers (17) | slightly, a little, a bit, somewhat, rather, moderately, etc |
| Minimizers (8) | hardly, scarcely, barely, almost, etc |

the words and used as features. For the CRF based system, each of the words in a tweet is marked with either positive or negative or neutral using all of the above lexicons. Whereas for the other system, we counted the total number of positive and negative words present in a tweet using the above lexicons.

## 4.5   Sentiment Abruptness (V)

Finally, in the present work, we proposed a special measure named as *Sentiment Abruptness*, which measures the variation of sentiment from positive to negative or vice versa in a tweet text. We plot each of the sentiment tokens on a graph with the help of SentiWordNet scores on the Y-axis and the token position of a tweet on the X-axis. Let us consider two sample tweets, $T_1$ and $T_2$.

$T_1$: *RT @TheeJesseHelton: "A Million Ways To Die In The West" looks about as **appealing** as **dysentery** (−4).*
$T_2$: *RT @TheeJesseHelton: "A Million Ways To Die In The West" looks like an **appealing** movie which I missed because of **dysentery** (0).*

If we consider '*appealing*' and '*dysentery*' as the only sentiment points on the above tweets, the two sentiment plots would look somewhat as shown Fig. 1.

Two sentiment words, the positive and negative sentiment values extracted from SentiWordNet in both of the tweets are same. But, we get different types of sentiment curves because of the difference in the vicinity of the two words. The "sharp turns" in the sentiment curve indicates higher level of sarcasm in the tweet.

Consider the following sentiment plot for a tweet in Fig. 2. Thus, the 'turn' or the degree of polarity can be detected by measuring the curvature of a circle passing through a given triplet of points ($P_1$ ($x_1$, $y_1$), $P_2$ ($x_2$, $y_2$), and $P_3$ ($x_3$, $y_3$)) in the sentiment curve (as a higher curvature indicates a sharper turn). We calculate the afore-mentioned curvature using the well-known Menger's curvature formula [7] indicated in Eq. 1. Thus we propose to detect the degree of the sentiment with the help of Menger's curvature and using the coordinates of the points, the abruptness score (K) is given by:

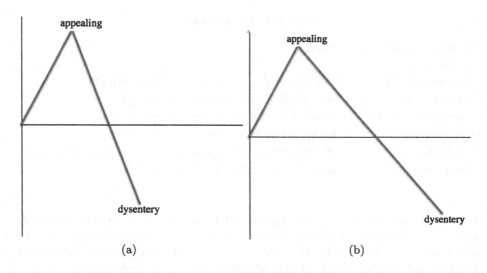

**Fig. 1.** Sentiment plots of $T_1$ (a) and $T_2$ (b)

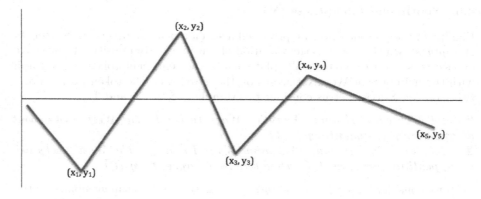

**Fig. 2.** Sample sentiment plot with five sentiment points.

$$\mathbf{K} = \frac{4 \times f(P_1, P_2, P_3)}{\sqrt{g(P_1, P_2, P_3)}} \tag{1}$$

where the area of triangle using the coordinates,

$$\mathbf{f(P_1, P_2, P_3)} = \frac{1}{2} \left| (x_2 - x_1)(y_3 - y_1) - (x_3 - x_1)(y_2 - y_1) \right| \tag{2}$$

The product of the three sides is $\sqrt{g(P_1, P_2, P_3)}$ and

$$\mathbf{g(P_1, P_2, P_3)} = ((x_2 - x_1)^2 + (y_2 - y_1)^2)((x_3 - x_1)^2 + (y_3 - y_1)^2)((x_3 - x_2)^2 + (y_3 - y_2)^2) \tag{3}$$

However, we have to normalize the sentiment abruptness score of a curve since a tweet with multiple sentiment points would score more than a tweet with a few sentiment points but with similar intensity. Therefore, we just divide the total score by the number of tokens present in the tweet. The algorithm used for calculating the sentiment abruptness score is given below.

**Algorithm (Sentiment abruptness):**

1: Initialize $Total_{abruptness} = 0$;
2: **for** each *tokens* in *tweet*:
   a. if *token* in *SentiWordNet* then put it in *SList*
3: **for** each triplet of points $P_1$, $P_2$, $P_3$ in *SList*:
   a. Calculate *sentiment abruptness score (K)*;
   b. $Total_{abruptness} = Total_{abruptness} + K$;
4: $Measured_{abruptness} = Total_{abruptness}/SList\_length$;
5: **return** $Measured_{abruptness}$;

We calculated the average sentiment abruptness score for each of the tweet classes. The values for the training dataset are as follows, 0.132039, 0.127876, 0.111354, 0.095409, 0.09699, 0.090894, 0.093322, 0.101701, 0.109143, 0.126745 and 0.266667 for $-5$, $-4$, $-3$, $-2$, $-1$, 0, $+1$, $+2$, $+3$, $+4$, and $+5$ tweet scores, respectively. We can observe that the sentiment abruptness score is higher for $-5$ and $+5$, whereas lower for the 0 tweet score. The sentiment abruptness score for each of the tweets is calculated using the above algorithm and this score is used as a feature for our experiments.

# 5    Multilevel Training and System Framework

The trial and training dataset are annotated with real valued scores whereas the test dataset is annotated with absolute values. Thus, we developed two basic systems; (a) a regression model followed by classification to cope up with real valued data and (b) a classification model capable to work on the absolute valued scores.

## 5.1    Evaluation Criteria

The performance of our model was evaluated based on the cosine similarity (CS) of the desired output with the system output as proposed in the Task 11 of SemEval-2015 [1]. We also evaluated performance of the systems using Mean Squared Error (MSE) metric. The equations for the CS and MSE are given in the Eqs. 4 and 5 respectively.

$$\mathbf{CS} = \frac{Actual.Predicted}{|Actual|\,|Predicted|} \tag{4}$$

$$\mathbf{MSE} = \frac{\sum_{i=1}^{n}(Actual_i - Predicted_i)^2}{n} \tag{5}$$

## 5.2   System 1

We developed our first system using the real valued scores of training and trial dataset. The test dataset contains one absolute scores, thus we have not used this for the first system. A total of 8906 tweets have been used to develop the first system. We used the CRF model first and then the Support Vector Machine Regression model of Weka[2] to build a multilevel classification framework.

Initially, to build the CRF model, we rounded off the scores of training and trail dataset to absolute values. We performed the 10-fold cross validation and in the first level, CRF classifier gives the maximum CS of 0.746 and minimum MSE of 3.096. The performance of the system according to each of the features and their combinations are provided in Table 3.

**Table 3.** Performance of the CRF based classifier in System 1.

| Features | CS | MSE |
|----------|------|-------|
| I | 0.736 | 3.270 |
| I+ II | 0.739 | 3.220 |
| I + II + III | 0.741 | 3.201 |
| I + II + III + IV | **0.746** | **3.096** |

The output of CRF, i.e. the absolute score of each tweet is then used as a feature in the second level regression model. We achieved the maximum CS of 0.823 and minimum MSE of 2.170 using the 10-fold cross validation. It was observed that CS did not vary even with the inclusion of CRF output as a feature, but the MSE reduced significantly. The CS and MSE with respect to the various features are given in Table 4.

## 5.3   System 2

We rounded off the scores from real values to integer for both trial and training datasets in order to consider the problem as a classification problem rather than regression. Therefore, we trained our second system using the absolute valued score of each tweet of trial and training dataset. We rounded off all the scores of tweets in the training and trial dataset and merged them together, i.e. we have a total of 8906 tweets for training. We used LibSVM classifier of Weka for classification purpose and tested the system on the test dataset of 4000 tweets.

Our second system achieves the maximum CS of 0.765 and minimum MSE of 2.973 as shown in Table 4. The performance obtained by this second system shows a marginal improvement of CS (0.007) over the best performing system of Task 11 at SemEval-2015 [5].

---

[2] http://www.cs.waikato.ac.nz/ml/weka/.

**Table 4.** Performance of the proposed systems.

| Features | System 1 | | System 2 | |
|---|---|---|---|---|
| | CS | MSE | CS | MSE |
| III + IV | 0.802 | 2.227 | 0.563 | 4.061 |
| V | **0.737** | **3.265** | **0.656** | **3.357** |
| III + IV + V | 0.822 | 2.230 | **0.765** | **2.973** |
| III + IV + V + CRF Class | **0.823** | **2.170** | X | X |

The noticeable improvement was found when we incorporated sentiment abruptness as a feature. This feature solely gives the maximum CS of 0.737 and 0.656 and minimum MSE of 3.265 and 3.357 for system 1 and system 2 respectively. Therefore, we added this measure into our existing feature set and the corresponding results are found in Table 4. One of the major problems faced during the experiment was to handle unequal number of tweet instances present in the datasets. The total count of the instances having scores $-2$ and $-3$ was larger than others and we observed the biasness of the classifiers towards these two classes.

## 6  Conclusion and Future work

In this paper we introduced a new measure, sentiment abruptness which achieves the maximum CS of 0.737 and 0.656 for the respective systems for identifying sentiment scores of the figurative texts. The system achieves the maximum CS of 0.823 with the help of multilevel classification along with other features.

In future, we plan to use the tweet dependency parsers to get the relations for different phrases. If the relations in a tweet are contradictory, then there may be a chance of tweet to be ironic or sarcastic. Another immediate goal is to develop some lexicons or ontology from the tweet data for sentiment analysis as developed in [5] and use this ontology to detect figurative language in social media texts. Moreover, we plan to consider different sentiment lexicons considering the abruptness measure.

**Acknowledgments.** The work reported in this paper is supported by a grant from the project "CLIA System Phase II" funded by Department of Electronics and Information Technology (DeitY), Ministry of Communications and Information Technology (MCIT), Government of India. The work of the fourth author is also supported by the SomEMBED TIN2015-71147-C2-1-P MINECO research project and by the Generalitat Valenciana under the grant ALMAPATER (PrometeoII/2014/030).

# References

1. Ghosh, A., Li, G., Veale, T., Rosso, P., Shutova, E., Reyes, A., Barnden, J.: Semeval-2015 task 11: sentiment analysis of figurative language in Twitter. In: 9th International Workshop on Semantic Evaluation (SemEval), Co-located with NAACL, Denver, Colorado, pp. 470–478. Association for Computational Linguistics (2015)
2. Reyes, A., Rosso, P., Veale, T.: A multidimensional approach for detecting irony in Twitter. Lang. Resour. Eval. **47**(1), 239–268 (2013)
3. Reyes, A., Rosso, P., Buscaldi, D.: From humor recognition to irony detection: the figurative language of social media. Data Knowl. Eng. **74**, 1–12 (2012)
4. Patra, B.G., Mandal, S., Das, D., Bandyopadhyay, S.: JU_CSE: a conditional random field (CRF) based approach to aspect based sentiment analysis. In: 8th International Workshop on Semantic Evaluation (SemEval), Co-located with COLING, Dublin, Ireland, pp. 370–374. Association for Computational Linguistics (2014)
5. Ozdemir, C., Bergler, S.: CLaC-SentiPipe: SemEval2015 subtasks 10 B, E, and task 11. In: 9th International Workshop on Semantic Evaluation (SemEval), Co-located with NAACL, Denver, Colorado, pp. 479–485. Association for Computational Linguistics (2015)
6. Strapparava, C., Valitutti, A.: Wordnet-affect: an affective extension of wordnet. In: 4th International Conference on Language Resources and Evaluation, pp. 1083–1086 (2004)
7. Léger, J.C.: Menger curvature and rectifiability. Ann. Math. **149**, 831–869 (1999)
8. Lafferty, J.D., McCallum, A., Pereira, F.C.N.: Conditional random fields: probabilistic models for segmenting and labeling sequence data. In: 18th International Conference on Machine Learning, pp. 282–289 (2001)
9. de Albornoz, J.C., Plaza, L., Gervas, P.: SentiSense: an easily scalable concept-based affective lexicon for sentiment analysis. In: 8th International Conference on Language Resources and Evaluation, pp. 3562–3567 (2012)
10. Taboada, M., Brooke, J., Tofiloski, M., Voll, K., Stede, M.: Lexicon-based methods for sentiment analysis. Comput. Linguist. **37**(2), 267–307 (2011)
11. Naveed, N., Gottron, T., Kunegis, J., Alhadi, A.C.: Bad news travel fast: a content-based analysis of interestingness on Twitter. In: 3rd International Web Science Conference. ACM (2011)
12. Owoputi, O., O'Connor, B., Dyer, C., Gimpel, K., Schneider, N., Smith, N.A.: Improved part-of-speech tagging for online conversational text with word clusters. In: NAACL. Association for Computational Linguistics (2013)
13. Mohammad, S., Turney, P.: Crowdsourcing a word-emotion association lexicon. Comput. Intell. **29**(3), 436–465 (2013)
14. Baccianella, S., Esuli, A., Sebastiani, F.: Sentiwordnet 3.0: an enhanced lexical resource for sentiment analysis and opinion mining. In: 7th Conference on International Language Resources and Evaluation, Valletta, Malta (2010)
15. Choi, Y., Wiebe, J.: +/−EffectWordNet: sense-level lexicon acquisition for opinion inference. In: EMNLP (2014)
16. Whissell, C., Fournier, M., Pelland, R., Weir, D., Makarec, K.: A dictionary of affect in language: IV. Reliability, validity, and applications. Percept. Mot. Skills **62**(3), 875–888 (1986)
17. Patra, B.G., Takamura, H., Das, D., Okumura, M., Bandyopadhyay, S.: Construction of emotional lexicon using potts model. In: International Joint Conference on Natural Language Processing (IJCNLP), pp. 674–679 (2013)

18. Pang, B., Lee, L.: Opinion mining and sentiment analysis. Found. Trends Inf. Retr. **2**, 1–135 (2008)
19. Vilares, D., Alonso, M.A., Gomez, C.: On the usefulness of lexical and syntactic processing in polarity classification of Twitter messages. J. Assoc. Inf. Sci. Technol. **66**(9), 1799–1816 (2015)
20. Barbieri, F., Ronzano, F., Saggion, H.: UPF-taln: SemEval 2015 tasks 10 and 11 sentiment analysis of literal and figurative language in Twitter. In: SemEval-2015, pp. 704–708 (2015)

# Determining Sentiment in Citation Text and Analyzing Its Impact on the Proposed Ranking Index

Souvick Ghosh[1](✉)(iD), Dipankar Das[1], and Tanmoy Chakraborty[2]

[1] Jadavpur University, Kolkata 700032, West Bengal, India
souvick.gh@gmail.com, dipankar.dipnil2005@gmail.com
[2] University of Maryland, College Park, MD 20742, USA
tanchak@umiacs.umd.edu

**Abstract.** Whenever human beings interact with each other, they exchange or express opinions, emotions and sentiments. These opinions can be expressed in text, speech or images. Analysis of these sentiments is one of the popular research areas of present day researchers. Sentiment analysis, also known as opinion mining tries to identify or classify these sentiments or opinions into two broad categories – positive and negative. Much work on sentiment analysis has been done on social media conversations, blog posts, newspaper articles and various narrative texts. However, when it came to identifying emotions from scientific papers, researchers used to face difficulties due to the implicit and hidden natures of opinions or emotions. As the citation instances are considered inherently positive in emotion, popular ranking and indexing paradigms often neglect the opinion present while citing. Therefore in the present paper, we deployed a system of citation sentiment analysis to achieve three major objectives. First, we identified sentiments in the citation text and assigned a score to each of the instances. We have used a supervised classifier for this purpose. Secondly, we have proposed a new index (we shall refer to it hereafter as M-index) which takes into account both the quantitative and qualitative factors while scoring a paper. Finally, we developed a ranking of research papers based on the M-index. We have also shown the impacts of M-index on the ranking of scientific papers.

**Keywords:** Sentiment analysis · Citation · Citation sentiment analysis
Citation polarity · Ranking · Bibliometrics

## 1 Introduction

Sentiment analysis of citation contexts is an unexplored field in the area of sentiment analysis, primarily because of the existing myth that most of the research papers are cited positively in general. Furthermore, the negative citations are hardly explicit and the criticisms are often veiled. This lack of explicit sentiment expressions poses a major challenge for successful polarity identification. However, sentiment analysis of citations in scientific papers and articles is a new and interesting problem which can open up many exciting new applications in bibliographic search and bibliometrics [1].

© Springer International Publishing AG, part of Springer Nature 2018
A. Gelbukh (Ed.): CICLing 2016, LNCS 9624, pp. 292–306, 2018.
https://doi.org/10.1007/978-3-319-75487-1_23

There are many linguistic differences between scientific texts and other genres [2] and the scientific community has to undertake new approaches for correct classification. In our work, we have used various existing and novel features. We have considered n-grams, specialized science-specific lexical features, dependency relations, various word lists, sentiment lexicons and negation features for our research [7, 17].

The importance of citation is due to the fact that it helps us in determining the impact of each cited paper. While most of the ranking indices rely solely on the number of citations each paper receives, we have added a qualitative measure to the ranking procedure. Our main goal is to use the sentiment information in each citation instance (qualitative) in addition to the number of citations (quantitative) to determine the worth of the paper. To the best of our knowledge, this is the first work that uses both the quantitative and qualitative indicators to determine the ranking of scientific papers.

This paper is structured in seven sections. First, we discuss the previous works that were done related to sentiment analysis of citations. Then, we explain how we prepared the corpus for this task. The next two sections explain the features used for classification and the classification procedure itself. The last three sections concentrate on the ranking algorithm, the results and the future work.

## 2 Related Work

Automated citation sentiment analysis has emerged as a new research topic in natural language processing over the last decade [1, 4–6, 15, 16]. An automated analysis would primarily take into account various linguistic cues, like tone of reference and any negative words and then make use of machine learning algorithms to evaluate the opinion of the author towards the cited paper.

Citation sentiment detection can also help researchers during search, by detecting problems with a particular approach. It can be used as a first step to scientific summarization [14], enable users to recognize unaddressed issues and possible gaps in the current research, and thus help them set their research directions [1]. Existing bibliometrics evaluation schemes, like H-index [8, 9], G-index [10], Impact Factor [11] and graph ranking algorithm like PageRank [12], focus mainly on the quantitative aspect of citations. However, for fair evaluation of a scientific paper, we need to consider the polarity of citation, or the qualitative aspect of citation. In many cases, we often find that the paper has been cited in a negative way, i.e., for the purpose of criticism. Bonzi et al. was one of the first proponents of this logic that if a cited work is criticized it should carry a lower or negative weight for bibliometric measures [13]. Abu-Jbara et al. worked on various linguistic analysis techniques for determining the purpose and polarity of citations [2].

Athar [3] explored various sentence structure-based features for automatic identification of sentiment polarity in scientific literature. Athar et al. also worked on context-enhanced citation sentiment detection and studied the effectiveness of the length of the context window.

## 3  Corpus Preparation

Most of the current work on citation sentiment detection focuses only on the citation sentences. The corpus has been obtained from Athar and groups [3]. It contains the source and the target paper, along with the citation sentence and its associated polarity (marked as 'o', 'p' or 'n' for *objective, positive* and *negative* respectively). This corpus has a total of 8736 sentences each of which is annotated manually with polarity. For our work, we selected 6736 instances for our training purposes and 2000 instances for our testing purposes.

Initially, we consider a baseline system with all the sentiments considered as neutral. As most of the citations are usually neutral (causing highly imbalanced classes), so the accuracy of the baseline system is quite well. It is also one of the reasons why we have to be careful while annotating them with a positive or negative tag as wrong tags might reduce the accuracy of the system to below baseline. We preprocessed the corpus to denote the polarity by three integers. +1 for *positive*, −1 for *negative* and 0 for *neutral*. We also used a list of polar words and phrases which are specific to citation texts in order to identify the opinion of the citing author (Table 1).

**Table 1.** Number of instances of each polarity in the dataset

|                | Positive | Neutral | Negative |
|----------------|----------|---------|----------|
| Entire dataset | 829      | 7627    | 280      |
| Training set   | 635      | 5888    | 213      |
| Test set       | 194      | 1739    | 67       |

## 4  Feature Identification

We evaluated the following features for identifying the sentiment polarity of the citation instances:

### 4.1  Automatic Sentiment (AS)

We have calculated an automatic sentiment score by splitting a sentence into a bag of words and then assigning score to each of the words. The words have been normalized before assigning scores. The score of individual words were formulated using SentiWordNet[1]. The sentiment score of the sentence is the sum of the scores of all the individual words multiplied by 100 (which helps in rounding the score). In the following example, the automatic score allocation by SentiWordNet is 43.0.

e.g.: *Dasgupta and Ng (2007) improves over (Creutz, 2003) by suggesting a simpler approach. (Citing paper id 'W09-0805', cited paper id 'N07-1020')*

---

[1] http://sentiwordnet.isti.cnr.it/download.php.

## 4.2   Positive Polarity Words (PPW)

We have used a list of words (unigrams, bigrams, trigrams and four-grams) with positive sentiment polarity. For each n-gram (up to n = 4) present in the sentence we have compared it with the collection of positive n-grams to determine if there is a match. The most frequent unigrams, bigrams and trigrams which are specific to citation texts and positive in polarity are illustrated in Tables 2, 3 and 4. The frequencies were obtained from the training dataset. The number of trigrams and 4-grams are significantly less than that of unigrams and bigrams. Phrases like '*improve performance of*', '*very high accuracy*', '*most widely used*' and '*state of the art*' are generally used in conjunction with citation texts which denote positive polarity.

**Table 2.**  Most frequent unigrams with positive polarity

| More (397) | Improvement (88) | Outperform (48) | Popular (35) |
|---|---|---|---|
| Most (308) | Important (86) | Correlate (47) | Efficient (31) |
| Improve (185) | High (82) | Higher (44) | Successful (30) |
| Best (153) | Effective (68) | Major (42) | Overcome (29) |
| Well (148) | Accurate (67) | Significant (39) | Consistent (23) |
| Better (141) | Development (67) | Highly (37) | Sophisticated (22) |
| Simple (110) | Useful (66) | Robust (36) | Benefit (20) |
| Good (100) | Successfully (56) | Considerable (36) | Simpler (19) |

**Table 3.**  Most frequent bigrams with positive polarity

| improvement in | success of | more efficient | most successful |
|---|---|---|---|
| good performance | can improve | very successful | most notable |
| good result | more accurate | best score | well known |
| development of | most important | widely used | effective at |
| high quality | achieve impressive | quite accurate | increase over |

**Table 4.**  Most frequent negative polarity words in citations

| However (125) | Unlike (26) | Worse (8) | Unrealistic (2) |
|---|---|---|---|
| While (119) | Restrict (14) | Unfortunately (7) | Insufficient (1) |
| Although (68) | Lack (13) | Complicated (5) | Inability (1) |
| Low (45) | Poor (10) | Daunting (4) | Lack of (12) |
| Without (40) | Unexplored (9) | Degrade (3) | Not well (2) |
| Difficult (36) | Little (8) | Burden (3) | Not able to (1) |

## 4.3   Negative Polarity Words (NPW)

Our next feature is obtained by using a collection of negative words. The set of negative words which are often encountered in scientific literature is relatively small in number. Owing to peer relations, criticisms of scientific papers are often hedged and implicit. Thus, we check the citation sentence to find any word which belongs to this collection of negatively polar words. Table 4 illustrates the negative words which usually occur in citation texts.

### 4.4 Presence of Specific Part-Of-Speech Tags (POS)

The output of the POS tagger is analyzed to check for the presence of specific tags like JJ, JJR, JJS, JJT (various forms of adjective), RB, RBR, RBT, RN, RT (forms of adverbs) and FW (foreign words). We also checked the occurrence of adverbs followed by adjectives (for example RB_JJ tag) as the presence of adverb along with adjective usually reflects subjectivity in sentence polarity.

   *e.g.: simpler/ JJR, well/RB, etc.*

Here, *simpler* and *well* are two subjective words which are tagged as JJR and RB respectively.

### 4.5 Presence of Specific Dependency Tags (DEP)

While obtaining dependency output, we check for the presence of tags *advmod* (adverb modifier), *acomp* (adverbial complement) and *amod* (adjectival modifier) in the sentence. These tags are also indicators of subjectivity in sentence.

   *e.g.: simpler approach, well known, etc.*

Here, *amod* (approach, simpler) and *advmod* (known, well) captures the polarity of citation. Similarly, *acomp* functions like an object of the verb and *amod* is any adjectival phrase that modifies the meaning of the noun phrase (NP). These relations are most frequent in sentences where polar sentiments are present.

### 4.6 Self Citation (SC)

We also check the presence of self citation in the citation sentence. This can be checked by verifying if the citing (source) paper refers to itself. We checked that there were no self citations in our dataset. When we constructed a graph representing citations, we found that it contained no self-loop. So we did not include this feature for classification purposes.

### 4.7 Opinion Lexicons (OL1 and OL2)

This feature is identified from a list of positive and negative opinion or sentiment words. The list was developed by Liu et al. [18] for comparing opinions on the web. We have used this list for identifying any sentiment words in the text. We have also used Vender Sentiment, which is another sentiment word list for identifying the sentiment words and determining their polarity. Both these lists have been split into positive and negative collections and then four features were introduced – the number of matches to each list - to train the classifier.

# 5   Classification

We used the machine learning software WEKA[2] [19]. We combined the above features to form a feature set and used the J48 classifier to generate a pruned C4.5 Decision Tree for three-way classification of the citation instances – *positive, negative* and *neutral*. The C4.5 algorithm generates a classification-decision tree for the given dataset by recursive partitioning of the data. It uses depth-first strategy and makes the selection based on highest information gain.

Individually, none of the features was able to detect positive or negative instances in citation. This was due to the large number of neutral instances present in the system and biasness of such neutral instances. We performed feature analysis by removing one feature at a time to determine if any feature was more important than the other. We also checked by adding one feature at a time.

The classification confidence score from WEKA and the number of matches to our citation specific lexicon were used to develop a post-processing algorithm. We added extra weight to the frequency of matches to the lexicon list. If the difference between frequency of positive and negative polarity words was more than $t_1$, we immediately assigned the instance as positive citation. If the number of negative polarity words was more than $n_1$, we assigned it as negative. Next we considered the confidence score of our WEKA classification. If it is more than $s_1$, we use the WEKA classification. Otherwise we use the polarity matches again to determine the polarity. The thresholds for this step are $t_2$ and $n_2$ respectively. This algorithm helped us to improve the accuracy of our result. Focusing on the best results obtained for different values of $t_1$ and $n_1$, ranging from one to five, and s1, ranging from 0.5 to 1.0, we settled for $t_1 = 3$, $n_1 = 2$, and $s_1 = 0.8$. Similarly best results were obtained by setting $t_2 = 2$ and $s_2 = 1$. Note that traditional accuracy measures are often not a good metric when the classes are imbalanced and/or cost of misclassification varies dramatically between the two classes.

## 5.1   Feature Analysis

See Tables 5 and 6

**Table 5.**  Impact of each feature calculated by eliminating one at a time

| Feature eliminated | Number of correct classifications | Number of incorrect classifications | Accuracy |
|---|---|---|---|
| SWN lexicon | 1740 | 260 | 0.87 |
| Citation specific lexicons | 1740 | 260 | 0.87 |
| Part of speech tags | 1731 | 269 | 0.8655 |
| Dependency tags | 1732 | 268 | 0.866 |
| Opinion lexicon 1 | 1740 | 260 | 0.87 |
| Opinion lexicon 2 | 1722 | 278 | 0.861 |

---

[2] http://www.cs.waikato.ac.nz/ml/weka/downloading.html.

**Table 6.** Impact of adding each feature iteratively to the last

| Feature added | Number of correct classifications | Number of incorrect classifications | Accuracy |
|---|---|---|---|
| SWN lexicon | 1740 | 260 | 0.87 |
| Citation specific lexicons | 1740 | 260 | 0.87 |
| Part of speech tags | 1746 | 254 | 0.873 |
| Dependency tags | 1744 | 256 | 0.872 |
| Opinion lexicon 1 | 1722 | 278 | 0.861 |
| Opinion lexicon 2 | 1736 | 264 | 0.868 |

## 5.2    Algorithm

See Table 7

**Table 7.** Algorithm to classify the citation instances

```
ALGORITHM : Program Classification
  begin
    L1 = citation specific lexicon list
    score = classification confidence score from WEKA
    C = Class assigned by WEKA
    posmatch = number of matches to positive polarity words
    negmatch = number of matches to negative polarity words
    if posmatch-negmatch > t1, class = "positive"
    else if negmatch > n1, class = "negative"
    else if score > s1, class = C
    else if posmatch – negmatch > t2, class = "positive"
    else if negmatch > n2, class = "negative"
    else class = "neutral"
  end
```

## 5.3    Results

Table 8 shows the confusion matrix for the polarity classification. The precision, recall and f-measure of the supervised and baseline systems are compared in Table 9.

**Table 8.** Confusion matrix for the classification result

| | Positive | Neutral | Negative |
|---|---|---|---|
| Positive | 33 | 159 | 2 |
| Neutral | 27 | 1704 | 8 |
| Negative | 3 | 51 | 13 |

**Table 9.** Precision, recall and F-measure of supervised system and the baseline

|                | Precision | Recall | F-measure |
|----------------|-----------|--------|-----------|
| Supervised system |       |        |           |
| Class positive | 0.524     | 0.17   | 0.257     |
| Class neutral  | 0.889     | 0.968  | 0.927     |
| Class negative | 0.545     | 0.179  | 0.27      |
| Baseline system |          |        |           |
| Class positive | 0         | 0      | 0         |
| Class neutral  | 0.864     | 1      | 0.927     |
| Class negative | 0         | 0      | 0         |

The baseline model considered all the instances to be of neutral polarity. So we can see that our supervised system shows improvement over the baseline model. However, the learning algorithm was slightly biased towards neutral classification which is evident from the confusion matrix. Most of the errors are due to positive and negative citations being identified as neutral.

In future works, we will need to fine tune our classification features so that the system can identify positive and negative citations more efficiently. Also using a larger dataset to train the system would eliminate the bias towards neutral classification of polarity.

# 6 Ranking Algorithm

For scientific literature, we generally use the H-index to find out the impact of an author. However, H-index is only a quantitative measurement. Also, it targets an author in particular. Our approach captures the importance of each paper based on the number and the opinion of the citing paper. We also capture both quantitative and qualitative factors in our index. The ranking index should help in identifying the paper which has not only been cited more often but also in a positive sense. By taking into account the criticism of the citing papers, we aim to make the ranking system more efficient and feasible.

## 6.1 Corpus

The dataset contains the following information:

**Citation sentence**
This is the sentence in a research paper where the paper refers to one or more scientific papers in order to explain, state, praise or criticize the claims made in the referred paper (s). While the first two cases would be of neutral polarity, the last two cases are of positive and negative polarity respectively.

   e.g.- *Dasgupta and Ng (2007) improves over (Creutz, 2003) by suggesting a simpler approach.*

### Source

This is the paper id of the source or citing paper. This paper quotes or borrows some idea or concept explained in the cited paper. In the previously mentioned example, the source paper id is '*W09-0805*', where we find the citation sentence.

### Target

This is typically the paper id of the cited paper. If we represent each citation instance using an edge of the graph, and the papers themselves as nodes of the graph, then we can find a directed edge from the source or citing paper to the target or cited paper.In the previously mentioned citation sentence, the target or cited paper id is '*N07-1020*'. This is the paper id for the paper by Dasgupta and Ng (2007).

### Polarity (calculated by our system)

The polarity is calculated by our sentiment classification system in the first place. We have considered only three types of polarity – positive, negative and neutral.

### 6.2  Naïve Algorithm

The naïve algorithm is the standard baseline which counts the number of times a particular paper is cited by other papers (Table 10).

### Algorithm

**Table 10.** Naïve algorithm to find the ranking of papers

| ALGORITHM |
|---|
| begin |
|    total_papers = Total number of papers in the collection |
|    total_instance = Total number of instances |
|    for (num=0; num<total_papers; num++) |
|      paper.count = 0; |
|    end for |
|    for each instance i |
|      for all the papers in collection |
|        if paper.id == target_paper.id |
|          paper.count +=1; |
|          break; |
|        end if |
|      end for |
|    end for |
|    sort all the papers by count to obtain ranked index |
| end |

## Ranking

The ranked list of papers was divided into buckets. Each bucket comprised 20% of the total number of cited papers. In our test set, there were 40 unique target (cited) papers for 2000 citation instances. So, we divided the ranked list into 5 buckets with each bucket consisting of 8 papers sorted by rank.

### 6.3    Proposed Algorithm (M-Index)

For scientific literature, we use the H-index to find out the impact of an author. However, H-index is only a quantitative measurement. We aim to capture both quantitative and qualitative factors in our index. So, we evaluated the impact of the paper based on two factors – the number of citations that the paper received (quantitative) and the polarity of the citation (qualitative). For the proposed algorithm, we have used three different kinds of scores – the reliability score, the polarity score and the M-Index Score.

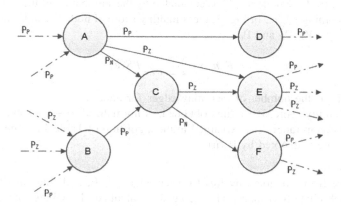

**Fig. 1.** Graphical representation of citations

We denote the citation dataset in the form of a directed graph $G = <V, E>$ where V is a set of all nodes and E is the set of directed edges over those nodes. Each node represents a research paper (cited or citing or both) and each outgoing edge represents a citation instance from the source (citing) node to the target (cited) node. The edges are marked with polarity scores of the instance.

### Polarity Score

Using our previous classification system, we judged the polarity of each citation instance. Polarity Score is denoted by PS(e) for any edge or instance $e \in E$. In Fig. 1, we have used $P_P$, $P_Z$, and $P_N$ to denote positive, neutral and negative polarity respectively. We have assigned a polarity of +1, 0.5 and −0.5 to papers of positive, neutral and negative polarity respectively for each citation instance. Therefore, $P_P = 1$, $P_Z = 0.5$ and $P_N = -0.5$. In Fig. 1, node C has two incoming edges with polarities $P_N$

and $P_P$ while node B has three incoming edges $P_Z$, $P_Z$ and $P_P$. We have tuned our system to reward positive citations by a larger weight. The weights of neutral and negative citations are kept the same. Neutral citations have been assigned with a positive score instead of zero to reward non-negative instances of citations. The polarity score is relevant only for the target or cited paper. It shows if the target paper has been referred subjectively and the polarity of that reference.

**Reliability Score**

For each paper, we can find out the extent to which we can rely on the paper for citations. If a paper has been cited negatively by most other papers, then we can assume that the paper lacks reliability. So we assign a score of $+2$, $+1$ and $-1$ to denote that the paper is *very reliable, fairly reliable* and *not reliable* respectively. In some cases, we may not be able to find the reliability of the source paper. This is due to the fact that the source paper may not appear as target in any citation instance. In such a situation, we will consider the source paper as *fairly reliable*, i.e., assign it a reliability score of $+1$. The reliability score is applicable only for the source or citing paper because we are concerned with the capacity of judgment of the citing paper. The reliability score, denoted by R (n) for a node n $\in$ V, is calculated by the summation of the polarity scores of all the incoming edges. In Fig. 1, the reliability score of node C can be calculated by finding the sum $P_Z$, $P_Z$ and $P_P$.

$$R(n) = \sum_{e \in En} PS(e) \tag{1}$$

Where $E_n$ is the number of incoming edges for node n.

The reliability score, R(n), thus obtained, is normalized to $+2$, $+1$ and $-1$ respectively for different ranges (as explained in the algorithm of Table 11). The normalized reliability score is denoted by R'(n).

**M-Index Score**

The Instance Score, denoted by I(e) for each edge e $\in$ E, is defined as the weighted score of each citation instance. This score was calculated by taking into account the reliability of the source or citing paper (or node) and the sentiment polarity of each reference. For a given node or paper n $\in$ V, we calculated MIS(n) as the sum of instance scores for all the incoming edges of node n.

$$I(e) = \left( PS(e) * R'^{(n)} \right) \tag{2}$$

$$MIS(n) = \sum_e I(e) \tag{3}$$

**Algorithm**

**Table 11.** Algorithm to determine ranking of papers by m-index

| ALGORITHM |
| --- |
| Begin |
|    total_papers = total number of papers in the collection |
|    total_instances = total number of instances in the collection |
|    for (num = 0; num < total_papers; num++) |
|      paper.relscore = 0 |
|    for each instance i |
|     for all papers |
|       if the paper.id = target.id |
|         if ( polarity > 0 ) paper.relscore += pp |
|         if ( polarity < 0) paper.relscore -= np |
|         if ( polarity == 0 ) paper.relscore += zp |
|       end if |
|     end for |
|    end for |
|    for (num=0; num< total_papers; num++) |
|      paper.mscore = 0; |
|      if (paper.relscore > 1)  paper.relscore = 2 |
|      else if (paper.relscore < 0)  paper.relscore = -1 |
|      else  paper_relscore = 1 |
|    end for |
|    for each citation instance i |
|      score1 = sourcepaper.relscore |
|      score2 = targetpaper.polarityscore |
|      instance_score = score1 * score2 |
|      targetpaper.mscore +=  instance_score |
|    end for |
|    sort cited papers by m-score to get the ranking index |
|   end |

**Ranking**

A total of 2000 citation instances had 40 unique cited papers. These papers were ranked by m-score. The ranked list was divided into 5 buckets, each bucket containing 8 papers.

## 7 Result Analysis

After ranking the papers based on naïve method and the modified algorithm, we divided them into 5 buckets and tried to evaluate the impact of the modified algorithm on each bucket. It did not present much variation in the overall ranking which was understandable owing to the limitations of the dataset. Only 3 papers out of 40 showed variations in the ranking and even then, the buckets were not altered.

The limited nature (imbalance and small size) of the corpus was one of the primary reasons as to why there was not much variation between the two ranked lists. If we look at the pie-chart of Fig. 2, we can notice that the top-ranked papers were clearly cited much more than the remaining ones. There was also a prominent gap in the number of citations for each paper. This resulted in fewer changes in ranking between the two methods (Fig. 3).

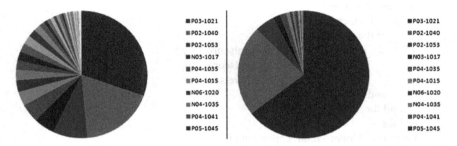

**Fig. 2.**  Score distribution of ranked papers (naïve score on the left and m-score on the right)

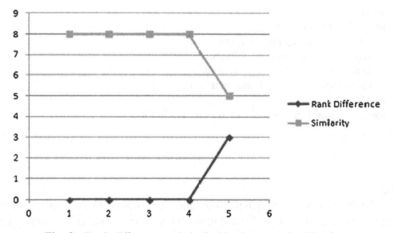

**Fig. 3.**  Rank difference and similarities between the 5 buckets

In future works, we aim to concentrate on preparing a larger corpus which will contain a larger proportion of subjective citations. This would help in reducing the bias of our sentiment analysis algorithm towards neutral classification. Qualitative factors would also have a higher impact in the overall ranking.

# 8   Conclusion and Future Work

In this paper, we focused on two aspects of citations, automatic detection of citation sentiment and ranking of scientific paper using a newly proposed index. First, we classified the polarity of citations using a statistical classifier C4.5 which made use of various sentence-based and linguistic features to generate decision trees. Our system achieved fairly accurate results with 87.5% accuracy.

Secondly, we proposed a new index, M-index, which takes into account the reliability of the citing paper and the type of polarity involved between the citing and cited paper. This index focuses mainly on a particular paper, unlike H-index, which is more author-specific. The ranked list of the cited papers was obtained using the new index. A similar ranked list was obtained using the naïve method which maintained a simple count of the number of times a paper was cited. We analyzed the impact of this new index by comparing the two ranked lists. Although the ranks did not show too much variation, yet the impact should be greater with a larger corpus.

For future work, we are working on a second corpus. This corpus is based on the ACL Anthology corpus[3] which has been annotated to take the dominant sentiment in the entire citation context into account. M-index based ranking uses both quantitative and qualitative information and its impact could be better analyzed by the larger corpus.

# References

1. Athar, A., Teufel, S.: Context-enhanced citation sentiment detection. In: Conference of the North American Chapter of the Association for Computational Linguistics: Human Language Technologies, Montreal, Canada, pp. 597–601 (2012)
2. Abu-Jbara, A., Ezra, J., Radev, D.: Purpose and polarity of citation: towards NLP-based bibliometrics. In: Proceedings of NAACL-HLT, Atlanta, Georgia, pp. 596–606 (2013)
3. Athar, A.: Sentiment analysis of citations using sentence structure-based features. In: Proceedings of the ACL-HLT 2011 Student Session, Portland, USA, pp. 81–87 (2011)
4. Tandon, N., Jain, A.: Citation context sentiment analysis for structured summarization of research papers. In: Poster and Demo Track of the 35th German Conference on Artificial Intelligence (KI-2012), pp. 98–102 (2014)
5. Athar, A.: Sentiment analysis of scientific citations. Technical report based on a dissertation submitted for the degree of Doctor of Philosophy. UCAM-CL-TR-856. University of Cambridge, Girton College (1994). ISSN 1476-2986
6. Yu, B.: Automated citation sentiment analysis: what can we learn from biomedical researchers. In: Proceedings of the 76th ASIS&T Annual Meeting: Beyond the Cloud: Rethinking Information Boundaries (2013)
7. Singh, S.K., Paul, S., Kumar, D.: Sentiment analysis approaches on different data set domain: survey. Int. J. Database Theory Appl. **7**(5), 39–50 (2014)
8. Hirsch, J.E.: An index to quantify an individual's scientific research output. In: Proceedings of the National Academy of Sciences, vol. 102, no. 46, pp. 16569–16572, November 2005
9. Hirsch, J.E.: An index to quantify an individual's scientific research output that takes into account the effect of multiple coauthorship. Scientometrics **85**(3), 741–754 (2010)

---

[3] http://clair.eecs.umich.edu/aan/index.php.

10. Egghe, L.: Theory and practise of the g-index. Scientometrics **69**, 131–152 (2006)
11. Garfield, E.: The Thomson reuters impact factor. In the current contents print editions, July 18, 1994. Institute for Scientific Information (1994)
12. Radev, D.R., Joseph, M.T., Gibson, B., Muthukrishnan, P.: A bibliometric and network analysis of the field of computational linguistics. J. Am. Soc. Inf. Sci. Technol. **1001**, 41092–48109 (2009)
13. Bonzi, S.: Characteristics of a literature as predictors of relatedness between cited and citing works. J. Am. Soc. Inf. Sci. **33**(4), 208–216 (1982)
14. Abu-Jbara, A., Radev, D.: Coherent citation based summarization of scientific papers. In: Proceedings of ACL (2011)
15. Athar, A., Teufel, S.: Detection of implicit citations for sentiment detection. In: Proceedings of the Workshop on Detecting Structure in Scholarly Discourse, Jeju Island, Korea, July, pp. 18–26. Association for Computational Linguistics (2012)
16. Teufel, S., Siddharthan, A., Tidhar, D.: Automatic classification of citation function. In: EMNLP, pp. 103–110. Association for Computational Linguistics (2006)
17. Pang, B., Lee, L., Vaithyanathan, S.: Thumbs up?: sentiment classification using machine learning techniques. In: EMNLP, pp. 79–86. Association for Computational Linguistics (2002)
18. Liu, B., Hu, M., Cheng, J.: Opinion observer: analyzing and comparing opinions on the web. In: Proceedings of the 14th International World Wide Web conference (WWW-2005), Chiba, Japan (2005)
19. Hall, M., Frank, E., Holmes, G., Pfahringer, B., Reutemann, P., Witten, I.H.: The WEKA data mining software: an update. SIGKDD Explor. **11**(1), 10–18 (2009)

# Combining Lexical Features and a Supervised Learning Approach for Arabic Sentiment Analysis

Samhaa R. El-Beltagy[(⊠)], Talaat Khalil, Amal Halaby,
and Muhammad Hammad

Center of Informatics Sciences, Nile University, Giza, Egypt
samhaa@computer.org, {t.maher, am.mahmoud}@nu.edu.eg,
mhammad@sci.cu.edu.eg

**Abstract.** The importance of building sentiment analysis tools for Arabic social media has been recognized during the past couple of years, especially with the rapid increase in the number of Arabic social media users. One of the main difficulties in tackling this problem is that text within social media is mostly colloquial, with many dialects being used within social media platforms. In this paper, we present a set of features that were integrated with a machine learning based sentiment analysis model and applied on Egyptian, Saudi, Levantine, and MSA Arabic social media datasets. Many of the proposed features were derived through the use of an Arabic Sentiment Lexicon. The model also presents emoticon based features, as well as input text related features such as the number of segments within the text, the length of the text, whether the text ends with a question mark or not, etc. We show that the presented features have resulted in an increased accuracy across six of the seven datasets we've experimented with and which are all benchmarked. Since the developed model outperforms all existing Arabic sentiment analysis systems that have publicly available datasets, we can state that this model presents state-of-the-art in Arabic sentiment analysis.

## 1 Introduction

Social media networks are playing an increasingly important role in the transmission of opinions about almost everything. Movies, products, actors, politicians, and events, are but a few examples of entities being targeted by opinionated posts. Because of this, social media has lately turned into a decision making tool, where decisions taken can vary from which political candidate to vote for, to which product to buy or which movie to watch. As a result, sentiment analysis has been the focus of many research studies in the past few years with sentiment Analysis in Arabic following the trend. Compared to the English language, the Arabic Language remains under-resourced with respect to annotated datasets and lexicons. However, more and more Arabic resources are starting to appear.

This paper presents a model for carrying out Arabic sentiment analysis by augmenting a machine learning approach with a set of features derived from an Arabic sentiment lexicon as well as from the text itself. To validate the model, it was applied to

© Springer International Publishing AG, part of Springer Nature 2018
A. Gelbukh (Ed.): CICLing 2016, LNCS 9624, pp. 307–319, 2018.
https://doi.org/10.1007/978-3-319-75487-1_24

all benchmarked datasets that the authors were able to acquire. The results of experimenting with these datasets, show that with the exception of one dataset, the model achieves higher polarity detection accuracy than all similar systems that have experimented with the same datasets.

The rest of this paper is organized as follows: Sect. 2 briefly describes related work, Sect. 3 overviews the proposed model and its preprocessing and feature extraction steps, Sect. 4 presents the experiments carried out to evaluate the presented model and their results, and finally Sect. 5 concludes this paper and presents future research directions.

## 2 Related Work

Work on Arabic Sentiment analysis has been gaining a lot of attention during the past couple of years. In [1], Abdulla et al., compared machine learning and lexicon based techniques for Arabic sentiment analysis on tweets written in the Jordanian dialect. The data set that was used for comparison consisted of 2000 tweets (1000 positive and 1000 negative). The preprocessing steps applied on the dataset included spelling correction, elongation and stop-word removal, and letter normalization. The authors experimented with a set of classifiers including: Support Vector Machines (SVM), Naive Bayes (NB), and K-Nearest Neighbor (KNN) with K = 9, using RapidMiner [2]. The best results were reported to be those of SVM and NB (using 5-fold cross validation and light stemming) with accuracies of 87.2% and 81.3% respectively.

The work presented in [3] targeted tweets written in the Egyptian dialect and was focused on examining the effect of different pre-processing steps on the task of sentiment analysis. The tweets that were used for training were chosen such that they contained only one opinion (positive or negative), were not sarcastic, and covered different topics. The tweet training set consisted of one thousand tweets (500 positive and 500 negative) manually annotated by two experts. Preprocessing included removing user-names, pictures, hash tags, stopwords, URLs, and all non-Arabic words. 10-fold cross validation was used for evaluating system performance. SVM was used for classification, and Weka [4] was used as the platform for experimentation. The results reported on that work were on raw data, normalized data, and light stemmed data using a modified version of El-Beltagy_Rafea Stemmer [5]. The best results were obtained by applying normalization, stemming, a combination of Unigrams and Bigrams, and stop word removal.

Salamah and Elkhlifi [6] developed a system for extracting sentiment from the Kuwaiti-Dialect. The system consisted of four components: a tweets collector, a preprocessing module, an opinionated terms extractor, and an opinion classifier. The preprocessing module consisted of a segmenter developed by the authors as well as the Stanford Arabic Tokenizer and was used to extract features from each tweet. The authors implemented their own set of resources of adjectival, nominal, verbal and adverbial indicators for the Kuwaiti dialect. They experimented with a manually annotated dataset comprised of 340,000 tweets, using SVM, J48, ADTREE, and Random Tree classifiers. The best result was obtained using SVM with a precision and a recall of 76% and 61% respectively.

In [7], Duwairi et al. present a sentiment analysis tool for Jordanian Arabic tweets. The authors created three lexicons to enhance the overall system accuracy. The first lexicon was used to map all dialect words to Modern Standard Arabic (MSA), the second lexicon was used to convert all Arabizi (Arabic words written in Roman Alphabet) words to MSA, and the third lexicon was used to convert emoticons to their respective meaning in the language. One thousand manually annotated Arabic Jordanian tweets were used for evaluating system performance. Data preprocessing steps involved: tokenization, stop words removal, links and elongation removal, letter normalization and stemming. RapidMiner [2] was used to train and test the model using NB, SVM and KNN. The NB classifier performed best in this experiment with an accuracy of 76% using 5-fold cross validation without stemming or stop words removal.

The goal of the work presented by Salameh et al. [8] was to investigate whether sentiment information was lost or persevered when translating from one language to another. In this study, the authors have used Arabic as a source language and English as a target and experimented with different configurations relating to how translation and sentiment annotation are carried out. In order to carry out their study, the authors adapted the NRC-Canada sentiment analysis system [9, 10], which can be considered as state of the art in English sentiment analysis[1], to work with Arabic. As such, and based on the results reported in their work, the system that they have developed for Arabic can be thought of as the state of the art in Arabic sentiment analysis. Tweet pre-processing included URL and mention normalization as well as character normalization, tokenization, part of speech tagging, and lemmatization. The main features used by the system were word and character n-grams, the count of each part-of-speech tag, the number of negated contexts, and a sentiment score. The sentiment score was calculated with the aid of a sentiment lexicon. The Arabic lexicon used in this work, was an automatically generated one. Evaluation of the Arabic system was carried out by testing it on two existing datasets: Mourad and Darwish [11] and Refaee and Rieser [12] as well as two other datasets that the authors have annotated (BBN and Syria). All four datasets have been used to test the work described in this paper and are described in detail in Sect. 4.1.

Shoukry and Rafea [13] present an approach that combines sentiment scores obtained using a lexicon with a machine learning approach and they apply it on Egyptian tweets. Pre-processing in their system involves character normalization, stemming, and stop word removal. Features are represented as a count vector of unigrams, bigrams, and tri-grams. The sentiment lexicon employed in this work, is one that was manually built by the authors. The lexicon contains 390 negative entries and 262 positive entries. To test their system, the authors annotated 4800 tweets[2] as positive, negative, or neutral (1600 positive, 1600 negative, 1600 neutral). The experiments conducted by the authors, show that adding the semantic orientation feature does in fact improve the result of the sentiment analysis task.

---

[1] The system was the best performer in SemEval 2013 and SemEval 2014 with respect to the message level polarity detection task [9, 10].

[2] The version provided to us by the authors had 4820 tweets.

# 3 Model Overview

## 3.1 Model Components

The model that we have adopted is one that employs statistical machine learning for evaluating the polarity of some input text. Our previous work presented in [14], has revealed that classifiers that perform best for the task of sentiment analysis are those that belong to the family of Naïve Bayes as well as SVM, but that the best performer amongst those, varies from one dataset to another. This conclusion is supported by the literature as presented in Sect. 2, where some authors have reported that they obtained their best results using an SVM classifier while others reported that Naïve Bayes classifiers performed better than SVM. In our model, we have chosen to use Complement Naïve Bayes (CNB) [15] as the default classifier. This choice is also based on experiments reported in [14], where CNB performed consistently well across experimented with datasets, even if it was not always the top performing classifier across these datasets. However, our system has been built in such a way that we can easily interchange CNB with any other classifier. The actual implementation of the system, was carried out in Java, with the Weka library [16] providing the machine learning functionality.

Like all supervised sentiment analysis systems, the first step to be carried out in our model is data pre-processing followed by feature extraction. Features extracted in the latter step, represent the actual inputs to the target classifier. The main contribution of the presented work is the introduction of the set of features that are extracted from input text and which we argue, improve classification accuracy. While some of these features are related to input text characteristics, such as the length of the text, whether it ends with a question mark or not, etc., other features are related to the occurrence of sentiment words or phrases within the input text. To extract these features a sentiment lexicon is needed. For the past few years, Arabic has been considered as an under-resourced language with respect to the task of sentiment analysis due to the almost non-existence of sentiment lexicons and training datasets annotated with sentiment. As time goes by, more and more annotated sentiment public datasets are starting to emerge [8, 12, 17]. Quality lexicons are still scarce, although efforts have been made to translate existing English lexicons to Arabic in order to fill this gap [8, 11]. To our knowledge, the largest Arabic sentiment lexicon with a manual like quality is NileULex which is presented in [18]. The lexicon has 5953 positive and negative entries and includes sentiments words and phrases from both Modern Standard Arabic and Egyptian. Many of the dialectical Egyptian entries, are also used in other Arabic speaking countries. In the presented work, we make use of this lexicon. The following two subsections, detail the pre-processing steps carried out in our model, and the features that are extracted and used by the sentiment classifier.

## 3.2 Preprocessing Steps

In this section, we present a detailed description of the preprocessing pipeline for our sentiment classifier.

## 1. Character Normalization

The first preprocessing step applied to input text, is character normalization. In this step, letters "آ", "إ" and "أ" are replaced with "ا" while the letter "ة" is replaced with "ه", and the letter "ى" is replaced with "ي". Diacritics are also removed in this step.

## 2. Elongation Removal

In this step, words that have been elongated, are reduced to their normal standard canonical form. In social media, elongation is a method for giving certain words more emphasis. An example of an elongated word is "رااااااااااائع" (magnificent). An English example is "yesssssss". The algorithm applied for elongation detection and removal is a simple one. A regular expression is used to detect if a character appears consecutively three or more times. If such a character is detected, the consecutive repetitions are replaced by a single instance of that character. As a result "yessssss" will be transformed to "yes" and "رااااااااااائع" to "رائع".

## 3. Emoticons Detection and Replacement

In this step, input text is matched against a predefined list of emoticons labeled as positive or negative. For this step, we have compiled a list comprised of 105 negative emoticons and 110 positive emoticons. All matched emoticons are replaced with a single term that is not actually part of the Arabic vocabulary depending on whether the match is with a positive or a negative emoticon. In the actual implementation of our system, any positive emoticon is replaced by the term "بجومنشومياي" while negative emoticons are replaced by the term"بلاسنشومياي". The number of emoticon matches encountered is stored as they it considered as part of the feature set.

## 4. Mention Normalization

In this step, any mention starting with @, is replaced with the English word "MENTION".

## 5. Named Entities Tagging

Although Named Entities identification may not seem to be directly related to sentiment analysis, as detailed in [19], in the absence of POS tags, many Arabic names can get confused with sentiment lexicon entries, which can have a negative impact over the overall accuracy of a sentiment analysis system. For example, without a named entity recognition (NER) system the word "طيبه" in the term "جامعه طيبه" (Teaba University), will match with lexicon entry "طيبه" which means "kind" or "kindness" depending on the context. There are many other examples given in [19]. We have used the system described in [20] to tag named entities. Through experimentation, we discovered that the effect of NER is not as great as we expected it to be, but that its inclusion still has a positive effect.

## 6. Matching with Lexicon Entries

In this step, input tweets/texts are matched against entries in the sentiment lexicon. Since Arabic is a morphologically complex language, having a lexicon containing all possible surface forms of its entries is an almost impossible task [8, 11]. On the other hand, straight forward stemming, even if just light, can alter the meaning of a word entirely. If we take for example the word "روعه" which means "magnificent", and stem it using any traditional Arabic light stemmer, the result will be the term "روع", which means "terrorized". While the first term is very positive, the second is very negative. This problem can be avoided through the use of lemmatization instead of stemming. To carry out lemmatization we have used a dictionary based tool [21] which is based on the work presented in [5]. The tool utilizes a large set of dictionaries built using large datasets as described in [5, 22]. In addition, the tool allows for the addition of manual entries. Base forms of lexicon entries were added to a "stem list" [5] to ensure that any word matching with those, does not get stemmed beyond its base form. Both the tweets/texts and lexicon entries are lemmatized and stemmed using this tool, prior to any matching steps. When a match occurs between text in the input, and a lexicon entry, a unique term is added to the input text immediately after the matching sentiment term depending on whether the match was with a positive or negative entry (we are using the English terms "pos" and "neg" as sentiment identifiers). Example:

النهايات السعيده pos دائما تاتي في المشمشneg

In the shown example, terms that matched with a lexicon entry are underlined and in bold. A count for positive and negative terms is also kept to be later used as part of the features. Negators are currently handled in a very simple way: encountering a negator before a sentiment term within a window $w$ results in the reversal of its polarity. We have observed that in some cases, this is not necessarily valid. For example, the term "حلو لا", in which the negator "no" appears before the word "nice", is actually used to affirm that something is nice.

## 7. Stemming

For stemming, we used the same tool described in step 6. However, here we used it in "stem enforce" mode. In this mode, any ta'a marbota (ة), ha'a (ه) or a trailing ya'a (ي) are removed even if they are part of a word that exists in the stemmer's dictionaries. However, other terms that are in the dictionaries and that have known suffixes or prefixes will be preserved. For example, the word حيوان will not be stemmed. The tool is also capable of reducing many broken plurals to their singular form.

### 3.3    Features

The following is the list of features used by our model:

- Stemmed word uni-grams and bi-grams represented by their idf weights. (Terms whose occurrence count is less than 2 are excluded from the feature vector).
- *startsWithLink:* a feature which is set to 1 if the input text starts with a link and to a 0 otherwise.
- *endsWithLink:* a feature which is set to 1 if the input text ends with a link and to 0 otherwise.
- *numOfPos:* a count of terms within the input text that have matched with positive entries in the sentiment lexicon. To give some extra weight to terms that are made up of more than one word (compound_terms) the following formula is used to set this feature:

$$numOfPos = \sum_{i=0}^{n} i + \sum_{j=0}^{c} j \times \alpha \qquad (1)$$

Where **n** is the number of positive single terms, **c** is the number of positive compound terms, and $\alpha$ is the boosting factor >1. We have set $\alpha$ to 1.5 based on empirical experimentation.

- *numOfNeg:* a count of terms within the input text that have matched with negative entries in the sentiment lexicon. The value of this feature is calculated in the same way the *numOfPos* is calculated.
- *length:* a feature that can take on one of three values {0, 1, 2} depending on the length of the input text. The numbers correspond to very short, short and normal. A tweet is categorized as "very short" if its length is less than 60 characters, "short" if it is less than 100, and normal otherwise.
- segments: a count for the number of distinct segments within the input text. We assume that segments are delimited by any of the following characters: "-|?.«؟,!:"
- *endsWithPostive:* a flag that indicates whether the last encountered sentiment word was a positive one or not.
- *endsWithNegative:* a flag that indicates whether the last encountered sentiment word was a negative one or not.
- *negPercentage:* a real number from 0 to 1 that represents the percentage of words in the text that are negative. For example, given the text "الوكيل ونعم الله حسبي", this number will be set to 1, as the entire text appears as a single entry in the lexicon. Given the text "ياه، ده وحش اوي", this number will be set to 0.25 as 1 word out of the four in the text is negative.
- *posPercentage:* a real number from 0 to 1 that represents the percentage of words in the text that are positive.
- *startsWithHashTag:* a flag that indicates whether the tweet starts with a hashtag.
- *numOfNegEmo:* the number of negative emoticons that have appeared in the tweet.
- *numOfPosEmo:* the number of positive emoticons that have appeared in the tweet.
- *endsWithQuestionMark:* a flag that indicates whether the tweets ends with a question mark or not.

# 4 Experiments and Results

The goal of the presented experiments was to determine whether the features introduced by this work do in fact improve sentiment analysis results or not. To determine this, we needed to compare between our model and existing benchmarks. Towards this end, we've collected as many benchmarked datasets as we were able to get. The description of the used datasets is provided in Sect. 4.1. The previously obtained results for these datasets can be found in [8, 11, 14]. Description of the experiments and their results, can be found in Sect. 5. All experiments were carried out using the WEKA workbench [4].

## 4.1   The Used Datasets

**The Talaat et al. dataset (NU)** [14]: The collection and annotation for this dataset is described in [14]. The dataset contains 3436 unique tweets, mostly written in Egyptian dialect. These tweets are divided into a training set consisting of 2746 tweets and a test set containing 683 tweets. The distribution of training tweets amongst polarity classes is: 1046 positive, 976 negative, and 724 neutral tweets. The distribution of the test dataset is: 263 positive, 228 negative and 192 neutral. This dataset is available by request from the author of this paper.

**The KSA_CSS dataset (NBI)** [14]: This dataset is one that was collected at a research center in Saudi Arabia under the supervision of Dr. Nasser Al-Biqami and which is also described in [14]. The majority of tweets in this dataset are in Saudi and MSA, but a few are written in Egyptian and other dialects. The tweets for this dataset have also been divided into a training set consisting of 9656 tweets and a test set comprised of 1414 tweets. The training set consists of 2686 positive, 3225 negative, and 3745 neutral tweets and the test set has 403 positive, 367 negative, and 644 neutral tweets.

**The Refaee_Rieser Dataset (RR2, RR3)** [12]: This dataset is a twitter subset of a dataset collected by Refaee and Rieser [23]. This particular dataset does not target a specific dialect. We re-constructed the dataset using the Twitter IDs provided by the authors, but some of those were deleted or not found. As a result the re-constructed dataset consisted of 724 positive tweets, 1565 negative tweets, and 3204 neutral tweets. In order to compare our model with existing benchmarks, this dataset was used in two configurations. In the first it was divided into a training set (4405 tweets) and a test set (1088 tweets) as described in [14]. The training dataset had 599 positive, 1241 negative, and 2565 neutral tweets, and the test dataset had 125 positive, 324 negative, and 639 neutral tweets. In the second configuration, the neutral class was omitted leaving 2289 tweets divided into 1563 negative tweets and 722 positive tweets.

**The Mourad_Darwish Dataset (MD)** [11]: Like the Refaee_Rieser Dataset, this dataset does not target a specific dialect. The dataset has a total of 1111 tweets of which 377 are classified as negative and 734 are classified as positive.

**The BBN Dataset (BBN)** [8]: This dataset consists of 1199 Levantine sentences, selected by the authors of [8] from LDC's BBN Arabic-Dialect–English Parallel Text. The sentences were extracted from social media posts. The polarity breakdown of the sentences in this dataset is as follows: 498 are positive, 575 are negative, and 126 are neutral.

**The Syria Dataset (SYR)** [8]: This dataset consists of 2000 Syrian tweets, so most of the tweets in this dataset are in Levantine. The dataset was collected by (Salameh and Mohammad) [8] and consists of 448 positive tweets, 1350 negative tweets and 202 neutral tweets.

**The Shoukry_Rafea Dataset (SR)** [13]: This dataset consists of 4820 Egyptian tweets divided into 1604 negative tweets, 1612 positive tweets and 1604 neutral tweets.

The authors of this dataset have kindly shared it with us, but the shared version is one that has been pre-processed by removing all mentions, hashtags and URLs.

Table 1 provides presents an overview of each of the used datasets described above.

**Table 1.** Summary of the size and distribution of used datasets among polarity classes

| Dataset | Total | Number of tweets | | | | | | | |
|---------|-------|------------------|---|---|---|---|---|---|---|
| | | *Training* | | | | *Testing* | | | |
| | | *Pos* | *Neg* | *Neu* | *Total* | *Pos* | *Neg* | *Neu* | *Total* |
| NU | 3436 | 1046 (38.1%) | 976 (35.5%) | 724 (26.4%) | 2746 | 263 (38.5%) | 228 (33.4%) | 192 (28.1%) | 683 |
| NBI | 11070 | 2686 (28.1%) | 3225 (33.7%) | 3745 (39.2%) | 9566 | 403 (28.5%) | 367 (26.0%) | 644 (45.5%) | 1414 |
| RR2 | 2285 | 722 (31.6%) | 1563 (68.4%) | – | 2285 | – | – | – | – |
| RR3 | 5493 | 599 (13.6%) | 1241 (28.2%) | 2565 (58.2%) | 4405 | 125 (11.5%) | 324 (29.8%) | 639 (58.7%) | 1088 |
| MD | 1111 | 734 (66.1%) | 377 (33.9%) | – | 1111 | – | – | – | – |
| BBN | 1199 | 498 (41.5%) | 575 (48%) | 126 (10.5%) | 1199 | – | – | – | – |
| Syria | 2000 | 448 (22.4%) | 1350 (67.5%) | 202 (10.1%) | 2000 | – | – | – | – |
| SR | 4820 | 1612 (33.4%) | 1604 (33.3%) | 1604 (33.3%) | 4820 | – | – | – | – |

## 5   Results

We tested our model which relies on the features presented in Sect. 3.2, on each of the datasets described in the previous subsection. Table 2, shows the accuracy results of applying 10 fold cross validation on the MD, RR2, BBN, SYR, and SR datasets. Each of these datasets has at least one previously published result as indicated in the table. Results published in Salameh et al. [8], were provided in two contexts that are relevant to this work. In the first (Ar Sys), polarity annotations were performed directly on Arabic text and were compared to system generated annotations. In the second (Eng Sys), automatically translated Arabic text was classified using an English sentiment analysis system and the labels were compared against original Arabic labels for the same text.

The results for our system are shown in two formats. The first format (default configuration) employs a Complement Naïve Bayes (CNB) classifier [15] with a smoothing factor of 1.0. Practical experience has shown us that smaller CNB smoothing factors result in better cross validation results, but poor test results.

As previously stated, work presented in [14] has shown that the best classifier to use for some given dataset, is often dataset dependent with Naïve Bayes classifiers and SVM usually yielding the best results. Accordingly, we have also included another

**Table 2.** Comparison between the accuracy (%) of various sentiment analysis systems on some of the used datasets

| Dataset | MD | RR2 | BBN | SYR | SR |
|---|---|---|---|---|---|
| Labels | Pos, neg | Pos, neg | Pos, neg, neu | Pos, neg, neu | Pos, neg, neu |
| Size | 1111 | 2285[a] | 1199 | 2000 | 4820 |
| Baseline | 67.87 | 66.78 | 56.88 | 75.4 | 73.84 |
| Mourad and Darwish [11] | 72.5 | – | – | – | – |
| Salameh et al. [8] (Ar Sys) | 74.62 | 85.23 | 63.89 | 78.65 | – |
| Salameh et al. (Eng Sys) | – | – | 62.49 | 78.11 | – |
| Shoukry and Rafea | – | – | – | – | 80.6 |
| Our system (default config) | 80.2 | 84.55[b] | **71.06** | 78.75 | 83.03[c] |
| Our system (best classifier) | **81** (MNBU) | **85.03** (SVM) | **71.06** (CNB) | **80.6** (SVM) | **83.13** (MNBU) |

[a]The version of the dataset that we have used is smaller than that used by [8], so our results and theirs are not directly comparable.
[b]In our system, features for this dataset were reduced using information gain.
[c]Features were reduced for this dataset using information gain.

format for our model, where the best results obtained by each of those classifiers is outlined. MNBU in the table, refers to a multi-nominal updateable Naïve Bayes classifier [24]. The baseline in this table, was obtained by representing raw unprocessed tweets using boolean vectors of unprocessed unigrams and classifying them using SVM (Linear Kernel with LibSVM's [25] default parameters).

The results shown in Table 2, illustrate that even when using the default configuration (which is not always the best), the presented model outperforms all other existing systems on all datasets except for RR2. The version of the RR2 dataset that we are using is smaller than that on which results in the table have been presented, so the results are not directly comparable. We've also found that using information gain (IG) to reduce the features for this dataset as well as for the SR dataset, improves the overall accuracy. The result for the SR dataset for example, would drop from 83.03% to 81.08% without using IG.

Table 3 shows the results of applying our model on datasets presented in [14]. In this table, the best results obtained in [14] are reported although it was often the case in this study that classifiers that scored best for 10 fold cross validation were not the ones that scored best for the test datasets. The baseline in this experiment has been calculated in the same way as for the experiment presented in Table 2. From the results it can be seen that the presented model performs quite well with respect to both the NU and NBI datasets. However, it performs rather badly with the RR3 dataset. When using CNB, which is the default classifier, the accuracy even drops below the baseline. We are not sure why the system performance drops like this with this particular dataset, but this will be subject to further analysis.

**Table 3.** Comparison between proposed model results and results presented in [14]

| Dataset | NU | | NBI | | RR3 | |
|---|---|---|---|---|---|---|
| Size | 10 FCV | Test | 10 FCV | Test | 10 FCV | Test |
| | 2746 | 683 | 9656 | 1414 | 4405 | 1088 |
| Baseline | 55.86 | 54.76 | 66.73 | 66.62 | 53.86 | 54.56 |
| Khalil et al. [14] | 70.84 | 57.25 | 77.34 | 69.81 | 66.7 | **59.45** |
| Our system (default config) | **73.53** | 59.23 | 79.06 | **75.05** | 63.8 | 53.7 |
| Our system (best classifier) | **73.53** (CNB) | **60.76** (MNBU) | 79.06 (CNB) | **75.05** (CNB) | **67.3** (SVM-RBF) | 58.8 (SVM-RBF) |

# 6  Conclusion and Future Work

In this paper we have presented a set of features that can be used with Arabic tweets in order to enhance the performance of a sentiment analysis system. We believe that the features that have the highest impact in enhancing the results, are those derived from a high quality sentiment lexicon. This hypothesis is supported by the work presented in [18]. By using these features within a sentiment analysis system, we have shown that our model outperforms all existing sentiment analysis systems on 6 out of the 7 datasets on which we have applied it.

In the future we would like to enhance the presented model by: handling negation in a better way, using elongation as an indicator for emphasis, and making use of intensifiers. We would also like to assign weights to various entries in the lexicon and use scores instead of counts. We have already taken initial steps towards this goal which are presented in [26]. We would also like to investigate augmenting the currently used lexicon with some or all of the currently available translated lexicons.

**Acknowledgements.** The authors would like to thank Amira Shoukry, Dr. Ahmed Rafea, and Dr. Kareem Darwish for kindly sharing their datasets.

# References

1. Abdulla, N.A., Ahmed, N.A., Shehab, M.A., Al-Ayyoub, M.: Arabic sentiment analysis: lexicon-based and corpus-based. In: 2013 IEEE Jordan Conference on Applied Electrical Engineering and Computing Technologies (AEECT), pp. 1–6. IEEE, Amman (2013)
2. Mierswa, I., Wurst, M., Klinkenberg, R., Scholz, M., Euler, T.: YALE: rapid prototyping for complex data mining tasks. In: Proceedings of ACM SIGKDD International Conference on Knowledge Discovery and Data Mining, pp. 935–940 (2006)

3. Shoukry, A., Rafea, A.: Preprocessing Egyptian dialect tweets for sentiment mining. In: Proceedings of 4th Workshop on Computational Approaches to Arabic Script-Based Languages, San Diego, California, USA, pp. 47–56 (2012)

4. Witten, I.H., Frank, E., Trigg, L., Hall, M., Holmes, G., Cunningham, S.J.: Weka: practical machine learning tools and techniques with Java implementations. In: Seminar, vol. 99, pp. 192–196 (1999)

5. El-Beltagy, S.R., Rafea, A.: An accuracy enhanced light stemmer for Arabic text. ACM Trans. Speech Lang. Process. **7**, 2–23 (2011)

6. Salamah, J.B., Elkhlifi, A.: Microblogging opinion mining approach for Kuwaiti dialect. In: International Conference on Computing Technology and Information Management (ICCTIM 2014), pp. 388–396 (2014)

7. Duwairi, R., Marji, R., Sha'ban, N., Rushaidat, S.: Sentiment analysis in Arabic tweets. In: 2014 5th International Conference Information and Communication Systems (ICICS), pp. 1–6. IEEE, Irbid (2014)

8. Salameh, M., Mohammad, S., Kiritchenko, S.: Sentiment after translation: a case-study on Arabic social media posts. In: Proceedings of 2015 Conference of the North American Chapter of Association for Computational Linguistics: Human Language Technologies, pp. 767–777. Association for Computational Linguistics, Denver (2015)

9. Mohammad, S.M., Kiritchenko, S., Zhu, X.: NRC-Canada: building the state-of-the-art in sentiment analysis of tweets. In: Proceedings of 7th International Workshop on Semantic Evaluation Exercises (SemEval-2013), Atlanta, Georgia, USA (2013)

10. Kiritchenko, S., Zhu, X., Mohammad, S.: Sentiment analysis of short informal texts. J. Artif. Intell. Res. **50**, 723–762 (2014)

11. Mourad, A., Darwish, K.: Subjectivity and sentiment analysis of modern standard Arabic and Arabic microblogs. In: Proceedings of 4th Workshop on Computational Approaches to Subjectivity, Sentiment and Social Media Analysis, pp. 55–64 (2013)

12. Refaee, E., Rieser, V.: Subjectivity and sentiment analysis of Arabic Twitter feeds with limited resources. In: Proceedings of Workshop on Free/Open-Source Arabic corpora and corpora processing tools, Reykjavik, Iceland, pp. 16–21 (2014)

13. Shoukry, A., Rafea, A.: A hybrid approach for sentiment classification of egyptian dialect tweets. In: 1st International Conference on Arabic Computational Linguistics (ACLing), Cairo, Egypt, pp. 78–85 (2015)

14. Khalil, T., Halaby, A., Hammad, M.H., El-Beltagy, S.R.: Which configuration works best? An experimental study on supervised Arabic twitter sentiment analysis. In: Proceedings of 1st Conference on Arabic Computational Liguistics (ACLing 2015), Co-located with CICLing 2015, Cairo, Egypt, pp. 86–93 (2015)

15. Rennie, J.D.M., Shih, L., Teevan, J., Karger, D.R.: Tackling the poor assumptions of Naive Bayes text classifiers. In: Proceedings of 20th International Conference on Machine Learning (ICML 2003), USA, vol. 20, pp. 616–623 (2003)

16. Frank, E., Hall, M., Holmes, G., Kirkby, R., Pfahringer, B., Witten, I.H., Trigg, L.: WEKA: a machine learning workbench for data mining. In: Maimon, O., Rokach, L. (eds.) Data Mining and Knowledge Discovery Handbook, pp. 1305–1314. Springer, Boston (2005). https://doi.org/10.1007/978-0-387-09823-4_66

17. ElSahar, H., El-Beltagy, S.R.: Building large Arabic multi-domain resources for sentiment analysis. In: Gelbukh, A. (ed.) CICLing 2015. LNCS, vol. 9042, pp. 23–34. Springer, Cham (2015). https://doi.org/10.1007/978-3-319-18117-2_2

18. El-Beltagy, S.R.: NileULex: a phrase and word level sentiment lexicon for Egyptian and modern standard Arabic. In: Proceedings of LREC 2016, Portorož, Slovenia (2016, to appear)

19. El-Beltagy, S.R., Ali, A.: Open issues in the sentiment analysis of Arabic social media: a case study. In: Proceedings of 9th International Conference on Innovations and Information Technology (IIT 2013), Al Ain, UAE (2013)
20. Zayed, O., El-Beltagy, S.R.: Named entity recognition of persons' names in Arabic tweets. In: Proceedings of Recent Advances in Natural Language Processing (RANLP 2015), Hissar, Bulgaria (2015)
21. El-Beltagy, S.R., Rafea, A.: LemaLight: a dictionary based Arabic lemmatizer and stemmer (2016)
22. El-Beltagy, S.R., Rafea, A.: A corpus based approach for the automatic creation of Arabic broken plural dictionaries. In: Gelbukh, A. (ed.) CICLing 2013. LNCS, vol. 7816, pp. 89–97. Springer, Heidelberg (2013). https://doi.org/10.1007/978-3-642-37247-6_8
23. Refaee, E., Rieser, V.: An Arabic Twitter Corpus for subjectivity and sentiment analysis. In: Proceedings of 9th Edition of Language Resources and Evaluation Conference (LREC 2014), Iceland (2014)
24. McCallum, A., Nigam, K.: A comparison of event models for Naive Bayes text classification. In: AAAI/ICML-1998 Workshop on Learning for Text Categorization, pp. 41–48 (1998)
25. Chang, C., Lin, C.: LIBSVM: a library for support vector machines. ACM Trans. Intell. Syst. Technol. **2**, 1–39 (2011)
26. El-Bletagy, S.R.: NileTMRG: deriving prior polarities for Arabic sentiment terms. In: Proceedings of SemEval 2016, San Diego, California (2014, submitted)

# Sentiment Analysis in Arabic Twitter Posts Using Supervised Methods with Combined Features

Rihab Bouchlaghem[1(✉)], Aymen Elkhelifi[2], and Rim Faiz[3]

[1] LARODEC, ISG, University of Tunis, Tunis, Tunisia
rihab.bouchlaghem@isg.rnu.tn
[2] Paris Sorbonne University, Paris, France
Aymen.Elkhlifi@paris.sorbonne.fr
[3] LARODEC, IHEC, University of Carthage, Tunis, Tunisia
Rim.Faiz@ihec.rnu.tn

**Abstract.** With the huge amount of daily generated social networks posts, reviews, ratings, recommendations and other forms of online expressions, the web 2.0 has turned into a crucial opinion rich resource. Since others' opinions seem to be determinant when making a decision both on individual and organizational level, several researches are currently looking to the sentiment analysis.

In this paper, we deal with sentiment analysis in Arabic written Twitter posts. Our proposed approach is leveraging a rich set of multilevel features like syntactic, surface-form, tweet-specific and linguistically motivated features. Sentiment features are also applied, being mainly inferred from both novel general-purpose as well as tweet-specific sentiment lexicons for Arabic words.

Several supervised classification algorithms (Support Vector Machines, Naive Bayes, Decision tree and Random Forest) were applied on our data focusing on modern standard Arabic (MSA) tweets. The experimental results using the proposed resources and methods indicate high performance levels given the challenge imposed by the Arabic language particularities.

**Keywords:** Sentiment analysis · Twitter · Modern standard Arabic
Supervised classification · Arabic sentiment lexicon

## 1 Introduction

The sentiment analysis task aims to detect subjective information polarity in the target text by applying Natural Language Processing (NLP), text analysis and Computational linguistics techniques. With the emergence of web 2.0, it becomes easy for Internet users to post their opinionated comments, reviews and share their thoughts and views via social networks, personal Blogs, forums, etc. Performing sentiment analysis on this raw data brings new challenges (for example: limited post length, misspellings, etc.) for which creative text mining techniques are needed in order to extract valuable information.

© Springer International Publishing AG, part of Springer Nature 2018
A. Gelbukh (Ed.): CICLing 2016, LNCS 9624, pp. 320–334, 2018.
https://doi.org/10.1007/978-3-319-75487-1_25

According to a study performed by Semiocast[1], Arabic was the fastest growing language on Twitter in 2011, and was the 6th most used language on Twitter in 2012. While a wide range of Arabic opinionated posts are broadcasted, research in the area of Arabic sentiment analysis remain sparse and show a very slow progress [15–17] compared to that being carried out in other languages mainly English [1–8].

This paper describes a supervised Sentiment analysis approach of Arabic written tweets to conclude the Twitter users' sentiments about a specific event or entity in a specific period. Our approach is using lexicon-based features for improving the detection of the sentiment polarity in Arabic tweets by integrating sentiment knowledge. Therefore, we introduce a general-purpose lexicon for Arabic sentiment words. The lexicon is created with large word-sentiment association lists which are automatically generated from an existed English polarity annotation corpus, manually filtered, and automatically expanded and translated. Furthermore, we propose a tweet-specific sentiment lexicon build with a gold seed words set manually annotated and extracted from the data set, and automatically expanded from 1.5 million tweets using co-occurrence and coordination computing methods. In addition to semantic features, the proposed data representation model includes: linguistic features (contextual Intensifiers, question keys, etc.), syntactic features (POS patterns, etc.), surface-form features (exclamation and question marks counts, etc.) and tweet-specific features (hashtag counts, tweet length, etc.)

We perform our experiments using three supervised classification algorithms: Support Vector Machines (SVM), Naïve Bayes (NB) and Random Forest (RF).

Experiments are carried out to asses both, the performance of the overall sentiment classification system as well as the quality and value of the proposed lexicons. The obtained results reveal that the SVM classifier conducts the highest F-score.

Furthermore, the ablation experiments allow exploring the impact of every feature group separately. Lexicon based features have positive impact on the classification performance since contribute to improve the F-score by over than 0.5%

The paper is organized as follows. We begin with a description of related work in Sect. 2. Next, we describe the data collection and annotation. Then, we detail the tweet's pre-processing performing tasks. Section 5 presents the sentiment lexicons used in our approach: general-purpose sentiment lexicon and tweet-specific sentiment lexicon. The detailed description of the used classification algorithms and the features set is presented in Sect. 6. Section 7 provides the results of the experiments evaluating the performance of the entire supervised sentiment analysis approach and examines the contribution of the features derived from our lexicons to the overall performance. Finally, we conclude and present future work in Sect. 8.

# 2  Related Works

Over the last decade, great effort has been devoted to the study of various aspects of sentiment analysis such as subjectivity classification and classifying positive and negative language. The proposed approaches used mainly machine learning techniques

---

[1]  http://semiocast.com/.

and have been applied to many different kinds of texts including customer reviews, newspaper headlines, novels, blogs, Twitter posts, etc. To classify sentiments conducted to single words, [1] used a supervised learning algorithm that automatically retrieves the adjectives' polarity from conjunction constraints collected from a large corpus. A specific unsupervised learning algorithm based on the mutual information statistical computing was applied in [2] to identify associations between sentences and the words "excellent" and "poor". In this work, only singletons "excellent" and "poor" were used as seed words having respectively positive and negative potential subjectivity.

While [3] developed a statistical approach for proposition opinion classification, comprising two methods. The first method used TREC 8, 9, and 11 text collections to compute, for each word, the difference between the frequency in subjective documents and the frequency in documents containing facts. The second method relied on co-occurrence score to extend a 1336 subjective adjectives seed list. [4] focused on subjective nested clauses classification and proposed a machine learning classification approach with wide range of syntactic features. [5] proposed an unsupervised classifier for subjective-objective classification. Various supervised methods were proposed in [6] for sentiment analysis in reviews extracted from travel blogs. In this work, the SVM classifier outperformed the Naive Bayes classifier.

Symbolic approaches such as proposed in [7–9] provided a better text analysis to represent the grammatical and semantic structure of analyzed text. However, such methods require an important manual work. In a similar context, [10] proposed an hybrid approach for sentiment classification applying different classifiers in series. The approach was tested on movie reviews, product reviews and *MySpace* comments. For the rule based classification, the authors combined existing rules with the ID3 and RIPPER induction algorithms to generate induced rule sets for classification. They also proposed a machine learning based method applying the SVM classifier for positive and negative sentiment classification.

Other studies aimed to build subjective lexicons. [11] presented a sentiment lexicon called *SentiFul*. The authors proposed methods to automatically generate the lexicon using an existing affect database. The generated lexicon was enlarged using direct synonymy relations and morphologic modifications. [12] developed two tweet-specific sentiment lexicons for English and proposed a supervised lexicon based system for sentiment classification in short messages (tweets and SMS).

Unfortunately, most of these systems are developed for the English language and are not directly usable on other language. Only a few works try to deal with sentiment analysis for morphologically rich languages such as Arabic. The most of the proposed works dealt with texts from the web and social media.

There are two main requirements to improve sentiment analysis effectively in any language and genres; high coverage sentiment lexicon and tagged corpora to train the sentiment classifier. [13] exploited web data collected from micro-blogs, forums and online market services to propose YADAC, a multi-genre dialectal Arabic corpus. In [14], the authors described AWATIF, which is a multi-genre corpus of Standard Arabic subjectivity and sentiment analysis. Three resources were exploited during the extraction process: the Penn Arabic Treebank collection, a set of Wikipedia user talk pages and conversations from online forums. In the same context, [15] developed a new annotated dataset for Arabic tweet subjectivity and sentiment analysis. Furthermore, [16] offered a manually labelled corpus for subjectivity and sentiment analysis, collected from Twitter. While [17] presented a sentence-Level sentiment analysis system for Modern standard Arabic (MSA) and Egyptian Dialect. They proposed a lexicon of Arabic sentiment words labelled as positive, negative and neutral. In addition to lexicon based features, they used many other features. However, the most of these resources are not publicly released yet.

Different approaches have applied language independent features selection methods to perform sentiment analysis in Arabic texts, such as genetic algorithms. [18] applied this method to identify discriminant features for both Arabic and English languages. The authors used many types of features, except semantic ones because they are language dependent. The proposed system was evaluated on movie reviews texts. Another way of performing sentiment analysis in Arabic texts consists in applying hybrid classifiers. For example, the work of [19] aimed to experiment sentiment analysis in a bilingual Arabic-English corpus using SVM and Naive Bayes (NB) classifiers. The applied features are both numeric and lexical.

Furthermore, the experiments performed by [15] used NB in order to prove the performance of the baseline they proposed, including various features. Other works used Arabic specific features in performing sentiment analysis. In this context, [20] proposed an approach of customers' Arabic comments mining which is based on new slang sentiment words extracted from web resources. The authors applied SVM classifiers to decide if a given comment conducts satisfaction or dissatisfaction.

In the same context, we propose an approach for sentiment analysis in Arabic written tweets. Our solution benefits from the use of tweet specific and general purpose sentiment lexicons which are semi-automatically built. A wide range of syntactic, linguistic and surface-form features are also applied for data representation.

## 3 Data

We started collecting data in June 2013 using our *Twitter API*[2] based collection module with specific key words. We have chosen to collect tweets related to recent major political events and social reforms in the Arabic world. Therefore, we usually had launched collect campaigns in specific time steps to retrieve relevant tweets which are subjective towards specific target entities, organizations, events, etc. We collect about 700 thousand MSA tweets.

From the collected corpus, we have selected a tweet subset to be manually annotated for subjectivity. The selection is performing according to specific conditions: holds one opinion, conducts a clear position, MSA, subjective, informative content, etc. We define a sentiment as positive, negative or neutral. Each data instance is marked with only a single annotation that tags the interpretation reflected by the tweet text, expressing the tweet owner's position. We also identify tweets non conducting opinion as "no_ sentiment" class. We annotated corpus contains 2000 MSA subjective tweets.

---

[2] http://twitter4j.org/en/index.html.

## 4  Pre-processing

The most of current NLP methods can't be directly applied to Arabic language and giving valid output. This is due to the peculiarities of the Arabic language. Therefore, our corpus requires much manipulations and pre-processing in order to be in a consistent form before applying these methods.

The first pre-processing task consists in Arabic tweets' text normalization. In fact, some Arabic script characters can cause confusion in texts. For example, the words "عشائه" and "عشاؤه" are different forms of the same word which means "his dinner". The normalization task consists, on the one hand, in converting all the various forms of a word to a common form. On the other hand, there are other Arabic characters which must be removed, being mainly the *shadda* ligature; a special symbol used to accentuate the consonant (e.g.: "عذّب", means "to torment"), and diacritics, representing short vowels in Arabic texts (e.g.: "عُذْبَ" , "عُذْب") [21]. We have used an existing normalizer to apply most of the Arabic language normalization rules such as: {ء،ؤ،ئ} → <ء>; {آ،أ،إ،ا} → <ا>.

Furthermore, we developed a regular expression based pre-processing module to remove all noisy elements (e.g.: useless whitespaces) and improve the tweet's form. In the same perspective, we perform the last pre-processing task allowing deleting URL, usermentions, and the '#' symbol, after extracting Twitter specific features.

The corpus is ready for specific NLP methods. We perform tokenization and POS (Part Of Speech) using the NLP tool Stanford Parser[3] [22].

## 5  Proposed Sentiment Lexicons

Sentiment lexicons are lists of words and expressions with prior associations to positive and negative sentiments. Some lexicons can additionally provide a sentiment score for a term to indicate its strength of evaluative intensity. Higher scores indicate greater intensity [12].

In our sentiment analysis approach, we built two sentiment lexicons from subjective corpus using several statistical methods and heuristics. The first one refers to an Arabic tweet-specific sentiment words lexicon generated by expanding a list of polarized words extracted from our dataset using more than 1.5 million subjective tweets. The next is an Arabic general-purpose sentiment words lexicon developed by exploiting a large polarized English corpus to identify the most subjective common terms and compute, for each candidate term, a PMI (*Pointwise Mutual Information*) based subjectivity score. The final MSA lexicon is obtained by translating the English selected subjective terms list.

We decided to use sentiment adjectives, nouns and verbs because we noticed that some opinion statements do not include any adjectives, but express negative or positive sentiment such as: " @جثث مسوخ #داعش بأرض العراق #باقية وتتعفن#نصرك_اللهم" means *"The bodies of monstrosities Daesh in land of Iraq remain and rot. Oh God, give us*

---

[3] http://nlp.stanford.edu/.

*victory*". There aren't any adjectives in this example, although it expresses the negative sentiment, the words "مسوخ" and "تتعفن" express the negative sentiment in the example.

## 5.1  General Purpose Arabic Sentiment Words Lexicon

Terms appearing in a subjective context tend to be polarized. We adapted that idea to create a general-purpose Arabic words lexicon by exploiting a large subjectivity annotation corpus for English, as described below.

We used the sentiment polarity corpus introduced in [23] holding 1000 positive and 1000 negative processed texts about movies reviews. This corpus consists of 1543347 words. We refer to this sentiment resource as basis to create a general purpose sentiment Arabic words lexicon. As can be seen from Fig. 1, our proposed lexicon generation process is divided into six steps.

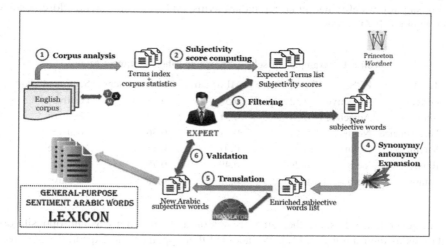

**Fig. 1.** General purpose lexicon building and expansion process

The first step consists in performing corpus analysis using the TXM[4] [24] software in order to carry out terms frequencies by part of speech and document class, in addition to useful statistics (total words number, distinct words number). From TXM output, three csv files are automatically generated (*Adjective_file, Noun_file, Verb_file*) per sentiment polarity class (positive, negative).

The next step is about computing sentiment score of all corpus terms, according to frequencies results given by TXM. We have chosen PMI [25] as a method for the association degree estimation of a term with a category because it is simple and robust and has been successfully applied in many NLP tasks [2, 12, 26].

The sentiment score for a term t in the corpus C was calculated according to the following equations:

---

[4] http://textometrie.ens-lyon.fr/?lang=en.

$$sentiment\_score(t) = PMI(t, positiveT) - PMI(t, negativeT) \qquad (1)$$

$$PMI(t, positiveT) = \log_2((freq(t, positiveT) * card(C)/(freq(t, C) * card(Cpos)) \qquad (2)$$

$$PMI(t, negativeT) = \log_2((freq(t, negativeT) * card(C)/(freq(t, c)) * card(Cneg)) \qquad (3)$$

Where: *freq (t, positiveT)* is the number of times a term *t* occurs in positive texts; *freq(t, negativeT)* is the number of times a term *t* occurs in negative texts; *freq (t, C)* is the total frequency of term *t* in the corpus; *card(Cpos)* is the total number of tokens in positive texts; *card(Cneg)* is the total number of tokens in negative texts; and *card(C)* is the total number of tokens in the corpus.

Table 1 illustrates examples of target terms and their sentiments scores. A positive sentiment score indicates greater overall association with positive sentiment, while a negative score indicates a greater association with negative sentiment. The next filtering step allows eliminating all redundant terms, ambiguous or having close positive and negative PMI scores.

**Table 1.** Examples of selected terms and their sentiment scores

| Term | POS | PMI-positive | PMI-negative | Sentiment score | Sentiment class |
|------|-----|--------------|--------------|-----------------|-----------------|
| Perfect | Adjective | 0.32 | −0.56 | 0.88 | Positive |
| Silly | Adjective | −0.38 | 0.31 | −0.69 | Negative |
| Performance | Noun | 0.15 | −0.21 | 0.36 | Positive |
| Victim | Noun | −0.06 | 0.07 | −0.13 | Negative |
| Raise | Verb | 0.29 | −0.49 | 0.79 | Positive |
| Annoy | Verb | −0.42 | 0.33 | −0.75 | Negative |

In general, words share the same orientation as their synonyms and opposite orientations as their antonyms [27]. We adopted this hypothesis to expand the new seed words list and assign correspondingly the subjectivity class to its synonyms/antonyms. We request Princeton Wordnet[5] [28] lexical database to get all synonyms and antonyms of each entry of the remaining words. Table 2 shows examples of expansion strategy results. A verification task is performed after every synonym/antonym adding, searching for redundancy or possible conflicts (e.g.: word having two polarity tags).

Finally, the enriched seed words set is sent to an online translator to get all Arabic equivalent seed words. After a last expert validation, the final lexicon first version; which we will refer to as general-purpose Arabic sentiment words lexicon; has 2513 entries for words.

---

[5] https://wordnet.princeton.edu.

**Table 2.** Examples of synonymy enrichment step output

| Example1 | Basic word | Excellent |
| --- | --- | --- |
| | Added synonyms | *First-class; fantabulous* |
| | Sentiment score | 1.24 |
| Example2 | Basic word | Neglect |
| | Added synonyms | *Miscarry; flunk; betray* |
| | Sentiment score | −1.26 |
| Example 3 | Basic word | Help |
| | Added synonyms | *Support; aid; assistance* |
| | Sentiment score | 0.055 |

## 5.2   Tweet Specific Arabic Sentiment Words Lexicon

We introduce an Arabic sentiment lexicon comprising 1462 words with various parts of speech (adjectives, nouns, verbs) which is manually created and automatically expansible (cf. Fig. 2).

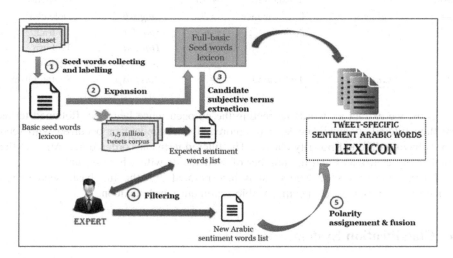

**Fig. 2.** Tweet-specific lexicon building and expansion process

Our starting point was a basic 560 seed word set manually collected from our dataset and labelled with "negative" or "positive" tags. Table 3 gives examples of collected seed words.

The lexicon expansion is performed in three levels. Firstly, we manually added, for each entry, the words sharing the same root with respect of the retained part of speech. As an example, the verb "نَكَّلَ" which means "to torment" is added to enrich the lexicon, when having the same root "نكل" of the seed noun "تَنْكِيل" ("harassment" in English).

**Table 3.** Examples of seed words collected from the data set

| Target word | Transliteration | Part of speech | English translation |
|---|---|---|---|
| هَجَرَ | haj ~ ara | Verb | To banish |
| مُتَطَرِّف | mutaTar ~ if | Adjective | Extremist |
| تَدْمِير | tadomiyr | Noun | Demolition |

The obtained set is expanded for the next time by applying a context based expansion strategy. We intend to investigate the seed words contexts in 0.5 million tweets corpus, specially collected for this purpose, and search for candidate subjective terms. We assume that if a word co-occurred with a given seed word in a tweet with reduced window, it tends to be subjective and share the same polarity class. The following table presents examples of new detected subjective terms (Table 4).

**Table 4.** Examples of co-occurred expected sentiment words

| Basic subjective word | | | Co-occurred words | | |
|---|---|---|---|---|---|
| Target word | Transliteration | English translation | Target word | Transliteration | English translation |
| غُلُوّ | guluw ~ | Extravagance | مَغلوب | maglwb | Defeated |
| كِذبَة | ki*bap | A lie | خَدَعَ | xadaEa | To trick |
| | | | حَرَقَ | Haraqa | To burn |
| | | | أَنجاس | >anojAs | Abusive |
| زَعَمَ | zaEama | To pretend | طَائِفِية | TaA}ifiyp | Sectarianism |

Thus, we extract all words respecting the co-occurrence heuristic. Retained terms are then filtered. A candidate term is ignored if is redundant or occurred with seed words having different polarity classes. The filtered terms are ordered according to the number of occurrences and the number of seed words with which occur.

Finally, the new seed words list is incorporated into the full basic seed words lexicon to form our tweet-specific Arabic sentiment words lexicon.

# 6   Classification System

In addition to lexicon based features, our proposed data representation includes tweet specific features; lexical, syntactic and surface-form features. We train four supervised classifiers on our dataset.

## 6.1   Classifiers

The classification step consists in performing multi-class categorization by mapping a set of tweets to the classes positive, negative, neutral or no-Sentiment. Our approach employs four supervised machine learning algorithms from different statistical approaches. The first one is Support Vector Machine (SVM) which is proved to be effective on text categorization tasks and robust on large feature spaces. In the preliminary experiments, a linear-kernel SVM showed better performance than an SVM with other used kernels (Gaussian, Polynomial and Sigmoid). Naïve Bayes (NB) is also applied, being yielded similar performance to SVM models [30]. The last applied classifiers are Random Forest which operates by constructing a multitude of decision trees, and the J45 classifier.

To built-in our classifiers, we have applied Weka API [29] using 10-fold cross validation. Experiments and results are presented and discussed in Sect. 7.

### 6.2   Features Extraction

We employ commonly used text classification features such as n-grams and part-of-speech tag counts, as well as common Twitter-specific features such as user mention and hashtag counts. Furthermore, we introduce several sentiment features that take advantage of the knowledge present in our proposed lexicons. Then, each tweet is represented as a features vector.

**Lexicon based features.** The sentiment lexicon features are derived from two newly created Arabic sentiment words lexicons.

*General-purpose lexicon.* Following [12], these features are generated from general purpose lexicon involving subjective words and their sentiment scores. For each token t occurring in a tweet and present in the lexicon, we use its sentiment score to compute:

- The number of tokens with score (t) $\neq 0$;
- The number of tokens with sentiment score (t) $> 0$;
- The number of tokens with sentiment score (t) $< 0$;
- The total sentiment score = $\sum_{t \in tweet} sentiment\_score(t)$;
- The maximal score = $max_{t \in tweet} sentiment\_score(t)$;

*Tweet specific lexicon.* We mainly employed two features inspired from the tweet specific lexicon:

- The number of tokens appearing in positive tweet specific lexicon;
- The number of tokens appearing in negative tweet specific lexicon;

**Linguistic features.** We explored a set of linguistic features able to handle several Arabic language structures such as negation, intensifiers, etc.

*Prayer and supplication.* Supplication means asking GOD for help in a humble way. It's generally expresses the supplicant's desperate wants and wishes. So, it can be used as sentiment marker. This talk style is often used in Arabic world. As an example, "اللهم العن داعش جميعا .. و انصرنا عليهم يا رب" means *"O Allah, curse Daesh all .. and give us victory over them"*; "لك الحمد يا الله" means *"praise be to you, o GOD"*. Therefore, we proposed a feature that counts, based on a list of Arabic supplication words, the number of supplication or wishful terms appearing in a given tweet.

*Negation.* The presence of Arabic negation term such as "لا", "ليس" and "لن" can reverse the sentiment polarity of words or change their evaluative intensity. We employed feature expressing how many negated terms that appear in the target tweet.

*Contextual intensifier.* Sentiment term polarity strength is fortified when being followed by an intensifier term. We employed a feature counting the number of Arabic intensifier ("كثير", "جدا") used in a given tweet.

*Question markers.* Questions are usually used to express: feeling, sadness, confusion, etc. That's why we considered the use of question in a tweet as sentiment indicator. In Arabic language, question types are multiples: "من", "ماذا", "كيف", "لماذا", (respectively "who", "what", "how", "why"). As an example, we give the following tweet "لا فلماذاالمتحمسين الشباب بعض يتعظ" which means *"Why do some young enthusiasts do not learn a lesson?"*

**Syntactic features.** We employed four features to indicate the number of occurrences of each part-of-speech tag. The authors in [17, 31] affirm that the POS tag bi-gram pattern *"noun+adjective"* typically represent a sentiment orientation. Thus, we used POS tag n-gram patterns frequencies in order to search the relevant patterns influencing the tweet sentiment polarity.

**Sentence level features**
*N-gram features.* We employed character n-gram, the presence or absence of contiguous sequences of 3 characters.

*Punctuation.* We employed the following punctuation specific features:

- The number of contiguous sequences of exclamation marks "!!!", question marks "???", and both exclamation and question marks "?!?!!".
- The total number of punctuation marks.

*Elongation.* We added a feature that counts the number of words with one character repeated more than two times, for example, "هههههههه".

**Tweet specific features.** As many existing works, we call Twitter-specific features the commands and conventions used by Twitter users in their posts. We used the following features:

- URL (or links): computes the number of links in tweet,
- User mentions: identifies the number of username mentions in a given tweets. It also indicates if the tweet replies to other users,
- Hashtag number,
- Presence of retweet symbol "RT",
- Tweet's Length.

# 7   Experiments

We trained the classifiers listed in the previous section on the set of 2000 annotated tweets using 10-fold cross validation. To evaluate the proposed classification process, we employ the precision, recall and F1 measurements presented in [28]. The obtained results are shown in Table 5.

**Table 5.** Classification results compared to baseline approach

|  | F-score | Precision | Recall |
|---|---|---|---|
| SVM | 70.64% | 69.66% | 72.28% |
| NB | 70.02% | 71.76% | 68.86% |
| Random forest | 66.84% | 66.2% | 76.02% |
| J48 | 64.59% | 63.49% | 65.85% |

As it can be seen, the SVM classifier achieves the best F1 (70.64%) superior than other classifiers. Where the Naïve Bayes classifier gives the second better result with 70.02% of F-score.

Table 6 and Fig. 3 show the results of the ablation experiments where we repeat the same classification process using SVM, but remove one feature group at a time.

**Table 6.** The F-scores obtained on the test set with one of the feature groups removed.

| Experiment | | F-score | Gain/loss |
|---|---|---|---|
| Run 1 | All features | 70.64% | – |
| Run 2-1 | All - general-purpose lexicon | 70.17% | −0.5% |
| Run 2-2 | All - tweet-specific lexicon | 69.7% | −0.94% |
| Run 3-1 | All - linguistic features | 70.59% | −0.05% |
| Run 3-2 | All - negation | 70.74% | +0.09% |
| Run 3-3 | All - intensifiers | 70.61% | −0.03% |
| Run 3-4 | All - supplication | 70.67% | +0.03% |
| Run 3-5 | All - question | 70.91% | +0.27% |
| Run 4 | All - syntactic features | 71.95% | +1.31 |
| Run 5 | All - sentence level features | 70.37% | −0.27% |
| Run 6 | All – N-gram features | 65.9% | −4.74% |
| Run 7 | All - tweet specific features | 70.94% | +0.3% |

Observe that the n-gram features were the most useful. In fact, removing just the character n-gram features results in a drop in performance (−4.74%). The sentiment lexicon features are the next most useful group. Removing Tweet specific lexicon based features leads to a drop in F-score of 0.94%. Where the use of general purpose lexicon based features enhances the F-score by 0.5%. It is interesting to note that sentence level features (elongation, punctuation sequences, etc.) contributed to improve the classification performance. Incorporating the linguistic tweets structures has also positive impact on performance, since the use of linguistic features improved performance by 0.05%. Removing the Tweet specific and syntactic features had almost no impact on performance, but this is probably because the discriminating information in them was also captured by some other features such as character n-grams.

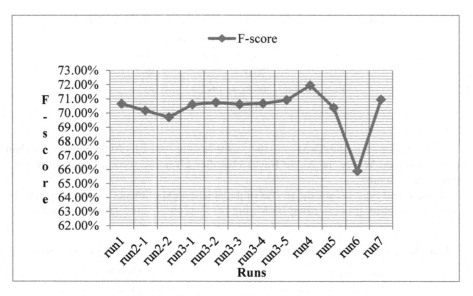

**Fig. 3.** The ablation experiments obtained F-scores

## 8   Conclusion and Perspective

In this paper, we presented a sentiment analysis approach for MSA and using a corpus of Arabic Twitter posts. Our goal was to deal with the complexity of Arabic language. That's why we employed various relevant and rich feature sets to handle Arabic specific structures (negation, question, etc.) to improve the classification performance. We show that exploiting general-purpose and tweet-specific sentiment lexicons has positive impact on classification performance. We also implemented a variety of features based on surface form and syntactic categories. The classification results are promising given the Arabic Natural Language Processing challenges, and this encourages us to continue working on this topic.

For future work, we will investigate other methods for association degree estimation to build general purpose lexicon. We also project to increase the SA training data size, and identify its impact on classifiers performance. In addition, we plan to improve the data representation by exploring other features groups that may be more discriminative, especially words embedding and vector representation.

## References

1. Hatzivassiloglou, V., McKeown, K.R.: Predicting the semantic orientation of adjectives. In: ACL 1997, Madrid, Spain, pp. 174–181 (1997)
2. Turney, P.D.: Thumbs up or thumbs down? Semantic orientation applied to unsupervised classification of reviews. In: ACL 2002, Philadelphia (2002)

3. Bethard, S., Yu, H., Thornton, A., Hatzivassiloglou, V., Jurafsky, D.: Automatic extraction of opinion propositions and their holders. In: Association for the Advancement of Artificial Intelligence (AAAI-2004), San Jose, California (2004)
4. Wilson, T., Wiebe, J., Hwa, R.: Just how mad are you? Finding strong and weak opinion clauses. In: Proceedings of Association for the Advancement of Artificial Intelligence (AAAI-2004), San Jose, California (2004)
5. Wiebe, J., Riloff, E.: Finding mutual benefit between subjectivity analysis and information extraction. IEEE Trans. Affect. Comput. 2(4), 175–191 (2011)
6. Ye, Q., Zhang, Z., Law, R.: Sentiment classification of online reviews to travel destinations by supervised machine learning approaches. Expert Syst. Appl. 36(3), 6527–6535 (2009). Part 2
7. Maurel, S., Dini, L.: Exploration de corpus pour l'analyse de sentiments. In: DEfi Fouille de Textes, Paris, France, pp. 11–23 (2009)
8. Vernier, M., Monceaux, L., Daille, B.L.: Catégorisation des évaluations dans un corpus de blogs multi-domaine. Revue des nouvelles technologies de l'information 25, 45–70 (2009)
9. Chardon, B., Muller, S., Laurent, D., Pradel, C., Séguéla, P.: Chaîne de traitement symbolique pour l'analyse d'opinion - l'analyseur d'opinions de Synapse Développement face à Twitter. In: Proceedings of DEfi Fouille de Textes, Caen, France (2015)
10. Prabowo, R., Thelwall, M.: Sentiment analysis: a combined approach. J. Inf. 3, 143–157 (2009)
11. Neviarouskaya, A., Prendinger, H., Ishizuka, M.: SentiFul: a lexicon for sentiment analysis. IEEE Trans. Affect. Comput. 2(1), 22–36 (2011)
12. Kiritchenko, S., Zhu, X., Mohammad, S.M.: Sentiment analysis of short informal texts. J. Artif. Intell. Res. Arch. 50(1), 723–762 (2014)
13. Al-Sabbagh, R., Girju, R.: YADAC: yet another dialectal Arabic corpus. In: 8th International Conference on Language Resources and Evaluation, Istanbul (2012)
14. Abdul-Mageed, M., Diab, M.: AWATIF: a multi-genre corpus for modern standard Arabic subjectivity and sentiment analysis. In: 8th International Conference on Language Resources and Evaluation, Istanbul (2012)
15. Mourad, A., Darwish, K.: Subjectivity and sentiment analysis of modern standard Arabic and Arabic microblogs. In: 4th Workshop on Computational Approaches to Subjectivity, Sentiment and Social Media Analysis, Atlanta, Georgia, pp. 55–64. Association for Computational Linguistic (2013)
16. Refaee, E., Rieser, V.: An Arabic Twitter Corpus for subjectivity and sentiment analysis. In: 9th International Conference on Language Resources and Evaluation (LREC 2014), Reykjavik, Iceland (2014)
17. Ibrahim, H.S., Abdou, S.M., Gheith, M.: Sentiment analysis for modern standard Arabic colloquial. Int. J. Nat. Lang. Comput. (IJNLC) 4(2), 95–109 (2015)
18. Abbasi, A., Chen, H., Salem, A.: Sentiment analysis in multiple languages: feature selection for opinion classification in web forums. ACM Trans. Inf. Syst. 26(3), Article ID: 12 (2008)
19. Rushdi-Saleh, M., Martin-Valdivia, M., Ureña-López, L., Perea-Ortega, J.: Bilingual experiments with an Arabic-English Corpus for opinion mining. In: Recent Advances in Natural Language, Hissar, Bulgaria, pp. 740–745 (2011)
20. Soliman, T.H.A., Elmasry, M.A., Hedar, A.R., Doss, M.M.: Mining social networks' Arabic slang comments. In: IADIS European Conference on Data Mining 2013 (ECDM 2013), Prague, Czech Republic (2013)
21. Bouchlaghem, R., Elkhelifi, A., Faiz, R.: Opinion mining in microblog texts using machine learning techniques. In: Knowledge Discovery and Data Analysis (KDDA 2015), Alger's, Algeria (2015)

22. Green, S., Manning, C.D.: Better Arabic parsing: baselines, evaluations, and analysis. In: COLING (2010)
23. Pang, B., Lee, L., Vaithyanathan, S.: Thumbs up? Sentiment classification using machine learning techniques. In: EMNLP (2002)
24. Heiden, S., Magué, J.-P., Pinceminb, B.: TXM: une plateforme logicielle open-source pour la textométrie conception et développement. In: JADT 2010, pp. 1021–1032 (2010)
25. Church, K.W., Hanks, P.: Word association norms, mutual information, and lexicography. Comput. Linguist. **16**(1), 22–29 (1990)
26. Turney, P., Littman, M.L.: Measuring praise and criticism: inference of semantic orientation from association. ACM Trans. Inf. Syst. **21**(4), 315–346 (2003)
27. Hu, M., Liu, B.: Mining and summarizing customer reviews. In: ACM SIGKDD Conference on Knowledge Discovery and Data Mining (KDD), pp. 168–177 (2004)
28. Fellbaum, C., Grabowski, J., Landes, S.: Performance and confidence in a semantic annotation task. In: Fellbaum, C. (ed.) WordNet: An Electronic Lexical Database, Language, Speech and Communication, pp. 216–237. The MIT Press, Cambridge (1998)
29. Hall, M., Frank, E., Holmes, G., Pfahringer, B., Reutemann, P., Witten, I.H.: The WEKA data mining software: an update. SIGKDD Explor. **11**(1), 10–18 (2009)
30. Morlane-Hondère, F., D'hondt, E.: Feature engineering for tweet polarity classification in the 2015 DEFT challenge. In: DEfi Fouille de Textes, Caen, France (2015)
31. Yi, J., Nasukawa, T., Bunescu, R., Niblack, W.: Sentiment analyzer: extracting sentiments about a given topic using natural language processing techniques. In: 3rd IEEE International Conference on Data Mining (ICDM), pp. 427–434 (2003)

# Interactions Between Term Weighting
# and Feature Selection Methods
# on the Sentiment Analysis of Turkish Reviews

Tuba Parlar[1]($\boxtimes$) ⓘ, Selma Ayşe Özel[2] ⓘ, and Fei Song[3] ⓘ

[1] Department of Mathematics, Mustafa Kemal University, Hatay, Turkey
tparlar@mku.edu.tr
[2] Department of Computer Engineering, Cukurova University, Adana, Turkey
saozel@cu.edu.tr
[3] School of Computer Science, University of Guelph, Guelph, Canada
fsong@uoguelph.ca

**Abstract.** Term weighting methods assign appropriate weights to the terms in a document so that more important terms receive higher weights for the text representation. In this study, we consider four term weighting and three feature selection methods and investigate how these term weighting methods respond to the reduced text representation. We conduct experiments on five Turkish review datasets so that we can establish baselines and compare the performance of these term weighting methods. We test these methods on the English reviews so that we can identify their differences with the Turkish reviews. We show that both $tf$ and $tp$ weighting methods are the best for the Turkish, while $tp$ is the best for the English reviews. When feature selection is applied, $tf * idf$ method with DFD and $\chi^2$ has the highest accuracies for the Turkish, while $tf * idf$ and $tp$ methods with $\chi^2$ have the best performance for the English reviews.

**Keywords:** Sentiment analysis · Feature selection · Term weighting

## 1 Introduction

Sentiment analysis classifies the sentiment of a review document, which is usually in the value of positive or negative, but can also be in a scale (e.g., 1 to 5, from most negative to most positive). Sentiment analysis has found many useful applications such as opinionated web search and automatic analysis of product reviews [1]. As a result, it is being actively studied by researchers particularly with the use of machine learning algorithms for various languages. However, not much work has been done for the Turkish language. The goal of this study is to analyze the effects of different term weighting methods along with different feature selection methods for text representation on the sentiment analysis of Turkish reviews.

Before classifying the sentiment of a review document, we need to find a suitable way to represent the text so that important features can be emphasized. In the vector space model, different features are assigned with appropriate weights so that more important features can receive higher weights for the text representation. Several term

© Springer International Publishing AG, part of Springer Nature 2018
A. Gelbukh (Ed.): CICLing 2016, LNCS 9624, pp. 335–346, 2018.
https://doi.org/10.1007/978-3-319-75487-1_26

weighting methods have been proposed in literature, especially in the field of Information Retrieval, such as term frequency ($tf$), term frequency and inverse document frequency ($tf * idf$), term presence ($tp$), and term frequency and relevance frequency ($tf * rf$).

For traditional text classification, the frequencies of words are usually important. For example, if a document repeatedly mentions certain sports terms and athlete names, it is more likely to be classified into the "sports" category. For sentiment analysis, however, people rarely repeat the sentiment-expressing words in the same review. As observed in [2], a consumer is unlikely to write: "This camera is great. It takes great pictures. The LCD screen is great. I love this camera." Rather, the consumer is more likely to write: "This camera is great. It takes breathtaking pictures. The LCD screen is bright and clear. I love this camera." In other words, the use of sentiment-expressing words tends to be less frequent within a review, but more frequent across different reviews. Thus, a feature selection method for sentiment analysis should accommodate the above observation as much as possible.

In this study, we want to investigate the interactions between term weighting and feature selection methods on the sentiment analysis of Turkish reviews. We also test these methods on the English reviews so that we can identify their differences with the Turkish reviews. To our knowledge, this is the first study that investigates the interactions between term weighting and feature selection methods on the sentiment analysis of Turkish reviews.

For the rest of this paper, we first introduce the related work on the sentiment analysis of Turkish reviews, feature selection methods, as well as machine learning algorithms. Then, we describe the methods that we use for text classification in terms of term weighting and feature selection. After that, we present the results and analyses for our experiments on five Turkish and five English review datasets, respectively. Finally, we conclude the paper and point out directions for future work.

## 2    Related Work

Many researchers have studied sentiment analysis, but its application to Turkish reviews is relatively new [3]. Çetin and Amasyalı [4] examine the term weighting methods for the feasibility of active learning and used a supervised term weighting method called delta1 in addition to the traditional term weighting methods: $tf$ and $tf * idf$. There are also researchers who use the term presence method [3, 5], both term frequency and term presence [6], or $tf * idf$ weighting method [4, 7].

Feature selection is an important task for text classification since it can not only reduce the number of features for improved efficiency, but also remove less important and noisy features for better accuracy. In addition to the popular feature selection methods such as Information Gain, Chi-square, there are also new methods such as Document Frequency Difference [2] and Optimal Orthogonal Centroid [8]. Nicholls and Song [2] also compare their proposed Document Frequency Difference with different feature selection methods using maximum entropy modeling classifier. Note that for the sentiment analysis of Turkish reviews, only Chi-square and Information Gain methods are used as feature selectors [5]. In addition, different machine learning algorithms such as Naïve Bayes, Naïve Bayes Multinomial, Support Vector Machine, and Maximum Entropy are used for sentiment analysis [9]. In particular, O'Keefe and Koprinska [10] compare three feature selection methods using six term weighting methods.

In our study, we investigate the interactions between term weighting and feature selections methods on the Turkish reviews and also compare the results against those for the English reviews. More specifically, we use four term weighting methods: term frequency, term presence, term frequency and inverse document frequency, term frequency and relevance frequency, and three feature selection methods: inverse document frequency cutoff, chi-square, and document frequency difference.

## 3 Methods

Sentiment analysis is essentially a classification task, and once the representation method is chosen for the review documents, we need to use a classifier to sort them into positive and negative ratings. In this study, we focus on the Naïve Bayes Multinomial (NBM) classifier since it tends to perform better for large vocabulary sizes [11]. NBM treats a document as a sequence of words, and models the distribution of words as a multinomial based on the word frequency information in a document. We propose that the word frequency information can be captured by different term weighting methods and intend to investigate their impacts on the sentiment analysis of Turkish reviews, particularly with reduced feature sizes for the text representation.

### 3.1 Term Weighting Methods

We represent each document $d$ by a feature vector $(w_1, ..., w_m)$, where $w_i$ corresponds to the $i$th term in the vocabulary and measures the importance of this term for the text representation.

**Term Frequency**

Term Frequency ($tf$) is the number of occurrences of a word in a document and is commonly used in a wide range of natural language applications. Intuitively, the higher the $tf$, the more important the word for a document. However, many stop words are highly frequent, but often do not carry much meaning for the related documents.

**Term Presence**

Term Presence ($tp$) is only concerned with whether a word appears in a document or not, and as a result, how frequent the word appears in the document makes no difference. As we mentioned in the Introduction, the use of sentiment-expressing words tends to be less frequent within a review, but more frequent across different reviews. Thus, $tp$ may capture what is stated in this observation.

**Term Frequency and Inverse Document Frequency**

One of the most popular term weighting methods is Term Frequency and Inverse Document Frequency ($tf * idf$), which has been widely used in information retrieval and text categorization [12]. Intuitively, $tf$ captures the importance of a term, while $idf$ emphasizes the specificity of the term, usually defined as follows:

$$idf = \log\left(\frac{N}{df^f}\right) \tag{1}$$

where $df^f$ is the number of documents that contains feature $f$ in the dataset and $N$ is the number of documents in the dataset.

When $tf$ and $idf$ are multiplied, we can favor the words that are both important and specific. In particular, the weights for the stop words will be reduced since they are not specific to a document even though they are popular in many documents.

**Term Frequency and Relevance Frequency**
Lan et al. [13] propose a supervised term weighting method called $tf * rf$, which replaces the $idf$ component by $rf$. This method assigns higher scores to terms that appear more frequently in the positive class, and thus can help improve the classification performance when the number of positive documents is much smaller than the number of negative documents. The $rf$ is defined as follows:

$$rf = \log\left(2 + \frac{df^f_+}{df^f_-}\right) \tag{2}$$

where $df^f_+$ is the number of documents in the positive class that contains feature $f$, while $df^f_-$ is the number of documents in the negative class that contains $f$.

## 3.2    Feature Selection Methods

Feature selection methods rank features according to certain measures so that the most valuable features can be chosen to improve the classification results. In this study we use Inverse Document Frequency Cutoff, Chi-square, and Document Frequency Difference methods so that we can compare their effectiveness with different term weighting methods for the sentiment analysis of Turkish reviews.

**Inverse Document Frequency Cutoff**
Inverse Document Frequency (IDF) is commonly used in information retrieval, which is not only easy to implement but also does not require training data for the calculation. Essentially, IDF Cutoff favors specific terms across a dataset so that it can be more discriminative for text classification.

**Chi-Square $(\chi^2)$**
Chi-square measures the lack of independence between a feature and a class. If a feature $f$ in the related class $c$ has a low score, it can be less informative, so it can be removed [2]. Using the 2-by-2 contingency table for feature $f$ and class $c$, where A is the number of documents in class $c$ that contains feature $f$, B is the number of documents in the other class that contains $f$, C is the number of documents in $c$ that does not contain $f$, D is the number of documents in the other class that does not contain $f$, and $N$ is the total number of documents, then the Chi-square score can be defined as follows:

$$\chi^2(f, c) = \frac{N(AD - CB)^2}{(A+C)(B+D)(A+B)(C+D)} \tag{3}$$

The Chi-square statistic can also be computed between each feature and each class in the dataset, and then combined across the class specific scores of each feature into following:

$$\chi^2(f) = \sum_{i=1}^{c} P(c_i)\chi^2(f, c_i) \tag{4}$$

**Document Frequency Difference (DFD)**

Nicholls and Song [2] propose the Document Frequency Difference feature selection method with the score normalized on a scale of 0 to 1. Intuitively, the scores should be proportional to other features, and thus can be used as the feature weight itself. More specifically, DFD is defined as follows:

$$score_f = \frac{\left| df_+^f - df_-^f \right|}{N} \tag{5}$$

where $df_+^f$ is the number of documents in the positive class that contains feature $f$, $df_-^f$ is the number of documents in the negative class that contains $f$, and $N$ is the number of documents in the dataset. DFD is designed to capture the observation that the use of sentiment-expressing words tends to be less frequent within a review, but more frequent across different reviews, and should be potentially helpful for sentiment analysis.

## 4 Experiments and Results

In order to examine the interactions between term weighting and feature selection methods on the sentiment analysis of Turkish reviews, we conduct experiments on five review datasets. We first apply the term weighting methods to all features of these datasets to establish the baseline results. After that, we try different feature selection methods, including IDF Cutoff, $\chi^2$, and DFD so that we can investigate how the reduced text representation interacts with different term weighting methods and whether we can further improve the accuracy for the sentiment analysis of Turkish reviews. Finally, we perform similar experiments on the English reviews as well so that we can identify their differences with the Turkish reviews.

We use Python with NLTK[1] for data preprocessing and feature vector construction, and the Weka[2] data mining tool for NBM classifier. All of our experiments are done with five-fold cross validation.

---

[1] http://nltk.org.
[2] http://www.cs.waikato.ac.nz/ml/weka.

### 4.1    Datasets

We use five Turkish review datasets in our experiments. The Turkish movie review dataset is collected from a publicly available website (beyazperde.com) by Sevindi [7]. It has 1057 positive and 978 negative reviews. Turkish multi-domain product reviews dataset[3] is introduced by Demirtas and Pechenizky [14]. This dataset is collected from a Turkish e-commerce website (hepsiburada.com) to conduct the training set expansion experiment with reviews from different domains, including books, DVD, electronics, and kitchen appliances. It has 700 positive and 700 negative reviews for each of the four categories.

We also use five English review datasets in our experiments. The English movie review dataset[4] is introduced by Pang and Lee [15]. This data consists of 1000 positive and 1000 negative movie reviews. The English product review dataset[5] is introduced by Blitzer et al. [16]. The dataset contains product reviews taken from Amazon.com in four categories: books, DVD, electronics, and kitchen appliances, each consisting of 1000 positive and 1000 negative reviews. In order to keep the same dataset sizes as the Turkish product reviews, we randomly select 700 positive and 700 negative reviews from each of the four categories for English product reviews.

### 4.2    Performance Evaluation

The performance of a classification system is typically evaluated by F measure, which is a composite score of precision and recall. Precision ($P$) is the number of correctly classified items over the total number of classified items with respect to a class. Recall ($R$) is the number of correctly classified items over the total number of items that belong to a given class. Together, the F measure gives the harmonic mean of precision and recall, and is calculated as [17]:

$$F = 2 \times \frac{P \times R}{P + R} \qquad (6)$$

### 4.3    Interactions Between Term Weighting and Feature Selection Methods for the Turkish Reviews

In order to investigate the effects of term weighting methods on the sentiment analysis, we conducted experiments on the five Turkish review datasets. After removing all punctuation patterns, we first apply the term weighting methods on these datasets. Then, we try different feature selection methods, including IDF cutoff, $\chi^2$, and DFD so that we can investigate how the reduced text representation interacts with different term weighting methods and whether we can further improve the accuracy for the sentiment analysis of Turkish reviews.

---

[3] http://www.win.tue.nl/~mpechen/projects/smm/#Datasets.

[4] http://www.cs.cornell.edu/people/pabo/movie-review-data/.

[5] http://www.cs.jhu.edu/~mdredze/datasets/sentiment/.

**Baseline Results for the Turkish Reviews**

We obtain the baseline results by applying the four term weighting methods on all five Turkish review datasets, where the best results for each dataset are shown in boldface in Table 1.

**Table 1.** Baseline results in F measures for the Turkish review datasets

|  |  | Term weighting methods | | | |
|---|---|---|---|---|---|
| Datasets | Features | tf | tf * idf | tp | tf * rf |
| Movie reviews | 18565 | 0.8259 | 0.7966 | **0.8337** | 0.8197 |
| Book reviews | 10500 | **0.8324** | 0.7887 | 0.8322 | 0.8299 |
| DVD reviews | 11334 | **0.7928** | 0.7511 | 0.7921 | 0.7795 |
| Electronic reviews | 10901 | **0.8148** | 0.7997 | 0.8133 | 0.8133 |
| Kitchen reviews | 9436 | 0.7767 | 0.7676 | **0.7807** | 0.7627 |

As can be seen in Table 1, the best performance is achieved with *tp* for the Turkish movie and kitchen review datasets, and with *tf* for the book, DVD, and electronics review datasets. In fact, the difference between *tp* and *tf* is much smaller than that between other term weighting methods. This confirms the observation we mentioned in the Introduction that the use of sentiment-expressing words tends to be less frequent within a review, but more frequent across different reviews.

**Feature Selection Results for the Turkish Reviews**

To see which term weighting method is the best fit for the reduced text representation obtained through feature selection, we tested the three feature selection methods introduced in this paper with the term weighting methods on all five review datasets. For each feature selection method, we try seven feature sizes from 500 to 3500, implying a total of $4 \times 3x7 = 84$ experiments for each dataset. The best results for each pair of term weighting and feature selection methods are shown in Tables 2, 3, 4, 5 and 6, respectively for each of the five review datasets. Along with the best results in F measures, we also show the feature sizes that help achieve the best results.

**Table 2.** The best results for the Turkish movie review dataset

|  | IDF | | DFD | | $\chi^2$ | |
|---|---|---|---|---|---|---|
|  | Features | F measure | Features | F measure | Features | F measure |
| tf | 3500 | 0.8240 | 3500 | 0.8854 | 1000 | 0.8849 |
| tf * idf | 1500 | 0.8147 | **3500** | **0.9002** | 1000 | 0.8957 |
| tp | 3500 | 0.8265 | 3500 | 0.8933 | 1000 | 0.8942 |
| tf * rf | 2000 | 0.8237 | 3500 | 0.8797 | 1500 | 0.8688 |

From these tables, we can see that the performance has been increased significantly over the baseline results. For example, the F measure for the movie review dataset is increased from 0.8337 to 0.9002. In addition, the best term weighting method is *tf * idf*

**Table 3.** The best results for the Turkish book review dataset

|  | IDF | | DFD | | $\chi^2$ | |
| --- | --- | --- | --- | --- | --- | --- |
|  | Features | F measure | Features | F measure | Features | F measure |
| *tf* | 3000 | 0.8371 | 2000 | 0.8764 | 1000 | 0.8864 |
| *tf * idf* | 3000 | 0.8171 | 3000 | 0.8949 | **1000** | **0.9093** |
| *tp* | 3000 | 0.8370 | 2500 | 0.8749 | 1000 | 0.8914 |
| *tf * rf* | 3000 | 0.8335 | 2000 | 0.8714 | 500 | 0.8799 |

**Table 4.** The best results for the Turkish DVD review dataset

|  | IDF | | DFD | | $\chi^2$ | |
| --- | --- | --- | --- | --- | --- | --- |
|  | Features | F measure | Features | F measure | Features | F measure |
| *tf* | 2000 | 0.7893 | 2500 | 0.8650 | 500 | 0.8656 |
| *tf * idf* | 2500 | 0.7649 | **2000** | **0.8893** | 500 | 0.8885 |
| *tp* | 2500 | 0.7842 | 2500 | 0.8679 | 500 | 0.8685 |
| *tf * rf* | 2500 | 0.7856 | 2500 | 0.8563 | 500 | 0.8635 |

**Table 5.** The best results for the Turkish electronics review dataset

|  | IDF | | DFD | | $\chi^2$ | |
| --- | --- | --- | --- | --- | --- | --- |
|  | Features | F measure | Features | F measure | Features | F measure |
| *tf* | 3000 | 0.8255 | 2000 | 0.8610 | 1000 | 0.8544 |
| *tf * idf* | 2500 | 0.8191 | **2500** | **0.8822** | 1000 | 0.8778 |
| *tp* | 3000 | 0.8234 | 1500 | 0.8626 | 1500 | 0.8624 |
| *tf * rf* | 1500 | 0.8283 | 2500 | 0.8675 | 1000 | 0.8652 |

**Table 6.** The best results for the Turkish kitchen review dataset

|  | IDF | | DFD | | $\chi^2$ | |
| --- | --- | --- | --- | --- | --- | --- |
|  | Features | F measure | Features | F measure | Features | F measure |
| *tf* | 2000 | 0.7815 | 2500 | 0.8262 | 1000 | 0.8392 |
| *tf * idf* | 1000 | 0.7803 | 2500 | 0.8593 | **500** | **0.8626** |
| *tp* | 2000 | 0.7824 | 1500 | 0.8305 | 1000 | 0.8458 |
| *tf * rf* | 500 | 0.7726 | 1000 | 0.8396 | 1000 | 0.8508 |

for all five datasets. This is indicating that with the reduced number of features, *tf * idf* can be more discriminating among the selected features. Finally, both $\chi^2$ and DFD help achieve the best performance for some datasets. DFD is the best performer for the movie, DVD, and electronics review datasets and $\chi^2$ is the best performer for the book and kitchen review datasets. While $\chi^2$ tends to work well with smaller feature sizes, DFD tends to favor bigger feature sizes.

## 4.4    Interactions Between Term Weighting and Feature Selection Methods for the English Reviews

For the English review datasets, we conducted similar experiments so that we can identify their differences with the Turkish review datasets.

**Baseline Results for the English Reviews**

Table 7 shows the best baseline results for each dataset among the four term weighting methods, along with the total number of features for each review dataset. As can be seen from Table 7, the best performance is achieved with *tp* for all five English review datasets. The classification results with *tf* method have only small differences compared with those for *tp* method. This provides an even stronger confirmation for the observation we commented about sentiment-expressing words in the Introduction.

**Table 7.** Baseline results in F measures for the English review datasets

|  | | Term weighting methods | | | |
|---|---|---|---|---|---|
| Datasets | Features | *tf* | *tf * idf* | *tp* | *tf * rf* |
| Movie reviews | 38863 | 0.8199 | 0.7868 | **0.8303** | 0.8066 |
| Book reviews | 18295 | 0.7597 | 0.7314 | **0.7768** | 0.7280 |
| DVD reviews | 17663 | 0.7880 | 0.7455 | **0.7981** | 0.7602 |
| Electronic reviews | 8998 | 0.7657 | 0.7421 | **0.7935** | 0.7547 |
| Kitchen reviews | 8066 | 0.8014 | 0.7714 | **0.8329** | 0.8214 |

**Feature Selection Results for the English Reviews**

For the English review datasets, we followed a similar process to test the four term weighting methods with the three feature selection methods and tried seven feature sizes from 500 to 3500 for a total of $4 \times 3x7 = 84$ experiments for each dataset. The best classification results for all pairs of term weighting and feature selection methods are shown in Tables 8, 9, 10, 11 and 12, respectively for each of the five review datasets, along with the corresponding feature sizes and the F measure scores.

**Table 8.** The best results for the English movie review dataset

|  | IDF | | DFD | | $\chi^2$ | |
|---|---|---|---|---|---|---|
|  | Features | F measure | Features | F measure | Features | F measure |
| *tf* | 3500 | 0.8139 | 3500 | 0.8614 | 2500 | 0.9185 |
| *tf * idf* | 3500 | 0.8119 | 3000 | 0.8679 | **2500** | **0.9565** |
| *tp* | 3000 | 0.8369 | 3500 | 0.8759 | 3000 | 0.9440 |
| *tf * rf* | 3500 | 0.8071 | 3000 | 0.8475 | 2500 | 0.9214 |

From Tables 8, 9, 10, 11 and 12, we see once again that the performance with feature selection methods has been increased significantly over that for the baseline results. As observed in Tables 8, 9, 10, 11 and 12, $\chi^2$ is the best performer for all five review datasets. The F-measure score of the English movie review dataset is increased

**Table 9.** The best results for the English book review dataset

|          | IDF | | DFD | | $\chi^2$ | |
|----------|----------|-----------|----------|-----------|----------|-----------|
|          | Features | F measure | Features | F measure | Features | F measure |
| *tf*       | 3500 | 0.7727 | 3500 | 0.8671 | 1000 | 0.8793 |
| *tf * idf* | 1500 | 0.7550 | 3500 | 0.8779 | 1500 | 0.9043 |
| *tp*       | 3500 | 0.7921 | 3500 | 0.8821 | **1500** | **0.9064** |
| *tf * rf*  | 1000 | 0.7633 | 2000 | 0.8399 | 1000 | 0.8769 |

**Table 10.** The best results for the English DVD review dataset

|          | IDF | | DFD | | $\chi^2$ | |
|----------|----------|-----------|----------|-----------|----------|-----------|
|          | Features | F measure | Features | F measure | Features | F measure |
| *tf*       | 2000 | 0.7968 | 2000 | 0.8538 | 1000 | 0.8999 |
| *tf * idf* | 1500 | 0.7925 | 3500 | 0.8676 | **1000** | **0.9186** |
| *tp*       | 1500 | 0.8098 | 3500 | 0.8688 | 1000 | 0.9171 |
| *tf * rf*  | 3500 | 0.7798 | 3000 | 0.8404 | 1500 | 0.8876 |

**Table 11.** The best results for the English electronics review dataset

|          | IDF | | DFD | | $\chi^2$ | |
|----------|----------|-----------|----------|-----------|----------|-----------|
|          | Features | F measure | Features | F measure | Features | F measure |
| *tf*       | 3000 | 0.7679 | 3000 | 0.8243 | 1000 | 0.8621 |
| *tf * idf* | 1500 | 0.7521 | 3000 | 0.8364 | 500 | 0.8729 |
| *tp*       | 3500 | 0.7963 | 2500 | 0.8550 | **500** | **0.8779** |
| *tf * rf*  | 1500 | 0.7686 | 1500 | 0.8199 | 500 | 0.8542 |

**Table 12.** The best results for the English kitchen review dataset

|          | IDF | | DFD | | $\chi^2$ | |
|----------|----------|-----------|----------|-----------|----------|-----------|
|          | Features | F measure | Features | F measure | Features | F measure |
| *tf*       | 2500 | 0.8107 | 2000 | 0.8571 | 500 | 0.8986 |
| *tf * idf* | 3500 | 0.7949 | 2000 | 0.8693 | 500 | 0.9114 |
| *tp*       | 3500 | 0.8371 | 2000 | 0.8814 | **500** | **0.9136** |
| *tf * rf*  | 3000 | 0.8329 | 2000 | 0.8721 | 500 | 0.8907 |

from 0.8303 to 0.9565 with *tf * idf* method. The *tp* and *tf* *idf* term weighting methods with $\chi^2$ method work reasonably well for all English datasets. For the English movie and DVD review datasets, *tf* *idf* method works better and for the English book, electronics, and kitchen review datasets *tp* method works better, even though with small differences. Overall, $\chi^2$ method achieves better classification results with small feature sizes compared with the other feature selection methods. Also observed is that IDF cutoff method lags behind $\chi^2$ and DFD methods. However, when compared with

the baseline results, it achieves better classification results with smaller feature sizes, ranging between 1500 and 3500, indicating that *idf* can still be helpful in discriminating among the selected features.

## 5   Conclusion and Future Work

We investigated the effects of four term weighting methods on the sentiment analysis of Turkish reviews. Furthermore, we examined the interactions between term weighting and feature selection methods for text representation using NBM classifier on five Turkish review datasets. Our experimental results show that different term weighting methods do affect the performance for the sentiment analysis of Turkish reviews and *tf\* idf* weighting method gives the best F measure values for the reduced text representation obtained through feature selection. In addition, both $\chi^2$ and DFD help produce the best results for some datasets, although $\chi^2$ tends to work well with smaller feature sizes, while DFD tends to favor bigger feature sizes.

We also investigated these weighting methods using NBM classifier on the English review datasets. Compared with Turkish reviews, *tp* works the best as the baselines and $\chi^2$ is the best performer for all five English datasets. Both *tp* and *tf\* idf* methods work reasonably well for these datasets with small differences in the performance, indicating that with a reduced number of features, *tf\* idf* can be effective in discriminating among the selected features.

For both Turkish and English review datasets, the IDF cutoff method lags behind the other two feature selection methods, but compared with the baseline results, it can achieve better classification results with smaller feature sizes, ranging between 1500 and 3500 features. Overall, the improvements are all significant when combining term weighting and feature selection methods, indicating that such a combination is both essential and important for the sentiment analysis of Turkish and English reviews.

For future work, it would be helpful to conduct similar experiments on aspect based sentiment analysis, which cannot only classify the overall sentiments for reviews, but also the individual ratings for the components mentioned within reviews.

**Acknowledgements.** This study was supported by Çukurova University Academic Research Project Unit under the grant no FDK-2015-3833 and by The Scientific and Technological Research Council of Turkey (TÜBİTAK) scholarship TUBITAK-2214-A.

## References

1. Pang, B., Lee, L.: Opinion mining and sentiment analysis. Found. Trends Inf. Retr. **2**(1–2), 1–135 (2008)
2. Nicholls, C., Song, F.: Comparison of feature selection methods for sentiment analysis. In: 23rd Canadian conference on Advances in Artificial Intelligence (AI 2010), pp. 286–289 (2010)
3. Erogul, U.: Sentiment analysis in Turkish. Master thesis, Middle East Technical University, Turkey (2009)

4. Çetin, M., Amasyali, M.F.: Active learning for Turkish sentiment analysis. In: IEEE International Symposium on Innovations in Intelligent Systems and Applications (INISTA) (2013)
5. Akba, F., Uçan, A., Sezer, E., Sever, H.: Assessment of feature selection metrics for sentiment analysis: Turkish movie reviews. In: 8th European Conference on Data Mining, pp. 180–184 (2014)
6. Kaya, M., Fidan, G., Toroslu, I.: Sentiment analysis of Turkish political news. In: IEEE/WIC/ACM International Conferences on Web Intelligence and Intelligent Agent Technology (WI-IAT), vol. 1, pp. 174–180 (2012)
7. Sevindi, B.I.: Comparison of supervised and dictionary based sentiment analysis approaches on Turkish text. Master thesis, Gazi University, Turkey (2013)
8. Yan, J., Liu, N., Zhang, B., Yan, S., Chen, Z., Cheng, Q., et al.: OCFS: optimal orthogonal centroid feature selection for text categorization. In: 28th Annual International ACM SIGIR Conference on Research and Development in Information Retrieval, pp. 122–129 (2005)
9. Pang, B., Lee, L., Vaithyanathan, V.: Thumbs up? Sentiment classification using machine learning techniques. In: Proceedings of Conference on Empirical Methods in Natural Language Processing, Morristown, pp. 79–86 (2002)
10. O'Keefe, T., Koprinska, I.: Feature selection and weighting methods. In: 14th Australian Document Computing Symposium on Sentiment Analysis, Sydney, Australia (2009)
11. McCallum, A., Nigam, K.A.: Comparison of event models for Naive Bayes text classification. In: Proceedings of AAAI (1998)
12. Robertson, S.E., Jones, K.S.: Relevance Weighting of Search Terms, pp. 143–160. Taylor Graham Publishing, London (1988)
13. Lan, M., Tan, C.L., Su, J., Lu, Y.: Supervised and traditional term weighting methods for automatic text categorization. IEEE Trans. Pattern Anal. Mach. Intell. **31**, 721–735 (2009). IEEE Computer Society
14. Demirtas, E., Pechenizkiy, M.: Cross-lingual polarity detection with machine translation. In: 2nd International Workshop on Issues of Sentiment Discovery and Opinion Mining (WISDOM 2013), vol. 9 (2013)
15. Pang, B., Lee, L.: A sentimental education: sentiment analysis using subjectivity summarization based on minimum cuts. In: Proceedings of Annual Meeting for the Association of Computational Linguists (2004)
16. Blitzer, J., Dredze, M., Pereira, F.: Biographies, bollywood, boom-boxes and blenders: domain adaptation for sentiment classification. Association of Computational Linguistics (ACL) (2007)
17. Han, J., Kamber, M.: Data Mining: Concepts and Techniques, 2nd edn. Morgan Kaufmann Publishing, San Francisco (2006)

# Developing a Concept-Level Knowledge Base for Sentiment Analysis in Singlish

Rajiv Bajpai$^{(\boxtimes)}$, Danyuan Ho, and Erik Cambria

School of Computer Science and Engineering,
Nanyang Technology University, Singapore, Singapore
{rbajpai,dyho,cambria}@ntu.edu.sg

**Abstract.** In this paper, we present Singlish SenticNet, a concept-level knowledge base for sentiment analysis that associates multiword expressions to a set of emotion labels and a polarity value. Unlike many other sentiment analysis resources, SenticNet is not built by manually labeling pieces of knowledge coming from general NLP resources such as WordNet or DBPedia. Instead, it is automatically constructed by applying graph-mining and multi-dimensional scaling techniques on the affective common-sense knowledge collected from three different sources. This knowledge is represented redundantly at three levels: semantic network, matrix, and vector space. Subsequently, the concepts are labeled by emotions and polarity through the ensemble application of spreading activation, neural networks and an emotion categorization model.

**Keywords:** Singlish · SenticNet · Sentiment analysis · Knowledge base

## 1 Introduction

Singapore Colloquial English (or Singlish), as the name suggests, refers to an English-based creole language, or rather the variety of English spoken in Singapore [1]. A product of colonial implantation, English as spoken by the British is gradually 'nativized' into Singlish due to extensive language borrowing and mixing with other language in the linguistic environment of Singapore. The end product of language contact is Singlish, an English variety that shows 'a high degree of influence from other local languages such as Hokkien, Cantonese, Mandarin, Malay and Tamil [19].

[11] noted that English in Singapore is diglossic in nature, which its domain of use is dependent on the degree of formality. The high variety is used in formal situations while the low variety used for informal occasions. The high variety of English is called Singapore Standard English and it is like other Standard English varieties (i.e., American English or British English). The low variety is called Singapore Colloquial English, or Singlish. It is distinct from Standard English because it demonstrates unique syntactical and lexical features [16]. [11], however, notes that there is no stark demarcation between Singlish and Singapore Standard English, but rather degree of variations in between.

© Springer International Publishing AG, part of Springer Nature 2018
A. Gelbukh (Ed.): CICLing 2016, LNCS 9624, pp. 347–361, 2018.
https://doi.org/10.1007/978-3-319-75487-1_27

Singapore is a multi-ethnic/multilingual country consisting of 77% Chinese, 14% Malay and 8% Indian. While Singlish serves as an inter-ethnic lingua franca [12], linguistic boundaries are present in the variety despite extensive contacts between different speakers. [26] observed that the speaker's linguistic background plays a part in influencing the linguistic repertoire of the user when using Singlish. For example a Chinese speaker may exclaim "John sibei hum sup one" whereas a Malay speaker may cry "John very buaya sia". While both phrases essentially mean "John is so lecherous" in Singlish, one's choice of lexicon in Singlish might be influenced by one's language background. The presence of ethnic differentiation therefore shows that Singlish, is not a 'monolithic' linguistic entity spoken by a homogenous Singaporean 'ethnicity'.

People in Singapore tend to speak and write in Singapore Colloquial English. Due to the informal nature of Singlish, it has found its way on the Internet as Singaporeans prefer to use Singlish in informal communication like online chatting, tweeting or interaction on social media like Facebook [27]. As such, large amount of data in Singlish is inaccessible because it is a variety that is highly distinct and incomprehensible to English speakers elsewhere. Researchers have been studying the linguistic features of Singlish extensively since the 1960s, but little research on Singlish has been done in natural language processing (NLP). A sub-branch of NLP research, [18] conducted a study on sentiment analysis in Singapore Colloquial English. Using 425 Singlish posts on a political issue in Singapore, the baseline of sentiment analysis was determined by manually annotating the target phrases with either a positive or negative polarity. The researcher then ran a phrasal sentiment analysis classifier to analyze the polarity of the dataset. An overall accuracy of 35.5% was achieved by this approach, in which a precision of 21% was achieved for positive posts and a precision of 94.4% was achieved for negative post.

## 2   AffectNet and ConceptNet

As an inventory of target labels and a source of training examples for the supervised classification, we used the emotion lists provided for the SemEval 2007 task 14: Affective text. According to the organizers of this task, the lists were extracted from WNA [25]. There are six lists corresponding to the six basic emotions: anger, fear, disgust, sadness, surprise, and joy. This dataset assigns emotion labels to synsets groups of words or concepts that are synonymous in the corresponding senses: e.g., a synset puppy love, calf love, crush, infatuation is assigned the label JOY. However, we ignored the synonymy information contained in the data and used the labels for individual words or concepts, i.e., puppy love→JOY, calf love→JOY, crush→JOY, infatuation→JOY. ConceptNet represents the information from the Open Mind corpus as a directed graph, in which the nodes are concepts and the labeled edges are common-sense assertions that interconnect them [6].

# 3   AffectiveSpace

AffectiveSpace [3] is a novel affective common-sense knowledge visualization and analysis system. The human mind constructs intelligible meanings by continuously compressing over vital relations [10]. The compression principles aim to transform diffuse and distended conceptual structures to more focused versions so as to become more congenial for human understanding. To this end, principal component analysis (PCA) has been applied on the matrix representation of AffectNet. In particular, truncated singular value decomposition (TSVD) has been preferred to other dimensionality reduction techniques for its simplicity, relatively low computational cost, and compactness. TSVD, in fact, is particularly suitable for measuring the cross-correlations between affective common-sense concepts as it uses an orthogonal transformation to convert the set of possibly correlated common-sense features associated with each concept into a set of values of uncorrelated variables (the principal components of the SVD).

By using Lanczos' method [14], moreover, the generalization process is relatively fast (a few seconds), despite the size and the sparseness of AffectNet. The objective of such compression is to allow many details in the blend of ConceptNet and WNA to be removed such that the blend only consists of a few essential features that represent the global picture. Applying TSVD on AffectNet, in fact, causes it to describe other features that could apply to known affective concepts by analogy: if a concept in the matrix has no value specified for a feature owned by many similar concepts, then by analogy the concept is likely to have that feature as well. In other words, concepts and features that point in similar directions and, therefore, have high dot products, are good candidates for analogies.

A pioneering work on understanding and visualizing the affective information associated with natural language text was conducted by Osgood et al. [17]. Osgood used multi-dimensional scaling (MDS) to create visualizations of affective words based on similarity ratings of the words provided to subjects from different cultures. Words can be thought of as points in a multi-dimensional space and the similarity ratings represent the distances between these words. MDS projects these distances to points in a smaller dimensional space (usually two or three dimensions). Similarly, AffectiveSpace aims to grasp the semantic and affective similarity between different concepts by plotting them into a multi-dimensional vector space [5]. Unlike Osgood's space, however, the building blocks of AffectiveSpace are not simply a limited set of similarity ratings between affect words, but rather millions of confidence scores related to pieces of common-sense knowledge linked to a hierarchy of affective domain labels [22]. Rather than merely determined by a few human annotators and represented as a word-word matrix, in fact, AffectiveSpace is built upon an affective common-sense knowledge base, namely AffectNet, represented as a concept-feature matrix. After performing TSVD on such matrix, hereby termed $A$ for the sake of conciseness, a low-rank approximation of it is obtained, that is, a new matrix $\tilde{A} = U_k \, \Sigma_k \, V_k^T$.

This approximation is based on minimizing the Frobenius norm of the difference between $A$ and $\tilde{A}$ under the constraint $rank(\tilde{A} = k$. For the Eckart–Young theorem [9], it represents the best approximation of $A$ in the least-square sense:

$$\min_{\tilde{A}|rank(\tilde{A}=k)} |A - \tilde{A}| = \min_{\tilde{A}|rank(\tilde{A}=k)} |\Sigma - U^*\tilde{A}V| = \min_{\tilde{A}|rank(\tilde{A}=k)} |\Sigma - S| \quad (1)$$

assuming that $\tilde{A}$ has the form $\tilde{A} = USV^*$, where $S$ is diagonal. From the rank constraint, i.e., $S$ has $k$ non-zero diagonal entries, the minimum of the above statement is obtained as follows:

$$\min_{\tilde{A}|rank(\tilde{A}=k)} |\Sigma - S| = \min_{s_i} \sqrt{\sum_{i=1}^{n} (\sigma_i - s_i)^2} \quad (2)$$

$$\min_{s_i} \sqrt{\sum_{i=1}^{n} (\sigma_i - s_i)^2} = \min_{s_i} \sqrt{\sum_{i=1}^{k} (\sigma_i - s_i)^2 + \sum_{i=k+1}^{n} \sigma_i^2} = \sqrt{\sum_{i=k+1}^{n} \sigma_i^2} \quad (3)$$

Therefore, $\tilde{A}$ of rank $k$ is the best approximation of $A$ in the Frobenius norm sense when $\sigma_i = s_i$ $(i = 1, \ldots, k)$ and the corresponding singular vectors are the same as those of $A$. If all but the first $k$ principal components are discarded, common-sense concepts and emotions are represented by vectors of $k$ coordinates. These coordinates can be seen as describing concepts in terms of 'eigenmoods' that form the axes of AffectiveSpace, i.e., the basis $e_0, \ldots, e_{k-1}$ of the vector space. For example, the most significant eigenmood, $e_0$, represents concepts with positive affective valence. That is, the larger a concept's component in the $e_0$ direction is, the more affectively positive it is likely to be.

Concepts with negative $e_0$ components, then, are likely to have negative affective valence. Thus, by exploiting the information sharing property of TSVD, concepts with the same affective valence are likely to have similar features – that is, concepts conveying the same emotion tend to fall near each other in AffectiveSpace. Concept similarity does not depend on their absolute positions in the vector space, but rather on the angle they make with the origin. For example, concepts such as gei_yan, see_buay, and cham_sheung are found very close in direction in the vector space, while concepts like act_blur, ah_long, and bo_chap are found in a completely different direction (nearly opposite with respect to the centre of the space).

## 4    The Emotion Categorization Model

The Hourglass of Emotions [7] is an affective categorization model inspired by Plutchik's studies on human emotions [20]. It reinterprets Plutchik's model by organizing primary emotions around four independent but concomitant dimensions, whose different levels of activation make up the total emotional state of the mind. Such a reinterpretation is inspired by Minsky's theory of the mind, according to which brain activity consists of different independent resources and that emotional states result from turning some set of these resources on and turning another set of them off.

This way, the model can potentially synthesize the full range of emotional experiences in terms of Pleasantness, Attention, Sensitivity, and Aptitude, as the different combined values of the four affective dimensions can also model affective states we do not have a specific name for, due to the ambiguity of natural language and the elusive nature of emotions. The primary quantity we can measure about an emotion we feel is its strength. But, when we feel a strong emotion, it is because we feel a very specific emotion. And, conversely, we cannot feel a specific emotion like fear or amazement without that emotion being reasonably strong. For such reasons, the transition between different emotional states is modeled, within the same affective dimension, using the function $G(x) = -\frac{1}{\sigma\sqrt{2\pi}}e^{-x^2/2\sigma^2}$, for its symmetric inverted bell curve shape that quickly rises up towards the unit value. In particular, the function models how valence or intensity of an affective dimension varies according to different values of arousal or activation ($x$), spanning from null value (emotional void) to the unit value (heightened emotionality). Justification for assuming that the Gaussian function (rather than a step or simple linear function) is appropriate for modeling the variation of emotion intensity is based on research into the neural and behavioral correlates of emotion, which are assumed to indicate emotional intensity in some sense. Nobody genuinely knows what function subjective emotion intensity follows, because it has never been truly or directly measured.

Each affective dimension of the Hourglass model is characterized by six levels of activation (measuring the strength of an emotion), termed 'sentic levels', which represent the intensity thresholds of the expressed or perceived emotion [8]. These levels are also labeled as a set of 24 basic emotions, six for each of the affective dimensions, in a way that allows the model to specify the affective information associated with text both in a dimensional and in a discrete form. The dimensional form, in particular, is termed 'sentic vector' and it is a four-dimensional *float* vector that can potentially synthesize the full range of emotional experiences in terms of Pleasantness, Attention, Sensitivity, and Aptitude. In the model, the vertical dimension represents the intensity of the different affective dimensions, i.e., their level of activation, while the radial dimension represents K-line that can activate configurations of the mind, which can either last just a few seconds or years. The model follows the pattern used in color theory and research in order to obtain judgments about combinations, i.e., the emotions that result when two or more fundamental emotions are combined, in the same way that red and blue make purple. Hence, some particular sets of sentic vectors have special names, as they specify well-known compound emotions. For example, the set of sentic vectors with a level of Pleasantness $\in$ [G(2/3), G(1/3)), i.e., joy, a level of Aptitude $\in$ [G(2/3), G(1/3)), i.e., trust, and a minor magnitude of Attention and Sensitivity, are termed 'love sentic vectors' since they specify the compound emotion of love. More complex emotions can be synthesized by using three, or even four, sentic levels, e.g., joy + trust + anger = jealousy. Therefore, analogous to the way primary colors combine to generate different color gradations (and even colors we do not have a name for), the primary emotions of the Hourglass model can blend to form the full spectrum of human emotional experience (Tables 1 and 2).

**Table 1.** Sensitivity cluster

| Rage | Vomit blood | Used to expressed something that is extremely aggravating, frustrating or difficult to endure, word-for-word translation. Loan translation from Chinese |
|---|---|---|
| Anger | Tu Lan | Used to express something that is extremely aggravating, frustrating or difficult to endure. Probably borrowed from Hokkien |
| Frustration | Pek Cek | To denote exasperation or frustration. Probably borrowed from Hokkien |
| Terror | Balls drop | To denote being frightened, scared or shocked |
| Fear | Khia-khia | To denote being afraid, worried or nervous. Literally "scared scared". Probably borrowed from Hokkien |
| Apprehension | Kan Cheong | To denote being nervous or tense. Probably borrowed from Cantonese |

**Table 2.** Attention cluster

| Vigilance | On the ball | To describe someone who is alert, hardworking and enthusiastic |
|---|---|---|
| Anticipation | Can't wait | Same as "Can't wait" or "looking forward to" in Standard English |
| Interest | Enthu | Showing or having eager enjoyment or interest. Truncation of the word 'Enthusiastic' |
| Amazement | Wah | Interjection to denote admiration or awe. Probably borrowed from Chinese |
| Surprise | Alamak | Interjection to denote surprise and shock. Probably borrowed from Malay |
| Distinction | Itchy finger | To denote restlessness or a restive person who does something disruptive out of boredom |

Beyond emotion detection, the Hourglass model is also used for polarity detection tasks. Since polarity is strongly connected to attitudes and feelings, it is defined in terms of the four affective dimensions, according to the formula:

$$p = \sum_{i=1}^{N} \frac{Pleasantness(c_i) + |Attention(c_i)| - |Sensitivity(c_i)| + Aptitude(c_i)}{3N}$$

$$(4)$$

**Table 3.** Pleasantness cluster

| Ecstasy | Shiok | To denote sheer delight. Probably borrowed from Malay "syok" or from Punjabi "shauk" |
|---|---|---|
| Joy | Happy like bird | To denote one who is overly happy or overjoyed |
| Serenity | Chill | To describe someone who is relaxed or easy-going. From North American informal usage but appears frequently in Singlish context |
| Grief | Sim Tia/Heart pain | Denotes heartache. Literally "heart pain". Probably borrowed from Hokkien |
| Sadness | Sad | Same as "sad" in Standard English |
| Pensiveness | Emo | Short form of 'emotional', to denote moodiness |

**Table 4.** Aptitude cluster

| Admiration | Suka | To like or be fond of. Probably borrowed from Malay |
|---|---|---|
| Trust | Trust | Same as "trust" in Standard English |
| Acceptance | Accept | Same as "accept" in Standard English |
| Loathing | Hate | Same as "hate" in Standard English |
| Disgust | Erxin | To denote disgust. Probably borrowed from Mandarin |
| Boredom | Sian | To denote boredom and frustration. Probably borrowed from Hokkien |

where $c_i$ is an input concept, $N$ the total number of concepts, and 3 the normalization factor (as the Hourglass dimensions are defined as float $\in [-1,+1]$). In the formula, Attention is taken as absolute value since both its positive and negative intensity values correspond to positive polarity values (e.g., 'surprise' is negative in the sense of lack of Attention, but positive from a polarity point of view) (Tables 3 and 4).

Similarly, Sensitivity is taken as negative absolute value since both its positive and negative intensity values correspond to negative polarity values (e.g., 'anger' is positive in the sense of level of activation of Sensitivity, but negative in terms of polarity). Besides practical reasons, the formula is important because it shows a clear connection between polarity (opinion mining) and emotions (sentiment analysis) [2].

## 5   Developing Singlish SenticNet

In this section, we elaborate the development of Singlish SenticNet based on the original English SenticNet [4] and its extensions [21,23]. We first explain the characteristics of Singlish language and concepts before going to the detail of the proposed method.

## 5.1   Syntactic Features of Singlish Language

This section contains a review of Singlish syntactical features that are relevant to the semantic analysis of the language. While these features characterize the syntax of Singlish, they are by no means compulsory in the grammatical construct of the language. The seemingly free variation of Singlish syntax could be attributed to the syntactical competition between British English, the suprastrate language that forms the grammatical base of Singlish, and substrate languages like Hokkien and Malay that influences the grammatical features of Singlish [24].

**Noun and Pronoun Omission.** Q: Why does Erik like to eat laksa?
A: Erik likes to eat laksa because it is delicious

1. Ø likes to eat laksa because it is delicious.
2. Erik likes to eat Ø because it is delicious.
3. Erik likes to eat laksa because Ø is delicious.

**Copula Dropping**

1. Eric Ø very clever (Eric is very clever)
2. Nick Ø going to church (Nick is going to church)
3. That one Ø Rachel husband. (That one is Rachel's husband)
4. My house Ø in Serangoon Gardens (My house is in Serangoon Gardens)
5. The dog Ø hit by a car (The dog is hit by a car)

The copular verb "to be" can be omitted in several instances in Singlish. It is usually omitted when the subject occurs with an adjective phrase containing the adverb "very" or "so" (2a), when it is before the present participle (-ing form) of a verb (2b), after a noun phrase (2c) or a locative word (2d), and in the passive construction (2e).

## 5.2   Lexical Features of Singlish

A review on the lexical features of Singlish is conducted. They could be categorized in three distinct forms, namely borrowing from non-English languages, English words that are "nativized" to the Singapore context, and English words that are similar to other English varieties.

The first source of Singlish originates from the borrowing of expressions from non-English languages spoken in Singapore. Singlish words of non-English origins include "shiok" that conveys sheer delight (from Malay "syok" or Punjabi "shauk"), "kan cheong" that denotes someone who is nervous or tense (from Cantonese), "sian" that refers to a state of boredom (from Hokkien) and more.

The second source of Singlish is English words that have different meaning in Singlish. The implantation of English in a different language setting has led to the idiosyncratic use of English words that are meaningful to the Singlish context. These English-origin Singlish words have meaning and usage that are different to Standard English, and they are created through the sharing of different ideals and expressions between cultures in a novel setting. These sources of words come from calques (word for word translation) like "vomit blood" from Chinese (literally "to vomit blood"), semantic shift (change in meaning) like "steady" which refers to someone who is calm and collected (rather than "firmly balanced" in Standard English) and the coining of idiosyncratic metaphors like "happy like bird" which denotes a state of great joy.

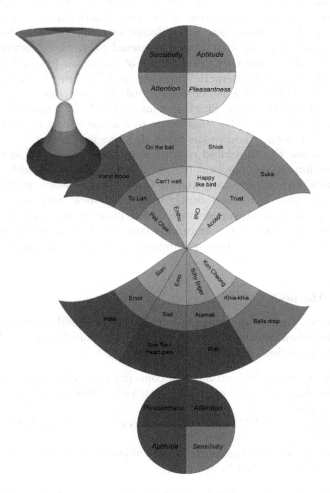

**Fig. 1.** Singlish concepts projected into Hourglass of Emotion

The third source of Singlish is from Standard English, and they have same meaning (like 'hate') like other varieties of English elsewhere. As much as the lexical composition of Singlish owes to the borrowing of other languages of Singapore, English still serves as the lexifier, or rather the dominant language where most of the vocabulary of Singlish originates. These English words in Singlish are either inherited from English, or borrowed from other English varieties, like "chill" from American informal usage.

### 5.3    Hourglass of Emotion Model and Singlish Emotion Expression

A survey on emotional expressions of Singlish is conducted by reviewing Singlish wordlists from the Coxford Singlish Dictionary and the Singlish Dictionary [15], and also the knowledge of the one of the researchers who is a native speaker of Singlish. Figure 1 shows how the different emotion terms fit into the emotional categories of the model. These emotional expressions commonly appear in Singlish discourse, and they are made up of loanwords from other languages, English words with 'idiosyncractic' meaning, and English words. The model, as far as possible, include Singlish emotion words that describe general emotional states, like 'tu lan' for anger. Singlish emotion words with specific emotion referents are excluded, like 'jelat' that refers to a sense of disgust that pertains to satiated by rich tasting food to the point that one is sick of it.

Figure 1 is not the translated equivalent of hourglass model as they contain Singlish emotion expressions that best fit into the model. As such, some of the emotion terms are not universally known. Chinese-origin words in Singlish like 'tu lan', 'pek chek' and 'erxin' are mostly used by Chinese Singaporeans and they might not be used by Singaporeans of other ethnicities. Some of the emotion terms might have restricted or emotional domains. For example the word 'sian' which usually refers to a state of boredom also has a secondary meaning of frustration.

### 5.4    The Method

In this section, we propose a simple but effective method to construct Singlish SenticNet based on the already developed English SenticNet as benchmark. This process involves following fundamental steps -

- Represent ConceptNet and Singlish concepts into AffectNet structure.
- Blend Singlish AffectNet and AffectNet with ConceptNet to obtain the new AffectiveSpace.
- After employing TSVD each Singlish concepts are represented by 100 dimensional vector.
- These vectors are used as features for a supervised classifier which is trained using English SenticNet benchmark.

Each Singlish concept has been labeled manually into one of the emotion categories (Anger, Disgust, Surprise, Joy, Sadness, Fear). Then we create the Singlish AffectNet graph in the following way -

- Take the emotions, i.e., Anger, Disgust Surprise, Joy, Sadness, Fear as the nodes in the graph.
- For each Singlish concept draw an edge from that concept to the emotion node in step 1. Each of such edge is labeled by "HasProperty".

ConceptNet is represented in the form of a labeled direct graph, with nodes being concepts such as, for example, spoon, eating, book, paper, and arcs being relations such as UsedFor (spoon–UsedFor→eating) and MadeOf (book–MadeOf→paper). Technically, a graph can be thought of as a matrix (Table 5).

**Table 5.** Cumulative analogy allows for the inference of new pieces of knowledge by comparing similar concepts. In the example, it is inferred that the concept special_occasion causes shiok as it shares the same set of semantic features with hari_raya and birthday

| Concepts | Semantic Features (*relationship+concept*) | | | | |
|---|---|---|---|---|---|
| | ... | *Causes* shiok | *IsA* event | *UsedFor* cooking | *MotivatedBy* celebration ... |
| ⋮ | | ⋮ | ⋮ | ⋮ | ⋮ |
| hari_raya | ... | x | x | − | x ... |
| makan | ... | x | − | − | − ... |
| bamboo_steamer | ... | − | − | x | − ... |
| **special_occasion** | ... | **x?** | x | − | x ... |
| birthday | ... | x | x | − | x ... |
| ⋮ | | ⋮ | ⋮ | ⋮ | ⋮ |

To perform inference on multiple matrices, blending is the most widely used technique. It allows multiple matrices to be combined in a single matrix, basing on the overlap between these matrices. The new matrix is rich in information and contains much of the information shared by the two original matrices. By means of the singular value decomposition on the new matrix, new connections are formed in source matrices based on the shared information and overlap between them. This method enables creation of a new resource, which is a combination of multiple resources representing different kinds of knowledge. In order to build a suitable knowledge base for affective reasoning, we applied the blending technique to ConceptNet, AffectNet and Singlish AffectNet. First, we represented AffectNet as a directed graph, similarly to ConceptNet. For example, the concept birthday party has the associated emotion joy; we considered birthday party and joy as two nodes, and added an assertion HasProperty on the edge directed from the node birthday party to the node joy. Next, we converted the three graphs, ConceptNet and AffectNet and Singlish AffectNet, to sparse matrices to blend them. After blending the two matrices, we performed TSVD on the resulting matrix, to discard those components that represent relatively small variations in the data. We kept only 100 components of the blended matrix to obtain a good approximation of the original matrix.

**Obtaining Polarity Values of Singlish Concepts.** The proposed framework is designed to receive as input a natural language concept represented according to an $M$-dimensional space and to predict the corresponding sentic levels for the four affective dimensions involved: Pleasantness, Attention, Sensitivity, and Aptitude. The dimensionality $M$ of the input space stems from the specific design of Singlish AffectiveSpace. As for the outputs, in principle each affective dimension can be characterized by an analog value in the range $[-1, 1]$, which represents the intensity of the expressed or received emotion (Fig. 2).

Indeed, those analog values are eventually remapped to obtain six different sentic levels for each affective dimension. The categorization framework spans each affective dimension separately, under the reasonable assumption that the various dimensions map perceptual phenomena that are mutually independent. As a result, each affective dimension is handled by a dedicated ELM [13], which addresses a regression problem.

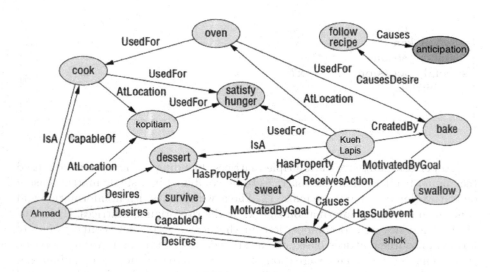

**Fig. 2.** Blending of Singlish AffectNet with ConceptNet and English AffectNet

Thus, each ELM-based predictor is fed by the $M$-dimensional vector describing the concept and yields as output the analog value that would eventually lead to the corresponding sentic level. Figure 3 provides the overall scheme of the framework; here, $g_X$ is the level of activation predicted by the ELM and $l_X$ is the corresponding sentic level. In theory, one might also implement the framework showed in Fig. 3 by using four independent predictors based on a multi-class classification schema. In such a case, each predictor would directly yield as output a sentic level out of the six available. However, two important aspects should be taken into consideration. First, the design of a reliable multi-class predictor is not straightforward, especially when considering that several alternative schemata have been proposed in the literature without a clearly established solution. Second, the emotion categorization scheme based on sentic levels stem from an inherently analog model, i.e., the Hourglass of Emotions. This ultimately motivates the choice of designing the four prediction systems as regression problems. In fact, the framework schematized in Fig. 3 represents an intermediate step in the development of the final emotion categorization system. One should take into account that every affective dimension can in practice take on seven different values: the six available sentic levels plus a 'neutral' value, which in theory correspond to the value $G(0)$ in the Hourglass model.

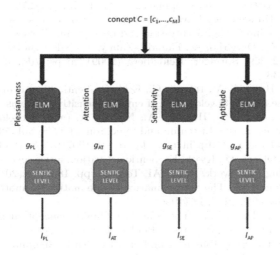

**Fig. 3.** The ELM-based framework for describing common-sense concepts in terms of the four Hourglass model's dimensions

In practice, though, the neutral level is assigned to those concepts that are characterized by a level activation that lies in an interval around $G(0)$ in that affective dimension. Therefore, the final framework should properly manage the eventual seven-level scale.

## 6    Conclusion

In this work, we proposed the construction of a concept-level Singlish sentiment lexicon using the English SenticNet framework. In future, we plan to extract 1 m Singlish concepts from various Singlish blogs and websites and employ the same framework. The proposed framework could also be improved by using different supervised classifiers.

# References

1. Brown, A.: Singapore English in a Nutshell: An Alphabetical Description of Its Features. Federal Publications, Singapore (1999)
2. Cambria, E.: Affective computing and sentiment analysis. IEEE Intell. Syst. **31**(2), 102–107 (2016)
3. Cambria, E., Fu, J., Bisio, F., Poria, S.: AffectiveSpace 2: enabling affective intuition for concept-level sentiment analysis. In: AAAI, pp. 508–514 (2015)
4. Cambria, E., Hussain, A.: Sentic Computing: A Common-Sense-Based Framework for Concept-Level Sentiment Analysis. Springer, Cham (2015). https://doi.org/10.1007/978-3-319-23654-4
5. Cambria, E., Hussain, A., Durrani, T., Havasi, C., Eckl, C., Munro, J.: Sentic computing for patient centered applications. In: ICSP, pp. 1279–1282. IEEE (2010)
6. Cambria, E., Hussain, A., Havasi, C., Eckl, C.: Common sense computing: from the society of mind to digital intuition and beyond. In: Fierrez, J., Ortega-Garcia, J., Esposito, A., Drygajlo, A., Faundez-Zanuy, M. (eds.) BioID 2009. LNCS, vol. 5707, pp. 252–259. Springer, Heidelberg (2009). https://doi.org/10.1007/978-3-642-04391-8_33
7. Cambria, E., Hussain, A., Havasi, C., Eckl, C.: Sentic computing: exploitation of common sense for the development of emotion-sensitive systems. In: Esposito, A., Campbell, N., Vogel, C., Hussain, A., Nijholt, A. (eds.) Development of Multimodal Interfaces: Active Listening and Synchrony. LNCS, vol. 5967, pp. 148–156. Springer, Heidelberg (2010). https://doi.org/10.1007/978-3-642-12397-9_12
8. Cambria, E., Olsher, D., Kwok, K.: Sentic activation: a two-level affective common sense reasoning framework. In: AAAI, Toronto, pp. 186–192 (2012)
9. Eckart, C., Young, G.: The approximation of one matrix by another of lower rank. Psychometrika **1**(3), 211–218 (1936)
10. Fauconnier, G., Turner, M.: The Way We Think: Conceptual Blending and the Mind's Hidden Complexities. Basic Books, New York (2008)
11. Gupta, A.F.: The Step-Tongue: Children's English in Singapore, vol. 101. Multilingual Matters, Bristol (1994)
12. Harada, S.: The roles of Singapore standard English and Singlish. Inf. Res. **40**, 70–82 (2009)
13. Huang, G.B., Cambria, E., Toh, K.A., Widrow, B., Xu, Z.: New trends of learning in computational intelligence. IEEE Comput. Intell. Mag. **10**(2), 16–17 (2015)
14. Lanczos, C.: An Iteration Method for the Solution of the Eigenvalue Problem of Linear Differential and Integral Operators. United States Governm. Press Office, Los Angeles (1950)
15. Lee, H., Grosse, R., Ranganath, R., Ng, A.Y.: Unsupervised learning of hierarchical representations with convolutional deep belief networks. Commun. ACM **54**(10), 95–103 (2011)
16. Low, E.L., Brown, A.: English in Singapore: An Introduction. McGraw-Hill, New York (2005)
17. Osgood, C.E., May, W.H., Miron, M.S.: Cross-cultural Universals of Affective Meaning. University of Illinois Press, Champaign (1975)

18. Phua, Y.L.: Social media sentiment analysis and topic detection for Singapore English. Technical report, DTIC document (2013)
19. Platt, J.T., Weber, H.: English in Singapore and Malaysia: Status, Features, Functions. Oxford University Press, Oxford (1980)
20. Plutchik, R.: The nature of emotions. Am. Sci. **89**(4), 344–350 (2001)
21. Poria, S., Gelbukh, A., Cambria, E., Das, D., Bandyopadhyay, S.: Enriching SenticNet polarity scores through semi-supervised fuzzy clustering. In: IEEE ICDM, Brussels, pp. 709–716 (2012)
22. Poria, S., Gelbukh, A., Cambria, E., Hussain, A., Huang, G.B.: EmoSenticSpace: a novel framework for affective common-sense reasoning. Knowl.-Based Syst. **69**, 108–123 (2014)
23. Poria, S., Gelbukh, A., Cambria, E., Yang, P., Hussain, A., Durrani, T.: Merging SenticNet and WordNet-affect emotion lists for sentiment analysis. In: ICSP, pp. 1251–1255. IEEE (2012)
24. Sato, Y., Kim, C.: Radical pro drop and the role of syntactic agreement in Colloquial Singapore English. Lingua **122**(8), 858–873 (2012)
25. Strapparava, C., Valitutti, A., et al.: WordNet affect: an affective extension of WordNet. In: LREC, vol. 4, pp. 1083–1086 (2004)
26. Tay, M.W.: The uses, users and features of English in Singapore. In: New Englishes, pp. 51–70 (1982)
27. Warschauer, M.: The internet and linguistic pluralism. In: Silicon Literacies: Communication, Innovation and Education in the Electronic Age, pp. 62–74 (2002)

# Using Syntactic and Semantic Features for Classifying Modal Values in the Portuguese Language

João Sequeira[1], Teresa Gonçalves[1(✉)] [iD], Paulo Quaresma[1,4] [iD],
Amália Mendes[2], and Iris Hendrickx[2,3]

[1] Department of Informatics, University of Évora, Évora, Portugal
tcg@uevora.pt
[2] Center for Linguistics of the University of Lisbon, Lisbon, Portugal
[3] Center for Language Studies, Radboud University Nijmegen,
Nijmegen, The Netherlands
[4] L2F – Spoken Language Systems Laboratory, INESC-ID, Lisbon, Portugal

**Abstract.** This paper presents a study made in a field poorly explored in the Portuguese language – modality and its automatic tagging. Our main goal was to find a set of attributes for the creation of automatic taggers with improved performance over the bag-of-words (bow) approach. The performance was measured using precision, recall and $F_1$. Because it is a relatively unexplored field, the study covers the creation of the corpus (composed by eleven verbs), the use of a parser to extract syntactic and semantic information from the sentences and a machine learning approach to identify modality values. Based on three different sets of attributes – from trigger itself and the trigger's path (from the parse tree) and context – the system creates a tagger for each verb achieving (in almost every verb) an improvement in $F_1$ when compared to the traditional bow approach.

## 1 Introduction

The automatic distinction between the factual and non-factual nature of events and the detection of the subjective perspective underlying texts is one of the concerns of the current trend in NLP that focuses on sentiment analysis and opinion mining. Modality is one such indicator of subjectivity and factuality in texts, as it is usually defined as the expression of the speaker's opinion and of his attitude towards the proposition [16]. It traditionally covers epistemic modality, which is related to the degree of commitment of the speaker to the truth of the proposition (whether the event is perceived as possible, probable or certain), but also deontic modality (obligation or permission), capacity and volition.

The present experiments are related to the automatic tagging of modality for the Portuguese language, a topic that has received little attention for languages other than English. In fact, one of our goals is to be able to create a tagger using a small corpus sample to (semi) automatically tag a larger corpus with modality information. For this purpose, we use a corpus of 158.553 tokens, manually

A. Gelbukh (Ed.): CICLing 2016, LNCS 9624, pp. 362–373, 2018.
https://doi.org/10.1007/978-3-319-75487-1_28

annotated with a modality scheme for Portuguese [8]. This paper restricts the experiments to eleven modal verbs: *arriscar* (chance/risk/dare), *aspirar* (aspire), *conseguir* (manage to/succeed in/be able to), *considerar* (consider/regard), *dever* (shall/might), *esperar* (wait/expect), *necessitar* (need/require), *permitir* (allow/permit), *poder* (may/can), *precisar* (need) and *saber* (know). These verbs were selected, out of the total set of triggers of the annotated corpus, based on their polysemy: indeed, they all express two, or even three, modal meanings. This increases the difficulty of the automatic annotation process and makes them an excellent object of study for our experiments.

Our experiments in the automatic annotation of modality will first identify the modal verbs (which we call the modal trigger) and then assign them a modal value (out of the possible set of modal meanings that each verb expresses). A modal verb may be ambiguous between several readings. For instance, the verb *poder* may be

- **Epistemic**, stating that something is possible, as in example (1)
- **Deontic**, denoting a permission, as in (2), or
- may express an **Internal Capacity**, the fact that someone is able to do something, as in (3).

(1) *O que é que vai fazer que julga <u>poder</u> ser marcante?*
    (What are [you] going to do that you believe may be significant?)
(2) *Nenhum atleta devia <u>poder</u> estar nos Jogos Olímpicos depois de uma deserção e creio que nenhum país deve sentir-se satisfeito por exibir medalhas conquistadas por estrangeiros.*
    (No athlete should be allowed to be (lit: should can be) at the Olympic Games after defecting and I believe that no country should be pleased to exhibit medals won by foreigners.)
(3) *Se a injustiça no mundo continuar, não sei até quando <u>poderemos</u> controlar as pessoas (...).*
    (If world injustice goes on, [I] don't know till when we will able to control people.)

To create the modality tagger, we first use a parser to get part-of-speech and syntactic information, then we identify modal triggers and, for those, we apply a machine learning approach to assign a modal value.

The paper is structured as follows: Sect. 2 presents related work on the automatic annotation of modality, Sect. 3 introduces the annotation scheme for Portuguese and presents statistics of the corpus used, Sect. 4 describes the features extracted and the system developed, and analyses the results obtained comparing them with a bag of words approach. Finally, in Sect. 5 we withdraw some conclusions and present some future work that would improve our system.

## 2   Related Work

The annotation schemes for modality apply mostly to the English language, as in Baker *et al.* [2], Matsuyoshi *et al.* [9], Nirenburg and McShane [14] and

Sauri *et al.* [19]. Two schemes have been devised for Portuguese: Hendrickx *et al.* [8] for written European Portuguese and Ávila *et al.* [21] for spoken Brazilian Portuguese. The existing schemes may focus on modality or integrate such information in a larger set of features that may cover, for instance, factuality, evidentiality, hedging, polarity and temporal information. These differences are discussed in detail in the work of Hendrickx *et al.* [8] and Nissim *et al.* [15]). Some of these schemes have been applied in experiments of automatic annotation and we focus here on reviewing the results that were obtained.

The experiments in Baker *et al.* [2] and Sauri *et al.* [19] report results with high success rates considering the complexity of the task at hand. Baker *et al.* [2] tested two rule-based modality taggers that identify both the modal trigger and its target and achieve results of 86% precision for a standard LDC data set. Sauri *et al.* [19] report, on the automatic identification of events in text and their characterization with modality features, to achieve accuracy values of 97.04% using the EviTA tool. Another experiment reported by Diab *et al.* [4] specifically addresses the annotation of belief by looking at auxiliary modal verbs. The authors do not consider the polysemy of modal verbs in their work and treat all auxiliary verbs as epistemic to avoid a source of noise in their system, although they acknowledge the fact that the verbs may have deontic meaning in certain contexts (note that our experiment, reported in Sect. 4, deals with the added complexity of multiple modal meanings). Prabhakaran *et al.* [17] extend the experiment on belief with tests on tagging different modality values (`Ability`, `Effort`, `Intention`, `Success` and `Want`); the authors report experiments on two very different annotated corpora: MTurk data composed of email threads and using only those examples for which at least two Turkers agreed on the modality and the target of the modality; and on a gold dataset, that contains sentences from newswire, letters and blogs in addition to emails. The results differ greatly according to the corpus used: the MTurk data achieves an overall 79.1% F-measure while the gold dataset presents 41.9% F-measure. Furthermore, there was a specific shared task at CoNLL2010 [5] on the detection of uncertainty and its linguistic scope by identifying hedging cues, which includes a broader set of lexical and syntactic clues when compared to modality as discussed in this paper. Finally, the annotation of events in the area of BioNLP includes in some cases values related to modality and factuality. The system described in [12] seeks to label events with the dimension 'level of certainty' and attains F-measures of 74.9% for 'low confidence' and 66.5% for 'high but not complete confidence'.

While our experiment focuses on auxiliary modal verbs but also main verbs with modal meaning, the work of Ruppenhofer and Rehbein [18] consider only the five auxiliary English verbs can/could, may/might, must, ought, shall/should. The authors predict the verb's modal value in context by training a maximum entropy classifier on features extracted from the corpus and improve the baseline for all verbs (but must), achieving accuracy numbers between 68.7% and 93.5%.

# 3   Modality Corpus

Our experiment applies over a corpus annotated with the annotation scheme for Portuguese presented in [8]. This scheme takes the concept of Modality as the expression of the speaker's attitude towards the proposition, so the concept of factuality is not included, contrary to approaches such as [15], who accounts for both values but in different layers of the annotation scheme. Tense and mood are also categories that are not taken into account, despite their relation with modality. The authors report that the approach is similar to the OntoSem ([10]) annotation scheme for modality [14]. Finally, the annotation is not restricted to modal verbs and covers also nouns, adjectives and adverbs, although for this experiment we only focus on a specific set of verbs.

Next subsections introduce the annotation scheme and corpus used for this work.

## 3.1   Annotation Scheme

Several modal values are included based not only on the modality literature but also on studies focused on annotation and information extraction (e.g. [1,2,16]). Seven main modal values are considered: Deontic, Effort, Epistemic, Evaluation, Participant-internal, Success and Volition. Some of these values are further classified into sub-values:

- Deontic modality has two subvalues: Obligation and Permission. This includes what is sometimes considered Participant-external modality, as in [1];
- Epistemic modality is further divided in Knowledge, Belief, Doubt, Possibility and Interrogative. Contexts traditionally considered of the modal type 'evidentials' (i.e., supported by evidence) are annotated as Epistemic Belief;
- Participant-internal modality has two sub-values: Necessity and Capacity.

The annotation scheme comprises several components: the trigger, which is the lexical element conveying the modal value; its target; the source of the event mention (speaker or writer); and the source of the modality (agent or experiencer). The trigger receives an attribute modal value, while both trigger and target are marked for polarity. For example, the modal verb *poder* in sentence (4) is underlined.

(4)   *Caso a avaliação seja positiva, a empresa de recauchutagem poderá salvar grande parte do equipamento que se encontra no interior das instalações, garantindo assim a laboração num curto espaço de tempo, que, segundo o administrador da empresa, António Santos, não poderá exceder os 15 dias.*
(If the evaluation is positive, the retreading production unit may save most of the equipment in the company premises, and so guaranty that the operation activity is done in a short span of time, that, according to the administrator of the company, António Santos, may not exceed 15 days.)

This sentence contains two other triggers: a first occurrence of the same verb *poderá* expresses the modal value Epistemic possibility, the trigger *garantido* (to guaranty) expresses Epistemic belief. We focus on the annotation of the second trigger *poder* in more detail. The target is discontinuous and we mark it here with the symbol @, although it is expressed in XML in our editor.

- Trigger: *poderá*
- Modal value: Deontic_permission. Polarity: negative
- Target: *a laboração@exceder os 15 dias*
- Source of the modality: *António Santos*
- Source of the event: writer
- Ambiguity: none

Full details on the annotation scheme and on the results of an inter-annotator experiment are provided in [8]. An enriched version with the interaction between Focus and Modality, specifically the case of exclusive adverbs, is presented in [11].

## 3.2 Corpus

The annotation scheme was applied to a corpus sample extracted from the written subpart of the Reference Corpus of Contemporary Portuguese (CRPC) [6]. Details about the selection of the sample are provided in [8]. The MMAX2[1] annotation software tool [13] was used for the manual annotation task. The elements of the annotation consist of markables that are linked to the same modal event, which is called a set.

For this study we used a subset of the annotated corpus by including the sentences from eleven verbs. Table 1 resumes the information about each verb.

**Table 1.** Corpus characterization: number of sentences per modal value for each verb.

| Verb | Modal values | Number of sentences | | | |
|------|--------------|-------|-------|-------|-------|
| | | total | val_1 | val_2 | val_3 |
| arriscar | 2 | 44 | 19 | 25 | |
| aspirar | 2 | 50 | 31 | 19 | |
| conseguir | 2 | 84 | 41 | 43 | |
| considerar | 2 | 29 | 18 | 11 | |
| dever | 2 | 108 | 37 | 71 | |
| esperar | 2 | 52 | 26 | 26 | |
| necessitar | 2 | 50 | 8 | 42 | |
| permitir | 2 | 78 | 60 | 18 | |
| poder | 3 | 236 | 42 | 154 | 40 |
| precisar | 2 | 54 | 45 | 9 | |
| saber | 2 | 103 | 93 | 10 | |

---

[1] The MMAX2 software is platform-independent, written in java and can freely be downloaded from http://mmax2.sourceforge.net/.

# 4    Developed System

The developed system works in two steps: first, it identifies the modal verbs and then it labels the appropriate modal value in its specific context. Modal verbs are identified by the automatic analysis of the output of the syntactic parser and the modal values are labelled using a Machine Learning approach.

The syntactic analysis is performed using the PALAVRAS parser [3] and the set of sentences that include modal verbs are selected to build the data for the Machine Learning algorithm.

## 4.1    Experimental Setup

Most machine learning algorithms use the vector space model to represent the input data. Using this approach each sentence needs to be transformed into a set of features that can be boolean, nominal or real valued. This work uses the output of PALAVRAS to build those features: we include information from the trigger itself, from the syntactic tree path and from the trigger's context. Besides this approach a bag-of-words representation of the sentences was also considered as a baseline representation. Next subsection describes in detail the information extracted from the syntactic tree and how it was represented.

The SVM (Support Vector Machine) algorithm [20] was chosen to label the modal value of each verb. Several initial experiments were conducted with different degrees of the polinomial kernel ($n \in 1, 2, 3$) and values of the C parameter ($C \in 0.0001, 0.001, 0.01, 0.1, 1, 10, 100, 1000, 10000$). Those experiments enabled us to chose the linear kernel with $C = 1$.

Different sets of extracted attributes were evaluated and compared with a typical bag-of-words approach. For the evaluation we used a 5-fold stratified cross-validation procedure (repeated twice) and computed average precision, recall and F1 performance measures. Appropriate statistical tests with 95% of significance were applied to analyse the differences between results.

These machine learning experiments were conducted using Weka framework [7].

## 4.2    Feature Extraction

The information extracted to build the attributes is inspired in the work by Ruppenhofer and Rehbein [18]. Their approach used three specific sets of attributes: information from the **trigger**, from the **path** (from the trigger to the root taken from the syntatic parse tree) and from the **context** around the trigger.

Using the PALAVRAS parser we also extracted attributes from the trigger, context and path including all possible information given by the parser output:

- for the trigger: besides the trigger itself we included info from the ancestral nodes (father, grandfather and great grandfather);
- for the path: besides collecting info from the trigger to the root, we also included info from the its left and right nodes;

– for the context: we collected info about the previous and following words using different size windows.

Table 2 sumarizes the attributes extracted and next sub-sections detail the extracted attributes and summarize them.

**Table 2.** Attributes extracted from trigger, path and context.

| Trigger | | Path | | Context | |
|---------|------------|---------------|------------|--------------------|------------|
| Source | Attributes | Source | Attributes | Source | Attributes |
| Trigger | POS | Siblings | POS | Left/right trigger | POS |
| | function | | function | | word |
| | role | | role | | lemma |
| | morphological | | morphological | | |
| | semantic | | semantic | | |
| Ancestors | POS | Trigger to root | POS | | |
| | function | | function | | |

**Trigger Related Attributes.** For each trigger word we extracted the POS tag, function, morphological and semantic information, and the role (if it exists). For the ancestral nodes (father, grandfather and great-grandfather) we extracted the POS tag and function. All information was represented as binary attributes (present/not present).

**Path Related Attributes.** For each trigger we extracted POS tags and functions from the tree's path (all nodes from the trigger word to the root) and also the POS tag, function, morphological and semantic information, and the role (if exists) from the path of sibling nodes (left and right). All information was represented as numerical attributes (counts over each possible value).

**Context Related Attributes.** For each trigger we extracted information about the POS tags, words and lemmas in the surrounding context with a size window equal to five words (with the trigger word in the middle). All information was represented as numerical attributes (counts over each possible value).

**Datasets Characterization.** As already mentioned, besides building a representation using the output of PALAVRAS, a traditional bag-of-words approach was also considered. Table 3 resumes, for each verb, the number extracted for each specific set of attributes.

**Table 3.** Datasets characterization: number attributes for bag-of-words, trigger, path and context sets.

| Verb | Bow | Trigger | Path | Context |
|------|-----|---------|------|---------|
| arriscar | 642 | 577 | 734 | 2548 |
| aspirar | 704 | 600 | 775 | 2736 |
| conseguir | 1093 | 649 | 838 | 4170 |
| considerar | 552 | 534 | 678 | 1918 |
| dever | 1578 | 645 | 845 | 5596 |
| esperar | 611 | 554 | 703 | 2358 |
| necessitar | 714 | 576 | 747 | 2732 |
| permitir | 1320 | 685 | 900 | 4966 |
| poder | 2736 | 715 | 940 | 9770 |
| precisar | 588 | 574 | 728 | 2320 |
| saber | 1237 | 696 | 912 | 4504 |

## 4.3 Experiments

In order to evaluate the discrimination power of each set of attributes, eight experiments were done: bag-of-words, trigger, path, context, trigger+path, trigger+context, path+context and trigger+path+context. Tables 4, 5 and 6 present the precision, recall and F1 values, respectively. Values statistically different (better or worse) from the corresponding bag-of-words experiment are boldfaced.

**Table 4.** Results: precision values

| | Bow | Trigger | Path | Context | tg+pth | tg+ct | pth+ct | All |
|------|-----|---------|------|---------|--------|-------|--------|-----|
| arriscar | .638 | .686 | .771 | .757 | .719 | .750 | .833 | .804 |
| aspirar | .741 | .853 | .795 | .694 | .778 | .756 | .778 | .779 |
| conseguir | .540 | .583 | .595 | .678 | .592 | .672 | .714 | .684 |
| considerar | .402 | .489 | .526 | .611 | .582 | .536 | .650 | .660 |
| dever | .700 | .662 | .568 | .626 | .636 | .692 | .611 | .602 |
| esperar | .745 | .610 | **.527** | .595 | .545 | .619 | .477 | .577 |
| necessitar | .708 | .723 | .735 | .732 | .709 | .701 | .690 | .698 |
| permitir | .593 | .666 | **.754** | **.785** | .702 | **.786** | **.812** | **.811** |
| poder | .530 | .486 | .529 | .522 | .544 | .520 | .484 | .536 |
| precisar | .698 | .700 | .669 | .757 | .700 | .788 | .736 | .736 |
| saber | .815 | **.906** | .833 | .861 | .881 | .903 | .843 | **.917** |

**Table 5.** Results: recall values

|            | Bow  | Trigger | Path | Context | tg+pth | tg+ct | pth+ct | All  |
|------------|------|---------|------|---------|--------|-------|--------|------|
| arriscar   | .614 | .672    | .726 | .740    | .669   | .718  | .800   | .765 |
| aspirar    | .710 | .820    | .76  | .710    | .770   | .730  | .770   | .760 |
| conseguir  | .530 | .565    | .59  | .671    | .577   | .647  | .696   | .660 |
| considerar | .587 | .510    | .513 | .647    | .547   | .513  | .657   | .623 |
| dever      | .700 | .652    | .601 | .645    | .611   | .690  | .626   | .607 |
| esperar    | .720 | .605    | .527 | .595    | .547   | .606  | .499   | .577 |
| necessitar | .840 | .780    | .73  | .830    | .740   | .790  | .720   | .770 |
| permitir   | .770 | .704    | .744 | .795    | .698   | .795  | .827   | .807 |
| poder      | .650 | .574    | .61  | .581    | **.574** | .568 | **.551** | .579 |
| precisar   | .835 | .739    | .732 | .826    | .750   | .827  | .799   | .817 |
| saber      | .903 | .898    | .869 | .913    | .903   | .937  | .908   | .942 |

## 4.4 Discussion of Results

For the bag-of-words approach we got results ranging between 0.402 (*considerar*) and 0.815 (*saber*) for precision; 0.530 (*conseguir*) and 0.903 (*saber*) for recall; and 0.472 (*conseguir*) and 0.857 (*saber*) for F1.

Looking at Table 2 we are able to find that *considerar*, *conseguir* and *saber* have 29, 84 and 103 examples, respectively (with 2 possible modal values); while *considerar* and *conseguir* constitute balanced datasets, for *saber* there is a ratio of 9:1. From this we can say that *saber* seems to be a easier verb to assign the modal value and that the unbalanced dataset does not seem hurt the performance.

**Table 6.** Results: F1 values

|            | Bow  | Trigger | Path | Context | tg+pth | tg+ct | pth+ct | All  |
|------------|------|---------|------|---------|--------|-------|--------|------|
| arriscar   | .605 | .660    | .712 | .736    | .658   | .708  | .794   | .758 |
| aspirar    | .693 | .823    | .759 | .694    | .770   | .730  | .763   | .757 |
| conseguir  | .552 | .554    | .581 | .670    | .563   | .640  | .689   | .650 |
| considerar | .472 | .470    | .510 | .611    | .543   | .489  | .634   | .618 |
| dever      | .668 | .644    | .566 | .617    | .609   | .685  | .610   | .597 |
| esperar    | .712 | .599    | .513 | .568    | .537   | .596  | .473   | .562 |
| necessitar | .768 | .746    | .724 | .775    | .718   | .742  | .699   | .728 |
| permitir   | .670 | .672    | .742 | .782    | .683   | .775  | **.807** | .794 |
| poder      | .540 | .520    | .535 | .537    | .548   | .538  | .510   | .550 |
| precisar   | .760 | .716    | .697 | .787    | .721   | .799  | .763   | .770 |
| saber      | .857 | .895    | .849 | .884    | .889   | .914  | .872   | **.923** |

From Table 4 its possible to state that, for most experiments, there are no statistical differences with the bag-of-words approach. Nevertheless, when using information extracted from PALAVRAS, we got improvements in precision values over the verbs *permitir* and *saber* when compared with the bag-of-words approach. For *permitir* we got better results when using path (0.754), context (0.785), trigger+context (0.786), path+context (0.812) and all (0.811) features; for *saber* better results were obtained with trigger (0.906) and all (0.917) attributes. If we look at the number of features of each setting (Table 3) we can conclude that for these verbs we are able to get higher precision using less attributes: for *permitir*, we have 747 vs. 1320 attributes for path vs. bag-of-words; for *saber*, we have 696 vs. 1237 attributes for trigger vs. bag-of-words. On the other hand, the verb *esperar* presents worse results when compared to bag-of-words for path (0.527) experiments.

For recall values (Table 5) we got even less statistical differences. Only verb *poder* got different results and they were worse (using trigger+path attributes with a recall of 0.574 and for path+context ones with a recall of 0.551).

Finally, and looking at Table 6, we can state that using information from the parse tree, we were able to improve the F1 values for 2 verbs: for *permitir* with the path+context setting (0.807) and for *saber* with all attributes (0.923).

# 5   Conclusions and Future Work

With this work, we tried to address a topic that has not been much explored in the Portuguese language – the automatic tagging of modality. The correct tagging of modality is important since it is linked to the current trend in NLP on sentiment analysis and opinion mining.

Due to this limited research it was necessary to implement a viable corpus for the work to be done: 11 verbs were chosen to be studied and the parser PALAVRAS was used to obtain morphological, syntatic and some semantic information from the sentences. Following a similar approach to Ruppenhofer and Rehbein [18] we defined three sets of attributes: trigger related, path related and context related ones.

Using Weka framework we conducted several experiments to study the effect on the usage of linguistic information to identify modal values. As baseline we used a bag-of-words approach and calculated precision, recall and $F_1$ measures. While we were able to get better results (precision and F1) with some settings for some verbs (*permitir* and *saber*), most experiments, even with higher performance values, are not significantly different from the bag-of-words approach (mainly because of the small number of training examples). Also, we obtained worse precision results for verb *esperar* using path attributes and worse recall values for verb *poder* using trigger+path and path+context ones.

With this work we were able to find a set of linguistic information related attributes that can be used to identify the modal values for the Portuguese language and from the results we can conclude that the use of information extracted from the parse trees does not harm the performance of the automatic taggers and can even, for some verbs and combinations enhance it.

Considering that our training corpus was relatively small and that we selected challenging verbs in our experiment, we believe that our goal, of creating a larger corpus with modal information by a (semi) automatic tagging process, could lead to positive results in the future. We plan to study the role played by each feature in our system and to observe in more detail the reasons why some verbs reach higher scores than others.

As we are currently applying a 'word expert' approach and training separate classifiers for different verbal triggers, it is clear that this approach will not be able to handle modal triggers that were not seen before. We intend to study this problem and try to train, for example, a general modal trigger classifier that is not dependent on the verb itself. Also, besides developing the tagger for the trigger we intend to build a system that is able to identify the target of the modality.

**Acknowledgements.** This work was partially supported by national funds through FCT – Fundação para a Ciência e Tecnologia, under project Pest-OE/EEI/LA0021/2013 and project PEst-OE/LIN/UI0214/2013.

# References

1. der Auwera, J.V., Plungian, V.A.: Modality's semantic map. Linguist. Typol. 1(2), 79–124 (1998)
2. Baker, K., Bloodgood, M., Dorr, B., Filardo, N.W., Levin, L., Piatko, C.: A modality Lexicon and its use in automatic tagging. In: Chair, N.C.C., Choukri, K., Maegaard, B., Mariani, J., Odijk, J., Piperidis, S., Rosner, M., Tapias, D. (eds.) Proceedings of the Seventh International Conference on Language Resources and Evaluation (LREC 2010). European Language Resources Association (ELRA), Valletta, Malta, May 2010
3. Bick, E.: The Parsing System PALAVRAS. Aarhus University Press, Aarhus (1999)
4. Diab, M.T., Levin, L.S., Mitamura, T., Rambow, O., Prabhakaran, V., Guo, W.: Committed belief annotation and tagging. In: Third Linguistic Annotation Workshop, pp. 68–73. The Association for Computer Linguistics, Singapore, August 2009
5. Farkas, R., Vincze, V., Móra, G., Csirik, J., Szarvas, G.: The CoNLL-2010 shared task: learning to detect hedges and their scope in natural language text. In: Proceedings of the Fourteenth Conference on Computational Natural Language Learning, pp. 1–12. Association for Computational Linguistics, Uppsala, Sweden, July 2010
6. Généreux, M., Hendrickx, I., Mendes, A.: Introducing the reference corpus of contemporary Portuguese on-line. In: Calzolari, N., Choukri, K., Declerck, T., Dogan, M.U., Maegaard, B., Mariani, J., Odijk, J., Piperidis, S. (eds.) LREC 2012, pp. 2237–2244. European Language Resources Association (ELRA), Istanbul (2012)
7. Hall, M., Frank, E., Holmes, G., Pfahringer, B., Reutemann, P., Witten, I.H.: The weka data mining software: an update. SIGKDD Explor. Newsl. 11(1), 10–18 (2009)
8. Hendrickx, I., Mendes, A., Mencarelli, S.: Modality in text: a proposal for corpus annotation. In: Chair, N.C.C., Choukri, K., Declerck, T., Doğan, M.U., Maegaard, B., Mariani, J., Moreno, A., Odijk, J., Piperidis, S. (eds.) Proceedings of the Eight International Conference on Language Resources and Evaluation (LREC 2012). European Language Resources Association (ELRA), Istanbul, Turkey, May 2012

9. Matsuyoshi, S., Eguchi, M., Sao, C., Murakami, K., Inui, K., Matsumoto, Y.: Annotating event mentions in text with modality, focus, and source information. In: Chair, N.C.C., Choukri, K., Maegaard, B., Mariani, J., Odijk, J., Piperidis, S., Rosner, M., Tapias, D. (eds.) Proceedings of the Seventh International Conference on Language Resources and Evaluation (LREC 2010). European Language Resources Association (ELRA), Valletta, Malta, May 2010

10. McShane, M., Nirenburg, S., Beale, S., O'Hara, T.: Semantically rich human-aided machine annotation. In: Proceedings of the Workshop on Frontiers in Corpus Annotations II: Pie in the Sky, pp. 68–75. Association for Computational Linguistics, Ann Arbor, Michigan, June 2005

11. Mendes, A., Hendrickx, I., Salgueiro, A., Ávila, L.: Annotating the interaction between focus and modality: the case of exclusive particles. In: Proceedings of the 7th Linguistic Annotation Workshop and Interoperability with Discourse, pp. 228–237. Association for Computational Linguistics, Sofia, Bulgaria, August 2013

12. Miwa, M., Thompson, P., McNaught, J., Kell, D.B., Ananiadou, S.: Extracting semantically enriched events from biomedical literature. BMC Bioinform. **13**, 108 (2012)

13. Müller, C., Strube, M.: Multi-level annotation of linguistic data with MMAX2. In: Braun, S., Kohn, K., Mukherjee, J. (eds.) Corpus Technology and Language Pedagogy: New Resources, New Tools, New Methods, pp. 197–214. Peter Lang, Frankfurt a.M., Germany (2006)

14. Nirenburg, S., McShane, M.: Annotating modality. Technical report, University of Maryland, Baltimore County, USA, March 2008

15. Nissim, M., Pietrandrea, P., Sanso, A., Mauri, C.: Cross-linguistic annotation of modality: a data-driven hierarchical model. In: Proceedings of IWCS 2013 WAMM Workshop on the Annotation of Modal Meaning in Natural Language, pp. 7–14. Association for Computational Linguistics, Postam, Germany (2013)

16. Palmer, F.R.: Mood and Modality. Cambridge Textbooks in Linguistics. Cambridge University Press, Cambridge (1986)

17. Prabhakaran, V., Bloodgood, M., Diab, M., Dorr, B., Levin, L., Piatko, C.D., Rambow, O., Van Durme, B.: Statistical modality tagging from rule-based annotations and crowdsourcing. In: Proceedings of the Workshop on Extra-Propositional Aspects of Meaning in Computational Linguistics, ExProM 2012, pp. 57–64. Association for Computational Linguistics, Stroudsburg, PA, USA (2012)

18. Ruppenhofer, J., Rehbein, I.: Yes we can!? Annotating English modal verbs. In: Chair, N.C.C., Choukri, K., Declerck, T., Doğan, M.U., Maegaard, B., Mariani, J., Odijk, J., Piperidis, S. (eds.) Proceedings of the Eight International Conference on Language Resources and Evaluation (LREC 2012). European Language Resources Association (ELRA), Istanbul, Turkey, May 2012

19. Sauri, R., Verhagen, M., Pustejovsky, J.: Annotating and recognizing event modality in text. In: FLAIRS Conference, pp. 333–339 (2006)

20. Vapnik, V.N.: Statistical Learning Theory. Wiley-Interscience, Hoboken (1998)

21. Ávila, L., Melo, H.: Challenges in modality annotation in a Brazilian Portuguese spontaneous speech corpus. In: Proceedings of IWCS 2013 WAMM Workshop on the Annotation of Modal Meaning in Natural Language. Association for Computational Linguistics, Postam, Germany (2013)

# Detecting the Likely Causes Behind the Emotion Spikes of Influential Twitter Users

Calkin Suero Montero[1(⊠)], Hatem Haddad[2], Maxim Mozgovoy[3], and Chedi Bechikh Ali[4]

[1] School of Computing, University of Eastern Finland, Joensuu, Finland
calkins@uef.fi
[2] Department of Computer and Decision Engineering,
Université Libre de Bruxelles, Brussels, Belgium
Hatem.Haddad@ulb.ac.be
[3] Active Knowledge Engineering Lab, University of Aizu,
Aizuwakamatsu, Japan
mozgovoy@u-aizu.ac.jp
[4] LISI Laboratory, INSAT, Carthage University, Tunis, Tunisia
chedi.bechikh@gmail.com

**Abstract.** Understanding the causes of spikes in the emotion flow of influential social media users is a key component when analyzing the diffusion and adoption of opinions and trends. Hence, in this work we focus on detecting the likely reasons or causes of spikes within influential Twitter users' emotion flow. To achieve this, once an emotion spike is identified we use linguistic and statistical analyses on the tweets surrounding the spike in order to reveal the spike's likely explanations or causes in the form of keyphrases. Experimental evaluation on emotion flow visualization, emotion spikes identification and likely cause extraction for several influential Twitter users shows that our method is effective for pinpointing interesting insights behind the causes of the emotion fluctuation. Implications of our work are highlighted by relating emotion flow spikes to real-world events and by the transversal application of our technique to other types of timestamped text.

**Keywords:** Emotion analysis · Emotion fluctuation explanation
Social media

## 1 Influence Through Twitter

Word of mouth (WOM) has long been considered one of the most efficient information sources that influence customers' decision when purchasing products or services [15]. Due to its strong implications when designing targeted advertising and marketing or political campaigns, the suggestive influence that consumers exert on each other through the WOM phenomenon has been extensively studied also in online social media contexts, as a new form of communication spread [15, 27]. Particular attention has been devoted to single out influential users in social media as inexpensive seed catalyzers in the dissemination of news, opinions, and products or services attractiveness [26]. Specifically, the Twitter microblogging platform has enticed great deal of

© Springer International Publishing AG, part of Springer Nature 2018
A. Gelbukh (Ed.): CICLing 2016, LNCS 9624, pp. 374–384, 2018.
https://doi.org/10.1007/978-3-319-75487-1_29

research on the analysis and identification of influential users due to ease that the platform presents in tracking the diffusion of information through its channels [4]. It is, therefore, not surprising that the analysis of the emotional state of influential users in Twitter reveals itself as a very important focal point of research.

Twitter, as a microblogging platform, receives over 500 million tweets worldwide every day as per 2016 [29]. This represents a well of information that is being exploited for journalism, business intelligence, monitoring natural and man-made disasters, terrorism and so forth [3, 10]. Hence, emotion analysis in Twitter has proven to be a valuable asset for marketing analysis including services consumers' satisfaction [7], and political candidate popularity [17], to name a few relevant examples. Since it has been shown that influential individuals in social media networks can positively diffuse their opinions and preferences to peers [2], this has implications for designing "intervention strategies, target advertising and policy making" [2]. However, the meaningful identification of the causes of fluctuation in the emotional flow dispositions of influential individuals through social media is a challenging and underexplored task.

Our application aims at identifying the likely explanations or causes of strong emotions fluctuations i.e., *emotion spikes*, within the temporal dimension of influential users' emotion flow in Twitter (see for example Kwak et al. [13], Twitter Counter [28] and Cha et al. [9] for influential users analysis and identification). These emotion spikes are presumably associated with a reaction to certain event. Hence, our system extracts *keyphrases,* phrases formed from linguistic patterns, adjacent to each identified emotion spike, and passes them to an analyst for subsequent examination. Our application goes beyond the detection of named-entities (e.g. "person", "company", "city", etc.) and events or topics identification since the extracted keyphrases are indicative of a change on user's sentiment, and represent the causes the emotion spike. This information is of value for business intelligence, designing targeted marketing campaigns and revamping brand image, for instance.

In this paper we present our contribution and ongoing work on the emotion analysis of influential Twitter users addressing two main issues:

- The visual representation and analysis of the temporal emotion flow of a user's tweets on which emotion spikes are identified.
- The extraction of the likely causes of the visualized emotion spikes using keyphrases for linguistic and statistical analyses.

The rest of the paper is organized as follows. Section 2 outlines existing work on emotional visualization and the identification of likely causes of emotion spikes. Section 3 describes our methodology. In Sect. 4 we describe our experimental evaluation. Finally, discussion and future work are covered in Sect. 5.

## 2 Related Work

### 2.1 Emotion Visualization

Emotion flow visualization in social media outlets has generated interest for a variety of applications. For instance, Mishne and De Rijke [19] developed a system called

MoodViews, a collection of tools for analyzing, tracking and visualizing moods and mood changes in blogs posted by LiveJournal users. Similarly, Kempter et al. [12] have proposed the EmotionWatch system to automatically recognize emotions and score tweets into 20 discrete emotion categories. Particularly relevant to our work is the TwitInfo tool proposed by Marcus et al. [16], a prototype system for monitoring events on Twitter, using a timeline graph to show major peaks of publication of tweets about a particular topic, the most relevant tweets, and the polarity of the opinions they express. These systems differ from our proposed approach in that their analysis is performed over the entire blogosphere instead of focusing on a singular user. Also, in our work we look at analyzing emotions alongside the likely explanations or causes that provoked the observed strong fluctuations over time in an emotion flow.

### 2.2   Likely Causes of Emotion Spikes

A relevant work to our research is that of Balog et al. [5] who proposed a method for identifying and explaining spikes in mood patterns in blogs. Balog et al. used empirical heuristics to identify spikes in users' reported moods within a LiveJournal blog corpus. From the identified spikes' period, a query formed by "overused words" was then extracted and used to search a news corpus in order to mine related news headlines. Similar to Balog et al., we effectively use empirical heuristics to detect and identify emotion spikes within the user's tweets timeline. However, our work differs from Balog et al. approach in that we use keyphrases instead of simple keywords to extract the likely causes of the spikes. By keyphrases we refer to a list of phrases formed from linguistic patterns that express the informative contents of tweets. Hence, in our approach the likely causes for the emotion fluctuation are extracted directly from the tweets themselves instead of an external source, which ensures that our system will return a likely explanation for the emotion spike.

   Our work also differs from events detection ED [22], topic detection and tracking TDT [1], and topic classification TC [31] research in social media. That is, given an emotion spike our approach aims at exploring its likely causes, answering questions such as 'why does the user have this strong emotion fluctuation?' and 'what are the likely causes that provoked the spike?'

## 3   Methodology

Our framework is outlined in Fig. 1. The processes involved are detailed here.

### 3.1   Twitter Data and Timeline

We download automatically tweets from a user's account using the Twitter API. The API crawls up to 3,200 most recent tweets and retweets of a user for further processing. Our application discards retweets from our timeline analysis. Once the tweets are downloaded, the system fragments them time-wise using the time-stamp of each tweet.

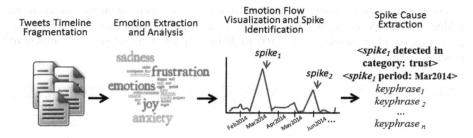

**Fig. 1.** Framework processes

## 3.2 Emotion Extraction and Analysis

After timeline fragmentation, the system extracts emotions from the input text using the NRC word-emotion association lexicon, EmoLex [20]. The lexicon has been manually annotated with the Plutchik's eight basic emotions i.e., joy, anger, sadness, fear, anticipation, surprise, disgust and trust [21]. An emotion score (*eScore*) is then calculated for each one of Plutchik's eight emotion categories represented in each text entry as follows:

$$eScore_{category} = \frac{eWords_{category}}{eWords_{all}} \tag{1}$$

where $eWords_{category}$ is the number of words in the analyzed text entry that have nonzero emotional score for the category according to the lexicon; and $eWords_{all}$ is the number of words in the text entry that have nonzero emotional score for any category according to the lexicon.

## 3.3 Emotion Spike Identification

The emotion flow visualization is given based on the calculated *eScore* on a monthly basis for easy access. A spike in the emotion flow is defined here as a sudden change in the *average* emotion flow of a user. That is, we calculate spikes in the emotion flow using the relative *eScore ratio* between adjacent months. A spike is detected if the ratio of eScore for an emotion category in one month to the eScore for the same emotion category in the month before and the month after is greater than an empirically predefined threshold. Formally:

$$Spike(c,m) = \left( \frac{eScore_{(c,m)}}{eScore_{(c,m-1)}} > \theta \right) AND \left( \frac{eScore_{(c,m)}}{eScore_{(c,m+1)}} > \theta \right) \tag{2}$$

where $c$ is an emotion category, $m$ is the month of the analysis and $\theta$ is the ratio threshold acceptable to define a spike. The value of $\theta$ was empirically set to *1.45* in our experiments, since it represented a balanced choice when identifying interesting spikes and filtering out noisy ones.

For our analysis, an emotion spike includes all the tweets from the month when the sudden change was detected. Figure 2 shows the emotion flows extracted from Dalai Lama tweets (1113 tweets, from 2010 to 2015), alongside highlighted examples of detected spikes.

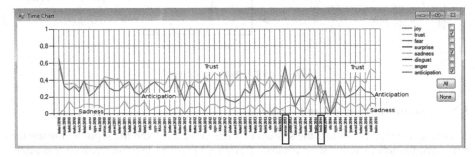

**Fig. 2.** Dalai Lama's emotion flow for *trust* (top), *anticipation* (middle) and *sadness* (bottom). Spikes were found for trust and anticipation in Nov 2013 and sadness in June 2014, among others

### 3.4    Extracting Likely Causes for Emotion Spikes

Aiming at uncovering the explanations or causes of a spike in the emotion flow, we extract keyphrases from the month period during which the spike is detected within the respective emotion category. Keyphrases have been used in several domains including text summarization [14], indexing [11] and searching [8], to describe and capture the main information enclosed in a given document [25]. Keyphrase extraction can be performed through linguistic analysis [8, 33], statistical analysis [18, 32] or through a combined linguistic and statistical analyses approach [6]. We use a combined linguistic and statistical analyses approach to keyphrase extraction as it achieves the best performance.

**Linguistic Analysis**
The system first performs part of speech tagging on the tweets in order to generate a tagged text where each word is labeled with a corresponding part-of-speech marker. Then, a set of manually defined syntactic rule patterns is used to extract keyphrases from the tagged text. Each rule is represented with a sequence of part-of-speech tags, describing the structure of the target keyphrase. Therefore, this method differs from n-gram filters and at the same time is general enough to allow creating patterns that describe keyphrases consisting of compound nouns, named-entities and events.

In our experiments, the length of the syntactic patterns varies from two to four part-of-speech tags, and all matching keyphrases are extracted. The syntactic patterns set consisted of twenty rules. For example, given the tweet *'The Senate already passed bipartisan immigration reform'*, its tagged text is: The/determiner Senate/noun already/adverb passed/verb bipartisan/noun immigration/noun reform/noun. Using two rule patterns "Noun Noun", and "Noun Noun Noun", for instance, the system extracted the following keyphrases:

- bipartisan_immigration (Noun Noun);
- immigration_reform (Noun Noun);
- bipartisan_immigration_reform (Noun Noun Noun)

**Statistical Analysis**

In order to select the most relevant keyphrases to represent the likely explanations or causes of an emotion spike, we use a two level filtering approach to reduce the number of candidates. Here we define relevancy based on the length as well as the frequency of the keyphrase. First, extracted keyphrases are filtered according to their length: longer keyphrases are preferred over shorter ones containing the same words [24]. Hence from the previous example, only the keyphrase *bipartisan_immigration_reform* is selected as a candidate.

Next, given that rare keyphrases are considered more informative than common ones [23], a second filter is applied to the spikes context based on a variation of the *tf-idf* weight. As a statistical measure, tf-idf provides scored weights to evaluate how important a keyphrase is to a spike within a set of detected spikes. We use tf-idf weights because this weighting scheme has been shown to achieve better performance than other algorithms like TextRank, SingleRank and ExpandRank [6]. With tf-idf we aim at scoring and ranking the extracted keyphrases in order to select *n*-best. In our application, the tf term analyzes the importance of a keyphrase against a spike, while the idf term analyzes the importance of the keyphrase against all the emotion spikes extracted. Hence, with $k$ representing a keyphrase (given by the first filter), $s$ an emotion spike (containing all the tweets in the analyzed month), and $S$ the set of all the emotion spikes, the tf-idf is calculated as:

$$tf\_idf(k, s, S) = tf(k, s) * idf(k, S) \tag{3}$$

where $k \in s$ and $s \in S$. We adapted the classic tf-idf weighting scheme [23] to assign the weight of keyphrase $k$ in a spike $s$ as:

$$tf\_idf(k, s, S) = tf(k, s) * log \frac{number\ of\ spikes\ in\ S}{number\ of\ spikes\ in\ S\ containing\ k} \tag{4}$$

where *tf(k, s)* is the number of times $k$ occurs in emotion spike s. Hence, considering the following $k_1$ (a rare keyphrase) and $k_2$ (a common keyphrase) from the Barack Obama spikes collection (3 spikes) found in the emotion category *anticipation:*

- $k_1$ = *anti-discrimination_law*, occurring once in the emotion spike period and only in this spike; then $tf\_idf(k_1) = 1 * log(3/1) = 0.47$.
- $k_2$ = *president_obama,* occurring 44 times in the emotion spike period and in 3 spikes; then $tf\_idf(k_2) = 44 * log(3/3) = 0$.

Therefore, $k_1$ has higher importance based on its tf-idf score.

## 4    Experimental Evaluation and Results

In order to test whether the automatic detection of emotion spikes is feasible, we set a preliminary experiment where we extracted tweets of three influential Twitter users in the period from 2010 to 2015: Barack Obama (2979 tweets), Bill Gates (1512 tweets) and Dalai Lama (1110 tweets) [28]. Table 1 shows the statistics of the detected spikes and extracted keyphrases for these 3 influential Twitter users. For Bill Gates and Barack Obama tweets, there were 9 spikes discovered, whereas 15 spikes were extracted from the Dalai Lama tweets. The average number of tweets per spike detected is highest for Bill Gates which in turn yields the highest number of extracted keyphrases.

**Table 1.** Spikes detection and keyphrases (KPs) extraction descriptions

| Influential user | Period | Tweets | Spikes | Tweets per spike | Extracted KPs | KPs per spike (avg) |
|---|---|---|---|---|---|---|
| Barack Obama | 8.11.2013 to 6.6.2015 | 2979 | 9 | 8,8 | 235 | 21,1 |
| Bill Gates | 19.1.2010 to 9.5.2015 | 1512 | 9 | 21,6 | 769 | 85,5 |
| Dalai Lama | 22.2.2010 to 12.5.2015 | 1110 | 15 | 12,6 | 680 | 15,6 |

Table 2 shows the extracted keyphrases corresponding to emotion spike cause candidates for two emotion categories for each user. From the extracted keyphrases we could speculate on the reasons for the respective spikes. For instance, the spike detected in the joy category in Dec 2014 on Bill Gates' emotion flow, could be related to a breakthrough in tuberculosis research, or to positive advances in the global discussion about inequality. Our system is also able to mine named-entities and events as per the syntactic patterns used. For instance, for the spike detected in trust, Nov 2013 within Bill Gates' tweets, the keyphrase 'conversation_with_bill_clinton' is extracted as one of the likely causes for the spike based on its tf-idf score. Conversely, the keyphrase 'wad2012_progress' was also extracted as a likely cause of this emotion spike, however it was discarded from the list of *3-best* candidates presented in Table 2 as its tf-idf score was low.

Although it is not possible to guarantee completely accurate results, our system can however serve as a significant aid to analysts who otherwise would have to identify the causes of emotion spikes without computer support. A comparative baseline could be created by the analyst reading every tweet of the target user and employing brainpower to analyze the data. Although this is certainly possible to achieve, such a manual process is still a very resource-consuming task. Our preliminary experiments show that our system is indeed able to extract blocks of text that convey likely explanations or causes for the emotional reaction of a user. In addition, even though the analysis of emotions, opinions and polarity among other features of texts has attracted an

**Table 2.** Extracted candidates of likely causes of emotion spikes for influential Twitter users

| User | Emotion and spike month | Samples of extracted likely causes *(keyphrases)* | tf idf |
|------|------------------------|--------------------------------------------------|--------|
| Barack Obama | Anticipation – Dec 2013 | `comprehensive_immigration_reform` | 0.97 |
| | | `bipartisan_immigration_reform` | 0.47 |
| | | `anti-discrimination_law` | 0.47 |
| | Anticipation – Aug 2014 | `economic_opportunity_for_all` | 2.38 |
| | | `private_sector_job` | 1.90 |
| | | `sector_job_creation` | 0.97 |
| Bill Gates | Trust – Nov 2013 | `conversation_with_bill_clinton` | 0.95 |
| | | `toilet_save_life` | 0.95 |
| | | `future_of_vaccination` | 0.95 |
| | Joy – Dec 2014 | `fantastic_global_discussion` | 0.95 |
| | | `discussion_about_inequality` | 0.95 |
| | | `breakthrough_in_tuberculosis` | 0.95 |
| Dalai Lama | Trust – Mar 2012 | `main_tibetan_temple` | 2.35 |
| | | `attitude_of_compassion` | 1.17 |
| | | `deep_inner_satisfaction` | 1.17 |
| | Sadness – June 2014 | `tibetan_childrens` | 1.17 |
| | | `village_school` | 1.17 |
| | | `condition_for_happiness` | 1.17 |

increasing amount of research, it is also very important to take into account the analysis of such features within the temporal dimension, i.e., how these features change over time. In our work, the change of emotions in the temporal dimension is capture through the analysis of identified spikes and the extraction of likely explanations. We further confirmed that time-stamped texts such as blogs and microblogs are convenient input sources for the type of analysis needed when reactions change over time.

## 5 Discussion and Future Work

While the mechanics of peer influence are complex and elusive to understand [2, 9], it has been long argued that influential people can catalyze peer behavior and opinions [30]. Hence, it is important to analyze influential users' emotion fluctuations in the temporal dimension, i.e., over a period of time, from the perspective of political voters' turnout, product demand or social unrest [2] and Twitter data shows value for these purposes.

Our contribution stands as the development and introduction of a very useful tool that works in detecting the possible reasons behind emotion spikes within a given user emotion flow timeline by using empirical methods heuristics. Our methodology works based on the combination of two algorithms: a lexicon-based algorithm that reacts to emotion-bearing keywords such as *'opportunity'*, and that can be considered as a coarse tool that shows likely places of emotion agitation or fluctuation; and a second algorithm that extracts complete keyphrases from where the emotion fluctuation was detected, such as *'economic opportunity for all'*.

It is important to observe that in our system the likely explanations for the emotion spikes are extracted in the form of keyphrases, only from within the tweets themselves. To further corroborate the validity of these extracted explanations or causes, external sources such as news corpora could also be utilized (see for example Balog et al. [5]). However, when using external sources of verification care should be taken since emotion spikes are subjective to the user and not necessarily have to be the result of an event reported elsewhere.

During the empirical testing for fine-tuning the value of the emotion spike identification threshold, we noticed that this value affects the successful retrieval of emotion spikes making the spikes noisier (if the threshold is lower) or more restricted (if the threshold is higher). This value can be dynamically set according to the analyst's needs and the individual user to be analyzed.

During our experiments we also observed that using our software the amount of some of the analyzed influential user tweets was rather small, against expectations. Noting this, we performed a meta-analysis on Bill Gates Twitter account and found that his tweeting frequency varied from one to three times a day during some weeks to no tweets at all for days during other weeks[1]. Following this pattern, it gave us a rough average of 360 tweets per year or 1800 tweets in a 5 years period available for retrieval. Hence, after discarding retweets, the amount of tweets from this user available for analysis in our application (1512) was made clear. Therefore, caution should be taken not to overestimate individual users' tweeting activity.

Our presented tool for emotion spike detection aims at assisting analysts when monitoring trends on the social media Twitter platform. The tool has been developed taking into account matters of simplicity and convenience of user interface and we are working towards releasing the software to the community as open source.

Our ongoing work includes performing a large-scale user evaluation of the system. Since there is no existing test collection for likely explanations or causes of strong emotions fluctuations, we plan to manually construct our test collection. Improvement of the spike detection procedure is also under consideration for future work, such that spikes are extracted on sliding averages instead of the current monthly basis. Also, while the core algorithm of the system for emotion spike detection is language-independent, emotions in documents are identified with English-based NRC lexicon. Thus, the adaptation of the system to languages other than English can be explored by using similar lexicons, customized for the given language.

## References

1. Allan, J. (ed.): Topic Detection and Tracking: Event-Based Information Organization, vol. 12. Springer Science & Business Media, Heidelberg (2002). https://doi.org/10.1007/978-1-4615-0933-2
2. Aral, S., Walker, D.: Identifying influential and susceptible members of social networks. Science **337**(6092), 337–341 (2012)

---

[1] Meta-analysis on Bill Gates tweeting activity, March 2016, https://twitter.com/BillGates.

3. Atefeh, F., Khreich, W.: A survey of techniques for event detection in Twitter. Comput. Intell. **31**(1), 132–164 (2013)
4. Bakshy, E., Hofman, J.M., Mason, W.A., Watts, D.J.: Everyone's an influencer: quantifying influence on Twitter. In: Proceedings of the Fourth ACM International Conference on Web Search and Data Mining, pp. 65–74. ACM (2011)
5. Balog, K., Mishne, G., De Rijke, M.: Why are they excited?: identifying and explaining spikes in blog mood levels. In: Proceedings of the Eleventh Conference of the European Chapter of the Association for Computational Linguistics: Posters and Demonstrations, pp. 207–210. ACL (2006)
6. Ali, C.B., Wang, R., Haddad, H.: A two-level keyphrase extraction approach. In: Gelbukh, A. (ed.) CICLing 2015. LNCS, vol. 9042, pp. 390–401. Springer, Cham (2015). https://doi.org/10.1007/978-3-319-18117-2_29
7. Bougie, R., Pieters, R., Zeelenberg, M.: Angry customers don't come back, they get back: the experience and behavioral implications of anger and dissatisfaction in services. J. Acad. Mark. Sci. **31**(4), 377–393 (2003)
8. Bracewell, D.B., Fuji, R., Kuriowa, S.: Multilingual single document keyword extraction for information retrieval. In: Proceedings of the IEEE NLP-KE 2005, pp. 517–522 (2005)
9. Cha, M., Haddadi, H., Benevenuto, F., Gummadi, P.K.: Measuring user influence in Twitter: the million follower fallacy. In: ICWSM 2010, pp. 10–17, 30 (2010)
10. Cheng, T., Wicks, T.: Event detection using Twitter: a spatio-temporal approach. PLoS ONE **9**(6), e97807 (2014)
11. Gutwin, C., Paynter, G., Witten, I., Nevill-Manning, C., Frank, E.: Improving browsing in digital libraries with keyphrase indexes. Decis. Support Syst. **27**(1), 81–104 (1999)
12. Kempter, R., Sintsova, V., Musat, C., Pu, P.: EmotionWatch: visualizing fine-grained emotions in event-related tweets. In: Eighth International AAAI Conference on Weblogs and Social Media (2014)
13. Kwak, H., Lee, C., Park, H., Moon, S.: What is Twitter, a social network or a news media? In: Proceedings of the 19th International Conference on World Wide Web, pp. 591–600. ACM (2010)
14. Litvak, M., Last, M.: Graph-based keyword extraction for single-document summarization. In: Proceedings of the Workshop on Multi-source Multilingual Information Extraction and Summarization. ACL (2008)
15. Litvin, S.W., Goldsmith, R.E., Pan, B.: Electronic word-of-mouth in hospitality and tourism management. Tourism Manag. **29**(3), 458–468 (2008)
16. Marcus, A., Bernstein, M.S., Badar, O., Karger, D.R., Madden, S., Miller, R.C.: Tweets as data: demonstration of TweeQL and Twitinfo. In: Proceedings of the ACM SIGMOD, pp. 1259–1262 (2011)
17. Melville, P., Gryc, W., Lawrence, R.D.: Sentiment analysis of blogs by combining lexical knowledge with text classification. In: Proceedings of the 15th ACM SIGKDD, pp. 1275–1284 (2009)
18. Mihalcea, R., Tarau, P.: TextRank: bringing order into texts. In: Proceedings of Conference on Empirical Methods in Natural Language Processing, pp. 404–411 (2004)
19. Mishne, G., De Rijke, M.: Moodviews: tools for blog mood analysis. In: AAAI Spring Symposium on Computational Approaches to Analysing Weblogs, pp. 153–154 (2006)
20. Mohammad, S.M., Turney, P.D.: Emotions evoked by common words and phrases: using mechanical turk to create an emotion lexicon. In: Proceedings of the NAACL HLT 2010 Workshop on Computational Approaches to Analysis and Generation of Emotion in Text. ACL (2010)
21. Plutchik, R.: A general psychoevolutionary theory of emotion. Theor. Emot. **1**, 3–31 (1980)

22. Psallidas, F., Becker, H., Naaman, M., Gravano, L.: Effective event identification in social media. IEEE Data Eng. Bull. **36**(3), 42–50 (2013)
23. Salton, G., Buckley, C.: Term-weighting approaches in automatic text retrieval. Inf. Process. Manag. **24**(5), 513–523 (1988)
24. Song, Y.I., Lee, J.T., Rim, H.C.: Word or phrase?: learning which unit to stress for information retrieval. In: Proceedings of the Joint Conference of the 47th Annual Meeting of the ACL and the 4th International Joint Conference on Natural Language Processing of the AFNLP: Volume 2–Volume 2, pp. 1048–1056. ACL (2009)
25. Turney, P.D.: Learning algorithms for keyphrase extraction. Inf. Retr. **2**(4), 303–336 (2000)
26. Trusov, M., Bodapati, A.V., Bucklin, R.E.: Determining influential users in internet social networks. J. Mark. Res. **47**(4), 643–658 (2010)
27. Trusov, M., Bucklin, R.E., Pauwels, K.: Effects of word-of-mouth versus traditional marketing: findings from an internet social networking site. J. Mark. **73**(5), 90–102 (2009)
28. Twitter counter - Twitter top 100 most followed (2015). http://twittercounter.com/
29. Twitter usage statistics, January 2016. http://www.internetlivestats.com/twitter-statistics/
30. Valente, T.W., Davis, R.L.: Accelerating the diffusion of innovations using opinion leaders. Ann. Am. Acad. Polit. Soc. Sci. **566**(1), 55–67 (1999)
31. Varga, A., Cano Basave, A.E., Rowe, M., Ciravegna, F., He, Y.: Linked knowledge sources for topic classification of microposts: a semantic graph-based approach. J. Web Semant. Sci. Serv. Agents World Wide Web **26**, 36–57 (2014)
32. Wang, R., Liu, W., McDonald, C.: How preprocessing affects unsupervised keyphrase extraction. In: Gelbukh, A. (ed.) CICLing 2014. LNCS, vol. 8403, pp. 163–176. Springer, Heidelberg (2014). https://doi.org/10.1007/978-3-642-54906-9_14
33. Zesch, T., Gurevych, I.: Approximate matching for evaluating keyphrase extraction. In: RANLP, pp. 484–489 (2009)

# Age Identification of Twitter Users: Classification Methods and Sociolinguistic Analysis

Vasiliki Simaki[1]([✉]), Iosif Mporas[2], and Vasileios Megalooikonomou[1]

[1] Multidimensional Data Analysis and Knowledge Management Laboratory,
Department of Computer Engineering and Informatics, University of Patras,
26500 Rio, Greece
{simaki, vasilis}@ceid.upatras.gr
[2] School of Engineering and Technology, University of Hertfordshire,
Hatfield, UK
i.mporas@herts.ac.uk

**Abstract.** In this article, we address the problem of age identification of Twitter users, after their online text. We used a set of text mining, sociolinguistic-based and content-related text features, and we evaluated a number of well-known and widely used machine learning algorithms for classification, in order to examine their appropriateness on this task. The experimental results showed that Random Forest algorithm offered superior performance achieving accuracy equal to 61%. We ranked the classification features after their informativity, using the ReliefF algorithm, and we analyzed the results in terms of the sociolinguistic principles on age linguistic variation.

**Keywords:** Text mining · Age identification · Text classification
Computational Sociolinguistics · Sociolinguistics

## 1 Introduction

The extensive expansion of the web and the plethora of options that social media provide, have resulted in the increase of the web users population. The daily use of the social media as a tool of communication and socialization with other users is a dominating reality, especially in the most developed countries. Twitter is one of the most popular media among men and women of different ages, with more than 255 million of active users per month, and 500 million Tweets sent per day. The automatic extraction of information from the everyday enormously growing volumes of data related not only to the twitter message itself but also to the gender, age and other demographic characteristics of the user are essential for e-government, security and e-commerce applications.

The age is the second most important social factor, after the author's gender, that can be easily extracted among the social media users, in the cases where the user declares his/her age explicitly (e.g. "Sharon, 33"), or when the user provides implicitly his/her age in his/her posts (e.g. "I was born in 1990"). The user's age is in strong correlation

© Springer International Publishing AG, part of Springer Nature 2018
A. Gelbukh (Ed.): CICLing 2016, LNCS 9624, pp. 385–395, 2018.
https://doi.org/10.1007/978-3-319-75487-1_30

with the user's language, as several sociolinguistic theories have proven [6]. Depending to the person's life stage, different linguistic attitudes and choices are observed, resulting the age linguistic variation. A basic principle that delimitates the language of adults from the "teens' language", is that adults use more standard types and normative structures than adolescents, who prefer neologisms (morphological, semantic, etc.), non-standard types and generally more unconventional language structures.

The age of social media users is an essential clue among the demographic information provided, which can be identified for several scientific, commercial and security purposes. Teens use Twitter often without being permitted to or supervised by adults, which could lead to unpleasant situations. It is thus important to identify the age category of social media users, in order to prevent harassment incidents.

Apart the minors' security, the detection of the user's age can be informative about the different trends, opinions, political and social views of each age group. This can enable social scientists to derive important clues about the anthropography among social media users, and how different age groups behave online. Market analysts and advertisers are also interested in such a study, in order to better promote a product or a service to an age group accordingly to their expressed interests and opinions.

In this article we attempt a multifaceted study including computational and theoretical aspects, bounded in the emerging field of Computational Sociolinguistics [11]. More specifically, we investigate the ability of several and dissimilar classification algorithms for the classification of Twitter users with respect to their age class. In order to evaluate them we relied on a large number of text-based features that have been reported in several publications related to age identification from text input. We perform then a feature ranking method, using the ReliefF algorithm, in order to detect the most efficient features during the classification process, and we try at the end, a sociolinguistic- driven interpretation of the most informative features in terms of the basic theories describing the age linguistic variation.

The rest of the paper is organized as follows: In Sect. 2 we describe the background work in the field of automatic age identification. In Sect. 3 we describe the evaluation methodology. Section 4 presents the experimental results. In Sect. 5 the most informative features are presented and discussed in analyzed according to existing sociolinguistic theories, and finally, in Sect. 6 we conclude this work.

## 2   Related Work

Several approaches have been proposed in the literature for the age class identification of documents. In [5] the evolution of the blogs' form during time and aims to predict the age of the users, based on their date of birth was studied. The authors observed that the size of the documents is a discriminative element and they also computed the percent of punctuation appearances, capital letters and spaces. Emoticons, acronyms, slang types, punctuation, capitalization, sentence length and part of speech tags were used, among others, as features in [18], where it was examined whether the online behavior of users can contribute effectively to the prediction of their age category. They observed that there are important and distinguishing changes in the writing style of bloggers before and after the social media expansion.

The stylistic differences in the writing style of bloggers according to their age were investigated in [8, 17, 19]. In [8, 19] they performed a stylometric analysis in terms of gender and age by using non-dictionary forms and the sentence length features. The slang, smileys, out-of-dictionary words, chat abbreviations, on the one hand, and the sentence length on the other, proved to be highly distinctive among different ages and gender, when combined with features proposed by earlier studies. The Naive Bayes classifier was used for the experiments. In [17] the authors focused on the use of slang words. They represented the co-occurrences of slang words that the bloggers use as a graph based model, where the slang words are the nodes and the edges stand for the number of co-occurrences.

Content-based and style-based characteristics were explored in [2, 3, 20]. In [2], one of the age-related conclusions was that the basic difference between older and younger bloggers is the extent to which their communication is outer or inner–directed. In [3], for the style-based features the researchers relied on parts of speech and function words. For the content-based features, they used the words that appear sufficiently enough in the dataset and that have a high value on the information gain measure. In [20] the authors created the "Blog Authorship Corpus", in order to identify the author's age and gender. They used style-related features (selected part-of-speech, function words, blog words and hyperlinks) and content-based characteristics in order to detect the gender and the age. They observed that specific forms and unigrams are more frequent in young bloggers, the blogging style and topics are different among 10's, 20's and 30's. In addition, lexical, syntactic and structural features were employed in [14], in order to identify the age and gender of the authors. Among other features, they relied on positive and negative words, stop words frequency, smiley word list and punctuation appearances. A decision tree classifier was used for classifying the author profile.

In [12] a study in language use among different age categories of Twitter users was performed. Their analysis proved that differences in style, references and conversation depended not only on the age category, but also on the life stage of the user.

Except as a categorical variable using age classes, age can be considered as a continuous variable also. In [13] the age identification task was performed using linear regression, using part of speech unigrams and bigrams and features from the LIWC [22], such as the percent of words having more than 6 letters.

## 3   Age Class Identification Methodology

For the evaluation of classification algorithms on the task of Twitter users' age identification we adopted a standard approach followed in most of the previous related work, i.e. preprocessing, feature extraction and classification structure was utilized, as illustrated in Fig. 1.

Specifically, each Twitter post is initially preprocessed. During pre-processing, each post is split into sentences and each sentence is split into words. Afterwards, three feature extraction methodologies are applied in parallel and independently to each other to each post. In detail, text mining, sociolinguistic based and context-based features are extracted constructing vectors $V_{TM}$, $V_{SL}$ and $V_{CT}$, respectively. These features are consequently concatenated to a super vector $V = V_{TM}||V_{SL}||V_{CT}$. This results to one

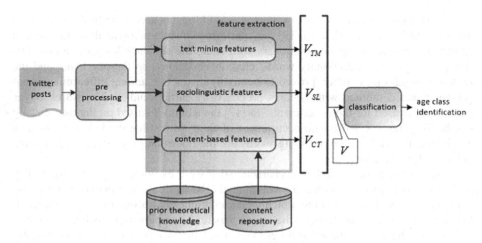

**Fig. 1.** Block diagram of the Twitter's posts age class identification scheme.

feature vector, $V$, per Twitter post, which is processed by a classification algorithm in order to label each post with an age class.

### 3.1    Feature Extraction

Three categories of features were computed for each Twitter post, namely the text mining features, the sociolinguistic-based features and the content-based features.

As considers the text mining features, they consist of statistical values in character and word level, used in several text classification tasks (authorship attribution, gender and age identification, genre classification), as presented in Table 1. These feature set is included in a vector $V_{TM}$, which has length equal to 40.

Regarding the second feature set, it consists of theoretical sociolinguistic markers of linguistic differentiation, which have been calculated as quantitative classification features, used in previous studies on gender classification [21]. The sociolinguistic features used in this evaluation are presented in Table 2, they are included in $V_{SL}$, which has length equal to 6.

For the third category, we implemented a number of features that are related to the age linguistic variation and could be useful in order to distinguish the writing style among the different age classes of the authors. As has been reported in [4, 15], with increasing age there is a more frequent use of future tense and fewer self-references. Therefore, these content-based features are the following: the *normalized number of future tense uses*, e.g. "will", "going to", "gonna"; the *normalized number of self-references* [16], e.g. "I", "me", "myself"; the *normalized number of hyperlink uses* [20]. The $V_{CT}$ feature vector contains the above content-based features and has length equal to 3.

The concatenation of the three feature vectors results to $V$, as described above. Thus, for each Twitter post one final feature vector, $V$, is constructed, which has length equal to $40 + 6 + 3 = 49$.

**Table 1.** The text mining classification features.

| Text mining features | |
|---|---|
| # of characters per tweet | standard deviation of the word length |
| normalized # of alphabetic characters | # of function words |
| normalized # of upper case characters | average # of sentences per paragraph |
| normalized # of digit characters | average # of characters per paragraph |
| normalized # of space characters | minimum word length |
| normalized # of tab ("\t") characters | normalized # of emoticons |
| # of occurrence of each alphabetic character | # of "hapax legomena" |
| # of adverbs | # of "hapax dislegomena" |
| total # of words | maximum word length |
| normalized # of words with length less than 4 characters | normalized # of words that start with a capital letter |
| normalized # of characters per word | normalized # of stop words |
| average word length | # of nouns |
| # of sentences | # of proper nouns |
| # of paragraphs | # of adjectives |
| # of lines | # of prepositions |
| average # of characters per sentence | # of verbs |
| average # of words per sentence | # of pronouns |
| normalized # of different words | # of interjections |
| # of articles | normalized # of occurrence of special characters ("@", |
| normalized # of words whose letters are all capital | "#", "$", "%", "&", "*", "~", "^", "–", "=", "+", ">", |
| # of punctuation symbols (".", ",", "!", "?", ":", ";", """, "\"") | "<", "[", "]", "{", "}", "\|", "\", "/") |

**Table 2.** The sociolinguistic classification features.

| Sociolinguistic features |
|---|
| syntactic complexity (normalized # of the sentence's words) |
| # of slang types per tweet |
| # of bad words per tweet |
| vocabulary richness (normalized # of different words per tweet without the stop words) |
| sentimental language (normalized # of positive and negative words per tweet according to SentiWordNet [7]) |

For the estimation of the above text features we used the NLTK [23] open-source toolkit.

### 3.2 Classification Algorithms

For the classification stage, we used a number of dissimilar machine learning algorithms, which are well studied and have extensively been used in several text classification tasks. In particular, we used:

- a multilayer perceptron neural network (MLP), using the back-propagation algorithm for training and three layers,
- the support vector machines (SVMs) using the sequential minimal optimization algorithm, which was tested using two different kernels, namely the radial basis kernel (rbf) and polynomial kernel (poly),
- four tree algorithms, namely the: pruned C4.5 decision tree (J48); the random tree (RandTree) constructing a tree that considers K randomly chosen attributes at each node; the random forest (RandForest) constructing a multitude of decision trees and the fast decision tree learner (RepTree) that builds a decision tree using information gain or variance and prunes it using reduced-error pruning with back-fitting.
- from the Bayesian classifiers, we used: the Bayes network learning (BayesNet) which is a probabilistic graphical model that represents a set of random variables and their conditional dependencies via a directed acyclic graph and the naive Bayes multinomial updateable (NBMU), in which feature vectors represent the frequencies with which certain events have been generated by a multinomial.
- two meta-classifiers: the Adaboost.M1, which is a boosting algorithm combined with Decision Stump and a bagging algorithm (Bagging) using a REPTree, aiming to reduce variance.

All classifiers were implemented using the WEKA toolkit [25]. For all algorithms, the free parameters that are not reported were kept in their standard values.

## 4    Age Class Identification Evaluation

The evaluation methodology for Twitter users' age class identification presented in the previous section was evaluated using the text-based features and the classifiers presented above. In order to avoid overlap between training and test subsets a 10-fold cross validation evaluation protocol was followed.

### 4.1 Dataset Description

For the present evaluation we collected and annotated posts from Twitter. Our dataset consists of 19,377 Twitter posts, written in English. The size of the corpus is 247,925 words. The number of the characters is equal to 1,486,681. The posts were divided in 6 age classes and each class corresponds to a different age range of the authors. The distribution of the age classes is tabulated in Table 3. The collection of the data includes Twitter posts from 46 different anonymous users.

**Table 3.** Distribution of the age classes.

| Age class | Age range | Number of posts |
|-----------|-----------|-----------------|
| A | 14–19 | 1,513 |
| B | 20–24 | 2,149 |
| C | 25–34 | 1,640 |
| D | 35–44 | 6,433 |
| E | 45–59 | 4,662 |
| F | >60 | 2,980 |

## 4.2  Results

The experimental results for the evaluated classification algorithms in terms of correctly classified web posts (i.e. accuracy), the proportion of actual positives which are correctly identified as such (i.e. sensitivity) and the proportion of negatives which are correctly identified as such (i.e. specificity) are tabulated in percentages in Table 4. The best performance for each of the above metrics is indicated in bold.

**Table 4.** Evaluation results for age class identification.

| Classifier | Accuracy | Sensitivity | Specificity |
|------------|----------|-------------|-------------|
| MLP | 57.63 | 57.60 | 86.40 |
| SVM-poly | 52.82 | 52.80 | 81.00 |
| SVM-rbf | 52.80 | 52.80 | 81.00 |
| J48 | 47.36 | 47.36 | 85.70 |
| RandForest | **61.00** | **60.60** | 84.40 |
| RandTree | 41.87 | 41.90 | 83.90 |
| Bayes Net | 42.61 | 42.60 | **86.50** |
| NBMU | 41.59 | 41.60 | 77.60 |
| REPTree | 50.65 | 50.78 | 82.90 |
| Bagging with REPTree | 56.56 | 56.60 | 84.30 |
| AdaBoost.M1 | 36.26 | 36.30 | 71.00 |

As can be seen in Table 4, the best performance both in terms of accuracy and sensitivity was achieved by the Random Forest algorithm. Specifically RandForest algorithm achieved 61% accuracy and 60.60% sensitivity, followed by the MLP neural network classifier with accuracy equal to 57.63% and the Bagging with accuracy equal to 56.56%. The worse performing algorithms are the NBMU, RandTree and BayesNet with accuracies equal to 41.59%, 41.87% and 42.61% respectively.

As can be observed from Table 4, the specificity performance is quite high for all the classifiers used and reaches up to 86.50% for the Bayes Net algorithm, i.e. there is a low percentage of false positive instances in the identification of the age class.

The discriminative algorithms that were evaluated, namely the multilayer perceptron neural network, the support vector machines using radial basis and polynomial kernel and the boosting algorithm, in average achieved an accuracy and sensitivity percent equal to 49.9% and a specificity percent equal to 79.8%. The tree algorithms, namely the J48, the random forest, the random tree, RepTree and Bagging, in average succeeded 51.5% in terms of accuracy and sensitivity and 84.2% in terms of specificity. The two probabilistic classifiers, Bayes Net and multinomial Naive Bayes, in average achieved accuracy and sensitivity equal to 42.1% and specificity equal to 82%. In average, the discriminative algorithms had the lowest value of specificity, while the tree algorithms had the highest specificity. Also, in average, as regards the accuracy and the sensitivity, the tree algorithms outperformed the rest of the algorithms, while the probabilistic classifiers had the lowest value in these metrics. The superior performance of the tree algorithms can be explained by the fact that they are robust, scalable and can perform well with large datasets, while the probabilistic classifiers are not as powerful when the dimensionality increases.

Finally, for the RandomForest and Bagging algorithms the results were quite good, while the AdaBoost algorithm performed poorly, since it returned the lowest values in all the three metrics that we examined. This can be explained by the fact that, although this algorithm may dramatically improve performance, sometimes over-fits. In contrast to this, over-fitting is not a problem for Random Forests, which are not very sensitive to outliers in training data. Their superiority is also owed to the fact that they do not need pruning trees and are easy to set parameters since accuracy and variable importance are generated automatically [25].

## 5    Feature Ranking and Sociolinguistic Analysis

After the classification experiments the competence of each feature was investigated, in order to highlight the most efficient and informative features and/or feature types for the age identification task. We used a Relief feature selection algorithm [9], which is heuristics-independent, noise-tolerant, robust to feature interactions and it runs in low-order polynomial time. In our case we used the updated ReliefF algorithm proposed by Koronenko [10], which improves the reliability of the probability approximation, it is robust to incomplete data, and generalized to multi-class problems. Our dataset was processed by the ReliefF algorithm, implemented using the WEKA machine learning toolkit [25], and feature ranking scores were estimated. The feature ranking results are tabulated in Table 5.

The ranking of the classification features demonstrates the importance and efficacy of the text mining features. Among the 20 first ranked features, 19 of them belong to the text mining category of characteristics. Only one content-based feature is highly ranked, in the third place, the normalized number of hyperlink uses. We observe though that the sociolinguistic features are not informative for the current task of the age identification. We may conclude that the characteristics based on theoretical markers of gender linguistic variation are not that significant for the age linguistic discrimination. Although the linguistic choice of slang types by teens appears to be a common ground among sociolinguists, in our experiments it is not detected as an important feature.

**Table 5.** The first 20-ranked classification features.

| Ranking | ReliefF score | Feature description |
|---------|---------------|---------------------|
| 1 | 0.0265742 | normalized # of words that start with a capital letter |
| 2 | 0.0153342 | normalized # of occurrence of special characters |
| 3 | 0.0131761 | normalized # of hyperlink uses |
| 4 | 0.0098577 | # of punctuation symbols |
| 5 | 0.0093385 | # of proper nouns |
| 6 | 0.0086022 | standard deviation of the word length |
| 7 | 0.007794 | normalized # of characters in capital |
| 8 | 0.007241 | hapax legomena |
| 9 | 0.0071615 | normalized # of short words |
| 10 | 0.0069192 | # of adjectives |
| 11 | 0.0065469 | # of verbs |
| 12 | 0.0063017 | maximum word length |
| 13 | 0.0062667 | # of nouns |
| 14 | 0.0060678 | hapax dislegomena |
| 15 | 0.0059304 | # of occurrence of each alphabetic character |
| 16 | 0.0056742 | average word length |
| 17 | 0.0054966 | # of acronyms |
| 18 | 0.0053539 | # of function words |
| 19 | 0.0053327 | # of pronouns |
| 20 | 0.0052823 | # of prepositions |

An important parameter to consider concerning the features selected is the text type of the dataset: our data collection consists of texts limited to 140 characters, a fact that influences the linguistic attitude of the user, by making different choices in morphological, lexical and syntactic level during formulating his message, in order to achieve the less semantic loss possible. This fact could explain the importance of features related to the differentiated use of proper nouns, acronyms and capitalized forms by the age classes, instead of other forms as stop words, articles, interjections, etc. that are probably eliminated as easily-meanings. The POS-tags features (verbs, adjectives, nouns, etc.) and their discriminative use by the different age categories, demonstrate that people of different ages produce different structures even in tweet level. The variation in lexical density (the contrast between function and content words in a text) proves that this marker is an important clue to be further investigated in age linguistic variation. Finally, the importance of the word length features evinces the theoretical markers discussed in [1], that teens use smaller lexical forms that adults, and short, semantic and other, neologisms.

Several of the most important features, according to their ReliefF scores, may be grouped in terms of their connection and the conclusions derived: the features that are related to the use of capitalized forms (1, 5, 7 according to their ranking), the features related to the sentence's lexical density (10, 11, 13, 17, 18, 19, 20) and the features related to the lexical forms' length (6, 9, 12, 16). There exists also a group of features that are related to the non-linguistic choices the user make: the use of special characters, punctuation, hyperlinks, and other characters, that may be connected and should be further examined. We observe though and differences in the use of hapax and dislegomena, but this might be a challenging task due to the ambiguity in interpreting the features: either it consists of neologisms and slang that the dictionaries used cannot recognize, which lead to a teen marker, or it consists of more complex forms that older people know and use.

## 6   Conclusion

In the present article we investigated the appropriateness of a number of machine learning algorithms for classification on the task of age class identification of web users. For the evaluation we relied on several text-mining, sociolinguistic and content-based features extracted from Twitter posts. The experimental results showed that Random Forest classification algorithm outperformed the rest of the evaluated algorithms in terms of accuracy and sensitivity, while the BayesNet had the best performance in terms of specificity. Moreover, the evaluation results showed that in general the decision tree algorithms perform well on the task of age class identification from web text. We used three different feature sets, each one inspired by different tasks, and we evaluated their efficacy during the classification experiments, using the ReliefF algorithm for the feature ranking process. The 20 most informative features are derived and analyzed, in terms of their efficiency in age linguistic variation. Sub-categories among the most important characteristics have arisen resulting to major differentiations between the different age groups in lexical density, capitalized forms, word length, and non-linguistic clues.

## References

1. Androutsopoulos, J.K., Georgakopoulou, A. (eds.): Discourse Constructions of Youth Identities, vol. 110. John Benjamins Publishing, Amsterdam (2003)
2. Argamon, S., Koppel, M., Pennebaker, J.W., Schler, J.: Mining the blogosphere: Age, gender and the varieties of self-expression. First Monday, **12**(9) (2007)
3. Argamon, S., Koppel, M., Pennebaker, J.W., Schler, J.: Automatically profiling the author of an anonymous text. Commun. ACM **52**(2), 119–123 (2009)
4. Barbieri, F.: Patterns of age-based linguistic variation in American English. J. Socioling. **12**(1), 58–88 (2008)
5. Burger, J.D., Henderson, J.C.: An exploration of observable features related to blogger age. In: AAAI Spring Symposium: Computational Approaches to Analyzing Weblogs, pp. 15–20, March 2006
6. Eckert, P.: Age as a sociolinguistic variable. In: The Handbook of Sociolinguistics, pp. 151–167 (1997)
7. Esuli, A., Sebastiani, F.: SentiWordNet: a publicly available lexical resource for opinion mining. In: Proceedings of LREC, vol. 6, pp. 417–422, May 2006
8. Goswami, S., Sarkar, S., Rustagi, M.: Stylometric analysis of bloggers' age and gender. In: Third International AAAI Conference on Weblogs and Social Media, March 2009
9. Kira, K., Rendell, L.: The feature selection problem: traditional methods and a new algorithm. In: AAAI (1992)

10. Kononenko, I.: Estimating attributes: analysis and extensions of RELIEF. In: Bergadano, F., De Raedt, L. (eds.) ECML 1994. LNCS, vol. 784, pp. 171–182. Springer, Heidelberg (1994). https://doi.org/10.1007/3-540-57868-4_57

11. Nguyen, D., Doğruöz, A.S., Rosé, C.P., de Jong, F.: Computational sociolinguistics: a survey. arXiv preprint arXiv:1508.07544 (2015)

12. Nguyen, D., Gravel, R., Trieschnigg, D., Meder, T.: "How old do you think i am?"; a study of language and age in Twitter. In: Proceedings of the Seventh International AAAI Conference on Weblogs and Social Media. AAAI Press (2013)

13. Nguyen, D., Smith, N.A., Rosé, C.P.: Author age prediction from text using linear regression. In: Proceedings of the 5th ACL-HLT Workshop on Language Technology for Cultural Heritage, Social Sciences, and Humanities, pp. 115–123. Association for Computational Linguistics, June 2011

14. Patra, B.G., Banerjee, S., Das, D., Saikh, T., Bandyopadhyay, S.: Automatic Author Profiling Based on Linguistic and Stylistic Features: Notebook for PAN at CLEF (2013)

15. Pennebaker, J.W., Stone, L.D.: Words of wisdom: language use over the life span. J. Pers. Soc. Psychol. **85**(2), 291 (2003)

16. Pfeil, U., Arjan, R., Zaphiris, P.: Age differences in online social networking–a study of user profiles and the social capital divide among teenagers and older users in MySpace. Comput. Hum. Behav. **25**(3), 643–654 (2009)

17. Prasath, R.R.: Learning age and gender using co-occurrence of non-dictionary words from stylistic variations. In: Szczuka, M., Kryszkiewicz, M., Ramanna, S., Jensen, R., Hu, Q. (eds.) RSCTC 2010. LNCS (LNAI), vol. 6086, pp. 544–550. Springer, Heidelberg (2010). https://doi.org/10.1007/978-3-642-13529-3_58

18. Rosenthal, S., McKeown, K.: Age prediction in blogs: a study of style, content, and online behavior in pre-and post-social media generations. In: Proceedings of the 49th Annual Meeting of the Association for Computational Linguistics: Human Language Technologies, vol. 1, pp. 763–772. Association for Computational Linguistics, June 2011

19. Rustagi, M., Prasath, R.R., Goswami, S., Sarkar, S.: Learning age and gender of blogger from stylistic variation. In: Chaudhury, S., Mitra, S., Murthy, C.A., Sastry, P.S., Pal, Sankar K. (eds.) PReMI 2009. LNCS, vol. 5909, pp. 205–212. Springer, Heidelberg (2009). https://doi.org/10.1007/978-3-642-11164-8_33

20. Schler, J., Koppel, M., Argamon, S., Pennebaker, J.W.: Effects of age and gender on blogging. In: AAAI Spring Symposium: Computational Approaches to Analyzing Weblogs, vol. 6, pp. 199–205, March 2006

21. Simaki, V., Aravantinou, C., Mporas, I., Megalooikonomou, V.: Using sociolinguistic inspired features for gender classification of web authors. In: Král, P., Matoušek, V. (eds.) TSD 2015. LNCS (LNAI), vol. 9302, pp. 587–594. Springer, Cham (2015). https://doi.org/10.1007/978-3-319-24033-6_66

22. Linguistic inquiry and word count. http://www.liwc.net/

23. http://www.nltk.org/

24. http://www.adweek.com/socialtimes/social-media-statistics-2014/499230

25. http://www.cs.waikato.ac.nz/ml/weka/

# Mining of Social Networks from Literary Texts of Resource Poor Languages

Pattabhi R. K. Rao and Sobha Lalitha Devi$^{(\boxtimes)}$

AU-KBC Research Centre, MIT Campus Anna University, Chennai, India
sobha@au-kbc.org

**Abstract.** We describe our work on automatic identification of social events and mining of social networks from literary texts in Tamil. Tamil belongs to Dravidian language family and is a morphologically rich language. This is a resource poor language; sophisticated resources for document processing such as parsers, phrase structure tree tagger are not available. In our work we have used shallow parsing for document processing. Conditional Random Fields (CRFs), a machine learning technique is used for automatic identification of social events. We have obtained an F-measure of 62% on social event identification. Social networks are mined by forming triads of the actors in the social events. The social networks are evaluated using graph comparison technique. The system generated social networks is compared with the gold network. We have obtained a very encouraging similarity score of 0.75.

**Keywords:** Social networks · Literary texts · Social event detection
Text mining · Resource poor languages · Tamil

## 1 Introduction

We describe our work on automatic identification of social events and mining of social networks from Tamil literary texts. Tamil is a Dravidian language and one of the popularly spoken languages in southern India and also in few other countries in South Asia such as Sri Lanka, Malaysia, and Singapore. It has a very rich heritage of literary texts and is considered as one of the old languages. In the present days mining of social networks from various sources such as social network websites, news articles, and micro blogs is one of the key research interests as well as business. Mining and analysis of social networks has been studied since early 20$^{th}$ century onwards. A social network consists of a set of social actors connected with each other. A social network is a social graph of people or organizations showing the connections or interactions or relationships between them. Social networks is an inter disciplinary field of study comprising sociology, psychology, anthropology, statistics, information sciences. Simmel [13] in his seminal work discussed about structural theory in sociology. He emphasized about groups and triads. Triads are a set of three social actors connected pair-wise to each other. In the last decade there has been greater research interest in the study of social networks analysis. Most of the research work towards social network mining has been from the perspective of social sciences where the sources of information were email conversations, speech conversational data, and interview questionnaires. One of the

© Springer International Publishing AG, part of Springer Nature 2018
A. Gelbukh (Ed.): CICLing 2016, LNCS 9624, pp. 396–405, 2018.
https://doi.org/10.1007/978-3-319-75487-1_31

widely used approaches was based on Name and word co-occurrence. A system called "Referral Web" was developed by Kautz et al. [9] which mines social networks from web. In their system the names in web pages are identified and the network of these names is mined using co-occurrence and links between names are identified. The co-occurrence information for names on the web is obtained using search engine queries results. Culotta et al. [5] present a system which extracts social networks from email messages and web documents. Their system identifies all unique person names in the email messages and finds person's web pages on the web. Connections are marked between the owner of the webpage and the names of person occurring in that web page.

The above described systems use co-occurrence information. Though we find most of the works are towards mining social networks using social media and mail information, there is huge treasure of social networks inside literary texts. Mining networks from literary texts help in better understanding of the literature and also aid in understanding of culture and interaction patterns in a discourse.

The first step towards mining social networks in literary texts is the detection of social events. Social events indicate interactions between actors. The basic idea is the more number of interactions or communications between persons indicate stronger connection between them. Event has been traditionally defined as an occurrence happening at a determinable time and place, with or without the participation of human agents. It may be a part of a chain of occurrences as an effect of a preceding occurrence and as the cause of a succeeding occurrence. Most of the works in event extractions have been to extract events such as appointments, calamities, accidents, crime etc. MUC series of information extraction events have been focused on extraction of such events. In this work we are interested in identifying social events, which describe events of interactions between two people. Identification of social events helps in mining social networks. Social networks as discussed earlier show connections or relations between persons. Hence in this work we focus on events involving interaction between two persons or groups of persons. Latest efforts have been to mine social networks from text by mining interactions between people expressed explicitly in the text [6, 8]. These approaches use only the interactions that are signaled by quotes. The other approaches used for event extraction are using pattern recognition, heuristic and linguistic rules as in traditional information extraction systems. Agarwal et al. [1–3] have worked on social networks extraction which is not restricted to interactions signalled by quoted speech.

The present work described in this paper uses the definition of social events as described in Agarwal et al. [2, 3]. The objective of our work is to automatically identify social events and their networks. In this work we differ substantially from the earlier works in the machine learning technique and in the feature space used for learning. We have worked on Tamil literary texts. In this paper we have the following major contributions:

(a) This work is a first of its kind in Indian languages.
(b) The methodology developed does not use parse tree structures and thus will be suitable for any less resource languages.
(c) We have used graph comparison technique from graph theory to evaluate the mined social networks. We are currently interested in measuring if the system has

mined the desired social network and not doing any deeper analysis of the network, we are not using other network analysis metrics.

In the literature we find that social networks are measured using various social network metrics such as graph density, connected components, degree centrality, and betweenness centrality [4, 7, 11, 12]. These metrics are useful when we do deeper social network analysis to identify which social actor is having prominent role, who is an influencer in the network. The goal of the present work is to automatically identify social events and thus mine social networks and have evaluation metrics which will give the performance of the automated system. Mining of social networks and their analysis has applications towards developing social network services targeting social network communities such as music groups, sports groups. The advent of easy affordable internet access has facilitated in lot of information exchange, sharing, formation of social network services. Social network services help in information exchange.

The paper is further organized as follows: In Sect. 2 we describe our work on social event identification, features used for training. In Sect. 3 we describe the work on mining social networks. In Sect. 4 we present the experiments and results. In Sect. 5 we conclude the paper with some future directions of research.

In our work there are two parts (a) Identifying Social Events (b) Mining of social networks using the detected social events. For the social event detection we have used precision, recall for evaluating and are described in Sect. 4.2. The maximal common sub-graph algorithm is used for formation of consolidated social network graph and not for evaluation (Sect. 4.3). The mined social network is evaluated using graph similarity technique (Sect. 4.3).

## 2    Social Events Identification

We have chosen a very popular contemporary Tamil literary work titled "Poniyin Selvan" authored by Kalki Krishnamoorthy during 1950. This is a historical novel. We have annotated our corpus following the schema described in Agarwal et al. [1, 2]. This defines a social event as an event in which two people interact such that at least one participant or social actor of the interaction is deliberate or conscious or fully aware of the interaction. A detailed description and formal backing on Social events is available in the [1]. Social network is a social graph of people or organizations showing the connections or interactions between them. Social events identify the interactions between entities. Thus by detecting the interactions we will be able to form grouping or network of these interactions, resulting in social networks. If we have to form social networks from literary texts identifying social events will be the first step.

We have used Conditional Random Fields (CRFs) [10], a machine learning technique to train and build our models. CRFs are feature based learning technique. We can provide linguistic based features and it is a discriminative and probabilistic modeling technique. This has been successfully used ML method. The input texts are preprocessed (shallow parsed) to obtain Morphological information, Part-of-Speech tag, Chunk information (Noun Phrases, Verb phrases etc.). Here we have made use of a

Named Entity Recognizer to identify Person entities (Individuals and groups). This task has two phases viz., training and testing. In training phase we use the manually annotated data with social events to build models. In the testing phase, an unseen text is given as input to the system for social event identification and markup. In the testing phase the models built during the training phase are used for the purpose of markup.

*Training Phase:* CRFs is a feature based learning algorithm. The feature identification and feature representation plays a very important role in learning by the system. Here we have used morphological properties such as case markers, tense markers along with POS, chunk information as features for learning and model files are built. Below we describe the feature space used for learning in detail. Features were identified based on the data analysis and exploiting the morphological richness of the Indian languages. We have come up with the feature space after performing various experiments with different feature combinations (feature engineering). The one described in the paper is most suitable feature space. The experiments for other texts or novels can be performed using the same feature space. The features and methodology are scalable for any literary texts. The feature space and the methodology used are not dependent on the Tamil language. These can be applied for any of the morphologically rich languages.

**Table 1.** Feature space used for learning

| Types | Features |
|---|---|
| Word level features | (a) Word – individual words or surface forms as they appear in the text |
| | (b) Root word – root form of the word in the text obtained from morphological analysis |
| | (c) POS tag – Part-of-Speech of the word |
| | (d) NE tag – Named entity tag of the word – whether Person or Non-person |
| | (e) Suffix – the suffix of the word |
| | (f) Suffix type – type of suffix such as accusative, dative, tense marker, PNG marker, clitic |
| Structural features | (a) Current word, its previous word, and its next word |
| | (b) Current Word's POS, previous word's POS, and next word's POS |
| | (c) Current Word, its POS, its NE tag, its suffix and suffix type |
| | (d) Current word, Previous words POS, Previous words NE |
| Positional features | (a) Number of words in the event text span |
| | (b) Number of chunks in the event text span |
| | (c) Event's start position – word count |
| | (d) Event's end position – word count |

Structural features and positional features help in the learning of the syntactic structure patterns in the sentences. These features help in improving the recall and maintaining good precision. In the section on experiments and results we will be seeing how the performance varies with respect to the different features used in learning.

*Testing Phase:* The input text is shallow parsed and the required features are extracted. The models built during the training phase are used to identify and markup the social events in the unseen input text.

## 3  Mining Social Networks

The social actors associated with each social event identified are formed as graphs as shown below:

[actor1] ----- >> (social event) ----- >> [actor2]

(Here the graph is represented in linear form, where nodes are represented inside square brackets [] and links are represented as parenthesis ().)

All such graphs are grouped or clustered based on the common actors in each such graph. We use a soft clustering methodology. The objective of using clustering for mining social networks is to group similar smaller actor-event-actor graphs based on the participating actors in an un-supervised manner. After the possible grouping are formed by the clustering algorithm, the complete social networks graph is constructed by merging the clusters formed based on the links or connections and the actors using graph operations of graph merging. We use Fuzzy C-means soft clustering algorithm. The main advantage of using soft clustering is that we are not going to have hard or rigid boundaries, the same data nodes can be in different groupings. One of the characteristics of social networks is that a person can be in associated with different communities or groups simultaneously. The associations are not mutually exclusive. Hence we need to identify all such network formations. Hence we go for soft clustering. For the purpose of mining complete social networks for a given literary text document, we merge different sub-graphs obtained in each cluster based on actors-edge overlap. Merging means common nodes and edges in each graph clusters are connected to form a single big network. The below graph example shows the merged output:

[actor1] ----- >> (social_event1) ----- >> [actor2]; [actor1] ----- >> (social_event2) ----- >> [actor4];
[actor1] ----- >> (social_event3) ----- >> [actor5];

After merging the output will be as follows:

[actor1] ----- >> (social_event1) ----- >> [actor2];
----- >> (social_even2) ----- >> [actor4];
----- >> (social_event3) ----- >> [actor5];

In the fuzzy C-means algorithm we have to specify the initial seed of the probable total number of clusters, "c". The "c" is set as (3* total number of social actors).

# 4 Experiments and Results

In this we present the experiments performed for the two tasks of social event detection and social networks mining separately. Before presenting the experiments and results a short description about the corpus is given below.

## 4.1 Corpus and Social Event Annotation

The corpus used for the experiments is "Ponniyin Selvan", a Tamil historical fiction novel written during 1950. This novel describes the political plot of events that have happened during the medieval period. The writing style of this novel is very unique; this is rich with humor and very simple. The narration of the story brings the readers live to those medieval eras of kings and describes the cultural and social aspects of that time in a very simple manner. There are mainly six characters or actors in this novel. Along with these six characters there are other nine characters. As said in the earlier section, we have annotated the text with social events following the annotation schema described in Agarwal et al. [2]. The original novel consists of six volumes. For the purpose of this work we have taken last 6 chapters in the last volume. The reason behind the choice of last six chapters is that these chapters contain very good number of interesting interactions between different characters and many of the events unfold in these chapters. This consists of 3500 sentences and 1155 social events. The corpus was annotated by a single annotator. The main challenge during the annotation was in determining the text span of the event. For example:

(1) Ta: muNNe   oru  murattu ciitaNai kaavarrkaaraNaaka vaiththirunthavar
         before    one  rough   disciple security+for        kept+3SM
         ippothu ippati       parinthu    upacariththu allaippathaRku yaarai
         now    this_manner reverence  courteously  call                who
         *amarththiyirukkiRaar* eNRa      eNNaththutaN   ulle
         position+past+3SM   that+3RP  thought+assoc  inside
         *piraveciththaarkall.*
         enter+past+3P

(Before a rough disciple was kept as a security person, now he has kept one who is calling with reverence and courteously who is that, with such thoughts they entered inside.)

Here there are two social events, triggered by "calling" and "thoughts". The span of the event triggered with "calling", could be "calling with reverence and courteously" or "now he has kept one which is calling with reverence and courteously". In this social event only one is aware, it is a perception event, where one of the actors perceives about other actor. In such instances we have considered the span of the event to be "calling with reverence and courteously" only. The choice of span of text is based on the optimum number of words required to express the event.

## 4.2   Social Event Detection

The annotated corpus is divided into training and testing partitions in the ratio of 80–20, where 80% is for training and 20% is for testing. The corpus is pre-processed using morphological analyser, part-of-speech tagger, chunker and named entity recognizer to extract the features for learning. Morphological analyser gives the morphological information such as root form, suffixes, and suffix type. Part-of-speech tagger gives the part-of-speech tag of each word, chunker mark ups the noun phrases, verb phrases. Named Entity recognizer identifies the named entities; here we are interested only in "Person" entities.

We have performed three experiments using different feature sets in training and thus built three different models. For each experiment we have performed 10-fold cross validation exercise. The baseline experiment is performed by considering the words and POS tags alone as features for learning. The Table 2, below shows the results of social event detection for different feature spaces.

**Table 2.**  Social event identification – evaluation results

| SNo | Features used | Precision | Recall | F-measure |
|-----|---------------|-----------|--------|-----------|
| 1 | Baseline – Words and POS tag are only used for learning | 36.54 | 74.65 | 49.06 |
| 2 | Word Level features (A) as described in Table 1 | 44.87 | 71.43 | 55.12 |
| 3 | All features (A), (B) and (C) as described in Table 1 | 54.34 | 70.42 | 61.34 |

The major challenge to the system is in the identification of social events where interactions include two different actors and not interactions with-in self or narrations. For example:

> (2) Ta: *kunthavaiyin  mukattil  kobam,*
>        kunthavai         face+loc  anger
>        *viyappu,      aiyam,    Attiram        mutaliya*
>        astonishment,   surprise, tension/anxiety etc
>        *vevveeRu  pAvankalY  toNRi           maRaintaNa.*
>        different   feelings      rise/show-up  vanished

(Kunthavai face had shown different feelings of anger, astonishment, surprise, anxiety, tension etc. and vanished.)

This appears as if it is a perception or cognitive event involving "Kunthavai" in cognition state, where the other actor is in the cognitive state of the other person. But actually this sentence is a narration by the author and is not an interaction between persons.

## 4.3  Mining Social Networks

Here we present the graph comparison algorithm used for comparing the mined social networks with the gold network graph. In the Sect. 3 we had described how the social networks are mined after the social events are detected. Let graph S1 be the social networks graph obtained by the system and let S2 be the gold network graph. The similarity between the two graphs is measured by the size of their maximal overlap graph or also called as maximum common sub-graph. The maximal common sub graph is obtained by building all possible common sub-graphs and then choosing the largest of all these sub-graphs. The problem of finding maximal common sub-graph is a well known NP-Hard problem and there is no known polynomial solution to this problem if we are finding exact solution. Generally all the algorithms for finding maximal common sub-graph provide approximate solution. The problem is modified to NP-Complete by modifying the problem to find maximal common induced sub-graphs. Here also we build the induced sub-graphs for both S1 and S2 and then identify the common sub-graphs. For this purpose, for graph S1 = (V, E), create a complete clique G1 = (V, E') such that E' = {(u, v) | u ! = v, for each u, v in V). Similarly for the gold network graph S2 = (V', E'), create a complete clique G2 = (V', E") such that E" = {(u, v) | u! = v, for each u, v in V'). Now the total number of vertices and edges that are common to G1 and G2 are identified to determine the similarity between the graphs S1 and S2. We have obtained 0.75 similarity score between the system obtained

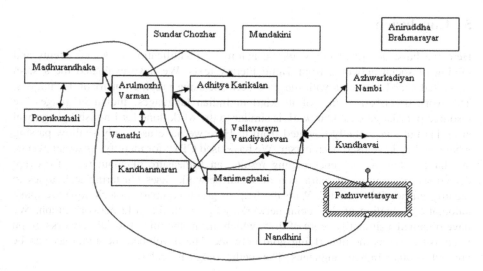

**Fig. 1.** System obtained social networks graph

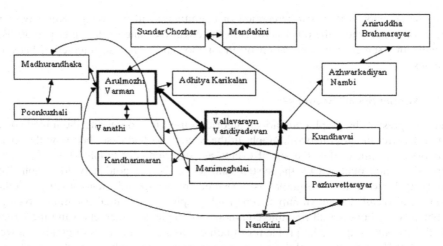

**Fig. 2.** Gold social networks graphs

graph and the gold network graph. The Figs. 1 and 2 show the network graph obtained by the system and the gold network respectively. It could be observed from the figures that the system could not identify the networks involving persons (or social actors) "Anirudha Brahmarayar", "Azhwarkadiyan Nambi", "Sundara Chozar" and "Mandakini". Since we are identifying connections based on the social events, the error in the social event detection has affected for few of the actors in the text. The system could not identify any interaction between these actors.

The thick arrows in the Figs. 1 and 2 indicate stronger connection between those two actors. The derived social network is an un-weighted, unlabeled graph.

## 5   Conclusion

Here we have described our work on automatic identification of social events and mining of social networks from Tamil literary texts. We have evaluated the system generated social networks with the gold network using graph comparison techniques. The work described is first of its kind in Indian languages. Indian languages are resource poor languages. Our work is similar to the work done for English by Agarwal et al. [1] but substantially deviated from it. Here we have used only shallow parsing where as they have used deep parser. We have used CRFs for identifying social events. We have obtained very encouraging results, an F-measure of about 62% for event detection. The mining or extraction of social networks uses clustering techniques to identify all possible "triads". With the help of graph comparison technique we make automatic evaluation of the social networks graph with the gold network graph. We have obtained a similarity score of 0.75, which means the mined social networks graph is almost same as the underlying gold network. The developed methodology can be applied to other Indian languages. The methodology is scalable.

In future we plan to apply this system for different Indian languages and build a multilingual system. We also plan to apply the models built for social event detection on other genres of Tamil texts such online news articles, blogs and micro blogs. Further we also aim to develop a detailed social networks analysis using different metrics to identify main actors in the text, their influence factor in the text. The larger goal is to create an application framework which would take a shallow parsed text as input and do all the necessary action to create a social networks graph automatically.

# References

1. Agarwal, A., Rambow, O.: Automatic detection and classification of social events. In: Proceedings of the 2010 Conference on Empirical Methods in Natural Language Processing, Cambridge, MA, pp. 1024–1034. Association for Computational Linguistics, October 2010
2. Agarwal, A., Rambow, O., Passonneau, R.J.: Annotation scheme for social network extraction from text. In: Proceedings of the Fourth Linguistic Annotation Workshop (2010)
3. Agarwal, A., Kotalwar, A., Zheng, J., Rambow, O.: SINNET: social network extractor from text. In: Proceedings of the 6th International Joint Conference on Natural Language Processing (IJCNLP 2013), Nagoya, Japan (2013)
4. Bavelas, A.: A mathematical model for group structures. Hum. Organ. **7**(3), 16–30 (1948)
5. Culotta, A., Bekkerman, R., McCallum, A.: Extracting social networks and contact information from email and the web. In: Proceedings of First Conference on Email and Anti-Spam (CEAS) (2004)
6. Elson, D.K., Dames, N., McKeown, K.R.: Extracting social networks from literary fiction. In: Proceedings of the 48th Annual Meeting of the Association for Computational Linguistics, pp. 138–147 (2010)
7. Freeman, L.C.: Centrality in social networks conceptual clarification. Soc. Netw. **1**(3), 215–239 (1979)
8. He, H., Barbosa, D., Kondrak, G.: Identification of speakers in novels. In: Proceedings of the 51st Annual Meeting of the Association for Computational Linguistics (2013)
9. Kautz, H., Selman, B., Shah, M.: Referral web: combining social networks and collaborative filtering. Commun. ACM **40**(3), 63–65 (1997)
10. Lafferty, J., McCallum, A., Pereira, F.: Conditional random fields: probabilistic models for segmenting and labeling sequence data. In: Proceedings of ICML, pp. 282–289 (2001)
11. Shaw, M.E.: Group structure and the behavior of individuals in small groups. J. Psychol. **38**(1), 139–149 (1954)
12. Shimbel, A.: Structural parameters of communication networks. Bull. Math. Biophys. **15**(4), 501–507 (1953)
13. Simmel, G.: The Sociology of Georg Simmel. The Free Press, New York (1950)

# Collecting and Annotating Indian Social Media Code-Mixed Corpora

Anupam Jamatia[1](✉) , Björn Gambäck[2] , and Amitava Das[3]

[1] National Institute of Technology, Agartala, Tripura, India
anupamjamatia@gmail.com
[2] Norwegian University of Science and Technology, Trondheim, Norway
gamback@idi.ntnu.no
[3] Indian Institute of Information Technology, Sri City, Andhra Pradesh, India
amitava.das@iiits.in

**Abstract.** The pervasiveness of social media in the present digital era has empowered the 'netizens' to be more creative and interactive, and to generate content using free language forms that often are closer to spoken language and hence show phenomena previously mainly analysed in speech. One such phenomenon is code-mixing, which occurs when multilingual persons switch freely between the languages they have in common. Code-mixing presents many new challenges for language processing and the paper discusses some of them, taking as a starting point the problems of collecting and annotating three corpora of code-mixed Indian social media text: one corpus with English-Bengali Twitter messages and two corpora containing English-Hindi Twitter and Facebook messages, respectively. We present statistics of these corpora, discuss part-of-speech tagging of the corpora using both a coarse-grained and a fine-grained tag set, and compare their complexity to several other code-mixed corpora based on a Code-Mixing Index.

**Keywords:** Social media text · Code-switching
Part-of-speech tagging

## 1 Introduction

In informal settings, such as in conversational spoken language and social media, and in regions where people are naturally bi- or multilingual (e.g., India), persons frequently alternate between the languages (codes) they have in common. When the code alternation/switching happens inside an utterance and below clause level, it is often referred to as *code-mixing*, while *code-switching* is the more general concept and most often refers to inter-clausal code alternation. We will here look at the tasks of collecting and annotating code-mixed English-Hindi and English-Bengali social media text. In contrast, most research on social media has concentrated either on completely monolingual text (in particular English tweets) or on text where code alternation occurs above the clause level.

© Springer International Publishing AG, part of Springer Nature 2018
A. Gelbukh (Ed.): CICLing 2016, LNCS 9624, pp. 406–417, 2018.
https://doi.org/10.1007/978-3-319-75487-1_32

Even though it previously was frown upon and regarded as dubious language usage, which in particular should be suppressed in language teaching, code-switching in conversational spoken language has been an acknowledged research theme in psycho- and socio-linguistics for half a century [13], and the ability to freely switch between languages and to build parallel language models is nowadays mostly seen as an asset for the individual, also in educational settings. However, code alternation in conventional text is not very prevalent, so even though the first work on applying language processing methods to code-switched text was carried out in the early 1980s [19], it was only with the increase of social media text that the phenomenon started to be studied more thoroughly within computational linguistics [22].

Here we will concentrate on the collection and annotation of these types of code-mixed social media texts. We have created three corpora consisting of Facebook chat messages and tweets that include all possible types of code-mixing diversity: varying number of code alternation points, different syntactic mixing, alternating language change orders, etc. The rest of the paper is organized as follows: in Sect. 2, we discuss the background and related work on social media text processing and code-switching. The collection and annotation of the code-mixed corpora are described in Sect. 3. Section 4 then discusses the issue of annotating the corpora with utterance breaks, while Sect. 5 targets annotation with part-of-speech tags. Section 6 compares the complexity of our corpora to several other code-mixed corpora. Section 7 then sums up the discussion.

## 2   Social Media and Code-Switching

The pervasiveness of social media—such as mails, tweets, forums, comments, and blogs—in the present digital era has empowered the 'netizens' to be more creative and interactive, and to generate content using free language forms that often are closer to spoken language and hence show phenomena previously mainly analysed in speech. In all types of social media, the level of formality of the language depends more on the style of the writer than on the media as such, although in general tweets (Twitter messages) tend to be more formal than chat messages in that they more often follow grammatical norms and use standard lexical items [18], while chats are more conversational [23], and hence less formal. Because of the ease of availability of Twitter, most previous research on social media text has focused on tweets; however, the conversational nature of chats tend to increase the level of code-mixing [6], so we have collected data both from Twitter and from Facebook posts.

Notably, social media in itself does not constitute a particular textual domain; we use the term 'social media text' as referring to the way these texts are communicated, rather than to a specific type of text. Indeed, there is a wide spectrum of different types of texts transmitted through social media, and the common denominator of social media text is not that it is 'noisy' or informal, but that it describes language in (rapid) change [1]. Although social media indeed often convey more ungrammatical text than more formal writings, the relative occurrence of non-standard English syntax tends to be fairly constant across several types of social media [2].

However, while the first works on social media concentrated on monolingual English texts, recent years has witnessed an increased interest in the study of non-English texts and of texts in a mix of languages, as shown by the shared task on word-level language detection in code-switched text [28] organized by the workshop on Computational Approaches to Code Switching at the 2014 Conference on Empirical Methods in Natural Language Processing (EMNLP), and the shared tasks on information retrieval from code-mixed text held at that the 2014 and 2015 workshops of the Forum for Information Retrieval Evaluation, FIRE [27]. Here we are in particular concerned with code-mixed social media text involving Indian languages. So though Diab and Kamboj [11] briefly explained the process of corpus collection and suggested crowd sourcing as a good method for annotating formal (non-social media) Hindi-English code-mixed data, the first Indian code-mixing social media text corpus (Bengali-Hindi-English) was reported by Das and Gambäck [7] in the context of language identification, while Bali *et al.* [3] argued that structural and discourse linguistic analysis is required in order to fully analyse code-mixing for Indian languages. Gupta *et al.* [17] discussed the phenomenon in the context of information retrieval (calling this 'mixed-script information retrieval'), applying deep learning techniques to the problem of identifying term equivalents in code-mixed text.

## 3   Data Collection

In order to create representative code-mixed corpora, we have collected text both from Facebook and Twitter: 500 raw tweets for English-Bengali (EN-BN), as well as 4,435 tweets and 1,236 Facebook posts for English-Hindi (EN-HI). The EN-BN tweets were mainly collected from celebrity twitter handles such as @monalithakur03, @sujoy_g, @rituparnas11, etc., and by using queries like

> "football" AND "khela"; "election" AND "kobe";
> "tumi" AND "chara"; "kichu" AND "ekta".

The EN-HI tweets were on various 'hot' topics (i.e., topics that are currently being discussed in news, social media, etc.) and collected with the Java-based Twitter API,[1] while the EN-HI Facebook posts were collected from campus-related university billboard postings on the Facebook "Confession" page[2] of the Indian Institute of Technology, Bombay (a predominantly Hindi-speaking university, since 95% of the students come from all over India). The posts on this page are mainly of the form of one longer story (a "confession" about something a student did on campus) followed by several shorter chat-style comments. The 'confession' posts tend to be written in more formal language and mainly in English with some Hindi mixed in, while the comments are more informal in style and freely mix English and Hindi.

All the 500 EN-BN tweets and 1,106 randomly selected EN-HI messages (552 Facebook posts and 554 tweets) were singled out for manual annotation.

---

[1] http://twitter4j.org/.
[2] www.facebook.com/Confessions.IITB.

**Table 1.** Token level language distribution (%)

| Source | Tokens | English | Hindi | Bengali | Univ | NE | Acro | Mixed | Undef |
|--------|--------|---------|-------|---------|------|-----|------|-------|-------|
| *English–Hindi* | | | | | | | | | |
| Facebook | 16,281 | 75.61 | 4.17 | – | 16.41 | 2.19 | 1.47 | 0.02 | 0.13 |
| Twitter | 10,886 | 22.24 | 48.48 | – | 21.54 | 6.70 | 0.88 | 0.08 | 0.07 |
| Total | 27,167 | 54.22 | 21.93 | – | 18.47 | 3.99 | 1.23 | 0.05 | 0.11 |
| *English–Bengali* | | | | | | | | | |
| Twitter | 38,223 | 40.45 | 2.63 | 34.05 | 18.96 | 2.86 | 0.83 | – | 0.22 |

Those messages were annotated automatically with language tags using Barman's system [4], and then checked manually using a customized GUI-based system. 230 (20.8%) of the messages were identified as monolingual whereas the rest were bilingual. The token level language distributions of the corpora are reported in Table 1, where 'Univ' stands for language independent symbols such as punctuation marks, 'NE' are named entities, and 'Mixed' are tokens showing code-mixing down at the character level (i.e., word internal). Most problematic for the annotation were tokens that are ambiguous between the languages, for example, words such as 'to', 'in', 'may/main' can be used in both Hindi and English. However, such ambiguities can normally be resolved by inspecting the context.

Note that the EN-HI Facebook posts are predominantly written in English, with 94.8% of the language specific tokens being English (making up 75.6% of all the tokens of the corpus), while the EN-HI tweets mainly are in Hindi (68.6% of the language specific tokens, and 48.5% of all the tokens). However, in the EN-BN Twitter corpus, English narrowly is the main language, represented by 52.4% of the language specific tokens and 40.5% of the total EN-BN corpus.

## 4    Tokenization and Utterance Boundary Insertion

Utterance boundary detection and tokenization can potentially be extra difficult in social media text due to its noisy nature. The CMU tokenizer [16], was used for the latter task; although it originally was developed for English, empirical testing showed this tokenizer to work reasonably well also for Indian languages.

Two annotators were employed for the task of manual utterance boundary insertion for English-Hindi corpus. At the beginning, the inter-annotator agreement on utterance breaks was 71%. In a second round, both annotators looked at the non-agreed cases and discussed those among themselves to reach an 86% agreement level. In addition, there were almost 8% cases where the annotators after discussion agreed on a third possibility. So finally, after discussions and corrections, the agreement between the annotators was 94%.

The following are two examples of tweets where the annotators disagreed. In both cases one of the annotators wanted to keep the original tweet, while the other wanted to insert an utterance boundary (after the URL in Tweet 1 and before *well* in Tweet 2).

**Tweet 1.** `I liked a @YouTube video http://t.co/Y9edo1yfRN Don`
`- khaike paan banaras wala old and new mix`

**Tweet 2.** `Aakir India south Africa KO world cup me jeet he`
`gaya well done India team`

The resulting EN-HI corpus has in total 2,583 utterances: 1,181 from Twitter and 1,402 from Facebook (compared to the 554 resp. 552 messages before boundary insertion). Notably, 876 of the original 1,106 selected messages (79.2%) were deemed multilingual, but after the utterance break insertion only 821 of the 2,583 utterances identified in these messages (31.8%) were judged to be multilingual. This sharp decrease in code-mixing when measured at the utterance level rather than message level shows the importance of the utterance boundary insertion. Tweet 3 is an example of this: initially, the entire tweet can be viewed as bilingual. However, after boundary insertion, only the first of the three resulting utterances contains code-mixing.

**Tweet 3.** `Yadav bhaiya good pace ! Bahut badiya` 😊 `aisa he`
`gola feko #IndvsSA #CWC15`

**U1** `Yadav bhaiya good pace !`
**U2** `Bahut badiya` 😊
**U3** `aisa he gola feko #IndvsSA #CWC15`

Utterance boundary detection for social media text is in general quite challenging and has not been discussed in detail previously. The main reason might be that much work on social media has been on tweets, that are limited to 140 characters and hence the whole tweet can be approximated to be one utterance. However, when working with Facebook chats, we found several long messages, with a high number of code alternation points.

## 5   Part-of-Speech Tagsets

Just as sentence boundary detection, part-of-speech (POS) tagging can be extra problematic in the context of social media. In order to create automatic POS taggers, annotated code-mixed data is needed. The English-Hindi corpora were thus part-of-speech tagged using both a coarse-grained and a fine-grained tagset. As can be seen in Table 2, the coarse-grained tagset is based on a combination of the eight Twitter specific tags introduced by Gimpel *et al.* [16] with the twelve tags in Google's Universal Tagset [24]. Google's Universal Tagset is a complete set by itself, but adding the Twitter specific tags [16] makes sense when addressing social media text, so we prefer to have a merged POS tagset.

**Table 2.** Part-of-speech tagsets

| Coarse-grained | | Fine-grained | |
| --- | --- | --- | --- |
| Tag | Description | Tag | Description |
| G_N | Noun | N_NN | Common Noun |
| | | N_NNV | Verbal Noun |
| | | N_NST | Spatio-temporal |
| | | N_NNP | Proper Noun |
| G_PRP | Pronoun | PR_PRP | Personal |
| | | PR_PRL | Relative |
| | | PR_PRF | Reflexive |
| | | PR_PRC | Reciprocal |
| | | PR_PRQ | Wh-Word |
| G_V | Verb | V_VM | Main |
| | | V_VAUX | Auxiliary |
| G_J | Adjective | JJ | Adjective |
| G_R | Adverb | RB_ALC | Locative Adverb |
| | | RB_AMN | Adverb of Manner |
| G_PRE | Demonstrative Adposition (Pre-/Postposition) | PSP | Pre-/Postposition |
| | | DM_DMD | Absolute |
| | | DM_DMI | Indefinite |
| | | DM_DMQ | Wh-word |
| | | DM_DMR | Relative |
| G_NUM | Quantifier | $ | Numeral |
| | | QT_QTF | General |
| | | QT_QTC | Cardinal |
| | | QT_QTO | Ordinal |
| G_PRT | Particle | RP_RPD | Default |
| | | RP_NEG | Negation |
| | | RP_INTF | Intensifier |
| | | RP_INJ | Interjection |
| G_SYM | Punctuation | SYM | Symbol |
| | | PUNC | Punctuation |
| G_CONJ | Conjunction | CC | Conjunction |
| G_DT | Determiner | DT | Determiner |
| G_X | Foreign Unknown Echo word Twitter-Specific (Gimpel et al., 2011) [16] | RDF | Foreign Word |
| | | UNK | Unknown |
| | | ECH | Echo Word |
| | | @ | At-mention |
| | | ~ | Re-Tweet/discourse |
| | | E | Emoticon |
| | | U | URL or email |
| | | # | Hashtag |

The mapping between our fine-grained tagset and the Google Universal Tagset is also shown in Table 2. The fine-grained tagset includes both the Twitter specific tags and a set of POS tags for Indian languages that combines the IL-POST tagset [5] with two tagsets developed, respectively, by the Indian

Government's Department of Information Technology (TDIL)[3] and the Central Institute of Indian Languages (LDCIL),[4] that is, an approach similar to that taken for Gujarati by Dholakia and Yoonus [10]. Combining all the three tagsets was necessary since some tags (e.g., 'numeral') are not in the TDIL tagset and were borrowed from the LDCIL tagset. The twitter-specific tags [16] are shown in the gray fields in the table and were thus used in both our tagsets.

To test the feasibility of using the tagsets, the Hindi-English corpora were annotated manually by one annotator using a custom GUI-based system. It was observed that specially for code-mixed text, the original lexical category of an embedded word often is lost in the context of the different languages of the corpus. So part-of-speech label prediction has to be based on the function of a token in a given context, as opposed to its de-contextualized lexical classification.

## 6   Measuring Corpora Complexity

An issue which is particularly interesting when comparing code-mixed corpora to each other, is the complexity of the code-mixing, that is, the level of mixing between languages. Both Kilgarriff [20] and Pinto *et al.* [25] discussed several statistical measures that can be used to compare corpora more objectively, but those measures presume that the corpora are essentially monolingual.

Debole and Sebastiani [9] analysed the complexity of the different subsets of the Reuters-21578 corpus in terms of the relative hardness of learning classifiers on the subcorpora, a strategy which does not assume monolinguality in the corpora. However, they were only interested in the relative difficulty and give no measure of the complexity as such. So, due to the mixed nature of our corpora, we will here instead adopt the Code-Mixing Index, $C$ of Gambäck and Das, first introduced in [8,14], but extended and detailed in [15].

This code-mixing measure is defined both at the utterance level ($C_u$) and over an entire corpus ($C_c$), and in short works as follows: if an utterance only contains language independent tokens, there is no mixing, so $C_u = 0$. For other utterances, $C_u$ is calculated by counting $N$, the number of tokens that belong to any of the languages $L_i$ in the utterance (i.e., all the tokens except for language independent ones) minus the ratio of tokens belonging to *the matrix language*, the most frequent language in the utterance, $\max_{L_i \in \mathbb{L}}\{t_{L_i}\}$, with $\mathbb{L}$ being the set of all languages in the corpus (and $1 \leq \max\{t_{L_i}\} \leq N$):

$$C_u(x) = \begin{cases} \dfrac{N(x) - \max\limits_{L_i \in \mathbb{L}}\{t_{L_i}\}(x)}{N(x)} & : N(x) > 0 \\ 0 & : N(x) = 0 \end{cases} \qquad (1)$$

Notably, for mono-lingual utterances $C_u = 0$ (since then $\max\{t_{L_i}\} = N$).

However, in addition to the number of tokens from the matrix language, the number of code alternation points ($P$) inside an utterance should also be taken

---

[3] www.tdil-dc.in/tdildcMain/articles/780732DraftPOSTagstandard.pdf.
[4] www.ldcil.org/Download/Tagset/LDCIL/6Hindi.pdf.

into account, since a higher number of language switches in an utterance arguably increases its complexity. In [15] we discuss how this additional information can weighted into the $C_u$ measure in general, but here we will assume equal weights assigned to the number of code alternation points per token and to the ratio of tokens belonging to the matrix language, giving Eq. 2:

$$C_u(x) = 100 \cdot \frac{N(x) - \max_{L_i \in \mathbb{L}}\{t_{L_i}\}(x) + P(x)}{2N(x)} \tag{2}$$

Again, $C_u = 0$ for monolingual utterances (since then $\max\{ti\} = N$ and $P = 0$).

**Table 3.** Code mixing and alternation points

| $C_u$ | English-Hindi | | | | | | English-Bengali | |
| | FB | | TW | | Total | | TW | |
| Range | (%) | $P_{avg}$ | (%) | $P_{avg}$ | (%) | $P_{avg}$ | (%) | $P_{avg}$ |
|---|---|---|---|---|---|---|---|---|
| $[\,0\,]$ | 84.59 | – | 48.18 | – | 67.94 | – | 79.85 | – |
| $(\,0, 10\,]$ | 4.07 | 1.74 | 2.88 | 1.44 | 3.52 | 1.63 | 1.37 | 1.64 |
| $(\,10, 20\,]$ | 4.99 | 2.06 | 15.41 | 1.82 | 9.76 | 1.89 | 5.59 | 2.13 |
| $(\,20, 30\,]$ | 3.57 | 2.28 | 14.90 | 2.43 | 8.75 | 2.40 | 6.42 | 2.41 |
| $(\,30, 40\,]$ | 1.57 | 2.14 | 11.18 | 2.67 | 5.96 | 2.60 | 4.14 | 3.19 |
| $(\,40, \infty\,)$ | 1.21 | 2.29 | 7.45 | 2.81 | 4.07 | 2.72 | 2.63 | 3.17 |

Table 3 shows the distribution of our corpora over ranges of code-mixing values, $C_u$ and average number of code alternation points, $P$. Interestingly, the EN-HI Twitter corpus has a higher percentage of mixed utterance ($C_u > 0$) than the Facebook one, while the number of code alternation points fairly steadily is around 2, also for utterances with a high level of mixing. The lower level of mixing in the Facebook data set might apparently contrast with the hypothesis that chat messages tend to increase the level of code-mixing. However, the explanation for this is quite certainly the nature of the posts on the IIT Bombay "Confession" page, as described at the beginning of Sect. 3.

Furthermore, $C_u$ only addresses code-alternation at the utterance level and does not account for code-alternation between utterances, nor for the frequency of code-switched utterances, that is, the number ($S$) of utterances that contain any switching divided by the total number ($U$) of utterances in the corpus. Incorporating these factors give the formula for calculating $C_c$, the $C$ measure at corpus level, as shown in Eq. 3:

$$C_c = \frac{100}{U}\left[\frac{1}{2}\sum_{x=1}^{U}\left(1 - \frac{\max_{L_i \in \mathbb{L}}\{t_{L_i}\}(x) - P(x)}{N(x)} + \delta(x)\right) + \tfrac{5}{6}S\right] \tag{3}$$

where $\delta(x)$ is 1 if a code-alternation point precedes the utterance, and 0 otherwise. The 5/6 weighting of $S$ (the number of utterances containing switching) comes from the classical 'Reading Ease' readability score [12], where Flesch similarly weighted the frequency of words per sentence as 1.2 times the number of syllables per word, based on psycho-linguistic experiments.

To evaluate the level of language mixing in our corpora, we compared their complexity to that of the English-Hindi corpus of Vyas et al. [29], the Dutch-Turkish of Nguyen and Doğruöz [21], and the corpora from the shared tasks at the EMNLP 2014 code switching workshop and at FIRE. The EMNLP corpora mix English with Spanish, Mandarin Chinese and Nepalese. A forth EMNLP corpus is dialectal: Standard Arabic mixed with Egyptian Arabic. The FIRE corpora mix English with Hindi, Gujarati, Bengali, Kannada, Malayalam and Tamil. However, the 2014 EN-KN, EN-TA and EN-ML corpora are small and inconsistently annotated, so those are not reliable as basis for comparison, and have thus been excluded here. The corpora from FIRE 2015 were not language tagged and thus not included either. The sizes and token level language distributions of the external corpora are shown in Table 4, where the values can be compared to the token distributions of our corpora, as given in Table 1.

**Table 4.** Token level language distribution of the external corpora (%)

| Languages | | Source | Tokens | Lang1 | Lang2 | Univ | NE | Mixed | Other |
|---|---|---|---|---|---|---|---|---|---|
| EN | HI | Vyas | 6,979 | 54.85 | 45.01 | – | – | – | 0.15 |
| EN | HI | FIRE | 23,967 | 44.11 | 38.58 | 17.27 | – | 0.04 | – |
| EN | BN | | 20,660 | 41.60 | 35.11 | 20.52 | – | 0.08 | 2.69 |
| EN | GU | | 937 | 5.02 | 94.98 | – | – | – | – |
| DU | TR | Nguyen | 70,768 | 41.50 | 36.98 | 21.52 | – | – | – |
| EN | ES | EMNLP | 140,746 | 54.78 | 23.52 | 19.34 | 2.07 | 0.04 | 0.24 |
| EN | ZH | | 17,430 | 69.50 | 13.95 | 5.89 | 10.60 | 0.07 | – |
| EN | NE | | 146,056 | 31.14 | 41.56 | 24.41 | 2.73 | 0.08 | 0.09 |
| ARB | ARZ | | 119,317 | 66.32 | 13.65 | 7.29 | 11.83 | 0.01 | 0.89 |

Table 5 shows the mixing for all the measured corpora, both over all utterances ($U$) and over only the utterances having a non-zero $C_u$ (i.e., those containing some code-mixing, $S$). The $C_u$, $P$ and $\delta$ columns give the average values. The final column ($C_c$) gives the total code-mixing value for each corpus. The values for the FIRE EN-HI corpus stand out in several of the column, but closer inspection reveals that that corpus contains too many errors and inconsistencies to useful for comparison. Instead, it is clear that the EMNLP English-Nepalese corpus exhibits a very high code-mixing complexity, as do our EN-HI Twitter corpus and the EMNLP English-Chinese corpus. More than half of the utterances in those three corpora also contain code-mixing. It is also interesting to note that the corpora from Vyas et al. [29] and from Nguyen and Doğruöz [21] show the highest level of inter-utterance code-switching.

**Table 5.** Code-switching levels in some corpora

| Language | | Source | utter. | switched | | $C_u$ | | $P$ | | $\delta$ | $C_c$ |
|---|---|---|---|---|---|---|---|---|---|---|---|
| Pair | | | (U) | (S) | (%) | (U) | (S) | (U) | (S) | (U) | |
| EN | BN | TW | 4,297 | 866 | 20.15 | 8.34 | 41.39 | 0.51 | 2.54 | 22.09 | 25.14 |
| EN | HI | TW | 1,181 | 612 | **51.82** | **21.19** | 40.89 | 1.19 | 2.30 | 30.99 | **64.38** |
| EN | HI | FB | 1,402 | 216 | 15.41 | 3.92 | 25.47 | 0.32 | 2.05 | 6.70 | 16.76 |
| EN | HI | FB + TW | 2,583 | 828 | 32.06 | 11.82 | 36.87 | 0.72 | 2.24 | 17.81 | 38.53 |
| EN | HI | Vyas | 671 | 160 | 23.85 | 11.44 | 47.98 | 0.53 | 2.24 | **53.50** | 31.31 |
| EN | HI | FIRE | 700 | 561 | 80.14 | 34.02 | 42.45 | 4.79 | 5.98 | 40.29 | 100.80 |
| EN | BN | | 700 | 165 | 23.57 | 11.44 | 48.53 | 1.70 | **7.19** | 44.14 | 31.08 |
| EN | GU | | 150 | 32 | 21.33 | 6.64 | 31.13 | 0.39 | 1.81 | 1.33 | 24.42 |
| DU | TR | Nguyen | 3,065 | 512 | 16.70 | 7.41 | 44.34 | 0.29 | 1.74 | **48.87** | 21.33 |
| EN | ES | EMNLP | 11,400 | 3,272 | 28.70 | 11.02 | 38.40 | 0.49 | 1.71 | 13.83 | 21.97 |
| EN | ZH | | 999 | 527 | **52.75** | 16.82 | 31.88 | 0.96 | 1.83 | 22.32 | **60.78** |
| EN | NE | | 9,993 | 7,274 | **72.79** | **31.03** | 42.63 | 1.95 | 2.67 | 35.18 | **91.69** |
| ARB | ARZ | | 5,839 | 1,005 | 17.21 | 5.21 | 30.29 | 0.21 | 1.22 | 13.29 | 19.56 |

# 7 Conclusion

The paper has reported work on collecting corpora of code-mixed English-Hindi and English-Bengali social media text (Twitter and Facebook posts), annotating them with languages at the word level, with utterance breaks, and with parts-of-speech tags, using both a coarse-grained and a fine-grained tagset.

The main contributions of this work are the creation of an annotated dataset of code-mixed Indian social media data. In addition to other the problems with annotating code-mixed text with language and part-of-speech tags, utterance boundary detection for social media text is a challenging task which has not been discussed in detail previously. Notably, the level of utterance-internal code alternation can decrease drastically if utterance boundaries are inserted into tweets and Facebook messages. This can make a major difference for the complexity of the code mixing, in particular for the often longer Facebook posts, and we have carried out some pilot experiments on training machine learners for automatic utterance boundary detection in our code-mixed corpora [26].

**Acknowledgements.** Thanks to the different researchers who have made their datasets available: the organisers of the shared tasks on code-switching at EMNLP 2014 and in transliteration at FIRE 2014 and FIRE 2015, as well as Dong Nguyen and Seza Doğruöz (respectively University of Twente and Tilburg University, The Netherlands), and Monojit Choudhury and Kalika Bali (both at Microsoft Research India). Thanks also to an anonymous reviewer for extensive and useful comments.

# References

1. Androutsopoulos, J.: Language change and digital media: a review of conceptions and evidence. In: Kristiansen, T., Coupland, N. (eds.) Standard Languages and Language Standards in a Changing Europe, pp. 145–159. Novus, Oslo (2011)
2. Baldwin, T., Cook, P., Lui, M., MacKinlay, A., Wang, L.: How noisy social media text, how diffrnt social media sources? In: Proceedings of the 6th International Joint Conference on Natural Language Processing, pp. 356–364. AFNLP, Nagoya, Japan, October 2013
3. Bali, K., Sharma, J., Choudhury, M., Vyas, Y.: "I am borrowing ya mixing?": An analysis of English-Hindi code mixing in Facebook. In: Proceedings of the 1st Workshop on Computational Approaches to Code Switching, pp. 116–126. ACL, Doha, Qatar, October 2014
4. Barman, U., Wagner, J., Chrupała, G., Foster, J.: DCU-UVT: word-level language classification with code-mixed data. In: Proceedings of the 1st Workshop on Computational Approaches to Code Switching, pp. 127–132. ACL, Doha, Qatar, October 2014
5. Baskaran, S., Bali, K., Bhattacharya, T., Bhattacharyya, P., Choudhury, M., Jha, G.N., Rajendran, S., Saravanan, K., Sobha, L., Subbarao, K.: A common parts-of-speech tagset framework for Indian languages. In: Proceedings of the 6th International Conference on Language Resources and Evaluation, pp. 1331–1337. ELRA, Marrakech, Marocco, May 2008
6. Cárdenas-Claros, M.S., Isharyanti, N.: Code switching and code mixing in internet chatting: between "yes", "ya", and "si" a case study. J. Comput.-Mediat. Commun. 5(3), 67–78 (2009)
7. Das, A., Gambäck, B.: Code-mixing in social media text: the last language identification frontier? Traitement Automatique des Langues 54(3), 41–64 (2013)
8. Das, A., Gambäck, B.: Identifying languages at the word level in code-mixed Indian social media text. In: Proceedings of the 11th International Conference on Natural Language Processing, pp. 169–178, Goa, India, December 2014
9. Debole, F., Sebastiani, F.: An analysis of the relative hardness of Reuters-21578 subsets. J. Am. Soc. Inf. Sci. Technol. 58(6), 584–596 (2005)
10. Dholakia, P.S., Yoonus, M.M.: Rule based approach for the transition of tagsets to build the POS annotated corpus. Int. J. Adv. Res. Comput. Commun. Eng. 3(7), 7417–7422 (2014)
11. Diab, M., Kamboj, A.: Feasibility of leveraging crowd sourcing for the creation of a large scale annotated resource for Hindi English code switched data: a pilot annotation. In: Proceedings of the 9th Workshop on Asian Language Resources, pp. 36–40. AFNLP, Chiang Mai, Thailand, November 2011
12. Flesch, R.: A new readability yardstick. J. Appl. Psychol. 32(3), 221–233 (1948)
13. Gafaranga, J., Torras, M.C.: Interactional otherness: towards a redefinition of codeswitching. Int. J. Biling. 6(1), 1–22 (2002)
14. Gambäck, B., Das, A.: On measuring the complexity of code-mixing. In: Proceedings of the 1st Workshop on Language Technologies for Indian Social Media, Goa, India, pp. 1–7, December 2014
15. Gambäck, B., Das, A.: Comparing the level of code-switching in corpora. In: Proceedings of the 10th International Conference on Language Resources and Evaluation. ELRA, Portorož, Slovenia, May 2016 (to appear)

16. Gimpel, K., Schneider, N., O'Connor, B., Das, D., Mills, D., Eisenstein, J., Heilman, M., Yogatama, D., Flanigan, J., Smith, N.A.: Part-of-speech tagging for Twitter: annotation, features, and experiments. In: Proceedings of the 49th Annual Meeting of the Association for Computational Linguistics, vol. 2, pp. 42–47. ACL, Portland, Oregon, June 2011

17. Gupta, P., Bali, K., Banchs, R.E., Choudhury, M., Rosso, P.: Query expansion for mixed-script information retrieval. In: Proceedings of the 37th International Conference on Research and Development in Information Retrieval, ACM SIGIR, Gold Coast, Queensland, Australia, pp. 677–686, July 2014

18. Hu, Y., Talamadupula, K., Kambhampati, S.: Dude, srsly?: The surprisingly formal nature of Twitter's language. In: Proceedings of the 7th International Conference on Weblogs and Social Media. AAAI, Boston, Massachusetts, July 2013

19. Joshi, A.K.: Processing of sentences with intra-sentential code-switching. In: Proceedings of the 9th International Conference on Computational Linguistics. ACL, Prague, Czechoslovakia, pp. 145–150, July 1982

20. Kilgarriff, A.: Comparing corpora. Int. J. Corpus Linguist. 6(1), 97–133 (2001)

21. Nguyen, D., Doğruöz, A.S.: Word level language identification in online multilingual communication. In: Proceedings of the 2013 Conference on Empirical Methods in Natural Language Processing, pp. 857–862. ACL, Seattle, Washington, October 2013

22. Paolillo, J.C.: Language choice on soc.culture.punjab. Electron. J. Commun./La Revue Electronique de Communication 6(3), n3 (1996)

23. Paolillo, J.: The virtual speech community: social network and language variation on IRC. J. Comput.-Mediat. Commun. 4(4), JCMC446 (1999)

24. Petrov, S., Das, D., McDonald, R.T.: A universal part-of-speech tagset. CoRR abs/1104.2086 (2011). http://arxiv.org/abs/1104.2086

25. Pinto, D., Rosso, P., Jiménez-Salazar, H.: A self-enriching methodology for clustering narrow domain short texts. Comput. J. 54(7), 1148–1165 (2011)

26. Rudrapal, D., Jamatia, A., Chakma, K., Das, A., Gambäck, B.: Sentence boundary detection for social media text. In: Proceedings of the 12th International Conference on Natural Language Processing, Trivandrum, India, pp. 91–97, December 2015

27. Sequiera, R., Choudhury, M., Gupta, P., Rosso, P., Kumar, S., Banerjee, S., Naskar, S.K., Bandyopadhyay, S., Chittaranjan, G., Das, A., Chakma, K.: Overview of FIRE-2015 shared task on mixed script information retrieval. In: Proceedings of the 7th Forum for Information Retrieval Evaluation, Gandhinagar, India, pp. 21–27, December 2015

28. Solorio, T., Blair, E., Maharjan, S., Bethard, S., Diab, M., Gohneim, M., Hawwari, A., AlGhamdi, F., Hirschberg, J., Chang, A., Fung, P.: Overview for the first shared task on language identification in code-switched data. In: Proceedings of the 1st Workshop on Computational Approaches to Code Switching, pp. 62–72. ACL, Doha, Qatar, October 2014

29. Vyas, Y., Gella, S., Sharma, J., Bali, K., Choudhury, M.: POS tagging of English-Hindi code-mixed social media content. In: Proceedings of the 2014 Conference on Empirical Methods in Natural Language Processing, pp. 974–979. ACL, Doha, Qatar, October 2014

# Turkish Normalization Lexicon
# for Social Media

Seniz Demir(✉), Murat Tan, and Berkay Topcu

TUBITAK-BILGEM, Kocaeli, Turkey
{seniz.demir,murat.tan,berkay.topcu}@tubitak.gov.tr

**Abstract.** Social media has its own evergrowing language and distinct characteristics. Although social media is shown to be of great utility to research studies, varying quality of written texts degrades the performance of existing NLP tools. Normalization of texts, transforming from informal to well-written texts, appears to be a reasonable preprocessing step to adapt tools trained on different domains to social media. In this study, we compile the first Turkish normalization lexicon that sheds light to the kinds of observed lexical variations in social media texts. A graphical representation acquired from a text corpus is used to model contextual similarities between normalization equivalences and the lexicon is automatically generated by performing random walks on this graph. The underlying framework not only enables different lexicons to be generated from the same corpus but also produces lexicons that are tuned to specific genres. Evaluation studies demonstrated the effectiveness of induced lexicon in normalizing Turkish texts.

**Keywords:** Text normalization · Graph-based lexicon · Turkish

## 1 Introduction

A growing body of research has been sought to unravel reasons behind the outbreak of social media usage for personal communication. Evidently, this outbreak mediates an upsurge in the amount of user generated social media content where the language significantly differs from formal languages. For instance, its very common to use non-standard words (e.g., "reallly baaaaaad"), slang words (e.g., "Y'all" for "you all"), abbreviations (e.g., "DAE" for "does anyone else"), phonetic substitutions (e.g., "4u" for "for you"), and emoticons in social media texts. Although some ill-formed words are typos or consequences of writing in a second language, some are intentionally produced in order to add extra pragmatic information to the content, save time, or meet the constraints of the media being used (e.g., text length limit in Twitter). Moreover, the language of a social media genre has its own characteristics which make it different from that of other genres (e.g., Blogs, Twitter, and SMS) [1]. It is widely accepted that social media language has been evolving with the addition of new words and language structures. Modeling linguistic changes and factors (e.g., gender, demographic background,

© Springer International Publishing AG, part of Springer Nature 2018
A. Gelbukh (Ed.): CICLing 2016, LNCS 9624, pp. 418–429, 2018.
https://doi.org/10.1007/978-3-319-75487-1_33

age, and time) that influence the way how people express themselves in social media is an active research area [2–4].

Everyday, millions of multilingual texts are produced in social media. This continuously growing real-time written data is a great resource for NLP studies. But, there is a major obstacle that hinders existing NLP tools (e.g., morphological parsers and named entity recognizers) from processing social media texts without a performance drop. NLP tools that are trained on formal and properly written texts often experience accuracy drops due to the use of non-standard words, improper capitalizations, ungrammatical phrases, and incomplete sentences [5,6]. To cope with the informal nature of language, these tools could be retrained with labeled data from social media domain. An alternative cost effective approach is to normalize texts by removing noise from their raw content before being processed. Normalization, as a preprocessing step, is of great importance to NLP tools where non-standard out of vocabulary (OOV) words are transformed into their standard in vocabulary (IV) forms (e.g., "u must be goin" to "you must be going").

In this work, we present the first publicly available Turkish normalization lexicon and describe the underlying framework that enables automatic generation of similar lexicons for different social media genres. The framework draws upon the work of Hassan and Menezes on unsupervised lexicon induction [7]. Words that are observed in a corpus of social media texts and their shared contexts are represented with a bipartite graph. Random walks are performed on this graph in order to explore normalization equivalences by utilizing their contextual similarities. A random walk starts from an OOV word and the IV word where this walk ends is considered as its normalization candidate. A metric that assesses contextual and lexical similarities between words is used to determine the entries of the induced lexicon from among all identified normalization candidates. The normalization lexicon contains entries in the form of a single OOV word along with its candidate IV forms (e.g., "aaba araba{car}|acaba{I wonder}")[1].

The underlying framework is designed to induce a Turkish normalization lexicon from unlabeled raw texts and labeled data is not required in any of the processing steps. Although a unique graph is built from a given text corpus, different lexicons could be produced from the same graph due to randomness in graph traversal. Moreover, the framework allows bipartite graphs to be adapted to informal writing characteristics of a specific social media genre by gathering input texts only from that medium. We envision that future normalization studies would benefit from our lexicon as an integral look-up table. In addition, the lexicon could be used to generate word confusion sets in a baseline system where the best normalized form is selected by a decoder. In our evaluation studies, we built such a normalization baseline using a Viterbi decoder along with a language model and assessed the effectiveness of our lexicon in normalizing Turkish

---

[1] Since the graph representation is for modeling contextual similarities between individual words, OOV words that contain more than one word due to omitted spaces (e.g., "şarkısözü" which is indeed "şarkı sözü{lyrics}") are manually removed. Automatic handling of these cases is left as future work.

tweets. The results showed that our baseline system achieved a normalization accuracy of 68–85%.

The rest of this paper is organized as follows. Section 2 discusses previous research on text normalization. Section 3 describes how a collected text corpus is represented as a bipartite graph along with the details of the lexicon induction framework. Section 4 presents the results of our evaluation studies and Sect. 5 concludes the paper.

## 2   Related Work

The literature of text normalization spans a variety of different techniques. Earlier approaches based on noisy channel model have used posterior probabilities to find the most probable standard form of an ill-formed word [8–11]. These approaches had some limitations such as the need for annotated data and proper categorization of non-standard words (e.g., abbreviation and spelling error). Approaches that addressed normalization as another form of machine translation from noisy to well-formed texts have also suffered from the requirement of hand annotated data [12,13]. Liu et al. [14] have developed a letter transformation approach where the generation of non-standard words from standard words is modeled at the character level. This work has been later enhanced with the incorporation of visual priming and string/phonetic similarity [15] in order to achieve a broader coverage. A recent approach of Han et al. [16] has addressed the normalization of short messages by first generating a confusion set of IV word candidates for an OOV word with the use of edit distance at phonemic transcription level. The best alternative is then selected by considering lexical string similarity and contextual features.

The motivation behind graph-based normalization approaches is that non-standard OOV words and their standard IV forms share similar contexts and these similarities can be captured via graphical representations. Hassan and Menezes [7] have tackled the normalization problem by building a bipartite graph from 73 million Twitter statuses and 50 million well-formed sentences and inducing a normalization lexicon from that graph. By utilizing the induced lexicon and a Viterbi decoder, their normalization approach has achieved a precision of 92.43% on English texts. Sönmez and Özgür [17] have used a word-association graph to represent relative positions of words to each other in written texts and their POS tags. Their unsupervised approach which takes contextual, grammatical, and lexical features of words into account has achieved a 94.1% precision on a well-known shared dataset of English texts.

Unfortunately, a few recent studies have focused on Turkish text normalization. In the cascaded approach [18], a non-standard OOV word is passed through seven components, each of which targets a specific kind of transformation (e.g., vowel restoration and accent normalization components). The normalization process ends when an IV word candidate is generated by any of these components. The approach has achieved an accuracy of 71% on a test set of 600 tweets. The most recent model [19] has comprised of a variety of techniques (e.g., lexical similarity and language model based contextual similarity) to handle different kinds of Turkish normalization problems. A precision of 80% on the average was reported on a set of Turkish tweets.

# 3   Lexicon Generation Framework

Previous normalization approaches have benefited from external resources such as slang word dictionaries and transliteration tables [10,17,18]. In this study, we developed a framework which produces a Turkish normalization lexicon from a given corpus of social media texts. The framework utilizes both contextual and lexical similarities between OOV words and IV normalization candidates in determining the entries of the lexicon.

For the corpus, a large amount of noisy and clean Turkish texts were collected from different resources. The corpus consisted of tweets (∼11 GB) retrieved via Twitter Streaming API[2] from April to October 2015, 20 million publicly available tweets[3], and clean Turkish texts (∼6 GB). The corpus was preprocessed by first discarding non-Turkish content as identified by a language identifier. Later, tweet-specific terms (e.g., # and RT), URLs, some punctuations, and repetitive characters were cleaned from the corpus. Finally, the remaining sentences of ∼9 GB were tokenized into 7,401,321 distinct words. For instance, the following shows a sentence before and after the preprocessing step:

*Before:* RT @iyiTweet: "Sen yeter ki iççinden de olsaaaa bir seni seviyorum de; benim kulaklarım çınlasın kafi...." /Cemal Süreya/

*After:* Sen yeter ki içinden de olsa bir seni seviyorum de; benim kulaklarım çınlasın kafi. Cemal Süreya

## 3.1   Graph-Based Representation

Given a text corpus, our framework represents the observed words and their contexts with a bipartite graph. In the graph, the first bipartite represents words (either IV or OOV) and the second bipartite represents all contexts where these words appear at least once. Any n-gram word sequence that has the word at its center is considered as a context of that word[4]. For instance, consider that 5-gram word sequences are extracted from the corpus (e.g., $w_1w_2w_3w_4w_5$). In this case, all center words (e.g., $w_3$) are used to build the first bipartite. The contexts, each of which consists of the two words on the left and the two words on the right of a center word (e.g., $w_1w_2w_4w_5$), are used to build the second bipartite. In the graph, a node that represents a center word (**word node**) is connected to all nodes (**context node**) that represent its contexts with undirected edges. The weight of an edge is the co-occurrence count of the corresponding word and

---

[2] twitter4j.org.

[3] http://www.kemik.yildiz.edu.tr/?id=28.

[4] The punctuation characters are omitted while identifying n-gram sequences.

context in the corpus. The degree of a context node hence represents the number of center words that appear within that context.

In this work, we constructed two graphs from the collected corpus with different word sequence lengths. Sequences of length three were used to generate the first graph (**Graph$_1$**) whereas 5-gram sequences were used for the second graph (**Graph$_2$**). The word sequences might contain one or more OOV words as determined by the Turkish morphological analyzer Zemberek[5]. As opposed to Hassan and Menezes [7], we used a word sequence even if it contains more than one OOV word or the only OOV word in the sequence is not at its center. Despite increasing the number of word and context nodes in the graph, this enables additional similarities between distinct word sequences to be captured. For instance, the similarity between 5-gram sequences ($IV_1$, $OOV_2$, $OOV_3$, $IV_4$, $IV_5$) and ($IV_1$, $OOV_2$, $IV_3$, $IV_4$, $IV_5$) where $IV_3$ is the normalized form of $OOV_3$ is encoded in our graphs.

The use of different sequence lengths resulted in two graphs with distinct characteristics as shown in Table 1. For instance, the highest degree of a word node was increased while that of a context node was dramatically decreased in Graph$_2$[6]. The highest edge weight in Graph$_1$ was 1,093,711 whereas was 57,984 in Graph$_2$. In addition, the majority of word and context nodes in both graphs had a degree of less than 5, which showed the sparsity of the collected data.

**Table 1.** Statistics about bipartite graphs

| | Bipartite | Number of nodes | Highest degree | Average degree | Average edge weight |
|---|---|---|---|---|---|
| Graph$_1$ (*3-gram*) | Word | 6,018,928 | 8,636,735 | 78.09 | 1.86 |
| | Context | 237,061,995 | 48,369 | 1.98 | |
| Graph$_2$ (*5-gram*) | Word | 5,122,873 | 14,445,174 | 114.56 | 1.18 |
| | Context | 567,078,012 | 2,547 | 1.03 | |

## 3.2 Lexicon Generation

We aim to produce a normalization lexicon by identifying IV word candidates that are contextually and lexically similar to OOV words. In our representation, this turns into the problem of finding which pairs of word nodes in the graph are similar in both respects. We describe contextual similarity between word nodes in terms of shared contexts, which correspond to context nodes with a degree of at least two in the graph. We assume that the more contexts are shared by the same word pair, the more contextually similar these words are. We also assume that words which are connected through a path of shared contexts show

---

[5] https://github.com/ahmetaa/zemberek-nlp.

[6] Stop words (e.g., "ve"{and}) and very frequent words (e.g.,"bir"{one}) were observed to have higher degrees.

more contextual similarities than those with no such paths. In order to explore contextually similar word pairs, our framework performs random walks over the bipartite graph as described in [7].

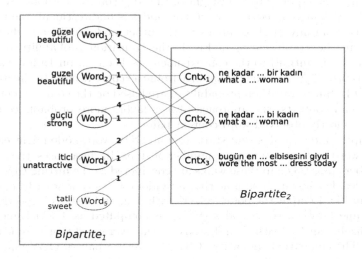

**Fig. 1.** A subgraph of Graph2 (5-gram word sequences)

A random walk starts from an OOV word node (a node that represents an OOV word) in the first bipartite. Each walk consists of an even number of sequential steps taken from a word node to a context node and vice versa[7]. At each step, the next node to be traversed is determined according to transition probabilities from the current node. Only one node is randomly selected from among the alternative nodes with highest transition probability. A random walk ends when either an IV word node (a node that represents an IV word) is reached or the maximum number of steps is taken without visiting an IV word node. Visited IV word node is identified as a candidate normalization node (NN(a)) of the OOV word node (node A) from where this walk starts.

For instance, consider the graph shown in Fig. 1 which has two shared contexts (i.e., $Cntx_1$ and $Cntx_2$) and two OOV word nodes (i.e., $Word_2$ and $Word_5$). Assume that, starting from the OOV word node "guzel", two random walks $Word_2$-$Cntx_1$-$Word_1$ (with two steps) and $Word_2$-$Cntx_2$-$Word_5$-$Cntx_2$-$Word_3$ (with four steps) are performed. After these walks, the IV word nodes $Word_1$ and $Word_3$ ("güzel" and "güçlü") are identified as candidate normalization nodes of the OOV word node "guzel".

The transition probability from node A to node B ($TP_{ab}$) is defined as the proportion of the co-occurrence count of these two nodes to all co-occurrences of node A with its neighbors in the graph:

---

[7] In a bipartite graph, a step cannot be taken between the nodes of the same bipartite.

$$TP_{ab} = Weight_{ab} \Big/ \sum_{\substack{x \in \\ Neighs(a)}} Weight_{ax} \tag{1}$$

$Weight_{ab}$ corresponds to the weight of the edge connecting nodes A and B in the graph. Although an edge between two nodes is undirectional, the transition probability from any of these nodes to the other node might be different. For instance, in the graph shown in Fig. 1, the transition probability from word node "güzel" (beautiful) to the context node "ne kadar ... bir kadın" (what a ... woman) is 0.78 (7/9) but the probability from the context node to the word node is 0.5 (7/14). Since transition probabilities are based on the co-occurrence counts of words and contexts, the approach always favors more probable connections and hence mostly observed word-context pairs.

Multiple random walks that start from an OOV word node might end at the same IV word node with a varying number of steps. In these cases, the average of the number steps taken in these walks are considered as the hitting time between these nodes. For instance, assume that 3 random walks that start from the node "guzel" in Fig. 1 end at the node "güzel" with 4, 2, and 6 steps respectively. The hitting time between these nodes ($HT_{ab}$) is computed as 4 which means that node "güzel" can be reached in 4 steps on the average starting from the node "guzel". The contextual similarity ($CS_{ab}$) between a pair of OOV and IV word nodes is computed as follows:

$$CS_{ab} = HT_{ab} \Big/ \sum_{\substack{x \in \\ NN(a)}} HT_{ax} \tag{2}$$

The lexical similarity between word nodes ($LS_{ab}$) is computed using two well-known metrics, namely the longest common subsequence ratio ($LCSR_{ab}$) and the edit distance ($ED_{ab}$):

$$LS_{ab} = LCSR_{ab} \Big/ ED_{ab} \tag{3}$$

$$LCSR_{ab} = LCS_{ab} \Big/ Max\langle Length(a), Length(b)\rangle \tag{4}$$

The normalization similarity score of word nodes ($NS_{ab}$) is the weighted sum of their contextual and lexical similarity scores. For each OOV word, our approach populates the lexicon with top N IV word candidates according to normalization similarity scores:

$$NS_{ab} = \lambda_1 \times CS_{ab} + \lambda_2 \times LS_{ab} \tag{5}$$

Randomness plays an important role while traversing the graph (i.e., choosing what to visit next). We observed that lexicons induced from the same graph in different trials coincide but might not be exactly the same. Even with a fixed number of random walks and a maximum step size, we observed differences between the identified IV word candidates for OOV words. The importance given to contextual and lexical similarities while computing the normalization similarity also affects the induced lexicon. Since our goal is to generate an effective (does not need to be the most comprehensive) normalization lexicon, these factors do not hinder what we want to achieve. However, as an attempt to visit at least some of the unvisited nodes in a single trial, two lexicons were generated from both Graph$_1$ and Graph$_2$[8]. In both cases, 100 random walks were performed from each OOV word node with a maximum of 6 steps. From among identified IV word candidates for the OOV word, the words with top 3 highest normalization similarity scores were selected. For each graph, the first lexicon was built by giving equal importance to contextual and lexical similarities (i.e., $\lambda_1 = 0.5$ and $\lambda_2 = 0.5$). However, in generating the second lexicon, the lexical similarity was favored (i.e., $\lambda_1 = 0.4$ and $\lambda_2 = 0.6$). Unfortunately, the approach didn't find IV word candidates for some OOV words in both trials[9].

The lexicons produced from the same graph were melded into a single lexicon, namely **Lexicon$_{3gram}$** for Graph$_1$ and **Lexicon$_{5gram}$** for Graph$_2$ by merging IV words candidates of the same OOV word. A lexical entry contains an OOV word along with one or more IV word candidates separated with a '|' character.

| | |
|---|---|
| berebere | berabere{*draw*} |
| insanlğın | insanlığın{*of humanity*} | insanların{*people's*} |
| fimlerin | dillerin{*of languages*} | filmlerin{*of films*} | isimlerin{*of names*} |

We observed that the entries in these lexicons differ in terms of coverage (i.e., the number of OOV words), the average number of IV word candidates per OOV word, and the identified IV word candidates for an OOV word. Although Lexicon$_{3gram}$ has significantly more lexical entries than Lexicon$_{5gram}$, some IV word candidates that appear in Lexicon$_{5gram}$ are not found in Lexicon$_{3gram}$.

Lexicon$_{3gram}$: aarştırma    araştırma{*research*} | araştırması{*his-her research*}
Lexicon$_{5gram}$: aarştırma    araştırma{*research*} | uygulama{*application*}

## 3.3   Lexicon Filtering

It was our observation that some OOV words and their IV word candidates, though being close in contextual use, are not quite similar lexically (e.g., the OOV word "aarştırma" and the IV word "uygulama"). Thus, we filtered from the lexicons all IV word candidates whose edit distance to corresponding OOV word is above a threshold[10]. As a final postprocessing step, we manually removed some lexical entries such as meaningless OOV words that consist of two characters (e.g., "cg") or more than one word (e.g., "geliyonmu"{are you coming?}).

---

[8] More than two trials could be made in order to reduce the effect of randomness.
[9] How these cases can be handled is indeed in our future work.
[10] In our evaluations, an edit distance of 2 was used.

## 4   Evaluation

We conducted two evaluation studies in order to assess the effectiveness of an induced lexicon in normalizing Turkish social media texts. In these studies, only one-to-one word mappings were evaluated and all OOV words were assumed to be identified beforehand. Since Lexicon$_{5gram}$ has less number of entries compared to Lexicon$_{3gram}$, it was used in both experiments.

In the first study, we evaluated the appropriateness and coverage of our IV word candidates as normalized forms. Our goal was to explore whether or not an IV word candidate from our lexicon is used to normalize corresponding OOV word by other normalizers. To the best of our knowledge, there is only one publicly available Turkish normalization system [18] that can be used for our comparisons in addition to a dictionary based spell checker. Neither the normalization system nor the spell checker utilizes the context of an OOV word during normalization. Thus, they are appropriate for comparing our look-up entries. For the study, we randomly selected 400 OOV words from our lexicon, half of which has only one IV word candidate whereas the remaining words have two or more candidates. We normalized these words (as a stand-alone word) using the Turkish normalization system and the MsWord spell checker. In a similar setting, we asked a native Turkish speaker to normalize these OOV words (without any given context) in order to obtain their normalized forms as our gold standard. Table 2 presents the percentage of agreement between parties on the normalized forms of OOV words. The first row represents the agreements for OOV words with a single IV word candidate whereas the second row represents those with more than one candidates.

**Table 2.** Results of the first evaluation

|  | Lexicon human | Lexicon norm. sys. | Lexicon spell C | Norm. sys. human | Norm. sys. spell C | Spell C. human |
|---|---|---|---|---|---|---|
| Single IV word | **88.5%** | 50% | 70% | 51.5% | 51% | 69% |
| Multiple IV words | **83.5%** | 57% | 62.5% | 36% | 43% | 37.5% |

The results showed that in at least half of the cases, an IV word from our lexicon was used by the Turkish normalization system to normalize the corresponding OOV word. On the other hand, our lexicon contained the IV word that was used by the human annotator in more than 83.5% of the cases. In at least 62.5% of the cases, the spell checker ranked an IV word from our lexicon as the most probable correction. Moreover, in majority of the remaining cases, IV word candidates from our lexicon were in the top 5 suggestions of the spell checker. These results are promising in terms of the appropriateness of IV word candidates as the normalized forms of OOV words in our lexicon.

In the second evaluation, we assessed whether an induced normalization lexicon can be used to generate word confusion sets in a baseline normalization system. This requires a look-up to the corresponding entry if the lexicon contains one IV word candidate for a non-standard word. However, in other cases, all IV word candidates should be used to fill in the confusion set and the most probable IV word candidate should be chosen as the best candidate using a language model and a Viterbi decoder. For this evaluation, we created such an in-house normalization system. For each of 340 randomly selected OOV words from our lexicon, we retrieved from Twitter a tweet which contains that word as the only OOV word. Our lexicon contained one IV word candidate for half of these OOV words and more than one IV words for the remaining OOV words. All tweets were normalized using our in-house normalization baseline. Two native Turkish speakers were given the original tweets and their normalized forms and asked to determine whether tweets are properly normalized or not. The participants were told to rate normalized tweets as 1 (properly normalized) or 0 (not properly normalized).

**Table 3.** Results of the second evaluation

|  | Participant₁ | Participant₂ | Both |
|---|---|---|---|
| Single IV word | 88% | 87% | 85% |
| Multiple IV words | 69% | 71% | 68% |

As shown in Table 3, the first participant agreed that our system properly normalized 88% of the sentences which contain an OOV word with one IV word candidate. However, once a selection has to be made using a language model (trained on well-written texts), the agreement was dropped to 69%. A similar drop (from 87% to 71%) was observed with the second participant. In 95% of the time, participants had an agreement on their ratings (both assigned 1 or 0 to normalizations). In those cases, participants approved at least 68% of the normalized tweets. The results showed that our induced lexicons could arguably be used as part of a normalization baseline for Turkish.

## 5  Conclusion

In this study, we developed a framework which produces a normalization lexicon from a given corpus of Turkish social media texts. A bipartite graph is used to represent contextual similarities between words that are observed in the corpus. The graph is explored using random walks in order to identify standard renderings of non-standard words. The framework offers flexibility in lexicon induction such as modeling a word context with varying sequence lengths, generating different lexicons from the same graph, and adapting lexicon to a specific social media genre. We provided two publicly available normalization lexicons that we believe would be a beneficial resource for future Turkish normalization studies. One or both of these lexicons could be integrated into existing normalization systems as a dictionary for easy and fast look-ups. Our evaluation studies showed the effectiveness of induced lexicons in normalizing ill-formed Turkish texts. As future work, we plan to expand the coverage of our lexicons by gathering large amount of ill-formed texts from different social media mediums. We will also study how randomness in lexicon induction can be reduced or totally eliminated. Finally, we plan to assess the effect of the size of lexicons on the performance of our in-house normalization baseline.

# References

1. Hu, Y., Talamadupula, K., Kambhampati, S.: Dude, srsly?: the surprisingly formal nature of Twitter's language. In: 7th International AAAI Conference on Weblogs and Social Media (ICWSM), pp. 244–253 (2013)
2. Eisenstein, J., O'Connor, B., Smith, N.A., Xing, E.P.: Diffusion of lexical change in social media. PLoS One **9** (2014)
3. Herdağdelen, A.: Twitter n-gram corpus with demographic metadata. Lang. Resour. Eval. **47**, 1127–1147 (2013)
4. Schwartz, H.A., Eichstaedt, J.C., Kern, M.L., Dziurzynski, L., Ramones, S.M., Agrawal, M., Shah, A., Kosinski, M., Stillwell, D., Seligman, M.E.P., Ungar, L.H.: Personality, gender, and age in the language of social media: the open-vocabulary approach. PLoS One **8** (2013)
5. Foster, J., Çetinoğlu, Ö., Wagner, J., Roux, J.L., Hogan, S., Nivre, J., Hogan, D., van Genabith, J.: # hardtoparse: POS tagging and parsing the twitterverse. In: The Workshop on Analyzing Microtext (AAAI), pp. 20–25 (2011)
6. Kucuk, D., Steinberger, R.: Experiments to improve named entity recognition on Turkish tweets. In: 5th Workshop on Language Analysis for Social Media, pp. 71–78 (2014)
7. Hassan, H., Menezes, A.: Social text normalization using contextual graph random walks. In: 51st Annual Meeting of the Association for Computational Linguistics, pp. 1577–1586 (2013)
8. Brill, E., Moore, R.C.: An improved error model for noisy channel spelling correction. In: 38th Annual Meeting on Association for Computational Linguistics, pp. 286–293 (2000)
9. Tautanova, K., Moore, R.C.: A pronunciation modeling for improved spelling correction. In: 40th Annual Meeting on Association for Computational Linguistics, pp. 144–151 (2002)
10. Choudhury, M., Saraf, R., Jain, V., Mukherjee, A., Sarkar, S., Basu, A.: Investigation and modeling of the structure of texting language. Int. J. Doc. Anal. Recogn. **10**, 157–174 (2007)
11. Cook, P., Stevenson, S.: An unsupervised model for text message normalization. In: 4th Workshop on Computational Approaches to Linguistic Creativity (CALC), pp. 71–78 (2009)
12. Aw, A., Zhang, M., Xiao, J., Su, J.: A phrase-based statistical model for SMS text normalization. In: 21st International Conference on Computational Linguistics/ACL, pp. 33–40 (2006)
13. Kaufmann, M., Kalita, J.: Syntactic normalization of Twitter messages. In: International Conference on Natural Language Processing (2010)

14. Liu, F., Weng, F., Wang, B., Liu, Y.: Insertion, deletion, or substitution?: normalizing text messages without pre-categorization nor supervision. In: 49th Annual Meeting of the Association for Computational Linguistics: Human Language Technologies (HLT), pp. 71–76 (2011)
15. Liu, F., Weng, F., Jiang, X.: A broad-coverage normalization system for social media language. In: 50th Annual Meeting of the Association for Computational Linguistics, pp. 1035–1044 (2012)
16. Han, B., Cook, P., Baldwin, T.: Lexical normalization for social media text. ACM Trans. Intell. Syst. Technol. (TIST) 4, 5(1)–5(27) (2013)
17. Sönmez, C., Özgür, A.: A graph-based approach for contextual text normalization. In: Conference on Empirical Methods on Natural Language Processing (EMNLP), pp. 313–324 (2014)
18. Torunoğlu, D., Eryiğit, G.: A cascaded approach for social media text normalization of Turkish. In: 5th Workshop on Language Analysis for Social Media (LASM), pp. 62–70 (2014)
19. Yıldırım, S., Yıldız, T.: An unsupervised text normalization architecture for Turkish language. In: 16th International Conference on Intelligent Text Processing and Computational Linguistics (CICLING) (2015)

# Text Classification and Categorization

# Introducing Semantics in Short Text Classification

Ameni Bouaziz[(⊠)], Célia da Costa Pereira, Christel Dartigues-Pallez,
and Frédéric Precioso

Laboratoire I3S (CNRS UMR-7271), Université Nice Sophia Antipolis, Nice, France
bouaziz@i3s.unice.fr,
{celia.pereira,christel.dartigues-pallez,precioso}@unice.fr

**Abstract.** To overcome short text classification issues due to shortness and sparseness, the enrichment process is classically proposed: topics (word clusters) are extracted from external knowledge sources using
Latent Dirichlet Allocation. All the words, associated to topics which
encompass short text words, are added to the initial short text content. We propose (i) an explicit representation of a two-level enrichment
method in which the enrichment is considered either with respect to
each word in the text or to the global semantic meaning of the short text
and (ii) a new semantic Random Forest kind in which semantic relations
between features are taken into account at node level rather than at tree
level as it was recently proposed in the literature to avoid potential tree
correlation. We demonstrate that our enrichment method is valid not
only for Random Forest based methods but also for other methods like
MaxEnt, SVM and Naive Bayes.

**Keywords:** Short text classification · Text enrichment · Semantics
LDA

## 1 Introduction

Applying the traditional classification algorithms, that were previously proven
very efficient in standard text classification, revealed lower performances on short
texts. This degradation is caused by shortness, word sparseness and lack of
contextual information, as explained by Phan *et al.* in [1]. Existing solutions
to overcome short text issues focus mainly on the pre-processing step. Text
representation is adapted by semantic enrichment or reduction of features [2–5].
The goal is first to maximize information contained in the features and second
to reduce noise and its impact on classification. However, except from a recent
but non-reproducible work from Bouaziz *et al.* [6], to the best of our knowledge,
none of the existing works propose to combine semantic enrichment both at the
level of individual words and of the whole text.

We present here our method for text enrichment in the pre-processing phase.
We rely on an external source of knowledge to fulfill the semantic enrichment of

© Springer International Publishing AG, part of Springer Nature 2018
A. Gelbukh (Ed.): CICLing 2016, LNCS 9624, pp. 433–445, 2018.
https://doi.org/10.1007/978-3-319-75487-1_34

short texts. We apply the Latent Dirichlet Allocation (LDA) [7] on the external data to group words that are semantically linked into common topics. Then, these topics are used in a two-level enrichment method. First, texts are considered as sets of words and each word is extended by the words from the nearest topic. Second, texts are taken as whole entities and we add to each text the words from the topic which is near to the overall context of the text. Bouaziz and colleagues propose in [6] a two-level approach for enriching short texts to be used in semantic Random Forests. However, they do not explicitly present their method and therefore, it would be hard to reproduce it.

In addition to the new enrichment method, we propose a new kind of semantic Random Forest in which the semantic relations between features are taken into account at node level rather than at tree level as proposed in [6]. In this case the whole features space is considered to build a tree like in standard RF.

Sumarizing, here, we propose a new semantic random forest method and a two-level enrichment method, which, besides being reproducible, allows to improve the accuracy of short-text classification using different techniques like semantic-based Random Forest methods, MaxEnt, SVM and Naive Bayes.

The paper is organized as follows. Section 2 presents an overview of related works, Sect. 3 introduces our enrichment method. Section 4 presents the usage of semantics in the learning algorithm of RF. Then, we present experimental results in Sect. 5. Finally, Sect. 6 concludes and presents some ideas for future work.

## 2   Related Work

Many learning algorithms have been developed for classifying text documents, such as Support Vectors Machine [8], Naive Bayes [9], Maximum Entropy [10] and Random Forests [11].

RF algorithm is known to be very efficient in text classification. This method is based on a set of decision tree classifiers. Each tree gives a decision and the final decision of the forest is taken through a majority vote among all the trees. Traditional RF are not really adapted to face issues raised by short texts classification. Indeed, this kind of texts has characteristics that have to be taken into account during classification.

Shortness [1] is one of the main characteristics of short texts. Usually, the size of a short text does not exceed few words. Thus, there is not enough information to measure semantically relevant similarities between several short texts.

Sparseness [1] is another short text issue. Indeed, vectors representing short texts are very sparse. Short texts lack also of contextual information, which implies a poor coverage of classes representing the dataset. The classification of new data becomes then complicated compared to standard texts.

Owing to these problems, bag-of-words representation combined with traditional machine learning algorithms is not efficient for short text classification. Works to overcome these issues try to propose alternative representations based on either enrichment or reduction of features. Reduction and enrichment use generally external sources of information to link features semantically. External knowledge can be taken from ontologies (wordNet, Wikipedia) [12,13] or from large scale datasets on which topic models techniques are applied (LDA, Latent Semantic Indexing [14]...).

Song et al. in [15] and Rafeeque and Sendhilkumar in [16] summarize these different techniques and show some of their usages in short text classification. Several reduction methods exist: feature abstraction, feature selection and LDA.

Yang et al. in [2] propose a new reduction approach, they use the topic model technique LDA to combine semantic and lexical aspects of features. Their method leads to a significant reduction of the feature space size. It provides good results, but it seems to be only adapted to equally distributed data. Sun in [17] propose to classify short texts using very few words. He selects the most important words of short texts and then submits them as a query to a local search engine in order to find the best matching label.

Regarding enrichment, Bouaziz and colleagues propose in [6], as Phan et al. in [1], to apply LDA on an external source of knowledge to generate topics and add them to the short text content. Phan et al. [1] classify the enriched texts using the Maximum Entropy algorithm and get high classification accuracy. Chen et al. [4] improve Phan's method by adding an algorithm that selects the best subset of topics for the enrichment step.

In [18] Vo and Ock propose to multiply the sources of external knowledge to have a better enrichment. Indeed, to classify the titles of scientific papers, they use three external enrichment background (DBLP, LNCS and Wikipedia). They apply LDA on these sources to generate a maximum of topics that are added to the paper titles.

To the best of our knowledge, existing enrichment methods consider a text as a set of words and enrich each word separately. None of these methods takes the text as a whole entity and enriches it based on its general meaning as we do in our works. Also all the solutions presented in the above works proposed to change the bag of words representation to solve the short texts issues. However, none of them propose to adapt machine learning algorithms to this kind of texts.

We found some studies on semantics introduction in RF, but they are mainly related to prediction tasks. Caragea et al. in [19] built a hierarchical ontology of features and combined it with four machine learning algorithms to predict friendships in social networks. Their best results were obtained with the RF. Chen and Zhang in [20] introduced a module guided RF in order to predict biological characteristics; they base the feature selection at tree nodes on a correlated network of features that they already built in advance.

Results obtained on prediction encourage us to introduce semantics in RF to solve short texts classification issues.

## 3 The Enrichment Method

To overcome the short text classification issues, as done in [1], we enrich short texts with words semantically related to their content. Thus the vector representing a short text becomes longer and contains more information. This additional information is found in external knowledge sources contained in documents that are closely related to the domain of our dataset and at the same time that are general enough to cover all the concerned domain. Figure 1 represents our method of enrichment, Figs. 2 and 3 illustrate through an example the mechanism of transforming the same dataset into a bag-of-word representation, without and with enrichment process respectively.

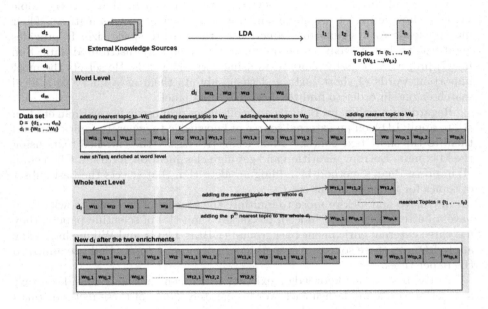

**Fig. 1.** Two-level enrichment of a short text.

In order to extract knowledge from those external sources, we apply LDA which is a generative probabilistic method that uses Dirichlet distribution to identify hidden topics of a dataset. Then, each word is associated to one or many of the generated topics based on similarity calculation. Each word has a given weight representing its importance in a topic. The external sources used to add more information in our dataset are then transformed in a set of $n$ topics containing $k$ words (Fig. 3b).

We start our enrichment process by considering a dataset $D = \{d_i, \ldots, d_m\}$ composed of $m$ short texts (Fig. 3a).

The first phase of our text enrichment method is local: it is a word level enrichment. During this first step, we add to each word contained in each of the $m$ short texts, all the words contained in the nearest topic that is the topic in which this word has the biggest weight (Fig. 3c).

Let us consider a document $d_i$ represented by a set of words $d_i = \{w_{i1}, \ldots, w_{il}\}$. Let $p_{zj}^i$ be the weight of the word $w_{iz}$ in topic $t_j$, with $z \in$

$\{1, \ldots, l\}$. The similarity between $w_{iz}$ and $t_j$ is given by $p_{zj}^i$ if $w_{iz}$ is in $t_j$ and 0 if it does not belong to the topic:

$$\text{sim}(w_{iz}, t_j) = \begin{cases} p_{zj}^i, & \text{if } w_{iz} \in t_j; \\ 0, & \text{otherwise.} \end{cases} \tag{1}$$

The topic $t^*$ in which the word $w_{iz}$ has the highest weight is given by:

$$t^* = \underset{j}{\text{argmax}}(\text{sim}(w_{iz}, t_j)). \tag{2}$$

This process transforms short texts in a set of general contexts. It then allows to build a generic model that is able to classify any text related to the domains of the initial short texts.

This first step is followed by a global enrichment, which is a whole text level enrichment. Here, the aim is not to consider each word of the short text but to try to capture the global meaning of the short text by considering all the contained words together. In order to catch this global meaning, we calculate the occurrence of a short text and we add to it all the words of the $p$ nearest topics to the short text (Fig. 3d). The occurrence is defined as the number of words in common between a text and a topic. It is calculated as follows:

The occurrence of a word $w_{iz}$ in a topic $t_j$, $\text{occ}(w_{iz}, t_j)$, is given by:

$$\text{occ}(w_{iz}, t_j) = \begin{cases} 1, & \text{if } w_{iz} \in t_j \\ 0, & \text{otherwise.} \end{cases} \tag{3}$$

The occurrence of a short text $d_i = \{w_{i1}, \ldots, w_{il}\}$ in a topic $t_j$, $\text{occ}(d_i, t_j)$, is given by:

$$\text{occ}(\{w_{i1}, \ldots, w_{il}\}, t_j) = \sum_{u=1}^{l} \text{occ}(w_{iu}, t_j). \tag{4}$$

The topic $t^{max}$ for which we get the best occurrence is given by:

$$t^{max} = \underset{j}{\text{argmax}}(\text{occ}(\{w_{i1}, \ldots, w_{il}\}, t_j)). \tag{5}$$

Once the initial short text has been locally (word level) and globally enriched (whole text level), a fusion of the two new texts is done in order to create a new set of words representing our initial short text (Fig. 3e). Finally, we generate a matrix corresponding to the new representation of our dataset from this new set of words. Each row of the matrix corresponds to a short text and each column corresponds to a word used after the two phases of enrichment. The value contained in a cell corresponding to the short text $i$ and the word $j$ represents how many times this word appears in the enriched short text. Comparing this last representation to the initial representation (Fig. 2f) we can see that vectors are larger and contain less null values. Thanks to this transformation, short text classification will be improved as confirmed by the next section.

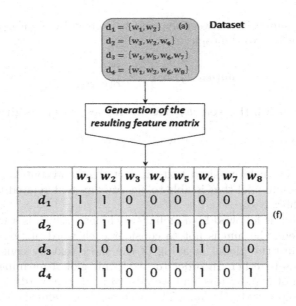

**Fig. 2.** Dataset before enrichment.

## 4   Semantic Random Forest

In order to compare our results with the ones obtained by Bouaziz in [6] by using a semantic Random Forest method, in this section, we first present such a method before presenting the semantic method we propose in this paper.

In RF algorithms, short text dataset is represented by a matrix where columns are all the words composing the dataset after preprocessing — the features. Each line of the matrix is an enriched short text. The values in the line represent the importance of the features in the text. There are several ways to compute the importance of a word in a text: number of occurrences, term frequency, tf-idf and so on. To build each decision tree of traditional RF, a subset of the matrix lines is randomly selected to form a bootstrap and the whole feature space is considered. Then, for each node of the tree a random subset of the features is selected, and a partitioning criterion (Gini, Entropy) is applied to find the best discriminant feature for splitting the node.

In [6], Bouaziz and colleagues introduce a new step: the trees are not anymore built using the whole features but a first selection determines the subset of features considered for each tree. This selection is done based on the semantic relations between features. They apply LDA to build a semantic network linking all the features. The features are then grouped into several topics and assigned weights according to their importance in the topics. To build a tree a small set of features are randomly selected and enlarged by adding from the nearest topics the features having weights bigger than a given threshold. The obtained set of feature is used to build a decision tree by applying on it the standard random feature selection at each node. The added step allowed us to obtain semantic trees.

This new type of RF showed significant improvements in short text classification (see [6]). Note that the initial small subset of feature remains chosen randomly for two reasons: first, to keep diversity and second to not lose features which were not assigned to any topic during LDA grouping. However, if unluckily the initial small subset is composed of words belonging to the same topic then the nodes of the built tree will be highly correlated. That's why we propose, in this paper, another Semantic Random Forest type in which semantic relations between features are taken into account at node level rather than at tree level. In this case the whole features space is considered to build a tree like in standard RF. The difference is at feature selection for nodes. Indeed, at each node the Random Feature Selection algorithm is replaced by a Semantic Feature Selection that starts by choosing randomly a small set of features (much smaller than the set used in standard RF), then, this set is enlarged by the semantically linked topics and the partitioning criterion applied on the enlarged set determines the feature to use in the current node. Figure 4 explains the principle of Semantic Random Forests at node level. Even though the trees obtained in this

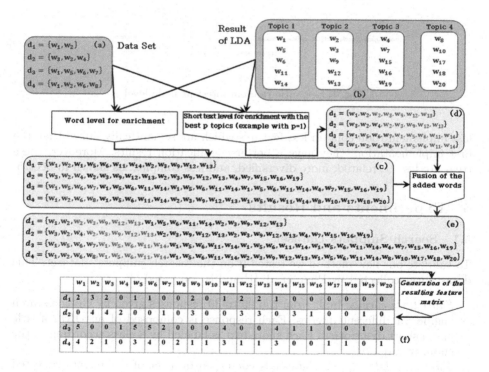

**Fig. 3.** Global enrichment process.

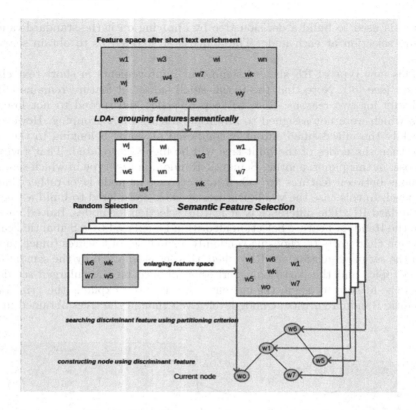

**Fig. 4.** Semantic random forest at node level.

last implementation contain less features which are semantically linked than the first implementation, they gave a better classification results. Moreover, these trees are less correlated, more diversified, and less complex, as shown in Sect. 5.

# 5    Experiments and Results

## 5.1    Search Snippets Dataset

To validate our new method we tested it on the "search-snippets" dataset. This dataset was collected by Phan et al. [1] and is composed of:

- *Short texts corpus*: built from top 20 to 30 responses given by Google search engine for different queries. Each response is composed of an URL, a title and a short description. Each short text is labeled by a class according to the submitted query.
- *Universal dataset*: this dataset is composed of a set of documents collected from Wikipedia as a response to queries containing some specific keywords.

The application of LDA on this document set generated 200 topics with 200 words each. We used those topics as an external source for short text enrichment.

To evaluate our algorithm, we used the accuracy defined as the ratio of correctly classified short texts to the total number of short texts.

## 5.2   Results and Interpretations

In this section we evaluate the added value of our two-level enrichment method. We present the results of the enrichment at text level and the improvement obtained thanks to word level enrichment. We ran our tests using RF classifier in a first step, then we confirm our results by running similar experiments considering MaxEnt, SVM and Naive Bayes.

**Enrichment at Whole Text Level.** In a first step we applied the enrichment of short text at whole text level to evaluate its contribution. We run classification experiments with traditional RF and an enrichment of 4 topics. This test was repeated with 10, 20 up to 100 trees.

**Enrichment at Word Level.** In order to evaluate the contribution of the second level of our enrichment model we run the following experiments: we take our short text dataset and we add to each text the nearest topics. Then, we apply the word level enrichment to enlarge the short text by topics following the Eq. 2. We build 10 RF classification models with respectively $10, 20 \ldots 100$ trees. Short text classification using the both text level and the two-level enrichment with the built RF allowed to obtain the results summarized in Table 1.

**Table 1.** Classification accuracy obtained by the two-level enrichment of the short texts.

|  | 10 | 20 | 30 | 40 | 50 | 60 | 70 | 80 | 90 | 100 |
|---|---|---|---|---|---|---|---|---|---|---|
| No enrichment | 0.57 | 0.592 | 0.598 | 0.589 | 0.589 | 0.592 | 0.594 | 0.595 | 0.6 | 0.59 |
| Text level enrichment | 0.703 | 0.721 | 0.724 | 0.724 | 0.73 | 0.732 | 0.733 | 0.733 | 0.728 | 0.733 |
| Two level enrichment | 0.728 | 0.745 | 0.753 | 0.755 | 0.758 | 0.758 | 0.76 | 0.761 | 0.76 | 0.766 |

Results show that enriching text with the first level is very interesting. The accuracy increased from 0.59 before enrichment to 0.733 after applying the first enrichment level. Results show also an additional improvement in accuracy when the word level enrichment is applieed. For the 100 tree forest for instance, the accuracy reaches 0.766, while it was 0.733 only with the text level enrichment alone. This improvement represents 4.5%. The results obtained for the experiences with different number of trees are quite similar.

**Enrichment Method with Other Classifiers.** To confirm the previous results, we tested our method with MaxEnt, SVM and Naive Bayes. Table 2 shows the results obtained with those classifiers. These results show that our two-level enrichment process contributes to the short text classification improvement with all the standard classifier types.

**Table 2.** Classification accuracy obtained with MaxEnt, SVM and Naive Bayes classifiers.

|  | MaxEnt | SVM | Naive Bayes |
|---|---|---|---|
| Traditional classifier | 0.657 | 0.611 | 0.493 |
| After enrichment | 0.734 | 0.694 | 0.761 |

**Semantic Random Forests.** We ran the same tests again on the short text enlarged with the full two-level enrichment model, but this time we replaced the learning algorithm of traditional RF by the SRF proposed in [6]. In a second experience we replaced RF again by our SRF Node level. As we can see in Table 3 and Fig. 5, our SRF Node level outperforms both the traditional RF and the SRF.

**Table 3.** Classification accuracy obtained by RF, enrichment only, SRF and SRF N.

|  | 10 | 20 | 30 | 40 | 50 | 60 | 70 | 80 | 90 | 100 |
|---|---|---|---|---|---|---|---|---|---|---|
| **RF** | 0.57 | 0.592 | 0.598 | 0.589 | 0.589 | 0.592 | 0.594 | 0.595 | 0.6 | 0.59 |
| **Enrichment only** | 0.728 | 0.745 | 0.753 | 0.755 | 0.758 | 0.758 | 0.76 | 0.761 | 0.76 | 0.766 |
| **Enrichment+SRF** | 0.73 | 0.761 | 0.771 | 0.776 | 0.776 | 0.78 | 0.784 | 0.784 | 0.786 | 0.789 |
| SRF/RF(%)[a] | 28.07 | 28.55 | 28.93 | 31.75 | 31.80 | 31.76 | 31.99 | 31.76 | 31.00 | 33.73 |
| SRF/Enrichment(%) | 0.27 | 2.15 | 2.39 | 2.78 | 2.41 | 2.90 | 3.16 | 3.02 | 3.42 | 3.00 |
| **Enrichment+SRF N** | 0.766 | 0.785 | 0.79 | 0.787 | 0.791 | 0.795 | 0.794 | 0.795 | 0.794 | 0.795 |
| SRF N/RF(%) | 34.39 | 32.60 | 32.11 | 33.62 | 34.30 | 34.29 | 33.67 | 33.61 | 32.33 | 34.75 |
| SRF N/Enrichment(%) | 5.22 | 5.37 | 4.91 | 4.24 | 4.35 | 4.88 | 4.47 | 4.47 | 4.47 | 3.79 |
| SRF N/SRF(%) | 4.93 | 3.15 | 2.46 | 1.42 | 1.89 | 1.92 | 1.28 | 1.40 | 1.02 | 0.76 |

[a]SRF/RF(%) is the percentage of accuracy improvement of SRF compared to RF

Indeed, with SRF Node level we obtained the best results of all our tests with an accuracy of more than 0.79, which represents more than 34% of global improvement. Results show also that the SRF Node level implementation outperforms with only 30 trees the best accuracy of the SRF implementation that is obtained for 100 trees. With more than 30 trees, accuracy of SRF Node level remains almost the same.

In [1], the usage of the Maximum Entropy algorithm of the same search snippets data set gave an accuracy of 0.657, which is better than the results of

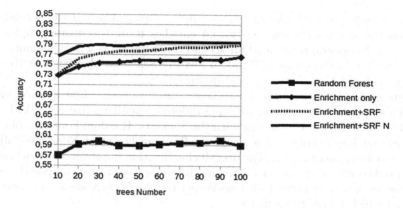

**Fig. 5.** Variation of short texts classification accuracy depending on trees number for short texts enriched using RF, RF with enrichment, SRF and SRF N.

standard RF. With our SRF Node level approach, we achieved a classification improvement of 21% compared to standard Maximum Entropy.

In addition to the classification accuracy amelioration our method presents an other advantage which lies in their trees size. Indeed, as shown in Table 4, traditional RF trees have 6272 nodes in average, with semantic RF, this average is reduced to less than the half (2826 nodes). In SRF Node Level, the number of nodes in each tree is even smaller (1911 nodes).

**Table 4.** Number of tree nodes for RF, SRF and SRF N.

| Tree | 1 | 2 | 3 | 4 | 5 | 6 | 7 | 8 | 9 | 10 | average |
|---|---|---|---|---|---|---|---|---|---|---|---|
| **Nodes in RF** | 6355 | 6451 | 6473 | 6119 | 5935 | 6143 | 6193 | 6339 | 6061 | 6657 | 6272 |
| **Nodes in SRF** | 2957 | 2883 | 2657 | 2653 | 2879 | 2851 | 2873 | 2717 | 3037 | 2753 | 2826 |
| **Nodes in SRF N** | 1883 | 1947 | 1921 | 1925 | 1899 | 1941 | 1851 | 1947 | 1867 | 1935 | 1911 |

The reduction of trees node number makes the classification algorithm quicker. The SRF Node level implementation is even faster since it requires fewer trees to achieve the same accuracy of SRF as we saw in the results above.

## 6   Conclusion

In this work, we proposed a new approach of short text classification based on semantics in the pre-processing step by a new enrichment method and in the learning step by making decision trees building dependent on semantic relations between features. We first considered an external source of knowledge to our dataset. We applied LDA to generate semantic topics. Then we used a two level enrichment process to enlarge short text. The first level takes a text word by word and adds to each one the most similar topic. Whereas the second level considers a text as a whole entity and enriches it by the nearest topics with respect to its general meaning. Our second contribution is a new Semantic Random Forest where the classic Random Feature Selection algorithm at the node level is replaced by a Semantic Feature Selection.

Our enrichment provides a significant improvement when applied on the "search snippets" dataset and by using different classifiers like RF, SVM, max-Ent and Naive Bayes. Combining with the Semantic Random Forest we obtained a significant improvement. The Semantic Random Forest at Node level allowed an even better classification. The overall classification improvement reached 34% compared to RF. Moreover our Semantic Random Forests are composed of small decision trees (an average of 1911 nodes per tree for SRF Node level versus 6272 for RF) which makes them faster.

For the future, we think that our semantic approach can be further improved by combining it with weighted Random Forest which are suitable for unbalanced dataset classification like "search snippets".

**Acknowledgments.** This work has been co-funded by Région Provence Alpes Côte d'Azur (PACA) and Semantic Grouping Company (SGC).

# References

1. Phan, X.H., Nguyen, L.M., Horiguchi, S.: Learning to classify short and sparse text and web with hidden topics from large-scale data collections. In: International Conference on World Wide Web, pp. 91–100. ACM (2008)
2. Yang, L., Li, C., Ding, Q., Li, L.: Combining lexical and semantic features for short text classification. Procedia Comput. Sci. **22**, 78–86 (2013)
3. Amaratunga, D., Cabrera, J., Lee, Y.S.: Enriched random forests. Bioinformatics **24**, 2010–2014 (2008)
4. Chen, M., Jin, X., Shen, D.: Short text classification improved by learning multi-granularity topics. In: IJCAI, pp. 1776–1781 (2011)
5. Song, Y., Wang, H., Wang, Z., Li, H., Chen, W.: Short text conceptualization using a probabilistic knowledgebase. In: IJCAI, pp. 2330–2336. AAAI Press (2011)
6. Bouaziz, A., Dartigues-Pallez, C., da Costa Pereira, C., Precioso, F., Lloret, P.: Short text classification using semantic random forest. In: Bellatreche, L., Mohania, M.K. (eds.) DaWaK 2014. LNCS, vol. 8646, pp. 288–299. Springer, Cham (2014). https://doi.org/10.1007/978-3-319-10160-6_26
7. Blei, D.M., Ng, A.Y., Jordan, M.I.: Latent dirichlet allocation. JMLR **3**, 993–1022 (2003)
8. Cortes, C., Vapnik, V.: Support-vector networks. Mach. Learn. **20**, 273–297 (1995)
9. Schneider, K.-M.: Techniques for improving the performance of Naive Bayes for text classification. In: Gelbukh, A. (ed.) CICLing 2005. LNCS, vol. 3406, pp. 682–693. Springer, Heidelberg (2005). https://doi.org/10.1007/978-3-540-30586-6_76
10. Berger, A.L., Pietra, V.J.D., Pietra, S.A.D.: A maximum entropy approach to natural language processing. Comput. Linguist. **22**, 39–71 (1996)
11. Breiman, L.: Random forests. Mach. Learn. **45**, 5–32 (2001)

12. Hu, X., Zhang, X., Lu, C., Park, E.K., Zhou, X.: Exploiting Wikipedia as external knowledge for document clustering. In: ACM SIGKDD International Conference on Knowledge Discovery and Data Mining, pp. 389–396. ACM (2009)
13. Hu, X., Sun, N., Zhang, C., Chua, T.S.: Exploiting internal and external semantics for the clustering of short texts using world knowledge. In: ACM Conference on Information and Knowledge Management, pp. 919–928 (2009)
14. Dumais, S., Furnas, G., Landauer, T., Deerwester, S., Deerwester, S., et al.: Latent semantic indexing. In: Proceedings of the Text Retrieval Conference (1995)
15. Song, G., Ye, Y., Du, X., Huang, X., Bie, S.: Short text classification: a survey. J. Multimed. **9**, 635–643 (2014)
16. Rafeeque, P., Sendhilkumar, S.: A survey on short text analysis in web. In: IEEE International Conference on Advanced Computing (ICoAC), pp. 365–371 (2011)
17. Sun, A.: Short text classification using very few words. In: ACM SIGIR Conference on Research and Development in Information Retrieval, pp. 1145–1146 (2012)
18. Vo, D.T., Ock, C.Y.: Learning to classify short text from scientific documents using topic models with various types of knowledge. Expert Syst. Appl. **42**, 1684–1698 (2015)
19. Caragea, D., Bahirwani, V., Aljandal, W., Hsu, W.H.: Ontology-based link prediction in the livejournal social network. In: SARA, vol. 9 (2009)
20. Chen, Z., Zhang, W.: Integrative analysis using module-guided random forests reveals correlated genetic factors related to mouse weight. PLoS Comput. Biol. **9** (2013)

# Topics and Label Propagation: Best of Both Worlds for Weakly Supervised Text Classification

Sachin Pawar[1,2], Nitin Ramrakhiyani[1(✉)],
Swapnil Hingmire[1,3], and Girish K. Palshikar[1]

[1] TCS Research, Tata Consultancy Services, Pune 411013, India
{sachin.pawar,nitin.ramrakhiyani,swapnil.hingmire,gk.palshikar}@tcs.com
[2] Department of CSE, Indian Institute of Technology Bombay, Mumbai 400076, India
[3] Department of CSE, Indian Institute of Technology Madras, Chennai 600036, India

**Abstract.** We propose a Label Propagation based algorithm for weakly supervised text classification. We construct a graph where each document is represented by a node and edge weights represent similarities among the documents. Additionally, we discover underlying topics using Latent Dirichlet Allocation (LDA) and enrich the document graph by including the topics in the form of additional nodes. The edge weights between a topic and a text document represent level of "affinity" between them. Our approach does not require document level labelling, instead it expects manual labels only for topic nodes. This significantly minimizes the level of supervision needed as only a few topics are observed to be enough for achieving sufficiently high accuracy. The Label Propagation Algorithm is employed on this enriched graph to propagate labels among the nodes. Our approach combines the advantages of Label Propagation (through document-document similarities) and Topic Modelling (for minimal but smart supervision). We demonstrate the effectiveness of our approach on various datasets and compare with state-of-the-art weakly supervised text classification approaches.

## 1 Introduction

Text classification is an important area of Natural Language Processing (NLP) with applications ranging from automatic request routing to text understanding. It has also been one of the most active and competitive areas of research in NLP. In this work, we propose a novel weakly supervised method to solve document classification.

We use the Label Propagation algorithm [24] which works on an undirected graph and involves iterative propagation of labels from a few labelled nodes to large number of unlabelled nodes. The algorithm stops when label distributions at all nodes have converged.

For Label Propagation, representation of documents in a graph and setting edge weights as similarity values among the documents is necessary. However, to

© Springer International Publishing AG, part of Springer Nature 2018
A. Gelbukh (Ed.): CICLing 2016, LNCS 9624, pp. 446–459, 2018.
https://doi.org/10.1007/978-3-319-75487-1_35

achieve high accuracy, providing a good number of labelled documents is necessary. Document labelling can be an expensive and time-consuming activity and would require domain expertise. To do away with the cumbersome labelling of documents we propose to add topics learned over the documents in the graph and solicit labels only for topic nodes. Hingmire et al. [12] and Razavi et al. [18] proposed labelling of topics instead of documents arguing that topic labelling invites lesser manual effort. We enrich the document similarity graph by adding labelled topics. We also introduce a topic influence parameter to control the topic enrichment process. Our algorithm LPA-TD (Label Propagation Algorithm - Topic Documents) constructs this topic enriched graph and runs Label Propagation on it to discover a classification of documents. The topic enriched graph is constructed for various configurations of the topic influence parameter and document similarities. Additionally, we experimented by constructing the topic enriched graph by dropping certain topic nodes which were incoherent and confusing to label. Closely seen, LPA-TD combines the power of topic modelling through smart manual tagging and iterative propagation of Label Propagation by harnessing document similarities.

We experiment on 4 public datasets from the 20Newsgroups (20NG) corpora and compare LPA-TD with multiple weakly supervised algorithms for text classification. We also compare LPA-TD with the performance of only Label Propagation (*OnlyLPA*) using some labelled documents. LPA-TD outperforms the OnlyLPA baseline on all datasets and also outperforms the other algorithms on two out of four 20 NG datasets. We also perform experiments on a real-world dataset comprising of about 4000 grievances raised by employees of a large IT organization. The grievance text needs to be analysed by classifying it into four classes related to appraisals, compensation, finance and administration. Based on a manually created gold standard, LPA-TD performs at an encouraging macro-F1 of 78% on this dataset.

The paper is organized as follows. In Sect. 2 we briefly describe the background of various techniques employed in the proposed LPA-TD algorithm. In Sect. 3, we describe the construction of the topic enriched graph and the topic influence parameter. Further in Sect. 4, we present details about the datasets, experimental setup, evaluation and analysis. Relevant related work is presented in Sect. 5. We finally detail some future work and conclude the paper.

## 2    Background

### 2.1    Label Propagation Algorithm

Zhu and Ghahramani [24] proposed the Label Propagation Algorithm which is a graph based semi-supervised method. It represents labelled and unlabelled instances as nodes in a graph with edges reflecting the similarity between nodes. The label information for any node is propagated to its nearby nodes through weighted edges iteratively and finally the labels of unlabelled examples are inferred when the propagation process is converged. The detailed version of the algorithm for transductive document classification is described in Algorithm 1.

**Data**: 1. $D_L$ (Set of labelled documents)
2. $D_U$ (Set of unlabelled documents)
3. $S$ ($n \times n$ Similarity matrix where $n = |D_L| + |D_U|$ and top $|D_L|$ rows
correspond to labelled documents)
4. $L = \{l_1, l_2, \cdots, l_m\}$ (Set of $m$ class labels)
**Result**: Label matrix $Y_{n \times m}$, where $Y_{ij}$ represents the probability of document
$d_i$ having label $l_j$

/* Begin Initialization                                            */

1 Define probability transition matrix $T$ such that $T_{ij} = \frac{s_{ij}}{\sum_k s_{kj}}$ which is the
probability of jumping from $l_j$ to $l_i$;

2 Define $\bar{T}$ as the row-normalized matrix of $T$ such that $\bar{T}_{ij} = \frac{T_{ij}}{\sum_k T_{ik}}$;

3 Set iteration index $t = 0$;

4 Let $Y^0$ be the label matrix for $0^{th}$ iteration and $Y_L^0$ be its top $|D_L|$ rows and $Y_U^0$
be its remaining rows;

5 Set $Y_{ij}^0 = 1$ if $d_i$ is labelled with $l_j$;

6 Set values of $Y_U^0$ arbitrarily;

/* End Initialization                                              */

7 Propagate the labels of any node to nearby nodes by $Y^{t+1} = \bar{T} Y^t$;

8 Replace the values of top $|D_L|$ rows of $Y^{t+1}$ with $Y_L^0$;

9 Set $t := t + 1$;

10 Repeat steps 7 to 9 until $Y$ converges;

11 **return** $Y^t$;

**Algorithm 1**: Label Propagation Algorithm for Transductive Document Classification

## 2.2   Topic Modelling

Topic modelling allows us to discover important and frequent themes or "topics" discussed in a large collection of text documents. The discovered topics provide an abstraction on the top of individual documents. Latent Dirichlet Allocation (LDA) [1] is the simplest topic model. LDA and its variants have numerous applications in natural language processing, image processing, social network analysis etc. It is widely used to browse a large corpus of documents using the most probable words of each topic and the distribution over topics for each document [3]. LDA assumes following generative process for generating documents.

1. Select word probabilities ($\phi_t$) for each topic $t$:
   $\phi_t \sim \text{Dirichlet}(\beta)$

2. Select topic proportions ($\theta_d$) for document $d$:
   $\theta_d \sim \text{Dirichlet}(\alpha)$

3. Select the topic for each word position ($z_{d,n}$):
   $z_{d,n} \sim \text{Multinomial}(\theta_d)$

4. Select the token for each word position $(w_{d,n})$:

$w_{d,n} \sim \text{Multinomial}(z_{d,n})$

($\alpha$ and $\beta$ are Dirichlet priors for document-topics and topic-words distributions respectively.)

## 2.3   Document Similarities

In order to assign weights to edges connecting various documents, we need to devise a way of computing similarity between any two documents. Various document similarity measures and their effect on document clustering performance are discussed in detail by Huang [13]. For all our experiments, we have employed "Cosine Similarity" measure.

Let $D = \{d_1, d_2, \cdots, d_n\}$ be a set of $n$ documents and $W = \{w_1, w_2, \cdots, w_m\}$ be the set of $m$ distinct words (excluding stop-words and all words occurring in only one document), constituting the vocabulary of the corpus $D$. We represent each document $d_i$ by a vector $V_i$ of length $m$ whose $j^{th}$ component ($V_i[j]$) corresponds to the $j^{th}$ word in $W$ and is computed as,

$$V_i[j] = TF(d_i, w_j) \cdot IDF(w_j)$$

where $TF(d_i, w_j)$ is number of times the word $w_j$ occurs in the document $d_i$ and $IDF(w_j)$ is computed using $ND(w_j)$, i.e. number of documents containing the word $w_j$ as $IDF(w_j) = \log\left(\frac{n}{ND(w_j)}\right)$.

For any two documents $(d_i, d_j)$ and their corresponding vector representations $(V_i, V_j)$, Cosine similarity between them is computed as follows:

$$CosSim(d_i, d_j) = \frac{V_i \cdot V_j}{|V_i||V_j|}$$

**Document Graph Construction:** We construct a document graph where each node represents a document and an edge between any two documents indicates that the documents are similar. The degree of similarity between two documents is represented by assigning appropriate proportional edge weight. Higher edge weight indicates that there is a high similarity between the two documents.

It was observed that each document generally has "low" cosine similarity with a large number of documents. In order to prevent label propagation among the dissimilar document nodes, we need to find some threshold on the document similarities such that if the similarity is below the threshold, then no edge will be added between such documents. We determine this threshold automatically for any set of documents. The threshold is determined such that at least 90% documents within the set are connected to at least $K$ other documents. All the similarities below this threshold are forced to 0 and hence no edge will be added in the document graph for a document pair with similarity below the threshold.

# 3    LPA-TD: Proposed Approach for Text Classification

In this section, we describe our novel approach for weakly supervised text classification.

## 3.1    Weak Supervision by Labelling Topics

The idea of manually obtaining labels for topics instead of instances was explored by Hingmire and Chakraborti [11,12] and Razavi et al. [18]. They observed that LDA topics are easily interpretable as they can be represented by their most probable words. A human annotator can provide the most suitable class label to each topic. The level of supervision in this case is quite low, as they found that a very few topics (typically twice the number of class labels) are generally enough. LDA topics uncover underlying semantic structure of the whole set of documents. Hence, even a few labelled topics, add significant information about most of the documents. Table 1 shows the discovered topics and corresponding labels by assigned by a human annotator for the MEDICAL vs SPACE classification problem in 20 newsgroups dataset.

**Table 1.** Examples of topics discovered in the MEDICAL vs SPACE classification of 20 newsgroups dataset and corresponding labels assigned by a human annotator

| Topic (most probable words) | Label |
| --- | --- |
| msg food doctor pain day problem read evidence problems doesn blood question case body dyer | Medical |
| space nasa science program system data research information shuttle technology station center based sci theory | Space |
| medical health water cancer disease number research information april care keyboard hiv center reported aids | Medical |
| launch earth space orbit moon lunar nasa high henry years spacecraft long cost mars pat | Space |

## 3.2    Affinity Between Topics and Documents

We use collapsed Gibbs sampling [9] to learn topics only using training documents. The collapsed Gibbs sampler for LDA gives topic-word distributions ($\phi_t$) for each topic and document-topic distribution ($\theta_d$) for each document. We use $\theta_{d,t}$ i.e. probability of generating document $d$ by topic $t$ as the affinity between a training document $d$ in corpus and topic $t$. In other words, affinity measures the belongingness of a topic to a particular document. We use $\phi_t$ to infer $\theta_d$ for an unseen document using collapsed Gibbs sampling method proposed by Heinrich [10]. For a particular document, its affinities with all the topics are normalized so that they sum to 1.

**Data:** $D = \{d_1, d_2, \cdots, d_{N_D}\} = D_{train} \cup D_{test}$ (Set of $N_D$ documents containing training and test documents), $N_T$ (Number of topics to be discovered and later used for enrichment), $L = \{l_1, l_2, \cdots, l_m\}$ (Set of $m$ class labels), $\tau$ (Topic influence parameter)

**Result:** $D_L = \{< d_1, l_1 >, < d_2, l_2 >, \cdots, < d_n, l_n >\}$ where $l_1, l_2, \cdots, l_n \in L$

1  $T :=$ Discover $N_T$ topics using LDA from documents in $D_{train}$;
2  $T_L := \{< t_1, l_1 >, < t_2, l_2 >, \cdots, < t_{N_T}, l_{N_T} >\}$;  /* Topic labels from human annotator */
3  $A_{dd} :=$ Compute document-document $N_D \times N_D$ similarity matrix;
4  $A_{td} :=$ Compute topic-document $N_T \times N_D$ affinity matrix;
5  **for** $i = 1$ to $N_D$ **do**
6  $\quad$ $\mu := 0$;        /* Total influence of the neighbouring documents */
7  $\quad$ **for** $j = 1$ to $N_D$ **do**
8  $\quad\quad$ $\mu := \mu + A_{dd}[i][j]$;
9  $\quad$ **end**
10 $\quad$ $c := \frac{\tau \cdot \mu}{1 - \tau}$;  /* Multiplier for topic-document affinities so that for each document, fraction of influence by topic nodes is $\tau$ */
11 $\quad$ **for** $j = 1$ to $N_T$ **do**
12 $\quad\quad$ $A_{td}[j][i] := c \cdot A_{td}[j][i]$;
13 $\quad$ **end**
14 **end**
   /* Similarity Matrix for Topic-enriched Graph                           */
15 $A := \begin{bmatrix} A_{td} & 0 \\ A_{dd} & A_{td}^T \end{bmatrix}_{(N_T+N_D) \times (N_T+N_D)}$ ;
16 $Y_{(N_T+N_D) \times m} := LPA(T_L, D, A, L)$;
17 $D_L := \Phi$;
18 **for** $i = N_T + 1$ to $N_T + N_D$ **do**
19 $\quad$ $j :=$ Index of maximum probability in $Y[i]$;
20 $\quad$ $D_L := D_L \cup \{< d_i, l_j >\}$
21 **end**
22 **return** $\underline{D_L}$

**Algorithm 2:** LPA-TD: Label Propagation Algorithm on a Topic-enriched Document Graph

## 3.3   Topic-Enriched Graph

We propose to enrich the document graph by adding a new node corresponding to each topic. As explained earlier, all these nodes are "labelled" nodes as the human supervision is provided at the topic level. All the other nodes representing documents are the "unlabelled" nodes. Each document node is connected to all topic nodes and the edge weight between a topic and a document is proportional to the "affinity" between them.

**Topic Influence Parameter ($\tau$):** During the iterations of Label Propagation Algorithm, each document receives label distributions from the topic nodes as well as document nodes it is connected with. In other words, there are two sources of label information for a document node, i.e. "labelled topics" and "similar documents". To control the flow of label information from the two sources, we define a Topic Influence Parameter $\tau$. Through this parameter, the influence of topic information on a particular node can be fixed to be a specific fraction of the total edge weight incident on that node. Consider a document at which sum of incident edge weights from document nodes is $\mu$ (sum of incident edge weights from topic nodes is 1 by definition as discussed in Sect. 3.2). To achieve the desired topic influence $\tau$, all the incident edge weights from topic nodes to the document are multiplied by a value $c$ changing the sum of incident edge weights from topic nodes to $c$. The value $c$ in turn can be expressed as a function of $\tau$ and $\mu$ as follows:

$$\tau = \frac{c}{\mu + c} \Rightarrow c = \frac{\tau * \mu}{1 - \tau} \tag{1}$$

Using a user-specified topic influence parameter, the topic-enriched document graph is constructed with appropriate edge weights. A classification of documents is then obtained by running Label Propagation Algorithm over this topic-enriched graph. Algorithm 2 describes the LPA-TD algorithm in detail.

## 4    Experimental Analysis

### 4.1    Datasets

We report the performance of our experiments on corpora from the **20Newsgroups (20NG)** dataset. This dataset contains messages across twenty different UseNet discussion groups, posted over a period of time. These twenty newsgroups are grouped into 6 major clusters. We use the *bydate* version of the 20Newsgroups dataset[1]. This version of the 20Newsgroups dataset contains 18,846 messages and it is sorted by the date of posting of the messages. The dataset is divided into training (60%) and test (40%) sets. We employ 4 different subsets of the 20NG dataset for our experiments, namely PC vs MAC, MEDICAL vs SPACE, POLITICS vs RELIGION and POLITICS vs SCIENCE. These subsets are fairly balanced in terms of representation of individual classes.

### 4.2    Experimental Setup

We start by learning double the number of topics as number of classes over the training documents. Here, it is important to note that class information of training documents is not used. Training documents are used in unsupervised way only to learn the topics. Also, we do not use test documents for learning the topics to ensure fair evaluation. Test documents are only used for reporting results.

The learned topics are labelled by a single human annotator. The annotator is asked to label a topic with only one of the most appropriate class. We use the learned topics to compute a topic-document affinity matrix ($A_{td}$) for both the training and test documents.

---

[1] http://qwone.com/~jason/20Newsgroups/.

Next, we construct the document to document similarity graph using two configurations: (i) K = 1 and (ii) K = 3 where K is minimum number of connected nodes for 90% nodes in the graph as discussed in the Sect. 2.3. We include both the training and test documents in the document graph. To form the topic enriched graph we introduce the learned topics as additional nodes in the document similarity graph, with labels as assigned by the human annotator. We create edges from each document to all topics and experiment with multiple values of the topic influence parameter ($\tau$) for assigning edge weights. Results on the three best values of $\tau$ each for K = 1 and K = 3 are reported.

As the process of learning topics is based on approximate inference, we carry out the topic learning, topic labelling and topic-document affinity computation processes 10 times. Hence, for a given configuration of the document similarity graph with a K and a $\tau$ value, the topic enriched graph is constructed 10 times. It is straightforward to see that the document-document similarity part remains same for all 10 runs, but topics and topic to document edges are added afresh for each run. We finally average the results over all 10 runs for a configuration and report them.

We compare the LPA-TD technique with multiple baselines presented below. In Table 2, we report the macro-F1 scores from the baselines and various configurations of LPA-TD. For the baselines requiring labelled documents, we provide them with the same number of labelled documents as number of topics to be labelled in LPA-TD.

- Expectation maximization with Naive Bayes for text classification proposed by Nigam et al. [16] with 4 randomly selected labelled documents
- GE-FL [5] with 10 labelled features, as reported by Hingmire and Chakraborti [11]
- TLC and ClassifyLDA, as proposed and reported by Hingmire and Chakraborti [11]
- Using only Label propagation: In this configuration, we only consider the document similarity graph without topic enrichment and label as many number of documents as topics in LPA-TD. We then run Label Propagation on this graph and obtain the classification results for evaluation. We ensure that the most connected documents for both classes in the graph are labelled in equal proportion to ensure fairness and getting the best from this baseline.

## 4.3   Incoherent Topics

As discussed earlier, to deal with approximate topic inference, we run all the experiments 10 times and report their average performance. However, for some particular runs, we observed that we get macro-F1 scores much below the average score for that configuration. Upon observing the learned topics in these runs, we found that some topics were incoherent and represent multiple classes. Hence, forcing them to one particular class label was introducing noise, in turn leading to poor performance. Table 3 shows examples of such incoherent topics.

**Table 2.** Experimental results

| | $\tau$ | pc-mac | med-space | politics-sci | politics-rel |
|---|---|---|---|---|---|
| TLC | | 0.68 | 0.943 | 0.911 | **0.922** |
| ClassifyLDA | | 0.641 | 0.926 | 0.899 | 0.892 |
| NB-EM | | 0.429 | **0.99** | 0.474 | 0.466 |
| GE-FL | | 0.666 | 0.939 | 0.618 | 0.765 |
| Only LPA (K = 1) | | 0.408 | 0.925 | 0.9407 | 0.8355 |
| Only LPA (K = 3) | | 0.37 | 0 | 0.9392 | 0.8325 |
| OnlyLPA (K = 1) | | 0.486 | 0.919 | 0.539 | 0.559 |
| OnlyLPA (K = 3) | | 0.374 | 0.953 | 0.601 | 0.657 |
| LPA-TD (K = 1) | 0.1 | 0.673 | 0.947 | 0.912 | 0.837 |
| | 0.05 | 0.682 | 0.949 | 0.912 | 0.836 |
| | 0.01 | 0.704 | 0.951 | 0.887 | 0.819 |
| LPA-TD (K = 3) | 0.2 | 0.661 | 0.948 | 0.918 | 0.840 |
| | 0.1 | 0.673 | 0.950 | 0.916 | 0.839 |
| | 0.05 | 0.671 | 0.949 | 0.903 | 0.823 |
| LPA-TD-Coh (K = 1) | 0.1 | 0.686 | 0.945 | 0.92 | 0.852 |
| | 0.05 | 0.696 | 0.950 | 0.918 | 0.854 |
| | 0.01 | **0.719** | 0.954 | 0.887 | 0.849 |
| LPA-TD-Coh (K = 3) | 0.2 | 0.671 | 0.945 | 0.904 | 0.858 |
| | 0.1 | 0.681 | 0.951 | **0.925** | 0.860 |
| | 0.05 | 0.674 | 0.953 | 0.918 | 0.853 |

**Table 3.** Examples of incoherent topics

| Noisy Topic (Most probable words) | Dataset |
|---|---|
| gun space nasa president launch weapon firearm science tax system job earth orbit clinton stephanopoulo | politics-sci |
| drive disk hard system controller floppy rom bios card port sound board power internal cable | pc-mac |
| msg food idea high pat money long read remember thought moon didn billion isn real | med-space |
| system children objective moral wrong fire fbi morality koresh opinions men doesn isn frank sex drugs | politics-religion |

In order to avoid adverse effect of incoherent topics on label propagation, we simply removed such topics from the topic-enriched graph while making sure that there is at least one topic mapped to each class. After removing incoherent topics from topic-enriched graph, we re-run LPA-TD algorithm. We refer this new approach as LPA-TD (Coherent) (*LPA-TD-Coh*).

## 4.4    Discussion

Table 2 shows experimental results of LPA-TD and LPA-TD (Coherent) along with other baselines. As we can observe from the results, LPA-TD outperforms both the configurations of the OnlyLPA baseline on all datasets. It also performs better than all other baselines on two datasets - PC vs MAC and POLITICS vs SCIENCE.

PC vs MAC is considered to be the most difficult dataset from the 20NG corpora due to significant overlap of words seen in both classes. This overlap results from high semantic similarity between the classes. On this dataset, LPA-TD outperforms all the baselines comfortably, re-iterating the merit of the new technique.

It however, doesn't perform as well as the TLC and ClassifyLDA techniques in the POLITICS vs RELIGION dataset. A look at the topics learned in the 10 runs reveals mostly complex and fuzzy topics not attributable to a single class, which brings down the overall performance. Also, on the MEDICAL vs SPACE dataset, NB-EM outperforms LPA-TD but it performs quite poorly on other datasets. On the other hand, LPA-TD demonstrates a consistent performance across all the datasets.

## 4.5    Case Study on Employee Grievances

We also carried out a case study to analyse a real-world industrial text dataset of grievances which were raised by employees of a major IT organization. The dataset contained about 4000 grievance descriptions related to areas like finance, compensation, appraisals and administration. However, no direct classification was available. So we got the dataset labelled from an HR executive for use as gold standard. Further, the grievances were sorted on the date of posting and we used first 70% grievances for training and the rest for testing. We tried out Naive Bayes and SVM classifiers using Weka[2] and obtained macro-F1 of about 80.9% and 71.2% respectively.

Here, it is important to note that both Naive Bayes and SVM are supervised classifiers and required 70% i.e. 2800 labelled grievances to achieve the above performance. Now, we employed our LPA-TD approach on this dataset. A few examples of the topics discovered and corresponding manual labels are shown in Table 4. The topic enriched graph comprised of the 4000 grievance nodes along with the 8 learned topics. From the various configurations we tried, we obtained the best macro-F1 of 77.8% for $K = 3$ and $\tau = 0.05$. This demonstrates a significant reduction in labelling effort (2800 grievances against only 8 topics) through use of the LPA-TD technique for comparable performance.

## 5    Related Work

Previous work in semi-supervised text classification can be broadly categorized into 4 different types based on the way supervision is provided: (i) Labelling a

---

[2] http://www.cs.waikato.ac.nz/ml/weka/.

**Table 4.** Examples of topics discovered in the Employees Grievances dataset and corresponding labels assigned by a human annotator

| Topic (Most probable words) | Label |
|---|---|
| basic salary grade compensation experience pay higher variable allowance months joined current designation letter mba | Compensation |
| rating project appraisal performance band work appraiser process team discussion client reviewer final disagreement worked | Appraisal |
| office bus working admin work food cab day provided facility service card transport canteen issue | Admin |
| salary amount month finance claim tax account months received deducted paid ticket allowance days payroll | Finance |

few documents, (ii) providing a list of features that are highly indicative of each class label, (iii) employing active learning and (iv) labelling topics.

### 5.1 Using Labelled and Unlabelled Documents

In this method, labelled documents and a large number of unlabelled documents are used for learning the classifier. While estimating the parameters of the classifier certain assumptions about the distribution of labelled and unlabelled documents will have to hold.

**Cluster assumption:** if instances are in the same cluster, they are likely to be of the same class. In other words, if the data are generated by a mixture model following a generative process and a mixture component represents one or more classes then the instances generated by a mixture component are likely to have the same class labels. Due to unlabelled data, the mixture model contains both observed and hidden variables and its parameters are estimated by Expectation-Maximization (EM) algorithm [4]. Nigam et al. [16] used EM Algorithm for semi-supervised text classification with a naive Bayes classifier.

**Low-density separation assumption:** the decision boundary of classification should lie in a low-density region. Using this assumption, Grandvalet and Bengio [8] proposed a maximum a posteriori (MAP) framework for learning a classifier using minimum entropy regularization. Another semi-supervised algorithm which makes this assumption for learning a text classifier using a small number of documents is *Transductive Support Vector Machines (TSVMS)* [14].

**Manifold assumption:** the high-dimensional data lie (roughly) on a low-dimensional manifold. Graph based semi-supervised methods make the manifold assumption to construct a graph in which nodes are both the labelled and unlabelled instances and edge weights represent similarity between instances. The Label Propagation Algorithm [25] discussed earlier falls in this category. Other graph based text classification approaches are by Subramanya and Bilmes [21] and Wang and Zhang [22].

**Multi-view assumption:** each instance has two or more "different" and "independent" views and each view is sufficient for good classification individually. Co-Training [2] algorithm is based on this assumption. The Co-Training process initially constructs a weak classifier for each view using labelled instances, then each weak classifier is bootstrapped using unlabelled instances.

## 5.2   Incorporating Labelled Features

Sometime it is easier for human annotators to describe a class of documents using a set of features than labelling large collections of documents. Liu et al. [15] proposed a text classification algorithm by labelling the most discriminative words. Eventually, these representative words are used to create a text classifier using the combination of naive Bayes classifier and the Expectation-Maximization (EM) algorithm. Other similar approaches are Schapire et al. [19] (based on AdaBoost), Wu and Srihari [23] (generalization of SVM, Weighted Margin SVM) and Druck et al. [5] (generalized expectation criteria based maximum entropy text classifier, i.e. GE-FL).

## 5.3   Using Active Learning

Active learning [20] systems attempt to overcome the labelling bottleneck by asking queries in the form of unlabelled instances to be labelled by a human annotator. Some important text classification approaches using active learning are Godbole et al. [7], Raghavan et al. [17] and Druck et al. [6].

## 5.4   Labelling Topics

Hingmire and Chakraborti [12] proposed the idea of obtaining labels for topics instead of documents. They propose the ClassifyLDA algorithm where a topic model is leaned using LDA and one class label is assigned to each topic. They use the Dirichlet distribution to aggregate all the same class label topics into a single topic and automatically classify unlabelled documents based on their similarity with the aggregated topics. Hingmire and Chakraborti [11] proposed the TLC algorithm which further improves the ClassifyLDA algorithm by allowing a topic to be labelled with multiple class labels instead of one.

# 6  Conclusions and Future Work

We proposed a weakly supervised text classification technique LPA-TD, based on Label Propagation and Topic Modelling. A topic enriched document graph is constructed for a set of documents where the only supervision is in the form of labelled topics. LPA-TD propagates labels over this topic enriched graph thereby exploiting benefits of both, document similarities and labelled topics. We evaluated LPA-TD on 4 datasets of the 20NG corpora and compared with multiple baselines. LPA-TD outperforms all the baselines on 2 out of these 4 datasets, including PC vs MAC which is considered to be one of the most difficult for text classification. Compared to other baselines, LPA-TD demonstrates a consistent performance across all the datasets. We also showed that the issue of incoherent topics can be handled by removing them from the topic enriched graph, without degrading LPA-TD's performance. Furthermore, removal of such incoherent topics resulted in better performance.

In future, we plan to extend LPA-TD by allowing fuzzy class labels to topics which naturally represent multiple classes. This will ease the restriction of assigning only one class per topic. Additionally, we plan to devise topic quality measures for automatic detection of incoherent topics.

# References

1. Blei, D.M., Ng, A.Y., Jordan, M.I.: Latent Dirichlet allocation. J. Mach. Learn. Res. **3**, 993–1022 (2003)
2. Blum, A., Mitchell, T.: Combining labeled and unlabeled data with co-training. In: Proceedings of 11th Annual Conference on Computational Learning Theory, pp. 92–100 (1998)
3. Chaney, A.J.B., Blei, D.M.: Visualizing topic models. In: ICWSM (2012)
4. Dempster, A.P., Laird, N.M., Rubin, D.B.: Maximum likelihood from incomplete data via the EM algorithm. J. Roy. Stat. Soc. B **39**(1), 1–38 (1977)
5. Druck, G., Mann, G., McCallum, A.: Learning from labeled features using generalized expectation criteria. In: SIGIR, pp. 595–602 (2008)
6. Druck, G., Settles, B., McCallum, A.: Active learning by labeling features. In: EMNLP, pp. 81–90 (2009)
7. Godbole, S., Harpale, A., Sarawagi, S., Chakrabarti, S.: Document classification through interactive supervision of document and term labels. In: PKDD, pp. 185–196 (2004)
8. Grandvalet, Y., Bengio, Y.: Semi-supervised learning by entropy minimization. In: NIPS (2004)
9. Griffiths, T.L., Steyvers, M.: Finding scientific topics. PNAS **101**(Suppl. 1), 5228–5235 (2004)
10. Heinrich, G.: Parameter estimation for text analysis. Technical report, University of Leipzig (2008)
11. Hingmire, S., Chakraborti, S.: Topic labeled text classification: a weakly supervised approach. In: SIGIR, pp. 385–394. ACM (2014)
12. Hingmire, S., Chougule, S., Palshikar, G.K., Chakraborti, S.: Document classification by topic labeling. In: SIGIR, pp. 877–880. ACM (2013)

13. Huang, A.: Similarity measures for text document clustering. In: Proceedings of 6th New Zealand Computer Science Research Student Conference (NZCSRSC 2008), pp. 49–56 (2008)
14. Joachims, T.: Transductive inference for text classification using support vector machines. In: ICML, pp. 200–209 (1999)
15. Liu, B., Li, X., Lee, W.S., Yu, P.S.: Text classification by labeling words. In: Proceedings of 19th National Conference on Artificial Intelligence, pp. 425–430 (2004)
16. Nigam, K., McCallum, A.K., Thrun, S., Mitchell, T.: Text classification from labeled and unlabeled documents using EM. Mach. Learn. - Special issue on Information Retrieval **39**(2-3), 103-134 (2000)
17. Raghavan, H., Madani, O., Jones, R.: Active learning with feedback on features and instances. JMLR **7**, 1655–1686 (2006)
18. Razavi, A.H., Inkpen, D., Brusilovsky, D., Bogouslavski, L.: General topic annotation in social networks: a latent Dirichlet allocation approach. In: Zaïane, O.R., Zilles, S. (eds.) AI 2013. LNCS (LNAI), vol. 7884, pp. 293–300. Springer, Heidelberg (2013). https://doi.org/10.1007/978-3-642-38457-8_29
19. Schapire, R.E., Rochery, M., Rahim, M.G., Gupta, N.K.: Incorporating prior knowledge into boosting. In: ICML, pp. 538–545 (2002)
20. Settles, B.: Active learning literature survey. Computer Sciences Technical report 1648, University of Wisconsin–Madison (2009)
21. Subramanya, A., Bilmes, J.: Soft-supervised learning for text classification. In: EMNLP, pp. 1090–1099. Association for Computational Linguistics (2008)
22. Wang, F., Zhang, C.: Label propagation through linear neighborhoods. IEEE Trans. Knowl. Data Eng. **20**(1), 55–67 (2008)
23. Wu, X., Srihari, R.: Incorporating prior knowledge with weighted margin support vector machines. In: Proceedings of 10th ACM SIGKDD International Conference on Knowledge Discovery and Data Mining, pp. 326–333 (2004)
24. Zhu, X., Ghahramani, Z.: Learning from labeled and unlabeled data with label propagation. Technical report, Citeseer (2002)
25. Zhu, X., Ghahramani, Z.: Learning from labeled and unlabeled data with label propagation. Technical report, Carnegie Mellon University (2002)

# Deep Neural Networks for Czech Multi-label Document Classification

Ladislav Lenc[1,2] and Pavel Král[1,2(✉)]

[1] Department of Computer Science and Engineering, Faculty of Applied Sciences,
University of West Bohemia, Plzeň, Czech Republic
{llenc,pkral}@kiv.zcu.cz
[2] NTIS - New Technologies for the Information Society, Faculty of Applied Sciences,
University of West Bohemia, Plzeň, Czech Republic

**Abstract.** This paper is focused on automatic multi-label document classification of Czech text documents. The current approaches usually use some pre-processing which can have negative impact (loss of information, additional implementation work, etc). Therefore, we would like to omit it and use deep neural networks that learn from simple features. This choice was motivated by their successful usage in many other machine learning fields. Two different networks are compared: the first one is a standard multi-layer perceptron, while the second one is a popular convolutional network. The experiments on a Czech newspaper corpus show that both networks significantly outperform baseline method which uses a rich set of features with maximum entropy classifier. We have also shown that convolutional network gives the best results.

**Keywords:** Czech · Deep neural networks · Document classification
Multi-label

## 1 Introduction

The amount of electronic text documents is growing extremely rapidly and therefore automatic document classification (or categorization) becomes very important for information organization, storage and retrieval. Multi-label classification is considerably more important than the single-label classification because it usually corresponds better to the needs of the current applications.

The modern approaches usually use several pre-processing tasks: feature selection/reduction [1]; precise document representation (e.g. POS-filtering, particular lexical and syntactic features, lemmatization, etc.) [2] to reduce the feature space with minimal negative impact on classification accuracy. However, this pre-processing has several drawbacks as for instance loss of information, significant additional implementation work, dependency on the task/application, etc.

Neural networks with deep learning are today very popular in machine learning field and it was proved that they outperform many state-of-the-art

© Springer International Publishing AG, part of Springer Nature 2018
A. Gelbukh (Ed.): CICLing 2016, LNCS 9624, pp. 460–471, 2018.
https://doi.org/10.1007/978-3-319-75487-1_36

approaches without any parametrization. This fact is particularly evident in image processing [3], however it was further showed that they are also superior in Natural Language Processing (NLP) including Part-Of-Speech (POS) tagging, chunking, named entity recognition or semantic role labelling [4]. However, to the best of our knowledge, the current published work does not include their application for multi-label document classification.

Therefore, the main goal of this paper consists in using neural networks for multi-label document classification of Czech text documents. We will compare several topologies with different number of parameters to show that they can have better accuracy than the state-of-the-art methods.

We use and compare standard feed-forward networks (i.e. multi-layer perceptron) and popular Convolutional Networks (CNNs). To the best of our knowledge, this comparison was never been done on this task before. Therefore, it is another contribution of this paper. Note that we expect better performance of the CNNs as shown for instance in the OCR task [5].

The results of this work should be integrated into an experimental multi-label document classification system. The system should be used to replace manual annotation of the newspaper documents which is very expensive and time consuming task and thus save the human resources in the Czech News Agency (ČTK)[1]

The rest of the paper is organized as follows. Section 2 is a short review of document classification methods with a particular focus on neural networks. Section 3 describes our document classification approaches. Section 4 deals with experiments realized on the ČTK corpus and then discusses the obtained results. In the last section, we conclude the experimental results and propose some future research directions.

## 2   Related Work

Document classification is usually based on supervised machine learning methods that exploit an annotated corpus to train a classifier which then assigns the classes to unlabelled documents. The most of works use Vector Space Model (VSM), which usually represents each document with a vector of all word occurrences weighted by their Term Frequency-Inverse Document Frequency (TF-IDF).

Several classification methods have been successfully used [6], for instance Bayesian classifiers, Maximum Entropy (ME), Support Vector Machines (SVMs), etc. However, the main issue of this task is that the feature space in the VSM is highly dimensional which decreases the accuracy of the classifier.

Numerous feature selection/reduction approaches have been introduced [1,7] to solve this problem. Furthermore, a better document representation should help to decrease the feature vector dimension, e.g. using lexical and syntactic features as shown in [2]. Chandrasekar and Srinivas further show in [8] that it is beneficial to use POS-tag filtration in order to represent a document more accurately.

---

[1] http://www.ctk.eu.

More recently, some interesting approaches based on Latent Dirichlet Allocation (L-LDA) [9] have been introduced. Another method exploits partial labels to discover latent topics [10]. Principal Component Analysis [11] incorporating semantic concepts [12] has also been used for the document classification.

Recently, "deep" Neural Nets (NN) have shown their superior performance in many natural language processing tasks including POS tagging, chunking, named entity recognition and semantic role labelling [4] without any parametrization. Several different topologies and learning algorithms were proposed.

For instance, the authors of [13] propose two Convolutional Neural Nets (CNN) for ontology classification, sentiment analysis and single-label document classification. Their networks are composed of 9 layers out of which 6 are convolutional layers and 3 fully-connected layers with different numbers of hidden units and frame sizes. They show that the proposed method significantly outperforms the baseline approaches (bag of words) on English and Chinese corpora. Another interesting work [14] uses in the first layer (i.e. lookup table) pre-trained vectors from word2vec [15]. The authors show that the proposed models outperform the state-of-the-art on 4 out of 7 tasks, which include sentiment analysis and question classification.

For additional information about architectures, algorithms, and applications of deep learning, please refer the survey [16].

On the other hand, classical feed-forward neural nets architectures represented particularly by multi-layer perceptrons are used rather rarely. However, these models were very popular before and some approaches for document classification exist. Manevitz and Yousef show in [17] that their simple feed-forward neural network with three layers (20 inputs, 6 neurons in hidden layer and 10 neurons in the output layer, i.e. number of classes) gives F-measure about 78% on the standard Reuters dataset.

Traditional multi-layer neural networks were also used for multi-label document classification in [18]. The authors have modified standard backpropagation algorithm for multi-label learning which employs a novel error function. This approach is evaluated on functional genomics and text categorization.

The most of the proposed approaches is focused on English and only few works deal with Czech language. Hrala and Král use in [19] lemmatization and Part-Of-Speech (POS) filtering for a precise representation of Czech documents. In [20], three different multi-label classification approaches are compared and evaluated. Another recent work proposes novel features based on the unsupervised machine learning [21]. To the best of our knowledge, no document classification approach using neural nets deals with Czech language.

## 3   Neural Nets for Multi-label Document Classification

### 3.1   Baseline Classification

The feature set is created according to Brychcín and Král [21] and is composed of words, stems and features created by S-LDA and COALS. They are used because the authors experimentally proved that the additional unsupervised features significantly improve classification results.

For multi-label classification, we use an efficient approach presented by Tsoumakas and Katakis in [22]. This method employs $n$ binary classifiers $C_{i=1}^n : d \rightarrow l, \neg l$ (i.e. each binary classifier assigns the document $d$ to the label $l$ iff the label is included in the document, $\neg l$ otherwise). The classification result is given by the following equation:

$$C(d) = \cup_{i=1}^n : C_i(d) \tag{1}$$

The Maximum Entropy (ME) model is used for classification.

## 3.2 Standard Feed-Forward Deep Neural Network (FDNN)

Feed-forward neural networks are probably the most commonly used type of NNs. We propose to use an MLP with two hidden layers which can be seen as a deep network[2]. As an input of our network we use the simple Bag of Words (BoW) which is a binary vector where value 1 means that the word with a given index is present in the document. The size of this vector depends on the size of the dictionary which is limited by $N$ most frequent words. The only preprocessing is the conversion of all characters to lower case and also replacing of all numbers by one common token.

The size of the input layer thus depends on the size of the dictionary that is used for the feature vector creation. The first hidden layer has 1024 while the second one has 512 nodes[3]. The output layer has size equal to the number of categories which is 37 in our case. To handle the multi-label classification, we threshold the values of nodes in the output layer. Only the values larger than a given threshold are assigned to the labels.

## 3.3 Convolutional Neural Network (CNN)

The input feature of the CNN is a sequence of words in the document. We use similar document preprocessing and also similar dictionary as in the previous approach. The words are then represented by the indexes into the dictionary.

The first important issue of this network for document classification is variable length of documents. It is usually solved by setting a fixed value and longer documents are shortened while shorter ones must be padded to ensure exactly the same length. The words that are not in the dictionary are assigned to a reserved index and the padding has also a reserved index.

The architecture of our network is motivated by Kim in [14]. However, we use just one size of the convolutional kernel and not the combination of several sizes. Our kernels have only 1 dimension (1D) while Kim have used larger 2 dimensional kernels. This is mainly due to our preliminary experiments where the simple 1 dimensional kernels gave better results than the larger ones.

---

[2] We have also experimented with an MLP with one hidden layer with lower accuracy.

[3] This configuration was set experimentally.

The input of our network is a vector of word indexes of the length $L$ where $L$ is the number of words used for document representation. The second layer is an embedding layer which represents each input word as a vector of a given length. The document is thus represented as a matrix with $L$ rows and $EMB$ columns where $EMB$ is the length of embedding vectors. The third layer is the convolutional one. We use $N_C$ convolution kernels of the size $K \times 1$ which means we do 1D convolution over one position in the embedding vector over $K$ input words. The following layer performs max pooling over the length $L - K + 1$ resulting in $N_C$ $1 \times EMB$ vectors. The output of this layer is then flattened and connected with the output layer containing 37 nodes.

The output of the network is then thresholded to get the final results. The values greater than a given threshold indicate the labels that are assigned to the classified document. The architecture of the network is depicted in Fig. 1.

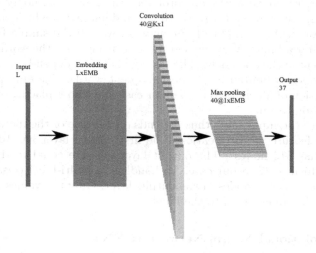

**Fig. 1.** Architecture of the convolutional network

## 4 Experiments

In this section we first describe the Czech document corpus that we used for evaluation of our methods. After that we describe the performed experiments and the final results. The results are compared with previously published results on the Czech document corpus.

### 4.1 Tools and Corpus

For implementation of all neural-nets we used Keras tool-kit [23] which is based on the Theano deep learning library [24]. It has been chosen mainly because of good performance and our previous experience with this tool. All experiments were computed on GPU to achieve reasonable computation times.

As already stated, the results of this work shall be used by the ČTK. Therefore, for the following experiments we used the Czech text documents provided by the ČTK. This corpus contains 2,974,040 words belonging to 11,955 documents. The documents are annotated from a set of 60 categories out of which we used 37 most frequent ones. The category reduction was done to allow comparison with previously reported results on this corpus where the same set of 37 categories was used. Figure 2 illustrates the distribution of the documents depending on the number of labels. Figure 3 shows the distribution of the document lengths (in word tokens). This corpus is freely available for research purposes at http://home.zcu.cz/~pkral/sw/.

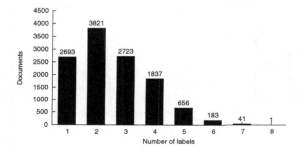

**Fig. 2.** Distribution of documents depending on the number of labels

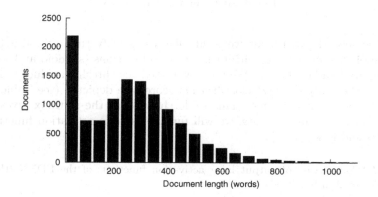

**Fig. 3.** Distribution of the document lengths

We use the five-folds cross validation procedure for all following experiments, where 20% of the corpus is reserved for testing and the remaining part for training of our models. For evaluation of the document classification accuracy, we use the standard F-measure (*F1*) metric [25]. The confidence interval of the experimental results is 0.6% at a confidence level of 0.95 [26].

## 4.2  Experimental Results

**FDNN.** As a first experiment, we would like to validate the proposition of thresholding applied to the output layer of the FDNN. For this task we use the Receiver Operating Characteristic (ROC) curve which clearly shows the relationship between the true positive and the false positive rate for different values of the *acceptance* threshold. We use 20,000 most common words to create the dictionary. The ROC curve is depicted in Fig. 4. According to the shape of this curve we can conclude that the proposed approach is suitable for multi-label document classification.

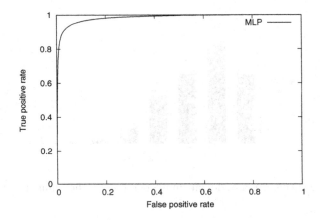

**Fig. 4.** ROC curve of the FDNN

In the second experiment we would like to identify the optimal activation function of the nodes in the output layer. Two functions (sigmoid and softmax) are compared and evaluated. We have evaluated the threshold values in interval $[0; 1]$, however only the best classification scores are depicted (see Table 1, best threshold values in brackets). This table shows that the softmax gives better results. Based on these results, we will further use this activation function and the threshold is set to 0.11.

**Table 1.** Comparison of output layer activation functions of the FDNN (threshold values depicted in brackets)

| Activation function | F1 [%] |
|---|---|
| Softmax | 83.8 (0.11) |
| Sigmoid | 82.3 (0.48) |

The third experiment studies the influence of the dictionary size on the performance of the FDNN. Table 2 shows the dependency of F-measure on the word number in the dictionary. This table shows that the previously chosen 20,000 words is a reasonable choice and further increasing the number does not bring any *significant* improvement.

**Table 2.** F-measure of FDNN with different numbers of words in the dictionary

| Word number | 1,000 | 2,000 | 5,000 | 10,000 | 15,000 | 20,000 | 25,000 | 30,000 |
|---|---|---|---|---|---|---|---|---|
| F1 [%] | 72.1 | 76.7 | 80.9 | 82.9 | 83.5 | 83.8 | 83.9 | 83.9 |

**CNN.** In all experiments performed with the CNN we use the same dictionary size (20,000 words) as in the case of FDNN to allow a straightforward comparison of the results. According to the analysis of our corpus we estimate that a suitable vector size for document representation is 400 words. As well as for the FDNN we first compute the ROC curve to validate the proposition of thresholding in the output. Figure 5 clearly shows that this approach is suitable for our task.

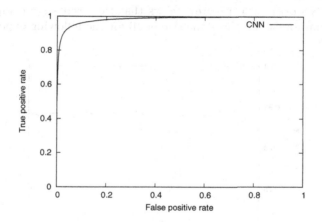

**Fig. 5.** ROC curve of the CNN

As a second experiment we identify an optimal activation function of neurons in the output layer. We compare the softmax and sigmoid functions. The achieved F-measures are depicted in Table 3. It is clearly visible that in this case the sigmoid function performs better. We will thus use the sigmoid activation function and the threshold will be set to 0.1 for all further experiments with CNN.

In this experiment, we will show the impact of the number of convolutional kernels in our network on the classification score. 400 words are used for document representation ($L = 400$) and the embedding vector size is 200. This experiment shows (see Table 4) that this parameter influences the classification score

**Table 3.** Comparison of output layer activation functions of the CNN (threshold values depicted in brackets)

| Activation function | F1 [%] |
|---|---|
| Softmax | 80.9 (0.07) |
| Sigmoid | 84.0 (0.10) |

**Table 4.** F-measure of CNN with different numbers of convolutional kernels

| Kernel no. | 12 | 16 | 20 | 24 | 28 | 32 | 36 | 40 | 44 | 48 | 52 | 56 | 60 | 64 |
|---|---|---|---|---|---|---|---|---|---|---|---|---|---|---|
| F1 [%] | 83.1 | 83.4 | 84.0 | 83.9 | 84.1 | 84.1 | 84.1 | 84.1 | 84.1 | 84.2 | 84.2 | 84.1 | 84.1 | 84.0 |

only very slightly ($\Delta F1 \sim +1\%$). All values from interval [20; 64] are suitable for our goal and therefore we chose the value of 40 for further experimentation.

The following experiment shows the dependency of F-measure on the size of convolutional kernels. We use 40 kernels and the size of the kernel varies from 2 to 40. The size of the kernels can be interpreted as the length of word sequences that the CNN works with. Figure 6 shows that the results are comparable and as a good compromise we chose the size of 16 for the following experiments.

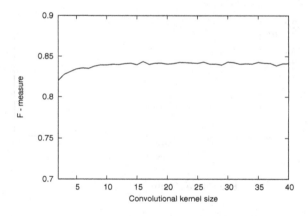

**Fig. 6.** Dependency of F-measure on the size of convolutional kernel

Finally, we tested our network with different numbers of input words and with varying size of the embedding vectors. Table 5 shows the achieved results with several combinations of these parameters. We can conclude that the 400 words that we chose at the beginning was a reasonable choice. However, it is beneficial to use longer embedding vectors. It must be noted that the further increasing of the embedding size has a strong impact on the computation time and might be not practical for real-world applications.

**Table 5.** F-measure of CNN with different word numbers and different embedding sizes [%]

| Word number Embedding length | 100 | 200 | 300 | 400 | 500 |
|---|---|---|---|---|---|
| 100 | 82.3 | 82.7 | 83.1 | 83.3 | 83.4 |
| 150 | 82.9 | 83.3 | 83.4 | 83.7 | 83.9 |
| 200 | 82.9 | 83.4 | 83.8 | 84.0 | 84.1 |
| 250 | 82.5 | 83.8 | 84.1 | 84.2 | 84.4 |
| 300 | 83.2 | 83.9 | 84.3 | 84.3 | 84.5 |
| 350 | 83.4 | 83.9 | 84.3 | 84.6 | 84.5 |
| 400 | 83.4 | 83.9 | 84.4 | 84.7 | 84.7 |
| 450 | 83.4 | 83.9 | 84.5 | 84.8 | 84.3 |

**Summary of the Results.** Table 6 compares the results of our approaches with another efficient method [21]. The results show that both proposed approaches significantly outperform this baseline approach that uses several features with ME classifier.

**Table 6.** Comparison of the results of our approaches with maximum entropy based method

| Method | Precision | Recall | F1 [%] |
|---|---|---|---|
| Brychcín and Král [21] | 89.0 | 75.6 | 81.7 |
| FDNN | 83.7 | 83.6 | 83.9 |
| CNN | 86.4 | 82.8 | 84.7 |

# 5  Conclusions and Future Work

In this paper, we have used two different neural nets for multi-label document classification of Czech text documents. Several experiments were realized to set optimal network topologies and parameters. An important contribution is the evaluation of the performance of neural networks using simple features. Therefore we have used the BoW representation for the FDNN and sequence of word indexes for the CNN as the inputs. Based on these experiments we can conclude:

- the two proposed network topologies together with thresholding of the output are efficient for multi-label classification task
- softmax activation function is better for FDNN, while sigmoid activation function gives better results for CNN
- CNN outperforms FDNN only very slightly ($\Delta$ F1 $\sim$ +0.6%)

– the most important is the fact that both neural nets with only basic pre-processing and without any parametrization significantly improve the baseline maximum entropy method with a rich set of parameters ($\Delta$ F1 $\sim$ +4%).

Based on these results, we want to integrate CNN into our experimental document classification system.

In this paper, we have used relatively simple convolution neural network. Therefore, our first perspective consists in designing a more complicated CNN architecture. According to the literature, we assume that more layers in this network will have a positive impact on the classification score. Our embedding layer was also not initialized by some pre-trained semantic vectors (e.g. word2vec or GloVe). Another perspective thus consists in initializing of the embedding CNN layer with pre-trained vectors.

**Acknowledgements.** This work has been supported by the project LO1506 of the Czech Ministry of Education, Youth and Sports. We also would like to thank Czech New Agency (ČTK) for support and for providing the data.

# References

1. Yang, Y., Pedersen, J.O.: A comparative study on feature selection in text categorization. In: Proceedings of the Fourteenth International Conference on Machine Learning. ICML 1997, pp. 412–420. Morgan Kaufmann Publishers Inc. San Francisco (1997)
2. Lim, C.S., Lee, K.J., Kim, G.C.: Multiple sets of features for automatic genre classification of web documents. Inf. Process. Manag. **41**, 1263–1276 (2005)
3. Krizhevsky, A., Sutskever, I., Hinton, G.E.: Imagenet classification with deep convolutional neural networks. In: Advances in Neural Information Processing Systems, pp. 1097–1105 (2012)
4. Collobert, R., Weston, J., Bottou, L., Karlen, M., Kavukcuoglu, K., Kuksa, P.: Natural language processing (almost) from scratch. J. Mach. Learn. Res. **12**, 2493–2537 (2011)
5. Peyrard, C., Mamalet, F., Garcia, C.: A comparison between multi-layer perceptrons and convolutional neural networks for text image super-resolution. In: International Conference on Computer Vision Theory and Applications (2015)
6. Della Pietra, S., Della Pietra, V., Lafferty, J.: Inducing features of random fields. IEEE Trans. Pattern Anal. Mach. Intell. **19**, 380–393 (1997)
7. Lamirel, J.C., Cuxac, P., Chivukula, A.S., Hajlaoui, K.: Optimizing text classification through efficient feature selection based on quality metric. J. Intell. Inf. Syst. **45**(3), 379–396 (2014)
8. Chandrasekar, R., Srinivas, B.: Using syntactic information in document filtering: a comparative study of part-of-speech tagging and supertagging (1996)
9. Ramage, D., Hall, D., Nallapati, R., Manning, C.D.: Labeled LDA: a supervised topic model for credit attribution in multi-labeled corpora. In: Proceedings of the 2009 Conference on Empirical Methods in Natural Language Processing, EMNLP 2009, vol. 1, pp. 248–256. Association for Computational Linguistics, Stroudsburg (2009)

10. Ramage, D., Manning, C.D., Dumais, S.: Partially labeled topic models for interpretable text mining. In: Proceedings of the 17th ACM SIGKDD International Conference on Knowledge Discovery and Data Mining, KDD 2011, pp. 457–465. ACM, New York (2011)
11. Gomez, J.C., Moens, M.F.: PCA document reconstruction for email classification. Comput. Stat. Data Anal. **56**, 741–751 (2012)
12. Yun, J., Jing, L., Yu, J., Huang, H.: A multi-layer text classification framework based on two-level representation model. Expert Syst. Appl. **39**(2), 2035–2046 (2012)
13. Zhang, X., LeCun, Y.: Text understanding from scratch. arXiv preprint arXiv:1502.01710 (2015)
14. Kim, Y.: Convolutional neural networks for sentence classification. arXiv preprint arXiv:1408.5882 (2014)
15. Mikolov, T., Chen, K., Corrado, G., Dean, J.: Efficient estimation of word representations in vector space. In: Proceedings of Workshop at ICLR (2013)
16. Deng, L.: A tutorial survey of architectures, algorithms, and applications for deep learning. APSIPA Trans. Signal Inf. Process. **3**, 1–29 (2014)
17. Manevitz, L., Yousef, M.: One-class document classification via neural networks. Neurocomputing **70**, 1466–1481 (2007)
18. Zhang, M.L., Zhou, Z.H.: Multilabel neural networks with applications to functional genomics and text categorization. IEEE Trans. Knowl. Data Eng. **18**, 1338–1351 (2006)
19. Hrala, M., Král, P.: Evaluation of the document classification approaches. In: Burduk, R., Jackowski, K., Kurzynski, M., Wozniak, M., Zolnierek, A. (eds.) CORES 2013. AISC, vol. 226, pp. 877–885. Springer, Heidelberg (2013). https://doi.org/10.1007/978-3-319-00969-8_86
20. Hrala, M., Král, P.: Multi-label document classification in Czech. In: Habernal, I., Matoušek, V. (eds.) TSD 2013. LNCS (LNAI), vol. 8082, pp. 343–351. Springer, Heidelberg (2013). https://doi.org/10.1007/978-3-642-40585-3_44
21. Brychcín, T., Král, P.: Novel unsupervised features for Czech multi-label document classification. In: Gelbukh, A., Espinoza, F.C., Galicia-Haro, S.N. (eds.) MICAI 2014. LNCS (LNAI), vol. 8856, pp. 70–79. Springer, Cham (2014). https://doi.org/10.1007/978-3-319-13647-9_8
22. Tsoumakas, G., Katakis, I.: Multi-label classification: an overview. Int. J. Data Warehous. Min. (IJDWM) **3**, 1–13 (2007)
23. Chollet, F.: Keras (2015). https://github.com/fchollet/keras
24. Bergstra, J., Breuleux, O., Bastien, F., Lamblin, P., Pascanu, R., Desjardins, G., Turian, J., Warde-Farley, D., Bengio, Y.: Theano: a CPU and GPU math expression compiler. In: Proceedings of the Python for Scientific Computing Conference (SciPy), Austin, TX, vol. 4, p. 3 (2010)
25. Powers, D.: Evaluation: from precision, recall and f-measure to roc., informedness, markedness & correlation. J. Mach. Learn. Technol. **2**, 37–63 (2011)
26. Press, W.H., Teukolsky, S.A., Vetterling, W.T., Flannery, B.P.: Numerical Recipes in C. vol. 2. Citeseer (1996)

# Turkish Document Classification with Coarse-Grained Semantic Matrix

İlknur Dönmez[(✉)] and Eşref Adalı

Computer Engineering Department, İstanbul Technical University,
Maslak, 34369 İstanbul, Turkey
buyukkuscu@itu.edu.tr

**Abstract.** In this paper, we present a novel method for Document Classification that uses semantic matrix representation of Turkish sentences by concentrating on the sentence phrases and their concepts in text. Our model has been designed to find phrases in a sentence, identify their relations with specific concepts, and represent the sentences as coarse-grained semantic matrix. Predicate features and semantic class type are also added to the coarse-grained semantic matrix representation. The highest success rate in Turkish Document Classification "97.12" is obtained by adding the coarse-grained semantic matrix representation to the data which has previous highest result in the previous studies about Turkish Document Classification.

## 1   Introduction

Representing words, sentences or text as condensed vectors or matrices has become crucial because of big data processing issues. In most natural language applications, sparseness is one of the important issues. Vector representation of the words is done via deep learning in 2013 [17]. The distance between the word vectors can show the semantic and syntactic relations between the words. But the best Pearson correlation of the semantic relatedness of word vectors is about 75% [16]. Meanings of larger units, calculated compositionally is still an issue for NLP and NLP deep learning applications [20].

In our study for generating the coarse-grained matrix representation for a sentence, first of all it is divided into sub-sentences. The number of sub-sentences depends on the gerunds (verbal nouns), participles (verbal adjectives) and converbs (verbal adverbs) in the sentence. Grammatical analysis involves the presence of argument phrases in the sub-sentences via parse tree outputs and case markers in a given sentence. Then the phrases of the sentence and the concept of each phrases is found. At the end, the predicate phrase category is also added to the matrix representation.

The coarse-grained semantic matrix representation of sentences (texts) can be used as an input for a great deal of semantic applications such as question answering, information extraction and text categorization. In this paper the utility of the target application is demonstrated with a Document Classification example.

© Springer International Publishing AG, part of Springer Nature 2018
A. Gelbukh (Ed.): CICLing 2016, LNCS 9624, pp. 472–484, 2018.
https://doi.org/10.1007/978-3-319-75487-1_37

A lot of approaches based on machine learning and intelligent agents have been applied to document classification task in recent years including K-nearest neighbor (KNN) [22, 26], Naive Bayes [12], Decision Trees [25], neural networks [15, 18], generative probabilistic classifiers [14], multivariate regression models [26] and support vector machines [11, 24]. There are some document classification application that uses statistical machine learning methods for Turkish [2, 6, 23, 27]. To the best of our knowledge the best result is "96.25" for text classification of five category in Turkish.

## 2    Phrases and Concepts

When a sentence is analysed semantically by a human, the sentences are separated into parts related to "where", "who", "when", "with whom", "why" and other questions. Then each answer of the question is connected to the related concept by the human so representing these answers as concept in a matrix can be a useful source for NLP semantic applications. In our study a Turkish sentence is represented with 51 concepts and 10 phrases. The presence of the arguments (phrase-content pairs) in the input sentence is used to represent sentences, not their locations because of free word order in Turkish. It is so free that sentences can be structured as all of the permutations of its phrases [10]. "Phrases" and "Concepts" are the important parts of our analysis and representation of sentences.

### 2.1    Phrase

In our study, the sentence is analysed as phrase and concept. The phrases that are seen on Table 1 are used in the study.

**Table 1.** Phrases that are used in application

| Phrases | Suffixes |
| --- | --- |
| P1: Subject phrase | - |
| P2: Nominative object phrase | - |
| P3: Accusative object phrase | -(y)ı, -(y)i, -(y)u, -(y)ü |
| P4: Destination Phrase (Dative Phrase) | -(y)a, -(y)e |
| P5: Location Phrase | -da, -de, -ta, -te |
| P6: Source Phrase (Ablative Phrase) | -dan, -den, -tan, -ten |
| P7: Instrument Phrase | -la, -le |
| P8: Adverb Phrase | - |
| P9: Preposition Phrase | - |
| P10: Predicate | - |

The four of the ten phrases are formed by case marker suffixes in Turkish. Besides the phrases that form via case marker suffixes, the suffix "-le" or "-la" is used for the instrument phrase. The phrases which do not take suffixes like adverb phrase, preposition phrase and predicate phrase are also used.

Turkish case markers and phrase relations are represented with formal languages on Table 2. This representation is formed as {P, N, T, S} so that P is generation rule, N is non terminal, T is terminal and S is starting symbol [3,7]. As seen on the Table 2, the suffixes "i", "e", "de", "den", "le", noun phrase, noun, adverb phrase, adverb and predicate are terminals. S, $P_1, P_2, P_3, P_4, P_5, P_6, P_7, P_8, P_9$, X and Z are non-terminals. $\lambda$ denotes the empty string. When we assume that the set of all permutations on the set

$$P = \{P_i : 1 <= i <= 9\} \tag{1}$$

is $\prod$, for the generation rule $p \in \prod$, S is denoted as $S \leftarrow pZ$.

**Table 2.** Formal representation of Turkish case markers and phrases relations

| Phrases | Suffixes |
| --- | --- |
| S ← p Z | S: Simple sentence: A sentence that includes, only one action, judgment or existence |
| P1 ← X \| $\lambda$ | P1: Subject phrase: Doer of the action or the thing that the predicate give knowledge directly. Subject answers "who" questions of the predicate |
| P2 ← X \| $\lambda$ | P2: Nominative Object Phrase: Nominative object is unspecific object that answers of "what" question of the predicate |
| P3 ← Xi \| $\lambda$ | P3: Accusative Object Phrase: it is specific object that answers of "what" question of the predicate |
| P4 ← Xe \| $\lambda$ | P4: Destination (Dative Phrase):it answers "to what", "to who", "to where", "to what time" questions of the predicate |
| P5 ← Xde \| $\lambda$ | P5: Location Phrase: it answers "in what", "on who", "in where","at what time" questions of the predicate |
| P6 ← Xden \| $\lambda$ | P6: Source (Ablative) Phrase: Ablative Phrase answers "from what", "from who", "from where", "from what time" questions of the predicate |
| P7 ← Xle \| $\lambda$ | P7: Instrumental Phrase: Instrumental Phrase answers "with what", "with who", "with where", "with what time" questions of the predicate |
| P8 ← adverb phrase \| adverb \| $\lambda$ | |
| P9 ← preposition phrase \| preposition \| $\lambda$ | |
| X ← noun phrase \| noun | |
| Z ← predicate | |

The phrase order is so flexible in Turkish that the sentence S can be formed with all of the permutations of $P_1, P_2, P_3, P_4, P_5, P_6, P_7, P_8, P_9$, non-terminals and Z. The computational analysis of the syntax and interpretation of "free" word order in Turkish is studied by Hoffman in 1995 [10]. In our study the dependency parser results and case markers are used to find the phrases within Turkish sentences.

## 2.2 Concepts

After the sentence phrases are found, the concept or concepts of each phrase are determined. The conceptual classifications of the entities are seen at ontologies. According to Lakoff [13], classification of entities is related with the physical environment, culture and is a comprehensive process that involves different variables.

In our study the most appropriate concepts are selected according to some critical issues. The first issue is which concepts are the basis for the entity categorization. The WordNet top hierarchical classes are examined for this purpose [9]. The second one is the distance between concepts that should be big enough to separate the entities and the last one is the term dictionaries that we maintained from Turkish Language Association. The concepts that are used in the study are shown in Table 3.

**Table 3.** Concepts that are used in application

| 1 Human | 10 Construction | 19 Biology | 28 Laws | 37 Health | 46 Commerce |
|---------|-----------------|------------|---------|-----------|-------------|
| 2 Anatomy | 11 Food | 20 Geography | 29 Linguistic | 38 Geology | 47 Theatre |
| 3 Animal | 12 Drink | 21 Physics | 30 Theology | 39 Marine | 48 Society |
| 4 Liquid | 13 Liquid | 22 Chemistry | 31 Literature | 40 Army | 49 Abstract |
| 5 Plant | 14 Gases | 23 Mining | 32 Education | 41 Cinema | 50 Reason |
| 6 Location | 15 Action | 24 Mathematics | 33 Economy | 42 Sport | 51 Unknown |
| 7 Time | 16 Organization | 25 Logic | 34 Philosophy | 43 History | |
| 8 Vehicle | 17 Informatics | 26 Geometry | 35 Astronomy | 44 Technical | |
| 9 Ware | 18 Measurement | 27 Music | 36 Psychology | 45 Television | |

After the concept titles are determined, noun phrases list are generated for each concept classes. Some of the concepts lists are obtained from TDK term dictionaries [4] but some of them were formed by us using Turkish dictionaries explanation parts and BalkaNet [5,21] which is an application of WordNet for Turkish.

## 2.3 Turkish Phrase-Concept as Thematic Roles Relation

The usage of phrases is essential for semantic analysis; the phrases are the answers of the "who", "what", "to what", "to who", "to where", "to what time",

**Table 4.** Equivalent phrase-concept pairs of VerbNet thematic roles

| Roles | Equivalent phrase-concept pair (in Turkish) | Example |
|---|---|---|
| Actor | Subject phrase-Person | |
| Agent | Subject phrase-Person; Subject phrase-Animal | "Susan whispered." Agent V |
| Co-agent | Instrument phrase-Person; Instrument phrase, Animal | "Brenda met with Molly." Agent V {with} Co-Agent |
| Asset | Subject phrase-Quantity | "$100,000 will build you a house." Asset V Beneficiary Product |
| Attribute | As equal "being" in Turkish, It start a new sub sentence in our work | "I fired him as my chief of staff." Agent V Theme {as} Attribute |
| Beneficiary | Dative phrase-Person; Dative phrase-Animal | "Carmen bought a dress for Mary." Agent V Theme {for} Beneficiary |
| Cause | Subject phrase-Abstract | "It freed him of guilt." Cause V Source {of} Theme |
| Destination | Dative phrase | "He came to Colorado." Theme V {+path} Destination |
| Source | Ablative phrase | "He converted from A to B." Patient V {from} Source {to} Goal |
| Location | Locative phrase | "Unicorns don't exist on Earth." Theme V {+loc} Location |
| Experiencer | Subject phrase-Person | "The tourists admired the paintings." Experiencer V Stimulus |
| Instrument | Instrument phrase | "Paula hit the ball with a stick." Agent V Patient {with} Instrument |
| Material | Subject phrase-Material; Object phrase-Material | "I kneaded the dough." Agent V Material |
| Product | Instrument phrase-Ware; Location phrase-Ware; Dative phrase Ware; Ablative phrase Ware | "The dough twirled into a pretzel." Material V {into} Product |
| Patient | Ablative phrase-Person; Ablative phrase-Animal | "Celia brushed the baby's hair." Agent V Patient |
| Stimulus | Dative phrase-Person; Dative phrase-Animal | "The tourists admired the paintings." Experiencer V Stimulus |
| Theme | Accusative phrase-All concepts Subject phrase-Not person and not animate concepts | "The thief stole the paint." Agent V Theme |
| Time | Time phrase | "John started the party at 5 oclock." Agent V Theme V Time |
| Topic | Ablative phrase-Abstract | "John ordered that she come ." Agent V Topic <+that_compt> |
| Predicate | Predicate phrase | |

"in what", "on who", "in where", "at what time", "from what", "from who", "from where", "from what time", "with what", "with who", "with where" and "with which quantity" questions.

When we find the concept of the phrases, we relate this question to the concepts. For the "Who did the action" question, The answers may be "human, animal or plant did the action". Or we can know the directly affected subject is a quantity, ware, vehicle or any of our concepts (Table 4).

If the source phrase has a location concept, the starting point of the action is a place, if the source phrase has a time concept; the starting point of the action is specific time. If the instrument phrase has a human concept, it means the action is done with a human (co-agent), if the instrument phrase has a ware concept; the action is done with a ware. It can be easily seen that the phrase concept pairs gives important semantic knowledge about the sentence.

In some cases Turkish phrase concept pairs overlap SRL thematic roles. For example in a sentence if the phrase is subject and its concept is person or animal, it correspond to "agent" thematic role in VerbNet. When the subject phrase has a quantity concept, it corresponds to "asset" thematic role in VerbNet [19].

## 3    Basic Model

In our study, the sentences are preprocessed by İTÜ NLP web service [8] and the sentences are parsed according to its dependencies form the inputs of our basic model.

As seen on the Fig. 1, the pre-processed sentence was given to the Sub-sentence Finder as input, each sub simple sentence of the sentence goes to the Phrase Finder, where all the nine phrases of the sub-sentence are found using the dependency parser results and case markers. Then the Concept Finder tool searches phrases in the concepts lists and forms $9 \times 51$ $X'$ matrix. Here searched noun phrases may be a polysemous terms and can be found in more than one concept list. All concepts are taken into account. The real related concept frequency will increase in the whole text according to the words related with the same true concept and will distinguish from the wrong one in text representation.

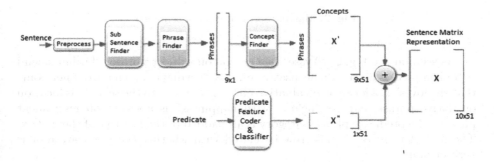

**Fig. 1.** Basic model

Simultaneously, the predicate is processed. If predicate has a noun root, it is searched in 51 noun phrase concept lists and predicate concepts list number is determined as binary form. Levin verb classes that are adapted to the Turkish are used if the verb has a verb root. The valence suffix, time, aspect modality (TAM) suffix and subject suffix of the predicate is also determined for forming $1 \times 51$ $X''$ matrix. For example if the predicate is "doldurdum (I filled)", it has a verb root, it is in "put verbs" class of Levin verb classes [19], it has past tense TAM suffix and it has first singular subject suffix. Turkish verbs are categorized manually according to Levin classes in this study. At the end the $X'$ matrix and $X''$ matrix are joined to form the X matrix that form a coarse-grained semantic representation of the sub-sentence.

## 3.1   Matrix Representation of Simple Sentences

In our study all types of sentences (compound and complex sentences) are converted to simple sentences. The elements of the matrix can take a value of 0 or 1. For the first nine rows, 1 means the sentence has the phrase-concept pair and 0 means the sentence does not have the phrase-concept pair.

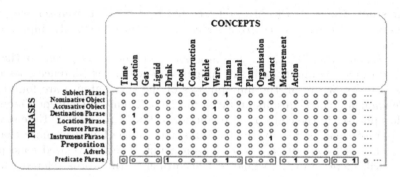

Ayşe kırılan kalemi sevdiği evinden okula sevinçle götürdü.
(Ayşe took her broken pen from her home that she loves to the school with happiness.)

**Fig. 2.** Matrix representation of sentence

As seen on the Fig. 2, "Ayşse" is a person on subject phrase, "kırılan kalemi = broken pen" is a ware on accusative phrase, "sevdiği evinden= from her home that she loves" is a location on ablative phrase, "okula = to the school" is location on dative phrase and "sevinçle = with happiness" is abstract on instrument phrase. The predicate phrase category is also added to the last row of the matrix. The first element of the last row is related with whether the predicate root is verb or noun.

If the predicate root is a noun, the second to fourth bits are related with suffixes types that transform the noun root into the predicate. If the predicate root is a noun and if the noun takes "-len" suffix that means "to have" in English, these three bits are "001"; If the predicate root is a noun and if the noun takes "-leş" suffix that means "to become" in English, these three bits are "010"; If

| Verb Root | - | Levin Class No | Valency Suffix Type | Time Suffix Type | Subject Suffix Type | The others |
|---|---|---|---|---|---|---|
| 0 | 000 | XXXXXXX | XXX | XXXXX | XXX | 00.. |
| | | | | | | |
| Noun Root | * | Concept No | Valency Suffix Type | Time Suffix Type | Subject Suffix Type | |
| 1 | 001 | XXXXXXX | XXX | XXXXX | XXX | 00.. |
| 1 | 010 | XXXXXXX | XXX | XXXXX | XXX | 00.. |
| 1 | 011 | XXXXXXX | XXX | XXXXX | XXX | 00.. |
| 1 | 100 | XXXXXXX | XXX | XXXXX | XXX | 00.. |

\* If the predicate root is noun and if noun takes "-len" suffix it is coded as "001", if noun takes "-leş" suffix it is coded as "010", if noun takes "-le" suffix it is coded as "011", if noun takes no suffix it is coded as "100".

**Fig. 3.** Predicate phrase data

the predicate root is a noun and if the noun takes "-le" suffix that means "to make" in English, these three bits are "011"; If the predicate root is a noun and if the noun takes no suffix, these three bits are "100". The fifth to eleventh bits are related with binary number of concept list as seen on Fig. 3.

For example the predicate "hüzünlendim" that means "I have sorrow" has the root word "hüzün" that means "sorrow", it is an element of abstract concepts list. In Turkish a predicate which has a noun root may have valance, TAM and subject suffixes. The twelfth to fourteenth bits are related with valence suffix type, fifteenth to nineteenth bits are related with TAM suffix type and twentieth to twenty-second bits are related with subject suffix type.

If the predicate root is a verb, the fifth to eleventh bits are related with the binary number of the Levin class (semantic verb class) the twelfth to fourteenth bits are related with valence suffix type, the fifteenth to nineteenth bits are related with TAM suffix type and the twentieth to twenty-second bits are related with subject suffix type.

## 3.2  Finding Sub Sentences of Compound and Complex Sentences

A compound sentence may have more than one complex sentence and each complex sentence may have more than one sub-sentence. In our model because of best representation of the sentence, the sentence is divided into sub sentences.

If the sentence is compound, the first complex sentence is taken and the reminder part is stored. For the complex part the number of the light verb gives the number of the sub sentences that we want to maintain. According to NLP parser result gerund (verbal noun), participle (verbal adjective) and verbal adverb and their dependent phrases are taken. For each light verb form and their related phrases the sub sentences are generated via using determined rules. After the entire sub sentences of the complex sentence is generated, the process goes on from the starting point, first complex sentence of the reminder part is found and algorithm goes on until all sub sentences are found. Here is an example (Fig. 4).

*"Mavi ceketli adam dün ofise geldi sonra devam eden işi ve işe dahil olan çalışanları yokladı". (The man with blue jacket came to the office yesterday then inspected the continued process and the employees that joined the process.)*

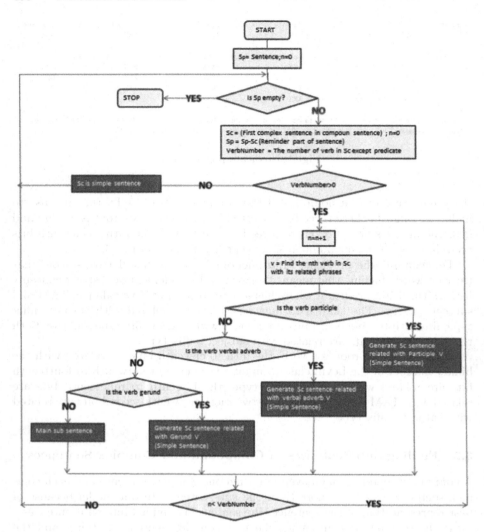

**Fig. 4.** Finding sub sentences

*First Complex (FC) and its Sub Sentence 1:* "*Mavi ceketli adam dün ofise geldi.*" *(The man with blue jacket came to the office yesterday.)*

*Second Complex (SC):* "*Devam eden işi ve işe dahil olan çalışanları yokladı.*" *(He inspected the continued process and the employees that joined the process.)*

*SC Sub Sentence 1:* "*İş devam eder.*"*(The process continues.)*

*SC Sub Sentence 2:* "*Çalışanlar işe dahil olur.*"*(The employees joined the process.)*

*SC Sub Sentence 3:* "*Çalışanları yokladı.*" *(He inspected the employees.)*

# 4   Approach

## 4.1   Problem Specifics

Our project focused on classifying articles for the list of documents that have "Politic", "Sport", "Health", "Economy" and "Magazine" categories. Supervised learning took place on an archive of 1150 articles and 230 articles from each of the five categories. This data has been used in Text documentation application and as far as we know this application has the best result in Turkish Document Classification [1].

## 4.2   Classification and Feature Selection Algorithms

In this work, WEKA Nave Bayes, LSVM, C 4.5 and K- Nearest Neighbor Classification Methods are used for the document classification problem. Chi square attribute evaluation model with ranker search is used for the feature selection algorithm. 10-fold cross validation was used for evaluation of success ratio.

## 4.3   Experimental Results

In our experiments, we showed whether the modelling of Turkish text with coarse-grained semantic matrix representation is successful approach or not in determining Document Classification.

Four different classifiers and three different Datasets are used for the classification. In the old best model F. Amasyalı used some equations with 39699 dimensions. This model is compared with our coarse-grained semantic representation model has 2550 dimensions and the mixture model. In our model, each dimention is related with the matrix elemtent of sub-sentence representation for 5 sub sentence. We also used the mixture model Default performance (dp) is the success ratio when all instances were classified as target class.

DataSet-I is the data set which was used in the model with the highest precision and recall rate in Turkish document classification. The highest success rate was 94.75% with SVM algorithm for this dataset. Dataset-II is the first 5 sub sentences of document which are represented in coarse-grained semantic representation as previously described. Each sub-sentence is represented with 510 dimensions thus the feature vector is 2550 dimensional vector. DataSet-I and DataSet-II are put together to form DataSet-III.

Success rate for different datasets and models are seen on Table 5.

**Table 5.** Success rate for different datasets and models

| db:16.6% | C 4.5% | NB% | SVM% | KNN% |
|---|---|---|---|---|
| DataSet-I (39699 feature) | 76 | 86.69 | 94.75 | 65 |
| DataSet-II (2550 feature) | 72.5 | 87.60 | 85.84 | 54.4 |
| DataSet-III (42249 feature) | 81 | 97.12 | 96.54 | 68.5 |

**Table 6.** Success rate for Dataset-II after feature selection

| db:16.6% | C 4.5% | NB% | SVM% | KNN% |
|---|---|---|---|---|
| DataSet-II (2550 feature) | 72.5 | 87.60 | 85.84 | 54.4 |
| DataSet-II (145feature) | 71.2 | 86.10 | 83 | 51.24 |

After applying Chi square attribute evaluation model with ranker search feature selection method to the DataSet-II, success rates are seen on Table 6.

After applying Chi square attribute evaluation model with ranker search feature selection method to the DataSet-III, success rates are seen on Table 7.

**Table 7.** Success rate for Dataset-III after feature selection

| db:16.6% | C 4.5% | NB% | SVM% | KNN% |
|---|---|---|---|---|
| DataSet-III (42249 feature) | 81 | 97.12 | 96.54 | 68.5 |
| DataSet-III (405 feature) | 79.5 | 96.60 | 93.7 | 67.2 |

## 5 Conclusion

In this study the WEKA packet is used for the classification and feature selection algorithms and the NLP tool is used for parsing the sentences. One of the difficulties of Text Classification Studies is the high number of dimensions in data representation. In traditional approaches, texts are represented with vectors with dimensions equal to the number of different words within the texts. The high number of dimensions in data representation can be reduced using the roots of the words in agglutinative languages like Turkish but the dimensions are still around one thousand. In this study, the dimension is reduced to 145 and 86.10 accuracy is maintained with these 145 features using coarse-grained semantic sentence representation. DataSet-III which is a combined form of DataSet-I and DataSet-II has the 97.12 accuracy with Naive Bayes which is a higher result than the best result so far.

## References

1. Amasyalı, M.F., Beken, A.: Türkçe kelimelerin anlamsal benzerliklerinin ölçülmesi ve metin sınıflandırmada kullanılması measurement of Turkish word semantic similarity and text categorization application
2. Amasyalı, M.F., Diri, B.: Automatic Turkish text categorization in terms of author, genre and gender. In: Kop, C., Fliedl, G., Mayr, H.C., Métais, E. (eds.) NLDB 2006. LNCS, vol. 3999, pp. 221–226. Springer, Heidelberg (2006). https://doi.org/10.1007/11765448_22
3. Backus, J.W.: The syntax and semantics of the proposed international algebraic language of the Zurich ACM-GAMM conference. In: 1959 Proceedings of the International Comference on Information Processing (1959)

4. Baytop, T.: Türkçe bitki adları sözlüğü, vol. 578. Turk Dil Kurumu, Ankara (1994)
5. Bilgin, O., Çetinoğlu, Ö., Oflazer, K.: Building a wordnet for Turkish. Rom. J. Inf. Sci. Technol. **7**(1–2), 163–172 (2004)
6. Çataltepe, Z., Turan, Y., Kesgin, F.: Turkish document classification using shorter roots. In: 2007 IEEE 15th Signal Processing and Communications Applications, SIU 2007, pp. 1–4. IEEE (2007)
7. Chomsky, N.: Syntactic Structures. Walter de Gruyter, Berlin (2002)
8. Eryigit, G.: ITU Turkish NLP web service. In: 2014 EACL, p. 1 (2014)
9. Fellbaum, C.: WordNet. Wiley Online Library, New York (1998)
10. Hoffman, B.: The computational analysis of the syntax and interpretation of "free" word order in Turkish. IRCS Technical reports Series, p. 130 (1995)
11. Joachims, T.: Text categorization with support vector machines: learning with many relevant features. In: Nédellec, C., Rouveirol, C. (eds.) ECML 1998. LNCS, vol. 1398, pp. 137–142. Springer, Heidelberg (1998). https://doi.org/10.1007/BFb0026683
12. Kim, S.-B., Rim, H.-C., Yook, D.S., Lim, H.-S.: Effective methods for improving naive Bayes text classifiers. In: Ishizuka, M., Sattar, A. (eds.) PRICAI 2002. LNCS (LNAI), vol. 2417, pp. 414–423. Springer, Heidelberg (2002). https://doi.org/10.1007/3-540-45683-X_45
13. Lakoff, G.: Women, Fire, and Dangerous Things: What Categories Reveal About the Mind, vol. 1. Cambridge University Press, Cambridge (1990)
14. Lewis, D.D.: Naive (Bayes) at forty: the independence assumption in information retrieval. In: Nédellec, C., Rouveirol, C. (eds.) ECML 1998. LNCS, vol. 1398, pp. 4–15. Springer, Heidelberg (1998). https://doi.org/10.1007/BFb0026666
15. Li, C.H., Park, S.C.: Text categorization based on artificial neural networks. In: King, I., Wang, J., Chan, L.-W., Wang, D.L. (eds.) ICONIP 2006. LNCS, vol. 4234, pp. 302–311. Springer, Heidelberg (2006). https://doi.org/10.1007/11893295_35
16. Marelli, M., Bentivogli, L., Baroni, M., Bernardi, R., Menini, S., Zamparelli, R.: Semeval-2014 task 1: evaluation of compositional distributional semantic models on full sentences through semantic relatedness and textual entailment. In: SemEval-2014 (2014)
17. Mikolov, T., Chen, K., Corrado, G., Dean, J.: Efficient estimation of word representations in vector space. arXiv preprint arXiv:1301.3781 (2013)
18. Nakayama, M., Shimizu, Y.: Subject categorization for web educational resources using MLP. In: ESANN, pp. 9–14 (2003)
19. Schuler, K.K.: Verbnet: a broad-coverage, comprehensive verb lexicon (2005)
20. Socher, R., Huval, B., Manning, C.D., Ng, A.Y.: Semantic compositionality through recursive matrix-vector spaces. In: Proceedings of the 2012 Joint Conference on Empirical Methods in Natural Language Processing and Computational Natural Language Learning, pp. 1201–1211. Association for Computational Linguistics (2012)
21. Stamou, S., Oflazer, K., Pala, K., Christoudoulakis, D., Cristea, D., Tufis, D., Koeva, S., Totkov, G., Dutoit, D., Grigoriadou, M.: BALKANET: a multilingual semantic network for the Balkan languages. In: Proceedings of the International Wordnet Conference, Mysore, India, pp. 21–25 (2002)
22. Tan, S.: An effective refinement strategy for KNN text classifier. Expert Syst. Appl. **30**(2), 290–298 (2006)
23. Torunoğlu, D., Çakırman, E., Ganiz, M.C., Akyokuş, S., Gürbüz, M.Z.: Analysis of preprocessing methods on classification of Turkish texts. In: 2011 International Symposium on Innovations in Intelligent Systems and Applications (INISTA), pp. 112–117. IEEE (2011)

24. Tsochantaridis, I., Hofmann, T., Joachims, T., Altun, Y.: Support vector machine learning for interdependent and structured output spaces. In: Proceedings of the Twenty-First International Conference on Machine Learning, p. 104. ACM (2004)
25. Wu, M.C., Lin, S.Y., Lin, C.H.: An effective application of decision tree to stock trading. Expert Syst. Appl. **31**(2), 270–274 (2006)
26. Yang, Y., Chute, C.G.: An example-based mapping method for text categorization and retrieval. ACM Trans. Inf. Syst. (TOIS) **12**(3), 252–277 (1994)
27. Yıldız, H., Gençtav, M., Usta, N., Diri, B., Amasyalı, M.: Metin sınıflandırmada yeni özellik çıkarımı. In: IEEE SIU 2007 15 Sinyal İşleme, İletişim ve Uygulamaları Kurultayı (2007)

# Supervised Topic Models for Diagnosis Code Assignment to Discharge Summaries

Mohamed Dermouche[1,2]($\boxtimes$), Julien Velcin[3], Rémi Flicoteaux[1,2,4],
Sylvie Chevret[1,2,4], and Namik Taright[1,5]

[1] INSERM, U1153 Epidemiology and Biostatistics Sorbonne Paris Cité
Research Center (CRESS), ECSTRA team, 75010 Paris, France
`mohamed.dermouche@inserm.fr`, {`remi.flicoteaux,namik.taright`}`@aphp.fr`,
`sylvie.chevret@univ-paris-diderot.fr`
[2] Paris Diderot University, Paris, France
[3] Université de Lyon (ERIC Lyon 2), Lyon, France
`julien.velcin@univ-lyon2.fr`
[4] Saint-Louis Hospital, AP-HP, 75010 Paris, France
[5] SIMAP/DOMU/AP-HP, 75004 Paris, France

**Abstract.** Mining medical data has significantly gained interest in the recent years thanks to the advances in data mining and machine learning fields. In this work, we focus on a challenging issue in medical data mining: automatic diagnosis code assignment to discharge summaries, i.e., characterizing patient's hospital stay (diseases, symptoms, treatments, etc.) with a set of codes usually derived from the International Classification of Diseases (ICD). We cast the problem as a machine learning task and we experiment some recent approaches based on the probabilistic topic models. We demonstrate the efficiency of these models in terms of high predictive scores and ease of result interpretation. As such, we show how topic models enable gaining insights into this field and provide new research opportunities for possible improvements.

**Keywords:** ICD code assignment · Topic models · Machine learning
Natural language processing · Text categorization · Text mining

## 1 Introduction

Health information systems are used at a large scale in the healthcare institutions and hospitals for various tasks, such as medical record management, medical prescription, and billing. As a result, increasing large volumes of healthcare data are regularly generated in the form of Electronic Medical Records (EMR). In this regard, textual data has a prominent place. Free text is actually a suitable form to describe a wide range of data related to patient's care including medical history, personal statistics, admission diagnosis, patient-caregiver exchange, etc.

© Springer International Publishing AG, part of Springer Nature 2018
A. Gelbukh (Ed.): CICLing 2016, LNCS 9624, pp. 485–497, 2018.
https://doi.org/10.1007/978-3-319-75487-1_38

However, despite of being an abundant and valuable resource, only low quantities of these data are actually used for specific mining tasks, e.g., [14,15].

One major issue that can be approached by capitalizing on the routinely generated textual data is the automation of diagnosis code assignment to medical notes [4,5,9–11,13–15,17,21]. The task involves characterizing patient's hospital stay (symptoms, diagnoses, treatments, etc.) by a small number of codes, usually derived from the International Classification of Diseases (ICD). Diagnosis codes provide a fast and easy understanding of patient's state evolution. The same codes are used as billing elements by the health insurance systems. Because of its importance, the task of code assignment is often performed manually by professional coders. However, manual coding is tedious and time-consuming: on average the coders spend about five minutes identifying only single code.

The main goal of this paper is to explore a new approach to automatic code assignment, based on probabilistic topic models. This approach has shown excellent performance in various text mining tasks, such as topic discovery, information retrieval, and sentiment analysis [8,20]. Moreover, topic models provide a natural way to error analysis based on topic description, for example using discriminant words. Though, apart from some sparse work, the application of topic models for medical text mining purposes remains relatively less explored than in the other fields [8,13,20].

In this work, we experiment some recently-proposed supervised topic models in the task of automatic code assignment to medical discharge summaries. Our contributions can be summarized as follows:

1. New benchmark data: we create two french datasets with discharge summaries and manually associated ICD codes.
2. New learning models: we experiment some recent probabilistic topic models in a supervised fashion [1,16].

Our evaluation setup allows a fair comparison of machine learning models in a more refined way than the traditional measures. Both code and data will be available for the community after anonymization.

A brief introduction to the ICD in given in Sect. 2. Then, an overview of prior work for diagnosis assignment is given in Sect. 3. Experiments (methods, data, and evaluation framework) are described in Sect. 4. Results and discussion are given in Sect. 5. Finally, the paper is concluded in Sect. 6.

## 2    International Classification of Diseases

According to the World Health Organization (WHO), the International Classification of Diseases (ICD) is the "standard diagnostic tool for epidemiology, health management and clinical purposes"[1]. This mainly includes diseases, but also symptoms, signs, procedures, and other content related to diseases. There exist a separate classification per language, that is regularly revised by the WHO.

---

[1] http://www.who.int/classifications/icd/.

**Table 1.** ICD code examples from CIM10 (top) and ICD9 (bottom).

| ICD | Language | Code | Label |
|---|---|---|---|
| CIM10 | French | C83.7 | *Lymphome de Burkitt* (Burkitt's lymphoma) |
| | | C88.0 | *Macroglobulinémie de Waldenström* (Waldenström's macroglobulinemia) |
| | | D30.0 | *Tumeur bénigne du rein* (benign kidney tumor) |
| ICD9 | English | 198.3 | Secondary malignant neoplasm of brain and spinal cord |
| | | 414.01 | Coronary atherosclerosis of native coronary artery |
| | | V34.01 | Other multiple birth (three or more), mates all liveborn, delivered by cesarean section |

Currently, the latest revision for English is ICD10 whereas for French it is called CIM10 (*Classification Internationale de Maladies*). However, ICD9 is the most widely-used classification for diagnosis coding, in particular ICD9-Clinical Modification (ICD9-CM) as it allows comparability and use of mortality and morbidity data.

For this paper, we use ICD9-CM (that we call ICD9) and CIM10 classifications for English and French respectively. In ICD9, diagnosis codes are 3–5 characters. The first character is numeric or alpha while characters 2–5 are numeric. In CIM10, diagnosis codes are far to 6 characters. The first character is always alpha and designs a high-level category, while the remaining are numeric. Table 1 shows some examples from the codes used in this paper. On the other hand, ICD can be structured in a tree hierarchy with edges representing "is-a" relationship between a parent code and its children. More details about ICD can be found on the WHO website (see footnote 1).

## 3   Related Work

The problem of diagnosis code assignment has been studied from both perspectives of machine learning and computational linguistics, leading to a number of significant works. A bulk of these works have been published with the Computational Medicine Center's 2007 medical NLP challenge involving ICD code assignment to radiology reports [15]. With 45 distinct codes, the best F-score from the challenge was 89% while the average F-score was 77%. Note that all these works used multi-label classification: a document is assigned to one or more diagnoses (which is outside the scope of this paper). Some of them focused on using the ICD hierarchy to improve classification accuracy [14, 21].

In [5, 21], statistical classifiers were learnt based on a bag-of-words representation. The works in [5] used BoosTexter, a boosting-like technique based on a weak classifier, to learn a set of classification rules. The best achieved F-scores were around 84%. In [21], a simple classifier was learnt based on the presence/absence of UMLS terms[2]. The achieved scores were around 86%.

Besides this challenge, there also were some significant work such as [9,13, 14,17]. In [9], both SVM and Ridge Regression classifiers have achieved a score of 68% on a dataset with 2,618 distinct codes and a large number of learning documents (nearly 100,000). In [14], SVM classifier has been tested under a flat and a hierarchical setting. On a dataset with 5,030 distinct codes, the achieved F-scores were around 27% under flat setting and 39% under hierarchical setting. In [17], the k-NN classifier has been tested on a French corpus of medical documents with more 10,000 distinct codes. The algorithm achieved about 74% precision score but very low recall levels.

The closest work to ours is by Perotte et al. [13]. The authors proposed a hierarchically-supervised topic model (HSLDA) combining LDA model [2] with the knowledge from ICD structure. The hierarchical structure of ICD codes was taken into account during the topic learning step. For this to happen, the final predicted code was constrained to derive from one branch of the tree (a code could not be assigned to a document if its parent were not). HSLDA have been tested on a dataset with 7,298 distinct codes, where it performed about 5% better than the non-hierarchical sLDA model [1]. Unfortunately, HSLDA source code is not publicly available which prevents us from including it in this study.

## 4    Experiments

### 4.1    Methods

We choose three traditional machine learning models: an example-based model (Decision Tree), a probabilistic model (Naive Bayes), and a kernel-based model (Support Vector Machines). We put these models against two others from the topic model family: sLDA [1] and labeledLDA [16]. To carry out the experiments, we rely on R rpart package that implements an efficient Decision Tree (DT) algorithm based on information-gain ratio as a splitting criterion [19]. Similarly, we use e1071 package to perform Naive Bayes (NB) and Support Vector Machines (SVM) classifiers [12]. For the latter one, the best performance is obtained with a linear kernel while all the remaining parameters are left to their default values.

sLDA and labeledLDA are both based of the well-known LDA topic model [2]. Both models implement supervised learning based on the hidden topic structures (latent variables). In fact, the topic modeling process can be assimilated to a fuzzy word clustering where the goal is to build semantically coherent clusters [2]. sLDA and labeledLDA rely on slightly different structures (see Fig. 1). In both cases, the supervised part is implemented through a response variable, depicted by the letter $c$, that gives the predicted class modality (here the ICD codes).

Despite their apparent similarities, sLDA and labeledLDA differ mainly on two key points:

---

[2] https://www.nlm.nih.gov/research/umls/.

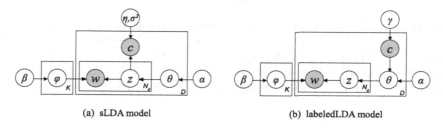

(a) sLDA model                    (b) labeledLDA model

**Fig. 1.** Plate notation of (a) sLDA and (b) labeledLDA topic models. In labeledLDA, the topics (variable $\theta$) are directly influenced by document's classes (variable $c$).

- Response formulation: in sLDA the response variable is derived from a Gaussian, while it follows a multinomial in labeledLDA. As such, labeledLDA would be more flexible and better fit with a multi-label classification, which is outside the scope of this paper.
- Topic supervision: in sLDA the response is calculated from empirical (learnt) topic distributions. For this, sLDA relies on a generalized linear model that maps the multinomial topic-document associations into a categorical response. In contrast, labeledLDA allows the knowledge from document's classes influencing topic construction. Thus, documents from the same classes are more likely to be linked with the same topics.

The parameters of sLDA and labeledLDA are fixed empirically in such a way to maximize the predictive scores on a held-out (test) sample[3]. For both models, the number of topics $K$ is set to the number of codes. For sLDA: $\alpha = 0.01$. For labeledLDA, $\alpha = 0.005, \beta = 0.07$. The remaining hyperparameters are learnt from data. In addition, to maintain low running time, the number of iterations is set to 10,000 for sLDA and 50,000 for labeledLDA.

## 4.2  Datasets

As a response to the need for benchmark datasets pointed out in the literature [14], we created two datasets by gathering discharge summaries from Saint-Louis university teaching hospital[4]: URO-FR and HEMATO-FR. These datasets were built by taking all the discharge summaries collected within urology and hematology services respectively, between 2009 and 2014. Apart from filtering out rare codes (with less than 10 documents), we did not make any restriction regarding data quality, such as the presence of noise and typos. The point was to create real-life issued data with more challenging analysis problems for the algorithms. In this work, we only focus on the primary diagnosis (the reason the patient came to therapy) to deal with a single-label classification problem. We leave secondary codes, including aftercare codes, to a future work that will rely on multi-label classification.

---

[3] Source codes from: http://www.cs.cmu.edu/~chongw/slda/ (sLDA) and https://github.com/myleott/JGibbLabeledLDA/ (labeledLDA).

[4] http://hopital-saintlouis.aphp.fr/.

**Table 2.** Dataset description.

| Dataset | ICD version | Lang. | #docs. | #unique words | #codes | Avg. #words/doc. | Avg. #docs./code |
|---------|-------------|-------|--------|---------------|--------|------------------|------------------|
| URO-FR | CIM10 | French | 4 690 | 11 143 | 60 | 46 | 78 |
| HEMATO-FR | CIM10 | French | 3 720 | 13 371 | 30 | 76 | 124 |
| MIMIC-EN | ICD9 | English | 7 956 | 12 951 | 252 | 59 | 32 |

The three datasets are highly imbalanced: about 70% of documents are assigned to 20% of codes. Once again, our goal is to experiment the models within the challenging real-world setup. Therefore, we choose to maintain the original imbalanced document distribution.

The third dataset MIMIC-EN is a subset of MIMIC-II Physiology database [18] using the following PostgreSQL query:

```
SELECT (subject_id, hadm_id, code, text)  FROM mimic2v26.icd9  JOIN
mimic2v26.noteevents USING (subject_id, hadm_id) WHERE (sequence='1'
AND category='DISCHARGE_SUMMARY' AND LENGTH(text) > 50);
```

The exact dataset used in this paper was obtained when discarding rare codes (a minimum of 15 documents has been chosen to make a trade-off between the total number of codes and the number of documents per code).

In order to mitigate the effects of high dimensionality, we systematically make the following text preprocessing:

1. Stemming
2. Filtering out the words occurring in less than 2 documents (3 for MIMIC-EN dataset because of its large size) or more than 300 documents
3. Removing stopwords and numerics.

The preprocessed text documents are then mapped into a bag-of-words representation where the words (unigrams) are weighted according to their presence/absence in the document. All the models are based on this representation. Table 2 gives an overview of the preprocessed datasets.

### 4.3   Evaluation Framework

Previous works in code assignment have mainly relied on automatic evaluation measures from information retrieval fields, specifically precision, recall, and F-score. The standard version of these measures, yet widely-used for many predictive tasks, is restrictive in that it only considers a single model response. For this paper, we suggest to use a more flexible version, called $F_k$-score, in order to take into account uncertainty of the predictive models. To this end, the evaluation of a given result is performed by considering all of the $k$ returned classes rather than one single class, as in [7,17]. This choice is motivated by the following observations:

– All of the models tested here return a set of ranked labels (NB, sLDA, and labeledLDA) or can easily be adapted to do so (DT and SVM) [3]. Our

evaluation enables retrieving the correct class in case it were not ranked first. This is particularly useful when some of the returned labels are ranked equally.
 – In practice, it is more prudent to make the task humanly-supervised rather than fully automatic. In this regard, a set of best ranked codes is returned by the model, from where the coder selects the appropriate ones.

$F_k$-score is calculated similarly to the standard F-score except that the correct class is fetched among the $k$ most probable classes returned by the model. That is, if the correct class is present within these classes, it is returned instead of the the most probable class.

## 5  Results and Analysis

Figures 2, 3, and 4 show the results from the three datasets. These are described in terms of micro (weighted average) and macro (average) $F_k$-scores, for $k$ ranging from 1 to 10. The results are obtained based on a 10-fold cross validation. Error bars give the standard deviations. In addition, Table 3 offers the exact scores for SVM and labeledLDA with $k \in \{1, 5, 10\}$. In the following, we discuss these results from three perspectives: (i) overall performance, (ii) performance w.r.t. $k$, and (iii) a comparison of topic models sLDA and labeledLDA.

*Overall performance:* as can be seen from these figures, SVM and labeledLDA always yield the best result compared to the other models. On URO-FR and HEMATO-FR, SVM has the best scores while on MIMIC-EN labaledLDA comes first. Based on a $t$-test throughout all the datasets, no evidence of any statistical difference could be observed between SVM and labeledLDA ($p$-value $> 0.05$). NB generally arrives third, followed by DT, then sLDA that performs comparably poor in this task, specifically on MIMIC-EN where the number of codes is large. The same trend can be observed with both micro and macro averaged F-scores.

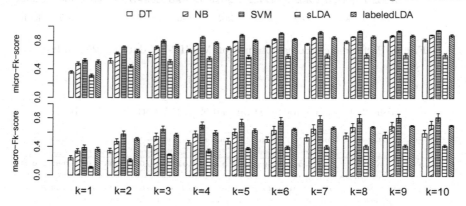

**Fig. 2.** Performance scores obtained on URO-FR dataset (60 codes).

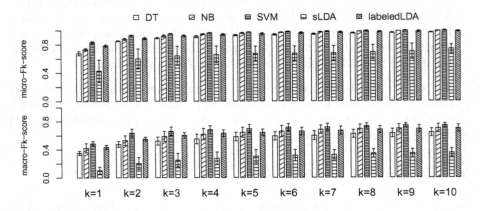

**Fig. 3.** Performance scores obtained on HEMATO-FR dataset (30 codes).

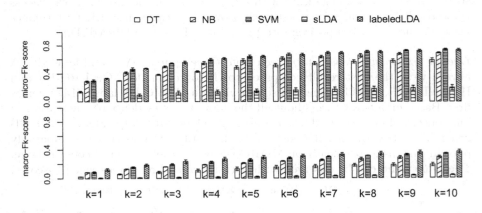

**Fig. 4.** Performance scores obtained on MIMIC-EN dataset (252 codes).

**Table 3.** $F_k$-scores for SVM and labeledLDA with $k \in \{1, 5, 10\}$.

| Dataset | Model | micro-$F_1$-score | macro-$F_1$-score | micro-$F_5$-score | macro-$F_5$-score | micro-$F_{10}$-score | macro-$F_{10}$-score |
|---|---|---|---|---|---|---|---|
| URO-FR | SVM | **0.52 ± 0.02** | **0.39 ± 0.03** | **0.89 ± 0.02** | **0.75 ±0.03** | **0.95 ± 0.01** | **0.82 ± 0.01** |
| | labeledLDA | 0.51 ±0.02 | 0.36 ±0.03 | 0.81 ±0.02 | 0.65 ±0.03 | 0.88 ±0.02 | 0.71 ±0.01 |
| HEMATO-FR | SVM | **0.83 ±0.01** | **0.49 ±0.03** | **0.95 ±0.01** | **0.69 ±0.02** | **0.99 ±0.01** | **0.74 ±0.03** |
| | labeledLDA | 0.78 ±0.01 | 0.44 ±0.03 | 0.92 ±0.01 | 0.64 ±0.02 | 0.98 ±0.01 | 0.70 ±0.05 |
| MIMIC-EN | SVM | 0.30 ±0.02 | 0.09 ±0.01 | 0.61 ±0.01 | 0.24 ±0.02 | **0.75 ±0.01** | 0.35 ±0.02 |
| | labeledLDA | **0.33 ±0.01** | **0.12 ±0.02** | **0.62 ±0.02** | **0.30 ±0.02** | 0.74 ±0.02 | **0.38 ±0.02** |

*Performance w.r.t. $k$:* better results are achieved when the value of $k$ grows up, giving more chance to the less probable codes to be selected. By averaging over all models and datasets, the gain in micro-$F_k$-score is equal to 14% when $k$ increases from 1 to 2. It is equal to 2% when $k$ increases from 5 to 6, and 1% when it increases from 9 to 10. To assess the statistical significance, we perform a paired $t$-test on the micro-$F_k$-scores obtained from MIMIC-EN dataset. The $p$-values result from comparing the means of micro-$F_{k+1}$-scores vs. micro-$F_k$-scores.

The difference is highly significant when $k$ increases from 1 to 2 or from 2 to 3 ($p$-value $< 10^{-6}$). In contrast, the difference is comparably much less significant for the greater values of $k$ ($p$-value $> 10^{-3}$).

*sLDA vs labeledLDA:* sLDA clearly achieves lower scores than the other topic model labeledLDA. For $k = 1$, the difference in micro-$F_k$-score is equal to 20% on URO-FR, 35% on HEMTAO-FR, and 30% on MIMIC-EN (see Table 3). We believe that this great difference in performance is due to the intrinsic difference in model structure. In labeledLDA, the knowledge from document's classes directly influences the topic construction. As such, documents from the same class (diagnosis code) are more likely to link to the same topics, which helps building more "diagnosis-based" topics. This feature, not shared by sLDA, is graphically depicted by the direction of the edge linking the variables $c$ and $\theta$ (see Fig. 1).

In addition to these quantitative results, we show the top 10 words characterizing the topics obtained with labeledLDA in Table 4. Each code is associated with the most likely topic based on empirical distributions [16]. We choose three examples from the best predicted scores and two examples with poorly predicted ones. The underlined words are manually annotated by medical experts as being semantically and clearly related to the associated diagnosis. Medical experts agree that these results are very informative. Most of the diagnoses are easily recognizable from their characterizing words. Moreover, a post hoc analysis of these results leads to the following observations:

- Topic's coherence is generally correlated to good predictive scores, as with the codes C81.9, C88.0, C91.1, N20.0 from the French data. Conversely, the codes with less coherent and/or mixed topics have poor predictive scores, such as C81.9, C83.0, N15.1, N20.1, N30.0. This observation may help explaining why certain codes are so easily-predicted by the model whereas others are not. In this regard, it is legitimate to believe that improving the topic's quality would lead to improve the predictive scores.
- A large number of codes can be characterized with medical concepts (n-grams, phrases) rather than single words, for example "arterial tension", "urinary tract infection", "blood test", etc. This observation should motivate the inclusion of medical concepts, either extracted statistically or based on medical ontology, into the vocabulary.

**Table 4.** Examples of codes and associated topics from URO-FR (top), HEMATO-FR (middle), and MIMIC-EN (bottom) datasets extracted with labeledLDA model.

| C61: *Tumeur maligne de la prostate* (Prostate cancer) | N39.3: *Incontinence urinaire d'effort* (Stress urinary incontinence) | Z52.4: *Donneur de rein* (Kidney donor) | N30.0: *Cystite aiguë* (Acute cystitis) | S30.2: *Contusion des organes génitaux externes* (Congestion of the external genitalia ) |
|---|---|---|---|---|
| *prostatectomie*[5] | *incontinent* | *prelèvement* (sample) | *pontage* (bypass) | *observer* (watch) |
| *radical* | *bandelette* (band) | *faveur* (favour) | *arterielle* (arterial) | *hospitalisé*(inpatient) |
| *laparotomie* (laparotomy) | *effort* (stress) | *manuel* (hand-operated) | *Ditropan* | *med* (medical) |
| *score* | *trans-obturatrice*[5] | *artère* (artery) | *post-mictionnel*[5] | *externe* (lateral) |
| *lobe* (lobus) | *urodynamique*[5] | *assisté* (assisted) | *Kardegic* | *motif* (cause) |
| *mini* | *toux* (cough) | *DFG* (GFR) | *diurne* (diurnal) | *chir* (surgery) |
| *capsulaire* (capsular) | *bud* (urodynam. test) | *laparoscopique*[5] | *surtout* (especially) | *ATCD* (med. history) |
| *élevé* (high) | *rééducation*[5] | *contre* (against) | *fonctionnel*(functional) | *clinique-uro*[5] |
| *extension* | *urgenturie*[5] | *apparenté* (related) | *impériosité* (urge) | *fam* (familial) |
| *curatif* (curative) | *position* | *min* (minute) | *hypertension*[5] | *suggérer* (suggest) |
| #documents=356 | #documents=47 | #documents=39 | #documents=16 | #documents=18 |
| F$_1$-score=0.68 | F$_1$-score=0.83 | F$_1$-score=0.96 | F$_1$-score=0.00 | F$_1$-score=0.22 |

| C81.9: *Lymphome de Hodgkin* (Hodgkin's lymphoma) | C88.0: *Macroglobulinmie de Waldenström* (Waldenström's macroglobulinemia) | D46.2: *Anémie réfractaire avec excès de blastes* (refractory anemia with excess of blasts) | C83.0: *Lymphome à petites cellules B* (small B-cell lymphoma) | E85.3: *Amylose généralisée secondaire* (secondary generalized amyloidosis) |
|---|---|---|---|---|
| *Hodgkin* | *Waldenström* | *senior* | *critère* (criterion) | *amylose* |
| *ABVD* | *IgM* | *multirésistant*(resistant) | *participer*(participate) | *troponine* (troponin) |
| *IVOX* | *lymphoplasmocytaire*[5] | *remise* (redelivery) | *accepter* (accept) | *formule* (formula) |
| *classique* (classical) | *macroglobulinémie*[5] | *blaste* (blast) | *consentement*(consent) | *proBNP* |
| *panoramique*(panoramic) | *monoclonal* | *AREB* (RAEB) | *aborder* (approach) | *VCD* |
| *escalade* (escalation) | *béta* (beta) | *leuco* | *attendu* (expected) | *évolution* (evolution) |
| *étoposide* (etoposide) | *créatininémie*[5] | *Vidaza* | *logistique* (logistics) | *dosage* (dose) |
| *BEAM* | *sup* (increased) | *myélodysplasique*[5] | *version* | *arriver* (reach) |
| *SPI* (IPS) | *stabilité* (stability) | *BHC* | *objectif* (goal) | *immunochimique*[5] |
| *nodulaire* (nodular) | *cérébral* (cerebral) | *mgX* (m.g.) | *contrainte*(constraint) | *physique* (physical) |
| #documents=168 | #documents=72 | #documents=37 | #documents=38 | #documents=85 |
| F$_1$-score=0.75 | F$_1$-score=0.74 | F$_1$-score=0.78 | F$_1$-score=0.38 | F$_1$-score=0.34 |

| 157.0: Malignant neoplasm of pancreas | 278.01: Morbid obesity | 430: Subarachnoid hemorrhage | 038.0: Streptococcal septicemia | 998.12: Hematoma complicating a procedure |
|---|---|---|---|---|
| duct | morbid | coil | vegetation | FFP |
| painless | roxicet | vasospasm | biliary | tube |
| biliary | elixir | nimodipine | streptococcus | yellow |
| bile | roux-en-y | fluent | surveillance | soften |
| whipple | crush | downgoing | endocardial | layer |
| CBD | laparoscopic | cistern | enterococcus | fiber |
| ERCP | actigall | angio | ductal | etc |
| endoscopic | bloated | pronation | cellular | colitis |
| duodenum | pill | sah | travel | everyday |
| cholangiopancreatography | program | EOM | medial | sleep |
| #documents=13 | #documents=13 | #documents=89 | #documents=26 | #documents=9 |
| F$_1$-score=1.00 | F$_1$-score=1.00 | F$_1$-score=0.69 | F$_1$-score=0.00 | F$_1$-score=0.00 |

[5] Term translation: *clinique-uro*: clinical-urological, *créatininémie*: creatininemia, *immunochimique*: immunochemical, *laparoscopique*: laparoscopic, *lymphoplasmocytaire*: lymphoplasmocytic, *macroglobulinémie*: macroglobulinemia, *myélodysplasique*: myelodysplastic, *post-mictionnel*: post-void, *prostatectomie*: prostatectomy, *rééducation*: reeducation, *trans-obturatrice*: transobturator, *urodynamique*: urodynamics, *urgenterie*: urge incontinence.

# 6    Conclusion

The work described in this paper is an example application of machine learning models to a real-world problem: diagnosis code assignment to discharge summaries. The models that we have chosen for experiments are issued from both classical machine learning research (DT, NB, and SVM) and modern NLP approaches (sLDA and labeledLDA). Despite the achieved results that are quite encouraging, the task would not allow a fully automatic coding because of the significant error rates. For example, based on labeledLDA model, 22% documents from hematology service would be miscoded. The rate rises to 51% on urology and 65% on intensive care medicine. A thorough analysis of prediction errors suggests that data quality (such as size, coverage, and specificity) is crucial for the success of the task. The code distribution is also an important factor as the codes with small sample sizes are generally hard to predict.

After a thorough discussion with medical experts, we believe that the automatic part of the coding process is very useful but cannot dispense with human supervision to make the conclusive choice. Nevertheless, as it has been shown in Table 3, the error rates are dramatically reduced when allowing larger values of $k$. For example, with $k=10$, the error rates are reduced by 25–45% points. In this way, the human coder can seek for the appropriate codes in a reduced space, which makes the coding task faster and easier.

Finally, in a semi-automatic approach the coder also has the ability to produce and express feedback for the learning algorithms. This can be done either by using the coder's choice among the proposed codes, or in a more specific way by asking the user to express a prior (e.g., "word-code" relation). Beyond the natural extension of this work to a multi-label task (including both primary and secondary codes), a challenging future work would be how to efficiently include user feedback, and more generally any type of prior knowledge. Conveniently, topic models are highly flexible for such purposes (see seededLDA model in [6]). Convinced by the utility of using prior knowledge and motivated by the promising results achieved with labeledLDA model, we consider extending our work to the embedding of this knowledge into labeledLDA model.

# References

1. Blei, D.M., Mcauliffe, J.D.: Supervised topic models. In: Advances in Neural Information Processing Systems (NIPS 2007), Vancouver, Canada, pp. 121–128. Curran Associates, Inc. (2007)
2. Blei, D.M., Ng, A.Y., Jordan, M.I.: Latent Dirichlet allocation. J. Mach. Learn. Res. (JMLR) **3**, 993–1022 (2003)
3. Cerri, R., De Carvalho, A.C.P.L.F., Freitas, A.A.: Adapting non-hierarchical multilabel classification methods for hierarchical multilabel classification. Intell. Data Anal. **15**(6), 861–887 (2011)
4. Farkas, R., Szarvas, G.: Automatic construction of rule-based ICD-9-CM coding systems. In: BMC Bioinformatics, vol. 9(Suppl. 3), p. S10 (2008)

5. Goldstein, I., Arzrumtsyan, A., Uzuner, O.: Three approaches to automatic assignment of ICD-9-CM codes to radiology reports. In: Proceedings of AMIA Symposium (AMIA 2007), pp. 279–283 (2007)
6. Jagarlamudi, J., Daumé III, H., Udupa, R.: Incorporating lexical priors into topic models. In: Proceedings of the European Chapter of the ACL (EACL 2012), Avignon, France, pp. 204–213. ACL (2012)
7. Krizhevsky, A., Sutskever, I., Hinton, G.E.: Imagenet classification with deep convolutional neural networks. In: Advances in Neural Information Processing Systems (NIPS 2012), Lake Tahoe, NV, USA, pp. 1106–1114. NIPS (2012)
8. Lin, C., He, Y., Everson, R., Ruger, S.: Weakly supervised joint sentiment-topic detection from text. IEEE Trans. Knowl. Data Eng. (TKDE) 24(6), 1134–1145 (2012)
9. Lita, L.V., Yu, S., Niculescu, S., Bi, J.: Large scale diagnostic code classification for medical patient records. In: Proceedings of the International Joint Conference on Natural Language Processing (IJCNLP 2008), Hyderabad, India, pp. 877–882. ACL (2008)
10. Medori, J., Fairon, C.: Machine learning and features selection for semi-automatic ICD-9-CM encoding. In: Proceedings of the NAACL HLT 2010 Second Louhi Workshop on Text and Data Mining of Health Documents (Louhi 2010), Los Angeles, CA, USA, pp. 84–89. ACL (2010)
11. Metais, E., Nakache, D., Timsit, J.-F.: Automatic classification of medical reports, the CIREA project. In: Proceedings of the 5th WSEAS International Conference on Telecommunications and Informatics (TELE-INFO 2006), Istanbul, Turkey, pp. 354–359. WSEAS (2006)
12. Meyer, D., Dimitriadou, E., Hornik, K., Weingessel, A., Leisch, F.: e1071: Misc Functions of the Department of Statistics, Probability Theory Group (2015)
13. Perotte, A., Bartlett, N., Wood, F., Elhadad, N.: Hierarchically supervised latent Dirichlet allocation. In: Advances in Neural Information Processing Systems (NIPS 2011), Granada, Spain, pp. 2609–2617 (2011)
14. Perotte, A., Pivovarov, R., Natarajan, K., Weiskopf, N., Wood, F., Elhadad, N.: Diagnosis code assignment: models and evaluation metrics. J. Am. Med. Inform. Assoc. (JAMIA) 21(2), 231–237 (2014)
15. Pestian, J.P., Brew, C., Matykiewicz, P., Hovermale, D.J., Johnson, N., Cohen, K.B., Duch, W.: A shared task involving multi-label classification of clinical free text. In: Proceedings of the Workshop on BioNLP 2007: Biological, Translational, and Clinical Language Processing (BioNLP 2007), Prague, Czech Republic, pp. 97–104. ACL (2007)
16. Ramage, D., Hall, D., Nallapati, R., Manning, C.D.: Labeled LDA : a supervised topic model for credit attribution in multi-labeled corpora. In: Proceedings of the 2009 Conference on Empirical Methods in Natural Language Processing (EMNLP 2009), August, Singapore, pp. 248–256. ACL, Singapore (2009)
17. Ruch, P., Gobeilla, J., Tbahritia, I., Geissbühlera, A.: From episodes of care to diagnosis codes: automatic text categorization for medico-economic encoding. In: Proccedings of the AMIA Symposium (AMIA 2008), Washington D.C., USA, pp. 636–640 (2008)
18. Saeed, M., Villarroel, M., Reisner, A.T., Clifford, G., Lehman, L., Moody, G., Heldt, T., Kyaw, T.H., Moody, B., Mark, R.G.: Multiparameter Intelligent Monitoring in Intensive Care II (MIMIC-II): a public-access intensive care unit database. Crit. Care Med. 39, 952–960 (2011)
19. Therneau, T., Atkinson, B., Ripley, B.: rpart: Recursive Partitioning and Regression Trees (2015)

20. Yi, X., Allan, J.: A comparative study of utilizing topic models for information retrieval. In: Boughanem, M., Berrut, C., Mothe, J., Soule-Dupuy, C. (eds.) ECIR 2009. LNCS, vol. 5478, pp. 29–41. Springer, Heidelberg (2009). https://doi.org/10.1007/978-3-642-00958-7_6
21. Zhang, Y.: A hierarchical approach to encoding medical concepts for clinical notes. In: Proceedings of the 46th Annual Meeting of the Association for Computational Linguistics on Human Language Technologies: Student Research Workshop (HLT-SRWS 2008), Columbus, OH, USA, pp. 67–72. ACL (2008)

# Information Extraction

# Identity and Granularity
# of Events in Text

Piek Vossen(✉) and Agata Cybulska

Computational Lexicology and Terminology Lab, Faculty of Humanities,
VU University Amsterdam, De Boelelaan 1105, 1081HV Amsterdam, Netherlands
{piek.vossen,a.k.cybulska}@vu.nl
http://cltl.nl

**Abstract.** In this paper we describe a method to detect event descriptions in different news articles and to model the semantics of events and their components using RDF representations. We compare these descriptions to solve a cross-document event coreference task. Our component approach to event semantics defines identity and granularity of events at different levels. It performs close to state-of-the-art approaches on the cross-document event coreference task, while outperforming other works when assuming similar quality of event detection. We demonstrate how granularity and identity are interconnected and we discuss how semantic anomaly could be used to define differences between coreference, subevent and topical relations.

**Keywords:** Event coreference · Event identity · Event relations

## 1 Introduction

News and blogs are common media that report on events that took place in the world. In the case of events with impact, we can expect that many different sources discuss the same event, partially providing the same information and partially differing from each other either in terms of the facts or perspective on these facts. Collections of news and blogs around the same topic therefore represent a challenging and natural task for cross-document event coreference. If done properly, event coreference resolution can be used to link event data across many different sources, resulting in deduplication and aggregation of data around events but also showing the different perspectives of these sources [32].

The task of cross-document event coreference is however far from trivial. Events exist within their temporal boundaries. The same type of event involving the same participants and the same action at a different point of time is not the same event, e.g. *John gave Mary the book on Tuesday* is a different event from *John gave Mary the book on Wednesday*. On the other hand, *Mary gave John the book on Tuesday* is also a different event, even though all event components are the same as for the first event but the roles differ. The precise semantics of these descriptions can be used to define identity across events. This becomes more

© Springer International Publishing AG, part of Springer Nature 2018
A. Gelbukh (Ed.): CICLing 2016, LNCS 9624, pp. 501–522, 2018.
https://doi.org/10.1007/978-3-319-75487-1_39

complicated when we underspecify and quantify the information. In the case of *John gave Mary several books this week* or *Mary often gets books from people* the precise details are not given and matching with the previous descriptions becomes more difficult. Also, it is not clear if the quantification should be seen as a quantification of the object *books* or of the event of *giving*. Furthermore, the action itself can be described in many different ways (*gets/takes/receives/borrows/buys/obtains*), exhibiting different manners or perspectives on the same event.

These examples clearly show that establishing identity across event descriptions is a hard problem to solve and on the one hand involves semantic components that play a role but on the other hand requires a robust matching across these components. In [12], we therefore described a model to measure identity across events as a function of the similarity of the event components. Such a model can be optimized on an annotated data set to weigh the contribution of the components for establishing event identity.

When implementing such a model for extracting and identifying events from raw text, another problem arises. The components necessary to compare events are spread over a complete article and are hardly found in a single sentence. Sentence-based approaches to event coreference cannot deal with the fact that event components are mentioned at different places of the text. In [15] a bag of events approach is suggested to group event components per source article and compare these across different articles. However, the implementation of the bag of events approach described in [15] was only tested on true mentions and did not consider the extraction of events from text. Furthermore, in the data set used in the experiments reported in [15] only a few sentences per news article are annotated, which means that only little information could be aggregated across these sentences.

In this paper, we describe a new implementation of the bag of events model proposed in [15] which completely processes news articles starting from raw text. First event instance representations are build from all mentions throughout the text within a single article and next these representations are compared across articles. The method employed here makes a distinction between mentions within an article and across articles on the one hand and event instances that are stable across these mentions on the other hand [17]. Likewise, we can compare event mentions but also event instance representations. Our system can be highly parameterized to establish identity. This can result in different levels of granularity of determined event instances. More loose constraints result in extreme aggregation and lumping, whereas very strict definitions result in each mention to be unique. We show that our instance-based approach performs close to state-of-the-art machine learning approaches and it out-performs these approaches when a similar quality of event detection is assumed.

In Sect. 2, we summarize previous work that is related to this task. In Sect. 3, we describe the event model and data set on which we evaluate our approach. Section 4 describes the bag of events approach that we proposed previously, which collects event information per document so that it can be used to solve event coreference. We present earlier implementation examples of the approach and their evaluation results achieved on true mentions in Sect. 5. Next in Sect. 6 we propose a new, door-to-door implementation of the bag of events approach. We evaluate the implementation and compare results with the state-of-the-art in Sect. 7. In Sect. 8, we discuss the results and speculate on the notion of granularity in relation to identity. We conclude in Sect. 9.

## 2    Related Work

Event coreference resolution is a difficult task of determining whether two event mentions refer to the same event instance. After the original focus on entity coreference, recently there is more and more interest in the field in event coreference resolution.

Some supervised approaches to event coreference resolution have been proposed by Humphreys et al. [19], Bagga and Baldwin [3], Ahn [1], or Chen and Ji [11]. These methods rely on supervised learning of rich linguistics features to determine coreference in a pairwise model and are strongly domain dependent. This is not the case for the method presented in Sect. 7 which is unsupervised. Furthermore, these works depend on local decisions without consideration of global event distribution at document level. The bag of events model implemented here allows us to overcome this deficiency.

More recently there are a few more works also using rich linguistic features to solve event coreference. One of them is the unsupervised approach of Bejan and Harabagiu [5] which relies on lexical, POS, event class and WordNet features as well as feature combinations. Another approach is the one of Liu et al. [24] and the most recent feature-rich model of Yang et al. [34]. Our method is similar to the approach of [24] in that it facilitates propagation of information about event participants between event mentions but our heuristic does this more globally by employing an instance level of event representation that aggregates this information at the document level as is also done [34]. We differ from [34] in that we use an RDF representation to do a logical comparison, whereas they do clustering using document level features. Our approach also differs from another recent approach to event coreference by Lee et al. [23], by making a distinction between specific entity types, whereas [23] disregard entity type information.

## 3    Event Model and Data

We model events as a combination of five slots. These five slots correspond to different elements of event information such as the action slot (or event trigger following the ACE [21] terminology) and four kinds of event arguments: time, location, human and non-human participant slots (see [13]). The next quote shows an excerpt from topic one, text number seven of the ECB corpus [5].

*The "American Pie" actress has entered Promises for undisclosed reasons. The actress, 33, reportedly headed to a Malibu treatment facility on Tuesday.*

Consider two event templates presenting the distribution of event information over the five event slots in the two example sentences (Tables 1 and 2). This event model has been employed to annotate the ECB+ data set [14] which is used in the experiments with event coreference described in the following sections. The approach to event coreference used in this work determines coreference between mentions of events through compatibility of slots of the five slot template. If event mentions are coreferent, they refer to the same event instance. Two or more event mentions are coreferent if they describe actions that happen or hold true in the same time and place and with involvement of the same participants.

**Table 1.** Sentence template ECB topic 1, text 7, sentence 1

| Action | *entered* |
|---|---|
| Time | N/A |
| Location | *Promises* |
| Human participant | *actress* |
| Non-human participant | N/A |

**Table 2.** Sentence template ECB topic 1, text 7, sentence 2

| Action | *headed* |
|---|---|
| Time | *on Tuesday* |
| Location | *to a Malibu treatment facility* |
| Human participant | *actress* |
| Non-human participant | N/A |

The EventCorefBank (ECB, [5]) was developed to test cross-document event coreference resolution. It consists of 43 different topics which correspond to seminal events, each topic with about 10 to 20 news articles reporting on a seminal event. Across articles from a topic, mentions of events are coreferent. The ECB+ corpus [14] is an extended and re-annotated version of the ECB. To each of the 43 ECB topics new texts were added about different event instances of the same event type. For example in addition to the topic of a particular celebrity checking into rehab covered in the ECB, to ECB+ descriptions were added of another event instance involving a different celebrity checking into another rehab facility. Likewise, the authors of ECB+ increased the referential ambiguity of event mentions. Table 3 shows some examples of seminal events represented in ECB+ with two different event instances per topic. Table 4 shows some statistics on the data set, most notably there are 6833 mentions of events annotated and 1958 coreference chains (instances). On average, 1.8 sentence per article was annotated for experiments on event coreference.

**Table 3.** Overview of seminal events in ECB and ECB+, topics 1–10

| Topic | Seminal Event type | Human participant ECB | ECB+ | Time ECB | ECB+ | Location ECB | ECB+ | Nr of docs ECB | ECB+ |
|---|---|---|---|---|---|---|---|---|---|
| 1 | rehab check-in | T. Reid | L. Lohan | 2008 | 2013 | Malibu | R. Mirage | 18 | 21 |
| 2 | Oscars host announced | H. Jackman | E. Degeneres | 2010 | 2014 | - | - | 10 | 11 |
| 3 | inmate escape | Brian Nicols 4 dead | A.J. Corneaux Jr | 2008 | 2009 | court house Atlanta | prison Texas | 9 | 11 |
| 4 | death | B. Page | E. Williams | 2008 | 2013 | LA | LA | 14 | 10 |
| 5 | head coach fired | Philadelphia 76ers M. Cheeks | Philadelphia 76ers J. O'Brien | 2008 | 2005 | - | - | 13 | 10 |
| 6 | "Hunger Games" sequel negotiations | C. Weitz | G. Ross | 2008 | 2012 | - | - | 9 | 11 |
| 7 | IBF, IBO, WBO titles defended | W. Klitchko H. Rahman | W. Klitchko T. Thompson | 2008 | 2012 | Germany | Switzerland | 11 | 11 |
| 8 | explosion at bank | - | - | 2008 | 2012 | Oregon | Athens | 8 | 11 |
| 9 | ESA changes | Bush | Obama | 2008 | 2009 | - | - | 10 | 13 |
| 10 | eigth-year offer | Angels M. Teixeira | Red Socks M. Teixeira | 2008 | 2008 | - | - | 8 | 13 |
| 45 | murder | S. Peterson L. Peterson | C. Simpson K. Flynn | | 2012 | - | - | 8 | 12 |

**Table 4.** ECB+ statistics

| ECB+ | # |
|---|---|
| Topics | 43 |
| Texts | 982 |
| Action mentions | 6833 |
| Location mentions | 1173 |
| Time mentions | 1093 |
| Human participant mentions | 4615 |
| Non-human participant mentions | 1408 |
| Coreference chains | 1958 |

# 4   Modelling Event Coreference: Bag of Events Approach

It is pretty much common practice to use information coming from event arguments for event coreference resolution ([5,9–12,19,23,24], among others). But using entities for event coreference resolution is complicated by the fact that event descriptions within a sentence often lack pieces of information. As pointed out by [19] it could be the case however that a lacking piece of information might be available elsewhere within discourse borders. This is a challenge for pairwise models comparing separate event mentions with one another on the sentence level. To be able to fully make use of information coming from event arguments, instead of looking at event information available within the same sentence, we propose to take a broader look at event descriptions surrounding the event mention in question within a unit of discourse. In this study we consider a document (a news article) as the unit of discourse.

The bag of events approach (for details see [15]) translates the structure of event descriptions into event templates for event coreference resolution. An event template can be created on different levels of information, such as a sentence, a paragraph or an entire document. The approach explicitly employs discourse structure to account for challenges following from the uneven distribution of event information across sentences of a document. Two templates are filled: a sentence and a document template. A sentence template collects event information from the sentence of an active action mention (Tables 1 and 2). By filling in a document template, one creates a "bag of events" for a document, that could be seen as a kind of document "summary" (Table 5).

**Table 5.** Document template ECB topic 1, text 7, sentences 1–2

| Action | *entered, headed* |
|---|---|
| Time | *on Tuesday* |
| Location | *Promises, to a Malibu treatment facility* |
| Human participant | *actress* |
| Non-human participant | N/A |

This heuristic employs clues coming from discourse structure and namely those implied by discourse borders. Descriptions of different event mentions occurring within a discourse unit, whether coreferent or related in some other way, unless stated otherwise, tend to share elements of their context. In our example text fragment the first sentence reveals that an actress has entered a rehab facility. From the second sentence the reader finds out where the facility is located (Malibu) and when the "American Pie" actress headed to the treatment center. It is clear to the reader of the example text fragment from the quotation that both events described in sentence one and two, happened on Tuesday. Also both sentences mention the same rehab center in Malibu. These observations are crucial for the approach.

The bag of events method can be implemented in different ways. In the following Sect. 5 we will briefly look at experiments with implementations as one- and as two-step classification tasks (the two-step implementation is described in detail in [15]). This implementation uses true mentions of event data and represents the document-based bag of events features as a loose set. In the one-step approach document-based bag of events features are combined with sentence-based features for pairwise mention comparison. In the two-step approach the bag of events features are first used for document clustering and the sentence-based features are used for pairwise mention comparison within a cluster. In Sect. 6, we describe another two-step implementation that extracts all event and entity data from the text and represents events as RDF instances, which aggregate all event data across the coreferential mentions in a single news article. These document-based RDF representations are more specific than the bag of events representations. In the second step, these RDF representations are compared semantically across all documents within a topic. The final instance results are mapped back to all the mentions for evaluation. We evaluate the RDF implementation in Sect. 7.

# 5    Experiments with Coreference on True Mentions

## 5.1    One- vs. Two-Step Classification

The specifics of the implementation of the two-step approach can be found in [15].

The two-step bag of events approach starts with filling in a document template, accumulating mentions of the five event slots: actions, locations, times, human and non-human participants from a document, as exemplified in Table 5. In a document template there is no distinction made between pieces of event information coming from different sentences of a document and no information is kept about elements being part of different mentions. A document template can be seen as a bag of events and event arguments. The template stores unique lemmas, to be precise a set of unique lemmas per event template slot. Document-based features from the template are used for preliminary document clustering. A supervised decision tree classifier (hereafter $DT$) determines whether two document templates share corefering event mentions. After all unique pairs of document templates from the test set have been classified by means of the DT classifier, "compatible" pairs are merged into document clusters based on pair overlap.

In the second step of the approach coreference is solved in a pairwise model between action mentions per document cluster created in step 1. For this task sentence templates are filled per true action mention from the corpus. Sentence templates collect event information from the sentence (consider examples in Tables 1 and 2). Pairs of sentence templates translate into features indicating compatibility across five template slots. A supervised classifier solves coreference between all unique pairs of action mentions per document cluster and finally pairs sharing common mentions are chained into equivalence classes.

In the one-step implementation of the approach all possible unique pairs of action mentions from the corpus are used as the starting point for classification. No initial document clustering is performed. For every action mention a sentence template is filled (see examples in Tables 1 and 2). Also, for every corpus document a document template is filled. Bag of events features indicating the degree of overlap between documents, from which two active mentions come from, are used for classification. In the one-step approach document features are used by a classifier together with sentence-based features; these combined create a feature vector per active action pair. One DT classifier is used to determine event coreference in a pairwise model. Pairs of mentions are classified based on a mix of information from a sentence and from a document. Finally, corefering pairs with overlap are merged into equivalence classes.

The one-step classification is implementation-wise simpler but it is computationally much more expensive. Ultimately every action mention has to be compared with every other action mention from the data set. This is a drawback of the one-step approach. On the other hand, it could be of advantage to have different types of information (sentence- and document-based) available simultaneously to determine event mention coreference.

## 5.2   Experiment Set-Up

For coreference experiments on true mentions we used a subset of ECB+ annotations (based on a list of 1840 selected sentences, [14]), that were additionally reviewed with focus on coreference relations. We divided the corpus into a training set (topics 1–35) and test set (topics 36–45).

The ECB+ texts are available in the XML format. The texts are tokenized, hence no sentence segmentation nor tokenization needed to be done. We POS-tagged (for the purpose of proper verb lemmatization) and lemmatized the corpus sentences. We used tools from the Natural Language Toolkit ([6], NLTK version 2.0.4): the NLTK's default POS tagger, Word-Net lemmatizer[1] as well as WordNet synset assignment by the NLTK[2]. For machine learning experiments we used scikit-learn [27].

**Table 6.** Features grouped into four categories: L-Lemma based, A-Action similarity, D-location within Discourse, E-Entity coreference and S-Synset based.

| Event slot | Mentions | Feature kind | Explanation |
|---|---|---|---|
| Action | Active mentions | Lemma overlap (L) <br> Synset overlap (S) | Numeric feature: overlap % <br> Numeric: overlap % |
| | | Action similarity (A) | Numeric: Leacock and Chodorow |
| | | Discourse location (D) | Binary: |
| | | - document | - the same document or not |
| | | - sentence | - the same sentence or not |
| | Sent. or doc. mentions | Lemma overlap (L) | Numeric: overlap % |
| | | Synset overlap (S) | Numeric: overlap % |
| Location | Sent. or doc mentions | Lemma overlap (L) | Numeric: overlap % |
| | | Entity coreference (E) | Numeric: cosine similarity |
| | | Synset overlap (S) | Numeric: overlap % |
| Time | Sent. or doc mentions | Lemma overlap (L) | Numeric: overlap % |
| | | Entity coreference (E) | Numeric: cosine similarity |
| | | Synset overlap (S) | Numeric: overlap % |
| Human participant | Sent. or doc mentions | Lemma overlap (L) | Numeric: overlap % |
| | | Entity coreference (E) | Numeric: cosine similarity |
| | | Synset overlap (S) | Numeric: overlap % |
| Non-human participant | Sent. or doc mentions | Lemma overlap (L) | Numeric: overlap % |
| | | Entity coreference (E) | Numeric: cosine similarity |
| | | Synset overlap (S) | Numeric: overlap % |

---

[1] www.nltk.org/modules/nltk/stem/wordnet.html.
[2] http://nltk.org/_modules/nltk/corpus/reader/wordnet.html.

In the experiments different features were assigned values per event slot (see Table 6). Note, that frequently one ends up with multiple entity mentions from the same sentence for an action mention (the relation between an action and involved entities is not annotated in ECB+). All entity mentions from the sentence (or a document in case of bag of events features) are considered. In case of document templates features referring to active action mentions were disregarded, instead action mentions from a document were considered. All feature values were rounded to the first decimal point.

We experimented with a few feature sets, considering per event slot lemma features only (L), or combining them with other features described in Table 6. Before fed to a classifier, missing values were imputed (no normalization was needed for the scikit-learn DT algorithm). All classifiers were trained on an unbalanced number of pairs of examples from the training set. We used grid search with ten fold cross-validation to optimize the hyper-parameters (maximum depth, criterion, minimum samples leafs and split) of the decision-tree algorithm.

## 5.3   Evaluation on True Mentions from ECB+

We will consider two baselines: a singleton baseline and a rule-based lemma match baseline. The singleton baseline considers event coreference evaluation scores generated taking into account all action mentions as singletons. The rule-based lemma baseline generates event coreference clusters based on full overlap between lemma or lemmas of compared event triggers (action slot) from the test set. Table 7 presents baselines' results in terms of recall (R), precision (P) and F-score (F) by employing the following metrics: MUC [31], B3 [2], CEAF [25], BLANC [29], and CoNLL F1 [28].

**Table 7.** Baseline results on the ECB+: singleton baseline and lemma match of event triggers evaluated in MUC, B3, mention-based CEAF, BLANC and CoNLL F.

| Baseline | MUC | | | B3 | | | CEAF | BLANC | | | CoNLL |
|---|---|---|---|---|---|---|---|---|---|---|---|
| | R | P | F | R | P | F | R/P/F | R | P | F | F |
| Singleton baseline | 0 | 0 | 0 | 45 | 100 | 62 | 45 | 50 | 50 | 50 | 39 |
| Action lemma baseline | 71 | 60 | 65 | 68 | 58 | 63 | 51 | 65 | 62 | 63 | **62** |

When discussing event coreference scores must be noted that some of the commonly used metrics depend on the evaluation data set, with scores going up or down with the number of singleton items in the data [29]. Our singleton baseline gives us zero scores in MUC, which is understandable due to the fact that the MUC measure promotes longer chains. B3 on the other hand seems to give additional points to responses with more singletons, hence the remarkably high scores achieved by the baseline in B3. CEAF and BLANC as well as the

CoNLL measures (the latter being an average of MUC, B3 and entity CEAF) give more realistic results. The lemma baseline reaches 62% CoNLL F1.

Table 8 evaluates final clusters of coreferent action mentions produced in the experiments by means of the one- and two-step classification when employing different features.

**Table 8.** Bag of events approach to event coreference resolution, evaluated on the ECB+ in MUC, B3, mention-based CEAF, BLANC and CoNLL F measures.

| Step1 | | | Step2 | | | MUC | | | B3 | | | CEAF | BLANC | | | | CoNLL |
|-------|---------|----------|-----|---------|----------|----|----|----|----|----|----|------|----|----|----|----|-------|
| Alg | Slot Nr | Features | Alg | Slot Nr | Features | R | P | F | R | P | F | F | R | P | F | F | F |
| - | - | - | DT | 5 | L | 61 | 76 | 68 | 66 | 79 | 72 | 61 | 67 | 69 | 68 | 70 | |
| - | - | - | DT | 5 | L+docL | 65 | 80 | 71 | 68 | 83 | 75 | 64 | 69 | 73 | 71 | 72 | |
| DT | 5 | docL | DT | 5 | L | 71 | 75 | 73 | 71 | 77 | 74 | 64 | 71 | 71 | 71 | 73 | |
| DT | 5 | docL | DT | 5 | LDES | 71 | 75 | 73 | 71 | 78 | 74 | 64 | 72 | 71 | 72 | **73** | |
| DT | 2 | docL | DT | 2 | LDES | 76 | 70 | 73 | 74 | 68 | 71 | 61 | 74 | 68 | 70 | 70 | |
| DT | 5 | docL | DT | 5 | LADES | 71 | 75 | 73 | 71 | 78 | 74 | 64 | 72 | 71 | 72 | 73 | |

When considering bag of events classifiers using exclusively lemma features L (row two and three), the two-step approach reached a 1% higher CoNLL F-score than the one-step approach with document-based lemma features (*docL*). The one-step method achieved in BLANC a 2% better precision but a 2% lower recall. This is understandable. In a two-step implementation when document clusters are created some precision is lost. In a one-step classification specific sentence information is always available for the classifier hence we see slightly higher precision scores (also in other metrics).

The best coreference evaluation scores with the highest CoNLL F-score of 73% and BLANC F of 72% were reached by the two-step bag of events approach with a combination of the DT document classifier using feature set L (document-based hence *docL*) across five event slots and the DT sentence classifier when employing features LDES (see Table 6 for a description of features). Adding action similarity (A) on top of LDES features in step two, does not make any difference on decision tree classifiers with a maximum depth of 5 using five slot templates. Our best CoNLL F-score of 73% is an 11% improvement over the strong rule based event trigger lemma baseline, and a 34% increase over the singleton baseline.

To quantify the contribution of document features, we contrast the results of classifiers using bag of events features with scores achieved when disregarding document features. The results reached with sentence template classification only (without any document features, row one in Table 8), give us some insights into the impact of the document features on our experiment. Note that one-step classification without preliminary document template clustering is computationally much more expensive than a two-step approach, which ultimately takes into account much less item pairs thanks to the initial document template clustering. The DT sentence template classifier trained on an unbalanced training set reaches 70% CoNLL F. This is 8% better than the strong baseline disregarding event arguments, but only 3% less than the two-step bag of events approach and 2% less than the one-step classification with document features. The reason for the relatively small contribution by document features could owe to the fact that in the ECB+ corpus not that many sentences are annotated per text. 1840 sentences are annotated in 982 corpus texts, i.e. 1.87 sentence per text. We expect that the impact of document features would be bigger, if more event descriptions from a discourse unit were taken into account than only the ground truth mentions.

We run an additional experiment with the two-step approach in which four entity types were bundled into one entity slot. Locations, times, human and non-human participants were combined into a cumulative entity slot resulting in a simplified two-slot template. When using two-slot templates for both, document and sentence classification on the ECB+ 70% CoNLL F score was reached. This is 3% less than with five-slot templates.

## 6    Door-to-Door Implementation

In the previous Sects. 4 and 5, true mentions have been used to evaluate the approach. In this section, we describe how we extract event data from raw text (system mentions) and then establish cross-document event coreference in a two-step-approach.

For extracting event data from text we use the NLP pipeline developed in the NewsReader project.[3] NewsReader applies a cascade of semantic modules among which: named-entity recognition and classification (NERC), named-entity-linking (NEL), semantic role labeling (SRL), time expression detection and normalization, time anchoring, event and entity coreference. The same entity, event or time expression can be mentioned several times in a document. The above modules will interpret each occurrence separately and annotate the tokens accordingly. The output of the NLP pipeline is thus mention-based.

From the mention-based annotation of tokens, we derive an instance-based representation of events, entities, time objects and relations between them. Our instance-based model follows the Simple Event Model (SEM, [18]). SEM is an RDF model for presenting event-instances through URIs (Unique Resource Identifiers) with triple relations to participants and dates, which also have unique URIs. We extended SEM with the Grounded Annotation Framework (GAF, [17]) to link each instance URI to mentions in the source documents. In Fig. 1, we show the SEM representations for the entities *Ka'loni Flynn* and *Christopher Simpson* resulting from processing topic 45 in ECB+. Each RDF structure has a URI representing an entity as the subject and various properties, such as the *rdfs:label* for the surface forms, *skos:prefLabel* for the most frequent surface form and *gaf:denotedBy* to link to char offsets of mentions. We can see that all mentions are offsets in different ECB+ files and there are no mentions in ECB files, which is correct.

---

[3] www.newsreader-project.eu.

For both persons, we see that not every mention was resolved to the same URI. For example, *Christopher Simpson* has one DBpedia URI, one representation as an entity without a DBpedia match and one representation as a so-called non-entity, i.e. phrases not detected as an entity by NERC but playing an important role in an event[4] In the case of *Ka'loni Flynn*, there is no DBpedia entry to match to and we see 3 entities and 1 non-entity. There are various reasons why the NLP modules did not match these mentions to the same entity. Such differences in URIs may also hamper the matching of events as we will see later.

```
1    dbp:Christopher_Simpson>
2          rdfs:label          "Christopher Simpson"  , "Simpson";
3          gaf:denotedBy
4             nwr:45_5ecbplus#char=167,186  ,   nwr:45_5ecbplus#char=306,325  ,
5             nwr:45_5ecbplus#char=607,626  ,   nwr:45_5ecbplus#char=1170,1189  ,
6             nwr:45_5ecbplus#char=2516,2535  ;
7          skos:prefLabel      "Christopher Simpson"  ;
8
9    nwr:/entities/ChristopherKenyonSimpson>
10         rdfs:label          "Christopher Kenyon Simpson"  ;
11         gaf:denotedBy
12            nwr:45_7ecbplus#char=567,593  , nwr:45_6ecbplus#char=405,431  ,
13            nwr:45_11ecbplus#char=476,502  ,   nwr:45_8ecbplus#char=476,502  ,
14            nwr:45_2ecbplus#char=491,517  ,  nwr:45_4ecbplus#char=371,397  ,
15            nwr:45_1ecbplus#char=532,558  ,  nwr:45_12ecbplus#char=287,313  .
16         skos:prefLabel      "Christopher Kenyon Simpson"  .
17
18   nwr:non−entities/purportedly+simpson
19         rdfs:label          "Purportedly Simpson"  ;
20         gaf:denotedBy       nwr:45_5ecbplus#char=2374,2393  ;
21         skos:prefLabel      "Purportedly Simpson"  .
22
23   nwr:entities/KaloniFlynn>
24         rdfs:label          "Ka'loni Flynn"  ;
25         gaf:denotedBy
26                     nwr:45_7ecbplus#char=639,652   ;
27         skos:prefLabel      "Ka'loni Flynn"  .
28
29   nwr:entities/KaLoniMarieFlynn>
30         rdfs:label          "ka'loni Marie Flynn"  ;
31         gaf:denotedBy
32                     nwr:45_2ecbplus#char=564,583  , nwr:45_4ecbplus#char=444,463  ,
33            nwr:45_9ecbplus#char=388,407,  nwr:45_3ecbplus#char=497,516  ;
34         skos:prefLabel      "ka'loni Marie Flynn"  .
35
36   nwr:entities/KaloniFlynn>
37         rdfs:label          "Ka'loni Flynn"  ;
38         gaf:denotedBy
39                     nwr:45_7ecbplus#char=639,652  ;
40         skos:prefLabel      "Ka'loni Flynn"  .
41
42   nwr:non−entities/flynn>
43         rdfs:label          "Flynn", 'Flynn's"  ;
44         gaf:denotedBy
45                     nwr:45_12ecbplus#char=584,589  , nwr:45_12ecbplus#char=733,738
46            nwr:45#11ecbplus#char=778,783
47         &word=w161&term=t161&sentence=8>  ;
48         skos:prefLabel      "Flynn"  .
```

**Fig. 1.** Entities in SEM format with gaf:denotedBy links to mentions

In the same way as for entities, also events are represented through unique URIs with properties as shown in Fig. 2. Since events are normally not stored in DBpedia and cannot be identified by their surface forms, we create meaningless unique identifiers (*nwr:45_12ecbplus#ev10* and *nwr:45_6ecbplus#ev16*). We further see similar properties as for the entities, such as rdfs:label, skos:prefLabel and gaf:denotedBy. We do not show the full list of mentions linked by the gaf:denotedBy property to save space but both events have mentions across

---

[4] We used FrameNet frame elements to decide on relevance of a participant.

various ECB+ files, implying that information has been aggregated from mentions in these files.

Other properties for events are the class information (property *a*), sem:Actor relations and a sem:hasTime relation. The class information consists of WordNet synsets [16][5], their hypernyms and the FrameNet frames associated with the events [4]. The WordNet synsets are selected from the highest scoring senses of all the mentions according to word-sense-disambiguation (WSD). As for the actors, we listed only the entities using their URIs with their surface forms between brackets.

```
 1   nwr:45_12ecbplus#ev10
 2           rdfs:label
 3                   murder , kill , assassination , execution , Killing ,
 4           Shooting , slaying ;
 5           skos:prefLabel  murder ;
 6      gaf:denotedBy
 7                   nwr:45_1ecbplus#char=1808,1815 , nwr:45_12ecbplus#char=109,115 ,
 8                   nwr:45_5ecbplus#char=3281,3287 , nwr:45_6ecbplus#char=99,107 ,
 9                   nwr:45_1ecbplus#char=1906,1913 , nwr:45_1ecbplus#char=5673,5686 ,
10                   etc... ;
11           a
12                   ili:i28310 ,  ili:i28306 ,  ili:i28311 ,  ili:i34133 ,
13           ili:i36562 , ili:i35417,  ili:i34134 ,  ili:i34139 ,
14           ili:i34130
15                   fn:Killing , fn:Attack ,fn:Execution  , , sem:Event , ;
16           sem:hasActor
17                   dbp:Jerome_Flynn (Flynn , Herbert Flynn , his , Ka'Loni Flynn ,
18                   Ka'Loni Flynn's , Ka'loni Flynn) ;
19           sem:hasTime   nwr:45_6ecb#tmx2 (time:20121112 , Nov. 12) .
20
21   nwr:45_6ecbplus#ev16
22           rdfs:label          charge , shooting , shoot ;
23           skos:prefLabel  shoot ;
24      gaf:denotedBy
25                   nwr:45_9ecbplus#char=640,644 , nwr:45_2ecbplus#char=633,637 ,
26                   nwr:45_4ecbplus#char=513,517 , nwr:45_7ecbplus#char=403,411 ,
27                   nwr:45_2ecbplus#char=359,366 , nwr:45_7ecbplus#char=69,77 ,
28                   etc... ;
29           a
30                   ili:i106612 ,  ili:i25451 ,  ili:i25858 ,  ili:i25860 ,
31           ili:i25976 ,  ili:i26598 ,  ili:i26600 ,  ili:i27206 ,
32           ili:i27278 ,  ili:i27293 ,  ili:i27599 ,  ili:i29722 ,
33           ili:i30898 ,  ili:i30954 ,  ili:i32022 ,  ili:i32053 ,
34           ili:i33338 ,  ili:i34100 ,  ili:i34141 ,  ili:i35084 ,
35                   ili:i36049 ,  ili:i36050 ,  ili:i36591 ,  ili:i40503 ,
36           ili:i70941 ,  ili:i27599 ,  ili:i32022 ,  ili:i26598 ,
37           ili:i33338 ,  ili:i36049 ,  ili:i30898 ,  ili:i106612 ,
38           ili:i27278 ,  ili:i26600 ,  ili:i25976 ,
39                   fn:Commerce_collect , fn:Motion , fn:Process_continue ,
40                   fn:Commerce_pay , fn:Killing , fn:Notification_of_charges ,
41                   fn:Hit_target , fn:Shoot_projectiles , fn:Use_firearm ;
42           sem:hasActor
43                   dbp:Electoral_division_of_Flynn ,
44                   dbp:Jerome_Flynn (Flynn , Herbert Flynn , his , Ka'Loni Flynn ,
45                   Ka'Loni Flynn's , Ka'loni Flynn) ,
46                   dbp:Oklahoma (okla , Oklahoma , Okla , Okla - man) ,
47           dbp:Robb_Flynn (Ka'loni Flynn , Flynn) ,
48                   dbp:Fort_Smith ,_Arkansas ,
49           nwr:entities/ChristopherKenyonSimpson ,
50                   dbp:Christopher_Simpson ,
51                   nwr:entities/Spiroman ,
52                   dbp:Arkansas ,
53                   dbp:O._J._Simpson (Purportedly Simpson , Simpson , his) ;
54           sem:hasTime   nwr:45_6ecb#tmx2 (time:2012 , 2012) .
```

**Fig. 2.** Events in SEM

Note that actors show some degree of lumping of mentions and in some cases have wrong URIs, e.g. *dbp:Jerome_Flynn (Flynn, Herbert Flynn, his, Ka'Loni*

---

[5] WordNet synsets are represented here as Inter Lingual Index (ili) records: www.globalwordnet.org/ili [33]. This enables us to compare events across different languages [32].

*Flynn, Ka'Loni Flynn's, Ka'loni Flynn)*, where references to both the murdered daughter *Ka'Loni Flynn* and her father *Herbert Flynn* got the same URI *dbp:Jerome_Flynn*, which is also to the wrong person. This lumping is mainly the result of wrong links coming from DBpedia spotlight [26] which gives preference to more popular entities. Wrong URIs will not harm our coreference matches as long as they are systematic, i.e. all mentions of *Simpson* get matched to the same URI *dbp:O._J._Simpson*. When different URIs are given, they can still be matched through surface forms (see below). Finally, we see that the events are linked to a date (sem:hasTime), which is a specific day for the first event and a year for the second event.

The above event instance representations are the result of a two-step approach that is implemented as follows:

1. Collect all event data for a single document and represent this as a SEM instance, giving access to all information on the action, the participants and the dates present in a document.
2. Compare the aggregated SEM representations across different documents to decide on identity. If identity is established then the SEM instance representations are merged.

For the first step, we collect all event mentions from a single document and then collect all participants and time anchors for these mentions into a single instance representation. We first group all mentions of the same lemma and next we determine the similarity across lemmas on the basis of the WordNet similarity scores [22] of their WordNet senses with the highest WSD score. Next, we check the Semantic Role structure created by the NewsReader pipeline to find all the roles for these mentions as well as all the time expressions to which these mentions are anchored. We thus already get aggregated instance representations with multiple surface forms across mentions in different sentences with their WordNet synsets, FrameNet frames, dates and actors.

In the second step compare these event structures across the different documents within a topic using the following heuristic:

1. the event actions need to match sufficiently (above a threshold) in terms of WordNet synsets or, if there are no synsets, in terms of the surface forms;
2. the time needs to match if present;
3. there must be overlap of actors in any role or the specified roles;

For comparing actions, we check the proportion of overlap across the WordNet synsets. In case two events have no synsets associated, we use the surface forms. If the overlap is mutually above a predefined threshold, we continue, otherwise the events do not match.

After passing the test for action similarity, we compare the overlap of the participants. We first check the URI. If the URI does not match, we check if the preferred form matches any of the surface forms. Participant matches can be required for specific semantic roles, e.g. PropBank [20] A0, A1, A2, or the role can be ignored. At least one participant needs to match in any role or, in the former case, per specified role.

If the above test fails, there is no coreference, otherwise we continue to compare the time anchors. Time is matched either per year, month or day, where a more specific time constraint also requires the more general ones, i.e. same month also implies same year, and same day also implies the same month and year. If one event has no time anchor and the other does, there is no coreference. If both events have no time anchor, they match.

In the case of Fig. 2, the two event structures did not get merged by our software. Their time anchors matched in terms of the year and there is an overlapping actor but none of the WordNet synsets overlap. Nevertheless, each of the event instances shows already quite some lumping of other mentions across the documents, as indicated by the mentions in different files in topic 45.

## 7    Experiments with Coreference on System Mentions

We evaluate the NewsReader system on system mentions on the same ECB+ data set. We compare the NewsReader results with Yang et al. [34], who report the best results for event coreference resolution system mentions for ECB+ and who also compare their results to other systems that have so far only been tested on ECB and not on ECB+. Yang et al. use a distance-dependent Chinese Restaurant Process (DDCRP [8]), which is an infinite clustering model that can account for data dependencies. They define a hierarchical variant (HDDCRP) in which they first cluster event mentions and data within a document and next cluster the within document clusters across documents. Their hierarchical strategy is similar to our approach using event components, in the sense that event data can be scattered over multiple sentences in a document and needs to be gathered first. Our approach differs from theirs in that we use a semantic representation to capture all event properties and do a logical comparison, while Yang et al. used machine learning methods (both unsupervised clustering and supervised mention based comparison). Yang et al. also report on a lemma-baseline as proposed by Cybulska and Vossen (2014) [14], where all event mentions with the same lemma within and across documents are simply joined in a single coreference set.

Yang et al. test their system on topics 24–43 while they used topics 1–20 as training data and topics 21–23 as the development set. They do not report on topics 44 and 45. To compare our results with theirs, we also used topics 24–43 for testing. In Table 9, we show Yang's lemma baseline (LEMMA), Yang's best results (HDDCRP), and the results for NewsReader (NWR). The NewsReader systems are not trained on ECB+ data and use logical comparison of event data. We tested the following variants of the NewsReader system, where systems starting with **NWR-X** (out-of-the-box), **NWR-T** (with TimeEval2013 training for event detection) or **NWR-G** (with true mentions of events). The remainder of the code expresses the time filter (Y = year, M = month, D = Day, N = none), the participant filter (A = any role, A1 = PropBank A1, N = none) and the action filter (c10, 30, 50, 70 = concept overlap, p10, 30, 50, 70 = phrase overlap):

**NWR-X-YAc30p30** NewsReader out-of-the-box, matching year (Y), any participant (A), concept overlap of 30% (c30), phrase overlap of 30% (p30).

**NWR-T-YAc30p30** NewsReader with event extraction using CRF trained on TemEval-2013 training data (T), matching year (Y), any participant (A), concept overlap of 30% (c30), phrase overlap of 30% (p30).

**NWR-G-YAc30p30** Newsreader with Gold event data (G), matching year (Y), any participant (A), concept overlap of 30% (c30), phrase overlap of 30% (p30).

**NWR-G-MAc30p30** Newsreader with Gold event data (G), matching month (M), any participant (A), concept overlap of 30% (c30), phrase overlap of 30% (p30).

**NWR-G-DAc30p30** Newsreader with Gold event data (G), matching day (D), any participant (A), concept overlap of 30% (c30), phrase overlap of 30% (p30).

**NWR-G-YAc10p10** Newsreader with Gold event data (G), matching year (Y), any participant (A), concept overlap of 10% (c10), phrase overlap of 10% (p10).

**NWR-G-YAc50p70** Newsreader with Gold event data (G), matching year (Y), any participant (A), concept overlap of 50% (c50), phrase overlap of 50% (p50).

**NWR-G-YAc70p70** Newsreader with Gold event data (G), matching year (Y), any participant (A), concept overlap of 70% (c70), phrase overlap of 70% (p70).

**NWR-G-YNc30p30** Newsreader with Gold event data (G), matching year (Y), no participant (N), concept overlap of 30% (c30), phrase overlap of 30% (p30).

**NWR-G-YA1c30p30** Newsreader with Gold event data (G), matching year (Y), A1 participant (A), concept overlap of 30% (c30), phrase overlap of 30% (p30).

**NWR-G-NAc30p30** Newsreader with Gold event data (G), no time (N), any participant (A), concept overlap of 30% (c30), phrase overlap of 30% (p30).

We first compare the NewsReader out-of-the-box system (NWR-X-YAc30p3) with Yang's results (LEMMA baseline and HDDCRP). The NewsReader system uses the following matching settings: year, a single participant in any role and 30% of the event concepts or, if not present, the event surface forms. We see that both Yang's *HDDCRP* and the lemma baseline outperform NWR-X-YAc30p3 by 14 and 9 points respectively in CoNLL $F_1$ score [28]. However, Yang et al. report that their system at first had an out-of-the-box accuracy for event detection of 56%. They therefore trained a separate Conditional Random Field (CRF) event detection system with event annotations of the first 20 topics (about half of the data set). They report an accuracy for this classifier of 95% on event detection and they used it as input for both the LEMMA baseline and HDDCRP. For comparison, the NewsReader system has an out-of-the-box accuracy of 67.99%, where events are detected by the MATE tool [7] which is trained on PropBank data. Clearly, what events have been annotated and how they were annotated

**Table 9.** Reference results macro averaged over ECB+ corpus as reported by Yang et al. [34] for state-of-the-art machine learning systems as compared to various News-Reader based systems. All NewsReader systems start with NWR-X (out-of-the-box), NWR-T (with TimeEval2013 training for event detection) or NWR-G (with true mentions of events). The remainder of the code expresses the time filter, the participant filter and the action filter. More explanation is given in the text.

| ECB+ | MUC | | | BCUB | | | CEAFe | | | CoNLL | Mention |
|---|---|---|---|---|---|---|---|---|---|---|---|
| Topics 24–43 | R | P | F$_1$ | R | P | F$_1$ | R | P | F$_1$ | F$_1$ | F$_1$ |
| LEMMA | 55.4 | 75.10 | 63.80 | 39.60 | 71.70 | 51 | 61.10 | 36.20 | 45.50 | 53.40 | 95 |
| **HDDCRP** | 67.10 | 80.30 | 73.10 | 40.60 | 73.10 | 53.50 | 68.90 | 38.60 | 49.50 | **58.70** | 95 |
| NWR-X-YAc30p30 | 44.85 | 50.16 | 47.35 | 46.88 | 45.3 | 46.08 | 47.45 | 34.89 | 40.22 | 44.55 | 67.99 |
| NWR-T-YAc30p30 | 48.99 | 58.5 | 53.33 | 45.37 | 55.48 | 49.92 | 41.37 | 45.56 | 43.36 | 48.87 | 75.03 |
| NWR-G-YAc30p30 | 64.12 | 72.03 | 67.85 | 65.21 | 74.89 | 69.72 | 66.35 | 57.39 | 61.55 | 66.37 | 99.84 |
| NWR-G-MAc30p30 | 64.12 | 72.03 | 67.85 | 65.21 | 74.89 | 69.72 | 66.35 | 57.39 | 61.55 | 66.37 | 99.84 |
| NWR-G-DAc30p30 | 62.12 | 70.99 | 66.26 | 61.93 | 75.69 | 68.12 | 66.57 | 56.52 | 61.14 | 65.17 | 99.84 |
| NWR-G-YAc10p10 | 64.81 | 70.6 | 67.58 | 65.57 | 72.84 | 69.02 | 63.75 | 57.1 | 60.24 | 65.61 | 99.84 |
| NWR-G-YAc50p50 | 63.49 | 72.55 | 67.72 | 64.63 | 75.84 | 69.79 | 67.48 | 57.29 | 61.97 | 66.49 | 99.84 |
| NWR-G-YAc70p70 | 62.61 | **72.81** | 67.33 | 63.8 | 76.92 | 69.75 | 67.9 | 56.61 | 61.74 | 66.27 | 99.84 |
| NWR-G-YNc30p30 | **77.4** | 69.68 | 73.34 | **72.92** | 64.24 | 68.31 | 54.99 | **65.39** | 59.74 | **67.13** | 99.84 |
| NWR-G-YA1c30p30 | 52.31 | 71.27 | 60.34 | 58 | **80.27** | 67.34 | **69.89** | 50.67 | 58.75 | 62.14 | 99.84 |
| NWR-G-NAc30p30 | 64.12 | 72.03 | 67.85 | 65.21 | 74.89 | 69.72 | 66.35 | 57.39 | 61.55 | 66.37 | 99.84 |

has a big impact on the results. To measure this impact on the actual event-coreference, we created two other variants of the NewsReader system: (1) NWR-T replaces the MATE event detection by a CRF classifier trained with SemEval 2013 - TempEval3 gold data [30] and (2) NWR-G uses the true mentions of the ECB+ annotation as the events (Gold). The event detection accuracy of NWR-T is 75.03% and the accuracy for NWR-G is 99.84%.[6]

In the last column of Table 9, we list the F$_1$ measures for the detection of event mentions by each variant system. It clearly shows that the differences in coreference results across systems are mainly due to the performance on the event detection. The NWR-G variants for example outperform HDDCRP by almost 10 points, while scoring only 5 points higher in event detection. NWR-T-YAc30p30 performs 7 points higher in event detection and 4 points higher in event coreference than NWR-X-YAc30p30. The NWR-X-YAc30p30, NWR-T-YAc30p30, NWR-G-YAc30p30 only differ in the extraction of the events with accuracies of 68%, 75% and 100% respectively. The other parameters for time, participant and action match are the same. Note that all NewsReader systems apply logical comparison and are not trained on the ECB+ data set. We thus can expect their performance to be relatively stable across data sets, whereas Yang et al's system is expected to perform significantly lower when applied to out-of-domain data.

---

[6] The reason that NWR-G is not 100% is because the NewsReader system could not process one of the evaluation files due to formatting problems.

Next, Table 9 lists the applied different variants of the NewsReader system using the true mentions of events (NWR-G) to see the impact on event coreference given the perfect event detection. Varying the settings for matching using the true mentions for events, shows only small differences. When we make the time more strict (year (Y), month (M), day (D)), we see that the month is as discriminative as the year but CoNLL $F_1$ is 1 point lower when the time needs to match at the level of the day. Next, we varied the threshold for overlapping concepts and surface forms (10%, 30%, 50%, 70%). We can see that precision for MUC and BCUB are higher when the thresholds are higher. This is in line with our expectation. For CEAFe, we see the same trend for recall. The last rows of Table 9 show the impact of the participants. In case of NWR-G-YNc30p30, no matching participant is required, for NWR-G-YA1c30p30 the PropBank A1 role should be identical, and for NWR-G-NAc30p30 we match any participant but we dropped the time constraint. Remarkably, dropping the participant match constraint gives best results, while requiring a matching A1 participant gives highest precision scores for BCUB and recall for CEAFe.

Overall, we can conclude that there is a slight tendency for more strict parameters to increase the precision but that we always loose more recall with slightly lower $F_1$ scores as a result. It is also clear that the event detection itself is the most important factor for improving event coreference.

Note that the best performance obtained with the true mentions of the events (67.13), scores only 6 points below our bag-of-events approach (73) using true mentions for all event components and it outperforms the lemma baseline on true mentions (63) reported in Sect. 5. Obviously, the NewsReader approach uses more data from the document than the annotated data (1.8 sentences) but on the other hand it also introduces more noise. We can expect that improving the linking of participants to entity mentions and improving the event detection is likely to bring the out-of-the-box system closer to 73 $F_1$ scores.

## 8   Discussion

The results have shown that cross-document event coreference is a hard task that is not completely solved despite the progress. We demonstrated that event components are critical but that they need to be collected from the complete document. Nevertheless, we have also seen that event detection as such is a major factor for the current performance starting from raw text.

With respect to the granularity of the matching of the components, we have used various ways to abstract from surface forms:

- different forms are matched to the same URI assigned by DBpedia Spotlight, created from surface forms or through nominal coreference;
- time expressions are normalised to the ISO dates;
- event mentions with different forms are matched through WordNet similarity and their word-senses;

Furthermore, we can parameterize the matching by setting loose or strict constraints:

- dates can be mapped by year or month instead of day;
- more or less participants to be shared, with or without their roles;
- degree of overlap of concepts and the range of concepts above a word-sense-disambiguation (WSD) threshold;

We have seen that making these constraints more tight results in more precision and lower recall. Making them more loose has the opposite effect.

There are still mentions that are not mapped to the same instances, e.g. "his girlfriend", "the daughter" to "Ka'loni Flynn", and there are many missed URIs as well as wrong URIs assigned. The quality of modules such as NERC, NED and WSD is crucial in this respect. Furthermore, time-anchors are very sparse and difficult to infer from the text as such.[7]

However, it is also the case that the ECB+ database is still too restricted to measure the true contribution of the component-based approach. We have seen that matching years instead of months makes no difference due to the fact that the events can already be distinguished by the year. Adding more seminal events to create more referential ambiguity for similar events around similar periods in time will require more precise analysis and component matching.

Defining the granularity of event descriptions provides an interesting view on event-coreference. How far can we go to lump together event data? In a way, we could lump all events that make up a story or a topic together and define a period of time in which the topic or story takes place with all the involved participants. This does not necessarily violate the idea of event-coreference since peoples' intuitions on decomposing events to smaller units are also not clear-cut. Obviously, at some point lumping of event data generates unclarity of scope relations between events and participants, such as more than one person murdering the same or different persons, or even semantic anomalies such as or the same person being at different places at the same time. This is where event coreference could set a hard border but this also means that annotation and evaluation of data sets may need to be different, e.g. assigning not only event-coreference relations but also subevent and topical relations.

A final aspect that still needs to be investigated is variation. Even though ECB+ has documents from various sources, events that are annotated as coreferential have mostly the same lemma and most events have no coreference relation at all (90% of all mentions). Annotation of event coreference is not an easy task and annotators tend to be conservative. More variation in reference to events is also expected to put higher demands on a semantic approach rather than approaches that are trained on mentions only.

## 9    Conclusion

We described a new method to detect event descriptions in text and to model semantics of events, their components and event coreference using RDF repre-

---

[7] We left out the document creation time as a baseline time-anchor because it may interfer with the task since the articles on each seminal event were published on different dates.

sentations. The proposed heuristic outperforms the state-of-the-art system when assuming equal quality in event detection. Our approach collects the component information from the complete document rather than from the local context of the event mentions that are compared. We have shown that event components play a role in obtaining precision and recall and that their matching needs to be adapted to the granularity of the task.

In future work, we want to investigate event-coreference in relation to topical structures and storylines. We believe that this also helps creating annotated data in which more variation and referential ambiguity is reflected. This will make both the annotation of data and the task more natural. Such data sets provide additional possibilities and challenges to obtain more precise temporal ordering of events for event coreference.

**Acknowledgments.** The NewsReader project was co-funded by the European Union as project number: 316404, FP7 Work Programme Call FP7-ICT-2011-8 Objective Cooperation Research theme "Information and Communication Technologies", challenge 4.4 - Area Intelligent Information Management.

# References

1. Ahn, D.: The stages of event extraction. In: Proceedings of the Workshop on Annotating and Reasoning About Time and Events (2006)
2. Bagga, A., Baldwin, B.: Algorithms for scoring coreference chains. In: Proceedings of the International Conference on Language Resources and Evaluation (LREC) (1998)
3. Bagga, A., Baldwin, B.: Cross-document event coreference: annotations, experiments, and observations. In: Proceedings of the ACL Workshop on Coreference and its Applications, p. 18 (1999)
4. Baker, C.F., Fillmore, C.J., Lowe, J.B.: The Berkeley framenet project. In: Proceedings of the 17th International Conference on Computational Linguistics, vol. 1, pp. 86–90. Association for Computational Linguistics (1998)
5. Bejan, C.A., Harabagiu, S.: Unsupervised event coreference resolution with rich linguistic features. In: Proceedings of the 48th Annual Meeting of the Association for Computational Linguistics, Uppsala, Sweden (2010)
6. Bird, S., Klein, E., Loper, E.: Natural Language Processing with Python. O'Reilly Media Inc., Sebastopol (2009). http://nltk.org/book
7. Björkelund, A., Bohnet, B., Hafdell, L., Nugues, P.: A high-performance syntactic and semantic dependency parser. In: Proceedings of the 23rd International Conference on Computational Linguistics: Demonstrations, COLING 2010, pp. 33–36. Association for Computational Linguistics, Stroudsburg, PA, USA (2010). http://dl.acm.org/citation.cfm?id=1944284.1944293
8. Blei, D.M., Frazier, P.I.: Distance dependent Chinese restaurant processes. J. Mach. Learn. Res. **12**, 2461–2488 (2011)
9. Chen, B., Su, J., Pan, S.J., Tan, C.L.: A unified event coreference resolution by integrating multiple resolvers. In: Proceedings of the 5th International Joint Conference on Natural Language Processing, Chiang Mai, Thailand, November 2011
10. Chen, Z., Ji, H.: Event coreference resolution: feature impact and evaluation. In: Proceedings of Events in Emerging Text Types (eETTs) Workshop (2009)

11. Chen, Z., Ji, H.: Graph-based event coreference resolution. In: TextGraphs-4 Proceedings of the 2009 Workshop on Graph-based Methods for Natural Language Processing, pp. 54–57 (2009)
12. Cybulska, A., Vossen, P.: Semantic relations between events and their time, locations and participants for event coreference resolution. In: Angelova, G., Bontcheva, K., Mitkov, R. (eds.) Proceedings of Recent Advances in Natural Language Processing (RANLP-2013), INCOMA Ltd., Hissar, Bulgaria, 7–14 September 2013. No. ISSN 1313–8502. http://aclweb.org/anthology//R/R13/R13-1021.pdf
13. Cybulska, A., Vossen, P.: Guidelines for ECB+ annotation of events and their coreference. Technical report NWR-2014-1, VU University Amsterdam (2014)
14. Cybulska, A., Vossen, P.: Using a sledgehammer to crack a nut? Lexical diversity and event coreference resolution. In: Proceedings of the 9th Language Resources and Evaluation Conference (LREC2014), Reykjavik, Iceland, 26–31 May 2014
15. Cybulska, A., Vossen, P.: "Bag of events" approach to event coreference resolution. Supervised classification of event templates. In: Proceedings of the 16th Cicling 2015 (Co-located: 1st International Arabic Computational Linguistics Conference), Cairo, Egypt, 14–20 April 2015
16. Fellbaum, C. (ed.): WordNet. An Electronic Lexical Database. MIT Press, Cambridge (1998)
17. Fokkens, A., Soroa, A., Beloki, Z., Ockeloen, N., Rigau, G., van Hage, W.R., Vossen, P.: NAF and GAF: linking linguistic annotations. In: Proceedings 10th Joint ISO-ACL SIGSEM Workshop on Interoperable Semantic Annotation, Reykjavik, Iceland, p. 9 (2014)
18. van Hage, W.R., Malaisé, V., Segers, R., Hollink, L., Schreiber, G.: Design and use of the Simple Event Model (SEM). J. Web Sem. 9(2), 128–136 (2011)
19. Humphreys, K., Gaizauskas, R., Azzam, S.: Event coreference for information extraction. In: ANARESOLUTION 1997 Proceedings of a Workshop on Operational Factors in Practical, Robust Anaphora Resolution for Unrestricted Texts (1997)
20. Kingsbury, P., Palmer, M.: From treebank to propbank. In: LREC. Citeseer (2002)
21. LDC: ACE (Automatic Content Extraction) English Annotation Guidelines for Events ver. 5.4.3 2005.07.01. In: Linguistic Data Consortium (2005)
22. Leacock, C., Chodorow, M.: Combining local context with wordnet similarity for word sense identification (1998)
23. Lee, H., Recasens, M., Chang, A., Surdeanu, M., Jurafsky, D.: Joint entity and event coreference resolution across documents. In: Proceedings of the 2012 Conference on Empirical Methods in Natural Language Processing and Natural Language Learning, EMNLPCoNLL 2012 (2012)
24. Liu, Z., Araki, J., Hovy, E., Mitamura, T.: Supervised within-document event coreference using information propagation. In: Proceedings of the International Conference on Language Resources and Evaluation, LREC 2014 (2014)
25. Luo, X.: On coreference resolution performance metrics. In: Proceedings of the Human Language Technology Conference and Conference on Empirical Methods in Natural Language Processing, EMNLP-2005 (2005)
26. Mendes, P.N., Jakob, M., García-Silva, A., Bizer, C.: Dbpedia spotlight: shedding light on the web of documents. In: Proceedings of the 7th International Conference on Semantic Systems, pp. 1–8. ACM (2011)
27. Pedregosa, F., Varoquaux, G., Gramfort, A., Michel, V., Thirion, B., Grisel, O., Blondel, M., Prettenhofer, P., Weiss, R., Dubourg, V., Vanderplas, J., Passos, A., Cournapeau, D., Brucher, M., Perrot, M., Duchesnay, E.: Scikit-learn: machine learning in Python. J. Mach. Learn. Res. 12, 2825–2830 (2011)

28. Pradhan, S., Ramshaw, L., Marcus, M., Palmer, M., Weischedel, R., Xue, N.: Conll-2011 shared task: modeling unrestricted coreference in ontonotes. In: Proceedings of CoNLL 2011: Shared Task (2011)

29. Recasens, M., Hovy, E.: Blanc: implementing the rand index for coreference evaluation. Nat. Lang. Eng. **17**(4), 485–510 (2011)

30. UzZaman, N., Llorens, H., Derczynski, L., Verhagen, M., Allen, J., Pustejovsky, J.: Semeval-2013 task 1: Tempeval-3: evaluating time expressions, events, and temporal relations (2013)

31. Vilain, M., Burger, J., Aberdeen, J., Connolly, D., Hirschman, L.: A model theoretic coreference scoring scheme. In: Proceedings of MUC-6 (1995)

32. Vossen, P., Agerri, R., Aldabe, I., Cybulska, A., van Erp, M., Fokkens, A., Laparra, E., Minard, A.L., Aprosio, A.P., Rigau, G., Rospocher, M., Segers, R.: Newsreader: how semantic web helps natural language processing helps semantic web. Special Issue Knowledge Based Systems, Elsevier (to appear)

33. Vossen, P., Bond, F., McCrae, J.: Toward a truly multilingual global wordnet grid. In: Proceedings of the 8th Global Wordnet Conference (2016)

34. Yang, B., Cardie, C., Frazier, P.I.: A hierarchical distance-dependent Bayesian model for event coreference resolution. CoRR abs/1504.05929 (2015). http://arxiv.org/abs/1504.05929

# An Informativeness Approach
# to Open IE Evaluation

William Léchelle[(✉)] and Philippe Langlais

RALI, University of Montreal, Montreal, Canada
{lechellw,felipe}@iro.umontreal.ca

**Abstract.** Open Information Extraction (OIE) systems extract relational tuples from text without requiring to specify in advance the relations of interest. Systems perform well on widely used metrics such as precision and yield, but a close look at systems output shows a general lack of informativeness in facts deemed correct.

We propose a new evaluation protocol, based on question answering, that is closer to text understanding and end user needs. Extracted information is judged upon its capacity to automatically answer questions about the source text. As a showcase for our protocol, we devise a small corpus of question/answer pairs, and evaluate available state-of-the-art OIE systems on it. Performance-wise, our results are in line with previous findings. Furthermore, we are able to estimate recall for the task, which is novel. We distribute our annotated data and automatic evaluation program.

**Keywords:** Open Information Extraction · Evaluation
Question answering

## 1 Introduction

OIE - information extraction without pre-specification of relations or entities to target - seeks to extract relational tuples from large corpora, in a scalable way and without domain-specific training [3,5]. Recently, there has been a trend of successful use of OIE output as a text understanding tool, for instance in [4,7,8].

However, a close look to systems output reveals that a large fraction of extracted facts, albeit correctly extracted from the text, are devoid of useful information. The main reason for this is lack of context: many extracted noun phrases, and facts, only have meaning in the context of their sentences. Once the source is lost, the remaining relation is empty, for factual purposes[1]. Figure 1 shows examples of uninformative facts. We discuss in Sect. 2 how state-of-the-art metrics of extraction performance fail to account for meaningless extractions.

We propose an evaluation procedure for open information extractors that more tightly fits downstream user needs. The most direct usage of information is

---

[1] For automatic language modelling purposes, on the other hand, extracted facts are a great source of learning material, as demonstrated in [8].

© Springer International Publishing AG, part of Springer Nature 2018
A. Gelbukh (Ed.): CICLing 2016, LNCS 9624, pp. 523–534, 2018.
https://doi.org/10.1007/978-3-319-75487-1_40

**Sentence :** In response, a group of Amherst College students held a patriotism rally in October, reciting the Pledge of Allegiance.
**Fact :** (a group of Amherst College students ; held ; a patriotism rally)
**Sentence :** That's a lot of maybes in a sport where the right thing seldom happens, but [given X, Y should Z if he wants T].
**Fact :** (That ; is ; a lot of maybes)

---

**Sentence :** The scandal has now forced resignations at Japan's fourth-largest bank and three of Japan's Big Four brokerages.
**Fact :** (Japan ; has ; fourth-largest bank)
**Sentence :** This leads to one of two inescapable conclusions : Either the president reads BioScope or I got lucky.
**Fact :** (the president ; reads ; BioScope or I got lucky)

---

**Fig. 1.** Extracted facts deemed correct by previous manual evaluations, respectively from ReVerb and ClausIE. In the first extraction, even though it is true in itself, crucial information from another sentence is missing to give the fact its meaning. In the second, the first argument it at best a vague idea. In the third, the fact holds true for most countries and holds little information by itself (it would be more adequate at another level of abstraction, e.g. *countries have banks*). The last extracted fact does not reflect the actual sentence meaning.

answering questions about it, so we evaluate extractors' output on their capacity to answer questions asked about the text at hand.

This procedure serves two purposes not previously addressed:

1. incorporate informativeness in the judging criterion for correct extractions;
2. estimate recall for the task.

Section 3 details the evaluation methodology we follow, and the guidelines that drive our annotations. Section 4 presents the dataset we built and our basic automatic evaluation procedure, and Sect. 5 exposes our results.

## 2 Related Work

Little work has directly addressed the issue of evaluating OIE performance. Up to now, performance of extractors mostly relied on 2 metrics: number of extracted facts, and precision of extraction, area under the precision-yield curve being a shorthand for both [5]. As the bulk of extractions are usually obtained with a precision in the 70–80% range, consecutive generations of extractors have mostly improved on the yield part:

"Ollie finds 4.4 times more correct extractions than ReVerb and 4.8 times more than WOE$^{parse}$ at a precision of about 0.75". [5]

"ClausIE produces 1.8–2.4 times more correct extractions than Ollie". [3]

---

**Sentence :** For the 2006-07 season, Pace played with the Nelson Giants in the New Zealand National Basketball League.

**4-ary fact :** (Pace ; played ; with the Nelson Giants ; for the 2006-07 season ; in the New Zealand National Basketball League)

**Questions :**
Who did Pace play with ?
When did Pace play with the Nelson Giants ?
In what league did Pace play with the Nelson Giants ?

---

**Fig. 2.** Some facts are intrinsically $n$-ary, and naturally answer many questions. An extractor capturing only binary relations would likely miss much context.

It is commonly lamented that absolute recall cannot be calculated for this task, because of the absence of a reference. We aim to address this issue.

Typically, precision is measured by sampling extractions and manually labelling them as correct or incorrect. As a rule of thumb, an extraction is deemed correct if it is implied by the sentence:

"Two annotators tagged the extractions as correct if the sentence asserted or implied that the relation was true." [5]

"We also asked labelers to be liberal with respect to coreference or entity resolution; e.g., a proposition such as ('he', 'has', 'office'), or any unlemmatized version thereof, is treated as correct." [3]

By contrast with previously employed criteria, we propose to incorporate the informativeness of extracted facts, as measured by their ability to answer relevant questions, in the judgement of their validity.

Figure 1 shows examples of previous facts manually labelled as correct, taken respectively from ReVerb[2] and ClausIE[3]. Except for the last, they pass the standard criteria for correctness. We seek to devise an evaluation protocol that would reject such extractions on the grounds that they are not informative.

As highlighted by [1], one major element of OIE performance is the handling of $n$-ary facts. Most extractors focus on binary relations, but many support $n$-ary relations to some extent. KrakeN [1] and Exemplar [6] are designed towards $n$-ary extractions. ClausIE [3] supports generation of $n$-ary propositions when optional adverbials are present and Ollie [5] can capture $n$-ary extractions by collapsing extractions where the relation phrase only differs by the preposition[4]. Though it is not our main focus, our evaluation protocol addresses this question in that $n$-ary facts that capture more information will answer more questions than binary facts that would leave out some of the arguments. Figure 2 shows an

---

[2] http://reverb.cs.washington.edu/reverb_emnlp2011_data.tar.gz.

[3] http://resources.mpi-inf.mpg.de/d5/clausie/.

[4] This was added to the distributed software since publication - see https://github.com/knowitall/ollie.

example of 4-ary fact, and the corresponding questions it answers. Information extractors will be evaluated on their ability to answer such questions.

In [6], the authors present an experimental comparison of several systems, over multiple datasets, on the very similar task of open relation extraction (that differs from OIE by only considering named entities as possible arguments). Their study focuses on the tradeoff between processing speed—depth of linguistical analysis—and accuracy. Much of their discussion stresses the difficulties of building a common evaluation methodology that is fair to various methods. In particular, their Sect. 3.3 is a good illustration of several evaluation difficulties.

# 3   Proposed Methodology

## 3.1   Evaluation Protocol

As hinted in Fig. 2, the evaluation protocol we propose for OIE is as follows:

> Given some input text, annotate all factoid questions that can be answered by information contained in that text, and the answers. Run the extractors on the input text. The evaluation metric is the number of questions that can be answered using only the output of each system.

The step of using extracted tuples to answer questions raises issues. Of course manual matching of extractions and questions would be most precise, but unrealistically expensive in human labor. An automatic scoring system also makes for more objective and easily replicable results, albeit less precise. Still, automatic question answering is a notoriously difficult problem that we would rather not tackle. In order to avoid this difficulty, we design our questions to be in the simplest possible form. Figure 3 shows a sample of our annotations. The questions are worded in very transparent ways.

We describe the automatic scoring system we use for this paper at greater length in Sect. 4.2, and release it along with our data.

## 3.2   Annotation Process

We wish to annotate all questions that can be answered by information contained in the input text, in a way that is easy to answer automatically.

Our primary goal being to evaluate OIE systems, we found useful to consider their output on the sentences at hand as a base for annotation. As stated before (Fig. 1), many extractions are not informative by themselves. Examples of correctly extracted facts that cannot answer real-world questions are showed in the first sentence of Fig. 4.

Given OIE output, all extracted relations that are factually informative are asked about, i.e. a question is added that is expected to be answered by it. Then, we also ask all other questions that can be answered with information contained in the sentence, without being overly specific.

---

**S**: Esaka and six other top executives will quit to take responsibility for 67.28 million yen in payoffs to corporate racketeer Ryuichi Koike, 54.

**Q**: Why will Esaka quit ?
**A**: to take responsibility for 67.28 million yen in payoffs to corporate racketeer Ryuichi Koike
**X**: 1

**YQ**: Will Esaka quit ?
**A**: Yes
**X**: 1

**Q**: What age is Ryuichi Koike ?
**A**: 54
**X**: 1

---

**S**: And he has eased up on team rules.

---

**S**: His teammates indeed loved the show.

---

**S**: Mrs. Yogeswaran was shot five times with a pistol near her Jaffna home on May 17, 1998.

**Q**: What was Mrs. Yogeswaran shot with ?
**A**: a pistol
**X**: 1

**Q**: How many times was Mrs. Yogeswaran shot ?
**A**: five
**X**: 1

---

**S**: Robert Barnard (born 23 November 1936) is an English crime writer, critic and lecturer.

**Q**: Who is Robert Barnard ?
**A**: an English crime writer
*(there is no X: annotation on this Q&A pair)*

**Q**: When is Robert Barnard born ?
**A**: 23 November 1936
**X**: 0

---

**Fig. 3.** Examples of annotated sentences, with questions and answers. Questions are worded following the original text so that the question answering step is simple to perform. Many sentences are embedded in so specific a context that they do not carry any extractable information, like the second and third sentences. Others usually yield a handful of facts. Lines are prefixed as **S**entences, **Q**uestions, and **A**nswers (**YQ** stands for yes/no questions). In our dataset, most questions (but not all) are tagged with an e**X**pected result of the Q&A system, given the extractions seen by the annotator (these are the **X**: lines – 1 if the answer will be found, 0 if it won't), for intrinsic evaluation purposes.

One could argue that the annotator seeing the output of the systems introduces a bias in the evaluation procedure. We do not believe this to be the case. As all OIE output is considered, the annotator is blind to specific extractors and the resulting dataset is fair to all systems. If proper attention is paid to asking questions about facts that are not correctly extracted, then the measure of recall

---

**Sentence:** While the Arab world is a rich prize in itself, Europe has been and remains the primary objective.
**Extraction:** (Europe ; remains ; the primary objective)
**Extraction:** (the Arab world ; is ; a rich prize)
*Important context is missing for these extractions to have meaning.*

---

**Sentence:** Wu worked as a reporter for United Press International from 1973 until 1978 when she joined WGBH-TV, Boston's public television station, as the Massachusetts State House reporter until 1983.
**Extraction:** (Wu ; worked as a reporter for ; United Press International)
**Sentence:** Daughter of the actor Ismael Sanchez Abellan and actress and writer Ana Maria Bueno, Gabriel was born in San Fernando, Cadiz ...
**Extraction:** (Gabriel ; was born in ; San Fernando)
*"Wu" is at the threshold of being sufficiently determined to ask questions about. "Gabriel" being a common first name, it is just on the other side of our threshold.*

---

**Sentence:** He was born in New York and died at Livonia, Michigan.
**Extraction:** (He ; was born in ; New York)
**Sentence:** He consults his family doctor for solution.
**Sentence:** Had I known then what I know now, I might have argued for a different arrangement.
*Coreference resolution is key in many sentences.*

---

**Fig. 4.** Ambiguity due to the loss of context is the main issue for annotation.

isn't biased towards technology performance either. In favor of using the information, it is easier to ask about useful extracted facts (e.g. in the very terms of the fact), which means all due credit is given to system successes. Additionally, annotation so helps reflecting on OIE capabilities and limitations.

### 3.3   Dealing with Ambiguity

The major issue in our search for informativeness is ambiguity due to lack of context. It is a problem on various levels, as exemplified in Fig. 4.

In some cases, each element of the relation is understandable, but the whole meaning cannot be understood out of context—for instance (*"the Arab world is a rich prize"*) and (*"Europe remains the primary objective"*) in Fig. 4. When we cannot understand what the text at hand is about, we naturally do not annotate anything. In the last sentence of Fig. 4, (*I; might have argued for; a different arrangement*) similarly relates to a lost specific context.

Often, and this is the key difficulty, the arguments themselves only make sense in context. In *"the scandal forced resignations ... "* (Fig. 1), what *"the scandal"* refers to maybe was obvious in the news context of the time, but is lost to us.

Therefore, there is a somewhat arbitrary line to draw regarding the amount of context we can assume a potential user has in mind when asking questions.

Amongst the many shades of ambiguity, *Albert Einstein* and *New York City* are utmostly self-explanatory[5]. *"I"*, to the contrary, is completely dependant on its particular utterance, and a clear definition of its reference will most often pertain to the metadata of the document at hand.

In between, consider *"Wu"* in the second sentence of Fig. 4. We consider this name to be on the threshold of acceptable ambiguity for our purpose. Using context, we can trace her to be Janet Wu, an American television reporter from the Boston area[6]. Also consider *"Gabriel"*[7] in the following sentence, which we consider to be on the other side of the threshold.

We envision two possible policies regarding context requirements.

What we did is the following: assume there is no word sense disambiguation. Every argument phrase used in our dataset should refer to a single entity. When an input sentence is about the most common use of its words (like *Europe* for the continent), we consider it well defined and annotate it, using its words in those senses. When sentences are about less common senses of their words (like *Gabriel* for Ruth Gabriel (See footnote 7) or *New York City* for the video game), we do not annotate them, on the grounds that further processing would be required to make use of such annotations, that is outside the scope of this work. It is normal for OIE to lose the context of extractions, and informativeness judgement calls shall take that into account.

Hence, by the fact that there is only one *"Wu"* in our corpus, it is a valid designation for an entity. *"Gabriel"*, on the other hand, is more often a common first name than refers to Ruth Gabriel, so we would not ask about her.

A looser policy would be to assume that the user that asks questions has in mind the same context as the writter of the original text. Within that view, some user may ask *"did the scandal force resignations ?"*, having in mind the Japanese banking public embarrassment of Fig. 1, or *"what is the primary objective ?"*, thinking of them who sell to the Arab world in Fig. 4.

### 3.4   Coreference

In a way, coreferential mentions are the extreme case of the ambiguity-without-context problem just mentioned.

Currently available OIE systems don't resolve coreferential mentions, and a significant portion of extracted facts have "it", "he" or "we" as arguments. As such, we consider that these extractions cannot answer questions, as the reference of such mentions is lost. On the Wu sentence, systems extract (**she**; *joined; WGBH-TV as the Massachusetts State House reporter*), but we lack the means of answering *"did Wu join WGBH-TV ?"* with that fact.

---

[5] Although they could refer to an American actor and a 1984 Atari video game.

[6] The sentence is from https://en.wikipedia.org/wiki/Janet_Wu_(WCVB). Incidentally, https://en.wikipedia.org/wiki/Janet_Wu_(WHDH) also is an American television reporter who worked in the Boston area. We consider this to be an ironic coincidence, but stand by our arbitrary line.

[7] Ruth Gabriel is a Spanish actress.

In our dataset, we ask such questions that would require coreference to be resolved at extraction time when it happens inside a sentence. As all sentences in our dataset were randomly picked from documents, all the cross-sentential references are lost, and we do not ask questions about them. In Fig. 4, we can't ask about the main theme of the last three sentences because of that.

It would be natural to extend the evaluation protocol to such facts, and it would enrich the measure of recall to examine how many facts are spread over several sentences, by annotating a whole document. Considering that the issue is not currently addressed by available systems (they treat all sentences independently), it would be a moot point for now.

# 4    Evaluation Data and Program

## 4.1    Question-Answer Dataset

As a proof of concept, we annotated slightly more than 100 Q&A pairs on a corpus previously employed for OIE evaluation. Rather than aiming for these to become a standard dataset, we encourage other researchers to write their own questions datasets, tailored to the needs of their particular OIE systems, and enrich the pool of available resources for evaluation.

The data we annotate is that distributed (See footnote 3) by [3]. Sentences were randomly picked from 3 sources:

- 500 sentences are the so-called "ReVerb dataset", obtained from the web via a Yahoo random-link service;
- 200 sentences come from Wikipedia;
- 200 sentences come from the New York Times.

A sample of annotations is shown in Fig. 3. To give an idea and as discussed in Sect. 3.3, about half of sentences are not suitable to ask meaningful questions about. On the other sentences, we typically find 2–4 questions that can be answered with their information (2.3 on average on the reverb dataset, 3 on the wikipedia sentences).

The data is available at http://www.CICLing.org/2016/data/92.

## 4.2    Question Matcher

With annotated Q&A pairs and extracted information in hand, the remaining step is to match and evaluate answers to the questions. We develop a very basic Q&A system, based on string matching. In short, it matches the text of extracted facts to questions, and assumes that a good match indicates the presence of an answer. It is not a very good Q&A system, but it is a decent evaluation script. Its most important features are to be fair to all systems, and easy to use, understand and replicate. Figure 5 illustrates how the evaluation system works.

As mentioned in Sect. 3.1, and by contrast with the task of open-domain question answering, e.g. studied in [4], we deliberately do not address the difficult

Q: What was Mrs. Yogeswaran shot with ?
Match words: <u>mrs</u>. <u>shot</u> <u>what</u> <u>yogeswaran</u>          **Threshold** $= 0.6$    **A:** <u>a pistol</u>

| Facts | Score | Returned answer | Evaluation metric |
|---|---|---|---|
| Esaka ; will quit ; to take responsibility for 67.28 million yen in payoffs | 0.0 | | |
| <u>Mrs</u>. <u>Yogeswaran</u> ; <u>was</u> <u>shot</u> ; five times with a pistol near her Jaffna home on May 17 1998 | 1.0 | five times with a pistol near her Jaffna home ... | Correct |
| <u>Mrs</u>. <u>Yogeswaran</u> ; <u>was</u> <u>shot</u> with ; a pistol | 1.0 | a pistol | Correct |
| Ashcroft ; said ; through <u>Mr</u>. Hilton <u>that</u> he had made the point <u>that</u> <u>there</u> <u>would</u> be no peace between him and the Governor until ... | 0.75 | Ashcroft | Wrong |

**Fig. 5.** Q&A script based on string matching. A fact that matches the words of a question past a given threshold is assumed to contain an answer to it. The part of that fact (arg1, rel, or arg2) that least matched the question is picked as the answer. A candidate answer is correct if it contains all the words of the reference answer.

problem of question understanding. Instead, questions were written in transparent ways so as to facilitate their automatic answering (see Fig. 3).

In order to answer a given question, the system attempts to match each extracted fact to it, at the word level. A fact that matches more than 60%[8] of a question's words is assumed to contain an answer to the question. Stopwords are excluded, using NLTK [2]. Edit distance is used to relax the words matching criteria[9], to make up for slight morphological variations between the words of interrogative questions and that of affirmative facts.

When a fact matches a question, we look for the part (first argument, relation, second argument) that least matches the question, and pick it as an answer (typically the second argument, arguments are favored in case of equality, as in the last line of Fig. 5). We consider an answer to be correct when all the gold answer words are contained in the returned answer.

Were we to build a true question-answering system, we would need to pick or order the various candidate answers gathered for each question. In practice, we seek to devise an evaluation script, and there are only a handful[10] of answers per question, so we consider that a question is correctly answered if any of its candidate answers is correct.

## 5  Results

### 5.1  Performance

The results of the evaluation procedure are showed in Table 1. On factoid questions, the automatic answering system finds candidate answers to 35–70% of

---

[8] We examine the impact of this factor in Sect. 5.2.

[9] Words differing by 1 or less of their characters are considered to match, as *Mrs.* and *Mr.* in Fig. 5.

[10] See Table 2 as the exact figure is directly dependant on the matching threshold.

questions, depending on the information extractor. When candidate answers are found by the answering system, at least one is correct in 20–50% of cases.

Most importantly, recall measures how much of the sentences' relevant information was captured by the extractions, and we can see that combining all the systems output, nearly 40% of the information is captured.

Our results fall in line with previous authors' findings. Pattern learning makes Ollie a significant improvement over the simplistic mechanism of ReVerb (at the computational price of dependency parsing), making it both more precise and yielding more facts, leading to a large increase in recall. ClausIE extracts more facts than Ollie at a similar level of precision, further boosting recall.

Practically though, both being open-source software, Ollie runs significantly faster than ClausIE, due to the difference in the embedded parsers they use.

**Table 1.** OIE systems performance results. Answered is the proportion of factoid questions for which at least one answer was proposed (correct or not); Precision is the proportion of answered questions to which one or more answers were correct; and Recall is the proportion of questions for which at least one candidate answer is correct. As a matter of fact, given the way metrics are computed, answered × precision = recall.

|        | Answered | Precision | Recall |
|--------|----------|-----------|--------|
| ReVerb | 35%      | 19%       | 6%     |
| Ollie  | 39%      | 43%       | 17%    |
| ClausIE| 68%      | 42%       | 29%    |
| All    | 71%      | 53%       | 38%    |

## 5.2  Analysis

In order to assess the quality of the evaluation script in terms of desired behaviour, manual assessment of whether a correct answer would be found or not was annotated on a sample of questions (80 out of 106). These are the **X:** lines in Fig. 3. This would be similar to a human-judged step of answering the questions given the facts, and comparison of the result of the automatic procedure with respect to the manual evaluation (but the annotator tagged it's expectation of the automatic procedure, rather than the desired behavior as in the manual judge case). On this sample, the evaluation system performed as predicted in upwards of 95% of cases, which is satisfying.

An important parameter of our approach, mentioned in Sect. 4.2, is the matching threshold past which a fact is assumed to contain an answer to the question it matched. Table 2 shows the impact of this parameter on our results, using extractions from all systems.

As expected, the lower the threshold, the looser the answers, and the higher the recall. We retained 0.6 as threshold for performance measures, for it has the highest precision, and above all maximises recall while keeping the average number of candidate answers reasonable (less than 5 rather than more than 20).

**Table 2.** Impact of matching threshold on evaluation metrics.

| Matching threshold | 0.5 | **0.6** | 0.7 | 0.8 |
|---|---|---|---|---|
| Questions answered | 95% | **71%** | 48% | 31% |
| Answers per question | 23 | **3.7** | 2.5 | 2.0 |
| Precision | 51% | **53%** | 46% | 38% |
| Recall | 48% | **38%** | 22% | 12% |

## 6    Conclusion and Future Work

We presented a new protocol for evaluation of OIE, that consists in annotating questions about the relevant information contained in input text, and automatically answering these questions using systems' output. Our performance metric more closely matches the usefulness of OIE output to end users than the previously employed methodology, by incorporating the informativeness of extracted facts in the annotation process. In addition, our protocol permits to estimate the recall of extraction in absolute terms, which to the best of our knowledge had never been performed. According to our results, about 40% of pieces of knowledge present in sentences are currently extracted by OIE systems.

We annotate a small dataset with Q&A pairs, and present our annotation guidelines, as well as the evaluation script we developed, in the form of a rudimentary Q&A system. We distribute our annotations and evaluation system to the community[11].

As directions for future work, we would like to annotate whole documents rather than isolated sentences, and measure the proportion of cross-sentential information. Our framework also naturally allows for evaluation of other text understanding systems, such as semantic parsers, or full-fledged question answering systems in the place of our own, which would be interesting to perform.

## References

1. Akbik, A., Löser, A.: Kraken: N-ary facts in open information extraction. In: Proceedings of Joint Workshop on Automatic Knowledge Base Construction and Web-Scale Knowledge Extraction, AKBC-WEKEX 2012, pp. 52–56. Association for Computational Linguistics, Stroudsburg (2012). http://dl.acm.org/citation.cfm?id=2391200.2391210
2. Bird, S., Klein, E., Loper, E.: Natural Language Processing with Python: Analyzing Text with the Natural Language Toolkit. O'Reilly, Beijing (2009). http://www.nltk.org/book
3. Del Corro, L., Gemulla, R.: ClausIE: clause-based open information extraction. In: Proceedings of 22nd International Conference on World Wide Web, WWW 2013, pp. 355–366. International World Wide Web Conferences Steering Committee, Republic and Canton of Geneva (2013). http://dl.acm.org/citation.cfm?id=2488388.2488420

---

[11] http://www.CICLing.org/2016/data/92.

4. Fader, A., Zettlemoyer, L., Etzioni, O.: Open question answering over curated and extracted knowledge bases. In: Proceedings of 20th ACM SIGKDD International Conference on Knowledge Discovery and Data Mining, KDD 2014, pp. 1156–1165. ACM, New York (2014). http://doi.acm.org/10.1145/2623330.2623677

5. Schmitz, M., Bart, R., Soderland, S., Etzioni, O.: Open language learning for information extraction. In: Proceedings of Conference on Empirical Methods in Natural Language Processing and Computational Natural Language Learning (EMNLP-CONLL) (2012)

6. Mesquita, F., Schmidek, J., Barbosa, D.: Effectiveness and efficiency of open relation extraction. In: Proceedings of 2013 Conference on Empirical Methods in Natural Language Processing, pp. 447–457. Association for Computational Linguistics, October 2013

7. Soderland, S., Gilmer, J., Bart, R., Etzioni, O., Weld, D.S.: Open information extraction to KBP relations in 3 hours. In: Proceedings of 6th Text Analysis Conference, TAC 2013, 18–19 November 2013, Gaithersburg, Maryland, USA. NIST (2013). http://www.nist.gov/tac/publications/2013/participant. papers/UWashington.TAC2013.proceedings.pdf

8. Stanovsky, G., Dagan, I., Mausam: Open IE as an intermediate structure for semantic tasks. In: Proceedings of 53rd Annual Meeting of the Association for Computational Linguistics and 7th International Joint Conference on Natural Language Processing, Short Papers, vol. 2, pp. 303–308. Association for Computational Linguistics, Beijing, July 2015. http://www.aclweb.org/anthology/P15-2050

# End-to-End Relation Extraction Using Markov Logic Networks

Sachin Pawar[1,2(✉)], Pushpak Bhattacharya[2,3], and Girish K. Palshikar[1]

[1] TCS Research, Tata Consultancy Services, Pune 411013, India
{sachin7.p,gk.palshikar}@tcs.com
[2] Department of CSE, Indian Institute of Technology Bombay, Mumbai 400076, India
pb@cse.iitb.ac.in
[3] Indian Institute of Technology Patna, Patna 801103, India

**Abstract.** The task of end-to-end relation extraction consists of two sub-tasks: (i) identifying entity mentions along with their types and (ii) recognizing semantic relations among the entity mention pairs. It has been shown that for better performance, it is necessary to address these two sub-tasks jointly [13,22]. We propose an approach for simultaneous extraction of entity mentions and relations in a sentence, by using inference in Markov Logic Networks (MLN) [21]. We learn three different classifiers: (i) local entity classifier, (ii) local relation classifier and (iii) "pipeline" relation classifier which uses predictions of the local entity classifier. Predictions of these classifiers may be inconsistent with each other. We represent these predictions along with some domain knowledge using weighted first-order logic rules in an MLN and perform joint inference over the MLN to obtain a global output with minimum inconsistencies. Experiments on the ACE (Automatic Content Extraction) 2004 dataset demonstrate that our approach of joint extraction using MLNs outperforms the baselines of individual classifiers. Our end-to-end relation extraction performance is better than 2 out of 3 previous results reported on the ACE 2004 dataset.

## 1 Introduction

Real world entities are referred in natural language sentences through *entity mentions* and these are often linked through meaningful *relations*. The task of end-to-end relation extraction consists of two sub-tasks: entity extraction and relation extraction. The sub-task of *entity extraction* deals with identifying entity mentions and determining their entity types. The other task of *relation extraction* deals with identifying whether any semantic relation exists between any two mentions in a sentence and also determining the relation type if it exists. In this paper, we refer to *entity extraction* and *relation extraction* tasks as defined by the Automatic Content Extraction (ACE) program [3] under the EDT (Entity Detection and Tracking) and RDC (Relation Detection and Characterization) tasks, respectively. ACE standard defined 7 entity types[1]: PER (person), ORG

---

[1] https://www.ldc.upenn.edu/sites/www.ldc.upenn.edu/files/english-edt-v4.2.6.pdf.

© Springer International Publishing AG, part of Springer Nature 2018
A. Gelbukh (Ed.): CICLing 2016, LNCS 9624, pp. 535–551, 2018.
https://doi.org/10.1007/978-3-319-75487-1_41

(organization), LOC (location), GPE (geo-political entity), FAC (facility), VEH (vehicle) and WEA (weapon). It also defined 7 coarse level relation types[2]: EMP-ORG (employment), PER-SOC (personal/social), PHYS (physical), GPE-AFF (GPE affiliation), OTHER-AFF (PER/ORG affiliation), ART (agent-artifact) and DISC (discourse).

Compared to the work (refer the surveys [18,19]) in Named Entity Recognition (NER), there are relatively few attempts [4,5,13,15] to address the more general entity extraction problem. NER extracts only named mentions (e.g. John Smith, Walmart) whereas entity extraction is expected to also identify common noun and pronoun mentions (e.g. company, leader, it, they) and their entity types. This task is more challenging than NER because entity type of mentions like leader or they may vary from sentence to sentence depending on which real life entity they are referring to in that sentence. For example, entity type of leader would be PER in the sentence John Smith was elected as the leader of the Socialist Party whereas its entity type would be ORG in the sentence Pepsi is a market leader in its segment.

There has been a lot of work for relation extraction like Zhou et al. [6], Jiang and Zhai [10], Bunescu and Mooney [1] and Qian et al. [20]. All of these approaches assume that the boundaries and the types of entity mentions are already known. Several features based on this information are used for relation prediction. In order to use such relation extraction systems, there should be separate entity extraction system whose output acts as an input for relation extraction. In such a "pipeline" method, the errors are propagated from first phase (entity extraction) to second phase (relation extraction) affecting the overall relation extraction performance. Another major disadvantage of the "pipeline" method is that it facilitates only one-way *information flow*, i.e. the knowledge about entities is used for relation extraction but not vice versa. However, the knowledge about relations can help in correcting some entity extraction errors.

In order to overcome these problems, we propose an approach which uses inference in Markov Logic Networks (MLN) for simultaneous extraction of entities and relations in a sentence. This approach facilitates two-way *information flow*. MLNs combine first-order logic and probabilistic graphical models in a single representation. An MLN contains a set of first-order logic rules, and each rule is associated with a weight. The fewer rules a world violates, the more probable it is. Also, higher the weight of a rule, greater is the probability of a world that satisfies the rule compared to the one that does not. In our approach, three separate classifiers are learned: a local entity classifier, a local relation classifier and a "pipeline" relation classifier which uses predictions of the local entity classifier. Predictions of these classifiers along with other domain knowledge are represented using weighted first-order logic rules in an MLN. Joint inference over this MLN is then performed to get a final output with least possible contradictions or inconsistencies among the individual classifiers.

---

[2] https://www.ldc.upenn.edu/sites/www.ldc.upenn.edu/files/english-rdc-v4.3.2.PDF.

The specific contributions of this work are: (i) a novel approach for joint extraction of entity mentions and relations using inference in MLNs and (ii) easy and compact representation of the domain knowledge using first-order logic rules in MLNs. The rest of the paper is organized as follows. Section 2 describes some background and necessary building blocks for our approach. Section 3 describes our approach in detail and Sect. 4 describes the working of our approach through an example. Experimental results are presented in Sect. 5. Related work is then described briefly in Sect. 6. Finally we conclude in Sect. 7 with brief discussion about the future work.

## 2   Building Blocks for Our Approach

### 2.1   Markov Logic Networks

Markov Logic Networks (MLN) which were proposed by Richardson and Domingos [21], combine first-order logic and probabilistic graphical models in a single representation. Formally, a Markov Logic Network $L$ is defined as a set of pairs $(F_i, w_i)$, where each $F_i$ is a formula in first-order logic with a real weight $w_i$. Along with a finite set of constants $C = \{C_1, C_2, \cdots, C_{|C|}\}$, it defines a Markov Network $M_{L,C}$ as follows:

1. $M_{L,C}$ contains one binary node for each possible grounding of each predicate appearing in $L$. The value of the node is 1 if the ground atom is true, and 0 otherwise.
2. $M_{L,C}$ contains one feature for each possible grounding of each formula $F_i$ in $L$. The value of this feature is 1 if the ground formula is true, and 0 otherwise. The weight of the feature is the $w_i$ associated with $F_i$ in $L$.

The probability distribution of random variable $X$ over possible worlds $x$ specified by Markov Network $M_{L,C}$ is given by,

$$P(X = x) = \frac{1}{Z} \exp\left(\sum_i w_i n_i(x)\right) \tag{1}$$

where $n_i(x)$ is the number of true groundings of $F_i$ in $x$ and $Z$ is the partition function. MLN can be used to find probability of a formula (say $F_1$) being true, given some other formula (say $F_2$) is true.

$$P(F_1|F_2, M_{L,C}) = \frac{P(F_1 \wedge F_2|M_{L,C})}{P(F_2|M_{L,C})} = \frac{\sum_{x \in X_{F_1} \cap X_{F_2}} P(X = x|M_{L,C})}{\sum_{x \in X_{F_1}} P(X = x|M_{L,C})}$$

where $X_{F_i}$ represents the set of worlds where $F_i$ holds and $P(X = x|M_{L,C})$ is computed using the Eq. 1.

## 2.2   Identifying Entity Mention Candidates

It is necessary to first identify the span (or boundaries) of each entity mention[3] in a given sentence. We model this as a sequence labelling problem. A sentence is a sequence of words and each word in a sentence is assigned a label indicating whether that word belongs to any entity mention or not. We use BIO encoding for this purpose.

- **O:** Label for the words which are not part of any entity mention
- **B:** Label for the first word of entity mentions
- **I:** Label for the subsequent words (except the first word) of entity mentions

We employ the Conditional Random Field (CRF) model [12], which is trained in a supervised manner. Given any new sentence, we use the trained CRF model to predict the 2 most probable label sequences as follows:

$S_1$ : A/O Palestinian/B Council/B member/B says/O anger/O is/O growing/O ./O

$S_2$ : A/O Palestinian/B Council/I member/B says/O anger/O is/O growing/O ./O

In this sentence, entity mention candidates from the topmost sequence are `Palestinian`, `Council` and `member`. Entity mention candidate `Palestinian Council` is generated from the second sequence. Generally, the candidates generated from the most probable sequence are more likely to be valid entity mentions. The candidates generated from the second most probable sequence are considered valid entity mentions only if they satisfy certain constraints. These constraints are applied in the form of first-order logic rules in MLNs and will be explained later. A special entity type *NONE* is assigned to a candidate entity mention if it is an invalid entity mention.

## 2.3   Local Entity Classifier

The local entity classifier is used to predict the most probable entity type for each candidate entity mention in a given sentence. This classifier is referred to as "local" as it takes an independent decision for each entity mention irrespective of its relation with other mentions. A Maximum Entropy Classifier is trained in a supervised manner which captures the characteristics of each entity mention $E$ using following features:

1. **Lexical Features:** Head word and other words in $E$, words preceding and succeeding $E$ in the sentence.
2. **Syntactic Features:** POS tags of the head word and other words in $E$, POS tags of the words preceding and succeeding $E$, parent of head word of $E$ in the dependency tree and also the dependency relation with the parent.

---

[3] We consider the "head" extent of a mention defined by ACE standard as the entity mention so that all the valid entity mentions are always non-overlapping.

3. **Semantic Features:** WordNet category (if any) of the head word of $E$. Some specific synsets in the WordNet (e.g. `person`, `location`, `vehicle`) are marked as possible "categories" and if any word is direct or indirect hypernym of such synsets, it is said to be falling in the corresponding "category".

As this classifier is trained using only the valid entity mentions in the training data, it always predicts one of the 7 ACE entity types and never predicts the $NONE$ type.

### 2.4  Local Relation Classifier

The local relation classifier is used to predict the most probable relation type for each pair of candidate entity mentions in a given sentence. This classifier is referred to as "local" as it takes an independent decision for each pair of entity mentions irrespective of their entity types.

In addition to ACE 2004 relation types, it considers two special relation types "NULL" (indicating that no semantic relation holds) and "IDN" (representing intra-sentence co-references). In the sentence `Pepsi is a market leader`, the entity mentions `leader` and `Pepsi` are co-references and hence we add the IDN relation between these mentions. With the help of IDN (identity) relation type, information about intra-sentence co-references can be incorporated in a principled way without using an external co-reference resolution system. Also, more number of entity mentions get involved in at least one relation, resulting in better entity extraction performance. For example, in the ACE 2004 dataset, there are 22718 entity mentions and 4328 relation instances resulting in only 7604 entity mentions involved in at least one relation. Considering the IDN relation, number of relation instances increases to 12060 covering 14930 entity mentions.

A Maximum Entropy Classifier is used which captures the characteristics of each entity mention pair $(E_1, E_2)$ with the help of following features:

1. **Lexical Features:** Head words and other words of $E_1$ & $E_2$, words preceding and succeeding $E_1$ & $E_2$ in the sentence.
2. **Syntactic Features:** POS tags of the head word and other words in $E_1$ & $E_2$, POS tags of the words preceding and succeeding $E_1$ & $E_2$, parents of head words of $E_1$ & $E_2$ in the dependency tree and also the dependency relations with the parents, path connecting $E_1$ & $E_2$ in the dependency tree, their common ancestor in the dependency tree.
3. **Semantic Features:** WordNet categories (if any) of the head words of $E_1$ & $E_2$, the common ancestor and other words on the path connecting $E_1$ & $E_2$ in the dependency tree of the sentence, syntactico-semantic structures identified in Chan and Roth [2].

### 2.5  Pipeline Relation Classifier

Unlike the local relation classifier, the "pipeline" relation classifier is dependent on the output of the local entity classifier. It uses following features in addition to the features used by the local relation classifier.

1. Entity types of $E_1$ and $E_2$ as predicted by the local entity classifier
2. Concatenation of entity types of $E_1$ and $E_2$
3. A binary feature indicating whether the types of $E_1$ and $E_2$ are same or not.

This classifier is referred to as a "pipeline" classifier because of unidirectional *information flow*. In other words, the knowledge about types of entity mentions is used by the relation classifier but not vice versa.

## 3    Joint Extraction Using Inference in MLNs

### 3.1    Motivation

As described in the previous section, we have 3 classifiers producing various predictions about entity types and relation types. These decisions may be inconsistent, i.e. relation type predicted by the local relation classifier may not be compatible with the entity types predicted by the local entity classifiers. Also, there may be contradiction in predictions of local relation classifier and "pipeline" relation classifier. Our aim is to take predictions of these classifiers as input and make a global prediction which minimizes such inconsistencies. MLN provides a perfect framework for this, where we can represent predictions of individual classifiers as first-order logic rules where weights of these rules are proportional to the prediction probabilities (soft constraints). Also, the consistency constraints among the relation types and entity types can be represented in the form of first-order logic rules with infinite weights (hard constraints). Now, the inference in such an MLN will provide a globally consistent output with maximum weighted satisfiability of the rules. The detailed explanation is provided in subsequent sections about how the first-order logic rules are created and how the corresponding weights are set.

### 3.2    Domains and Predicates

We specify one MLN for a sentence, i.e. for all candidate entity mentions and possible relation instances in a sentence. The software package used for inference in MLN is Alchemy[4]. We define 3 domains: *entity*, *etype* and *rtype*. The *entity* domain represents entity mentions where an unique ID is assigned to each entity mention. It is specified as follows in Alchemy for a sentence with $n$ entity mentions having IDs from 1 to $n$:

$$entity = \{1, 2, \cdots, n\}$$

The next domain *etype* represents the set of all possible entity types and another domain *rtype* represents the set of all possible relation types. These domains are specified in Alchemy as follows:

$$etype = \{PER, ORG, LOC, GPE, WEA, FAC, VEH, NONE\}$$

---

[4] http://alchemy.cs.washington.edu/.

$$rtype = \{EMPORG, GPEAFF, OTHERAFF, PERSOC, PHYS, ART, NULL, IDN\}$$

We define following predicates which are used for writing various first-order logic rules. The arguments for these predicates come from the above domains.

1. *ET(entity, etype)*: $ET(i, E)$ is true only when entity type of the entity mention $i$ is equal to $E$. It is true for one and only one entity type. It represents the entity type prediction of the local entity classifier and used as an *evidence* during inference.
2. *RTP(entity, entity, rtype)*: $RTP(i, j, R)$ is true only when type of relation between entity mentions $i$ and $j$ is equal to $R$. It is true for one and only one relation type. It represents relation type prediction of "pipeline" relation classifier. It is also used as an *evidence*.
3. *RTL(entity, entity, rtype)*: Similar to *RTP* but represents relation type prediction of local relation classifier.
4. *ETFinal(entity, etype)*: Similar to *ET* but represents global entity type prediction and is used as a *query* predicate during inference.
5. *RTFinal(entity, entity, rtype)*: Similar to *RTP* but represents global relation type prediction and is used as a *query* predicate.

During the inference in MLN, the probabilities of all possible groundings of *query* predicates are computed, conditioned on the specific groundings of the *evidence* predicates.

## 3.3 Generic Rules

Although one MLN is created for each sentence, some first-order rules are common and they are added to MLNs of all the sentences. We refer to these rules as *Generic Rules*. These rules represent some universal truths about the domain and hence the weight associated with each of these rules is set to *infinity*. In other words, any world that violates any of these rules, is practically impossible. These rules provide an easy and effective way of incorporating the domain knowledge about entity types and relation types. For each valid combination of relation type and entity type of one of its argument, we write rules to constrain the possible entity types for the other argument. Such rules can be easily devised by going through the ACE 2004 labelling guidelines. Following are some representative examples[5]. Note that the variables $x, y$ are universally quantified at the outermost level.

1. If there is an *EMPORG* relation between two entity mentions and entity type of any mention is *PER*, then entity type of other mention can only be one of: *ORG* or *GPE*.

   $RTFinal(x, y, EMPORG) \wedge ETFinal(x, PER) \Rightarrow (ETFinal(y, ORG) \vee ETFinal(y, GPE))$.

   $RTFinal(x, y, EMPORG) \wedge ETFinal(y, PER) \Rightarrow (ETFinal(x, ORG) \vee ETFinal(x, GPE))$.

---

[5] All the rules can't be listed because of the space constraints.

2. For the "identity" relation type $IDN$, the constraint is that the entity types of both the mentions should be same.

$$RTFinal(x, y, IDN) \wedge ETFinal(x, z) \Rightarrow ETFinal(y, z).$$
$$RTFinal(x, y, IDN) \wedge ETFinal(y, z) \Rightarrow ETFinal(x, z).$$

### 3.4  Sentence-Specific Rules

These rules are specific to each sentence and represent the predictions by the individual baseline classifiers. Unlike the *Generic Rules*, these rules are added with finite weights.

**Weight Assignment Strategies:** In order to learn the weights of various first-order logic rules, historical examples of predictions of 3 base classifiers along with gold-standard predictions would be required. Instead we chose to compute these weights by using some functions of the corresponding prediction probabilities. The work by Jain [9] discussed various ways of weight assignments to represent knowledge in MLNs. In another work, Heckmann et al. [8] adjusted the rule weights experimentally for citation segmentation using MLNs. On the similar lines, following two strategies are adopted for weight assignments.

1. **Log of Odds Ratio (LOR):** Richardson and Domingos [21] states that the weight of a formula $F$ is log odds between a world where $F$ is true and a world where $F$ is false. For a prediction with probability $p$, we set the weight of corresponding formula as $\log\left(\frac{p}{1-p}\right)$. Here, the penalty for violating any formula will increase logarithmically with its probability.
2. **Constant Multiplier (CM):** As per this strategy, for a prediction with probability $p$, we set the weight of corresponding formula as $K \cdot p$. Here, the penalty for violating any formula will increase linearly with its probability. We have used $K = 10$ in all our experiments.

**Rules induced by the Local Entity Classifier:** For each candidate entity mention, the entity type predicted by the local entity classifier acts as an *evidence* for the MLN inference. The classifier also assigns some probability to each possible entity type. For each entity mention id $i$, for each possible entity type $E$, following rule is added with the weight proportional to the probability of prediction.

$$ET(i, E_{max}) \Leftrightarrow ETFinal(i, E)$$

Here, $E_{max}$ is the entity type predicted by the local entity classifier. The weights assigned to this rule as per above strategies would be $\log\left(\frac{P_e(E|i)}{1-P_e(E|i)}\right)$ and $K \cdot P_e(E|i)$, where $P_e(E|i)$ is the probability assigned to entity type $E$ for the entity mention id $i$ by the local entity classifier.

**Rules induced by the Pipeline Relation Classifier:** For each pair of entity mentions, the relation type predicted by the "pipeline" classifier acts as an *evidence* for the MLN inference. For each pair of candidate entity mentions $(i, j)$, for each possible relation type $R$, following rule is added with the weight proportional to the probability of prediction.

$$RTP(i, j, R_{max}) \Leftrightarrow RTFinal(i, j, R)$$

Here, $R_{max}$ is the relation type predicted by the "pipeline" relation classifier. The weights assigned to this rule would be $\log\left(\frac{wt_p \cdot P_r^P(R|i,j)}{1 - P_r^P(R|i,j)}\right)$ and $K \cdot P_r^P(R|i, j) \cdot wt_p$, where $P_r^P(R|i, j)$ is the probability assigned to the relation type $R$ for the pair $(i, j)$ by the "pipeline" relation classifier. And $wt_p$ is the reliability of prediction of "pipeline" classifier, which indicates how confident the local entity classifier was in predicting entity types for entity mentions $i$ and $j$. We set $wt_p = P_e(E_{max}^i|i) \cdot P_e(E_{max}^j|j)$.

**Rules induced by the Local Relation Classifier:** For each pair of candidate entity mentions, the relation type predicted by the local classifier acts as an *evidence* for the MLN inference. For each pair entity mentions $(i, j)$, for each possible relation type $R$, following rule is added with the weight proportional to the probability of prediction.

$$RTL(i, j, R_{max}^L) \Leftrightarrow RTFinal(i, j, R)$$

Here, $R_{max}^L$ is the relation type predicted by the local relation classifier. The weights assigned to this rule would be $\log\left(\frac{P_r^L(R|i,j)}{1 - P_r^L(R|i,j)}\right)$ and $K \cdot P_r^L(R|i, j)$, where $P_r^L(R|i, j)$ is the probability assigned to the relation type $R$ for the pair $(i, j)$ by the local relation classifier.

**Rules for identifying valid/invalid entity mentions:** We generate candidate entity mentions using top 2 most probable BIO sequences. In general, we have a high confidence that candidate mentions from the topmost sequence are valid and have a lower confidence for candidates from the second sequence. This intuition is captured by addition of following rules. For each candidate $i$ from the topmost sequence, we add $!ETFinal(i, NONE)$ with the weight $\log\left(\frac{p}{1-p}\right)$ or $K \cdot p$, based on the weighing strategy employed. Also for each candidate $i$ from the second sequence, we add $ETFinal(i, NONE)$ with the weight $\log\left(\frac{1-p}{p}\right)$ or $K \cdot (1 - p)$. In both the cases, $p$ is the highest probability for any entity type predicted for that mention by the local entity classifier. As we are generating candidate entity mentions by using top 2 most probable BIO sequences, there may be some overlapping entity mentions. For each pair of such overlapping candidate entity mentions (say $i$ and $j$), following rules are added so that at most one of them is a valid entity mention.

$$!ETFinal(i, NONE) \Rightarrow ETFinal(j, NONE).$$
$$!ETFinal(j, NONE) \Rightarrow ETFinal(i, NONE).$$

We assume candidate mentions generated from the second BIO sequence to be valid, only if they are involved in some valid relation other than $NULL$. Also, an invalid entity mention should not be involved is any non $NULL$ relation with any other mention. To ensure this desired consistency, following rules are added for each pair of candidate mentions $(i, j)$ where one of them (say $i$) is generated from second BIO sequence.

$$!RTFinal(i, j, NULL) \Rightarrow !ETFinal(i, NONE).$$
$$ETFinal(i, NONE) \Rightarrow RTFinal(i, j, NULL).$$

After the inference, if the probability of $ETFinal(i, NONE)$ is the highest for any candidate mention $i$, then it is identified as an invalid mention. And because of above rules ensuring consistency, such mentions are never involved in any non $NULL$ relation.

### 3.5   Additional Semantic Rules

We explored the possibility of incorporating some domain knowledge by exploiting the easy and effective representability of the first-order logic. In order to incorporate the additional rules, we define following new predicates:

1. $CONS(entity, entity)$: $CONS(i, j)$ is true only when there is no other entity mention occurring in between the mentions $i$ and $j$ in a sentence.
2. $CONJ(entity, entity)$: $CONJ(i, j)$ is true only when there is a conjunction (i.e. connected through the dependency relations "conj:and" or "conj:or" in the dependency tree) between the two mentions $i$ and $j$.

**Using the knowledge of conjunctions:** When two entity mentions are connected through a conjunction (like **and, or**) and one of them is connected to a third entity mention with PHYS (i.e. located at) relation, then the other entity mention is also very likely to be connected to the third mention with PHYS relation. E.g. in the sentence fragment `troops in Israel and Syria`, a PHYS relation between `troops` and `Israel` implies another PHYS relation between `troops` and `Syria`. To incorporate this knowledge, following generic rules are added in MLNs of all sentences.

$$RTFinal(x, y, PHYS) \wedge ((CONJ(y, z) \wedge CONS(y, z)) \vee (CONJ(z, y) \wedge CONS(z, y)))$$
$$\wedge ET(y, t) \wedge ET(z, t) \Rightarrow RTFinal(x, z, PHYS).$$

$$RTFinal(x, y, PHYS) \wedge ((CONJ(w, x) \wedge CONS(w, x)) \vee (CONJ(x, w) \wedge CONS(x, w)))$$
$$\wedge ET(w, t) \wedge ET(x, t) \Rightarrow RTFinal(w, y, PHYS).$$

**Linking entity mentions with same types:** The entity mentions linked through certain dependency relations tend to share the same entity type. E.g. in the sentence fragment `companies such as Nielsen`, the mentions `companies` and `Nielsen` are very likely to have the same entity type. This is one of the Hearst patterns [7] to automatically identify hyponyms from text. If entity mentions $i$ and $j$ follow such a pattern, we add following rule to their sentence's MLN : $ETFinal(i, x) \Leftrightarrow ETFinal(j, x)$.

**Using knowledge about relation types:** If an entity mention of type $PER$ is involved in a $EMPORG$ relation, then it is highly unlikely that the same person will be connected to any other mention with the $EMPORG$ relation. This is because any person can have at most one employer mentioned in a single sentence. To impose this constraint, we add following rule.

$$RTFinal(x, y, EMPORG) \wedge (y \neq z) \wedge !RTFinal(y, z, IDN) \wedge !RTFinal(z, y, IDN)$$
$$\wedge ETFinal(x, PER) \Rightarrow !RTFinal(x, z, EMPORG) \wedge !RTFinal(z, x, EMPORG).$$

### 3.6 Joint Inference

As described above, an MLN is created for a sentence using some *Generic Rules* with infinite weights and some sentence-specific rules. Given such an MLN, we are interested to know the most probable groundings of the *query* predicates given some specific groundings of *evidence* predicates. In our case, $ETFinal$ and $RTFinal$ are the *query* predicates and $ET$, $RTP$, $RTL$, $CONS$ and $CONJ$ are the *evidence* predicates. Inference over this MLN gives the probability of each possible grounding of the *query* predicates, conditioned on the given values of the *evidence* predicates. We used the default inference algorithm in Alchemy named "Lifted Belief Propagation" [26]. For each candidate entity mention $i$, grounding of the predicate $ETFinal(i, E)$ with the highest probability is chosen and corresponding value of $E$ is its final entity type except the case when $E = NONE$. In that case, we do not identify the corresponding candidate mentions as a valid entity mention. Similarly, for each entity mention pair $(i, j)$, grounding of the predicate $RTFinal(i, j, R)$ with the highest probability is chosen and corresponding $R$ value is its final relation type.

## 4 Example

In this section, we describe an example sentence where the joint inference helps in correcting the prediction errors by the individual classifiers. Consider the sentence from the ACE 2004 dataset: `she is the new chair of the black caucus`. In order to identify the candidate entity mentions, top 2 label sequences predicted by the CRF model are considered.

1. `she/B is/O the/O new/O chair/O of/O the/O black/O caucus/B ./O`
2. `she/B is/O the/O new/O chair/B of/O the/O black/O caucus/B ./O`

**Table 1.** Candidate entity mentions identified in the example sentence

| ID | Entity mention | From first BIO sequence? | Predicted type | Actual type |
|----|----------------|--------------------------|----------------|-------------|
| 1  | she            | Yes                      | PER            | PER         |
| 2  | chair          | No                       | PER            | PER         |
| 3  | caucus         | Yes                      | PER            | ORG         |

Table 1 shows all the candidate entity mentions identified along with their IDs and predictions of the local entity classifier. It can be observed that mention ID 2 is generated from the second best BIO sequence and hence will be considered a valid mention only if it is involved in a relation with some other mention. Moreover, the entity type predicted for the mention ID 3 (caucus) is incorrect. This error propagates to the relation classification with "pipeline" classifier predicting relation between chair and caucus to be IDN instead of EMP-ORG. But the local classifier predicts the correct relation type EMP-ORG for this pair as it is not using the entity type features. The first-order logic rules for this sentence's MLN are shown in the Table 2. The LOR (log of odds ratio) weights assignment strategy is used. In case of soft constraints, the number preceding each rule indicates its weight. No weight is explicitly specified for the hard constraints and they always end with a period.

**Table 2.** First-order logic rules for the MLN of example sentence

| Rules induced by the local entity classifier | Rules for identifying valid/invalid entity mentions |
|---|---|
| $6.13 \quad ET(1, PER) \Leftrightarrow ETFinal(1, PER)$ | $6.13 \quad !ETFinal(1, NONE)$ |
| $-0.93 \quad ET(2, PER) \Leftrightarrow ETFinal(2, LOC)$ | $0.71 \quad ETFinal(2, NONE)$ |
| $-0.89 \quad ET(2, PER) \Leftrightarrow ETFinal(2, ORG)$ | $0.15 \quad !ETFinal(3, NONE)$ |
| $-0.71 \quad ET(2, PER) \Leftrightarrow ETFinal(2, PER)$ | $ETFinal(2, NONE) \Rightarrow RTFinal(1, 2, NULL)$ |
| $-0.53 \quad ET(3, PER) \Leftrightarrow ETFinal(3, ORG)$ | $!RTFinal(1, 2, NULL) \Rightarrow !ETFinal(2, NONE)$ |
| $0.15 \quad ET(3, PER) \Leftrightarrow ETFinal(3, PER)$ | $ETFinal(2, NONE) \Rightarrow RTFinal(2, 3, NULL)$ |
| | $!RTFinal(2, 3, NULL) \Rightarrow !ETFinal(2, NONE)$ |

| Rules induced by the local and pipeline relation classifiers |
|---|
| $3.37 \quad RTL(1, 2, IDN) \Leftrightarrow RTFinal(1, 2, IDN)$ |
| $2.99 \quad RTP(1, 2, IDN) \Leftrightarrow RTFinal(1, 2, IDN)$ |
| $1.52 \quad RTL(1, 3, NULL) \Leftrightarrow RTFinal(1, 3, NULL)$ |
| $-1.66 \quad RTL(1, 3, NULL) \Leftrightarrow RTFinal(1, 3, IDN)$ |
| $0.35 \quad RTP(1, 3, NULL) \Leftrightarrow RTFinal(1, 3, NULL)$ |
| $-1.63 \quad RTP(1, 3, NULL) \Leftrightarrow RTFinal(1, 3, IDN)$ |
| $-1.80 \quad RTL(2, 3, EMPORG) \Leftrightarrow RTFinal(2, 3, PHYS)$ |
| $-1.09 \quad RTL(2, 3, EMPORG) \Leftrightarrow RTFinal(2, 3, IDN)$ |
| $0.24 \quad RTL(2, 3, EMPORG) \Leftrightarrow RTFinal(2, 3, EMPORG)$ |
| $-0.46 \quad RTP(2, 3, IDN) \Leftrightarrow RTFinal(2, 3, IDN)$ |

**Table 3.** MLN inference output for entity types

| she (ID 1) | chair (ID 2) | caucus (ID 3) |
|---|---|---|
| $ETFinal(1, PER) = \textbf{0.99}$ | $ETFinal(2, PER) = \textbf{0.92}$ | $ETFinal(3, PER) = 0.35$ |
| $ETFinal(1, GPE) = 0.01$ | $ETFinal(3, GPE) = 0.03$ | $ETFinal(3, ORG) = \textbf{0.39}$ |
| | $ETFinal(2, GPE) = 0.02$ | $ETFinal(3, NONE) = 0.14$ |
| | $ETFinal(2, NONE) = 0.01$ | $ETFinal(3, FAC) = 0.03$ |

**Table 4.** MLN inference output for relation types

| (she,chair) | (she, caucus) |
|---|---|
| $RTFinal(1, 2, IDN) = \textbf{0.92}$ | $RTFinal(1, 3, EMPORG) = 0.02$ |
| $RTFinal(1, 2, PHYS) = 0.01$ | $RTFinal(1, 3, PERSOC) = 0.02$ |
| $RTFinal(1, 2, ART) = 0.01$ | $RTFinal(1, 3, OTHERAFF) = 0.02$ |
| $RTFinal(1, 2, OTHERAFF) = 0.02$ | $RTFinal(1, 3, NULL) = \textbf{0.90}$ |
| $RTFinal(1, 2, NULL) = 0.01$ | $RTFinal(1, 3, IDN) = 0.02$ |
| **(chair,caucus)** | |
| $RTFinal(2, 3, EMPORG) = \textbf{0.33}$ | |
| $RTFinal(2, 3, PHYS) = 0.10$ | |
| $RTFinal(2, 3, GPEAFF) = 0.10$ | |
| $RTFinal(2, 3, OTHERAFF) = 0.09$ | |
| $RTFinal(2, 3, IDN) = 0.20$ | |

The joint inference combines the evidence from the above three classifiers and generates a globally consistent output. The outputs for the query predicates *ETFinal* and *RTFinal* are shown in the Tables 3 and 4, respectively. The predicate groundings which have negligible probability are not shown. Here, it can be observed that the entity mention `chair` (ID 2) has been correctly identified as a valid mention and the type of entity mention `caucus` has been correctly predicted as ORG. Also the correct relation type of EMP-ORG between `chair` and `caucus` has been chosen as the global prediction.

## 5  Experimental Analysis

In order to demonstrate the effectiveness of our approach, we compare its performance with other approaches which have reported their results for end-to-end relation extraction on ACE 2004 dataset[6]. For fair comparison, we follow the same assumptions made by Chan and Roth [2] and Li and Ji [13], i.e. ignoring the DISC relation, not treating implicit relations as false positives and using

---

[6] We have not yet acquired a more recent ACE 2005 dataset.

**Table 5.** Results on the ACE 2004 dataset (Micro-averaged, 5-fold cross-validation)

| Approach | Entity extraction | | | Relation extraction | | | Entity+Relation | | |
|---|---|---|---|---|---|---|---|---|---|
| | P | R | F | P | R | F | P | R | F |
| Local classifiers | 80.9 | 77.6 | 79.2 | 53.2 | 43.9 | 48.1 | 46.2 | 38.1 | 41.8 |
| Pipeline classifier | | | | 53.3 | 46.4 | 49.6 | 48.7 | 42.5 | 45.4 |
| Chan and Roth [2] | | | | 42.9 | 38.9 | 40.8 | | | |
| Li and Ji [13] | 83.5 | 76.2 | 79.7 | 64.7 | 38.5 | 48.3 | 60.8 | 36.1 | 45.3 |
| Miwa and Bansal [16] | 83.3 | 79.2 | **81.2** | | | | 56.1 | 40.8 | **47.2** |
| MLN (LOR) | 79.3 | 79.9 | 79.6 | 56.2 | 45.2 | 50.1 | 50.6 | 40.8 | 45.2 |
| MLN (LOR)+Rules | 79.3 | 80.0 | 79.6 | 56.6 | 45.1 | 50.2 | 51.0 | 40.6 | 45.2 |
| MLN (CM) | 78.9 | 80.1 | 79.5 | 57.2 | 45.2 | 50.5 | 51.6 | 40.8 | 45.6 |
| MLN (CM)+Rules | 79.0 | 80.1 | 79.5 | 57.9 | 45.6 | **51.0** | 52.4 | 41.3 | 46.2 |

coarse entity and relation types. All the results are obtained by 5-fold cross-validation on ACE-2004 data. Note that the actual folds used by each algorithm may differ.

Comparative performances of all the approaches are shown in the Table 5. A true positive for the task of entity extraction means that an entity mention has been correctly identified as the valid mention and also its type has been identified correctly. A true positive for the task of relation extraction means that for a pair of valid entity mentions, its relation type (except for special relation types $NULL$ and $IDN$) has been identified correctly. For entity+relation extraction, a stricter criteria is used where a true positive means that for a pair of valid entity mentions, not only its relation type is identified correctly but types of both the mentions are also identified correctly. Even if any one of these predictions is incorrect, we consider it as a false positive for the predicted combination of entity types and relation type and also as a false negative for the true combination of entity types and relation type.

It can be observed that MLN inference with CM (Constant Multiplier) weights assignment strategy performs better that the LOR (Log of Odds Ratio) in case of relation extraction whereas for entity extraction LOR strategy is better. Addition of semantic rules (discussed in the Sect. 3.5) results in better performance for both the strategies. Also, we can observe that MLN (CM) with semantic rules comfortably outperforms the individual classifiers: local entity classifier, local relation classifier and "pipeline" relation classifier. In case of end-to-end relation extraction, our approach outperforms the approaches of Chan and Roth [2] and Li and Ji [13] on the ACE 2004 dataset and also achieves a comparable performance as compared to Miwa and Bansal [16]. We also achieve comparable performance in case of entity extraction as compared to Li and Ji [13] but underperform in comparison with Miwa and Bansal [16].

# 6    Related Work

Previous work on joint extraction of entities and relations can be broadly classified into 5 categories: (i) Integer Linear Programming (ILP) based approaches [22,24], (ii) Probabilistic Graphical Models [23,25], (iii) Card-pyramid parsing [11], (iv) Structured Prediction [13,14,17] and (v) Recurrent Neural Network (RNN) based model [16]. Our approach is similar to ILP based approaches, but we use MLNs for joint inference which provide much better representation to incorporate complex domain knowledge as compared to ILP. For example, the rules defined in the Sect. 3.5 are quite easy to incorporate using first-order logic but the same would be cumbersome in ILP. The approaches by Singh et al. [25] and Li and Ji [13] not only carry out joint "inference" but also create a joint "model" where the parameters for both the tasks are learned jointly.

Zhang et al. [27] used Markov Logic rules to perform *Ontological Smoothing*. The concept of *Ontological Smoothing* is to find a mapping from a user-specified target relation to a background knowledge base. This mapping is then used to generate extra training data for distant supervision. Similar to our approach, they also use Markov logic rules to ensure consistency between relation types and entity types. One major difference is that the relation types used by them were quite specific and not as general as ACE 2004 relation types. Zhu et al. [28] also used MLNs but they addressed a relation extraction problem which is bit different from the ACE 2004 RDC task. It requires the explicit mention of relation in the form of words other than the words inside entity mentions. This is not always true for ACE 2004 relations. For example, EMP-ORG relation holds between Indian and soldiers in the sentence Indian soldiers attacked the terrorists.

# 7    Conclusion and Future Work

We described the problem of end-to-end relation extraction and the need to jointly address its sub-tasks of entity and relation extraction. We proposed a new approach for joint extraction of entity mentions and relations at the sentence level, which uses joint inference in Markov Logic Networks (MLN). We described in detail about the domains, predicates and first-order logic rules used to create an MLN for a sentence. We also explored how the effective representability of first-order logic can be used to incorporate various semantic rules and domain knowledge. Finally, we demonstrated better than the state-of-the-art end-to-end relation extraction performance on the standard dataset of ACE 2004.

In future, we plan to analyze the two weights assignment strategies (CM and LOR) in detail and develop deeper understanding of pros and cons of each one. Also, we have tried only a small number of additional semantic rules. In future, we wish to take advantage of the first-order logic framework to incorporate deeper semantic knowledge. Another important direction to explore is about learning the weights of first-order logic rules automatically.

# References

1. Bunescu, R.C., Mooney, R.J.: A shortest path dependency kernel for relation extraction. In: Proceedings of Conference on Human Language Technology and Empirical Methods in Natural Language Processing, pp. 724–731. ACL (2005)
2. Chan, Y.S., Roth, D.: Exploiting syntactico-semantic structures for relation extraction. In: Proceedings of 49th Annual Meeting of the Association for Computational Linguistics: Human Language Technologies, vol. 1, pp. 551–560. ACL (2011)
3. Doddington, G.R., Mitchell, A., Przybocki, M.A., Ramshaw, L.A., Strassel, S., Weischedel, R.M.: The Automatic Content Extraction (ACE) program-tasks, data, and evaluation. In: LREC, vol. 2, p. 1 (2004)
4. Florian, R., Jing, H., Kambhatla, N., Zitouni, I.: Factorizing complex models: a case study in mention detection. In: Proceedings of 21st International Conference on Computational Linguistics and 44th Annual Meeting of the Association for Computational Linguistics, pp. 473–480. ACL (2006)
5. Florian, R., Pitrelli, J.F., Roukos, S., Zitouni, I.: Improving mention detection robustness to noisy input. In: Proceedings of 2010 Conference on Empirical Methods in Natural Language Processing, pp. 335–345. ACL (2010)
6. GuoDong, Z., Jian, S., Jie, Z., Min, Z.: Exploring various knowledge in relation extraction. In: Proceedings of 43rd Annual Meeting on Association for Computational Linguistics, pp. 427–434. Association for Computational Linguistics (2005)
7. Hearst, M.A.: Automatic acquisition of hyponyms from large text corpora. In: Proceedings of 14th Conference on Computational Linguistics, vol. 2, pp. 539–545. ACL (1992)
8. Heckmann, D., Frank, A., Arnold, M., Gietz, P., Roth, C.: Citation segmentation from sparse & noisy data: an unsupervised joint inference approach with Markov logic networks (2013)
9. Jain, D.: Knowledge engineering with Markov logic networks: a review. Evolv. Knowl. Theory Appl. **16** (2011)
10. Jiang, J., Zhai, C.: A systematic exploration of the feature space for relation extraction. In: HLT-NAACL, pp. 113–120 (2007)
11. Kate, R.J., Mooney, R.J.: Joint entity and relation extraction using card-pyramid parsing. In: Proceedings of 14th Conference on Computational Natural Language Learning, pp. 203–212. ACL (2010)
12. Lafferty, J., McCallum, A., Pereira, F.C.: Conditional random fields: probabilistic models for segmenting and labeling sequence data (2001)
13. Li, Q., Ji, H.: Incremental joint extraction of entity mentions and relations. In: ACL (2014)
14. Li, Q., Ji, H., Hong, Y., Li, S.: Constructing information networks using one single model. In: EMNLP (2014)
15. Lu, W., Roth, D.: Joint mention extraction and classification with mention hypergraphs. In: Proceedings of Conference on Empirical Methods in Natural Language Processing (EMNLP 2015) (2015)
16. Miwa, M., Bansal, M.: End-to-end relation extraction using LSTMs on sequences and tree structures. arXiv preprint arXiv:1601.00770 (2016)
17. Miwa, M., Sasaki, Y.: Modeling joint entity and relation extraction with table representation. In: EMNLP, pp. 1858–1869 (2014)
18. Nadeau, D., Sekine, S.: A survey of named entity recognition and classification. Lingvisticae Investigationes **30**(1), 3–26 (2007)

19. Palshikar, G.K.: Techniques for named entity recognition. In: Bioinformatics: Concepts, Methodologies, Tools, and Applications, p. 400 (2013)
20. Qian, L., Zhou, G., Kong, F., Zhu, Q., Qian, P.: Exploiting constituent dependencies for tree kernel-based semantic relation extraction. In: Proceedings of 22nd International Conference on Computational Linguistics, vol. 1, pp. 697–704. ACL (2008)
21. Richardson, M., Domingos, P.: Markov logic networks. Mach. Learn. **62**(1–2), 107–136 (2006)
22. Roth, D., Yih, W.: A linear programming formulation for global inference in natural language tasks. In: CoNLL, pp. 1–8 (2004)
23. Roth, D., Yih, W.T.: Probabilistic reasoning for entity & relation recognition. In: Proceedings of 19th International Conference on Computational Linguistics, vol. 1, pp. 1–7. ACL (2002)
24. Roth, D., Yih, W.T.: Global inference for entity and relation identification via a linear programming formulation. In: Introduction to Statistical Relational Learning, pp. 553–580 (2007)
25. Singh, S., Riedel, S., Martin, B., Zheng, J., McCallum, A.: Joint inference of entities, relations, and coreference. In: Proceedings of 2013 Workshop on Automated Knowledge Base Construction, pp. 1–6. ACM (2013)
26. Singla, P., Domingos, P.: Lifted first-order belief propagation. In: AAAI, vol. 8, pp. 1094–1099 (2008)
27. Zhang, C., Hoffmann, R., Weld, D.S.: Ontological smoothing for relation extraction with minimal supervision. In: AAAI (2012)
28. Zhu, J., Nie, Z., Liu, X., Zhang, B., Wen, J.R.: StatSnowball: a statistical approach to extracting entity relationships. In: Proceedings of 18th International Conference on World Wide Web, pp. 101–110. ACM (2009)

# Knowledge Extraction with NooJ Using a Syntactico-Semantic Approach for the Arabic Utterances Understanding

Chahira Lhioui[1,2,3,4](✉), Anis Zouaghi[1,2,3,4], and Mounir Zrigui[1,2,3,4]

[1] ISIM of Medenine, Gabes University, Road Djerba, 4100 Medenine, Tunisia
Chahira_ml983@yahoo.fr, Anis.Zouaghi@gmail.com,
Mounir.Zrigui@fsm.rnu.tn
[2] ISSAT of Sousse, Sousse University, Taffala city (Ibn Khaldoun),
4003 Sousse, Tunisia
[3] FSM of Monastir, Monastir University, Avenue of the environnement,
5019 Monastir, Tunisia
[4] LATICE Laboratory, ESSTT, Tunis, Tunisia

**Abstract.** Regarding the amelioration of NLP field, knowledge extraction has become an interesting research topic. Indeed, the need to an improvement through the NLP techniques has become also necessary and advantageous. Hence, in a general context of the construction of an Arabic touristic corpus equivalent to those of European projects MEDIA and LUNA, and due to the lack of Arabic electronic resources, we had the opportunity to expand the EL-DicAr of [11] by knowledge hinging on Touristic Information and Hotel Reservations (TIHR). Thus, in the same manner of [11], we have developed local grammars for the recognition of essential knowledge in our field of study. This task facilitates greatly the subsequent work of understanding user utterances interacting with a dialogue system.

**Keywords:** Arabic · Named Entity · Compound words · NooJ

## 1 Introduction

Knowledge Extraction is considered as an important preprocessing step for many tasks such as document classification or clustering, machine translation (MT), information retrieval (IR), automatic language understanding and other text processing applications. Our concern in this study is to examine the symbolic automatic understanding utterances. In our work, knowledge extraction refers to the recognition and categorization of Named Entity (NE) [3, 4, 15–17], compound words (CW) and frozen expressions (FE).

Knowledge Extraction in Arabic language represents many interesting challenges. In fact, Arabic language is characterized by the lack of digital resources and mainly by the absence of the uppercase and lowercase distinction which is a very useful indicator to extract the NE such as the proper names in many languages dealing with Latin alphabet. Furthermore, Arabic is a morphologically rich and complex language, its automatic processing is complicated with the absence of the voyellation in most of

© Springer International Publishing AG, part of Springer Nature 2018
A. Gelbukh (Ed.): CICLing 2016, LNCS 9624, pp. 552–564, 2018.
https://doi.org/10.1007/978-3-319-75487-1_42

written texts and the existence of many spelling variants multiplying unknown forms in these texts.

In order to process users' requests addressed to a language understanding system, we have established a syntactico-semantic parsing for the extraction of knowledge required to their understanding.

During this parsing step, we exhibited some linguistic concepts, then, we checked their validity and attested their belonging to particular classes such as named entities, compound words, frozen expressions, etc. To do this, we used the CFG (context-free-grammar) grammars provided within the NooJ development platform. These grammars are called local grammars [8, 14] that serve to precisely locate local frequently repeated phenomena in users' utterances such as dates, numerical determinants, compound words, frozen expressions, etc. These grammars are also considered as lexicalized graphs which refer to dictionaries of simple and composed words as well as equivalent to Recursive Transition Networks (RTN) or even Augmented Recursive Networks (ARTN).

It is in this context that the present work is situated. The main objective is to follow a linguistic approach of Arabic knowledge extraction occurring in users' requests. To reach our objective, we have to ameliorate the El-DicAr of the author [11] with complementary dictionaries that represent our studied domain (touristic domain). We also have to establish a set of transducers resolving morphological and syntactical phenomena related to the Arabic knowledge and implemented with the linguistic platform NooJ.

In this paper, we present, firstly, a brief overview of the state-of the art. Then, we exposit our linguistic approach. After that, we detail our resources construction and their implementation in the linguistic platform NooJ. Finally, the paper concludes with some perspectives.

## 2   Knowledge Extraction

Knowledge Extraction is a current research topic with regard to the amelioration of Arabic Natural Language Processing (NLP) field. The need to such improvement through NLP techniques has become necessary and interesting. Moreover, in our work, knowledge extraction involves three kinds of subdivision:

- Named Entity extraction (NE)
- Compound words extraction (CW)
- Frozen expressions extraction (FE).

For the extraction of compound words and frozen expressions, few works are recorded covering such a phenomenon despite its importance [18, 19]. In fact, the treatment of compound words and frozen expressions facilitates enormously the Arabic corpus processing given their frequent redundancy.

According to MUC[1] and to the notation of [11], we distinguish three NE categories which are respectively:

---

[1] http://en.wikipedia.org/wiki/Message_Understanding_Conference.

– ENAMEX: This category regroups the proper names. It can be subcategorized as follow:
  • Persons: famous proper names such as presidents names, doctors names, etc.
  • Organizations: Banks, associations, youth club, etc.
  • Localizations: towns, villages, beaches, ridges, mountains, rivers, addresses,
  • Entertainment means: Galleries, public gardens, hotels, museums, shops, restaurants, cinemas, clubs, festivals, shopping centers, etc.
  • Itinerary: highway, runs great, streets, avenues, etc.
  • Locations: cars, houses, marine tools, etc.
  • Transport Means: plane, train, metro, buses, Taxis, renting, etc.
  • Specialties: foods, sports activities, products, trend marks, etc.
  • Categories: star ratings for hotels, etc.
  • Contacts: phone number, email, vocal servers, etc.
– TIMEX: numerical expressions of percentage, numbers, sizes, distances, areas measures, volumes, monetary expressions, amount, etc.
– NUMEX: temporal expressions of time, date, or period.

An entity relation may be established between two or more NE, such as a person (محمد صالح, Mohammed Salah), an organization (البنك الفلاحي, agriculture bank), a location (أم العرايس, Om La'Arayes), or a specific time (منتصف النهار, midday). In this case, we talk about compound words (CW). The relationship between NE can be binary, or may involve more entities [9].

In addition, the third type of knowledge that can be factorized is the frozen expressions. Indeed, in Arabic language, there are a lot of expressions that are not modifiable in interlocutors' utterances. Among these frozen expressions we can cite:

– Greeting expressions: such as صباح الخير (good morning)
– Reference expressions: such as هنا (there), هناك (there)
– Excuses expressions: such as عفوا (sorry), لو سمحت (if you please)
– Thanks expressions: such as شكرا (thanks), بارك الله فيك (thank you)
– Affirmation expressions: such as بلى (of course), حسنا (ok), وهو كذلك (That is right)
– Negation expressions: such as لا أظن ذلك (I do not think so)
– Temporal expressions: such as الجاري (current), المقبل (following), الماضي (precedent)
– Expressions for pray: الله المستعان (God bless you).

## 3  Related Work About Knowledge Extraction

The knowledge extraction is a very active research area for many years in several languages. Three fundamental approaches exist in literature:

– Rule-based approach: This approach exploits the advances in NLP and is especially based on the use of formal grammars manually built by an expert linguist. It is particularly based on the description of NE with rules that operate lexical markers, proper names dictionaries and sometimes by a syntactic labeling. The works of [1, 11, 13] are very well illustrations for such approach.

- Learning-based approach: It uses stochastic techniques and learns the specificities on a large learning corpus where targets NE were labeled. Learning algorithms are then applied automatically to develop a knowledge base using several statistical models (such as HMM, MEMM, SVM, CRF, etc.). This approach was considered to have some intelligence when making decisions. [7] has developed an SVM learning technique for the implementation of their system for the recognition of Arabic NE. This system has introduced an overall F-measure of 82.7%. Other researches follow also to this approach [2, 6, 7, 12].
- Hybrid approach: This approach combines the two above-mentioned approaches for their complementarity. It uses manually written rules but also builds a part of its rules on the use of both syntactic information and information extracted from the data through learning algorithms. [5] adopts a hybrid approach to the extraction of Arabic NE taking advantage of the symbolic and based learning approach. [5, 9] follow also this approach.

In this paper, we opt for a linguistic approach. The choice of the rule-based approach in our treatment is guided by the fact that our studied domain (touristic field) is restricted and limited. Besides, stochastic approach requires a huge amount of learning data which is quasi-absent for Arabic resources. That is why a syntactico-semantic method is appreciated.

## 4 Difficulties Hindering Arabic Knowledge Extraction

The extraction of Arabic knowledge is a complex task whose complexity is manifested in:

- The nesting and the coordination of NE, compound words and frozen expressions
- The complexity related to specific Arabic aspects which are:

### 4.1 The Agglutination Phenomenon

Unlike Romance languages, Arabic is an agglutinative language. Articles, prepositions and pronouns stick to adjectives, nouns, verbs; which requires to proceed with cutting words before lemmatization task. Most of Arabic words are composed by the aggregation of elementary lexical items. The agglutinated form corresponds to a sequence of forms "glued". For example, the determination may be expressed by:

- Agglutination article ال (the) before the word. For example, الصيدلية (the pharmacy).
- Agglutination of a clitic at the end of the word. For example, صيدليته (his pharmacy).

### 4.2 The Absence of Capitalization

Unlike Indo-European languages, Arabic language does not have the concept of capitalization. This represents a major obstacle for the Arabic language during the extraction of NE. In fact, capital latter is very effective in the proper names recognition process for certain languages as English or French. So its absence in the Arabic language imposes an urgent necessity to find alternatives and, ultimately, to use other conventional means such as lexicons, triggers words and grammatical rules.

### 4.3 The Absence of Vowels

A non-vowelized word has many ambiguities in meaning or syntactic function. For example, the word "ذهب" can be a verb (go) to a voyellation or a proper name (Gold) for another. Thus, this example illustrates the impact of the vowels lack in words recognitions.

### 4.4 The Morphological Complexity

Arabic is a highly-inflected language. It uses an agglutinative strategy to form a word. If NE appears with agglutinative form, then this poses a difficulty for the identification of this entity.

### 4.5 The Segmental Ambiguity

The agglutination requires a segmentation of the morphological units, which will generate multiple ambiguities, knowing that no vowelized statements may have a significant number of possible segmentations, they did not all have the same meaning. In its vowelized form, the same word المهم accepts five potential segmentations as it is shown in the table below:

Table 1. Different divisions and interpretations of the word 'المهم'

| Possible cutting | Translation into English |
|---|---|
| المُهِمُّ | interesting |
| اَلَّم + هَمِّ | sake + pain |
| اَلَمَ+هُمْ | they + is what |
| اَلَّم +هُمْ | their + pain |
| أ+لِمَ+هُمْ | they + it + is |

## 5 The Preconized Approach for the Arabic Knowledge Extraction

To overcome all these problems, we decide to build a recognition system of named entities; compound words and frozen expressions for the Arabic language. The construction of such a system requires firstly the collection of a corpus collecting a sufficient number of texts. These texts serve not only as an observation corpus (to constitute rules) but also as a corpus test for validation.

After collecting the maximum information for all forms of utterances recognized by morphological analysis or by consultation of electronic dictionaries, this information will be used in local syntactic-semantic grammars, to locate the relevant sequences (Fig. 1).

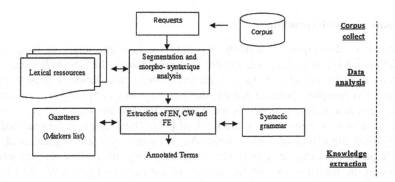

**Fig. 1.** Steps for knowledge extraction

## 5.1 Corpus Collect

In the context of this contribution, we were guided to the collection of dialogues transcriptions realized with hundreds of people speaking the Modern Standard Arabic (MSA). These conversations are related to tourist information and hotel reservations for the Tunisia. The corpus contains entirely 4000 queries. We also note that in addition to the units written in standard Arabic, the corpus includes also 2771 sequences asking for train schedules, planes and buses reservations dates in different Tunisian hotels names.

## 5.2 Segmentation and Morpho-Syntactic Analysis

In this phase, we segment interlocutors queries in words based on spaces delimiter. Then we proceed by a morphological analysis of the corpus to extract useful information that will be used in the knowledge extraction system.

Given the agglutinative structure that has the majority of Arabic words, our morphological analyzer NooJ [14] can separate and identify morpheme shapes entry and associate all information necessary to ongoing treatment.

Given the agglutinative structure that has the majority of Arabic words, our morphological analyzer NooJ can separate and identify morpheme and associate all information necessary to the current treatment. These forms are decomposed to recognize affixes (conjunctions, prepositions, pronouns, etc.) attached to it. These morphological possibilities in this analyzer can facilitate the identification of trigger words, names of people or locality, compound words and frozen expressions even when clustered. As a matter of fact, each of these forms is associated with a set of useful linguistic information for the following step: lemma, grammatical labels, gender and number, syntactic information (such as: + Transitive) distributional information (such as: + Human), etc. Consequently, instead of listing all inflected forms (singular, dual, plural, masculine, feminine) of cities names considered as lexical marker names of locations, for example, we use the regular expression syntax of NooJ where the grammatical symbol refers to all potential inflected vocalize, partly vocalized, as well as the unvoiced form for this lemma.

### 5.3   Knowledge Extraction

We began by developing a lexicon or what is called gazetteers to recognize the knowledge in our study. Then, the information provided by the morphological analysis is directly used by our knowledge recognition system. In addition to these gazetteers and collected morpho-syntactic information, our system is based on recognition of another linguistic resource which is local grammars. For the construction of our linguistic rules, we were inspired by the work of [11]. These rules are represented in the form of NooJ augmented transitions networks. They can represent sequences of words. These sequences are described through written manually rules. They produce results in some linguistic information such as the type of the identified NE, CW and FE.

The rules described in these grammars are used to group all the elements of a knowledge entity. They are generally formed of trigger, forms derived from gazetteers, and occasionally unknown words. These words sequences can be perfectly labeled when they appear in a certain context or when they contain a word or a trigger input of our markers list. Finally, it remains to explore all potential phrasal constructions since the knowledge entity can be accompanied by:

- only a right context
- only a left context
- both right and left contexts.

The preponderance of unknown forms in knowledge entities implies a lack of information which widens the boundary of error probabilities. The NooJ syntactic grammar respects certain heuristics when applying recognition rules. They locate the "Longuest Matches" for the only grammar and "All Match" for all grammars.

In the recognition phase, we have solved the problems related to the Arabic language (e.g., agglutination) establishing morphological grammars built into the NooJ platform. This phase contains 17 graphs respecting the local grammars identified in the studied corpus.

### Temporal expressions recognition.

Our focus in this part of study is to identify fragments of sentences that express the time in interlocutors queries wishing to learn about train schedules, buses and planes or to book a room in hotels for the nights. This work is based on the construction of local grammars recognitions represented as an ATN using the graphical editor in NooJ. The transducer of Fig. 2 defines the "Time Recognition Grammar" and contains three sub-graphs. Each sub-graph represents a category identified in the Time hierarchy. This grammar allows Time requests recognition. Each path of each sub-graph point out to a rule extracted in the study corpus.

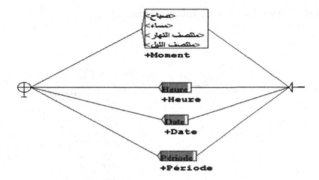

**Fig. 2.** Temporal expressions recognition

More specifically, the description of certain phenomena is typically equivalent to the construction of dozens of graphs. Within these graphs, many more or less complex elementary phenomena occur in different contexts. For instance, in this case, the list of day and month names can be used in a graph to outline both dates and period supplements. So we associate with corresponding sequences a single graph that will be called several times in the overall sequence.

In practical terms, this type of representation has at least three major advantages:

- Recursion calls between graphs and sub-graphs;
- The possibility of imposing conditions on transitions;
- The ability to associate actions and outputs to the performed transitions.

For example, the sub-graph "Time" of the main graph "Time Recognition Grammar" is defined as another transducer embedded in the main transducer (see Fig. 3).

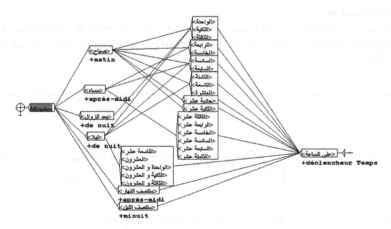

**Fig. 3.** Time expressions recognition

The "Minutes" transducer is as follows (Fig. 4):

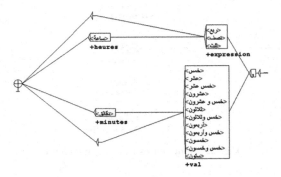

**Fig. 4.** Minute expressions recognition

Thus, these graphs allow recognition of queries containing phrases such as:

أريد تذكرة القطار المنطق من قابس المتجه نحو سوسة على الساعة الثامنة صباحا والنصف

*I want a ticket from the train starting from Gabes going towards Sousse at eight thirty in the morning.*

أريد حجز غرفة في نزل القنطاوي بداية من 17 أفريل إلى 24 من نفس الشهر

*I want to book a room at El-Kantaoui Hotel from 17 April to 24 of the same month.*

**Price recognition.**
We are also interested in identifying fragments of sentences that express prices in tourist queries wanting to learn about the prices of public transport or the room prices in hotels (Fig. 5).

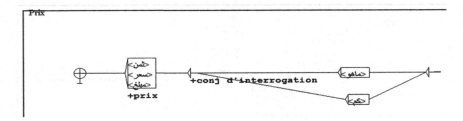

**Fig. 5.** Price expressions recognition

The following graph recognizes the cardinality room, seats in transport means and the number of individuals (Fig. 6).

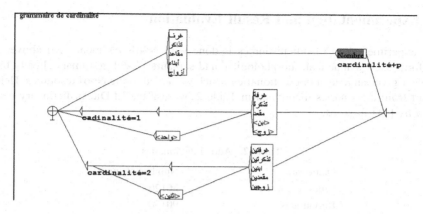

**Fig. 6.** Cardinality expressions recognition

Thus, these graphs allow recognition queries containing phrases such as:

ماهو سعر التذكرة للدخول إلى المعرض الدولي للكتاب بصفاقس بالنسبة للفرد الواحد

*What is the price of the entrance ticket to the international book fair in Sfax for a person.*

**Compound word and frozen expression recognition.**
The following NooJ transducer explicit a non-exhaustive and modifiable list of the different compound words that can be found in the interlocutor queries (Fig. 7).

**Fig. 7.** Compound word expressions recognition

We note that each word of this list should be in the undefined and singular form.
The frozen expressions that are frequently used in the users' requests are also represented in a NooJ transducer. This later contains hundreds of frozen expressions.

## 6    Experimentation and Result Evaluation

The experimentation of our resources is done with NooJ. As mentioned above, this platform uses (syntactical, morphological and semantic) local grammars already built. Table 1 gives an idea about dictionaries which we added to the NooJ resources. Hence, a part from dictionaries mentioned in Table 2, we use the El-DicAr dictionary developed by [11] in NooJ.

**Table 2.** Added dictionaries

| Dictionaries | Number of inputs |
|---|---|
| Cities and towns | 30120 |
| Restaurants | 90130 |
| Itineraries appellations | 30100 |
| Locations | 60125 |
| Organizations | 90125 |
| Persons names | 40125 |
| Entertainments | 60150 |
| Localizations | 80125 |
| Transport fields | 60120 |
| Specialties | 10130 |
| Hotel and restaurant categories | 11125 |
| Contacts | 60125 |

To El-DicAr, we include entries related to our touristic domain. We also involve morphological grammars of all entries in all mentioned dictionaries taken into account their inflected and agglutinated forms.

To experiment and evaluate our work, we have applied our resources to three types of corpus: Touristic Information Retrieval corpus, Hotel Reservation corpus and Transport information corpus.

To assess the recognition phase, we have applied our resources to a corpus formed by 4000 requests in the touristic domain (different of the study corpus). This corpus is collected from a hundred interviewed tourists. Table 3 shows detailed statistics for each category names.

**Table 3.** Statistics about added knowledge to El-DicAr

| | El-DicAr | Our dictionary |
|---|---|---|
| Nouns | 19504 | 8789 |
| Verbs | 10162 | 345 |
| NEs | 3686 localizations + 11860 Proper names | 622 500 |
| CWs | 720 | 13 098 |
| FEs | – | 2943 |

Our syntactico semantic recognizer yields an F-score of 91% which is a satisfying measure compared to [5, 7, 12].

# 7    Conclusion

In this paper, we have proposed an approach for recognition of Arabic NE, CW and FE based on a syntactico-semantic method. Besides, we have enriched the El-DicAr by knowledge related to TIHR domain. We have also given an experimentation and evaluation on this studied domain. The experimentation and the evaluation are done in the linguistic platform NooJ. The obtained results are satisfactory. As perspectives, we try to prove the cover of our local grammars in the recognition of all structure types of Arabic phrases.

# References

1. Abdallah, S., Shaalan, K., Shoaib, M.: Integrating rule-based system with classification for Arabic named entity recognition. In: Gelbukh, A. (ed.) CICLing 2012. LNCS, vol. 7181, pp. 311–322. Springer, Heidelberg (2012). https://doi.org/10.1007/978-3-642-28604-9_26
2. Algahtani, S.: Arabic named entity recognition: a corpus-based study. Ph.D. thesis, The University of Manchester, UK (2011)
3. Alkharashi, I.: Person named entity generation and recognition for Arabic language. In: Proceedings of the Second International Conference on Arabic Language Resources and Tools, Cairo, pp. 205–208 (2009)
4. Attia, M., Toral, A., Tounsi, L., Monachini, M., van Genabith, J.: An automatically built named entity lexicon for Arabic. In: Proceedings of the Seventh International Conference on Language Resources and Evaluation (LREC 2010), Valletta, pp. 3,614–3,621 (2010)
5. Aubulei, S.: Hybrid system for extracting and classifying Arabic proper names. In: Proceedings of the WSEAS International Conference on Artificial Intelligence, Knowledge Engineering and Data Bases, Madrid, Spain, pp. 205–210 (2006)
6. Benajiba, Y., Rosso, P.: Arabic named entity recognition using conditional random fields. In: Proceedings of the Workshop on HLT & NLP within the Sixth International Conference on Language Resources and Evaluation (LREC 2008), Marrakech, pp. 143–153 (2008)
7. Benajiba, Y., Diab, M., Rosso, P.: Using language independent and language specific features to enhance Arabic named entity recognition. Int. Arab J. Inf. Technol. (IAJIT) 6(5), 463–471 (2009)
8. Traboulsi, H.: Arabic named entity extraction: a local grammar-based approach. In: Proceedings of the International Multi-conference on Computer Science and Information Technology, pp. 139–143 (2009). ISBN 978-83-60810-22-4
9. Zribi, I., Hammami, S.M., Belguith, L.H.: L'apport d'une approche hybride pour la reconnaissance des entités nommées en langue arabe. In: TALN 2010, Montréal, 19–23 juillet 2010
10. Shalaan, K.: A survey of arabic named entity recognition and classification. Assoc. Comput. Linguist. 40, 469–510 (2014)
11. Mesfar, S.: Named entity recognition for Arabic using syntactic grammars. In: Kedad, Z., Lammari, N., Métais, E., Meziane, F., Rezgui, Y. (eds.) NLDB 2007. LNCS, vol. 4592, pp. 305–316. Springer, Heidelberg (2007). https://doi.org/10.1007/978-3-540-73351-5_27

12. Mohit, B., Schneider, N., Bhowmick, R., Oflazer, K., Smith, N.: Recall-oriented learning of named entities in Arabic Wikipedia. In: Proceedings of the 13th Conference of the European Chapter of the Association for Computational Linguistics (EACL) (2012)

13. Boulaknadel, S., Talha, M., Aboutajdine, D.: Amazighe named entity recognition using a A rule based approach. In: 2014 IEEE/ACS 11th International Conference on Computer Systems and Applications (AICCSA), pp. 478–484 (2014). https://doi.org/10.1109/aiccsa.2014.7073237

14. Silberztein, M.: Formalizing Natural Languages: The NooJ Approach, p. 346. Wiley-ISTE, Totnes (2016). ISBN 978-1-84821-902-1

15. Zaghouani, W., Pouliquen, B., Ebrahim, M., Steinberger, R.: Adapting a resource-light highly multilingual named entity recognition system to Arabic. In: Proceedings of LREC 2010, Valletta, Malta (2010)

16. Benajiba, Y., Diab, M., Rosso, P.: Using language independent and language specific features to enhance Arabic named entity recognition. Int. Arab J. Inf. Technol. 6(5), 464–472 (2009)

17. Zayed, O., El-Beltagy, S.: Person name extraction from modern standard Arabic or colloquial text. In: Proceedings of the 8th International Conference on Informatics and Systems Conference (INFOS 2012), NLP track, Cairo, pp. 44–48 (2012)

18. Boujelben, I., Mesfar, S., Hamadou, A.B.: Methodological approach of terminological extraction applied to biomedical domain (2011)

19. Hammouda, F.K., Haddar, K., Abdelwahed, A.: Construction d'un dictionnaire de noms composés en arabe. In: 3rd International Conference on Arabic Language Processing (CITALA 2009), Rabat, Morocco, 4–5 May 2009

# Adapting TimeML to Basque: Event Annotation

Begoña Altuna[✉], María Jesús Aranzabe, and Arantza Díaz de Ilarraza

Universidad del País Vasco/Euskal Herriko Unibertsitatea,
Manuel Lardizabal pasealekua 1, 20018 Donostia, Spain
{begona.altuna,maxux.aranzabe,a.diazdeilarraza}@ehu.eus

**Abstract.** In this paper we present an event annotation effort following EusTimeML, a temporal mark-up language for Basque based on TimeML. For this, we first describe events and their main ontological and grammatical features. We base our analysis on Basque grammars and TimeML mark-up language classification of events. Annotation guidelines have been created to address the event information annotation for Basque and an annotation experiment has been conducted. A first round has served to evaluate the preliminary guidelines and decisions on event annotation have been taken according to annotations and inter-annotator agreement results. Then a guideline tuning period has followed. In the second round, we have created a manually-annotated gold standard corpus for event annotation in Basque. Event analysis and annotation experiment are part of a complete temporal information analysis and corpus creation work.

**Keywords:** EusTimeML · Events · Event extraction
Event classification · Temporal information · Corpus creation

## 1 Introduction

Events—situations that happen or occur—can only be perceived through time, by means of the changes in the reality or the absence of those. Humans conceptualise time as points and intervals which are employed to locate those events in a chronology, through past, present and future, and events express an action or state located in a certain time or period. Apart from the events, there are time expressions that express those points and intervals in the temporal continuum and some structures, implicit or explicit, which convey temporal relation information such as "before", "after" or "simultaneous". These all help the speakers to situate the events in time, with a direct reference or one relative to another.

Event analysis is a major issue on natural language processing (NLP) as a part of temporal information analysis and processing, since they are the core of the discourse: the actions and situations we are talking about. Many evaluation challenges such as SemEval 2015 Task 4 [1] make us aware of its relevance. In order to take advantage of event information, experimentation (*e.g.* extraction,

© Springer International Publishing AG, part of Springer Nature 2018
A. Gelbukh (Ed.): CICLing 2016, LNCS 9624, pp. 565–577, 2018.
https://doi.org/10.1007/978-3-319-75487-1_43

analysis, annotation) on events has to be done. Event information has to be made machine readable by means of a mark-up language. For example, TimeML [2] is a mark-up language to code events and temporal expressions, their features and the relations among them based on XML. The information will then be represented in temporally annotated corpora like TimeBank [3] or WikiWars [4], which may be used to train machine-learning based tools such as TIPSem [5]. The information extracted from events and saved in corpora can be useful in many NLP tasks such as event forecasting [6] or timeline creation [7].

This paper has two main parts. First, an adaptation of TimeML temporal mark-up language [8] for event annotation in Basque. This has been done through an analysis of the event expressions in Basque of which we have analysed the main features. Second, the annotation of events following EusTimeML [9].

The article is structured as follows: in Sect. 2 we give a definition of event and we classify events according to their lexical content to give a theoretical basis for the annotation of two most relevant event features. In Sect. 3, we present the linguistically based decisions on event annotation for the experiment presented in Sect. 4. To conclude, we sum up our main ideas and propose further research on the temporal information processing field in Sect. 5.

## 2    Definition of Events

Event is a cover term for situations that happen, occur, hold, or take place and states and circumstances in which something obtains or holds true [10]. This definition already shows a difference between actions and states, but a more thorough classification can also be done.

Apart from their meaning, events convey different linguistic information. Some features such as the class are semantic, whereas features like the part of speech category are grammatical. These features can be normalised through attributes and a set of values.

### 2.1    Event Categorisation

Events can be classified according to their semantic features. For the annotation of Basque events, the event classification described in TimeML annotations have been followed:

- **Occurrence:** these are dynamic events that happen or occur, *e.g. salto egin* (to jump), *dantzatu* (to dance) or *ibili* (to walk).
- **State:** these are events describing circumstances in which something obtains or holds true and do not vary over time, *e.g. egon* (to be) or *geratu* (to remain).
- **Reporting:** reporting events describe the utterance, narration, description, etc. of an event, *e.g. esan* (to say) or *iragarri* (to announce).
- **Aspectual:** aspectual events indicate the beginning, continuity or end of an event: *e.g. hasi* (to begin), *jarraitu* (to continue) or *amaitu* (to end).

- **Perception:** these events describe the physical perception of another event, *e.g. ikusi* (to see), *entzun* (to hear) or *sumatu* (to perceive).
- **Intensional action:** are dynamic events that select for an event-denoting argument which is explicitly in the text, *e.g. saiatu* (to try), *agindu* (to order) or *aztertu* (to analyse).
- **Intensional state:** these are states that, as intensional actions do, select for an event-denoting argument which is explicitly in the text, *e.g. pentsatu* (to think), *gorrotatu* (to hate) or *prest egon* (to be ready).

This categorisation fulfils two major objectives. First it serves to determine whether an event may be taken as an argument for another event. Secondly, the difference between dynamic and stative events offers a preliminary view on how events happen in time, that is to say, whether they are punctual or last through a period of time.

The categorisation presented in this section is intended to cover all the different event types and give relevant semantic information of them. The different events described in this section can be represented by means of the expressions described in Sect. 3.

## 3    Event Expressions in Basque: A Syntactic Perspective

Events can be expressed by more than one grammatical category. As in many other languages, mainly verbs (1), nouns (2), adjectives (3) and adverbs (4) (in bold) can express events in Basque:

(1)    Hor    ez **dira sartuko**      Edesako      langileak.
       There no **AUX enter.FUT** Edesa.REL workers.ABS
       'Edesa workers **will** not **enter** in there'.

(2)    Fagor Etxetresnak enpresak        **konkurtsora** joko      du.
       Fagor Etxetresnak company.ERG **tender.ALL**  go.FUT AUX
       'Fagor Etxetresnak company will go out to a **tender**'.

(3)    Sartu    den          emakumeak **gaztea** dirudi.
       Come.in AUX.REL woman.ERG **young** looks
       'The woman who has come in looks **young**'.

(4)    **Txaloka**    egin dute ibilaldi guztia.
       **Clapping** do   AUX walk    all.DET
       'They have done all the walk **clapping**'.

The events in bold in (1–4) express a single event: a single action or state. For the annotation experiments, only the lexical head of each event expression will be marked, although all morpho-syntactic information contained in the phrase (auxiliaries, demonstratives, etc.) will be taken into account. For events expressed by nouns, adjectives and adverbs, those will be considered the lexical heads. For events expressed by verbs, instead, only the lexical head, *sartuko* (1), will get an event tag.

The linguistic analysis and decisions we have taken have been extracted from Basque grammars [11–13], a classification of complex predicates [14] and the decisions taken for temporal annotation in other languages [8,15]. These expressions are described and examples for each are given below.

### 3.1  Events Expressed by Verbs

Events in Basque are mainly expressed by verbs. In the following sections the verb forms that may express an event are presented.

**Synthetic Forms.** Synthetic forms are one-word units. The lexical root conveys the semantic information and a compound of morphemes add aspect, tense, person and mood information. Only a handful of verbs possess synthetic forms and their extension in the verbal paradigm is also restricted to some tenses. As a consequence, the majority of verbal events in Basque will be expressed by means of periphrastic forms.

(5)  Zientzialariek    urteak    **daramatzate** fusiozko    energia
     Scientists.ERG years.ABS **have.been**    fussion.INS power
     merkearen   bila.
     cheap.GEN looking.for.
     'Scientists **have** long **been** looking for cheap fusion power'.

**Periphrastic Forms.** For periphrastic forms (also called analytical) we stick to the traditional Basque definition. These forms are formed by a lexical head which bears aspectual information and a mood, person and tempus carrying auxiliary. All verbs in Basque have periphrastic forms. The main formal variation happens in auxiliaries, which drastically reduces the mechanisms of morphological creation since lexical heads do not largely vary. This phenomenon makes the creation of periphrastic forms an easy language resource and leads to the reduction of synthetic form use.

(6)  8etan      **atera dira** mendizaleak mendi      tontorrerantza.
     8.PL.LOC **leave AUX** hikers.ABS   mountain summit.DIR
     'Hikers **have left** at 8 towards the summit'.

**Non-finite Forms.** Verbal expressions may also appear in Basque texts as non-finite forms (radical, participles and verbal nouns). These forms are used on their own as sentence heads in contexts such as fossilised expressions, exclamatory sentences and questions.

(7)  Akordiorik          **lortu** ezean, grebari    eutsiko   diote.
     Agreement.PART **reach** NEG, strike.DAT continue AUX
     'If no agreement **is reached**, (they) will continue on strike'.

## 3.2  Events Expressed by Nouns

Some nouns may express events. These can be verb nouns (8), common nouns (9) or proper nouns denoting a particular event (10).

(8)  Derrigorrezkoa da Greziari     zorraren     zati     bat
     Compulsory   is  Greece.DAT debt.SG.GEN part.ABS a
     **barkatzea.**
     **condone.ABS**

     'It is compulsory to **condone** a part of the Greek debt'.

(9)  Lau  **eskaera** nagusi     egin dituzte.
     Four **request** major.ABS do   AUX

     '(They) have done four major **requests**'.

(10)  2016an     **Olinpiar Jokoak**     Rio de Janeiron ospatuko  dira.
      2016.LOC **Olympic Games**.ABS Rio de Janeiro   held.FUT AUX

      'In 2016 **Olympic Games** will be held in Rio de Janeiro'.

We follow a test proposed in [8] to decide whether a noun refers to an event. A noun may express an event if it fits at least in two of the presented settings.

- NOUN lasted for several minutes/days/years/...
- NOUN was very fast/immediate/...
- NOUN took/takes/will take place *in temporal expression*.
- NOUN began/continued/ended *in temporal expression*.

## 3.3  Events Expressed by Adjectives

Adjectives express the qualities of the entity they refer to. Although they may appear in many contexts, we only consider events the adjectives acting as predicate adjectives.

(11)  Zer     egin behar  da     enpresa **bideragarria** egiteko?
      What do   have.to AUX business **profitable**     make.FIN

      'What has to be done to make the business **profitable**?'

## 3.4  Events Expressed by Adverbs

Adverbs will be considered event expressions when they accompany verbs to create a more complex event construction (Sect. 3.6). These will mainly be adverbs of manner.

(12)  Arrakasta     **harro** egoteko modukoa  da
      Success.ABS **proud** be.FIN likely.ABS is

      'Success is big enough to be **proud** of it'.

## 3.5   Events Expressed by Pronouns

Pronouns themselves do not express events, but may have a deictic value when they corefer with another event in the text.

(13)   Bihar      egingo    da          mozorro desfilea.    **Horretarako**
       Tomorrow do.FUT AUX.3.SG costume parade.ABS .PUN
       erdigunea  itxiko      dute.
       **That.FIN** centre.ABS close.FUT AUX

       'Costume parade will be done tomorrow. For **that** the centre will be closed'.

## 3.6   Complex Structures

Complex predicates conform a non-homogeneous yet gradated group [14]. We will now present Basque complex predicate structures, based on Jedzrejko's [14] classification for Polish complex predicates. We have adapted this list to accommodate Basque predicate features as can be seen in the following lines:

– **Standard nominal predicates** are constructions with a basic auxiliary verb (*izan* (to be), *ukan* (to have)). The verbs in these constructions are semantically or referentially empty and the nominal or adjectival predicate carries all the predicative information.

   (14)   Ordenagailua   **geldoa da.**
          Computer.ABS   **slow.ABS is**

          'The computer **is slow**'.

– **Modal predicates** are complex constructions formed by a lexical verb in its participle form and a conjugated modal verb (*nahi* (to want), *behar* (to have to, must) and *ahal* (to can)) or derived nouns expressing modality (15) (*nahi* (wish), *behar* (need, obligation), *ahal* (possibility)) and a participle.

   (15)   Jende  nagusiak     noizbehinka **jesarri beharra du.**
          People elderly.ERG sometimes   **sit.down need.ABS has**

          'Elderly people **has the need to sit down** sometimes'.

– **Aspectual predicates** are formed by an aspectual verb or noun and a verbal or nominal predicate. The aspectual expression in the construction marks the phase of its argument.

   (16)   **Beherapenak amaitu dira**
          **Sales.ABS end AUX**

          'Sales have ended'.

– **Generic verb constructions.** Generic verbs are those which are used to give predicative properties to nouns. Therefore, generic verb constructions are formed by a noun that carries the lexical meaning of the event and a verb that provides the syntactic information.

(17)  Ona   da lagunek       elkarri        **musu ematea.**
Good is   friends.ERG each.other.DAT **kiss give.ABS**
' It is good for friend to **kiss** each other.,

– **Metaphors.** In these constructions a fully predicating verb is used next to a noun phrase with a meaning other than its main meaning. The verb, apart from adding the grammatical and syntactic information, modifies the conceptual information of the noun it accompanies.

(18)  Entzuleak    **barrez lehertu ziren.**
Hearers.ABS **laugh.INS explode AUX**
'Hearers **laughed** a lot'.

– **Idiomatic expressions** are formed by a noun and a verb which carries the grammatical information of the construction. Nevertheless, idiomatic constructions cannot be seen as a simple sum of the meanings of its parts and can only be understood as a single meaning unit.

(19)  Adierazpenek    **hautsak harrotu zituzten.**
Statements.ERG **dusts.ABS raise AUX**
'The statements **caused a commotion**'.

Although these complex constructions express a single complex event, they may contain more than one event expressing form and all those forms will be annotated as single events according to the EusTimeML guidelines to show that complexity.

# 4    Experimentation

In order to prove the correctness and universality of the EusTimeML mark-up language and annotation guidelines, we have conducted a two-round annotation experiment on event identification and feature extraction. The first was a preliminary experiment to evaluate and discuss the guidelines [16]. There was a guideline tuning period following this first round in which the annotating team added or corrected annotation features. Once the new guidelines [9] were finished, a second annotation round was used to annotate a gold standard corpus of verbal event expression in Basque.

Both annotation efforts have been done using the CELCT Annotation Tool [17], which is easily customizable and offers a range of interesting features for textual annotation such as inter-annotator agreement metrics.

## 4.1   First Annotation Round

For this first experiment about 172 events[1] were annotated. The annotated documents are part of a 25 article corpus that contains news on the closure of a

---

[1] The amount of events varies among the annotations.

company extracted from a Basque newspaper. The events were annotated according to the EusTimeML guidelines [16], a set of guidelines for Basque temporal annotation based on the TimeML annotation scheme. Three annotators (A, B and C) took part in this annotation effort.

In this annotation round the agreement on event identification and extension were evaluated. The annotations of the three annotators were evaluated in pairs. Agreement levels ranged between 0.864 and 0.947 in weighted Dice's coefficient [18] depending on the annotator pair. The agreement level on the part of speech category, modality and whether events are aspectual were also evaluated.

We found that events expressed by a single token were unanimously annotated in most of the cases. We also discovered an unexpectedly high agreement on events expressed by nouns and adjectives. However, although agreement in general was high, some annotation features were troublesome; we list them below:

- Some tokens were incorrectly considered events; mainly verbs taking part in time expressions and discourse markers.
- Some events on complex structures were neglected.
- Event expressions derived from verbs, were not consistently given the same part of speech category.

In order to overcome these disagreement issues in the forthcoming annotation experiment, discussion on the annotation and guidelines among annotators was crucial; mainly in what referred to obscure annotation guidelines and ambiguous categories (namely, grammatical categories). Then we revisited Basque grammars and we updated the annotation guidelines adding more accurate information.

## 4.2    Second Annotation Round

After the grammatical reanalysis, a second annotation round has been conducted. This second time, four annotators have taken part; three of them were familiar with EusTimeML and the CAT annotation tool and the fourth one had a deep knowledge on temporal annotation as well as the guidelines and the annotation tool. The annotation has been done on 15 documents of the Basque version of the MEANTIME corpus [19] used in NewsReader project [20]. The first three annotators have annotated 115 sentences and their annotations have been compared to the fourth annotator's.

The number annotations for each annotator (A, B, C) and super-annotator (fourth annotator) and a counting of unanimously annotated events is given in Table 1. The numbers already show a relatively high agreement.

The main reason for disagreement has been the difficulty to class some entities as events. In example (20) there is a linguistic form which expresses an event in the MEANTIME corpus and in example (21) there is the same form not expressing any event. This phenomenon has been more pronounced in the cases in which a form refers to a process and the final product of that process.

**Table 1.** Annotated events by each annotator and agreed events

| Document sets | Annotator | Super-annotator | Agreed events |
|---|---|---|---|
| First (Ann. A) | 96 | 74 | 69 |
| Second (Ann. B) | 394 | 418 | 358 |
| Third (Ann. C) | 95 | 99 | 84 |

(20)  **Ekoizpena**          AEBra      ekartzeko asmoa          du.
      **Production.DET.ABS** USA.ALL bring.FIN intention.ABS has
      '(He/She) intends to bring the **production** to the USA'.

(21)  Nekazariek    euren **ekoizpena**         salgai    jarriko    dute.
      Farmers.ERG their  **production.ABS** to.be.sold put.FUT AUX
      'Farmers will put their **production** on sale'.

State denoting events have also been a disagreement point. It has been sometimes difficult to decide whether they are events as they do not always express
an ongoing state but a very generic situation.

(22)  Oso  **desengainatuta** gaude
      Very **disappointed**     are
      '(We) are very **disappointed**'.

In Basque the verb *egin* (to do) is used to focus events expressed by verbs.
This verb, does not offer any event information and, although it was stated not
to annotate it, it has sometimes been annotated.

**Table 2.** Event extent agreement results

| Annotator pairs | Micro-average (markable) | Micro-average (token) | Macro-average (markable) | Macro-average (token) |
|---|---|---|---|---|
| A – SA | 0.812 | 0.812 | 0.819 | 0.819 |
| B – SA | 0.877 | 0.877 | 0.875 | 0.875 |
| C – SA | 0.866 | 0.866 | 0.883 | 0.883 |

Results in Table 2 show a high agreement [18] on markable extent between
annotators (the first three and the super-annotator). Markable extent agreement
refers to the perfect overlap of the tags of two different annotators. Token extent
agreement, instead, refers to the markable extent considering only the overlapping tokens. In our case both, markable and token extent, agreement results
get the same values as markables have always a single-token extent. One may
consequently deduce that all annotators have respected the single-token rule for
event annotation in EustimeML guidelines.

**Table 3.** Unanimously annotated POS

| Event annotation | A-GS | B-GS | C-GS |
|---|---|---|---|
| Verbs | 45 | 242 | 51 |
| Nouns | 14 | 58 | 33 |
| Adjectives | 1 | 3 | 1 |
| Adverbs | 0 | 9 | 2 |
| Pronouns | 0 | 1 | 0 |
| Other | 0 | 0 | 0 |
| TOTAL | 60 (87%) | 311 (87%) | 77 (92%) |

As shown in Table 3, a rather high agreement on the grammatical category of events has been reached. Most of the disagreement is due to one of the annotators not giving any value to an event or forgetting to change the default value. However some other disagreement is due to grammatical reasons.

– Verbal nouns ended with *-tea/-tzea* have been annotated as nouns and verbs.
– Participles with a relative mark *-tako/-riko* have been annotated as adjectives and verbs.
– Some adverbs have been considered part of the verb form and have been given a verb value or an "other" value.

**Table 4.** Modality agreement results

| Modal event annotation | A-GS | B-GS | C-GS |
|---|---|---|---|
| BEHAR | 0 | 5 | 2 |
| NAHI | 0 | 3 | 1 |
| AHAL | 3 | 4 | 0 |
| TOTAL | 3 | 12 | 3 |

The modal verbs unanimously annotated by the first three annotators and the gold standard can be seen in Table 4. Modal events have been easy to identify, since there is little variation on the modality expressing forms. Moreover, there is virtually no possibility of confusedly giving a wrong value to a modal event expression as they have very distant meanings. Although the number of modal events is low, the result analysis has shown that mistakes in the annotation were due to annotators' mistakes during the annotation; not to wrong perceptions of those events.

Finally we have measured the agreement on event category. The results are not as high as expected (A-SA: 58%, B-SA: 56% and C-SA: 49%), however, it is to mention that the agreement strongly varies between categories. Reporting and aspectual events have been easily identifiable, despite the fact that some have been incorrectly annotated presumably by mistake in many of the cases. Occurrence and intensional actions, instead, have been a major matter for disagreement.

From a thorough analysis of the agreement, we have noticed that the event documents that have been annotated later get higher agreement in event categorisation. Therefore, one may deduce that the more the training the better results in categorisation.

### 4.3   Final Guideline Tuning

Once we have analysed the annotation results, we have dropped some conclusions and have made some decisions:

- The more trained and familiar with the task an annotator is, the less mistakes will make and the higher agreement will achieve.
- In order not to forget filling or saving the attribute values, a means for it will be designed.
- Event identification and part-of-speech categorisation do not seem difficult to master.
- Although modal events have been correctly annotated in general, we expect further discussion and training on them to improve the results.
- Event categorisation agreement has been lower than expected. Although some categories seem easier to assign, we will set a new analysis guideline tuning period for the most conflictive.

After the corrections to the 15 annotated documents are done, the trained annotators will continue enlarging the gold standard corpus, as well as annotating more temporal structures such as time expressions, temporal linking constructions and temporal relations.

## 5   Conclusions and Future Work

The analysis and processing of event information is a very relevant task in text processing. Some information can be extracted, analysed and processed language-independently, but other needs a previous analysis on the forms of a certain language. In this paper, we have offered a summary of the linguistic forms in Basque that can express event information based on Basque grammars and we have also highlighted their main features. We have also classified the different eventive forms according to their semantics following the classification proposed in TimeML.

This information has been made explicit in EusTimeML, a temporal mark-up language for Basque, and the manual annotation guidelines have been written. A series of experiments to evaluate our linguistic decisions and create a corpus annotated with temporal information have been run. First annotation effort served to train annotators and tune the annotation guidelines. The second has led to the creation of a gold standard corpus of event information.

This work is part of a complete analysis of temporal information and corpus creation for Basque. After having analysed temporal expressions and events, our ongoing research is focused on the analysis of the temporal relations between those events and time expressing constructions. With each analysis and annotation effort, we are building a gold standard corpus for temporal information in Basque. This corpus is expected to be used for the evaluation of automatic information extraction tools in a first instance.

**Aknowledgement.** This research is funded by the Basque Government PRE_2015_2_0284 grant.

# References

1. Minard, A.-L., Speranza, M., Agirre, E., Aldabe, I., van Erp, M., Magnini, B., Rigau, G., Urizar, R.: SemEval-2015 Task 4: TimeLine: cross-document event ordering. In: Proceedings of 9th International Workshop on Semantic Evaluation (SemEval 2015), Denver, Colorado, pp. 778–786. Association for Computational Linguistics, June 2015
2. Pustejovsky, J., Castaño, J.M., Ingria, R., Saurí, R., Gaizauskas, R.J., Setzer, A., Katz, G., Radev, D.R.: TimeML: robust specification of event and temporal expressions in text. N. Dir. Quest. Answ. **3**, 28–34 (2003)
3. Boguraev, B., Pustejovsky, J., Ando, R., Verhagen, M.: TimeBank evolution as a community resource for TimeML parsing. Lang. Resour. Eval. **41**(1), 91–115 (2007). https://doi.org/10.1007/s10579-007-9018-8
4. Mazur, P., Dale, R.: WikiWars: a new corpus for research on temporal expressions. In: Proceedings of 2010 Conference on Empirical Methods in Natural Language Processing, EMNLP 2010, Stroudsburg, PA, USA, pp. 913–922. Association for Computational Linguistics (2010). http://dl.acm.org/citation.cfm?id=1870658.1870747
5. Llorens, H., Saquete, E., Navarro, B.: TIPSem (English and Spanish): evaluating CRFs and semantic roles in TempEval-2. In: Proceedings of 5th International Workshop on Semantic Evaluation, SemEval 2010, Stroudsburg, PA, USA, pp. 284–291. Association for Computational Linguistics (2010). http://dl.acm.org/citation.cfm?id=1859664.1859727
6. Radinsky, K., Horvitz, E.: Mining the web to predict future events. In: Proceedings of 6th ACM International Conference on Web Search and Data Mining, pp. 255–264. ACM (2013)
7. Bauer, S., Clark, S., Graepel, T.: Learning to identify historical figures for timeline creation from Wikipedia articles. In: Proceedings of HistoInformatics 2014 - 2nd International Workshop on Computational History, Barcelona, Spain (2014)
8. TimeML Working Group: TimeML Annotation Guidelines Version 1.3., Manuscript. Technical report, Brandeis University (2010)
9. Altuna, B., Aranzabe, M.J., de Ilarraza, A.D.: Euskarazko denbora-egiturak etiketatzeko gidalerroak v2.0. Technical report, Lengoaia eta Sistema Informatikoak Saila, UPV/EHU. UPV/EHU/LSI/TR;01-2016 (2016). https://addi.ehu.es/handle/10810/17305
10. Saurı, R., Batiukova, O., Pustejovsky, J.: Annotating events in Spanish. TimeML annotation guidelines. Technical report, Version TempEval-2010, Barcelona Media-Innovation Center (2009)
11. Altuna, P., Salaburu, P., Goenaga, P., Lasarte, M.P., Akesolo, L., Azkarate, M., Charriton, P., Eguskitza, A., Haritschelhar, J., King, A., Larrarte, J.M., Mujika, J.A., Oyharabal, B., Rotaetxe, K.: Euskal Gramatika Lehen Urratsak (EGLU) I. Euskaltzaindiko Gramatika Batzordea, Euskaltzaindia, Bilbao (1985)

12. Altuna, P., Salaburu, P., Goenaga, P., Lasarte, M.P., Akesolo, L., Azkarate, M., Charriton, P., Eguskitza, A., Haritschelhar, J., King, A., Larrarte, J.M., Mujika, J.A., Oyharabal, B., Rotaetxe, K.: Euskal Gramatika Lehen Urratsak (EGLU) II. Euskaltzaindiko Gramatika Batzordea, Euskaltzaindia, Bilbao (1987)
13. Hualde, J.I., de Urbina, J.O.: A Grammar of Basque, vol. 26. Walter de Gruyter, Boston (2003)
14. Jędrzejko, E.: The problematics of describing periphrastic predication between word and image. Stud. Pol. Linguist. **6**(1), 27–44 (2011)
15. Caselli, T., Lenzi, V.B., Sprugnoli, R., Pianta, E., Prodanof, I.: Annotating events, temporal expressions and relations in Italian: the It-TimeML experience for the Ita-TimeBank. In: Proceedings of 5th Linguistic Annotation Workshop, pp. 143–151. Association for Computational Linguistics, Portland (2011)
16. Altuna, B., Aranzabe, M.J., de Ilarraza, A.D.: Euskarazko denbora-egiturak etiketatzeko gidalerroak. Technical report, Lengoaia eta Sistema Informatikoak Saila, UPV/EHU. UPV/EHU/LSI/TR;01-2016 (2014). http://ixa.si.ehu.es/Ixa/Argitalpenak/Barne$_t$xostenak/1414871293/publikoak/Denbora-egiturak%20etikeatzeko%20gidalerroak
17. Lenzi, V.B., Moretti, G., Sprugnoli, R.: CAT: the CELCT annotation tool. In: Calzolari, N., (Conference Chair), Choukri, K., Declerck, T., Doğan, M.U., Maegaard, B., Mariani, J., Odijk, J., Piperidis, S. (eds.) Proceedings of 8th International Conference on Language Resources and Evaluation (LREC 2012), Istanbul, Turkey, pp. 333–338. European Language Resources Association (ELRA) (2012)
18. Dice, L.R.: Measures of the amount of ecologic association between species. Ecology **26**, 297–302 (1945)
19. Minard, A.-L., Speranza, M., Urizar, R., Altuna, B., van Erp, M., Schoen, A., van Son, C.: MEANTIME, the NewsReader multilingual event and time corpus. In: Proceedings of LREC 2016 (2016)
20. Agerri, R., Agirre, E., Aldabe, I., Altuna, B., Beloki, Z., Laparra, E., de Lacalle, M.L., Rigau, G., Soroa, A., Urizar, R.: NewsReader project. Procesamiento del Lenguaje Natural **53**, 155–158 (2014)

# Applications

# Deeper Summarisation:
# The Second Time Around
## An Overview and Some Practical Suggestions

Simone Teufel[(✉)]

Computer Laboratory, University Cambridge,
JJ Thomson Avenue, Cambridge CB3 0FD, UK
Simone.Teufel@cl.cam.ac.uk

**Abstract.** This paper advocates deeper summarisation methods: methods that are closer to text understanding; methods that manipulate intermediate semantic representations. As a field, we are not yet in a position to create these representations perfectly, but I still believe that now is a good time to be a bit more ambitious again in our goals for summarisation. I think that a summariser should be able to provide some form of explanation for the summary it just created; and if we want those types of summarisers, we will have to start manipulating semantic representations.

Considering the state of the art in NLP in 2016, I believe that the field is ready for a second attempt at going deeper in summarisation. We NLP folk have come a long way since the days of early AI research. Twenty-five years of statistical research in NLP have given us more robust, more informative processing of many aspects of semantics – such as semantic similarity and relatedness between words (and maybe larger things), semantic role labelling, co-reference resolution, and sentiment detection. Now, with these new tools under our belt, we can try again to create the right kind of intermediate representations for summarisation, and then do something exciting with them. Of course, exactly how is a very big question. In this opinion paper, I will bring forward some suggestions, by taking a second look at historical summarisation models from the era of Strong AI. These may have been over-ambitious back then, but people still talk about them now because of their explanatory power: they make statements about which meaning units in a text are always important, and why.

I will discuss two 1980s models for text understanding and summarisation (Wendy Lehnert's Plot Units, and Kintsch and van Dijk's memory-restricted discourse structure), both of which have recently been revived by their first modern implementations. The implementation of Plot Unit-style affect analysis is by Goyal et al. (2013), the KvD implementation is by my student Yimai Fang, using a new corpus of language learner texts (Fang and Teufel 2014). Looking at those systems, I will argue that even an imperfect deeper summariser is exciting news.

© Springer International Publishing AG, part of Springer Nature 2018
A. Gelbukh (Ed.): CICLing 2016, LNCS 9624, pp. 581–598, 2018.
https://doi.org/10.1007/978-3-319-75487-1_44

# 1   Introduction

The task of text summarisation is the production of a much shorter piece of text that still expresses the essence, the most important statements, of the full input text. From the early days of computer science and AI, the ambitious vision behind summarisation has always been that in order to produce a summary of a text, that text should have been digested by whatever automatic method created it: consumed, understood, rearranged, and then turned into a summary containing just the gist.

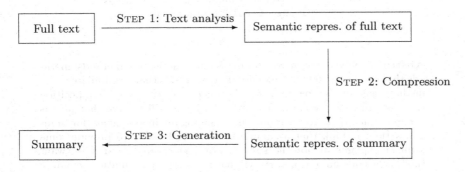

**Fig. 1.** Summarisation by text comprehension.

This ambitious model implies that some semantic representation is created from the text – maybe in first order logic, maybe in some other form. Once that representation exists (after Step 1 in Fig. 1), we can move to the content selection step (Step 2). We can now manipulate the semantic representation, and start separating the wheat from the chaff. We might do this with no claim at cognitive plausibility at all, or alternatively, we might want to follow the reasoning steps that a human reading this text would perform. There are many appealing models of how a summary can be achieved. For instance, we might zoom in on the main characters of a story, and only report what happens to them. Alternatively, we might look at all the events, choose the most surprising or newsworthy one according to our world knowledge, and report mainly about this event. Or we can follow a discourse-based strategy, looking at how the different information units in the story belong together, and choosing information that connects the existing textual material best. Under this model, summarisation becomes a testbed for text understanding.

Much of the emphasis of AI in the 1970s and 1980s was on Step 2. Several researchers proposed plausible, appealing, general summarisation strategies (e.g., Schank and Abelson 1977; Lehnert et al. 1983; DeJong 1982; Kintsch and van Dijk 1978; Lehnert 1981). And because many compaction operations are concerned with connections between sentences, researchers interested in discourse studies or even argumentation theories had something to contribute too (e.g., Cohen 1984).

Producing the text itself, the task natural language generation (NLG), according to this model, is Step 3. Given the semantic representation of the outcome of the content selection process, it has to be verbalised to produce a smooth new text. Verbalisation could be seen as mere convenience to make the output of the content selection step more accessible to a human, but it is not really necessary to make this summariser operate. In principle, the correctness or plausibility of Step 2 could be evaluated on basis of the semantic representations the method chose, rather than on newly created text.

What I have sketched above is the ambitious deep text-understanding model of Strong AI. Needless to say, no such system was ever built, and it is 2016 now. Most computational linguists would agree that it is mainly Step 1 that has been the stumbling-stone: the transformation of an arbitrary text into a semantic representation, along with the need to model an arbitrarily large amount of world knowledge possibly required during Step 2. Many subtasks are involved: all material would have to be syntactically parsed correctly, abstracting away from the particular surface form of the material in the text; one would have to deal with synonymy and polysemy, resolve figurative language, interpret negation, identify the objects that the text talks about and resolve these onto each other, and deal with all the other language vehicles such as omissions, irony, and creative on-the-spot neologisms. Even if the explicit message was decoded perfectly, we next would have resolve pragmatic effects in order to fill in additional, implicit facts that are assumed or presupposed in the text even if they are not explicitly said, using inference.

We humans decode all of this instinctively and with ease. In fact, when we assume the role of writer, we are able to produce non-redundant texts because we know that our listeners can do the same. The fact that writers don't have to spell out everything makes communication efficient for humans, but turns Step 1 into a near-impossible decoding exercise for a machine.

The other steps are less hopeless. In Step 3, we have all relevant information available to us and "only" have to verbalise it. It is likely that a halfway decent language generation system for Step 3 could probably be built with a lot of work; there already exist good solutions for domain-specific subtasks (e.g., Reiter 2005; McKeown et al. 1995). If the range of linguistic impression of this hypothetical system was initially boring and repetitive, it could be step-wise improved towards more naturalness.

Researchers in the tradition of Strong AI zoomed in on Step 2. For instance, Lehnert (1981, 1981a) proposed a model of text understanding that relies on recognising how each participant in the story is affected by any event that happens – is it positive or negative for them? She hypothesised that humans prefer stories of a predictable shape – symmetric stories, or stories containing certain plots such as "revenge" or "a good thing turned sour". When summarising a text, the semantic representation should consist of plots and part of plots (called *plot units*), which represent a set of affect-states and transitions between them. Summarising a text means recognising higher-level plot units from lower-level ones and the affect-states, and expressing them as summary text.

Kintsch and van Dijk (1978) propose that we can summarise a text by keeping track of how clause-like meaning units called *propositions* are connected. The only type of connections that is needed in their model is *argument overlap*. Two propositions are connected if they share arguments, i.e., if a participant occurs in both propositions. They argue that an incremental representation of the entire text based on this principle, combined with a recurrent process of "forgetting" less connected propositions, can simulate human memory under resource limitations. The detail is forgotten after each sentence is processed and its meaning has been incorporated into a memory tree. By this process, what will stick in a human's mind is the essence. They validate this theory by hand-simulating it on one text, and by measuring memory recall in human experiment, in comparison to predictions of their model.

Both models are general and elegant solutions. Both add a layer of explanation to the text that wasn't there before, and are thus clear examples of machine intelligence. However, neither the KvD nor the Plot Unit model have ever been fully implemented. In Sects. 3.3 and 3.2, I will discuss two current attempts at changing this.

## 2   Shallow, General Summarisation Models

Much has happened in summarisation since the early AI models were proposed. The field is too large to mention all relevant achievements; I will therefore concentrate more on content selection rather than on presentation. For instance, I will say nothing about sentence shortening (e.g., Knight and Marcu 2000; Dorr et al. 2003; Unno et al. 2006).

In multi-document summarisation, it is possible to use the idea of *information redundancy* as a model of importance: if a fact or a concept has been repeated in several documents about the same topic, they are likely to be highly relevant to the topic. There are many increasingly sophisticated ways to measure similarity of information, including paraphrasing methods (Barzilay and Lee 2003; Androutsopoulos and Malakasiotis 2010) and various versions of high-dimensional semantic spaces (Landauer and Dumais 1997; Steyvers and Griffiths 2007; Baroni and Lenci 2010). Another early idea is MMR (maximal marginal relevance), the combination of maximising importance and simultaneously minimising redundancy with other summary sentences (Carbonell and Goldstein 1998). However, leaving multi-document summarisation and the useful trick of redundancy aside, I am interested here in single–document, general summarisation models.

Radev and McKeown (1998) avoid the "Step 1 problem" by using domain-specific information extraction (IE) as their semantic representation, for instance a MUC-template for terrorist attacks. This allowed Radev and McKeown to move directly to Steps 2 and 3. Their system produces smooth summaries from scratch. However, the restriction to a narrow domain makes the system applicable to only a very small number of pre-coded topics, as well as dependent on the IE technology. I will from now on discuss general, robust summarisers only, which impose no restrictions on the allowable content of their texts.

Most of the content selection methods I will consider use connections between lexical items as their main content selection criterion. Over the years, these have been mostly based on the idea of lexical entities' relative importance (from which the relative importance of entire sentences can be induced), and the connections between lexical items (where generally the more connected items are deemed to be more important). The history of lexical and sentence importance metrics is very long and started with Luhn (1958), who proposed simple word frequency (without inverse document frequency), and includes the idea of centroid-based summarisation (Radev et al. 2004). Also, two summarisers based on eigenvector centrality have been developed, Textrank (Mihalcea and Tarau 2004) and Lexrank (Erkan and Radev 2004).

Nishikawa et al. (2014) introduce a model that uses a HSMM, an HMM that can has states of non-uniform "length", together with a model of lexical coherence and a model for statistical sentence shortening. Using the knapsack algorithm, they train optimal combinations of sentence segments which are lexically overall maximally coherent. This model requires a large training base, but is able to produce very short, highly informative summaries.

Often, methods such as these creates very usable summaries, for instance for tasks such as relevance decision in an information retrieval situation under time pressure, or to remind a user whether they already came across a particular text earlier in their life. Nevertheless, it is the lack of understanding and explanatoriness behind these methods that motivates the current paper. I am of course happy if a summarisation system does deliver decent output, but I am far more interested in what that summary reveals about how much of the text has been "understood" in the process of creating it. For instance, I demand a justification for the summary, i.e., an answer to the question "OK, if this is the summary that you created, can you defend it? Why did you include this particular fact then?" I will not accept as an explanation something along the lines of "this lexical items is the most important one according to eigenvector centrality, and together with the other lexical items in that (part of the) sentence, they form the best compromise, so I extracted that sentence". In my opinion, lexical items are too uninformative (and short) a meaning unit to serve as a justification, and sentences are too long, containing irrelevant material that had nothing to do with the reason why it was chosen. A better, more human-like explanation is needed, which should be based on more informative information units and non-trivial connections between them.

# 3 Recent Implementations of Deep Summarisation Models

## 3.1 Stories and Scripts

Another interesting recent development in text understanding is the unsupervised learning of narrative schemes proposed by Chambers and Jurafsky (2008, 2009), and similar work on narrative recognition (Kasch and Oates 2010; Elson and McKeown 2009). Chamber and Jurafsky's work is a reimplementation of Schank's (1975) scripts and story grammars on a web-size unsupervised scale. Schank's definition of a script is a knowledge representation of recurrent, default processes, which include sub-events and participants. For instance, the RESTAU-RANT script might contain certain actions such as *entering, being assigned a seat, ordering, eating, paying, leaving*, and agents such as *customer, maitre-de, busboy*, and *waiter*. An early summariser (DeJong's (1982) FRUMP system) was based on manually-written scripts, but despite some successes in some domains, this system was unable to fully scale up to unlimited domains. Chambers and Jurafsky harvested frequently occurring events involving the same participants in a large corpus of web text using bootstrapping and decided which of these sequences could quality as frames.

Farrow et al. (2015) consider the task of creating stories, rather than recognising them. The stories they create are personalised to a particular user of social media, and expressed in photos that follow a common theme. The themes are recognised from the captions submitted with the photos, and an extensive human experiment is performed to validate the system output.

## 3.2   Lehnert's Plot Units

Lehnert (1981) proposed a conceptual knowledge structure called *Plot Units* to represent the affect states of characters in narrative stories. She argued that affect states and the relations between them are central to story understanding and summarisation. Stories written by humans are predictable to a certain degree because of what we perceive as pleasant or natural in terms of story lines – for instance, an exchange of wares is a cooperative act that maintains symmetry between two people, whereas revenge is an example of how symmetry can be restored after it was disturbed by a hostile act. In Lehnert's model, the meaning of a story is represented as positive and negative effects of events happening to each participant in a story, and their intentions: i.e. the three affect states positive $(+)$, negative $(-)$, and mental (M) states. The story emerges as patterns of change between these states.

Primitive plot units consist of two affect states associated with the same character and a causal link between them (motivation (m), actualisation (a), termination (t), and equivalence (e)). Events that affect more than one characters provide a connection between their characters' respective affect states. The pattern in Fig. 2 represents the concept of a "serial exchange", an example of reciprocation of cooperative behaviour. The person represented by the left column requests something from the person on the right; both their affect states turn to positive when that thing is being given (subplot: Success). Then the person on the right requests something in turn, a request that is equally honoured.

Plot units are built chronologically from the text, as the plot unfolds, in a bottom-up fashion. This is different from top-down representations such as story grammars or scripts (e.g., Schank 1975; Rumelhart 1977).

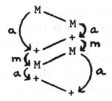

**Fig. 2.** Serial exchange, from Lehnert (1981)

Importantly, Lehnert restricted herself to a finite number of mini-plots (called plot units), which compose a full story. The graph structure allows her to analyse a story at different levels of granularity. It is this rich semantic representation, with negative, positive and mental states per participant, and causal links between them, that serves as the meaning representation for general story understanding (and thus question-asking ability), but also for summarisation at different abstraction levels (and lengths).

Lehnert's mental states are closely related to intentions, but in her scheme, all mental states are simply represented as "M" and left undifferentiated. This has the advantage that an automatic process would not have to recognise or classify them – it just needs to infer that some mental state is now taking place (a task that also bears some similarity to Goldberg et al.'s (2009) task of "wish detection"). This definition makes intention detection more tractable.

Goyal et al. (2013) present a model called AESOP that learns a core aspect of Lehnert's (1981) story representations – affect analysis. AESOP produces plot unit representations by exploiting existing sentiment resources combined with rules that assign affect states to characters and create links between them. They enrich the lexica with the syntactic and semantic information they need and provide a rule-based coreference system dealing with the special phenomena in fables. They added a bootstrapping module which automatically acquires verbs that impart polarity on their patients (called patient polarity verbs or PPV). Their final best classification system successfully classify events into positive, negative and mental affect states, and achieves some acceptable results even for link classification.

Goyal et al. distinguish between direct plan/goal expressions ("the lion wanted to find food"), direct speech acts, and indirect plan/goal expressions ("the lion hunted deer"), where the goal of finding food is treated as if it was a presupposition. Often, indirect plan/goal expressions involve a form of inference, e.g., the fact that the serpent spat poison into the mans water indicates that the serpent has a plan to kill the man. Of these, speech acts and indirect plans/goals were empirically found to be the most frequent affect states.

Goyal et al. used the following tools: FrameNet (Baker et al. 1998, the MPQA Lexicon (Wilson et al. 2005b), OpinionFinder (Wilson et al. 2005), the Semantic Orientation Lexicon (Takamura et al. 2007), and Wierzbicka's (1987) speech act list. The best classification results were achieved with FrameNet and the

Semantic Orientation Lexicon ($F = 0.39$). They were also able to show that their list of patient polarity verbs significantly improved classification results in combined with the lexical resources (best system in combination with Framenet: $F = 0.45$). This is interesting as they harvested these from the web in a bootstrapping manner, assuming that "evil" characters do things that are bad news to the patients of these actions, and "angelic" characters in turn do things which result in a positive affect state in the patients of the actions. For the task of link prediction, lower recognition rates were achieved ($F = 0.05$ to $0.25$), but this is a much harder task, and the result was measured in pipeline with automatic affect recognition (i.e., with $F = 0.45$).

Overall, this is an exciting system that combines a limited form of intention recognition with sentiment detection and speech act detection, and provides an affect analysis that could serve as the semantic representation of a deeper summariser. On the basis of this representation, such a summariser would have the possibility of formulating a justification for its content selection. It also shows that current NLP resources can be exploited and built together to achieve something that is larger than its parts.

### 3.3    Kintsch and Van Dijk's Memory Restricted Summariser

One summarisation model that allows manipulation of semantic structures of texts was proposed by Kintsch and van Dijk (1978). It is fundamentally a cognitive account of human text comprehension, where the text is turned into propositions and processed incrementally, sentence by sentence. In this model, the reader constructs from the text a meaning representation based on propositions. Propositions are then incrementally organised into a coherence tree, which represents the reader's working memory. After each sentence, the memory is reorganised by attaching the sentence's propositions to the existing ones. This is called a *memory cycle*. At the end of a cycle, as a simulation of limited memory, only a few important propositions are carried over to the next cycle. The forgotten propositions can be recalled to provide bridging connections, if a new proposition cannot be otherwise attached. The final summary is based on those propositions whose semantic participants (arguments) are well-connected and hence likely to be remembered by a human reading the text, under the assumption of memory limitations.

For its explanatory power and simplicity, the model has been well-received not only in the fields of cognitive psychology (e.g., Paivio 1990; Lave 1988) and education (Gay 1976), but also in the summarisation community (Hahn 1984; Uyttendaele et al. 1998; Moens et al. 2003).

My student Yimai Fang has been working on a reimplementation of a KvD summariser. It uses approximate propositions constructed from a syntactic parse. It is not perfect, but it produces what I consider deeper explanations, and it is also competitive with current summarisers when evaluated with current summarisation evaluation methodology. KvD's experiments show that three factors control the reproduction probability of a proposition in recall or summarisation tasks: (a) the frequency of it being selected to remain in memory, (b) whether it is a generalisation, and (c) whether it is a meta-statement (or macro-proposition). Of these factors, we can only model the first, robust inference being beyond current NLP capability.

The central mechanism of the KvD model is *argument overlap* of propositions. Two arguments overlap if they co-refer to one concept, which can be an entity, an event, or a class of things. Argument overlap is a linguistic phenomenon that is conceptually simple, but computationally complex as it could be modelled using many different combinations of methods. A wide range of relevant linguistic phenomena are connected to KvD's argument overlap, including but not limited to coreference, paraphrasing, polysemy, and genericity. In KvD's manual demonstration of the algorithm, the resolution of textual expressions to concepts relies on human intelligence. When automating this process, computing an effective value of argument overlap is the key to successful content selection, because a new proposition is attached to the old proposition that is most attractive to its arguments, i.e. has the maximal argument overlap.

We have performed an in-detail comparison of three models of argument overlap (lexical chains (Galley et al. 2003), Lin-type lexical similarity (Lin 1998) and Word2Vec-type lexical similarity (Mikolov et al. 2013)). Of these, the Lexical Chain-based argument overlap is by far superior in our setting. All our models are complemented by coreference chain information.

When the input texts are chosen from the news domain (the DUC-2 corpus), our summariser's performance is in the ballpark of the best freely available lexical-item attraction type summarisers, with MEAD being overall the strongest summariser, and with no summarisers statistically better than the n-best baseline. But we propose to move away from news texts for summarisation for various reasons, and instead advocate English learner texts. On this data, the proposition-based summariser significantly outperforms the lexical-attraction based summarisers (on all ROUGE measures considered, namely ROUGE-1, ROUGE-2, ROUGE-S and ROUGE-SU4). It also produces, as a side-effect, data structures that can be interpreted as an explanation of why a certain content unit was chosen for the summary. It thereby adds a new dimension of explanatoriness to summaries, even if many of its decisions are necessarily based on imperfect pre-processing.

## 4  Evaluation Matters

Should "deep summaries" be evaluated any differently from summaries derived in a shallower way? Or should all summaries be evaluated by methods that place more stress on equality of meaning, not surface string? De facto summarisation evaluation for the past 15 years or so has been by ROUGE (Lin 2004), a string-based similarity metric. ROUGE is a recall-based metric that counts how many string-units (many kinds are supported) are shared between the $n$ model summaries (gold standard summaries written by $n$ different humans) and the system summary. Because ROUGE is recall-based, it is extremely important to control for length of summaries when setting up an experiment with ROUGE. In recall-based metrics, longer summaries always have a higher chance of producing higher numerical values, with no disincentive working the other way.

Is ROUGE good enough? It is a compromise, and it is a generally accepted measure that can deliver an automatic evaluation. It is of course shallow, and it therefore runs into all the problems with synonymy and polysemy and everything else associated with shallow string-based mechanisms. It is not designed to capture deep meaning equivalences between two texts; no current automatic method can do that. However, high ROUGE overlap is hard to reach, and is therefore not meaningless. ROUGE incorporates several variations of string units, including ngrams, shared sub-sequences, and skip-grams. Some of these are better at capturing the shared meaning between the two texts than others, but nothing can guarantee in advance which ROUGE metric will turn out to be more meaningful for any particular input text and systems compared. That is the reason why researchers generally report more than one ROUGE variant.

Could it be done any better? The pyramid method (Nenkova and Passonneau 2004) and the factoid method (van Halteren and Teufel 2003) are expensive but more informative alternatives. They use human judgement to determine information units of variable sizes, and then count overlap in information units between gold standard and system summaries. Humans, once they are trained, are surprisingly consistent in annotating information units in text and have many shared intuitions that cannot be simulated automatically in advance. Of course information-unit based evaluation results in higher quality evaluation, which is more convincing overall and more likely to persuade critics of any particular summarisation method. But it is very expensive.

The core problem that makes it so expensive is that the manual annotation process cannot be done in an *off-line* manner. What I mean by "off-line" here is that it is not possible to create a gold-standard annotation in advance, as a one-time effort. Off-Line-evaluation, where it is possible to perform all manual annotation once only (*before* the creation of summaries by systems) is attractive because it can be reused indefinitely and for free, for as many systems and system variants as one wishes. On-line summary evaluation, in contrast, is performed *after* automatic summary creation, and thus requires human annotation effort for each new summary that a system creates. Pyramid/factoid annotation is impossible to perform off-line because one cannot predict all the different ways how an automatic summariser might choose to verbalise a given information unit. Similarly, and more obviously, subjective evaluation is also always on-line – a human cannot judge a summary before it has been created.

The cost of such human evaluation might therefore be justifiable for a one-shot, project-final evaluation that settles all accounts. When we look at day-to-day evaluation during development of a method, the cost would almost always prohibit the use of this deep evaluation method, resulting in a ROUGE-type automatic evaluation. If one is working on deeper summarisation methods for general text, on a large scale, then this evaluation is a bit like "working in the dark"; and it definitely impedes the development of better summarisation methods.

I was recently involved in the evaluation of a timeline summariser (Bauer and Teufel 2015), where we developed a system for off-line automatic deep evaluation. This was possible because we were able to "anchor" information units ("Historical content units"; HCUs) in the input text itself, and then made the working assumption that each summariser would select from the pieces of input text annotated with HCUs. (If an automatic summariser hypothetically were to produce real deep-generated summaries in such a way that it could not, for each event it named, point to a particular place in the text, it would not be evaluatable with our framework.) We use the TimeML framework (Pustejovsky 2003; Saurí et al. 2006), which has become a de-facto standard in all types of processing to do with temporal information. The fact that in history events and time descriptions (such as dates) are so central and closely connected to meaning units provided a further helping hand. Even so, the processes of annotation scheme definition and annotation itself were very time-intensive. But the huge advantage of this method is that the associated test corpus[1] can be reused by us and by other research groups again and again.

Let us now return to the situation of evaluating deeper summaries. I think it is possible to develop a fully automatic evaluation of summarisation based on propositions. HCUs, SCUs, and factoids are related to propositions, but there is not necessarily a one-to-one mapping between them. In particular, the timeline based anchoring of HCUs into the text was a solution that worked in this particular type of summary and domain, but would not necessarily work in the general case. The method I am proposing would also be off-line like the HCU-evaluation, i.e., require only a one-time effort in human annotation before summary creation time, and it would be usable forever for free afterwards. However, it would not be designed to evaluate the quality of the entire deep summarisation pipeline (Steps 1 through 3 in Fig. 1), but only Step 2 (content selection). Even so, it would require some agreement in the community to make such an effort worthwhile:

- **Propositions:** There would need to be some definition of the propositions used as semantic representation that at least a sub-community in the field of summarisation agrees on. This definition of propositions would have to be automatically derivable. It might be based on DMRS (Copestake 2009) or a similar semantic representation, or it might be based on more ad-hoc propositions based on dependency parsing, such as the ones that we use in our previous work (Fang and Teufel 2014), or that Stanovsky et al. (2014) propose. Guidelines describing the meaning of a wide range of propositions, covering many semantic and syntactic phenomena, will have to be written.
- **Annotator Training:** It is crucial that annotators have a shared and well-defined image of what a proposition is, before we can allow them to the selection annotation. Otherwise, if we ask them to select from propositions they don't understand well, the task becomes meaningless. Annotators would have to demonstrate a high level of understanding of the meaning of propositions, according to the definition in the guidelines. This could take the form

---

[1] http://www.cl.cam.ac.uk/~smb89/form.html.

of annotators being asked to produce propositions in a training setup, which would be compared to the parser's propositions, or a test where annotators have to describe what some preposition means.

- **Selection/Ranking Goldstandard:** They will then be asked to provide a ranking of the propositions they deem most important for a summary. Because it is well-known that humans do not agree much on content selection, it is important that, similar to the philosophy behind ROUGE, there is a high enough number of annotators that a mathematically stable ranking scheme can be developed.
- **Development of Agreement Metric:** Once all these pieces are in place, a metric for summary quality in terms of content selection could be developed, in analogy to the Pyramid scheme. In contrast to the original Pyramid scheme, which is on-line, the proposed methodology would be a one-time effort, and result in a reusable off-line gold-standard. Each summarisation system would be able to start from the same, automatically derived, propositions, and be instantly comparable to all other systems as well as to the humans.

Such an evaluation scheme would enable the creation of a growing repository of content selection gold standards, shared across the community. It would allow for experiments that establish which content selection schemes works best, or at least most robustly. The repository would also be able to cover many different text types, writing styles and difficulty levels.

## 5    NLP Task Interfaces

Decades of work on methodology allowed the creation of more robust and generally much better solutions to hard problems in all areas of NLP – for instance parsing, semantic role labelling, and the modelling of semantic similarity at different granularities of linguistic units. However, when we started using standard systems for supposedly well-defined tasks such as parsing and co-reference, in the framework of proposition-based summarisation, we found that there is a gap between how NLP tasks are currently defined and how our system would have needed them.

Our problems weren't only to do with the actual noise in the performance – it is rather understandable that for a system trained on news texts, various things will go wrong when it is asked to process general text. Instead, it had to do with how the tasks are *defined*. This is not to be a criticism of the developers of these schemes – I am extremely happy that these tools exist and that I can use them. It is more a call for more of a dialogue between text understanding and more engineering-based disciplines.

For instance, the task of coreference seems so simple and clear – look at every referring expression in text, and decide which discourse referent it refers to. Discourse referents can be real and fictional, abstract and concrete, things and kinds of things in the world, instances or types. Once this has been done, return equivalence classes of referring expressions that refer to the same entity in the real word. When we ran two state of the art coreference resolvers (Björkelund and Farkas 2012; Lee et al. 2011), we were pleased to see that they do find many of the links that improve our summariser. Dropping coreference resulted in a large drop in ROUGE scores. However, there are also still many useful links/overlaps (essential to text coherence) that they are not yet able to identify. Some of the missing useful links are anaphoric (which the current coreference guideline recognises (with some limits)), but others are of a more difficult kind, such as bridging.

The guidelines for OntoNotes annotation (Weischedel et al. 2011) gave us possible answers for this problem – annotators are free not to annotate certain entities, eg., abstract entities, generics and event references. I do admit that all three are difficult phenomena, both philosophically and practically, but some sort of practical answer is needed for deeper summarisation and for text understanding tasks. Generic, event references and abstract concepts are a frequent phenomenon in the texts we process, and if annotators are allowed to avoid the annotation of difficult instances, real-world tasks will suffer as a consequence. OntoNotes-trained coreference systems, of course, can only mimic what is annotated. I feel these are examples of areas were a joint definition of phenomena and annotation instructions could be extremely fruitful.

Part of the problems we experienced come from the overuse of news text in NLP. The Penn Treebank has fostered much research since 1993, when it was released, and has helped develop many a great NLP tools. At some point in time, however, not moving on to other domains will hurt us as a field (and many argue that we have long reached this point). News text has many properties that make it less interesting for researchers interested in text understanding, deeper summarisation, question answering and similar areas. The texts in the Wall Street Journal are written for financial experts, and much world knowledge is assumed in the reader. This creates serious obstacles for most deeper NLP tasks.

First, if a reader is reading a text which assues they already know much of the relevant information, they will have to perform much inference in order to connect the explicit information; much more than they would have to do in a simpler text. Inference is of course a problem for us as a field; much work has been spent on it and not much progress has been made.

Secondly, for researchers of summarisation, news texts have another toxic disadvantage – there is the n-first baseline, a "baseline" that is anything but. Journalists are highly trained experts who carefully produce text that starts with a one or two sentence summary. On the one hand, one could argue this lucky situation lessens the need for automatic summarisation of news overall drastically. However, if one evaluates sentence extraction based summarisers against this supposed baseline, it will create an entirely wrong impression about the actual quality of the summarising systems, and can actively impede progress in the field.

My conclusion from all this is that it is high time we move away from texts that have been specifically written only for experts (in the case of the WSJ, experts in finance and politics), and towards simpler texts. I am not advocating that we artificially dumb down our texts, and I am certainly not advocating a return to the toy stories used in the 1970s and 1980s. If decades of corpora work have taught us anything, it is that we should only ever use naturally occurring texts, and nothing else. Naturally occurring texts provide a reality check – they are unpredictable and contain unexpected linguistic expressions we would never have been able to create ourselves. We should search for other kinds of texts that have a satisfactory level of difficulty, that do not insult our intelligence, but that do not assume an unrealistic level of world knowledge either.

Goyal et al. (2013) use unadulterated fables, an inspired choice. They needed stories written according to sociologically accepted "plot" norms; fables, often containing a moral message, are of this type. They may be simple, but they *are* naturally occurring texts, and they *are not* news. (Unfortunately, they turned out to have other peculiarities that required special attention, such as the unfamiliar use of capitalisation and coreference).

The texts that me and my student Yimai Fang found are English Language Learners texts from the International English Language Testing System (IELTS; www.ielts.org). These texts are aimed at speakers of the language of a certain level of sophistication. They are edited but not excessively so. They usually concentrate on a random cluster of facts from the sciences and humanities, or cover societal topics that can be seen as being of general interest. They are carefully written so as not to assume too much from the readers. The texts need to deliver the relevant facts, so that the texts' readers can later be tested for their understanding of the texts. We concluded that these circumstances were perfect for our purposes. Quite unsurprisingly, the n-best baseline is beatable in this domain by any intelligent automatic summariser - and this is obviously so because the texts do not routinely start with a summary. The corpus (4 summaries each for 31 test texts) will be made available for download in early 2017.

There are likely to be many other exciting text genres that have advantages for deeper text understanding approaches. We should waste no time in starting to looking for them, and move away from the practice of using news texts for any and all NLP tasks.

## 6   Conclusions

Recently there has been a lot of interest in revisiting ambitious AI problems, such as narrative scheme creation (Chambers and Jurafsky 2008), affect analysis for plot units (Goyal et al. 2013) and memory-limited proposition-based summarisation (Fang and Teufel 2014) – a series of second attacks on problems which had previously been thought unimplementable. I attribute this new-found interest to a swing of the pendulum, a craving for meaning representations after years of methodology-driven research. I for one am curious about which AI/language problems our field will turn to in the near future.

I have in this opinion paper argued why now is a good time for deeper summarisation's "second chance": lower-level NLP methodology has reached a level of maturity where we can once again be ambitious. But this time, unlike the first time around, the field has learned valuable lessons about robustness, about lexical sparsity, about empirical methodology. As a field, we have grown to expect large-scale demonstration of applicability. One of the important preconditions for creating convincing evaluations is our continued work on gold standards – the question of which semantic components a good summary should contain – which ones can it not be without, which ones would be nice to have, and which ones are definitely too detailed to make it into a summary. We need gold standards because that means off-line evaluation (thus cheaper and reusable); we need *deep* gold standards because deep means higher quality compared to string-based evaluation.

The emerging breed of AI-flavoured text understanding tasks profits from what I have maybe unfairly described as "subsidiary" NLP tasks – parsing, coreference resolution, and semantic relatedness. It also has something to offer to these subsidiary tasks, namely real-world applicability. The interface between the two areas consists in the intermediary results – the dependencies, coreference equivalence classes, distributed representations of semantic relatedness, sentiment aspects of different semantic units, all the way up to full semantic representations. By actively collaborating, the lower-level tasks can gain by getting more realistic and usable for a real-task, whereas the deeper summarisation, by communicating which information it actually needs, can gain by receiving more accurate and robust intermediary results. In order to achieve progress right now, both sides would profit from more a deeper dialogue about the exact nature of this interface.

# References

Androutsopoulos, I., Malakasiotis, P.: A survey of paraphrasing and textual entailment methods. J. Artif. Intell. Res. **38**, 135–187 (2010)

Baker, C.F., Fillmore, C.J., Lowe, J.B.: The Berkeley framenet project. In: Proceedings COLING, pp. 86–90 (1998)

Baroni, M., Lenci, A.: Distributional memory: a general framework for corpus-based semantics. Comput. Linguist. **36**(4), 673–721 (2010)

Barzilay, R., Lee, L.: Learning to paraphrase: an unsupervised approach using multiple-sequence alignment. In: Proceedings of HLT, pp. 16–23 (2003)

Bauer, S., Teufel, S.: A methodology for evaluating timeline generation algorithms based on deep semantic units. In: Proceedings of ACL, p. 834 (2015)

Björkelund, A., Farkas, R.: Data-driven multilingual coreference resolution using resolver stacking. In: Joint Conference on EMNLP and CoNLL-Shared Task, pp. 49–55 (2012)

Carbonell, J., Goldstein, J.: The use of MMR, diversity-based reranking for reordering documents and producing summaries. In: Proceedings of SIGIR, pp. 335–336 (1998)

Chambers, N., Jurafsky, D.: Unsupervised learning of narrative schemas and their participants. In: Proceedings of ACL, pp. 602–610 (2009)

Chambers, N., Jurafsky, D.: Unsupervised learning of narrative event chains. In: Proceedings of ACL, pp. 789–797 (2008)

Cohen, R.: A computational theory of the function of clue words in argument understanding. In: Proceedings of COLING, pp. 251–255 (1984)

Copestake, A.: Slacker semantics: why superficiality, dependency and avoidance of commitment can be the right way to go. In: Proceedings of EACL, pp. 1–9 (2009)

DeJong, G.F.: An overview of the FRUMP system. In: Lehner, W.G., Ringle, M.H. (eds.) Strategies for Natural Language Processing, chap. 5. Lawrence Erlbaum, Hillsdale (1982)

Dorr, B.J., Zajic, D., Schwartz, R.: Hedge: a parse-and-trim approach to headline generation. In: Proceedings of the HLT Text Summarization Workshop, pp. 1–8 (2003)

Elson, D.K., McKeown, K.: Extending and evaluating a platform for story understanding. In: AAAI Spring Symposium: Intelligent Narrative Technologies II, pp. 32–35 (2009)

Erkan, G., Radev, D.R.: Lexrank: graph-based lexical centrality as salience in text summarization. J. Artif. Intell. Res. **22**, 457–479 (2004)

Fang, Y., Teufel, S.: A summariser based on human memory limitations and lexical competition. In: Proceedings of EACL, pp. 732–741 (2014)

Farrow, E., Dickinson, T., Aylett, M.P.: Generating narratives from personal digital data: using sentiment, themes, and named entities to construct stories. In: Abascal, J., Barbosa, S., Fetter, M., Gross, T., Palanque, P., Winckler, M. (eds.) INTERACT 2015. LNCS, vol. 9299, pp. 473–477. Springer, Cham (2015). https://doi.org/10.1007/978-3-319-22723-8_41

Galley, M., McKeown, K., Fosler-Lussier, E., Jing, H.: Discourse segmentation of multiparty conversation. In: Proceedings of ACL, pp. 562–569 (2003)

Gay, L.R., Mills, G.E., Airasian, P.W.: Educational Research: Competencies for Analysis and Application. Merrill, Columbus (1976)

Goldberg, A.B., Fillmore, N., Andrzejewski, D., Xu, Z., Gibson, B., Zhu, X.: May all your wishes come true: a study of wishes and how to recognize them. In: Proceedings of HLT/NAACL, pp. 263–271 (2009)

Goyal, A., Riloff, E., Daume III, H.: A computational model for plot units. Computational Intelligence **29**(3), 466–488 (2013)

Hahn, U., Reimer, U.: Computing text constituency: an algorithmic approach to the generation of text graphs. In: Proceedings of SIGIR, pp. 343–368 (1984)

van Halteren, H., Teufel, S.: Examining the consensus between human summaries: initial experiments with factoid analysis. In: Proceedings of the HLT Text Summarization Workshop (2003)

Kasch, N., Oates, T.: Mining script-like structures from the web. In: Proceedings of the NAACL HLT 2010 First International Workshop on Formalisms and Methodology for Learning by Reading, pp. 34–42 (2010)

Kintsch, W., van Dijk, T.A.: Toward a model of text comprehension and production. Psychol. Rev. **85**(5), 363–394 (1978)

Knight, K., Marcu, D.: Statistics-based summarization – step one: sentence compression. In: Proceeding of AAAI-2000, pp. 703–710 (2000)

Landauer, T.K., Dumais, S.T.: A solution to Plato's problem: the latent semantic analysis theory of acquisition, induction, and representation of knowledge. Psychol. Rev. **104**(2), 211 (1997)

Lave, J.: Cognition in Practice: Mind, Mathematics and Culture in Everyday Life. Cambridge University Press, Cambridge (1988)

Lee, H., Peirsman, Y., Chang, A., Chambers, N., Surdeanu, M., Jurafsky, D.: Stanford's multi-pass sieve coreference resolution system at the CoNLL-2011 shared task. In: Proceedings of the Fifteenth Conference on Computational Natural Language Learning: Shared Task, pp. 28–34 (2011)

Lehnert, W.G.: Plot units: a narrative summarisation strategy. In: Lehnert, W.G., Ringle, M.H. (eds.) Strategies for Natural Language Processing, chap. 4, pp. 223–244. Lawrence Erlbaum, Hillsdale (1981a)

Lehnert, W.G.: Plot units and narrative summarization. Cogn. Sci. **4**, 293–331 (1981)

Lehnert, W.G., Dyer, M.G., Johnson, P.N., Yang, C., Harley, S.: BORISan experiment in in-depth understanding of narratives. Artif. Intell. **20**(1), 15–62 (1983)

Lin, D.: Using collocation statistics in information extraction. In: Proceedings of ACL/COLING 1998, Montreal, Canada (1998)

Lin, C.Y.: Rouge: a package for automatic evaluation of summaries. In: Proceedings of Workshop "Text Summarization Branches Out" at ACL 2004 (2004)

Luhn, H.P.: The automatic creation of literature abstracts. IBM J. Res. Dev. **2**(2), 159–165 (1958)

McKeown, K., Robin, J., Kukich, K.: Generating concise natural language summaries. Inf. Process. Manag. **31**(5), 703–733 (1995)

Mihalcea, R., Tarau, P.: Textrank: bringing order into texts. In: Proceedings of EMNLP (2004)

Mikolov, T., Chen, K., Corrado, G., Dean, J.: Efficient estimation of word representations in vector space. arXiv preprint arXiv:1301.3781 (2013)

Moens, M.F., Angheluta, R., De Busser, R.: Summarization of texts found on the world wide web. In: Abramowicz, W. (ed.) Knowledge-Based Information Retrieval and Filtering from the Web, vol. 746, pp. 101–120. Springer, Boston (2003). https://doi.org/10.1007/978-1-4757-3739-4_5

Nenkova, A., Passonneau, R.J.: Evaluating content selection in summarization: the pyramid method. In: Proceedings of NAACL/HLT 2004, Boston, MA (2004)

Nishikawa, H., Arita, K., Tanaka, K., Hirao, T., Makino, T., Matsuo, Y.: Learning to generate coherent summary with discriminative hidden semi-Markov model. In: Proceedings of COLING, pp. 1648–1659 (2014)

Paivio, A.: Mental Representations. Oxford University Press, Oxford (1990)

Pustejovsky, J., Castano, J.M., Ingria, R., Sauri, R., Gaizauskas, R.J., Setzer, A., Katz, G., Radev, D.R.: TimeML: robust specification of event and temporal expressions in text. In: New Directions in Question Answering, vol. 3, pp. 28–34 (2003)

Radev, D., Allison, T., Blair-Goldensohn, S., Blitzer, J., Celebi, A., Dimitrov, S., Drabek, E., Hakim, A., Lam, W., Liu, D., et al.: MEAD - a platform for multi-document multilingual text summarization. In: Proceedings of LREC (2004)

Radev, D.R., McKeown, K.R.: Generating natural language summaries from multiple on-line sources. Comput. Linguist. **24**(3), 469–500 (1998)

Reiter, E., Sripada, S., Hunter, J., Yu, J., Davy, I.: Choosing words in computer-generated weather forecasts. Artif. Intell. **167**(1), 137–169 (2005)

Rumelhart, D.E.: Understanding and summarizing brief stories. In: Laberge, D., Samuels, S. (eds.) Basic Processes in Reading, Perception and Comprehension. Lawrence Erlbaum, Hillsdale (1977)

Saurí, R., Littman, J., Gaizauskas, R., Setzer, A., Pustejovsky, J.: TimeML Annotation Guidelines, Version 1.2.1 (2006)

Schank, R.C.: Conceptual Information Processing. North-Holland, Amsterdam (1975)

Schank, R.C., Abelson, R.P.: Scripts, Goals, Plans and Understanding. Lawrence Erlbaum, Hillsdale (1977)

Stanovsky, G., Ficler, J., Dagan, I., Goldberg, Y.: Intermediary semantic representation through proposition structures. In: Proceedings of ACL, p. 66 (2014)

Steyvers, M., Griffiths, T.: Probabilistic topic models. In: Handbook of Latent Semantic Analysis, vol. 427, no. 7, pp. 424–440 (2007)

Takamura, H., Inui, T., Okumura, M.: Extracting semantic orientations of phrases from dictionary. In: Proceedings of HLT-NAACL, vol. 2007, pp. 292–299 (2007)

Unno, Y., Ninomiya, T., Miyao, Y., Tsujii, J.: Trimming CFG parse trees for sentence compression using machine learning approaches. In: Proceedings of COLING/ACL, pp. 850–857 (2006)

Uyttendaele, C., Moens, M.F., Dumortier, J.: Salomon: automatic abstracting of legal cases for effective access to court decisions. Artif. Intell. Law 6(1), 59–79 (1998)

Weischedel, R., Consortium, L.D., et al.: OntoNotes Release 4.0. Linguistic Data Consortium (2011)

Wierzbicka, A.: English Speech Act Verbs: A Semantic Dictionary. Academic Press, Cambridge (1987)

Wilson, T., Hoffmann, P., Somasundaran, S., Kessler, J., Wiebe, J., Choi, Y., Cardie, C., Riloff, E., Patwardhan, S.: OpinionFinder: a system for subjectivity analysis. In: Proceedings of HLT/EMNLP, pp. 34–35 (2005)

Wilson, T., Wiebe, J., Hoffmann, P.: Recognizing contextual polarity in phrase-level sentiment analysis. In: Proceedings of the Conference on HLT and EMNLP, pp. 347–354 (2005b)

# Tracing Language Variation for Romanian

Daniela Gîfu$^{(\boxtimes)}$ (iD) and Radu Simionescu

Faculty of Computer Science, Alexandru Ioan Cuza University,
16 General Berthelot St., 700483 Iaşi, Romania
{daniela.gifu, radu.simionescu}@info.uaic.ro

**Abstract.** This paper illustrates a pilot study on two collections of publications, written at the middle of the 19th century in two countries, Romania and Republic of Moldavia. The corpus includes articles from the most important Romanian and Bessarabian publications, categorized in three periods: 1840–1917, 1918–1940, and 1941–1991. The research conducted on these resources focuses on the lexical evolution of words. We use a machine learning approach to explore the patterns that govern the lexical differences between two lexicons. The model is used for automatically correlating different forms of a word. The approach is suitable for bootstrapping, in order to increase the quantity and quality of the training data. The presented approach is language independent. By using the contemporary language as a pivot, the data is analyzed and compared from various perspectives.

**Keywords:** Diachronic texts · Lexical evolution · Comparable corpora
Lexical approach · Statistics

## 1 Introduction

For many decades, historical and comparative linguistics were fascinated about the connection between language acquisition and diachronic change [1]. The diachronically and synchronically comparative studies of the Romance languages expose the presence of many similarities. Especially in diachronic studies, the similarities of the Romance languages are more numerous [2]. Latin was the starting point, but issues about substratum, superstratum and adstratum, which contributed to differentiate languages, were not set aside. Contributions assigned to this dimension are closely related to the previous ones, as many of the traits of Romance linguistics are also found in diachronic study on the Romanian language [3, 4].

In this paper, the differences between the lexicons of two versions of a language, from the perspective of transformation patterns were analyzed. We consider that it is highly important that research on language evolution cover the patterns, which govern the changing units of a lexicon. However, it is difficult to model the rules, which govern the lexical changes of a language. Most such rules apply only on some words, based on various aspects such as etymology, declination and even length. There are usually exceptions, which apply to such rules. Creating a comprehensive inventory with the transformations of a lexicon of a language requires high effort and profound linguistic knowledge.

© Springer International Publishing AG, part of Springer Nature 2018
A. Gelbukh (Ed.): CICLing 2016, LNCS 9624, pp. 599–610, 2018.
https://doi.org/10.1007/978-3-319-75487-1_45

The methodology described in this paper can be used for extracting transformational rules from the differences between two lexicons, in a unsupervised manner. By maintaining an inventory of these rules, automatic correspondence between the words of different lexicons can be traced.

The lexicons of two versions of a language can be observed as two sets of distinct word forms. Based on word-to-word association, one can present the intersection of the two lexicons, and emphasize the transformation rules, which lead to the observable surface differences in the lexical forms of the same words.

To overcome the lack of a comprehensive resource, which would indicate word-to-word association between lexicons of different dialects or periods, we developed a mechanism for automatic correlation of different forms of the same words. A statistical model is trained on a list of known word-to-word correlations between the two lexicons. Even though the size of the training data is not necessarily big, it must expose a wide variety of lexical evolution phenomena. This aspect is difficult to achieve in a single iteration, when developing such a resource. The development can be structured around a bootstrapping methodology, and the diversity of the exposed phenomena can be increased, by analyzing the words for which a correlation cannot be made automatically.

The paper is organized as follows: Sect. 2 shortly reviews the relevant literature on various diachronic approaches, Sect. 3 discusses a new method using the MaxEnt (Maximum Entropy), results are then described in Sect. 4 and, finally, our conclusions are given in Sect. 5.

## 2   Diachronic Studies in Computational Linguistics

Since the '80s, the construction language corpora have been a subject of real interest in research in natural language processing field, starting with English language corpora, and later with European languages (French, German, Romanian, Italian, Portuguese etc.) and Asian languages[1].

Currently, diachronic corpora of an increasing number of languages are constructed and annotated with syntactically relevant information in order to address them and related issues. Up to the 16th century, almost all-scientific writing in Europe was conducted in Latin. The construction and annotation of historical corpora is challenging in many ways [5–10].

Reading the studies published by our predecessors helped us to better perceive the differences occurring in the Romanian language, especially in the diachronic and diatopic variation. Taking over the way to interpret the language facts from them, our approach is developed based on lexical analysis of the words found in analyzed ancient texts.

---

[1] Here are some corpus resources that involve diverse languages: ELDA (*Evaluations and Language Resource Distribution Agency*), TRACTOR (*TELTRI Research Archive of Computational Tools and Resources*), OTA (*Oxford Text Archive*), LDC (*Linguistic Data Consortium*), COROLA (COrpus of ROmanian Language).

The vast literature tells its own story regarding the usefulness of technology and information services [11–18]. The development and use of software for natural language processing (NLP) highlight the defining aspects of the text (morphological and syntactic analysis, semantic analysis and, more recently, pragmatic analysis).

The similarities between languages are interesting for historical and comparative linguistics, as well as for language acquisition. We will show details about this in the next chapter. Scannell [19] and Hajič et al. [20] argue for the possibility of obtaining a better quality in translation using simple methods for very closely related languages. Koppel and Ordan [21] studied the impact of the distance between languages on the translation product and conclude that it is directly correlated with the ability to distinguish translations from a given source language from non-translated text.

It has been established that some genetically related languages have a high degree of similarity to each other, and its speakers are able to communicate without prior instructions [22, 23]. Various aspects present relevance when investigating the level of relatedness between languages, for example orthographic, phonetic, syntactic and semantic differences.

The approach for the study of the evolution of Romanian language is focusing only on the lexical similarity. The basis for this approach consists of the idea that phonetic alterations have a lexical correspondent [24].

Different approaches have been used in previous case studies in order to assess the orthographic distance similarity between related words. Their accuracy has been investigated and compared [25, 26], but a clear conclusion could not be drawn with respect to which method is the most appropriate for a given task.

Metrics will be used to determine the orthographic similarity between related words. For instance, the syllabic similarities of the Romanian language in different geographic areas and periods of time [27]. They used orthographic metrics like: the edit distance, the longest common subsequence ratio, and the rank distance.

## 3 Research Methodology

The diachronic corpus is a valuable resource for investigating linguistic variation over time. But to properly study the evolution of words, a word to word association between old forms and current forms of words, is required. This would be a highly valuable resource for linguists and NLP technicians alike. Unfortunately, such resources are very sparse for most languages when it comes to dialect or diachronic lexicon correlation. We present our method of developing such a resource, based on a modest sized list of known word-to-word associations.

In this section, we describe methodological preliminaries, starting with the lexicons of the two languages and a dictionary, which maps some words from one lexicon with words from the other.

For each word pair from the input list, a set of textual differences is extracted that arise from the linguistic phenomena as well as the data involved. These are modeled as substring replacement operations (referred as "REP"), at particular character positions inside the words.

By applying these operations on the first word of the pair, the other is obtained.

Example: a noun educației (En: education) in the oblique case:

educatiunei - **educației**: {REP(5, t, ț), REP(7, un,)}

where educatiunei is the "source" word, and educației is the "destination" word.

The lexicons from which the two word forms belong are also referred as the "source" and "destination" lexicons, respectively.

To avoid excessive segmentation of the replacement operations and to model transformation rules based on longer patterns, we impose a minimum distance of two characters between two REP operations. By applying this regulation on the previous example, only one operation is extracted:

educatiunei - **educației**: {REP(5,tiun,ți)}

We use a MaxEnt (*Maximum Entropy*) model to classify all the REPs extracted from the known correlations list, based on the character-level context in which they are applied in the source word. This model handles successfully data, which is classified by a large amount of different outcomes. The training context contains the characters to the left and to the right of the replaced substrings, up to a distance of 3 characters, in the source words.

The trained model is used for predicting REPs, which can be applied on a previously unseen word from the source lexicon. Based on these REPs, a set of fictive/candidate words are generated, each having a trust score attached. If a candidate word is found in the destination lexicon, the two words are marked as a corresponding pair.

For instance, for the old Romanian word **imperatesci**[2] (Current form: "împărătești" - En: kingly/royal) the model indicates the following transformations that could be applied at different positions:

REP(0,i,î), REP(3,e,ă), REP(5,a,ă), REP(8,sc,șt)

A trust score threshold has been used at this point to filter REPs with scores, which have empirically proved of no use. To generate candidate destination words, all possible combinations of REPs are applied in turn on the source/unknown word.

In this example, the candidate word generated by applying all the indicated REPs is the form used in today's language. The list of generated destination candidates:

```
imperatești,   imperătesci,   impăratesci,   împeratesci,
imperătești,   impăratești,   împeratești,   impărătesci,
împerătesci,   împăratesci,   impărătești,   împerătești,
împăratești,   împărătesci,   împărătești
```

By looking up candidates forms in the destination lexicon, which should be a comprehensive resource, word form associations can be inferred automatically. In this case:

---

[2] The word was extracted from *Foaia pentru minte, anima si literature*, 1854.

`imperatesci -` **împărătești**

When more candidate forms are found in the destination lexicon, the one with the highest trust score is considered.

Another example: **cetatianii** (En: citizens). REPs indicated by the trained model:

```
REP(3, atia, ăţe), REP(4, t, ţ)
```

The two REPs overlap so they cannot be applied at the same time. The current form of the word is obtained by applying the first REP only. The candidate destination forms:

```
cetaţianii, cetăţenii
```

By looking up candidates forms in the destination lexicon, the final word form associated is:

`cetatianii -` **cetăţenii**

The source words, for which none of the candidate words is found in the destination lexicon, are subject of further human analysis. These words might have not been correlated automatically because some lexical evolution phenomena have not been observed in the training data. Alternatively, they might be words, which are part of only one lexicon. On the other hand, they might be foreign words. These situations require human supervision to establish the nature of the event. At this point, the initial training data set can be extended to expose more transformation operations, under human supervision.

Since the candidate words are looked up in the destination lexicon, it is generally required to use a morphologic dictionary (e.g. DEX on-line[3]) as the destination lexicon, while the source lexicon can be anything smaller than that.

To compare the word form of a particular text with the current language, the source lexicon can contain only the words from the given text, while the destination lexicon should be based on a comprehensive morphologic dictionary of the words currently in language use, to ensure proper overlap.

When comparing two texts written in the same language, but marked by different temporal or geographical traits, usually there are insufficient resources available. Morphologic dictionaries of dialects or of diachronic versions of languages (e.g. eDTLR (*Dictionary Thesaurus of the Romanian Language in electronic form*) are resources, which are difficult to develop. However, morphologic dictionaries of contemporary lexicons are much more common. To compare texts marked by different traits (in our case, a collection of journalistic articles after 1992), a contemporary lexicon can be used as a pivot destination lexicon, to correlate words from the two texts automatically.

---

[3] https://dexonline.ro/.

## 3.1    Corpora

While preparing the preliminary conclusions in the configuration model, we decided to include in our corpora different journalistic texts containing almost 60.000 lexical tokens (see Table 1). All documents were converted from PDF into plain text using the software package Apache PDFBox[4]. Furthermore, the raw texts were manually corrected.

We intend to enlarge our corpora with numerous texts from the other fields and thus improving discovering rates and information extraction patterns.

**Table 1.**  Statistical data

| Region | Period | N words | Sources |
|---|---|---|---|
| Bessarabia | 1840–1917 | 2661 | Basarabia reinoită, Candela, Curierul, Cuvânt moldovenesc, Democratul Basarabiei, Deșteptarea |
| | 1918–1940 | 11189 | Basarabia, Basarabia Chișinăului, Curierul, Democratul Basarabiei, Dreptatea, Glasul Basarabiei, Luminătorul, Sfatul țării, România nouă |
| | 1941–1991 | 8267 | Basarabia, Curierul, Deșteptarea, Literatură și artă |
| Romania | 1840–1917 | 8152 | Organul Luminarei, Foaia pentru Minte, Anima și Literatura, Literatorul, Pressa, Telegraful roman, Albina romanescă, Familia, Gazeta Transilvaniei |
| | 1918–1940 | 18701 | Deșteptarea, Adevereul, Curentul, Universul, Clujul, Drepteta, Moldova Socialistă |
| | 1941–1991 | 9594 | Deșteptarea, România literară, Scânteia, Convorbiri literare, Moldova socialistă, Vatra românească |
| Total | | 58564 | |

## 4    Statistics and Interpretation

From the entire set of data presented above, our corpus highlights the values for the three periods from two countries, Romania (Ro) and Republic of Moldavia (Mo). Using the mechanism of automatic correlation of unknown words, various experiments were conducted on the available corpora. An evaluation of the model is presented in Table 2. The training data for each period/region consists of a correlation list, which associates approx. 50% of the unknown words from the corpus with their forms in contemporary language use. The precision for various training scenarios is presented. Given the relatively small data, a 10 fold cross validation was used. The value indicated for each dataset represents the successful automatic correlation percent of the evaluation data.

---

[4] http://pdfbox.apache.org/.

**Table 2.** Evaluation of the automatic correlation of unknown words mechanism

| Ro 1840–1917 | Ro 1918–1940 | Ro 1941–1991 | Mo 1840–1917 | Mo 1918–1940 | Mo 1941–1991 | All texts |
|---|---|---|---|---|---|---|
| 85.71% | 74.3% | 67.5% | 80% | 76.03% | 65.3% | 71.59% |

Table 3 presents various detailed aspects of the data and the percent of correlations that were automatically performed by the presented mechanism.

**Table 3.** Unknown words in corpus and their automatic correlations coverage

| Region/Time | Ro 1840–1917 | Ro 1918–1940 | Ro 1941–1991 | Mo 1840–1917 | Mo 1918–1940 | Mo 1941–1991 |
|---|---|---|---|---|---|---|
| Total distinct words | 7622 | 17784 | 9150 | 2436 | 10685 | 7928 |
| Total unknown distinct words | 3408 | 6162 | 3826 | 1210 | 4219 | 3320 |
| Percent of known words (%) | 78.26 | 96.51 | 98.37 | 74.51 | 96.49 | 97.58 |
| Total automatic correlations | 746 | 189 | 48 | 181 | 138 | 50 |
| Total words covered by automatic correlations | 1189 | 267 | 56 | 240 | 173 | 61 |
| Percent of automatic correlations of unknown words (%) | 71.75 | 43.06 | 37.58 | 38.64 | 46.13 | 31.77 |

The percent of correlated words decreases in more recent periods, while the number of known words percent increases. The closer the period of time, the more words are found in the contemporary morphologic dictionary. The decreased correlation percent in the more recent texts is explained by the fact that the remaining unknown words are mostly foreign words, or proper names, while the quantity of words, which have deformed based on some observable lexical evolution phenomenon is reduced.

The chart presented below (Fig. 1) shows the most frequent REP operations extracted from texts, clustered by the different periods of time and regions.

For instance, the REP, "u ->", deletes a character from words, usually in the suffix. It represents almost 30% of all the REPs founded in the Romanian texts from 1840–1917. A synthesis about the rules that have increased the phonological character of the Romanian orthography [28].

Nowdays, the character "u" at the end of the words have disappeared from words or grammatical forms which is ending in diphthongs as: *ai, îi, ei, oi, ui* and in the consonants [k'], [g'] and [č]followed by *i* non-syllabic: *luai* (En: take), *lămâi* (En: lemon), *mei* (En: my), *auzii* (En: hear), *pietroi* (En: rock), *cui* (En: nail); *ochi* (En: eye), *vechi* (En: old), *unghi* (En: angle), etc.

The words of this type were written with "u" at the end, although it had been vanished long time ago from the pronunciation. "U" at the end of the words appears in

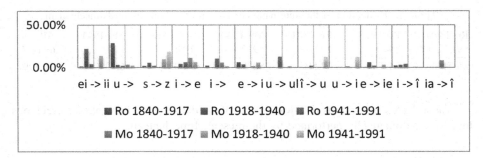

**Fig. 1.** The distributions of the 12 most frequent REPs per time frame and region

the current spelling only in the contexts of a phonetic reality and it must be pronounced *ambulatoriu* (En: ambulatory), *beneficiu* (En: ambulatory), *contradictoriu* (En: ambulatory), *interogatoriu* (En. interrogation), *obligatoriu* (En: obligatorily), *onorariu* (En: honorarium), *salariu* (En: salary), *serviciu* (En: service), *teritoriu* (En: territory), and so on [28].

In Table 4, is given one example for each REP's relevant for our corpus.

**Table 4.** Example with frequent REPs distribution per time frame and region

| REPs/Ex. | Romania (old word form – present word form) | | English term | Republic of moldavia (old word form – present word form) | | English term |
|---|---|---|---|---|---|---|
| ei -> ii | apunerei | apunerii | dawn's | inimei | inimii | heart's |
| u -> | introdusu | introdus | introduced | comisu | comis | commit |
| s -> z | sburatu | zburat | flown | desvoltarea | dezvoltarea | development |
| i -> e | cari | care | which | aducire | aducere | bringing |
| i -> | eminentia | eminența | eminence | priimi | primi | receive |
| e -> i | stenge | stinge | extinguish | aceeași | aceiași | same |
| u -> ul | minutu | minutul | minute | anulu | anul | year |
| î -> u | sîntem | suntem | (we) are | sînt | sunt | (I) am |
| u -> i | priimitu | priimiti | recieved | simțu | simți | feel |
| e -> i.e. | trebuesc | trebuiesc | must | mângăeri | mângâieri | caresses |
| i -> î | imperatu | împăratul | emperor | intre | între | between |
| ia -> î | – | – | – | ian | în | in |

One can observe that some REP operations present a higher than average frequency for both analyzed regions, in the same period of time, like the first and the third REPs, for the first period and the second period respectively. Some of the REPs are occurring in similar morphological situations for both regions. The first REP, for example occurs most frequently in the suffix of oblique nouns, for both regions. Other REPs have a more generic spectrum, like *s->z*, which occurs without any correlation to morphological traits.

Part of the motivation of this research is related to developing a diachronic POS tagger for Romanian. The presented model could be used to add enhanced support for unknown words, for a POS tagger for contemporary Romanian. Syntax and orthography play a very important role in POS tagging, and Gold corpora is required for training such a model. However, since this is not available, being able to associate unknown words with their known forms would still be highly beneficial to POS tagging, because that would greatly reduce the ambiguity of unknown words in the process of POS tagging diachronic texts.

The table (Table 5) and charts (Fig. 2) below present a very promising perspective. The "Identifiable words" represents the percent of words from the texts, which are known words or which can be correlated automatically with a known form. The "Automatic correlations of unknown words" indicates the amount of the unknown words, which can be correlated automatically. As observed earlier, this indicator decreases with more periods that are recent.

**Table 5.** Known words vs. identifiable words comparison

| Region/time | Ro 1840–1917 | Ro 1918–1940 | Ro 1941–1991 | Mo 1840–1917 | Mo 1918–1940 | Mo 1941–1991 |
|---|---|---|---|---|---|---|
| Known words (%) | 78.26 | 96.51 | 98.37 | 74.51 | 96.49 | 97.58 |
| Identifiable words (%) | 93.86 | 98.02 | 98.98 | 84.36 | 98.11 | 98.35 |
| Automatic correlations of unknown words (%) | 71.75 | 43.06 | 37.583 | 38.64 | 46.13 | 31.77 |

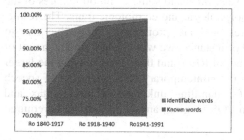
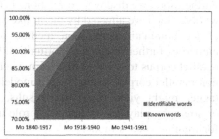

**Fig. 2.** The percentage of known words vs. identifiable words over time in the two regions (left side – Ro and right side - Mo)

In the case of old Romanian texts before 1917, 93% of the words can be recovered, where 71% of these are automatically correlated with a precision of 85%. In the case of Moldavian texts before 1919, only 83% of the words can be recovered. However, the ratio with the volume of known words is maintained.

The lower percent of the known words in old Moldavian texts, before 1917, is explained by the fact that Moldavia has undergone a "Russification" period. Romanian was gradually forbidden in Moldavia by Soviet authorities: in 1829 in administration, in 1833 in churches, in 1842 in secondary schools, in 1860 in elementary schools. After the First World War, in 1920, a Cyrillic Moldovan alphabet was introduced and adopted as the official alphabet for the Moldavian language.

In 1989, the contemporary Romanian version of the Latin alphabet was adopted as the official script of the Moldavian language, and in 1991, the official language was identified as Romanian. Interestingly, this sequence of historical events has impact on the presented statistics, when using the contemporary Romanian as a baseline lexicon.

# 5    Conclusions and Future Work

When computing the variation between two lexicons belonging to different versions of the same language, there are many aspects, which can be taken into, account, some of which being subjective in nature. The preliminary conclusion drawn from the previous chapters, having scrutinized various substring replacement operations at explaining the relationship between language acquisition and diachronic change in the lexical domain, confirms that language variations have a crucial role in explanations of lexical change. Moreover, the semi-automatic recognition of language deviations represents the basis for defining the link between them, exploring the changes in the lexical rules, while comparing two versions of a language from different periods of time and areas.

Out of this context, it is also important for aligned journalistic corpora with the most important Romanian language resources as WebDEX (*Romanian Academy Explanatory*), DEX on-line, and eDTLR (*Dictionary Thesaurus of the Romanian Language in electronic form*).

On all the REPs extracted from the input pairs various statistics were performed. A few rules were observed which govern the differences between the two lexicons. Furthermore, we discover correlations between rules and some common features of the words, which are affected by them, like morphologic and etymologic traits. The results of the automatic correlation method presented are promising. They give rise to numerous further research opportunities. For instance, we will create a small amount of parallel corpus to use it to first learn a kernel of REPs and then grow it using the larger non-parallel corpus. Using such a model in a contemporary Romanian part of speech tagger might yield interesting results, for handling unknown words. Syntax and orthography play a very important role in POS tagging, and Gold corpora is required for training such a model.

**Acknowledgments.** We thank to all students from 1st and 3rd year of Journalism at the Faculty of Letters of the "Alexandru Ioan Cuza" University of Iași have participated in the initial phases of the acquisition of the collection of texts.

# References

1. Baron, N.S.: Language Acquisition and Historical Change. North-Holland, Amsterdam (1977)
2. Densusianu, O.: Filologia Romanică în universitatea noastră. Bucuresci, J. V. Socecu Editeur, p. 23 (1902)

3. Maiden, M., Smith, J.C., Ledgewav, A. (eds.): The Cambridge History of the Romance Languages. Cambridge University Press, Cambridge (2011)
4. Saltarelii, M., Wanner, D. (eds.): Diachronic Studies in Romance Linguistics at the Conference on Diachronic Romance Linguistics, University of Illinois, April 1972, Janua Linguarum. Series Practica 207 (1972)
5. Lüdeling, A., Poschenrieder, T., Faulstich, L.C., et al.: DeutschDiachronDigital - Ein diachrones Korpus des Deutschen. Jahrbuch für Computerphilologie 6, 119–136 (2005)
6. Chiarcos, C., Dipper, S., Götze, M., Leser, U., Lüdeling, A., Ritz, J., Stede, M.: A flexible framework for integrating annotations from different tools and tag sets. Traitment Automatique des Langues 49, 271–293 (2008)
7. Claridge, C.: Historical corpora. In: Lüdeling, A., Kytö, M. (eds.) Corpus Linguistics: An International Handbook, vol. 1, pp. 242–259. De Gruyter, Berlin (2008)
8. Rissanen, M.: Corpus linguistics and historical linguistics. In: Lüdeling, A., Kytö, M. (eds.) Corpus Linguistics: An International Handbook, vol. 1, pp. 53–68. Walter de Gruyter, Berlin and New York (2008)
9. Kytö, M.: Corpora and historical linguistics. Rev. Bras. Linguística Apl. 11(2), 417–457 (2011)
10. Kytö, M., Pahta, P.: Evidence from historical corpora up to the twentieth century. In: Nevalainen, T., Traugott, E.C. (eds.) The Oxford Handbook of the History of English, pp. 123–133. Oxford University Press, Oxford (2012)
11. Carstensen, K.-U., Ebert, C., Ebert, C., Jekat, S., Langer, H., Klabunde, R. (eds.): Computerlinguistik und Sprachtechnologie: Eine Einführung. Spektrum Akademischer Verlag, Heidelberg (2009). https://doi.org/10.1007/978-3-8274-2224-8
12. Jurafsky, D., Martin, J.H.: Speech and Language Processing, 2nd edn. Prentice Hall, Upper Saddle River (2009)
13. Manning, C.D., Schütze, H.: Foundations of Statistical Natural Language Processing. MIT Press, Cambridge (1999)
14. Cole, R., Mariani, J., Uszkoreit, H., Varile, G.B., Zaenen, A., Zampolli, A. (eds.) Survey of the State of the Art in Human Language Technology. Cambridge University Press, Cambridge (1998)
15. Tufiş, D., Filip, F.Gh. (coord.): Limba română în Societatea informaţională – Societatea Cunoaşterii. Expert, Bucureşti (2002)
16. Cristea, D., Butnariu, C.: Hierarchical XML representation for heavily annotated corpora. In: Proceedings of the LREC 2004 Workshop on XML-Based Richly Annotated Corpora, Lisbon, Portugal (2004)
17. Rehm, G., Uszkoreit, H. (eds.): The Romanian Language in the Digital Age. White Paper Series. Springer, Berlin (2012). https://doi.org/10.1007/978-3-642-30703-4
18. Gîfu, D.: Contrastive diachronic study on romanian language. In: Cojocaru, S., Gaindric, C. (eds.) Proceedings FOI-2015 Institute of Mathematics and Computer Science, Academy of Sciences of Moldova, pp. 296–310 (2015)
19. Scannel, K.: Statistical models for text normalization and machine translation. In: Proceedings of the First Celtic Language Technology Workshop, Dublin, Ireland, pp. 33–40, 23 August 2014
20. Hajič, J., Hric, J., Kuboň, V.: Machine translation of very close languages. In: Proceedings of the 6th Applied Natural Language Processing Conference, pp. 7–12. Association for Computational Linguistics (2000)
21. Koppel, M., Ordan, N.: Translationese and its dialects. In: Proceedings of the 49th Annual Meeting of the Association for Computational Linguistics (ACL 2011), Portland, Oregon (2011)

22. Gooskens, C.: Linguistic and extra-linguistic predictors of Inter-Scandinavian intelligibility. In: Van de Weijer, J., Los, B. (eds.) Linguistics in the Netherlands, vol. 23, pp. 101–113. John Benjamins, Amsterdam (2006)

23. Gooskens, C., Beijering, K., Heeringa, W.: Phonetic and lexical predictors of intelligibility. Int. J. Humanit. Arts Comput. 2(1–2), 63–81 (2008)

24. Delmestri, A., Cristianini, N.: String similarity measures and PAM-like matrices for cognate identification. Bucharest Work. Pap. Linguist. 12(2), 71–82 (2010)

25. Frunza, O., Inkpen, D., Nadeau, D.: A text processing tool for the Romanian language. In: Proceedings of the EuroLAN 2005 Workshop on Cross-Language Knowledge Induction (2005)

26. Rama, T., Borin, L.: Comparative evaluation of string similarity measures for automatic language classification. In: Mikros, G.K., Macutek, J. (eds.) Sequences in Language and Text. De Gruyter Mouton (2014)

27. Ciobanu, A., Dinu, L.: An etymological approach to cross-language orthographic similarity. In: Application on Romanian in "Proceedings of EMNLP-2014", Doha, Quatar, 25–29 October 2014, pp. 1047–1058 (2014)

28. Hristea, T.: Sinteze de limba română, 2nd edn, pp. 100–102. Didactică şi Pedagogică, Bucureşti (1981)

# Aoidos: A System for the Automatic Scansion of Poetry Written in Portuguese

Adiel Mittmann[1(✉)], Aldo von Wangenheim[1], and Alckmar Luiz dos Santos[2]

[1] Graduate Program in Computer Science (PPGCC),
National Institute for Digital Convergence (INCOD),
Federal University of Santa Catarina (UFSC), Florianópolis, Brazil
`adiel@inf.ufsc.br, aldo.vw@ufsc.br`
[2] Center for Research in Informatics, Literature and Linguistics (NUPILL),
Federal University of Santa Catarina (UFSC), Florianópolis, Brazil
`alckmar@cce.ufsc.br`

**Abstract.** Scansion is the activity of determining the patterns that give verses their poetic rhythm. In Portuguese, this means discovering the number of syllables that the verses in a poem have and fitting all verses to this measure, while attempting to pronounce syllables so that an adequate stress pattern is produced. This article presents Aoidos, a rule-based system that takes a poem written in the Portuguese language and performs scansion automatically, further providing an analysis of rhymes. The system works by making a phonetic transcription of a poem, determining the number of poetic syllables that the verses in the poem should have, fitting all the verses according to this measure and looking for verses that rhyme. Experiments show that the system attains a high accuracy rate (above 98%).

## 1 Introduction

Poetry scansion is the task of correctly determining the metrical structure of a verse. The precise rules that must be taken into consideration when scanning a verse depend on the language, but it usually involves placing syllables with specific properties in specific positions. In languages such as Portuguese, this task can be complicated because a verse can be pronounced in many ways, depending on how one chooses to join adjacent vowels.

The objective of this article is to describe Aoidos, a system that automatically performs the scansion of poetry written in the Portuguese language. The system takes a poem, transcribes its words phonetically, discovers the poem's metrical pattern, fits all verses to the pattern and finds rhyming verses. Experiments with works from three different poets show that Aoidos achieves high accuracy and that it can be extended to other poets without difficulty. Although Aoidos has been developed with Brazilian Portuguese in mind, an experiment is presented where verses of a Portuguese poet are analyzed as well. The system was named after the Ancient Greek word ἀοιδός, which means "singer", "bard".

© Springer International Publishing AG, part of Springer Nature 2018
A. Gelbukh (Ed.): CICLing 2016, LNCS 9624, pp. 611–628, 2018.
https://doi.org/10.1007/978-3-319-75487-1_46

Aoidos is a rule-based system, a fact which provides several advantages. First, unknown words, unknown meters and unknown stress patterns are handled gracefully, since the system requires no training data. Second, the system keeps track of the evolution of a verse, from the raw text all the way to phonemes, which provides accountability for the system's behavior. Finally, the system uses high-level, declarative rules that can be used to produce stylistic information.

With the help of a system like Aoidos, the automatic analysis of large poetry corpora becomes possible. This is an important step towards distant reading [1], where one would like to extract relevant features from verses in order to draw conclusions from looking at a large number of poems.

This article is organized as follows: Sect. 2 presents related work; Sect. 3 provides relevant background information concerning poetry written in Portuguese; Sect. 4 describes the Aoidos system in detail; Sect. 5 reports three experiments that intend to validate the system; finally, Sect. 6 contains discussion and conclusions.

## 2   Related Work

There have been numerous proposals for the automatic analysis of several elements of poetry. Due to the unique features of each language and poetic tradition, such proposals are usually specific to one language, though more general approaches can be found, for example, for rhyme analysis [2].

For English poetry, a connectionist model [3] has been used early on to study differences in style among 10 poets. A program named AnalysePoems [4] has been presented that is capable of identifying the dominant meter of poems in English as well as locating rhymes. More recently, a tool called ZeuScansion [5] has been introduced that robustly discovers the global meter of poems in English and achieves a per-syllable accuracy of 87%. Algorithms have also been proposed for counting the number of syllables in English [6].

Proposals for analyzing several poetic elements exist in other languages too. Poetry written in Ancient Greek hexameter [7] has been subject to automatic scansion. There has been a proposal for the automatic analysis of Czech verse [8]. Several features of Classical Arabic poetry [9] have been used to extract poems from web pages. A system named SPARSAR [10] has been proposed to recognize both rhythm and rhyme in Italian and English poetry. Metrical annotation has been added to a corpus of Old Occitan poetry [11] by using an automatic system. Metricalizer [12] is a tool capable of analyzing the prosody, meter and rhyme of German poetry. A tool named Anamètre [13] has been proposed to extract the meter and rhymes of French verse. An automatic scansion system has been used to annotate a large corpus of Spanish poetry [14].

To the best of our knowledge, the only similar research on the Portuguese language has been conducted by Araújo and Mamede [15, 16]. They proposed a system to classify various features of poems written in Portuguese. They use an external tool that provides the phonetic transcription of verses, but they do not attempt to fit verses to the meter; the examples they give show that the system does not find the correct number of syllables in verses where, e.g., a synaloepha is required. They did perform rhyme analysis, but they only conducted experiments to evaluate the system's response time, not its accuracy in any respect.

Research on speech synthesis has produced a number of systems that are related to the early stages of poetry analysis. A rule-based system for the phonetic transcription of European Portuguese has been proposed as early as 1992 [17]. A more complete description of such a system has also been described [18] and work more focused on syllabification can also be found [19]. Statistical approaches to phonetic transcription have also been proposed [20].

The present article introduces and evaluates Aoidos, the first complete system for the automatic scansion of poetry written in the Portuguese language. As most approaches in the literature, Aoidos is rule-based, which allows for unsupervised poetry scansion.

## 3   Background

Portuguese is the official language of several countries, including Portugal and Brazil. This article is primarily concerned with poetry written in the context of Brazilian Literature. The system herein described was designed to process poetry written according to the orthography rules currently practiced in Brazil since the 1970s, although poetry from other times can also be processed as long as it is encoded in current orthography. In this article, the adjective *Portuguese* when applied to nouns such as poetry, verse, orthography, etc., refers to the Portuguese language, not to Portugal.

The current orthography of Portuguese is phonetic to a certain degree. In particular, the position of the stress in a word can be found deterministically in the vast majority of words. Syllabification can also be performed in most cases without errors. Phonetic transcription can be carried out with a certain accuracy, but there are cases where the precise pronunciation of a word cannot be deduced from orthography; such cases, however, do not interfere much, if at all, with the scansion of poetry.

Metrical poems written in the Portuguese language follow syllabic patterns. Scansion of a poem, therefore, requires determining the number of syllables its verses have and fitting every verse to the metric. Vowels are commonly put in contact with each other, forming groups of two, three or more; there are many possible ways of resolving such groups, so that an inexperienced person who begins to read a poem might face difficulties determining the number of syllables of the first verse in isolation. Even when the number of syllables is known, there is frequently more than one way to divide the verse into syllables. Stress patterns are also important in Portuguese poetry. For example, it is very common for verses containing 10 poetic syllables to be stressed on their 6th; such verses are called heroic decasyllables. This kind of stress pattern is important for the correct scansion of a verse.

The current practice in both Brazil and Portugal dictates that syllables in a verse are only counted up to and including the last stressed syllable. The remaining syllables are to be pronounced regularly, but they are not included in the total number of poetic syllables a verse has. Thus, a verse with 10 poetic syllables might have 10, 11 or 12 actual syllables—but usually not more, since the stress in Portuguese can only go back to as far as the third syllable from the end of a word. In general, starting from a conservative, unhurried pronunciation of a verse, one is looking to diminish the number of syllables in a verse by joining vowels or converting them into semivowels. Portuguese poetry also features the so-called broken verses. These are verses that are shorter than the full-length verses that appear in the same poem, but the number of poetic syllables in a broken verse is counted in the same fashion as in full verses.

## 4    Methods

This section describes Aoidos in detail. The first body of poems fully analyzed by Aoidos was the complete works of Brazilian poet Augusto dos Anjos (1884–1914), which comprises 284 poems and 6,580 verses. This choice was made because prior experience indicated that his poetry would provide rich material for analyzable poetic elements. All examples in this section are taken from his works, except those provided in the discussion about phonetic transcription. The latter ones were not taken from any particular source.

### 4.1    Overview

The complete scansion system is composed of several modules. Before going into the details of each one of them, this overview subsection attempts to provide a general idea of how the system works. An example of the evolution of a verse across all modules in the systems is given by Fig. 1.

| Input | <l>Com a cara hirta, tatuada de fuligens</l> |
|---|---|
| | *With his stiff face, tattooed with soot* |
| Pre-processing | com a cara hirta tatuada de fuligens |
| Transcription | /kõ/ /a/ /'ka.ra/ /'ir.ta/ /ta.tu'a.da/ /dɪ/ /fu'li.ʒẽs/ |
| Naïve prosody | /kõ.a'ka**'rir**.ta.ta**'twa**.da.dɪ.fu'li.ʒẽs/ |
| *Rhythm* | 10/6 |
| Informed prosody | /**kwa**'ka'rir.ta.ta'twa.da.dɪ.fu'li.ʒẽs/ |
| *Rhyme* | AABCCB |

**Fig. 1.** An example of a verse going through the pipeline of the modules that make up the scansion system. The two highlighted modules (rhythm and rhyme) take into consideration the poem as a whole. Highlighted syllables are the ones that changed from the previous step.

The system takes as input a Text Encoding Initiative (TEI) XML file. Each verse in that file is extracted and undergoes *pre-processing*, which involves regularizing the text, removing all punctuation and converting it to lower–case letters. In the next stage a broad *phonetic transcription* is produced for every word in each verse. Words are then joined and processed by a *prosody* module that applies rules ranging from simple ones like obligatory voicing of a word–final sibilant before a vowel or voiced consonant to more advanced rules such as crasis and synaloepha.

The prosody module is used twice. It is employed to perform a scansion based on a *naïve prosody*, one that has no access to information regarding the overall metric of the poem. The *rhythm* module then analyzes the result produced by naïve prosody for all verses in the poem and discovers the metric of the poem (10 syllables in Fig. 1), as well as secondary stress patterns (the 6th should be stressed). Now scansion can be performed again, this time with an *informed prosody* that knows the metrical patterns that must be obeyed and will do everything in its power to fit the verse to those patterns.

Finally, the *rhyme* module analyzes the final phonetic transcription of all verses in a poem and assigns them letters which correspond to the rhyme scheme. In the example, the verse being analyzed corresponds to the first "C"—the other "C" corresponding to the rhyming word in the next verse, *origens* "origins".

## 4.2   Pre-processing

Aoidos takes as input a file in the TEI XML format. TEI defines elements for many types of texts, including prose, drama and poetry. By using this format, the same source file can be used for Aoidos to perform its analyses and for a publishable HTML version to be generated by a stylesheet. Within TEI, the lg element is a stanza, the l element is a verse and div can be used for poems and poem sections.

The system looks for l elements within an lg and considers lg elements with the same parent as composing a poem. Furthermore, the input file can use the choice element to aid the system in two cases when phonetic transcription would produce wrong results or fail: foreign words, which are not written according to Portuguese orthography, and symbols, abbreviations, numbers, etc., which are not written out as words. Thus choice can be used to specify that the English word *clown* should be read as if it were written *cláun* in Portuguese, and that D. "Mrs." should be read as *dona*.

Once raw text has been produced for every l element, the system proceeds to resolve apostrophes. Portuguese orthography does not require apostrophes the way, for example, English does. Common contractions are written without them, as in *do* "of the" or *daquele* "of that". In poetry, however, the author or publisher may choose to indicate certain poetic contractions by using an apostrophe, as in *minh'alma* "my soul" instead of the full *minha alma*. Such apostrophe usage is neither required nor applied consistently, so that the system must be ready to perform such contractions on its own and to accept text that indicates the contractions by using apostrophes; the former is accomplished by using suitable phonetic rules in later stages, the latter by removing the apostrophe and joining words together.

Finally, punctuation is removed and letters are converted to lower case. It may be strange that punctuation is deleted at this stage, since it could convey information regarding the pronunciation of verses. For example, it could be that a full stop or an exclamation mark would strongly separate words and prevent them from being closely joined together phonetically. However, this is not at all the case with Portuguese poetry. The punctuation of verses carry important information, but words are to be scanned as if they did not exist. Consider this example:

> *Meu Deus! E este morcego! E, agora, vede:*
> /ˈmewˈdewˈzjes.tɪ.morˈse.gjaˈgɔ.raˈve.dɪ/
> "My God! And this bat! And, now, consider:"

Here, the vowel before the second exclamation mark and the two vowels that follow it must be coalesced into one syllable for the verse to be correctly scanned.

### 4.3   Phonetic Transcription

The system's phonetic transcription is rule-based and is divided into three stages: location of the stress position, syllabification and mapping of letters to phonemes. The three stages are represented in Fig. 2.

| Original | com | a | cara | hirta | tatuada | de | fuligens |
|---|---|---|---|---|---|---|---|
| **1. Stress position** | com | a | ca̱ra | hi̱rta | tatua̱da | de | fuli̱gens |
| **2. Syllabification** | com | a | ca̱-ra | hi̱r-ta | ta-tu-a̱-da | de | fu-li̱-gens |
| **3. Transcription** | /kõ/ | /a/ | /ˈka.ra/ | /ˈir.ta/ | /ta.tuˈa.da/ | /dɪ/ | /fuˈli.ʒẽs/ |

*With his stiff face, tattooed with soot*

Fig. 2. The three stages of phonetic transcription.

Stress location, in this article, refers to the process of finding the letter in a written word that corresponds to the phonetic vowel that bears the stress. In Portuguese, the stress can be deterministically found in the vast majority of cases. The algorithm used by Aoidos has four rules for stress location. The first rule deals with exceptions in the writing system, that is, words that cannot be fully and correctly represented in the current official orthography; that's the case with *com* "with", which is unstressed but would otherwise be considered stressed, as *to̱m* "tone". The second rule deals with words with a stress-determining diacritic, as in *á̱cido* "acid", *â̱mbar* "amber", *maçã̱* "apple". The third rule finds the stress in words that end in certain consonants, in which case the stress falls on the vowel letter that precedes them, as in *alugue̱l* "rent", *come̱r* "to eat", *duple̱x* "duplex" and *acide̱z* "acidity". The fourth rule examines the final vowel letters and either finds the letter within that group that bears the stress, *torneio* "tournament", *uruguaio* "Uruguayan", *caju* "cashew", or finds a letter further back in the word, as in *imagem* "image", *ainda* "yet", *caieira* "lime kiln".

Syllabification proceeds by classifying letters into consonants, vowels and semivowels. In Portuguese, the main difficulty at this stage is to correctly determine the status of the letters *u* and *i*, which can be vowels or semivowels. The system first finds the consonants, then some vowels (e.g., *a* is always a vowel; *e* and *o* are in most positions vowels; the stressed letter in a word is always a vowel) and finally resolves groups of *u* and *i*. For each such group, the algorithm classifies letters by alternating between vowel and semivowel. The first letter in the group is a semivowel if the letter that precedes the group is a vowel; otherwise it is a vowel. Thus, the *i* in *migalha* "crumb" is a vowel, and the group in *buiuçu* "horse-eye bean" is a vowel, a semivowel and a vowel. Once letters are thus classified, syllabification becomes a matter of correctly splitting consonant groups. For most cases, splitting can be accomplished by keeping the first consonant in one syllable and placing the remaining ones in the next. Exceptions are groups containing one of *b*, *p*, *c*, *g*, *t*, *d*, *f* or *v*, followed by either *l* or *r*. Such groups are split immediately before these groups.

The final stage is mapping letters to phonemes. Most consonants can be directly mapped, such as *b*, *d*, *ç* and others; some consonants can be part of digraphs, as *lh* and *nh*, which become, respectively, /ʎ/ and /ɲ/. Most vowels can also be mapped without further difficulties, as is the case with nasal vowels or the stressed vowels in proparoxytone words.

By far the two biggest sources of errors in the phonetic transcription module arise when determining whether a stressed *e* or *o* is open or closed (that is, whether they are /ɛ/ or /e/, /ɔ/ or /o/), and which consonant an *x* corresponds to. Both are difficult problems that cannot be gracefully handled by a rule-based system. For example, *olho* "eye" and *olho* "I look" are written exactly the same, but the former is pronounced /ˈo.ʎʊ/ and the latter /ˈɔ.ʎʊ/; *fixo* "fixed" is pronounced /ˈfik.sʊ/, but *lixo* "garbage" is pronounced /ˈli.ʃʊ/.

## 4.4  Rule Engine

The rule engine is used by the prosody module in order to apply *rules* to *utterances*. In the context of this article, an utterance is a data structure that holds four main items:

- A string of phonemes, i.e., symbols from the International Phonetic Alphabet (IPA);
- Stress information on each phoneme, i.e., a boolean array specifying whether the corresponding phoneme is stressed or not;
- Syllabic information, such as boundaries and whether a given syllable marks the beginning or the end of a word (or both);
- A score, which is an integer that the system uses to attempt to keep track of how far-fetched the pronunciation of the utterance is.

Rules, in Aoidos, specify how to modify utterances. They are completely declarative, that is, they contain no code, only data. The prosody module works by applying rules to utterances, attempting to stick to the principle that the original utterance is pronounceable and that the modified utterance produced by them should also be.

A rule has a pattern, which the engine matches against the utterance string. If no match is found, the rule is not applied. The pattern is divided into three consecutive sections: the middle one specifies the phonemes or letters that are directly involved in the change, while the two surrounding sections provide context, as shown in Fig. 3. A rule also specifies the total number of syllables (even if partial ones) that the pattern must span. This condition is important to make sure that the rule is not applied in situations which it was not designed for.

| Original | /'ka.ra'ir.ta/ | /ta.tu'a.da/ |
|---|---|---|
| Matching | /'ka.[r]**[a]**'[i]r.ta/ | /ta.[t]**[u]**'[a].da/ |
| Replacing | /'ka'[r][]**[]**[i]r.ta/ | /ta'[t]**[w]**[a].da/ |
| Final | /'ka'rir.ta/ | /ta'twa.da/ |
|  | *stiff face* | *tattooed* |

**Fig. 3.** Application of two rules, an elision and a synaeresis. The brackets indicate the portion of the string that was matched. The middle brackets, in bold, indicate the text that is changed; the other brackets define the context.

Rules may specify whether the syllables matched by the pattern must follow a certain scheme concerning word boundaries. For example, elision rules match two vowels, the first of which must be at the final syllable of a word and the second must be at the beginning of the next word. Rules may also indicate whether the syllables matched by the pattern must follow a stress scheme. For example, an /i/ after an /a/ can become the semivowel /j/, but only if it is unstressed—or, at any rate, a stressed /i/ becoming a /j/ is an entirely different matter. In Fig. 3, it should be noted that, though the IPA stress mark changes position, internally the stressed vowel remains the same.

A rule specifies a replacement string, which is used when it has matched all conditions. The part of the original string that gets replaced is the middle section of the rule pattern; the other sections remain unchanged. The rule engine, as it applies the changes, adjusts the stress and syllabic information of the utterance.

Rules further contain a delta, that is, an integer that gets added to the utterance score if the rule is applied. Such scores are crucial in the prosody module, where many utterances are generated and only one of them will be chosen as the final version. Therefore, the delta is a measure of how unlikely or how drastic the change specified by the rule is. For example, elision is fairly common, and elision rules should therefore have a low delta; on the other hand, a rule that transforms the stressed /u/ in /'su.a/ "his" into the semivowel /w/ should have a higher delta, since, although this is not a drastic change, other rules should be tried first.

## 4.5  Prosody

The prosody module takes the phonetic transcription of individual words, joins them and produces a series of possible utterances, each with its own score, by applying rules. The application of a rule produces a new utterance, which is kept alongside the original. The rules in this module are divided into two groups: basic and advanced. The difference between the two groups is that rules in the former use criteria concerning word boundaries and are only applied in the beginning of the process; rules in the latter group do not use word boundaries as criteria and might be allowed to combine indefinitely. For example, elision rules are located in the basic group, since such rules are based on word boundaries and should not be compounded; crasis rules, as performed by Aoidos, are applied regardless of word boundaries and may be compounded, and therefore are placed in the advanced group.

The prosody module is used twice during the scansion process: once before rhythm analysis, when the meter is not known (naïve prosody), and once afterwards, when such information is available (informed prosody). During the naïve prosody analysis, rules are applied to all matching instances in a verse at once. For example, an elision rule that matches two vowels in naïve prosody would elide both of them at once, while in informed prosody a total of *three* new derivations will be created from the original utterance: one with only the first elision applied, another with only the second elision, and a third with both. This ensures that the space of possible ways to read a verse is thoroughly explored in informed prosody. Another difference between naïve and informed prosody is that only rules with a low delta are applied during the naïve prosody analysis, while informed prosody is allowed access to the full rule set.

Naïve prosody, after applying all rules, chooses the utterance with the lowest score as its final result. Informed prosody is iterative: it begins by using rules with a low delta and rules with increasingly higher deltas are applied until at least one satisfactory utterance has been produced. A satisfactory utterance is one whose number of poetic syllables matches either the verse length or one of the secondary stresses prescribed by the rhythm analysis module. If more than one satisfactory utterance is found, then they are sorted according to an index based on their score and on an evaluation of how well they fit the stress pattern found by the rhythm analysis, and the best utterance is selected.

**Basic rules.** The basic rules are responsible for joining words together in one unified utterance and taking care of phenomena which rely on word boundary information. The majority of rules in this module can be divided into:

- Elision rules. These rules remove the final vowel of a word when the next one begins with a vowel too; examples are *esfera opaca* "opaque sphere" and *escapava entre* "escaped between", which become /es'fɛ.ro'pa.ka/ and /es.ka'pa've̜.trɪ/.
- Syncope rules, which remove certain unstressed phonemes from the interior of a word. Examples are *símbolo* "symbol" and *século* "century", which become /'sĩ.blʊ/ and /'sɛ.klʊ/.

– Apheresis rules. They remove an initial unstressed vowel when followed by certain consonant clusters, as in *espírito* "spirit" and *escaldante* "scalding", which become /'spi.ri.tʊ/ and /skaw'dã.tɪ/.

Each item in the list above talks about *rules*, in the plural, because a given phonetic phenomenon, such as elision, manifests itself in different contexts, which must be individually considered if a meaningful delta is to be assigned to each case.

**Advanced rules.** The rules in this group deal with phenomena that take place regardless of word boundaries. The majority of the rules in this group are:

– Crasis rules. These rules mix two similar vowels into one, thus reducing the number of syllables in the verse by one. Examples are *vêem-se* "they are seen" and *a alma* "the soul", which from /'ve.ẽ.sɪ/ and /a'aw.ma/ are converted into /'vẽ.sɪ/ and /'aw.ma/.
– Synaloepha rules, which compress vowels and semivowels from different syllables into one syllable, forming diphthongs, triphthongs and, in more extreme cases, dropping phonemes. A simple example is *e o sol* "and the sun", which is converted from /ɪ.ʊ'sɔw/ to /jʊ'sɔw/. Such rules are also responsible for synaeresis, which is similar to synaloepha but happens inside a word, as in *afluem* "they flow", which is transformed from /a'flu.ẽ/ into /a'flwẽ/ by shifting the stress from one vowel to the other.

The system currently has 9 crasis rules and 20 synaloepha rules. Such a number is required to cover a great number of contexts while keeping each context

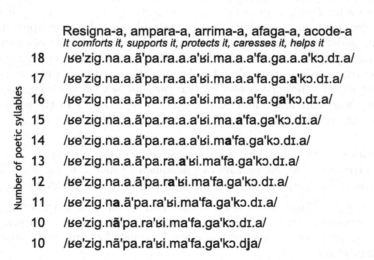

Fig. 4. Sequence of rules applied in the prosody module, reducing the number of poetic syllables from 18 to 10.

as specific as possible. When rules are too general, the risk is high that they will end up being applied in circumstances they were not planned for.

Figure 4 shows an example where crasis rules compound to greatly reduce the number of syllables in a verse. The steps shown in the figure were produced by recursively tracing back each utterance's origin. Because rules are applied to each match separately, there were many more possible combinations; this is just one of them.

## 4.6  Rhythm Analysis

The rhythm analysis module is responsible for determining the number of poetic syllables that verses in a poem should have and the syllables that are commonly stressed in those verses. If the module fails to determine the number of poetic syllables for the poem as a whole, it resorts to a more flexible, stanza-based analysis.

The module takes the result of the prosody analysis and looks at the distribution of the number of poetic syllables among verses, as shown in Table 1. The poems in that table were chosen because they provide a good source of examples, but they are not necessarily typical of poet A. dos Anjos.

**Table 1.** Distribution (%) of the number of poetic syllables found by the prosody module in the verses of several poems. The bold values indicate the correct number of syllables in the full length verses.

| Poem | 6 | 7 | 8 | 9 | 10 | 11 | 12 | 13 |
|---|---|---|---|---|---|---|---|---|
| Canto Íntimo | 10.7 | **85.7** | 3.6 | | | | | |
| Numa Forja | 47.7 | 2.3 | | | **45.3** | 4.7 | | |
| Vênus Morta | 3.1 | | | 12.5 | **81.2** | 3.1 | | |
| O Morcego | | | | 7.1 | **78.6** | 7.1 | 7.1 | |
| Agonia de um Filósofo | | | | | **92.9** | 7.1 | | |
| Asas de Corvo | | | | | **71.4** | 21.4 | 7.1 | |
| Guerra | | | | | **64.3** | 28.6 | | 7.1 |
| Súplica num Túmulo | | | | | | 7.1 | **85.7** | 7.1 |

Finding the correct number of syllables in a poem can be problematic because, even in the context of metric poetry, not all of its verses may have the same number of poetic syllables. In particular, a common feature in Portuguese poetry are the so-called broken verses, which are shorter than the full verses in a given poem. The system, in such cases, finds the length of full verses and the informed prosody attempts to fit broken verses according to the commonly stressed syllables, which generally include the exact poetic length of the broken verses.

The algorithm that detects the full length of verses begins by trying to locate a pair of consecutive lengths whose presence in the verses of a poem add up to at least 75%. The idea is that, although the naïve prosody may not generate very precise results, it will not be wrong by a wide margin, so that two adjacent lengths are usually enough to reach the threshold. Among poems in Table 1, this strategy finds the right answer (shown there in bold) in all cases but the second. The second poem contains a good amount of broken verses (exactly half of them), so that the lengths are distributed around the full length of the verse (10 syllables) and the broken length (6 syllables). The second poem is also an example of why picking the largest value does not work: there are more verses with length 6 than with length 10. It should be noted that the third poem in that table also contains broken verses, but in a much smaller proportion, so that a more sophisticated strategy is not required for analyzing that poem.

If there is no consecutive group of two lengths whose proportions add up to at least 75%, the system then considers the hypothesis that perhaps there are two main lengths instead of just one. In this case, the system looks for two groups of consecutive lengths. The combined percentage of each group should reach at least 37.5%, for a total combined percentage of 75%. If such groups are found, then the system considers that the full length of the verses in the poem is the largest length among the two groups. This strategy takes care of the case of the second poem in Table 1.

If the system fails to find the full length of verses using these two strategies, then it considers that there might not be a very discernible pattern at the poem level and that it should instead examine the poem stanza by stanza. Stanzas vary in length but they can be very short, as in two or three verses, so that a complex analysis of frequencies is not possible. The system, therefore, simply chooses the length with higher frequency. This was added to the system as a fall back strategy, since there were no cases in the works of A. dos Anjos that required it.

Once the full length of verses has been determined, the system looks for secondary stress patterns. The strategy here is straightforward: all lengths that are stressed in at least 60% of verses are considered to be secondary stresses. The final poetic syllable of broken verses is usually found among the secondary stresses.

## 4.7   Rhyme Analysis

This module attempts to identify rhymes at the end of verses. An uppercase Latin letter starting from "A" is attributed to each verse; those verses that rhyme are assigned the same letter. An example is given in Fig. 5.

The rhyme analysis algorithm extracts the final syllables from verses, starting from the last stressed syllable and going up to the very last phoneme. Consonants are then dropped from the beginning of the stressed syllable, leaving what is called, in this article, the rhyming segment. As can be seen in Fig. 5, the rhyming segments can be compared to each other in order to establish the rhymes that exist between verses.

| Verse | Ending | Rhyming Segment | Result |
|---|---|---|---|
| Aí vem sujo, a coçar chagas plebeias, | ple'bɛj.as | ɛj.as | **A** |
| Trazendo no deserto das ideias | zi'dɛj.as | ɛj.as | **A** |
| O desespero endêmico do inferno, | dwĩ'fɛr.nʊ | ɛr.nʊ | **B** |
| Com a cara hirta, tatuada de fuligens | fu'li.ʒẽs | i.ʒẽs | **C** |
| Esse mineiro doido das origens, | zo'ri.ʒẽs | i.ʒẽs | **C** |
| Que se chama o Filósofo Moderno! | mo'dɛr.nʊ | ɛr.nʊ | **B** |

*There he comes, dirty, scratching plebeian wounds,*
*Dragging through the desert of ideas*
*Hell's endemic despair,*
*With his stiff face, tattooed with soot*
*This mad miner of origins,*
*Whom we call the Modern Philosopher!*

**Fig. 5.** Rhyme analysis for a complete stanza. The ending displayed here is composed of the last three *phonetic* syllables.

When the rhyming segments for each verse have been found, the algorithm assigns a letter to each verse. To do so, the algorithm keeps a dictionary of the 10 most recently seen rhyming elements and the letters originally attributed to them. If the current verse's rhyming element is found in the dictionary, it is assigned the corresponding letter; otherwise, the next available letter is chosen and a new rhyming element is added to the dictionary.

## 5 Experiments

This section describes three experiments, in decreasing order of significance. In each experiment, poems by different poets were evaluated. The first experiment evaluates the phonetic transcription, scansion and rhyme analysis produced by Aoidos; the second evaluates one component of scansion (number of poetic syllables); and the third provides evidence that Aoidos can be applied without difficulty to poems from other times and cultures.

### 5.1 Augusto dos Anjos

The complete works of Brazilian poet Augusto dos Anjos comprise 284 poems, among which 167 are sonnets (poems with four stanzas and fourteen verses). A total of 111 sonnets (1,554 verses) were randomly chosen and were subject to manual evaluation by experts. Each sonnet was independently assessed by two experts. The experts that participated in the experiment were not involved in the definition of the rules used by the prosody module. Figure 6 shows the output of Aoidos as seen in the evaluation screen. Experts had the option to toggle between phonetic transcription and regular text, and between highlighting or unhighlighting the stressed syllables. Experts could further add comments to each verse, explaining their decisions or providing comments.

| | | | | | | | | | | |
|---|---|---|---|---|---|---|---|---|---|---|
| **A** No | tem- | po | de | meu | Pai, | sob | es- | tes | ga- | lhos, |
| **B** Co- | mo u- | ma | ve- | la | fú- | ne- | bre | de | ce- | ra, |
| **B** Cho- | rei | bi- | lhões | de | ve- | zes | com a | can- | sei- | ra |
| **A** De i- | ne- | xo- | ra- | bi- | li- | ssi- | mos | tra- | ba- | lhos! |
| | | | | | | | | | | |
| **A** Ho- | je, es- | ta ár- | vo- | re, | de am- | plo- | s a- | ga- | sa- | lhos, |
| **B** Guar- | da, | co- | mo u- | ma | cai- | xa | de- | rra- | dei- | ra, |
| **B** O | pa- | ssa- | do | da | Flo- | ra | Bra- | si- | lei- | ra |
| **A** E a | pa- | leon- | to- | lo- | gi- | a | dos | Car- | va- | lhos! |
| | | | | | | | | | | |
| **C** Quan- | do | pa- | ra- | rem | to- | do- | s os | re- | ló- | gios |
| **C** De | mi- | nha | vi- | da, e a | voz | dos | ne- | cro- | ló- | gios |
| **D** Gri- | tar | nos | no- | ti- | ciá- | rios | que eu | mo- | rri, | |
| | | | | | | | | | | |
| **E** Vol- | tan- | do à | pá- | tria | da ho- | mo- | ge- | nei- | da- | de, |
| **E** A- | bra- | ça- | da | com a | pró- | pria E- | ter- | ni- | da- | de |
| **D** A | mi- | nha | som- | bra há | de | fi- | ca- | r a- | qui! | |

*In the time of my father, under these branches,*   *Today this wide-branched tree*
*Like a funeral wax candle,*   *Keeps, as an ultimate strongbox,*
*I cried billions of times with the fatigue*   *The past of the Brazilian Flora,*
*Of most inexorable chores!*   *And the paleontology of Oaks!*

*When all clocks of my life*   *And returned to the land of homogeneity,*
*Stop, and the voice of obituaries*   *Embraced with eternity itself*
*Shouts in the paper I have died,*   *My shadow will remain here!*

**Fig. 6.** Final result produced by Aoidos for A. dos Anjos' poem *Under the Tamarind.*

Experts were asked to evaluate three criteria for each verse:

- *Phonetic transcription:* is the phonetic transcription acceptable for the purposes of scansion?
- *Scansion:* is the final scansion performed by the system acceptable?
- *Rhyme analysis:* is the rhyme analysis provided by the system acceptable?

Because the definition of "acceptable" may vary from one expert to another, a criterion for a given verse was considered unacceptable if any of the two experts deemed it so. Table 2 summarizes the results; all criteria were judged acceptable for at least 98% of verses.

**Table 2.** Results of the evaluation by experts of 111 poems automatically scanned by Aoidos.

| | Total | Acceptable | | Unacceptable | |
|---|---|---|---|---|---|
| Phonetic transcription | 1,554 | 1,549 | 99.68% | 5 | 0.32% |
| Scansion | 1,554 | 1,523 | 98.01% | 31 | 1.99% |
| Rhyme analysis | 1,554 | 1,550 | 99.74% | 4 | 0.26% |

Scansion was found unacceptable in 31 cases, generally because the rules applied by the system produced a pronunciation that, though independently valid, was not the adequate choice. For example, in the following verse, a different scansion was indicated by the expert, who explained that it is better to keep *que* "that" in a syllable of its own in order to avoid two consecutive stressed syllables:

|        | 1    | 2    | 3     | 4   | 5   | 6       | 7   | 8   | 9    | 10   | 11  |
|--------|------|------|-------|-----|-----|---------|-----|-----|------|------|-----|
| Aoidos | Ven- | cen- | do o  | a-  | zul | que an- | te  | si  | s'er | gue  | ra. |
| Expert | Ven- | cen- | do o a| zul | que | an-     | te  | si  | s'er | gue  | ra. |
| | | *Overcoming the blue sky that had risen.* | | | | | | | | | |

There were also cases when the experts were impressed by the results produced by the system. For example, there was a case when the same word was used in two adjacent verses, but in one verse the system produced a synaeresis within that word, and in the other verse it did not. Such behavior is required for the correct scansion of these two verses and is perceived as "intelligent". There were also cases when the experts commented that they reached the end of a verse only to find out that their scansion was not appropriate and that the system had produced the scansion that they should have used. This is a testimony to the fact that verses in Portuguese can be quite difficult to be scanned right the first time: an early decision to make a synaloepha or not might cause an extra or missing syllable once the reader has finished the verse.

## 5.2   Gustavo Teixeira

Two books by Brazilian poet Gustavo Teixeira (1881–1937), *Notebook* and *Lyric Poems*, were read by an expert, who classified poems and verses according to their number of poetic of syllables. The two books contain a total of 112 poems and 3,178 verses. The automatic scansion system was applied to these poems and the final scansions were compared to those indicated by the expert; results are shown in Table 3.

The system found the correct number of poetic syllables in 99.8% of cases. It mostly failed to correctly scan broken verses, which are more prominent in the poetry of G. Teixeira than in that of A. dos Anjos.

## 5.3   Luís de Camões

In order to test the system on a more diverse body of poetry, the automatic scansion system was applied to *The Lusiads*, an epic poem written by Portuguese poet Luís de Camões (1524–1580), which comprises 10 cantos and a total of 8,816 verses. A total of 8 new rules had to be added to the system so that all verses could be scanned. These new rules were mostly synaloephas not found in the works of A. dos Anjos, as in the following verse:

**Table 3.** Proportion of correctly identified lengths of verses in poems by G. Teixeira.

| Full length | Actual length | Count | Correct | |
|---|---|---|---|---|
| 12 | 12 | 1,523 | 1,523 | 100.0% |
| 12 | 8 | 10 | 8 | 80.0% |
| 12 | 6 | 17 | 17 | 100.0% |
| 10 | 10 | 1,031 | 1,031 | 100.0% |
| 10 | 6 | 30 | 29 | 96.7% |
| 9 | 9 | 42 | 42 | 100.0% |
| 9 | 4 | 14 | 14 | 100.0% |
| 8 | 8 | 172 | 172 | 100.0% |
| 7 | 7 | 303 | 303 | 100.0% |
| 7 | 4 | 36 | 32 | 88.9% |
| | | 3,178 | 3,171 | 99.8% |

Responde **ao em**baixador, que tanto estima:
*He replies to the ambassador, whom he highly cherishes:*

The system correctly identified that all verses contain 10 poetic syllables, and for all cantos the 6th syllable was correctly considered a secondary stress. This needs not be manually evaluated: it is known that all verses in the *The Lusiads* are decasyllables and that most of them are heroic, that is, have their 6th syllable stressed. The results produced by the system indicate that the sixth syllable was stressed in 97,03% of verses. The system was able to process the poetry of L. de Camões significantly faster than that of either A. dos Anjos and G. Teixeira. The main reason is that the number of pronunciation alternatives considered by the system before settling on its final version was larger for the two latter poets. Whereas the system considered an average of 35.79 alternatives for each verse in the poetry of A. dos Anjos and 26.03 for G. Teixeira, the average for L. de Camões was only 9.30. This is certainly a reflection of stylistic features.

# 6    Discussion and Conclusions

This article introduced Aoidos, a rule-based system that extracts information regarding scansion and rhyming from poetry written in Portuguese. Its results can be used for several purposes. The system can be employed to process poetic corpora that are too large to be analyzed by humans; at a rate of about 40 verses per second in an ordinary computer, processing even millions of verses becomes feasible. The analysis of verses and poems can be used to understand the style of different poets and times; and not only the final, scanned verse can be used for such comparisons, but also the very set of phonetic rules employed during the analysis. It might be that certain poets favor certain types of synaloephas, or that they use them more sparingly. Finally, as a robotic minstrel, Aoidos does not get emotionally involved when scanning poetry and therefore spots mistakes in the source material that render verses impossible to fit the metric; indeed, quite a few mistakes have been found in our digital editions thanks to Aoidos.

The system was evaluated by experiments, which show that it is capable of scanning verses with great accuracy (above 98%). Aoidos applies phonetic rules to verses and attempts to produce a pronounceable, metrified transcription of poems. Although its rules were designed initially for a 20th-century Brazilian poet, the system proved capable of analyzing poems by another poet from the same place and time without modifications as well as an 8,816-verse poem written three centuries earlier by a Portuguese poet, this time with the addition of only 8 new rules.

Aoidos can still be improved in several aspects. Its current treatment of broken verses, an important feature of Portuguese poetry, is not yet very robust, as evidenced by the second experiment. Its understanding of stress patterns, i.e., rhythm, works well for most cases, but the few mistakes it does commit are evidence that it can be improved. In particular, it currently has no conception of a rhythmic stress, so that all stress information used in its analyses is derived from each word's primary stress. Its pre-processing stage currently does not handle foreign words or abbreviations (unless annotations are provided), which would certainly cause mistakes to be made in larger, untreated corpora.

The possibility of extending Aoidos to support other languages is currently being investigated. One important challenge to take into account when considering other languages is obtaining a suitable phonetic transcription from the source text: whereas the orthography of Portuguese is fairly phonetic, this is not the case with, e.g., English (where the same letter can correspond to different phonemes) and Russian (where stress position cannot be determined from the written text). Preliminary results of applying Aoidos to poetry written in Spanish are encouraging.

**Acknowledgments.** We would like to thank Isabela M. B. Sandoval, Lívia Guimarães and Samanta Maia for their contribution in the experiments. This work was partially supported by the Brazilian National Council for Scientific and Technological Development (CNPq) and by the Santa Catarina Foundation for the Support of Research and Innovation (FAPESC).

# References

1. Moretti, F.: Distant Reading. Verso Books, New York (2013)
2. Reddy, S., Knight, K.: Unsupervised discovery of rhyme schemes. In: Proceedings of the 49th Annual Meeting of the Association for Computational Linguistics: Human Language Technologies: short papers, vol. 2, pp. 77–82. Association for Computational Linguistics (2011)
3. Hayward, M.: Analysis of a corpus of poetry by a connectionist model of poetic meter. Poetics **24**, 1–11 (1996)
4. Plamondon, M.R.: Virtual verse analysis: analysing patterns in poetry. Literary Linguist. Comput. **21**, 127–141 (2006)

5. Agirrezabal, M., Arrieta, B., Astigarraga, A., Hulden, M.: ZeuScansion: a tool for scansion of English poetry. In: Proceedings of the 11th International Conference on Finite State Methods and Natural Language Processing, St Andrews, Scotland, pp. 18–24 (2013)
6. Hammond, M.: Calculating syllable count automatically from fixed-meter poetry in English and Welsh. Literary Linguist. Comput. **29**, 218–233 (2014)
7. Papakitsos, E.C.: Computerized scansion of Ancient Greek hexameter. Literary Linguist. Comput. **26**(1), 57–69 (2010)
8. Ibrahim, R., Plecháč, P.: Towards the automatic analysis of Czech verse. In: Formal Methods in Poetics, pp. 295–305. RAM, Lüdenscheid (2011)
9. Almuhareb, A., Alkharashi, I., AL Saud, L., Altuwaijri, H.: Recognition of classical Arabic poems. In: Proceedings of the Workshop on Computational Linguistics for Literature, Atlanta, Georgia, Association for Computational Linguistics, pp. 9–16 (2013)
10. Delmonte, R.: A computational approach to poetic structure, rhythm and rhyme. In: Proceedings of the First Italian Conference on Computational Linguistics CLiC-it 2014 and of the Fourth International Workshop EVALITA 2014, pp. 144–150. Pisa University Press (2014)
11. Rainsford, T., Scrivner, O.: Metrical annotation for a verse treebank. In: The 13th International Workshop on Treebanks and Linguistic Theories (TLT13), pp. 149–159 (2014)
12. Bobenhausen, K., Hammerich, B.: Métrique littéraire, métrique linguistique et métrique algorithmique de l'allemand mises en jeu dans le programme metricalizer. Langages **3**, 67–88 (2015)
13. Delente, E., Renault, R.: Traitement automatique des formes métriques des textes versifiés. In: 22ème Traitement Automatique des Langues Naturelles (2015)
14. Navarro, B., Ribes-Lafoz, M., Sánchez, N.: Metrical annotation of a large corpus of Spanish sonnets: representation, scansion and evaluation. In: Language Resources and Evaluation Conference (2016)
15. de Araújo, P.A.M.: Classificação de poemas e sugestão das palavras finais dos versos. Universidade Té cnica de Lisboa (2004)
16. de Araújo, P.A.M., Mamede, N.J.: Classificador de poemas. In: Conferência Científica e Tecnológica em Engenharia (2002)
17. Oliviera, L.C., Viana, M., Trancoso, I.M.: A rule-based text-to-speech system for Portuguese. In: 1992 IEEE International Conference on Acoustics, Speech, and Signal Processing, ICASSP 1992, vol. 2, pp. 73–76. IEEE (1992)
18. Braga, D., Coelho, L., Resende Jr., F.G.V.: A rule-based grapheme-to-phone converter for TTS systems in European Portuguese. In: 2006 International Telecommunications Symposium, pp. 328–333. IEEE (2006)
19. Neto, N., Rocha, W., Sousa, G.: An open-source rule-based syllabification tool for Brazilian Portuguese. J. Braz. Comput. Soc. **21**, 1–10 (2015)
20. Couto, I., Neto, N., Tadaiesky, V., Klautau, A., Maia, R.: An open source HMM-based text-to-speech system for Brazilian Portuguese. In: 7th International Telecommunications Symposium (2010)

# Author Index

Printed in the United States
By Bookmasters